RODD'S CHEMISTRY OF CARBON COMPOUNDS

ELSEVIER PUBLISHING COMPANY
335 JAN VAN GALENSTRAAT
P.O.BOX 211, AMSTERDAM, THE NETHERLANDS

ELSEVIER PUBLISHING CO. LTD.
BARKING, ESSEX, ENGLAND

AMERICAN ELSEVIER PUBLISHING COMPANY, INC.
52 VANDERBILT AVENUE, NEW YORK, N.Y. 10017

LIBRARY OF CONGRESS CARD NUMBER: 64-4605

ISBN 0-444-40878-9

WITH 44 TABLES AND 12 ILLUSTRATIONS

PRINTED IN THE NETHERLANDS

RODD'S CHEMISTRY OF CARBON COMPOUNDS

ADVISORS

Professor Sir ROBERT ROBINSON, O.M., M.A. (Oxon.), D.SC. (Manc.), HON.D.SC. (Lond., Liv., Wales, Dunelm, Sheff., Belfast, Bris., Oxon., Nott., Strath., Delhi, Sydney, Zagreb), HON. SC.D. (Cantab.), HON. LL.D. (Manc., Edin., Birm., St. Andrews, Glas., Liv.), HON. D. PHARM. (Madrid and Paris), HON. F.R.S.E., F.R.S., *London*

Chairman

Professor A. R. BATTERSBY, M.SC. (Manc.), PH.D. (St. Andrews), D.SC. (Bris.), F.R.S., *Cambridge*

Professor R. N. HASZELDINE, M.A., PH.D., SC.D. (Cantab.), PH.D., SC.D. (Birm.), F.R.I.C., *Manchester*

Professor R. D. HAWORTH, D.SC., PH.D. (Manc.), B.SC. (Oxon.), F.R.I.C., F.R.S., *Sheffield*

Professor Sir EDMUND HIRST, M.A., PH.D. (St. Andrews), D.SC. (Birm.), HON. LL.D. (St. Andrews, Aberdeen, Birm., Strath.), HON. D.SC. (Dublin), F.R.I.C., F.R.S., *Edinburgh*

(The late) Dr. E. H. Rodd, D.SC. (Lond.), F.C.G.I., F.R.I.C., *Newton Abbot, Devon*

Professor Lord TODD, M.A. (Cantab.), D.SC. (Glas.), D.PHIL. (Oxon.), DR. PHIL. NAT. (Frankfurt), HON. LL.D. (Glas., Edin., Melb., Calif.), HON. DR. RER. NAT. (Kiel), HON. D. MET. (Sheff.), HON. D.SC. (Oxon., Dunelm, Lond., Exe, Leic., Liv., Adel., Alig., Madrid, Stras., Wales, Strath.), F.R.I.C., F.R.S., *Cambridge*

RODD'S CHEMISTRY OF CARBON COMPOUNDS

VOLUME I

GENERAL INTRODUCTION

ALIPHATIC COMPOUNDS

★

VOLUME II

ALICYCLIC COMPOUNDS

★

VOLUME III

AROMATIC COMPOUNDS

★

VOLUME IV

HETEROCYCLIC COMPOUNDS

★

VOLUME V

MISCELLANEOUS

GENERAL INDEX

★

RODD'S CHEMISTRY OF CARBON COMPOUNDS

A modern comprehensive treatise

SECOND EDITION

Edited by

S. COFFEY

M.Sc. (London), D.Sc. (Leyden), F.R.I.C.

formerly of

I.C.I. Dyestuffs Division, Blackley, Manchester

VOLUME III PART A

AROMATIC COMPOUNDS, GENERAL INTRODUCTION

Mononuclear hydrocarbons and their halogeno derivatives, and derivatives with nuclear substituents attached through nonmetallic elements from group VI of the Periodic Table

ELSEVIER PUBLISHING COMPANY

AMSTERDAM LONDON NEW YORK

1971

CONTRIBUTORS TO THIS VOLUME

N. CAMPBELL, O.B.E., PH.D., D.SC.
Department of Chemistry, The University, Edinburgh, 9.

G. S. CHANDLER, B.SC., PH.D.
Research School of Chemistry, Australian National University, Canberra, A.C.T. 2600 (Australia).

N. B. CHAPMAN, M.A., PH.D., F.R.I.C.
Department of Chemistry, The University, Hull.

D. P. CRAIG, PH.D., D.SC., F.R.I.C., F.R.A.C.I., F.R.S.
Research School of Chemistry, Australian National University, Canberra, A.C.T. 2600 (Australia).

A. G. DAVIES, D.SC., F.R.I.C.
Department of Chemistry, University College, London, W.C.1.

P. D. B. DE LA MARE, PH.D., D.SC.
Department of Chemistry, The University, Auckland (New Zealand).

J. D. DOWNER, B.SC., PH.D., F.R.I.C.
Texaco Trinidad, Inc., Pointe-a-Pierre (Trinidad, W.I.).

W. J. FEAST, B.SC., PH.D.
Department of Chemistry, The University, Durham City.

A. R. FORRESTER, PH.D.
Department of Chemistry, The University, Old Aberdeen, Scotland.

D. H. HEY, PH.D., D.SC., F.R.I.C., F.R.S.
Department of Chemistry, King's College, London, W.C.2.

D. R. HOGG, B.SC., PH.D.
Department of Chemistry, The University, Old Aberdeen, Scotland.

W. K. R. MUSGRAVE, PH.D., D.SC.
Department of Chemistry, The University, Durham City.

J. L. WARDELL, B.SC., PH.D.
Department of Chemistry, The University, Old Aberdeen, Scotland.

G. H. WILLIAMS, B.SC., PH.D.
Bedford College, London, N.W.1.

R. E. FAIRBAIRN, B.SC., PH.D., F.R.I.C.
formerly of Research Department, Dyestuffs Division, I.C.I. Ltd., Manchester, 9 (Index).

PREFACE TO VOLUME IIIA

In presenting Volume III of this second edition of *Rodd's* "Chemistry of Carbon Compounds" the general plan used in Volumes I and II is again followed. The subject matter deals with the chemistry of carbocyclic aromatic compounds (the latter term to be understood in its classical sense), *i.e.*, with benzenoid hydrocarbons, their homologues and polynuclear analogues, and their derivatives. During the last two decades since the original edition was compiled, our ideas on aromaticity and aromatic behaviour have become more accurately defined and meaningful and it is now realised that aromatic character depends on physical properties rather than chemical reactivity and varies in degree depending on the structure of the compound in question. The early work indicating the tentative beginnings of this realisation was discussed in the original edition in a chapter on so-called pseudo-aromatic compounds. In the present edition this chapter has been deleted. The subject of aromaticity and aromatic behaviour in its contemporary setting is discussed by Dr. G. S. Chandler and Professor D. P. Craig in the introductory chapter to this volume, and the relevant classes of "aromatic" compounds not derived from benzene and its analogues are discussed systematically in appropriate places in Volume II. One other change from the original arrangements should be mentioned. In the original Volume III, in the order in which substituted derivatives of the various classes of parent hydrocarbons were described, compounds with substituents linked through nitrogen preceded those linked through oxygen and other elements of group VI of the Periodic Table, which was contrary to the arrangement adopted in the first two Volumes. In the present edition the order is consistent in all three volumes.

It is planned that Volume III shall appear in six sub-volumes dealing respectively with (A) a general introduction to aromatic compounds, benzenoid hydrocarbons and their derivatives carrying substituents linked through elements of groups VI, VIB, VII and VIIB of the Periodic Table; (B) derivatives with substituents linked through elements of the remaining groups of the Table; (C) compounds with substituted side-chains; (D) substances with unsaturated side-chains, and poly-aryl compounds; (E) and (F) aromatic compounds with condensed carbocyclic rings.

In Volume IIIA the General Introduction to the chemistry of aromatic compounds has sub-chapters on aromatic character and the benzene nucleus, electrophilic, nucleophilic and homolytic aromatic substitutions, and the formation and fission of the benzene nucleus. Subsequent chapters deal systematically with benzene and its homologues, their halogeno derivatives, and compounds containing groups attached to the nucleus through oxygen, sulphur, selenium or tellurium.

May 1971 S. COFFEY

CONTENTS

VOLUME IIIA

Aromatic Compounds; General Introduction; Mononuclear Hydrocarbons and their halogeno derivatives; and derivatives with nuclear substituents attached through nonmetallic elements from Group VI of the Periodic Table

Chapter 2. Mononuclear Hydrocarbons: Benzene and its Homologues

by J. D. DOWNER

Chapter 3. Halogen Derivatives of Benzene and its Homologues
by W. J. FEAST AND W. K. R. MUSGRAVE

Chapter 4. Nuclear Hydroxy Derivatives of Benzene and its Homologues
by A. R. FORRESTER AND J. L. WARDELL

Chapter 5. Mononuclear hydrocarbons carrying substituents attached through sulphur: Thiophenols, Sulphides, etc.

by A. R. Forrester and J. L. Wardell

Chapter 6. Mononuclear hydrocarbons carrying nuclear substituents containing sulphur: Sulphonic Acids, Sulphinic Acids and Sulphenyl Compounds of the Benzene Series

by D. R. Hogg

Chapter 7. Mononuclear hydrocarbons carrying nuclear substituents containing Selenium or Tellurium

by A. G. DAVIES

VOLUMES IIIB, C, D, E AND F

Vol. IIIB Derivatives with substituents linked through elements of the remaining groups of the Periodic Table

Vol. IIIC Compounds with substituted side-chains

Vol. IIID Substances with unsaturated side-chains, and poly-aryl compounds

Vol. IIIE and IIIF Aromatic compounds with condensed carbocyclic rings.

OFFICIAL PUBLICATIONS

B.P.	British (United Kingdom) Patent
F.P.	French Patent
G.P.	German Patent
Sw.P.	Swiss Patent
U.S.P.	United States Patent
U.S.S.R.P.	Russian Patent
B.I.O.S.	British Intelligence Objectives Sub-Committee Reports, H.M. Stationery Office, London.
C.I.O.S.	Combined Intelligence Objectives Sub-Committee Reports
F.I.A.T.	Field Information Agency, Technical Reports of U.S. Group Control Council for Germany
B.S.	British Standards Specification
A.S.T.M.	American Society for Testing and Materials
A.P.I.	American Petroleum Institute Projects
C.I.	Colour Index Number of Dyestuffs and Pigments

SCIENTIFIC JOURNALS AND PERIODICALS

With few obvious and self-explanatory modifications the abbreviations used in references to journals and periodicals comprising the extensive literature on organic chemistry, are those used in the World List of Scientific Periodicals.

LIST OF COMMON ABBREVIATIONS AND SYMBOLS USED

A	acid
Å	Ångström units
Ac	acetyl
a	axial
as, *asymm.*	asymmetrical
at.	atmosphere
B	base
Bu	butyl
b.p.	boiling point
C, mC and μC	curie, millicurie and microcurie
c, *C*	concentration
conc.	concentrated
crit.	critical
D	Debye unit, 1×10^{-18} e.s.u.
D	dissociation energy
D	dextro-rotatory; dextro configuration
DL	optically inactive (externally compensated)
d	density
dec. or decomp.	with decomposition
deriv.	derivative
E	energy; extinction; electromeric effect
E1, E2	uni- and bi-molecular elimination mechanisms
E1cB	unimolecular elimination in conjugate base
e.s.r.	electron spin resonance
Et	ethyl
e	nuclear charge; equatorial
f	oscillator strength
f.p.	freezing point
G	free energy
g.l.c.	gas liquid chromatography
g	spectroscopic splitting factor, 2.0023
H	applied magnetic field; heat content
h	Planck's constant
Hz	hertz
I	spin quantum number; intensity; inductive effect
i.r.	infrared
J	coupling constant in n.m.r. spectra
K	dissociation constant
k	Boltzmann constant; velocity constant
kcal.	kilocalories
L	laevorotatory; laevo configuration
M	molecular weight; molar; mesomeric effect
Me	methyl
m	mass; mole; molecule; *meta*-
ml	millilitre
m.p.	melting point
Ms	mesyl (methanesulphonyl)

[M]	molecular rotation
N	Avogadro number; normal
n.m.r.	nuclear magnetic resonance
n	normal; refractive index; principal quantum number
o	*ortho-*
o.r.d.	optical rotatory dispersion
P	polarisation; probability; orbital state
Pr	propyl
Ph	phenyl
p	*para-;* orbital
p.m.r.	proton magnetic resonance
R	clockwise configuration
S	counterclockwise config.; entropy; net spin of incompleted electronic shells; orbital state
S_N1, S_N2	uni- and bi-molecular nucleophilic substitution mechanisms
S_Ni	internal nucleophilic substitution mechanisms
s	symmetrical; orbital
sec	secondary
soln.	solution
symm.	symmetrical
T	absolute temperature
Tosyl	*p*-toluenesulphonyl
Trityl	triphenylmethyl
t	time
temp.	temperature (in degrees centigrade)
tert	tertiary
U	potential energy
u.v.	ultraviolet
v	velocity
α	optical rotation (in water unless otherwise stated)
$[\alpha]$	specific optical rotation
α_A	atomic susceptibility
α_E	electronic susceptibility
ε	dielectric constant; extinction coefficient
μ	microns (10^{-4} cm); dipole moment; magnetic moment
μ_B	Bohr magneton
μg	microgram (10^{-6} g)
λ	wavelength
v	frequency; wave number
χ, χ_d, χ_μ	magnetic, diamagnetic and paramagnetic susceptibilities
$\sim$	about
(+)	dextrorotatory
(—)	laevorotatory
$\ominus$	negative charge
$\oplus$	positive charge

Chapter 1

General Introduction

The aromatic compounds, as was stated in Volume IA (p. 21) when discussing the classification of carbon compounds, form the second subdivision of the carbocyclic class. They are conveniently again subdivided into

A. mononuclear compounds, *i.e.*, benzene and its homologues and their substitution derivatives;

B. polynuclear compounds, which include
   1. compounds containing two or more benzene rings linked directly, as in biphenyl (phenylbenzene);
   2. di- and poly-arylalkanes and their derivatives;
   3. compounds containing condensed rings in which one or more is benzenoid, as in tetralin, naphthalene, anthracene, etc.

Compounds in which a benzenoid ring is attached to or fused with a heterocyclic ring are included with heterocyclic compounds in Volume IV, with certain obvious exceptions such as some cyclic lactones, cyclic ethers of *o*-dihydroxy compounds and derivatives of *o*-dicarboxylic acids such as their anhydrides and imides.

In the study of aromatic compounds fascinating problems of a kind different from those met with in aliphatic chemistry are encountered, to the solution of which a succession of the most eminent organic chemists have applied themselves during the course of more than a century. The great stability of the benzene ring, the isomerism of its derivatives, the effect of substituents in directing, accelerating or retarding attack by substituting agents, and the genesis of intense colour in certain types of compound were among the effects calling for explanation. These and the practical uses found for many compounds stimulated purely chemical investigations. Although the description "aromatic" is traditionally applied to benzenoid compounds, that is, to those containing a cyclohexa-1,3,5-triene ring, modern research has revealed compounds of marked aromatic character arising from other conjugated cyclic structures [tropylium salts and tropolone (Vol. IIB, p. 359 *et seq.*), having a seven-membered ring, are noteworthy examples]. The introduction of modern physicochemical techniques and theoretical conceptions is enabling us to elucidate and evaluate the fundamental basis of "aromaticity", which is assuming much wider significance than it did earlier. In this chapter these theoretical studies, dealing in particular with the development of ideas on the structure of benzene and the nature of aromatic character in a wider connotation, are discussed by Dr. G. S. Chandler and Professor D. P. Craig. Orientation in electrophilic substitution is dealt with by Professor P. D. B. de la Mare, whilst nucleophilic substitution and homolytic substitution in aromatic compounds are discussed, respectively by Professor N. B. Chapman, and Professor D. H. Hey and Professor G. H. Williams. Finally, an account of the formation of compounds containing the benzene ring from non-aromatic progenitors and of the reverse process, the scission of the benzene ring, is given by Professor Neil Campbell.

In the following chapters benzene and its homologues and their simpler substitution products and substances derived from them, are described and discussed in considerable detail, since their chemistry is to a great extent typical of that of the more complex polycyclic aromatic substances. They serve, moreover, as intermediates for the synthesis not only of more complex homocyclic compounds but also of many heterocyclic compounds to be described in Volume IV. Within the limits of this work it would be impossible to deal in such detail with the higher aromatic hydrocarbons and their derivatives, for every additional ring multiplies the number of possible isomeric derivatives (there are, for example, ten nitronaphthylamines against three nitroanilines) while at the same time introducing new factors influencing their chemical behaviour. The chapters dealing with these classes of compounds must therefore be far more compressed and restricted to a description of what is important fundamentally for their understanding and as a guide to the directions of modern research in their fields.

As regards the *nomenclature* of aromatic compounds, historically it has long been customary to use the substitutive rather than the radico-functional method, which is frequently employed in naming aliphatic compounds; thus C_6H_5Br is bromobenzene, not phenyl bromide. When the first edition of this book was published (1953) strenuous efforts were being made to achieve universal acceptance of an agreed scheme of nomenclature for organic chemistry; however, such agreement had not then been reached and alternative ways of numbering substituent groups were in use, the one favoured in the original Edition of *Rodd*'s Chemistry of Carbon Compounds being that then accepted by the Chemical Society, London. With the publication and general acceptance of the I.U.P.A.C. Rules for Nomenclature of Organic Chemistry (Butterworths, London, Sections A and B, 2nd Edition, 1966, Section C, 1965), definite rules, which are followed here, are set out for the selection of the principal group, the order of priority of other substituents and for the numbering of the various substituents.

1. Aromatic character and the benzene nucleus

G. S. CHANDLER and D. P. CRAIG

(a) Early developments

In 1825 Faraday isolated from coal gas a liquid to which he gave the name "bicarburetted hydrogen". In 1834 Mitscherlich obtained the same substance by distillation of benzoic acid with lime. Their substance, benzene, was later recognised as the parent of the most numerous and industrially important series of compounds in chemistry. The great extensions of this branch of chemistry in later decades was dependent on the success achieved in the development of a theory of the structure of benzene.

In the early 1860's organic chemistry was emerging from the type theory: structural formulae were being increasingly employed. Frankland, Williamson, Kekulé and others had made clearer the rules for formulating structures and the laws of valency largely with reference to organic compounds of the aliphatic series. There also existed a group of "aromatic" substances, not yet numerous, which in composition and behaviour differed from aliphatic compounds, and seemed to form a notable exception to the rules of structure and valency. In 1865 *A. Kekulé* showed how this anomaly might be resolved, pointing out that aromatic compounds could be brought within the established rules, if it were assumed that a nucleus of six carbon atoms was common to all. He then proceeded to his famous hypothesis that the atoms formed a ring of alternate single and double bonds, leaving each atom with one unit of combining power. He believed, though he admitted that he could not prove, that the six combining positions were equivalent (Bull. Soc. chim. Fr., 1865, **3**, [ii] 98; Ann., 1866, **137**, 169; Lehrbuch der organischen Chemie, 1867, II, 493). By 1874 Ladenburg had completed a proof of Kekulé's postulate of six-fold equivalence: the demonstration was improved and simplified by Wroblewsky in 1878, and by Hubner in 1879 (*A. Ladenburg*, Ber., 1869, **2**, 140; Ann., 1874, **172**, 344; *E. Wroblewsky*, Ber., 1872, **5**, 30; Ann., 1873, **168**, 147; 1878, **192**, 196; *H. Hubner* and *A. Petermann*, *ibid.*, 1869, **149**, 129; *Hubner*, *ibid.*, 1879, **195**, 1). Another major problem solved in principle in the same period was to determine the relative positions of two substituents in aromatic compounds. Kekulé suggested that all compounds whose substituents were considered to be in the same relative positions as those of an agreed standard "ortho compound" should be called "ortho", and similarly for other such designations. The designations were used at first without any implication of positional relations. Later it was found that most compounds which had been classified as *ortho* had their substituents in

1,2-positions. Therefore, the prefix *ortho* was reserved for 1,2-compounds, and *meta* and *para* for the 1,3- and 1,4-compounds, respectively.

The third of the major problems with which aromatic chemistry was faced was that of the disposition of the six carbon valencies not required to maintain the ring or hold the attached atoms. This was of a different order of difficulty. The problem of the disposal of the six "spare" valencies was attacked just as energetically as the other problems up to the end of the century, and it was in some respects clarified; but it was not then solved, and we now know that it could never have been solved solely by organic chemistry. Only in the present century, following general scientific advances in physics and chemistry, has this become incorporated into a rational physical theory.

Kekulé's assignment, in 1865, of the six valencies to three double bonds was provisional; and in 1872 he issued a supplementary hypothesis that an oscillation occurs between the two arrangements of the double bonds (*ibid.*, 1872, **162**, 77). His reason was that, as Ladenburg had already pointed out, the original formula admitted unrealised possibilities of isomerism in benzene derivatives. It soon became clear that the additional symmetry introduced by the oscillation hypothesis was demanded by the observed extent of isomerism. Accordingly the oscillation postulate as shown in (I) has since been regarded as an integral part of Kekulé's theory:

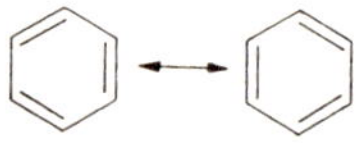

(I)

Quite the most advanced of the early attempts to interpret aromatic stability, while still satisfying the symmetry condition, was that made in 1899 by *J. Thiele* (Ann., 1899, **306**, 125). As a result of a general study of unsaturation he had been led to assume the reactive valencies of double bonds to exist in most circumstances largely bound, but partly also free, and he had shown how in a chain of alternate single and double bonds, there could be *conjugation* between the double bonds, *i.e.*, a spread of affinity over the double bonds and the intervening single bonds, so that at the ends of the chain free affinity was available (II). Thus he interpreted terminal addition to conjugated unsaturated systems in the aliphatic series. He noted that in Kekulé's benzene structure the conjugation is cyclic: that there can be little additive reactivity (III). The symmetry condition was fulfilled by starting with either of the alternative Kekulé structures and allowing the affinity distribution to proceed equally (IV). The outstanding merit of this theory, as C. K. Ingold has emphasised, was that it dealt with aromatic stability on the basis of concepts developed outside the aromatic series.

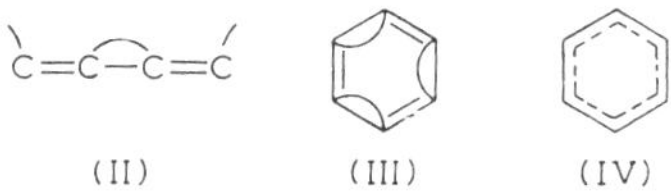

(II) (III) (IV)

The general characteristics of aromatic systems with which a structural theory should be concerned are symmetry, stability, and chemical transformations. Any consideration of the modes in which the benzenoid system suffers disruption indicates appeal in some form to Kekulé's structure. Transformations which destroy the aromatic system without opening the ring, produce either addition products, or *ortho*-quinonoid, or *para*-quinonoid compounds and all such conversions are more easily understood in terms of Kekulé's formula than of the others. But Kekulé's formula can be entertained only in a dynamical system which will provide the necessary symmetry. At a very early period *J. Dewar* had proposed V as a possible formula for benzene (Proc. roy. Soc., Edinburgh, 1866, p. 82). As a static formula it was quite unacceptable, lacking the requisite symmetry. In 1922 *Ingold* revived it (J. chem. Soc., 1922, **121**, 1133), interpreting the *para*-bond as qualitatively comparable to the reactive component of a double bond and he proposed to incorporate Dewar's formula with Kekulé's, each with the number of orientations (2 or 3) needed to fulfil the symmetry condition, in the complex dynamical system VI. The argument focussed attention on the transformations of the aromatic system, the primary object of including the two types of structure being to permit parallel interpretations of *ortho*- and *para*-quinonoid conversions. The lack in the theory is that the stability problem was not dealt with otherwise than through the undeveloped idea, not then capable of theoretical support, that the rapid interconversion of forms might

(V) (VI)

itself be of energetic significance. Lastly among the dominant ideas in this period there is that of the *aromatic sextet* deriving from *E. Bamberger* (Ber., 1891, **24**, 1758; 1893, **26**, 1946; Ann., 1893, **273**, 373), whose argument referred to a stable set of six *valencies* as exhibited in the Armstrong centric formula (VII):

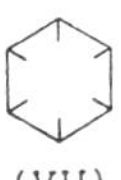

(VII)

In electronic terms (a specially stable set of six *electrons*) it was to be incorporated into the electronic theories of structure and reactions developed by Robinson and Ingold in the mid-1920's. The attempt to represent the chemical properties of benzene in a formula that would conform with the quadrivalency of carbon without the then unknown idea of delocalization had produced Dewar's formula (V), either for the ground state or, in Ingold's refinement, as one component in a mixed representation in which this formula would account for the *para* reactivity of the molecule. The same attempt had many years before also produced Ladenburg's prism formula (VIII) (*Ladenburg*, Ber., 1869, **2**, 140, 272):

(VIII)

Valence isomers of benzene with structures V and VIII have recently been prepared as stable or metastable compounds. The first to be prepared is an isomer of 1,2,4-tri-*tert*-butylbenzene (*S. P. Pappas* and *E. E. Van Tamelen*, J. Amer. chem. Soc., 1962, **84**, 3789). Under ultraviolet irradiation this gives a non-planar hydrocarbon with an ultraviolet spectrum and other properties supporting the formulation as a derivative of Dewar benzene (V). It is reconverted into the parent benzene by heating for a short time at 200°. Subsequently, the Dewar valence isomer of benzene itself has been made (*idem, ibid.*, J. Amer. chem. Soc., 1963, **85**, 3297). This hydrocarbon is metastable with respect to benzene and is converted to it quantitatively at 90°.

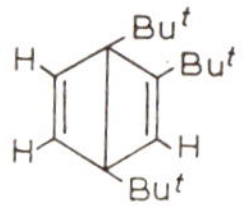

As to "Ladenburg benzene" one of the trimeric polymerization products from *tert*-butylfluoroacetylene is obtained in a pure state by simple gas chromatography of the reaction mixture (*H. G. Viehe et al.*, Angew. Chem., 1964, **76**, 922). The compound is thermally stable up to 250°. It shows no C = C infrared absorption, and support is given to the prism structure by the observation that the n.m.r. spectrum has only one *tert*-butyl proton resonance and one fluorine resonance.

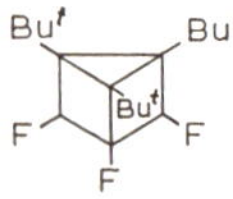

(b) Electronic description of aromatic systems

It will be clear that a well-defined concept of what is characteristic of the chemical behaviour of aromatic molecules was arrived at empirically, and to a degree rationalized, before the discovery of the quantum theory of electrons. The idea of aromaticity grew both before and after this time in relation to the chemical behaviour of particular compounds and also in relation to physical properties in which aromatic molecules were exceptional, such as diamagnetic susceptibility. Aromatic character was at first associated with particular types of reactivity rather than with the properties of molecules in their ground states. Ingold's theory of mesomerism, the earliest successful theory, emphasised this typically chemical viewpoint. The connections between mesomerism and the aromatic sextet and the quantum physics of electrons were shown by *E. Hückel, L. Pauling* and others in a short period around 1930. Two main forms of approximate description of aromatic systems were developed at this time, the valence-bond (VB) and the molecular-orbital (MO) methods; both were physically on secure foundations, and both enabled the calculation of the thermochemical resonance energy, which was a measurable property of the molecular ground state (*Hückel*, Z. Phys., 1931, **70**, 204; 1931, **72**, 310; 1932, **76**, 628; *Pauling* and *G. W. Wheland*, J. chem. Phys., 1933, **1**, 362).

This change in emphasis from chemical manifestations of aromatic character in reactivity, to physical properties, is typical of more recent developments and reflects the greater success of theoretical calculations of ground state properties than those of reactivity and reaction rates. There is a strong tendency to confine the term *aromaticity* to ground state properties (for example, energetics and magnetic properties) and to include qualities seen in reactions in the term *aromatic reactivity*. The real point is that widening knowledge on several fronts makes clear that there is no single measurable property which allows the *degree* of aromatic character to be dealt with in an acceptable way. The common underlying physical characteristic is a cyclic system of delocalized electrons occupying a closed shell of bonding orbitals; the problem is to recognise where such systems exist and to measure how they contribute to chemical and physical properties. In this chapter aromatic character refers to any aspect of behaviour conferred on a system by cyclic delocalization. It is necessarily a vague descriptive term.

A general discussion of aromatic character in the present state of knowledge must go beyond the benzene nucleus. Historically benzene was the first discovered example of stabilisation by electron delocalization. However, aromaticity is by no means confined to the benzene nucleus and its electronic nature is illuminated by the properties of other systems, some of them not even containing carbon atoms, in which similar electronic situations occur.

It was early recognised that the six-electron configuration of the aromatic sextet could occur in both 5-membered and 7-membered rings, in the former as a negatively charged, and in the latter as a positively charged ion. Further extensions, for example in azulene, were accepted, and work over the last 15 years has done much to provide examples of cyclic molecules with numbers of electrons other than six. Two-electron systems are represented by the cyclopropenyl cations (Vol. IIA, p. 36 *et seq.*); cyclodecapentaene with 10 π-electrons has been reported (cf. Vol. IIB, p. 399; *S. Masamune* and *R. T. Seidner*, Chem. Comm., 1969, 542); heptalene (Vol. IIC, p. 73) has 12 π-electrons, and the annulenes (Vol. IIB, p. 390) have upwards of 14. These molecules display varied behaviour, in some cases and in some respects aromatic, but the problem of how to define aromatic character becomes more and more difficult as examples multiply in which characteristics occur together in new and perplexing ways.

The possession of aromatic character by molecules other than benzenoid is one matter. The further extension to include examples in which the electronic nature of cyclic delocalization is different in more fundamental ways from that of benzene has to be more carefully examined. Much interest has been aroused by new knowledge of the phosphonitrilic ring systems and others like them in having rings made of alternating first and second-row elements of the periodic table. These may be formulated in electronic terms that have important points of resemblance to benzenoid systems. They will be discussed later in this chapter (see p. 20).

(c) π-Electron theories

In the π-electron theory of cyclic and conjugated systems each carbon atom provides one electron in a $2p\pi$-type atomic orbital; the neighbour

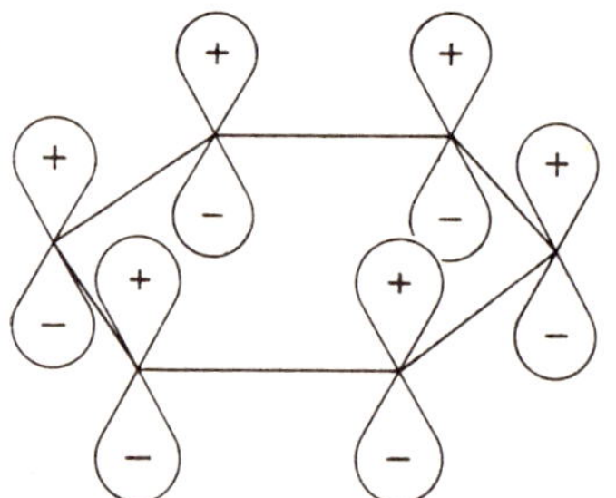

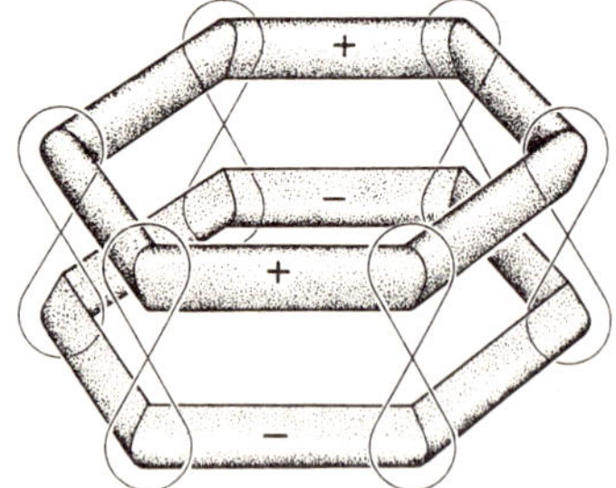

Figure 1
Localized $2p\pi$-carbon orbitals in benzene (left-hand diagram) and delocalized molecular orbital (right-hand diagram).

atomic orbitals overlap and the electrons are delocalized over the whole system of connected carbon atoms (Figure 1).

The atomic orbitals are combined together in molecular orbitals, which are occupied by electrons under the restrictions of the Pauli principle, namely that in each molecular orbital there can be no more than two electrons. The energies of the molecular orbitals of benzene are shown in Figure 2 (a).

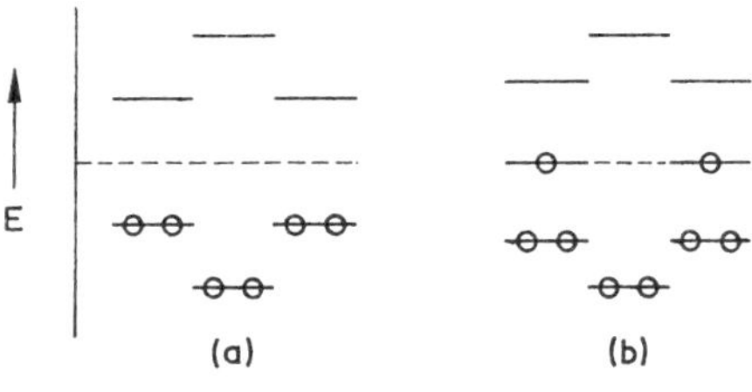

Figure 2
Molecular-orbital energies for (a) benzene and (b) planar cyclo-octatetraene. Electrons occupying the orbitals in the molecular ground state are shown by the circles.

At the lowest energy there is a single molecular orbital. Then there are two degenerate pairs, and at the highest energy a single nondegenerate orbital. In the most stable state of benzene the six π-electrons occupy the three lowest orbitals, giving a closed shell of electrons, in a similar sense to that in which the atomic octet makes a closed shell. A seventh electron in the benzene negative ion would have to occupy a higher orbital and in the benzene positive ion, with five electrons, there must be a vacancy in the topmost occupied orbital. According to the theory both of these ions are chemically unstable in either readily losing or gaining an electron in a chemical reaction. The neutral benzene molecule on the other hand is stable towards electron loss or gain and its six π-electrons are more stable in this configuration than they would be in a structure in which there were three double bonds. Calculations can be made showing that the amount of the extra stability is equal to 2β, where β is the *resonance integral* characteristic of the overlap of two adjacent carbon $2p\pi$-orbitals. The aromatic sextet is thus accounted for in π-electron theory on the ground of the special stability of a closed-shell configuration. Other closed shells in benzene belong to configurations of two electrons and ten, in which either the lowest only, or the lowest five, of the orbitals are occupied. These configurations correspond to benzene carrying four positive charges, or four negative charges, and are physically unrealised.

The molecular orbitals for a cyclic system of eight carbon atoms, as in the planar model of cyclo-octatetraene, are arranged as in Figure 2 (b) differing from the benzene arrangement of orbitals only by the addition of a further pair at the energy zero. Occupation by the eight electrons in a planar model of cyclo-octatetraene would not, however, make a closed shell because there are now only two electrons occupying the upper two molecular orbitals, instead of the four required. According to the π-electron theory in its simplest form the lack of a closed shell in cyclo-octatetraene accounts for the absence of aromatic stability.

Cyclic systems of ten and fourteen electrons again give closed shells. For cycles of even numbers of carbon atoms this alternating pattern of configurations, with and without closed shells, underlies the Hückel rule (*Hückel, loc. cit.*) according to which monocyclic systems of $4n+2$ carbon atoms give stable aromatic-type delocalized sets of electrons, and sets of $4n$ electrons do not. In polycyclic conjugated systems Hückel's rule is of uncertain force. Both the linearly annelated hydrocarbons, (the acenes), and the angularly annelated, (the phenes), have numbers of $4n + 2$ carbon atoms, but for reasons that may have little to do with Hückel's rule. These numbers are determined by the common building block of the six-membered ring system; when rings are joined together with two carbon atoms in common the total number belongs to the set $4n + 2$. The reason for the prime place occupied by the six-membered ring is simply that the 120^0 bond angle at carbon is strainless. We find that when six-membered rings are joined together other than by annelation, as in biphenylene (IX) the number of carbon atoms

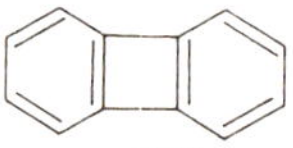

(IX)

may not belong to the $4n+2$ series. In biphenylene the number is 12, and there are many similar examples in which Hückel's rule, if it were thought to apply, would be disobeyed. On the other hand, in the *acenes* and *phenes* all the carbon atoms do lie on the perimeter of the molecule, and Hückel's rule may retain some vestigial force. In molecules with interior carbon atoms, such as pyrene and perylene, there is no basis for the application of the rule and the occurrence of stable $4n$ systems is common. Thus in cases other than the simple monocyclic systems other criteria of stability and of potential aromaticity must be sought. These may depend on the symmetry of the atomic arrangement alone, in which case they may be applied simply by inspection of the supposed molecular structure, or they may be quantitative, depending on calculations of π-electron energies and comparison of the energies of stabilisation with known standards.

An extension can be made to the picture of aromaticity, to allow the concept to go beyond overlapping π-orbitals on adjacent carbon atoms in the cyclic system. If one of the bonds in cycloheptatrienylium cation is broken and replaced by a methylene chain to give the homotropylium ion (X), cyclic delocalization can still occur if the unsaturated carbons at the ends of the saturated chain are close enough for significant overlap of their $2p\pi$-orbitals. Aromatic properties in molecules of this type would be expected to be governed by Hückel's rule. Such systems, extensively investigated by *S. Winstein et al.* (in "Aromaticity", Chemical Society special publication No. 21, London, 1967), are *homoaromatic* and will be discussed more fully later.

$(CH_2)_n$

(X)

π-Electron theory has also recently been used to recognise aromaticity in non-planar bicyclic systems. Bicycloaromaticity has been defined, for example, as the enhanced thermodynamic stability of the hydrocarbon XI, in comparison with an appropriate reference compound XII which possesses the same number of trigonally hybridized carbons, *i.e.*, in which u + v = x + y + z, (*M. J. Goldstein*, J. Amer. chem. Soc., 1967, 89, 6357). Theory has yielded a rule similar to Hückel's for predicting the occurrence of the phenomenon. An ion from XI will be bicycloaromatic if it consists of one

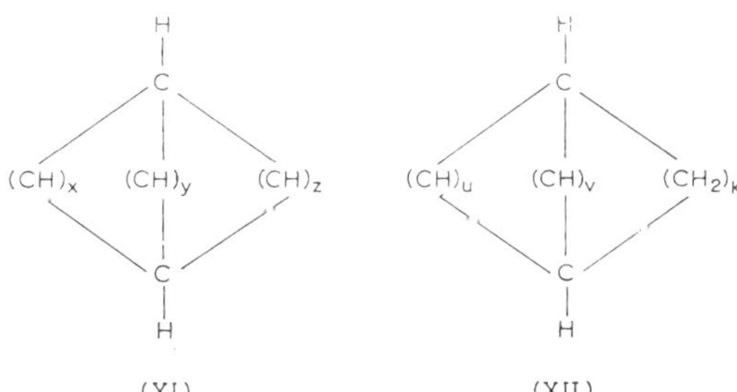

(XI) (XII)

odd-membered bridge and two even-membered bridges in which the total number of π-electrons in the unique odd bridge and one even bridge (the longer one, if they differ) is $4p + 2$ ($p = 0, 1, 2\ldots$). In addition the reference ion must be relatively destabilized, which will occur if the total number of π-electrons is 4n (n = 1, 2, 3). An extension of the idea leads to the prediction of antibicycloaromaticity, where the total number of π-electrons in the ion from XI is $(4n + 2)$ making it destabilized relative to XII. Evidence for bicycloaromaticity is scant as yet, being based on the stability of the 7-norbornadienyl cation (*Winstein* and *C. Ordronneau, ibid.*, 1960, **82**, 2084), and the lack of succes shown by attempts to prepare a number of ions predicted to be antibicycloaromatic (*R. A. Finnegan* and *R. S. McNees*, J. org. Chem. 1964, **29**, 3234; *Goldstein* and *B. G. Odell*, J. Amer. chem. Soc., 1967, **89**, 6356; *A. S. Kende* and *T. L. Bogard*, Tetrahedron Letters, 1967, 3383).

Extension of the term "aromaticity" to include certain inorganic compounds is suggested by a similarity in the electronic delocalization that may take place. Their importance in the present discussion is in showing that the conditions in carbon compounds are not the only ones that may lead to cyclic delocalization and aromatic-type stability (*D. P. Craig* and *N. L. Paddock*, Nature, 1958, **181**, 1052). Figure 3 illustrates how a cyclic set of alternating first- and second-row elements, as in the phosphonitrilic compounds, may have electron delocalization by the overlapping of adjacent $p\pi$- and $d\pi$-orbitals. This in itself is capable of several variations depending on whether one or both of the $d\pi$-orbitals at the second-row element is involved. Delocalization of this kind is different in important respects from that in carbon compounds, notably in the fact that there is no alternation of properties with ring size, and aromatic-type properties may be expected to appear in rings of any even number of elements. This is confirmed experimentally. There is no reason to believe, however, that in terms of resonance energies or other static properties the degree of stabilisation in these cases will be so great as that in aromatic carbon compounds, but it is the case that such cyclic inorganic molecules are remarkably more stable chemically than their open chain relatives.

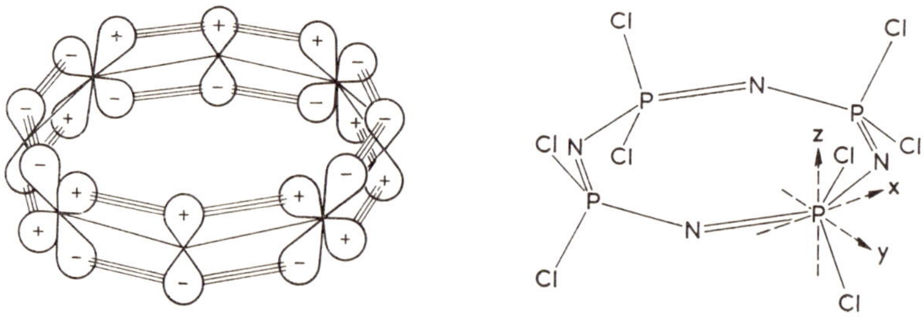

Figure 3

Overlapping $p\pi$- and $d\pi$-orbitals in $(NPCl_2)_4$. Each phosphorus atom also has a second $d\pi$-orbital related to the one illustrated by a rotation of 90° about an axis normal to the plane.

(d) *Molecular symmetry and the aromatic-pseudoaromatic classification*

Early studies of π-electron energies of cyclic and other conjugated systems were made both by the molecular-orbital method and by the valence bond method. Both have undergone important refinements since. In the valence-bond method the molecular wave function is supposed to be built by using only those atomic electron configurations with one electron assigned to each carbon $2p\pi$-orbital. This leads to a basis of "structures" representing extreme modes in which the electrons on different carbon atoms may be paired together to form bonds. The two Kekulé structures (I) of benzene symbolise pairing schemes of this kind, to be supplemented by pairing schemes according to the three Dewar structures (XIII), with two bonds between adjacent atoms and one long bond between *para*-atoms.

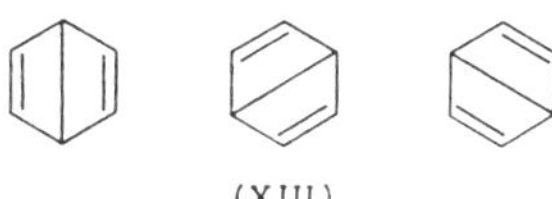

(XIII)

The ground state of benzene is described in this approximation as a linear combination of the bond eigenfunctions describing each one of the structures I and XIII. When this method of calculation is applied to other hydrocarbons the results fall into two classes. In one, represented by benzene and the common aromatic molecules, the electronic wave function for the ground state is totally symmetrical in the sense that it is unchanged under all the symmetry operations that send the nuclear framework of the molecule into itself. In some other (pseudoaromatic) molecules the valence-bond ground state of the molecule is non-totally symmetrical in that it changes sign under some of the operations of the nuclear framework. In these cases the electrons are coupled together differently from the standard aromatic molecules, and the mere fact that they are cyclic and that they may be written in structures with alternating double and single bonds in more than one way does not imply aromatic-type stabilisation. According to this classification the molecules XIV–XVI illustrated below are not expected to be aromatic in type.

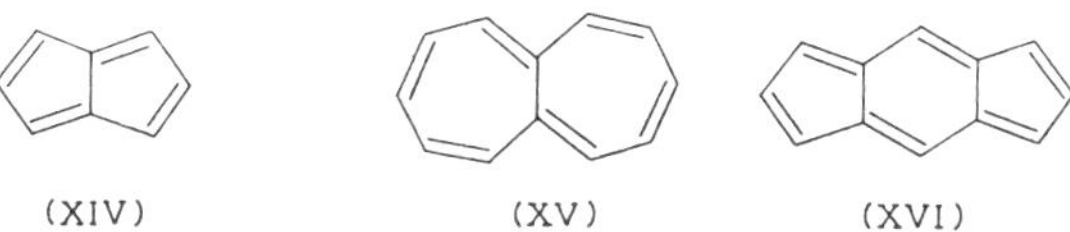

(XIV) (XV) (XVI)

The properties of heptalene (XV) (*H. J. Dauben* and *D. J. Bertelli*, J. Amer.

chem. Soc., 1961, **83**, 4659) *s*-indacene (XVI) (*K. Hafner et al.*, Angew Chem., 1963, **75**, 35) and the recently synthesized 1,3-bis(dimethylamino)pentalene (*Hafner, K. F. Bangert* and *V. Orfanos*, Angew. Chem. internat. Edn., 1967, **6**, 451) confirm these predictions insofar as the molecules are polyolefinic in behaviour with proton resonances showing no evidence of ring currents. Simple dynamical considerations suggest that a molecule not substantially stabilised by aromatic-type delocalization will not have equal carbon-carbon bond lengths. In the absence of such stabilisation the molecular energy will be less if the carbon-carbon bonds adopt the lengths of alternating double and single bonds. There is evidence that this occurs.

Rules for applying the valence-bond criterion have been given (*Craig*, J. chem. Soc., 1951, p. 3175; "Non-Benzenoid Aromatic Hydrocarbons", Interscience, New York, 1959, Chap. 1.). In some molecules in which the carbon atoms do not all lie on the perimeter as in pleiapentalene (XVIII), and aceheptylene (XVII), azulenoheptalene (XIX) and azupyrene (XX), application of the symmetry criterion may give ambiguous results, and the symmetry of the valence-bond ground state can only be found by an energy calculation:

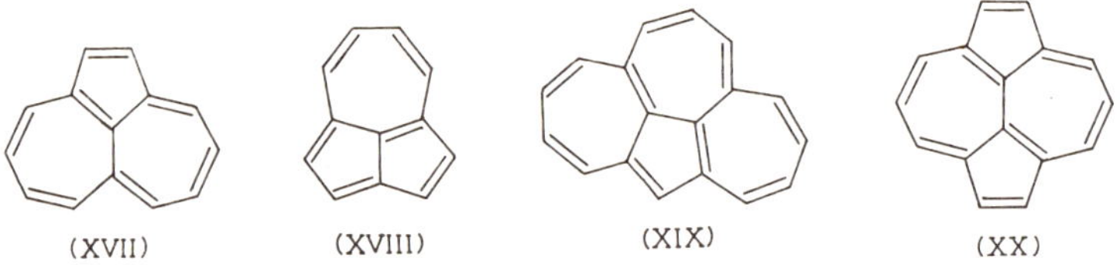

In the these examples the lowest symmetric and nonsymmetric states fall closely together, a result which itself suggests electronic behaviour different from that in normal aromatics.

Energy calculations using the valence-bond method and molecular-orbital investigations have indicated that XVII should be aromatic having a totally symmetric ground state and XVIII, with a non-totally symmetric ground state, should be pseudoaromatic (*M. A. Ali* and *C. A. Coulson*, Mol. Phys., 1961, **4**, 65). The matter is not settled, since more sophisticated calculations using self consistent field techniques with configuration interaction (*Ali* and *N. K. DasGupta*, Theoret. Chim. Acta, 1966, **4**, 101) indicate that both XVII and XVIII should be aromatic, having totally symmetric ground states. Methylated derivatives of aceheptylene and pleiapentalene were first synthesized by *Hafner* and *J. Schneider* (Ann., 1959, **624**, 37). 3,5,8,10-Tetramethylaceheptylene has a planar structure with peripheral bond lengths between 1.451 Å and 1.364 Å (*E. Carstensen-Oeser* and *G. Habermehl*, Angew. Chem. internat. Edn., 1968, **7**, 543), the dimethyl derivative behaves like an azulene with a conjugated diene bridge which participates in the resonance of the bicyclic system (*Hafner, ibid.*, 1964, **3**, 165) and has been considered to be aromatic (*D. Lloyd*, "Carbocyclic Non-Benzenoid Aromatic Compounds", Elsevier, Amsterdam, 1966, p. 210), while the pleiapentalene has the characteristics of a polyolefin (*Hafner, loc. cit.*). By way of contrast the diamagnetic susceptibility exaltation of 5,7-dimethyl-2-phenylpleiapentalene (30 units) and 3,5-dimethylaceheptylene (0 units) (*Dauben, J. L. Laity* and *J. D. Wilson*, J. Amer. chem. Soc., 1969, **91**, 1991) indicate the opposite conclusion as do the proton magnetic resonances (see *D. E. Jung*, Tetrahedron, 1969, **25**, 129). Of the other two compounds azuleno[8,8a,1,2-*def*]heptalene, has high field ring proton resonances (*Hafner, G. Hafner-Schneider* and *F. Bauer*, Angew. Chem. internat. Edn., 1968, **7**, 808), but azupyrene has the low field resonances typical of an aromatic compound (*A. G. Anderson, A. A. MacDonald* and *A. F. Montana*, J. Amer. chem. Soc., 1968, **90**, 2993). Thus conclusions concerning the aromaticity of some members of this class of molecules must await further investigation. In such manifestly borderline cases a graded sequence in the degree of aromatic character is expected rather than a hard division into distinct classes of aromatics and others.

(e) Molecular-orbital calculations

Molecular-orbital calculations of pseudoaromatic systems like pentalene and heptalene made with the assumption of a symmetrical structure lead to results in conflict with those of valence-bond theory, in that different electronic states are held to be the lowest in the two theories (*Craig* and *A. Maccoll*, J. chem. Soc., 1949, p. 964). In normal aromatics no such difference occurs. The conflict has been resolved by the demonstration that the symmetrical nuclear arrangement (D_{2h} symmetry) is unstable in the ground state, and that a lower symmetry corresponding to bonds alternating in length in the manner of a single Kekulé structure is the stable one for heptalene and for a number of other systems (*H. C. Longuet-Higgins* and *L. Salem*, Proc. roy. Soc. London, 1959, **A251**, 172; *ibid.*, 1960, **255**, 435; *P. C. den Boer-Veenendaal et al.*, Tetrahedron, 1962, **18**, 1325; *L. C. Snyder*, J. phys. Chem., 1962, **66**, 2299; *T. Nakajima* and *S. Katagiri*, Mol. Phys., 1963, **7**, 149). Moreover, by allowing for bond alternation it is possible to get improved agreement between calculated and experimental resonance energies for a wide variety of cyclic conjugated systems both aromatic and pseudoaromatic.

The calculations depend upon a *bond alternation parameter k*, given as the ratio $k = \beta_s/\beta_d$ of resonance parameters β_s and β_d for single and double bonds in a Kekulé structure for the molecule. A trial value of k is taken and the Hückel secular equations solved. The solutions give bond orders, which are used to deduce a set of bond lengths r according to the linear relation

$$r = 1{\cdot}520 - a\rho$$

where 1·520 Å is the length of a C–C bond of zero π-bond order, a is a constant, and ρ the bond order. After *Nakajima* ("Molecular Orbitals in Chemistry Physics and Biology", Academic Press, New York, 1964, p. 451) a is taken to be 0·186 Å. The resonance integral β for the bond is found for the length r by using an exponential relation,

$$\beta(r) = \beta^0 \exp[b(1{\cdot}397 - r)]$$

where β^0 is the value in benzene ($r = 1{\cdot}397$ Å) and $b = 4{\cdot}599$ Å^{-1} (*Nakajima, loc. cit.*). Then the total molecular energy V, including both π- and σ-electrons, is found from the expression

$$V(k) = (2/ab) \sum_{i<j} \beta_{ij} + \sum_i q_i\alpha_i + \text{const.}$$

where β_{ij} is the resonance integral for the bond between atoms i and j. q_i is the electron density and α_i the coulomb parameter for the i-th atom,

$$\alpha_i = \alpha^0 + (1 - q_i)\beta^0$$

The calculation is repeated for several different k values, and an optimum $k = k_{min}$ is found from the minimum of the curve of V against k. For benzene $k = 1$, corresponding to equal lengths for all C–C bonds, and for butadiene $k = 0{\cdot}55$ (bond lengths 1·34 and 1·47 Å). Calculations for k for nonbenzenoids are given in Table 1 taken from *Nakajima* (*loc. cit.*)

TABLE I

BOND-ALTERNATION PARAMETERS FOR NON-BENZENOIDS
(*Nakajima*, 1964)

Molecule	k_{min}	Molecule	k_{min}
Fulvene	0.56	*s*-Indacene	0.59
Pentalene	0.55	Acepentylene	0.54
Heptalene	0.58	Pleiaheptalene	0.60

The Table includes two unknown molecules, acepentylene (XXI) and pleiaheptalene (XXII). In all six systems a degree of bond alternation similar to that of butadiene is expected.

(XXI) (XXII)

Such studies may be further generalised (*G. Binsch* and *E. Heilbronner*, "Structural Chemistry and Molecular Biology", Eds. *A. Rich* and *N. Davidson*, Freeman, San Francisco, 1968, p. 816) to remove the assumption that alternation is the favoured pattern of length variation in a conjugated molecule and to allow the pattern itself to appear from the theory. It is also desirable to distinguish causes of bond-length differences that are allowed for in the bond order from others which are not, such as those affecting lengths in extended polymers. In the more general theory of *Binsch et al.* (*loc. cit.*) the total energy is the sum of independent terms for σ- and π-electrons,

$$E = E_\sigma + E_\pi$$

The molecular framework is built up from a σ-bonded skeleton with all C–C distances equal to a common value R_0. The system of overlapping carbon $2p$-orbitals is then filled with its complement of electrons, which distort the bond lengths by individual increments $\Delta R_{\mu v}$. The most favoured distortion is calculated as the one giving the lowest energy. The energy is expanded in a Taylor series, about R_0, where the sums are over bonded atoms.

$$E = E(R_0) + \sum_{\mu < v} \left(\frac{\partial E}{\partial R_{\mu v}}\right)_{R_0} \Delta R_{\mu v} + \tfrac{1}{2} \sum_{\mu < v} \sum_{x < \lambda} \left(\frac{\partial^2 E}{\partial R_{\mu v} \partial R_{x\lambda}}\right)_{R_0} \Delta R_{\mu v} \Delta R_{x\lambda} + \text{higher terms}$$

The first-order terms lead to the usual bond-order bond-length relation (see before). By including second-order terms a criterion can be found for the relative importance of second-order effects in the π-energy and the σ-compression energy, so giving a criterion for the appearance of second-order distortions. These will in general cause the molecule to lose its original symmetry.

Application of the theory to pentalene and heptalene show that second-order bond distortions will occur, and that in these examples bond alternation is the most favourable distortion. Bond alternation is not however favoured for all systems (*Binsch* and *Heilbronner*, Tetrahedron, 1968, **24**, 1215).

(f) Delocalization in "Mcbius strip" molecules

The overlap of $p\pi$-orbitals depicted in Figure 1 and occurring in aromatic carbon compounds, and that of alternating $p\pi$- and $d\pi$-orbitals (Figure 3) in certain cyclic inorganic molecules belong to delocalized systems with different characteristics. In a classification according to the types of atomic orbital able to participate in delocalized π-bonding they are the only types possible. However, a chain of π-orbitals may be connected end-to-end in two different ways, defined in relation to the symmetry planes of successive orbitals. In an ordinary carbon aromatic such as benzene, or a $p\pi$-$d\pi$ system such as $(PNCl_2)_3$, the atoms of the ring lie in one plane, and this plane is the common symmetry plane of all the atomic π-orbitals. The ring may be thought of as formed from an open chain by joining the local symmetry planes of the π-orbitals to form an ordinary two-sided surface.

Figure 4
Schematic diagram of π-atomic orbitals in a planar set of ten carbon atoms, with local planes folded to form a "Mobius molecule".

The alternative (*Heilbronner*, Tetrahedron Letters, 1964, p. 1923; *S. F. Mason*, Nature, 1965, **205**, 495) is that the local symmetry planes are connected to form a Mobius single-sided surface as in Figure 4, so that the local axes of the π-orbitals are twisted through 180^0 in passing once round the closed cycle of atoms. In a large-ring compound of the type of the annulenes C_nH_n, and the larger phosphonitrilic rings $(PNCl_2)_n$, conformations of this type are possible so far as steric considerations go, and π-electron energetics are on that account of interest. If only nearest neighbour interactions count, as usual in simple π-electron theories, it is unimportant whether the ring atoms are coplanar. The connection of local atomic symmetry-planes is topologically equivalent, and the delocalization behaviour similar to one or another of the two types.

If the two *symmetry* types (represented as examples of $p\pi$-$p\pi$ and $p\pi$-$d\pi$ systems) are combined with the two *topological* types there are in all, four kinds of delocalization behaviour. The classification by symmetry types is into homomorphic and heteromorphic classes according to whether the atomic orbitals all belong to the same local symmetry species (*e.g.*, $p\pi$-$p\pi$ or $p\pi$-d_{yz}) or to different species ($p\pi$-d_{xz}) (*Craig*, J. chem. Soc., 1959, p. 997). The classification by topological species is into normal and Mobius molecules. The differences between the types may be brought out in simple molecular-orbital theory with the assumption of equal electronegativities for the atoms concerned, and equal resonance integrals β. Following *Mason (loc. cit.)* we find:

(1) *Normal homomorphic delocalization.* The total π-electron energy for the n electron system where n/2 is odd is given by expression [1]

$$E_\pi = \text{n}\alpha + 4\beta \operatorname{cosec} (\pi/\text{n}) \ldots \qquad [1]$$

and for n/2 even by the smaller quantity [2]

$$E_\pi = \text{n}\alpha + 4\beta \cot (\pi/\text{n}) \ldots \qquad [2]$$

(2) *Normal heteromorphic delocalization.* The example of this type is the phosphonitrilic ring system with d_{xz} orbitals. The energies of the π-electrons are given for n/2 either even or odd by the smaller of the two expressions under (1) namely [2]. The delocalization energy per π-electron does not oscillate as in case (1) according to the parity of n/2, but changes steadily.

(3) *Mobius homomorphic delocalization.* To make an example of this type it is only necessary to suppose that a cyclic polyene of type (1) is broken at any one bond, and the ends rejoined after twisting one of them through 180°. The behaviour again differs with the parity of n/2, but in the opposite sense from case (1). Now systems of π-electrons where n/2 is odd have the lower energies given by expressions [2] and those with n/2 even the higher energies [1]. Of course in this case the neighbour resonance integral has a smaller value than before, because the axes of neighbour orbitals are inclined to one another by π/n, and the appropriate resonance integral is $\beta \cos \pi/\text{n}$.

(4) *Mobius heteromorphic delocalization.* The case is exemplified by a phosphonitrilic ring system broken and rejoined after a 180° twist. For all even numbers of π-electrons the energy is given by the greater of the two expressions, namely (2). The conclusions are summarised in Table 2, in which the energy-dependence on electron number n is expressed as "cosec" or "cot" according to whether formula [1] or [2] applies.

The use of the forms of expression for molecular-orbital energies for

TABLE 2

MOLECULAR-ORBITAL ENERGIES IN THE FOUR DELOCALIZATION TYPES

Local Symmetry Type \ Topological Type		Normal	Mobius
Homomorphic	n/2 odd	cosec	cot
	n/2 even	cot	cosec
Heteromorphic		cot	cosec

classification does not imply that the energies given should themselves be taken very seriously. Allowance would certainly have to be made for variations of bond length and for differences of electronegativity.

Although Mobius molecules have no symmetry the distinctions between the four cases (1)–(4) may be made starting from a basis of linear polyenes built on the two site-symmetry species and imposing cyclic boundary conditions adapted to the two topological types. An n-atom polyene folded into a normal cyclic system (Figure 4) has its cyclic symmetry covered by the group C_n, with representations

$$\chi_l = e^{2\pi i l/n} = 0 \pm 1, \pm 2 \ldots n/2 \ldots \qquad [3]$$

These are representations to which molecular orbitals must conform. If n/2 is odd one half of the molecular orbitals are doubly filled, giving a closed shell configuration. If n/2 is even two molecular orbitals (belonging to l = n/4 in [3]) are incompletely filled, giving an open shell configuration. The energy sums over occupied orbitals in these two situations lead to expressions [1] and [2].

In the Mobius molecules the linear polyene to be closed into a cycle is in the form of a helix of period 2n. The symmetry group of the infinite helix has as a factor group the cyclic group of order 2n, and the Mobius molecule can be simulated as to symmetry by applying boundary conditions at points in the helix n atoms apart. The representations of the cyclic group of period 2n are

$$\chi_l = e^{2\pi i l/2n} = e^{\pi i l/n} \qquad [4]$$

and only those representations can occur with a change of sign in the repeat period of atoms, namely $l = \pm 1, \pm 3, \pm n-1$. Molecular orbitals transforming according to these representations fall in degenerate pairs throughout, and are thus different from those transforming as expression [3].

In adopting a similar point of view for heteromorphic systems one takes the structural unit to be a pair of atoms. In a system of n atoms there are n/2 structural units. In the Mobius heteromorphic molecules the boundary condition to be imposed along the helix is that for n/2 odd only representations with the same sign at points n/2 units apart occur, and for n/2 even only representations with opposite sign. Thus, with reference to the representations [4] the permitted values of l are:

$$n/2 \text{ odd: } l = 0, \pm 2, \pm 4 \ldots$$

$$n/2 \text{ even: } l = \pm 1, \pm 3, \ldots$$

The former give the same representations as those in [3] with the same pattern of non-degenerate and degenerate states, and the latter are the same as those for the homomorphic Mobius case. It must be noticed that in the normal molecules the symmetry analysis is based on cyclic symmetry with respect to the principal n-fold axis of the molecule itself. In the Mobius molecules the analysis uses the helix axis of the parent linear helix.

(g) Physical criteria of aromatic character

(i) Resonance energy

Benzene and the condensed aromatic molecules are thermochemically more stable and less reactive than molecules with alternate double and single bonds are expected to be. Broadly benzene is more stable by 36 kcal per mole than if it had three double bonds of the cyclohexene type; or as defined by *M. J. S. Dewar, A. J. Harget* and *N. Trinajstić* (J. Amer. chem. Soc., 1969, 91, 6321) more stable than 1,3,5-hexatriene. The resonance energy goes up in the larger aromatics roughly in proportion to the number of π-electrons, but at the same time they are generally more reactive, emphasising the distinction between the chemical and physical viewpoints and showing that inertness and resonance stabilisation do not always go together.

Most measurements of resonance energies depend on heats of combustion; the more accurate minority on heats of hydrogenation. In the combustion method the resonance energy is the difference between the heat of combustion found by burning the compound, and a calculated value equal to the sum of contributions by the atoms in the molecule according to their state of combination in the most stable classical structure. The bond heat contributions themselves are found by analysis of combustion results. There has been controversy over whether conventional methods of calculating resonance energy over-estimate its magnitude by assigning to delocalization some terms really assignable to changes in single bond energy associated with hybridisation and the relief of strain in a single bonded system (*Dewar* and *H. N. Schmeising*, Tetrahedron, 1959, 5, 166; 1960, **11**, 96; *R. S. Mulliken*, *ibid.*, 1959, 6, 68). The balance of these arguments does not yet seem finally settled. Several different sets of bond heat contributions have been proposed leading to different resonance energies for the same molecule. *H. D. Springall*, *T. R. White* and *R. C. Cass* have proposed a set of averaged values (Trans. Faraday Soc., 1954, **50**, 815) and *F. Klages* a set of environment-dependent values (Ber., 1949, **82**, 358) for the analysis of combustion results. Both sets give resonance energies for benzene equal to that found from hydrogenation, but diverge in other molecules. A comparison is given in Table 3.

TABLE 3

RESONANCE ENERGIES OF BENZENOID AROMATICS
(kcal/mole)

Molecule	From averaged bond constants	From environment-dependent bond constants
Benzene	36	36
Naphthalene	67	61
Anthracene	99	86
Phenanthrene	112	99
Biphenyl[a]	81	74
Bibenzyl	78	70

[a] Based on a value of 1510·7 kcal/mole for the gaseous heat of combustion.

Comparison of the two columns in the table shows how sensitive calculations of resonance energies are to small changes in bond constants. This sensitivity explains why the resonance energy concept has remained incompletely defined in spite of efforts at final specification and why much attention, often misplaced, has been given to seeming anomalies that have no physical origin in π-electron effects, but are the result of the accumulation of small differences in bond constants. A fuller discussion of these points is given elsewhere ("Non-benzenoid Aromatic Compounds", *loc. cit.*). Among non-benzenoid compounds resonance energies have been measured in several examples (Table 4).

TABLE 4

RESONANCE ENERGIES OF NON-BENZENOID MOLECULES (kcal/mole)

Molecule	Resonance energy
Dimethylfulvene[a]	12
Azulene[a]	31
Cyclo-octatetraene[a]	4
Biphenylene[a]	20
Tropolone[a]	21
Heptafulvene[b]	13
Heptafulvalene[b]	28

[a] Combustion values based on environment-dependent bond constants.
[b] From heats of hydrogenation (*R. B. Turner*, "Theoretical Organic Chemistry", Academic Press, New York, 1959, p. 67).

Where the single-bonded framework is strained, as in five- and seven-membered rings, the values are differences between the π-resonance energy and the strain energy. If, in these cases, an allowance of 10 kcal/mole is made for the strain energy, the net resonance energy applicable to π-electron effects is about 3 to 5 kcal/mole per electron compared with 5 to 7 for benzenoid hydrocarbons. Cyclo-octatetraene has a resonance energy calculated by these methods of about 4 kcal/mole, which is less than twice that of butadiene, and less than one kcal/mole per π-electron, a small value in agreement with its polyolefinic chemical character. In this case the calculated energy can have little relevance to cyclic conjugation because in the now accepted tub structure the orbitals at the ends of each single bond are at right angles and cannot overlap.

(ii) Diamagnetic susceptibilities

The diamagnetic susceptibility of a molecule is approximately an additive property of its constituent atoms and bonds with corrections for conjugation. These corrections, which are differences between measured susceptibilities and the summed contribution by atoms and bonds, are in part contributed by delocalized electrons and form a useful auxiliary criterion of aromatic character.

In a magnetic field atomic electrons acquire a small angular momentum and a magnetic moment associated with it in a way analogous to that in which current and an associated magnetic field are induced in a circular wire perpendicular to a changing magnetic field. The induced field opposes the applied field and is proportional to the cross-sectional area, that is, to the square of the "radius" of the orbital of the electron. In molecules the simple extension of the spherical atom picture is made complicated by alterations to atomic electrons as the result of bonding. An atom in a molecule is no longer spherically symmetrical and the induced magnetic fields are not anti-parallel to the external field, but point in other directions fixed by the spatial relation of the molecule to the external field. In cyclically conjugated molecules, relatively large magnetic moments are induced in the delocalized system which can be ascribed to circulation of the π-electrons round the ring, embracing an area much larger than that of the atomic orbitals themselves — the ring currents (see *L. Salem*, "The Molecular Orbital Theory of Conjugated Systems", Benjamin, New York, 1966, p. 177). Although criticized by *J. I. Musher* (J. chem. Phys., 1965, **43**, 4081) this view leads to values in agreement with experiment (*J. Nowakowski*, Theor. Chim. Acta, 1968, **10**, 79). The ring current gives two measurable effects, a contribution to the susceptibilities averaged over all field directions and a large magnetic anisotropy, where the susceptibility perpendicular to the plane of the ring is greater than the susceptibilities measured in the plane of the ring. The first effect can be approached using empirical methods based on a set of constants obtained by Pascal and modified by later writers (see for example, *C. K. Ingold*, "Structure and Mechanism in Organic Chemistry", Cornell U.P., Ithaca, New York, 1953, p. 188; *W. Haberditzl*, Angew. Chem. internat. Edn., 1966, **5**, 288). The measured average diamagnetism of aliphatic and unsaturated hydrocarbons can be fitted remarkably closely to the sums of contributions by atoms and bonds, there being only a small though definite exaltation by conjugated double bonds in, for example, polyenes. Gross discrepancies are found in aromatic hydrocarbons. For example, the exaltation in benzene arises from the spread of electron orbitals round all six carbon atoms which include an area many times larger than that of the atomic orbitals of carbon itself.

In condensed aromatic systems the exaltation increases roughly linearly with the number of benzenoid rings, a result that may be interpreted in terms of an increase in molecule area bounded by the electron "circuit". Typical values of the exaltation of diamagnetism ($-\Delta\chi$) and its value per benzenoid ring are given in Table 5. Two of the nonaromatic systems, 9,10-dimethyldibenzopentalene and [16]annulene, show significant negative exaltation. These data support other theoretical and p.m.r. evidence (see

TABLE 5

SUSCEPTIBILITY EXALTATIONS IN HYDROCARBONS*
(c.g.s., units times 10^{-6})

Molecule	$-\chi$ (Expt.)	$-\chi$† by summing	$-\Delta\chi$	$-\Delta\chi/n$
Benzene	54.8	41.1	13.7	13.7
Naphthalene	91.9	61.4[a]	30.5	15.3
Anthracene	130.3	81.7[a]	48.6	16.2
Chrysene	167	102[a]	65	16.3
Azulene	91.0	61.4	29.6	
1,6-Methano[10]annulene	111.9 ± 0.4	75.1	36.8	
9,10-Dimethyldibenzopentalene	132	146[b]	−14	
[16]Annulene	105 ± 2	110	−5	
3,5-Dimethylaceheptalene	112 ± 3	112	0	
Cyclo-octatetraene	53.9	54.8	−0.9	

* Taken from *H. J. Dauben, J. D. Wilson* and *J. L. Laity*, J. Amer. chem. Soc., 1969, **91**, 1991.

† Based on constants compiled by *W. Haberditzl*, Angew. Chem. internat. Edn., 1966, **5**, 288.

(a) Includes correction for annelation effects.

(b) Benzene-ring exaltation included.

p. 28) that annulenes containing 4n π-electrons (*J. A. Pople* and *K. G. Untch*, J. Amer. chem. Soc., 1966, **88**, 4811) and heptalene and pentalene can sustain paramagnetic ring currents.

Magnetic anisotropy has been very difficult to measure in the past, requiring single crystals in which the orientation of the molecules is known, for its determination. Developments in the observation of the Zeeman effect in rotational spectra have made the effect more accessible (*W. Hüttner* and *W. H. Flygare*, J. chem. Phys., 1967, **47**, 4137; *Hüttner, M. K. Lo* and *Flygare, ibid.*, 1968, **48**, 1206). In aromatic systems the susceptibility perpendicular to the ring is indeed the largest, whereas in other unsaturated cyclic molecules the effect is considerably smaller. Values for the diamagnetic anisotropy are shown in Table 6. Anisotropies are large, as are exaltations, compared with the total susceptibility and they both should, therefore, be a sensitive measure of cyclic conjugation and of aromatic character, in cases where there is no confusion from competing diamagnetic and paramagnetic effects. Thus the first clear physical evidence of lack of aromatic character in cyclo-octatetraene came from an application of average susceptibilities (*R. C. Pink* and *A. R. Ubbelohde*, Trans. Faraday Soc., 1948, **44**, 708). The very small exaltation is incompatible with any large peripheral circulation of electrons, a result in harmony with other indications of the polyolefinic character of cyclo-octatetraene.

TABLE 6

SUSCEPTIBILITY ANISOTROPY IN CONJUGATED MOLECULES*
(c.g.s., units times 10^{-6})

Molecule	$-[\chi_{zz} - {}^1/_2(\chi_{yy} + \chi_{xx})]$
Benzene	59.7
Thiophene	50.1
Pyrrole	42.4
Furan	38.7
Cyclopropene	17.0
1,3-Cyclohexadiene	7.4

The z axis is perpendicular to the molecular plane.

* Values taken from *J. M. Pochan* and *W. H. Flygare,* J. Amer. chem. Soc., 1969, **91**, 5928.

(iii) Nuclear magnetic resonance

The circulation of π-electrons induced by a magnetic field has another consequence that can be exploited in measurements of proton magnetic resonance. As illustrated in Figure 5 the induced circulation of π-electrons produces a field which, when measured at points within the hexagon, is opposed to the external field, but at points outside it gives an increment so that the total field is greater. The proton magnetic resonance is a sensitive

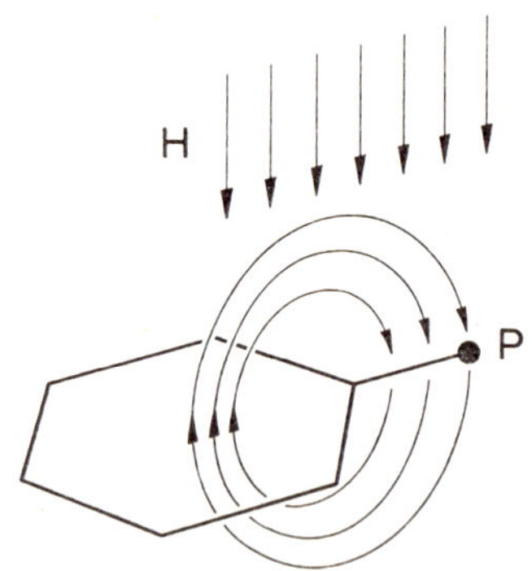

Figure 5
External field **H** and induced field of the aromatic ring current, showing the increased net field at the proton of an attached hydrogen atom.

index of net field and allows a change produced by aromatic electron circulation to be detected. In benzene the protons are external to the ring and give n.m.r. signals at lower applied fields. The protons lying inside the ring system of, for example, [18]annulene give signals at higher applied fields. Proton resonance has been widely and effectively used to detect cyclic delocalization, especially in the annulenes (see p. 38).

The magnitude of the ring current contribution to the shift in proton resonance is estimated by comparing the resonances in benzene and in compounds with localized double bonds in an environment as similar as possible, such as cyclohexa-1,3-diene. This gives the down-field shift as 1·48 p.p.m. (*J. S. Waugh* and *R. W. Fessenden*, J. Amer. chem. Soc., 1957, **79**, 846). Comparison with the mean of cyclo-octatetraene and cyclo-octatriene gives 1·55 p.p.m. (*J. A. Elvidge* and *L. M. Jackman*, J. chem. Soc., 1961, p. 859). These values are acceptably in agreement with theoretical estimates (see *Pople, W. G. Schneider* and *H. J. Bernstein*, "High Resolution Nuclear Magnetic Resonance", McGraw Hill, New York, 1959, p. 181). A magnetic field $\mathbf{H}_0$, perpendicular to a planar aromatic molecule causes the π-electrons to circulate about the lines of force with cyclic frequency $e\mathbf{H}_0/4\pi mc$. The magnetic effect of the corresponding current may be simulated by a magnetic dipole at the centre of the ring. If R is the distance from the centre of the ring to the proton, and a the radius of the path of the circulating electrons the ring current contributions to the mean screening constant can be shown to be, approximately,

$$\frac{e^2a^2}{12mc^2\mathrm{R}^3} \qquad [5]$$

for each π-electron. For benzene, with six π-electrons, this gives a shift of 1·75 p.p.m.

Since the ring current contribution is often one of several comparable terms in a measured chemical shift, its evaluation may be a delicate matter. Induced local atomic currents, which differ in aromatic and non-aromatic compounds because of differences in hybridization, make a considerable contribution to the down field shift in aromatics (*A. F. Ferguson* and *Pople*, J. chem. Phys., 1965, **42**, 1560). Electron densities also have an effect, thus the chemical shift values for cyclo-octatetraene and its dianion are almost identical (*T. J. Katz*, J. Amer. chem. Soc., 1960, **80**, 3784). However, *Elvidge* and *Jackman* (*loc. cit.*) analysed a number of measured spectra in support of the proposal that the ring current contribution should be taken as a quantitative measure of aromaticity, assigning for example to γ-pyridone 35% of the aromaticity of benzene.

Extension of this method to a measurement of the aromaticity of furan and thiophene resulted in a controversy which indicates the difficulty of choosing suitable model compounds to compare with the aromatic molecule in question (*R. J. Abraham et al.*, Chem. Comm., 1965, p. 43; J. chem. Soc., B, 1966, p. 127; *Elvidge*, Chem. Comm., 1965, p. 160). *Abraham et al.* (*loc. cit.*) concluded that it is in the detection of aromaticity rather than in the measurement of the degree of aromatic character that the aromatic ring current is most useful.

Whether the three criteria (resonance energy, and the two magnetic criteria) are accepted as measures of aromatic character depends upon convention. They are criteria of aromatic character interpreted to mean that cyclic delocalization of electrons is energetically of importance in the molecular ground state. In this sense they are static criteria. Their connection with aromatic-type reactivity and chemical stability is only secondary; there are indeed no generally accepted standards of aromaticity in this chemical sense. In many cases molecules or ions with aromatic character, such as the tropylium ion, do display stability in chemical reactions that is not expected from experience of olefinic double and single bonds, but problems arise as soon as the attempt is made to put such vague connections on a quantitative scale. Aromatic character can only be measured quantitatively by selecting a single criterion and accepting it as the unique index. The criterion might be thermochemical resonance energy per π-electron, or ring current contribution to proton magnetic resonance, it being accepted that these static measures may give conflicting results. There seems little point in seeking quantitative precision. The underlying concept, that aromatic character goes along with cyclic delocalization of electrons, is the nearest to a satisfactory definition so far given, and its occurrence in a particular case may be recognised in one or more of several different ways. It may occur to a greater or less degree, again according to a chosen criterion, in a given molecule, but not in a way that can so far be reduced to allow simple comparisons between one molecule and another.

Vicinal coupling constants have also been used to recognise aromatic character. Theoretical work by *M. Karplus* (J. Amer. chem. Soc., 1963, 85, 2870) indicates that vicinal coupling constants should decrease linearly with increasing length of the intervening C–C bond, provided that the internal bond angles and hybridization remain constant. Satisfactory agreement with this conclusion is shown in five- and six-membered rings (*W. B. Smith, W. H. Watson* and *S. Chiranjievi, ibid.*, 1967, 89, 1438) and polycyclic benzenoid hydrocarbons (*M. S. Cooper* and *S. L. Manatt, ibid.*, 1969, 91, 6326). The large coupling constant $J_{1,2}$ in 6,6-dimethyl-, 6,6-diphenyl-, and 6,6-dibenzyl-fulvene ($\sim$ 5·3 Hz) compared to azulene (3·5 Hz) and the pentalenyldianion (3·0 Hz) caused *Smith* and *B.A. Shoulders* (*ibid.*, 1964, 86, 3118) to suggest that these fulvenes should be regarded as cyclic dienes.

One other magnetic effect of interest has been recognised. There is a systematic difference between molecules with cyclic delocalization and others in their rotation of the plane of polarization of light by molecules in a magnetic field (the Faraday effect) (*J. F. Labarre* and *F. Crasnier*, J. Chim. phys., 1967, 64, 1664).

(h) Aromatic character in systems of two π-electrons: cyclopropenium

According to Hückel's rule a cyclic system possessing two π-electrons is the simplest in which aromatic character appears. Cycles of three and four carbon atoms could have two π-electrons in a unipositive cyclopropenium cation and in a cyclobutadiene dication. Examples of the former are now well known, and examples of the latter have been reported, the tetraphenylcyclobutenium dication (XXIII) (*F. A. Olah, J. M. Bollinger* and *A. M. White*, J. Amer. chem. Soc., 1969, 91, 3667), being obtained in solution.

R. Breslow (*ibid.*, 1957, 79, 5318) made a salt of triphenylcyclopropenium (XXIV) (cf. Vol. IIA, p. 36 *et seq.*) and since then a number of other similar compounds have been prepared (see, for example, *D. N. Kursanov* and *M. E. Vol'pin*, Zhur. Vsesoyuz. Khim. obshch. im D. I. Mendeleeva, 1962, 7, 282):

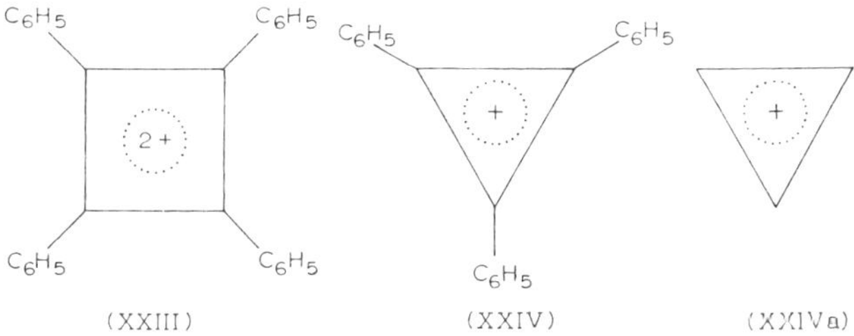

The parent ion, XXIVa, has been made (*Breslow, J. T. Grover* and *G. Ryan*, *ibid.*, 1967, 89, 5048; *D. G. Farnum, G. Mehta* and *R. G. Silberman*, *ibid.*, 1967, 89, 5049). According to simple molecular-orbital theory the pattern of molecular-orbital energies for three- and four-atom systems includes a non-degenerate lowest orbital with energy 2β below the reference zero. Many other factors must be taken into account in assessing stability, such as total charge and non-neighbour interactions, but evidence continues to accumulate in support of the aromatic character of cyclopropenium, both from the viewpoint of ground-state properties and of reactivity (see *A. W. Krebs*, Angew. Chem. internat. Edn., 1965, 4, 10).

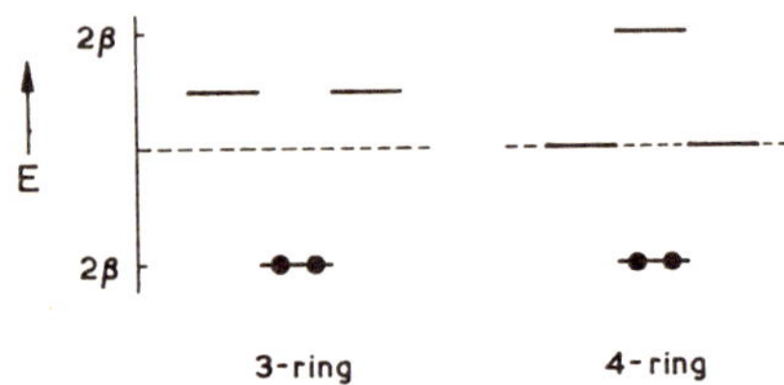

The dipropylcyclopropenium cation (*Breslow* and *H. Höver*, J. Amer. chem. Soc., 1960, **82**, 2644) seems to be as stable as the diphenylcyclopropenium cation, suggesting that the stability does not derive from the attached phenyl groups but is a property of the three-atom two-electron grouping. The parent cation precipitates as the hexachloroantimonate on mixing 3-chlorocyclopropene and antimony pentachloride in dichloromethane. It is stable at $-20°$ and for some time at room temperature, but decomposes on warming. Diphenylcyclopropenone (XXV) has the particular interest of a close analogy with tropone (XXVI), which displays properties suggesting partial expulsion of an electron on to the carbonyl oxygen, leaving an aromatic sextet in the seven-membered ring. Diphenylcyclopropenone (*Breslow et al., ibid.*, 1959, 81, 247; *Kursanov* and *Vol'pin, loc. cit.*) despite its highly strained ring, is a relatively stable system. It fails to give many reactions of ketones and has a dipole moment of 5·08*D* suggesting partial expulsion of a ring electron to leave a stable aromatic electron pair. The evidence is thus easily reconciled to the view that the double bond in the classical formula, XXV, is delocalized:

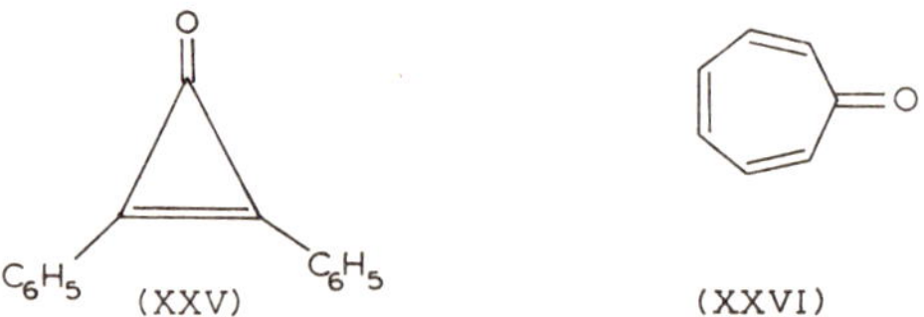

The available physical evidence of delocalization includes the X-ray structure analysis of triphenylcyclopropenylium perchlorate, showing that the bonds in the three-membered ring are equal, and 1·40 Å long (*M. Sundaralingam* and *L. H. Jensen*, J. Amer. chem. Soc., 1963, 85, 3302). Also proton magnetic resonances(τ — 0·31) (in 5% $CF_3{\cdot}CO_2H$), for the ring proton of the dipropylcyclopropenylium cation (*Breslow, Höver* and *H. W. Chang, ibid.*, 1962, **84**, 3168) and the ring protons in the cyclopropenylium cation (τ −0·87) (in FSO_3H) (*Breslow, Groves* and *Ryan, loc. cit.*) are in close agreement with expectation for an aromatic system, as discussed in section (*j*) (p. 35). The tetramethylcyclobutenium dication has been obtained only in solution being prepared in sulphur dioxide–antimony pentafluoride. The structure was inferred from the p.m.r. of the methyl protons τ 6.32 (in SO_2–SbF_5) and the ^{13}C chemical shifts $\delta = -14{\cdot}4$ p.p.m. (in SbF_5–SO_2) compared with CS_2. The ^{13}C value fits the linear relation between π-electron densities and ^{13}C chemical shifts shown for $C_5H_5^{\ominus}$, C_6H_6, $C_7H_7^{\oplus}$ and $C_8H_8^{2\ominus}$ by *H. Spiesecke* and *W. G. Schneider* (Tetrahedron Letters, 1961, p. 468) if the dication is assumed to have the delocalized π system XXIII.

There is also a series of oxygenated dianions which are considered to be aromatic. Squaric acid (XXVII) is a strong acid which gives the aromatic squarate dianion (XXVIII) (*R. West et al.*, J. Amer. chem. Soc., 1960, **82**, 6204). If as has been suggested XXIX makes a significant contribution to the structure (*H. E. Sprenger* and *W. Ziegenbein*, Angew. Chem. internat. Edn., 1968, **7**, 530; compare with *West* and *D. L. Powell*, J. Amer. chem. Soc., 1963, **85**, 2577) this ion may be considered as a 2π-electron system. Evidence for cyclic delocalization in the dianion is found in replacement of the C=O and C=C stretching band in the infrared spectrum of the acid by a broad band from 1540 to 1490 cm^{-1} in the ion (*S. Cohen, J. R. Lacher* and *J. D. Park, ibid.*, 1959, **81**, 3480). The infrared and Raman spectra are consistent with a planar structure having D_{4h} symmetry (*M. Ito* and *West, ibid.*, 1963, **85**, 2581):

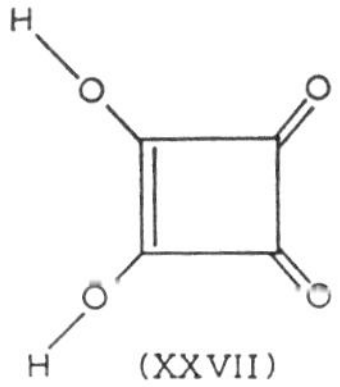

(XXVII) 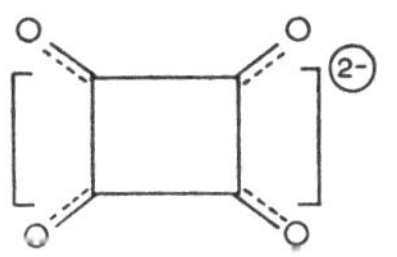 (XXVIII) 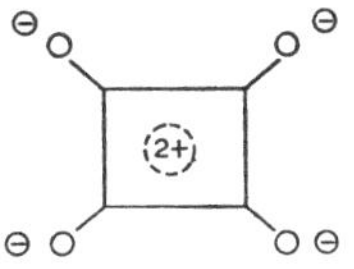 (XXIX)

(i) The four-electron system: cyclobutadiene

The search for cyclobutadiene and its simple derivatives (see Vol. IIA, p. 84 *et seq.*), though so far unsuccessful, bears on the subject of aromatic character. Moreover the search is increasingly hopeful as unstable or metastable species, like the valence isomers of benzene, [section (*a*)] which might have seemed incapable of preparation, are made.

Theoretical studies of cyclobutadiene by molecular-orbital methods (see for example, *Craig*, Proc. roy. Soc., 1950, **A202**, 498) confirm the expectation from Hückel's rule that the π-electron system is not stabilised by delocalization. It might easily be destabilised, as it appears that the cyclopropenyl anion is (*Breslow* and *M. Douek*, J. Amer. chem. Soc., 1968, **90**, 2698), in the sense of being made less stable than expected of a system of localized double bonds. Also the σ-bond system is strained. This leads to the conclusion that cyclobutadiene has nothing to gain from cyclic delocalization, and so is non-aromatic, but has little bearing on the question whether it would be chemically stable, metastable or unstable. The Dewar valence isomer of benzene [section (*a*)] is metastable, and attempts to prepare cyclobutadiene have continued in the hope that it would similarly be metastable with respect to dimerisation and other chemical transformations.

The present position, reviewed by *P. J. Garratt* and *M. V. Sargent* (Advances in organic Chemistry, Interscience, London 1969, Vol. 6, p. 19) is that no derivative of cyclobutadiene has yet been isolated, although there is evidence of short-lived species which can be distilled (*L. Watts, J. D. Fitzpatrick* and *R. Pettit*, J. Amer. chem. Soc., 1965, **87**, 3253). In a number of reactions in solution plausible reaction courses have been proposed in which cyclobutadiene derivatives appear as intermediates either free or in transient complexes. On the other hand, stable metal complexes have been made in which substituted cyclobutadienes are attached to nickel, iron and palladium; but attempts to liberate free cyclobutadienes from them by pyrolysis or through displacement in favour of another complexing group have failed.

Typical of the metal complexes is the first-prepared of them, tetramethylcyclobutadienenickelous chloride (*R. Criegee* and *G. Schroder*, Ann., 1959, **623**, 1) made by reaction in benzene of nickel tetracarbonyl with dichlorotetramethylcyclobutene. The crystal structure (*J. D. Dunitz et al.*, Helv., 1962, **45**, 647) shows a dimeric molecular unit, the two nickel atoms joined by a pair of bridging chlorine atoms. Each nickel is attached to a third chlorine and to one cyclobutadiene ring, the four ring carbons being equidistant. The electronic structure may be rationalised in terms of the rare gas rule, according to which the valence shell of the metal attains a krypton configuration by the addition of bonding electrons from the ligands. Nickel, with ten electrons, completes the krypton configuration with one electron from each bridging chlorine, two from the third (non-bridging) chlorine, and four from cyclobutadiene. The complex as a whole including the cyclobutadiene gains its stability by molecular-orbital binding within the whole ligand-plus-metal array. Cyclobutadiene itself is present as a ligand group in the complex of cyclobutadieneirontricarbonyl, prepared by *G. F. Emerson, Watts* and *Pettit* (J. Amer. chem. Soc., 1965, **87**, 131).

(*j*) *Six- and ten-electron monocyclic systems*

Two six-electron systems of high symmetry, other than benzene, are known. Both the tropylium cation $C_7H_7^{\oplus}$ (Vol. IIB, p. 359) and the cyclopentadienide anion $C_5H_5^{\ominus}$ (Vol. IIA, p. 119) are stable in ways suggesting aromatic chemical stability. The ring bond lengths are equal, and measurements of the proton magnetic resonance spectra can very reasonably be interpreted in favour of a ring current contribution to the chemical shift of roughly equal magnitude in each case. According to the simplest theory of the ring current contribution (*J. A. Pople,* Mol. Phys., 1958, **1**, 175) (see section *g iii*, p. 28) the contribution by ring currents to the chemical shift in the tropylium and cyclopentadienide rings should differ from that in benzene by —0.66 and +0.09 p.p.m., respectively on account of the small changes in ring dimensions.

The observed shifts (*G. Fraenkel et al.,* J. Amer. chem. Soc., 1960, **82**, 5846) are near to —1·9 p.p.m. and +1·85 p.p.m. These large shifts, corrected for the small ring-current differences, can be explained as the result of the changes in π-electron charge on the carbons to which the protons are attached. The tropylium protons are de-shielded by the +1/7 charge on each carbon, and their resonances shifted to lower field; the cyclopentadienide protons shifted to higher field by the shielding of the —1/5 surplus charge. The approximately linear relation between shift and surplus charge, corresponding to a slope of about 10 p.p.m. per electron, enables a consistent description of charge displacements in other systems, for example, in azulene (*Spiesecke* and *Schneider, loc. cit.,*) and is plausible on theoretical grounds. Any large difference in ring-current contribution from that of benzene (~ —1·5 p.p.m.) would be at once apparent.

Two ten-electron systems offer similar possibilities. They are both anions. In the cyclo-octatetraenyl dianion (XXX) (cf. Vol. IIB, p. 392 *et seq.*) there is

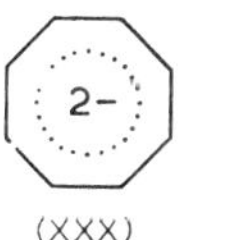

(XXX)

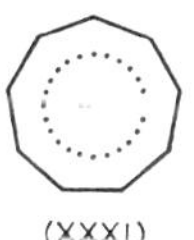
(XXXI)

a variety of evidence in support of substantial delocalization stability, in contrast with the lack of it in cyclo-octatetraene itself (see, *e.g. T. J. Katz,* J. Amer. chem. Soc., 1960, **82**, 3784). Thus the dianion is flat; it is produced from cyclo-octatetraene in a single two-electron polarographic reduction, and the proton resonance gives evidence of a ring current contribution to the chemical shift. The n.m.r. spectra of dipotassium and dilithium cyclo-octatetraenide consist of a single sharp absorption very close to that of cyclo-octatetraene itself. Now each (equivalent) carbon has a π-electron charge excess of $-\frac{1}{4}$ units; this should give a large shielding equal to about 2·5 p.p.m. on the basis of the slope of 10 p.p.m. per electron found for the system. This shift must in this case be offset by a ring current contribution giving a low-field displacement of the resonance of the same magnitude. This is indeed reasonable from equation [5] (p. 29), which gives a ring current shift of about 1·8 times that of benzene, or 2·7 p.p.m. allowing for the large ring, and the complement of ten electrons.

The other ten-electron monocyclic system is that of the anion of cyclononatetraenyl (XXXI) (Vol. IIB, p. 399) reported by *Katz* and *P. J. Garratt* (J. Amer. chem. Soc., 1963, 85, 2852) and by *E. A. LaLancette* and *R. E. Benson* (*ibid.*, p. 2853) as the $Li^{\oplus}$ and $NH_4^{\oplus}$ salts. The n.m.r. spectrum again provides important evidence. The excess charge per carbon is $-^1/_9$ units; the shift from XXX produced by the shielding difference is 1·4 p.p.m. so that, on the assumption of equal ring current contributions, this should be the measured difference in proton resonances. The observed difference is 1·3 p.p.m. The ^{13}C n.m.r. chemical shifts are also in agreement with a planar conjugated system (*idem, ibid.*, 1965, 87, 1941). It may be concluded that both of these ten-electron systems have delocalized π-electron systems and are aromatic. Figure 6 shows the proton resonances [relative to $(CH_3)_4Si$ as 10] in the six- and ten-electron systems, as a function of the π-electron excess on the carbon atoms. The ten-electron systems fall in a line nearly 1 p.p.m. lower than that of the six-electron systems because of their greater ring current shift.

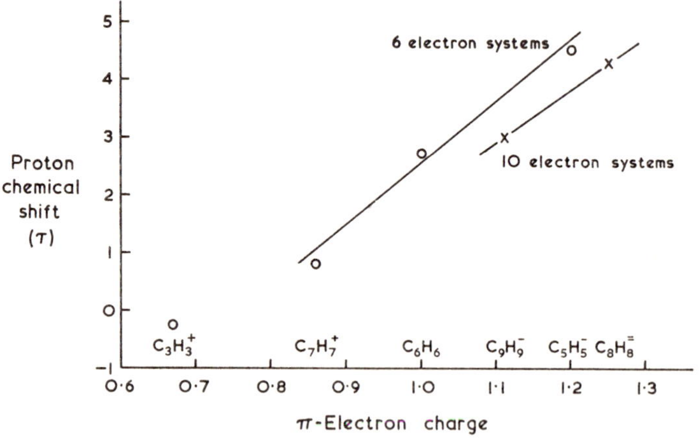

Figure 6

Proton chemical shifts for monocyclic hydrocarbon ions and molecules of formula C_nH_n as a function of net π-electron charge per carbon.

Consideration of cyclodecapentaene, ([10]annulene), isomers (Vol. IIB, p. 399) indicate two factors against stabilization by cyclic conjugation; bond-angle strain is large in the all *cis* configuration XXXII, and in the di-*trans* configuration XXXIII angle strain is minimized but the hydrogen atoms inside the ring have large non-bonded interactions. The non-bonded interactions may be removed by replacing the internal hydrogen atoms by a bridge (XXXIV), allowing the C_{10} skeleton to become nearly planar.

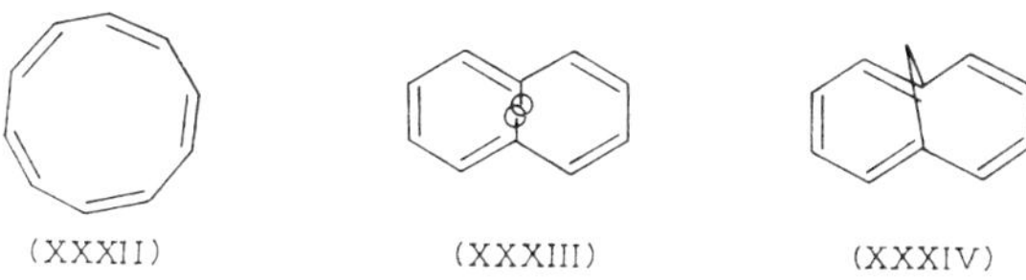

(XXXII) (XXXIII) (XXXIV)

1,6-Methano[10]annulene (XXXIV) was the first of these bridged compounds to be made (*E. Vogel* and *H. D. Roth*, Angew. Chem. internat. Edn., 1964, **3**, 288). Evidence for the cyclic delocalization of the π-electrons in this system comes from the p.m.r. spectrum and the X-ray structure of a derivative. 1,6-Methano[10]annulene-2-carboxylic acid has an almost planar structure with bond lengths between 1·38Å and 1·42Å (*M. Dobler* and *Dunitz*, Helv., 1965, **48**, 1429). The p.m.r. spectrum of XXXIV shows the vinylic protons as an A_2B_2 system at low field (centred at τ 2·9) and a sharp singlet (τ 10·5) for the methylene protons. The compound also exhibits "aromatic reactivity". No addition occurs when it is heated with maleic anhydride in refluxing benzene, and with bromine it forms a substituted monobromo-compound.

Other [10]annulenes with a variety of bridging groups have been prepared (for a review see *Vogel*, Chimia, 1968, **22**, 21). These 1,6-bridged compounds are valence tautomers of the corresponding 9,10-dihydronaphthalene. There is evidence that if there is more than one bridging atom the non-aromatic form is favoured (*J. J. Bloomfield* and *J. R. S. Irelan*, Tetrahedron Letters, 1966, 2971).

Irradiation of *cis*-9,10-dihydronaphthalene at low temperatures is claimed to give two isomers of [10]annulene (*S. Masamune* and *R. T. Seidner*, Chem. Comm., 1969, 542), the *trans*-isomer (XXXV) and the *cis*-isomer (XXXII). Both compounds readily isomerize thermally to 9,10-dihydronaphthalenes, and their p.m.r. spectrum does not show evidence for a significant ring current.

10π-electron systems have also been made in bridged ions. Both bicyclo-[5.4.1]dodecapentaenium cation (XXXVI) (*W. Grimme, H. Hoffmann* and *Vogel*, Angew. Chem. internat. Edn., 1965, **4**, 354) and the bicyclo[4.3.1]-

decapentaenyl anion (XXXVII) (*Grimme et al., ibid.*, 1966, 5, 604; *P. Radlick* and *W. Rosen*, J. Amer. chem. Soc., 1966, 88, 3461) have been prepared:

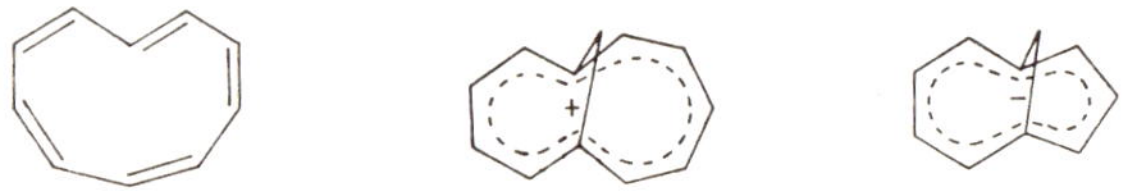

(XXXV) (XXXVI) (XXXVII)

Evidence for the structure and for the delocalization of the π-electrons in these compounds is provided by their p.m.r. spectra. In the cation the ring protons are at low field (τ 0·4–1·7) and the bridge protons are shielded (τ 10·3–11·8), in the anion the ring protons are at higher field as is expected because of the negative charge, but the bridge protons are still shielded (τ 10·7–11·2).

(*k*) *The annulenes*

The preparation by *F. Sondheimer* and his colleagues of large monocyclic polyenes C_nH_n (see Vol. II B, p. 390 *et seq.*) has allowed the study of aromaticity to be carried a stage further (for reviews with references see *Sondheimer et al.*, in "Aromaticity", Chem. Soc. Special Publication No. 21, 1967; *P. J. Garratt* and *M. V. Sargent*, in Advances in org. Chem., Interscience, New York, 1969, Vol 6). [18]Annulene (XXXVIII) and [16]annulene (XXXIX) are of particular interest being representative of the 4n + 2 and 4n π-electron systems. In the larger cyclic polyenes the problem of the interaction of internal hydrogens is still present in structures in which bond-angle strain has been minimized. Accordingly a number of dehydroannulenes where the problem is less have been made.

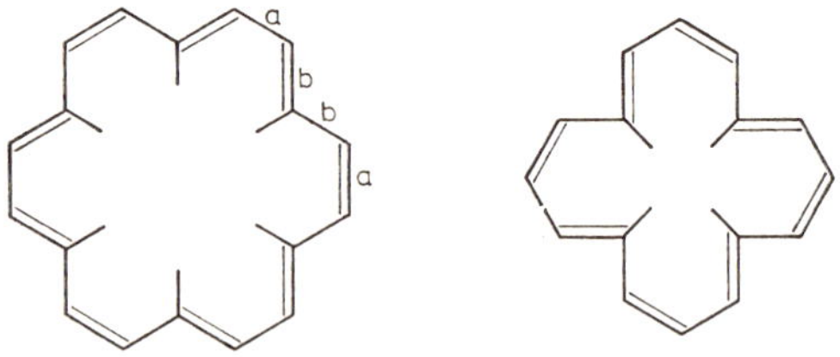

a b b a

(XXXVIII) (XXXIX)

An X-ray structure analysis shows that the molecule [18]annulene has a centre of symmetry and that its carbon atoms deviate from a mean plane by less than 0·1Å. The C–C bond lengths fall into two sets shown in the illustrated structure by *a*, mean length = 1·419 Å, and *b*, mean length = 1·382 Å (*J. Brigman et al.*, Acta Cryst., 1965, **19**, 227; *F. L. Hirschfeld* and *D. Rabinovich*, *ibid.*, p. 285). The bonds do not alternate in length as they would in a Kekulé structure and the lengths themselves are well within a range spanned by the bonds in condensed benzenoid aromatics. The structural evidence of planarity and near-equality of bond lengths is strongly in favour of cyclic electron delocalization and, to that extent, of aromatic character. The proton magnetic resonance spectrum (*Jackman et al.*, J. Amer. chem. Soc., 1962, **84**, 4307), shows two broad bands, one at low field and one at high field (τ 1·1 and τ 11·8) with relative areas showing that they are to be assigned respectively to the twelve outer protons and the six inner protons; the former showing de-shielding and the latter shielding, as expected from the influence of a large aromatic ring current. The molecule has the large stabilisation energy of 100 ± 6 kcal per mole (*A. E. Beezer et al.*, J. chem. Soc., 1965, p. 216), only a little less per π-electron (5·6 kcal per mole) than benzene, naphthalene and anthracene (6·0–6·2 kcal per mole). Aromatic character in the physical sense is thus established. In chemical behaviour [18]annulene shows no special stability, readily giving addition products with bromine and maleic anhydride, although it does nitrate with cupric nitrate in acetic anhydride and acetylate with acetic anhydride – boron trifluoride etherate (*I. C. Calder et al.*, quoted by *Garratt* and *Sargent*, *loc. cit.*).

Tetradehydro- and tridehydro-[18]annulenes related to the parent by the loss of four and three molecules of hydrogen, respectively, have also been isolated. The proton resonances give evidence of a ring current. Only two of the four π-electrons in each triple bond take part in delocalization, giving an 18π-electron system as in [18]annulene itself; both types of molecule are Hückel 4n + 2 systems and their aromatic properties are in agreement with expectation. An 18π-electron system is also present in the [16]annulenyl

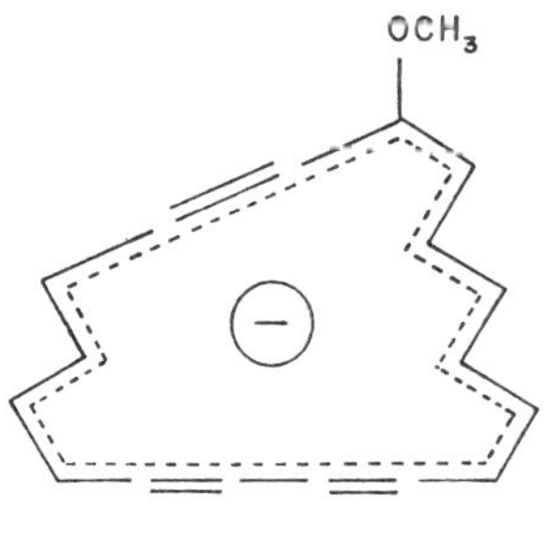

(XL)

dianion (*G. Anthoine, J. M. Gilles* and *J. F. M. Oth*, Tetrahedron Letters, 1968, 6265) and the 1-methoxy-2,8,10-tridehydro-[17]annulenyl anion (XL) (*J. Griffiths* and *Sondheimer*, J. Amer. chem. Soc., 1969, 91, 7518). The p.m.r. spectra show low field resonances for the outer protons in these anions and exceptionally high field resonances for the inner protons, indicating the delocalized nature of the π-electrons.

The behaviour of Hückel 4n systems as exemplified by [16]annulene is quite different. X-ray structure analysis shows that the bonds in [16]-annulene alternate, the average "single-bond" distance being 1·46 Å and the average "double-bond" distance 1·34Å (*S. M. Johnson* and *I. C. Paul*, *ibid.*, 1968, 90, 6555). The carbon atoms lie up to 0.57Å away from the best plane through four carbon atoms. These deviations occur in a way which would relieve the interactions between the internal hydrogens. The p.m.r. spectrum at room temperature of [16]annulene is a time-averaged one, brought about because bond rotation causes all protons to spend some time both inside and outside the ring. At low temperatures this process slows and at -120^0 the spectrum shows one multiplet at low field (τ $-0{\cdot}32$) and one at high field (τ 4·8). The relative areas of the peaks indicate that in this case the inner protons have the low field resonance, suggesting a paramagnetic ring current. Bond alternation in 4n electron systems would split the degeneracy of the half filled pair of non-bonding orbitals of molecular-orbital theory. The one-electron excitation between these split orbitals has a magnetic moment perpendicular to the ring, thus the excitation makes a paramagnetic contribution to the ring current (*H. C. Longuet-Higgins* in "Aromaticity", Chem. Soc. Special Publication No. 21, 1967; *Pople* and *K. G. Untch* J. Amer. Chem. Soc., 1966, 88, 4811).

In systems smaller than [16]annulene internal hydrogen interactions are severe. Nevertheless, the p.m.r. criterion shows [14]annulene as well as a number of its dehydro-derivatives to be aromatic. The same standard shows that the Hückel 4n systems [12], [16], [20] and [24]annulene or their dehydro-derivatives are not aromatic.

Some interesting bridged 14π-electron systems have been proposed. *trans*-15,16-Dimethyldihydropyrene (XLI) (*V. Boekelheide* and *J. B. Phillips*, Proc. natl. Acad. Sci., U.S., 1964, 51, 550) has proton resonances

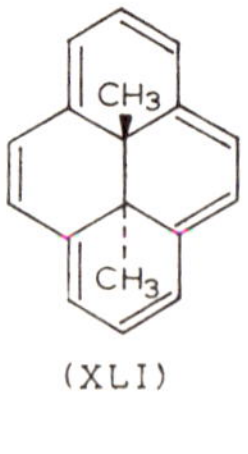

CH_3 CH_3

(XLI)

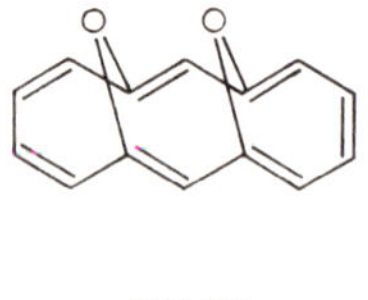

O O

(XLII)

for the ring protons at τ 1·33–2·02 and a high field singlet for the methyl protons at τ 14.25 and also exhibits aromatic reactivity, undergoing electrophilic substitution on acylation, bromination and nitration.

syn-1,6; 8,13-Bisoxido-[14]annulene (XLII) which is analogous with the bridged [10]annulenes has also been prepared (*Vogel, loc. cit.,*) and its p.m.r. spectrum shows it to be aromatic. It too shows aromatic reactivity undergoing bromination, nitration and Friedel-Crafts acylation to give substitution products. *anti*-Bridged [14]annulenes in which the bridging groups lie on opposite sides of the ring are not expected to show aromatic behaviour because models show that in these systems some of the carbons are twisted so that the $p\pi$-orbitals are no longer parallel and the cyclic delocalization is interferred with. As confirmation of this, *anti*-1,6;8,13-bismethano-[14]annulene has a p.m.r. spectrum typical of a polyolefin.

Another possibility with large ring compounds of this type is that a configuration with equal or nearly equal bond lengths may not be the most stable even if Hückel's 4n + 2 rule is obeyed (*Longuet-Higgins* and *Salem, loc. cit.*). Energy of stretching and compression is expended in transforming a system of alternating long and short bonds into one with equal lengths, and whether this structure is stable depends upon the gain in π-delocalization energy compared with the σ-bond stretching and compression. Calculations of the balance of these energy terms have been described in section (*e*); the delocalization energy per electron diminishes in the larger rings, so that with increasing ring size the energy balance tends in favour of unequal bond lengths. If we use the energy formulae [1] and [2] (p. 21) for cyclic systems belonging to the 4n + 2 and 4n systems, respectively (in the nearest neighbour approximation it is not important that the annulenes do not have n-fold cyclic symmetry) we find for the delocalization energy per electron the two formulae [6] and [7]:

$$E_\pi/\mathrm{n} = 4\beta/\mathrm{n}\ \mathrm{cosec}\ (\pi/\mathrm{n}) - \beta \qquad [6]$$

$$E_\pi/\mathrm{n} = 4\beta/\mathrm{n}\ \cot\ (\pi/\mathrm{n}) - \beta \qquad [7]$$

In Hückel systems E_π/n decreases steadily as the ring size increases with values 0·33 (n = 6, benzene), 0·29 (n = 10), 0·28 (n – 14) to a limiting $(4/\pi-1)$ 0·28. In the non-Hückel systems [7] the energy steadily increases to the same limit, with values of 0 (n = 4), 0·22 (n = 8), 0·24 (n = 12), 0·26 (n = 16). It is apparent from these values that the stabilisation of the larger systems is only marginally different in Hückel and non-Hückel molecules, the distinction between classes becoming quickly blurred. In Hückel systems a change from uniform to alternating lengths is expected in molecules of large enough size. It has been predicted (*M. J. S. Dewar* and

G. J. Gleicher, J. Amer. chem. Soc., 1965, **87,** 685) that bond alternation would occur in systems larger than [22]annulene. It is not evident in [18]-annulene, but the proton magnetic resonance spectra of trisdehydro-[26]-annulene and two dehydro-[30]annulenes, showing broad multiplets which are unchanged on cooling, has been interpreted as indicating the loss of a ring current through bond alternation (*C. C. Leznoff* and *Sondheimer, ibid.*, 1967, **89,** 4247). In non-Hückel systems alternating lengths are expected in all cases.

(l) Homoaromaticity

The central idea in homoaromaticity is that some delocalization in a system of π-electrons is possible even where the cycle of overlapping orbitals has to be completed across an intervening group, such as a methylene group, by the overlapping of $p\pi$-orbitals not on adjacent carbon atoms, as in the homotropylium cation (XLV). For the delocalization to be energetically significant these orbitals must be oriented to have a large overlap. The possibilities are illustrated in the homoallyl cation (XLIII) (*S. Winstein,* Quart. Reviews, 1969, **23,** 141). If the axes of the p-orbitals lie in the plane formed

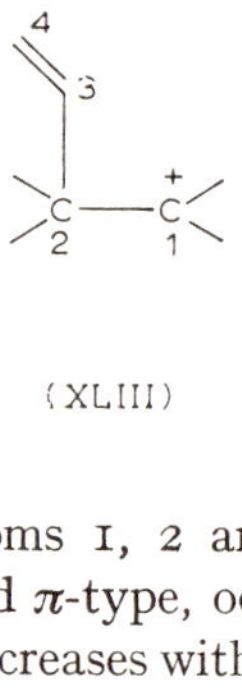

(XLIII) 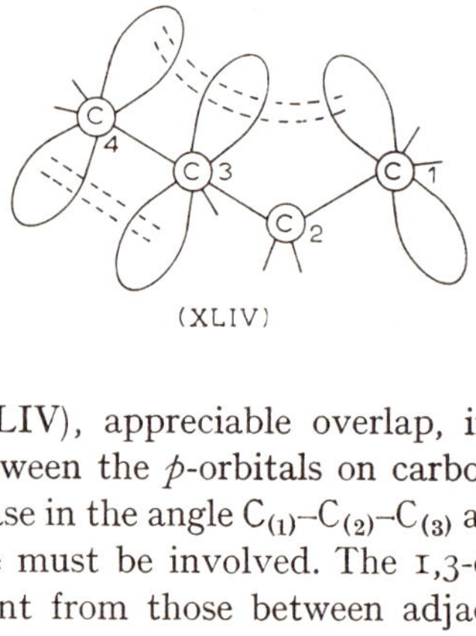 (XLIV)

by carbon atoms 1, 2 and 3 (XLIV), appreciable overlap, intermediate between σ- and π-type, occurs between the p-orbitals on carbons 1 and 3. This overlap increases with a drecease in the angle $C_{(1)}$–$C_{(2)}$–$C_{(3)}$ at the cost of increased strain and some balance must be involved. The 1,3-overlap and resonance integrals will be different from those between adjacent carbon atoms, but molecular-orbital calculations suggest that this type of interaction can give substantial stabilization by cyclic delocalization in favourable circumstances, as for instance in the monohomotropylium ion (XLV) where, even with the 1,3-type resonance integral assigned only one half of the normal C–C value, the delocalization energy is still 80% of that for the tropylium cation. Such cyclic delocalization in homoaromatic compounds should confer properties to some degree similar to those of normal aromatics.

The subject has been reviewed by *Winstein* (*loc. cit.*, and in "Aromaticity", Chem. Soc. Spec. Publ. No. 21, 1967, p. 1).

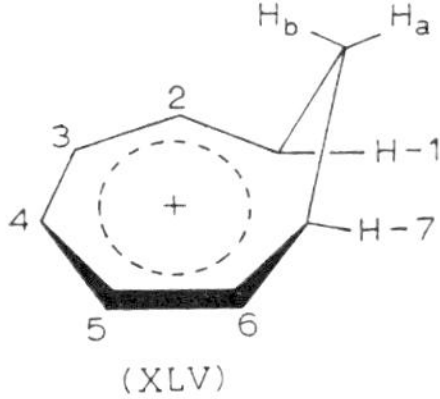

(XLV)

The existence of homoaromatic ions as intermediates in solvolysis reactions has been inferred from kinetic studies. In contrast to these short lived species a number of longer lived homoaromatic ions have been isolated as salts or identified in solution. The best studied is the homotropylium cation (XLV), obtained as the hexachloroantimonate by treatment of cyclo-octatetraene with hydrogen chloride and antimony pentachloride (*J. E. Mahler, R. Pettit* and *J. L. Rosenberg*, J. Amer. chem. Soc., 1962, 84, 2482). The proton magnetic resonances for the salt show a five-proton signal (τ 1·5), assigned to the protons on carbons 2–6, a signal at τ 3·58 for protons 1 and 7, a low field signal (τ 4·90) for the outer proton H_a and a high field signal (τ 10·67) for the inner proton H_b. These resonances strongly support the structure XLV and the existence of a ring current, leading to the low τ values of the protons on $C_{(2)}$–$C_{(6)}$ and the large difference in chemical shift between inner and outer protons on the methylene carbon atom. Further support for this comes from the unusually large diamagnetic susceptibility exaltation (20 c.g.s. units $\times$ 10^{-6}). A number of substituted monohomotropylium ions have been made and show p.m.r. spectra consistent with homoaromaticity.

The nature of the postulated homoaromatic delocalization carries the implication that Hückel's rule should apply to restrict the cases to those of $4n + 2$ π-electrons. This is found in practice. Homotropylium is a six-electron system, and homoallyl a two-electron system. Contrariwise, an eight-electron system, the homocycloheptatrienide anion appears to show antihomoaromatic behaviour (*Winstein, loc. cit.*).

The bishomocyclopentadienide ion (XLVII), a six-electron ion, has been prepared by shaking the allylic ether XLVIa in tetrahydrofuran or 1,2-dimethoxyethane with sodium-potassium alloy at $\sim 0^0$ (*J. M. Brown*, Chem. Comm., 1967, 639; *Winstein et al.*, J. Amer. chem. Soc., 1967, 89, 3656). Evidence for cyclic delocalization in this ion rests on a comparison of the p.m.r. spectra of the parent hydrocarbon XLVIb and the ion.

There is a significant upfield shift of the resonance signal of the protons on $C_{(6)}$ and $C_{(7)}$ indicating that delocalization has spread the negative charge to embrace these carbons. The protons on $C_{(8)}$, lying above the ring, also have signals at considerably higher field compatibly with a ring current effect.

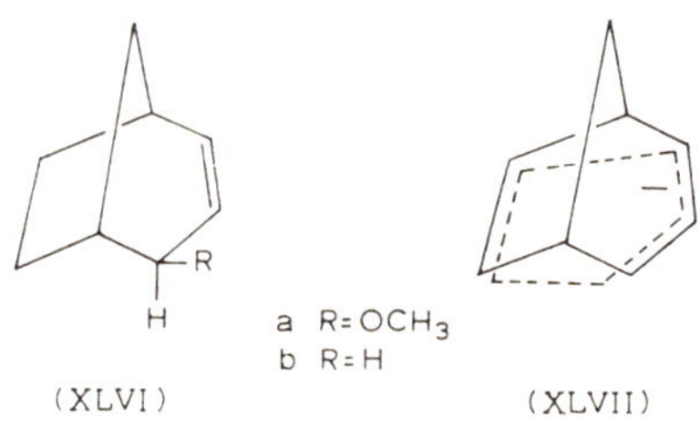

(XLVI) (XLVII)

The ten π-electron monohomocyclo-octatetraene dianion (XLVIII) (*M. Oglianuso, R. Riecke* and *Winstein, ibid.*, 1966, **88,** 4731) and a number of the two π-electron homocyclopropenium cation derivatives (XLIX) (*T. J. Katz* and *E. H. Gold, ibid.*, 1964, **86,** 1600) have been prepared. The p.m.r. spectra support the view that there is homoaromatic character.

(XLVIII) (XLIX)

2. Electrophilic aromatic substitution

P. B. D. DE LA MARE

(a) Historical introduction

One of the very early results of the establishment of sound methods for the determination of the positions of substituents in the benzene nucleus was the discovery of regularities concerning the orientation of substitution. It appeared that there were two broadly contrasting types of orientation, *ortho* and *para* collectively on the one hand, and *meta* on the other. It also appeared that the nature of the substituent already present determined the position or positions which a new substituent must take up. Already in the 1870's attempts were being made to reduce the known facts to a rule.

The notion that atoms and groups have an electrochemical character had survived from the teachings of Berzelius, although the time for the great revival of such concepts was yet to come. The first three of the proposed rules of orientation in aromatic substitution, those of Körner, Hübner and Noelting, were essentially forms of the same rule: it expressed the conviction that the significant property of an orienting group was its electropolar nature (*G. Körner*, Gazz., 1874, **4**, 305, 446; *H. Hübner*, Ber., 1875, **8**, 873; *E. Noelting, ibid.*, 1876, **9**, 1797). Electronegative groups (*i.e.* those which confer or enhance acidic properties), said this rule, will, provided their electronegativity is strong enough, direct a new substituent to the *meta*-position; and electropositive groups (*i.e.*, those which confer or enhance basic properties), as well as neutral, and even weakly electronegative groups, will direct new substituents into *ortho*- and *para*-positions. It was appreciated that a number of the substituents which, in being introduced in place of hydrogen, are subject to these rules of orientation, themselves have an electronegative character.

These early investigators made a real contribution to the subsequently developed theory of the subject, inasmuch as they foreshadowed the part played by the inductive effect in the modern theory of orientation in aromatic substitution. Electronegative groups are those whose atomic nuclei are under-screened: they are the groups which, for this electrostatic reason, attract electrons, reducing the electron-density in an attached residue; they exert what is now termed the negative inductive ($-I$) effect*. Elec-

* Groups which carry a positive ionic charge (*e.g.* $\cdot N^{\oplus}Me_3$) are *ipso facto* electronegative; and groups which carry a negative ionic charge (*e.g.* $\cdot O^{\ominus}$) are electropositive. It is not necessary to evade this situation. To escape from it, as some authors have, by reversing electrochemical terminology is to break a continuity in the designation of electropolar character which has prevailed since the foundations of chemical theory.

tropositive groups electrostatically raise electron-density in an attached residue; that is, they exert a positive inductive ($+I$) effect. The rules of Körner, Hübner and Noelting, then, are equivalent to a statement that groups which attract electrons strongly ($-I$ effect) direct substituents to the *meta*-position, whilst other groups direct them to the *ortho*- and *para*-positions; and this is an approximately correct description of the actual situation.

In the succeeding decades, rules were advanced which directly or indirectly connect orientation with the saturation or unsaturation of the directing substituents. In 1887, *H. E. Armstrong* pointed out that substituents attached to the benzene ring by an atom involved in a multiple bond were *meta*-directing, whereas other substituents were *ortho*- and *para*-directing (J. chem. Soc., 1887, **51**, 258). The same correlation was later stressed by *D. Vorländer* (Ann., 1902, **320**, 122). In 1892, *A. Crum Brown* and *J. Gibson* announced their famous rule to the effect that a substituent X would be *meta*-directing if HX could be directly oxidized to HOX, and *ortho*- and *para*-directing otherwise (J. chem. Soc., 1892, **61**, 367). Neither rule is completely accurate, but the second is more accurate than the first.

Although the connection was never made clear, these rules are related; and they foreshadow the part played by the mesomeric effect in the modern theory of aromatic orientation. A substance HX will be the more easily oxidisable to HOX if the latter has some special factor of stability denied to the former, such as a mesomeric system in which the unshared electrons of oxygen can participate; and this condition is fulfilled when ·X is of a form, ·Y : Z, having a multiple bond, provided also that Z will readily accept unshared electrons:

$$\mathrm{H{-}\overset{\frown}{O}{-}Y\overset{\frown}{=}Z}$$

These are the groups which become *meta*-directing, if the power of the system to withdraw aromatic electrons by an analogous mechanism is sufficient:

$$\mathrm{\overset{\frown}{Ar}{-}Y\overset{\frown}{=}Z}$$

Armstrong and Vorländer's rule provides unsaturation as the cause of *meta*-orientation, but in fact unsaturation alone is not enough. Crum Brown and Gibson's rule provides electron-attraction accompanying unsaturation, and this will give *meta*-orientation; therefore their rule is the more accurate. But it is still not satisfactory, because it does not reflect the fact that strong electron-attraction, even without unsaturation, can produce the same result.

In 1902 the first serious theory (as distinct from a rule) of orientation was advanced by *B. Flürscheim* (J. pr. Chem., 1902, **66**, 321; 1905, **71**, 497; Ber., 1906, **39**, 2015; Chem. and Ind., 1925, **44**, 246; J. chem. Soc., 1926,

1562). He used Werner's idea that chemical affinity, even apart from unsaturation, is continuously divisible, and is partly in the bonds, and partly free. He assumed that an atom (*e.g.* bivalent oxygen) able to increase its valency, when linked to an aromatic carbon atom, would make a large affinity-demand on the latter, leaving it with relatively little affinity for its other bonds, and little free affinity. This atom would accordingly make a small affinity-demand on the next atom; and thus the disturbance would be propagated, and an alternating distribution of bound and of free affinity would be set up, which would lead to much *ortho* and *para* free-affinity (I), and thus to *ortho*- and *para*-orientation. A substituent atom in its highest valency state (*e.g.* nitrogen in a nitro group) would make a small affinity-demand on the nuclear atom to which it was bound, and this would produce a different affinity distribution (II), leading to *meta*-orientation. A carbon atom linked to the aromatic ring would, if unsaturated, make a high affinity demand; whilst, if it were saturated, its character would be determined by the affinity-demand of the other atoms to which it was combined.

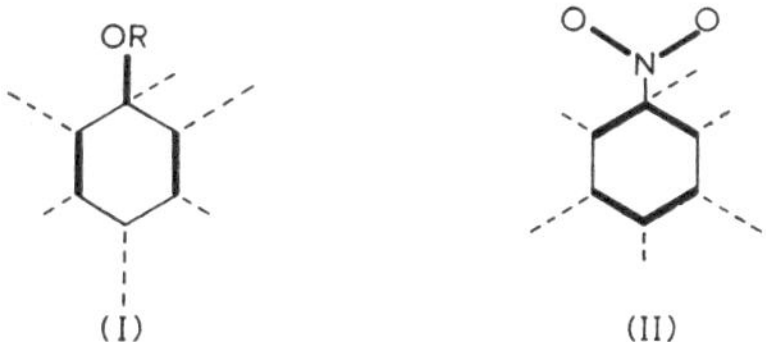

(I) (II)

Flürscheim's theory fulfilled a notable role in the development of the modern theory. For, first, it embodied a widened idea of unsaturation, recognising in the power of an atom to increase its valency (to share its unshared electrons, as we should now say), a condition which could, similarly to a double or triple bond, disturb the distribution of affinity in its neighbourhood. Secondly, it described the affinity redistribution resulting from this kind of unsaturation quite correctly in its formula covering *ortho*- and *para*-orientation. Expression I could be (and was) directly translated into electronic terms; and thus it foreran the part played by the electromeric effect in the modern theory of aromatic orientation. However, difficulties arose in the interpretation of *meta*-directive effects. Already in 1902, one *meta*-orienting substituent was known, the nitromethyl group, for which the theory could not account; and Flürscheim was forced into a special and highly unconvincing explanation of the case. In the 1920's many more such examples were demonstrated, and could not be explained away. The cause of the trouble was partly that Flürscheim's theory, as a pre-electronic theory, treats electrochemical character as a property separable from valency distribution; and partly that, based as the theory is on Werner's view of valency, it does not recognise the major difference of mobility that exists between unsaturated and saturated valencies.

A. F. Holleman gave an important criterion by which all theories of orientation must be judged. In 1910 his book appeared ("Die direkte Einführung von Substituenten in den Benzolkern", Verlag von Veit & Co., Leipzig). In it he set forth the known facts of orientation, including numerous quantitative findings of his own. He also offered a theoretical treatment, in which he assumed that the first step in substitution is addition to a Kekulé double-bond, and that substituents orient by modifying the additive reactivity; but he did not attempt to complete a theory of orientation by correlating these presumed effects with the constitutions of the substituents. The valuable element in his discussion is his insistence on the connexion between orientation and the reactivity of the aromatic nucleus. He points out that *ortho*- and *para*-orienting substituents usually increase the rate at which substitution takes place, whilst *meta*-orientation is definitely associated with a decreased rate of substitution. Therefore no theory can be satisfactory if it modifies *meta*-carbon atoms in *meta*-orientation exactly as it modifies *ortho*- and *para*-carbon atoms in *ortho*- and *para*-orientation. Flürscheim's theory would suggest activated *meta*-carbon atoms in *meta*-orientation (II), whereas the rate relations show that all nuclear positions are deactivated.

Over the next fifteen years, a group of closely related theories, often known, individually or collectively, as the "theory of alternate polarities", secured a very notable amount of attention (*H. S. Fry*, J. Amer. chem. Soc., 1912, **34**, 664; 1914, **36**, 248, 262, 1038; 1915, **37**, 855; "The Electronic Conception of Valence and the Constitution of Benzene", Longmans, London, 1912; *D. Vorländer et al.*, Ber., 1919, **52**, 263 *et seq.*; 1925, **58**, 1893; *A. Lapworth*, Mem. Manchester phil. Soc., 1920, **64**, iii, 2; J. chem. Soc., 1922, **121**, 416; *W. O. Kermack* and *R. Robinson*, *ibid.*, p. 427; *T. M. Lowry*, *ibid.*, 1923, **123**, 822). The first form of the theory, due to Fry, assumed that the atoms of benzene had alternating charges in two equivalent, interconvertible forms, or "electromers" (III and IV). In most derivatives of benzene, one electromer predominated (V and VI). Substitution preferentially replaced positive hydrogen, and therefore a favoured electromer corresponded to a favoured mode of orientation; thus expression V represents *ortho*- and *para*-orientation, and VI *meta*-orientation. Which arrangement of charges would be favoured was deduced from Crum Brown and Gibson's rule, as follows. Oxidation, by definition, involves electron release; and therefore a facile reaction HX $\rightarrow$ HOX signifies that X tends to charge an attached atom negatively. Thus, the reaction $HNO_2 \rightarrow HNO_3$ being facile, a nitro group will repel electrons to an adjacent nuclear carbon atom (as in VI). It is interesting that so plausible an argument should have yielded the wrong answer (owing to a neglected effect of unsaturation). But given this false start, an equally false assumption of alternating polarities produced the right orientation (*meta;* cf. VI):

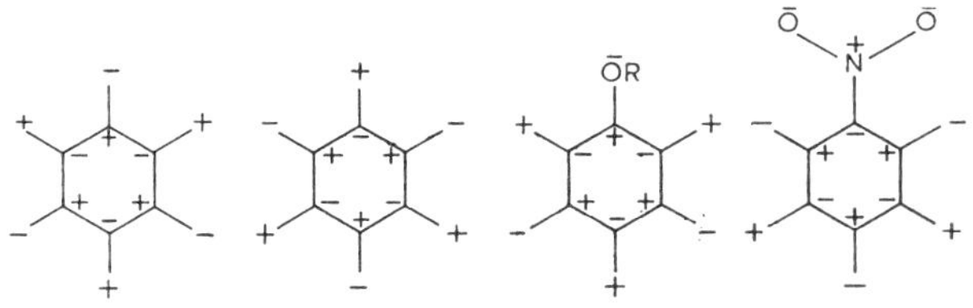

(III) (IV) (V) (VI)

Vorländer and Lapworth obtained the same preferred sign distributions by allowing the most electrochemically extreme atom, called by Lapworth the "key atom", to determine the form of the alternations. Thus O in ·OR or $\cdot NO_2$ had to be negative (so that N in $\cdot NO_2$ became positive), and H in $\cdot CH_3$ had to be positive (so that the C atom became negative). Vorländer added the proposition that the sign of an ionic charge carried by an atom directly linked to the benzene ring could be accepted as determining the polarities. Vorländer made a permanent contribution to the subject of orientation when he showed experimentally that cationic substitutents, such as $\cdot N^{\oplus}Me_3$, are *meta*-orienting, whilst it was known that the anionic substituent $\cdot O^{\ominus}$ is *ortho,para*-orienting. As to mechanism, Lapworth and Lowry insisted that the assumed charges arise only during reaction, and Lowry that the carbon atoms, but not the hydrogen atoms of the benzene molecule, receive charges; whilst Kermack and Robinson offered an electronic picture of the production of the charges by the alternate contraction and expansion of octets.

This theory encountered many difficulties. Some were of an essentially practical nature, as when Vorländer disclosed an inconsistency, that *tert*-butylbenzene, like toluene, was *ortho,para*-orienting (*loc. cit.*). It was found also that an α-styryl ether was *ortho,para*-orienting; evidently, the O atom of $\cdot C(OR){:}CH_2$ was not behaving properly as "key atom" (*C. K. Ingold* and *E. H. Ingold,* J. chem. Soc., 1925, **127**, 870). More serious, perhaps, was the implication which can be drawn from Holleman's discussion (*loc. cit.*; Chem. Reviews, 1925, **1**, 187) namely that to treat *meta*-orientation as implying *meta*-activation, and accordingly to modify *meta*-positions in *meta*-orientation exactly as *ortho*- and *para*-positions are modified in *ortho,para*-orientation, cannot be correct.

By 1926 there existed much of the material required for the construction of a general theory of constitutional effects on reactivity. *G. N. Lewis* ("Valence and the Structure of Atoms and Molecules", Chemical Catalog

Co., N.Y., 1923) had established his form of electron displacement, as a permanent molecular condition, now termed the inductive effect; and Lowry had proposed and illustrated his form of electron displacement, as an activation phenomenon, now called the electromeric effect (cf. p. 56). In a discussion of addition to olefins, *H. J. Lucas* had combined the inductive and electromeric effects, showing how the former could assist and give direction to the latter (J. Amer. chem. Soc., 1924, **46**, 2475; 1925, 47, 1459, 1462). Reagents had been classified as either seekers of electrons (oxidants, acids, etc.; *i.e.* electrophilic reagents, as they are usually called today), or seekers of atomic nuclei (reductants, bases; *i.e.* nucleophilic reagents as they are now commonly termed)*. It was clear that the reagents participating in the aromatic substitutions to which the usual orientation rules apply belong to the category of electrophilic reagents (*Fry*, "Electronic Conception of Valence", Longmans, London, 1921; *J. Stieglitz*, J. Amer. chem. Soc., 1922, **44**, 1293; *Lapworth*, Mem. Manchester phil. Soc., 1925, 69,, p. xviii; Nature, 1925, **115**, 625). It remained to combine these ideas and some others into a coherent theory of aromatic orientation, to which major contributions were made by *R.* (*Sir Robert*) *Robinson* and his associates (cf. *Robinson et al.*, J. chem. Soc., 1926, 401; *H. R. Ing* and *Robinson*, *ibid.*, 1655) and by *C. K.* (*Sir Christopher*) *Ingold* and his coworkers (cf. *Ingold* and *Ingold*, *ibid.*, 1310; *Ingold*, Annual Reports, 1926, **23**, 129). At about the same time, *P. Pfeiffer* and *R. Wizinger* (Ann., 1928, 461, 132) formulated the intermediates involved in determining orientation in these aromatic substitutions as aromatic cations, using analogies from their experimental studies of the reactions of arylethylenes with bromine. The stage was thus set not only for the development of the qualitative electronic theory, but also for more detailed consideration of the reaction paths which may lead from reactants to products.

(b) The qualitative electronic theory of aromatic substitution

The theory was outlined in a general form in a historic review in 1934 (*Ingold*, Chem. Reviews, 1934, **15**, 225). Its applicability to aromatic substitution was already being tested by crucial experiments. Fundamental to these was Ingold's description of relative reactivities in terms of *partial rate factors*, the relative rates of substitution at a particular position in an

* Lapworth called the two classes of reagents "kationoid" and "anionoid"; and these names are still sometimes used. But they tend to obscure Brønsted's teaching that fundamental similarity of function cuts across all differences of total charge. Brønsted illustrated this by reference to acids and bases, but it is just as true of the broader classes, *i.e.* of electrophilic reagents, which include acids, and nucleophilic reagents, which include bases.

aromatic nucleus as compared with the corresponding rate of substitution at a single position in benzene. To determine these quantities, both relative rates of reaction and orientational information are necessary. The main results are summarised (*Ingold*, "Structure and Mechanism in Organic Chemistry", 2nd Edn., G. Bell, London, 1969) and can be used to illustrate the essential features of the qualitative electronic theory as applied to the reactivities of aromatic compounds with electrophilic reagents.

(*i*) *The electron-withdrawing inductive effect* (*symbol*: —I)

We note first that the electron-withdrawing inductive effect, shown in its purest form by the $\cdot N^{\oplus}Me_3$ substituent, deactivates every position in the aromatic nucleus, but deactivates the *meta*-positions least (VII; *J. H. Ridd et al.*, J. chem. Soc., B, 1968, 528, 534, 1063, 1068). It appears, therefore, that the electronic distribution, as the reagent encounters the molecule, can be represented approximately* by the partial charges in formula VIII. Part of the transmission of charge from the substituent to the ring must act through the bonding system, diminishing with distance and damped by each successively intervening atom. Part also may (especially for the *ortho*-position) act through space by the direct effect, which is sometimes separately specified, but is difficult to separate from other inductive influences. It is assumed, however, because of the alternation in incidence of the inductive effect (the *meta*-position being the least affected), that the positive charges can in part reach the *ortho*- and *para*-positions conjugatively; the relevant electronic movements are those shown by the curved arrows in VIII:

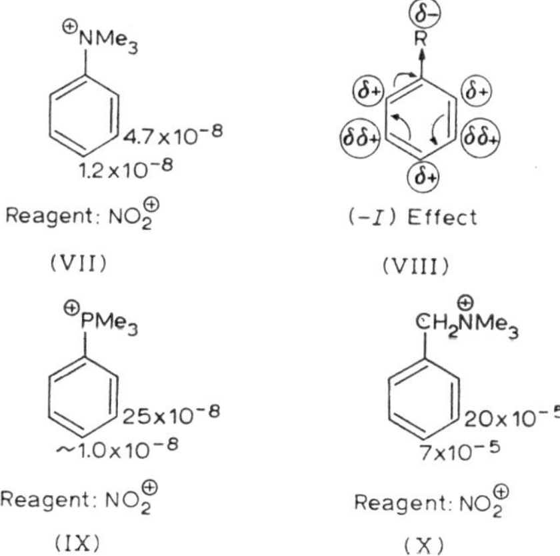

* The balancing partial negative charge is sometimes not shown, it being assumed that the arrow on the Ar–R bond implies this electronic movement; *i.e.* that Ar→—R means $Ar^{\delta+}$→—$R^{\delta-}$.

The charges are considered to reach the *meta*-position in part by the less direct route of inductive relay from the *ortho*- and *para*-positions. If the positive charge is dispersed over a larger atom (IX), or is moved further away (X), the effect of the charge on the rate of reaction is diminished (*J. H. Ridd et al., loc. cit.*).

This work, together with more recent studies (*Ridd et al.*, J. chem. Soc., B, 1969, 684), establishes that the selectivity of the inductive effect in determining the orientation of aromatic substitution has until recently been over-emphasised. In particular, the difference between the *meta*- and the *para*-positions as far as the incidence of the inductive effect is concerned may be exaggerated in formula VIII, which might better have the *para*-position labelled $\delta\delta+$. There is, however, no doubt that an alternation in reactivity is usually observed. An approach based on electrostatic considerations (cf. *Ridd et al., loc. cit.*) suffers from the disadvantages of a rather over-simplified model, and also (in the writer's view) from a neglect of hyper-conjugative influences. These may be small in absolute magnitude, but, superposed on the large and apparently only slightly discriminatory inductive effect, may be significant differentially.

(ii) The electron-releasing inductive effect (symbol: +I)

The converse charge distribution resulting from the electron-releasing inductive effect is shown by the partial charges in XI; the arrows indicate the related electronic movements. All positions should be activated for electrophilic substitution, but the *meta*-position should be activated the least. It is difficult to choose a group which shows this effect free from any other; the methyl substituent (XII; *L. M. Stock*, J. org. Chem., 1961, **26**, 4120; cf. *Ingold, loc. cit.*) probably is as good as any, but we shall need to return later to consider more fully the effects of alkyl groups (section xi, p. 60):

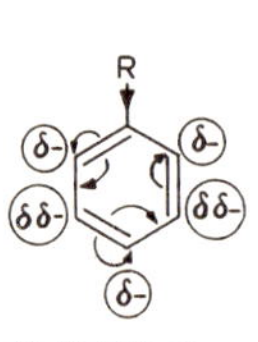

R
(δ-) (δ-)
(δδ-) (δδ-)
(δ-)
(+*I*) Effect
(XI)

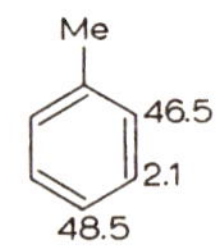

Me
46.5
2.1
48.5
Reagent: $NO_2^{\oplus}$
(XII)

(iii) The electron-releasing mesomeric effect (symbol: +M)

It can readily be shown that inductive effects cannot be the only structural influences important in determining orientation. The methoxyl group, for example, is known to be inductively electron-withdrawing; but it selectively activates the *ortho*- and *para*-positions very markedly. The arrows in structure XIII indicate the nature of the conjugative effects concerned here:

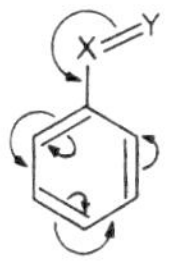

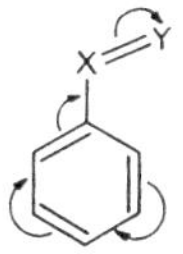

(+*M*) Effect (−*M*) Effect

(XIII) (XIV) (XV)

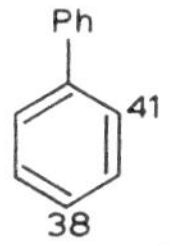

Reagent: $NO_2^{\oplus}$ (XVI)

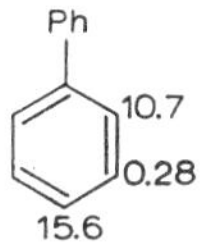

Reagent: '$Br^{\oplus}$' (XVII)

Unsaturated groups (·X:Y) conjugated with the aromatic nucleus also can have (+*M*) effects (structure XIV); they differ from groups like methoxyl in that in general the reverse mode of electron movement (XV) is also possible, so that all these groups in principle should be classified as (±*M*). Depending on the relative electron-withdrawing power of X and Y, one or other of these modes is expected to be dominant. The phenyl group itself, for example, markedly activates the *ortho*- and *para*-positions (cf. XVI, XVII; *O. Simamura* and *Y. Mizuno*, Bull. chem. Soc. Japan, 1957, 30, 196; *P.B.D. de la Mare* and ***M. Hassan***, J. chem. Soc., 1957, 3004), and so is (+*M*) for electrophilic aromatic substitution; its deactivation of the *meta*-position (XVII) shows that it is also (−*I*) in character.

(iv) Steric requirements for conjugation: steric inhibition of mesomerism

Such conjugative effects are subject to geometrical restrictions. In derivatives of biphenyl, for example, the formal structure XVIII will make its maximum contribution to the stability of the molecule only if the two rings are in the same plane, so that the geometry about the central bond can be similar to that of ethylene. Experimentally, it is known that bulky groups in the *ortho*-position of a biphenyl system can modify the rate and orientation of electrophilic substitution. Quantitative measurements, (XIX, XX), which show the diminution in reactivity *ortho* and *para* to the Ar–Ar bond in such compounds as 2-methylbiphenyl, have recently been reported for chlorination (*de la Mare et al., ibid.*, 1962, 3784):

(XVIII) 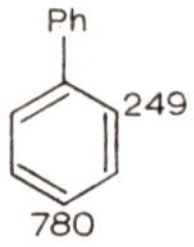 (XIX) (XX)

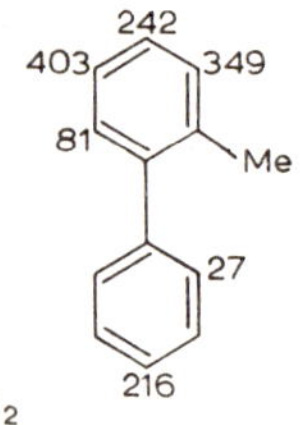

Reagent: Cl_2

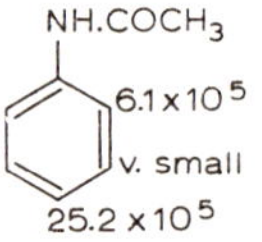

(XXI) 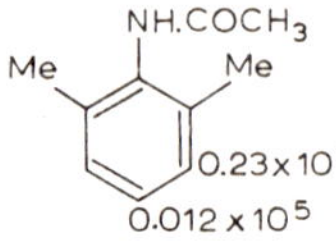 (XXII)

Reagent: Cl_2

Related steric requirements are imposed on the conjugating power of groups such as ·OR and ·NHR, in which electron-release can be at its maximum only when the lone pairs of electrons can be most favourably oriented. The partial rate factors in diagrams XXI and XXII illustrate how adjacent methyl groups can interfere with conjugative electron-release from the acetamido group (*de la Mare* and *Hassan, ibid.*, 1958, 1519); similar illustrations can be given for the methoxyl group (*G. Baddeley, N. H. P. Smith* and *M. A. Vickars, ibid.*, 1956, 2455).

(*v*) *The electron-withdrawing mesomeric effect* (*symbol*: −M)

Substituents like the ethoxycarbonyl and nitro groups are deactivating and *meta*-orienting. Partial rate factors are shown in diagrams XXIII (*Ingold, loc. cit.*) and XXIV (*J. G. Tillett,* J. chem. Soc., 1962, 5142):

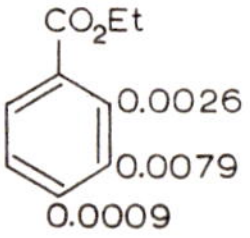

Reagent: $NO_2^{\oplus}$

(XXIII)

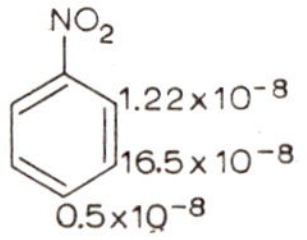

Reagent: $NO_2^{\oplus}$

(XXIV)

These groups have the form Ar·X:Y shown in structures XIV and XV: but the atom (Y) further from the ring is substantially more electron-withdrawing than that (X) attached to the ring. An electron-withdrawing (−*I*) effect would therefore be expected, and *Ingold* (*loc. cit.*; cf. *Rodd,* C.C.C.,

1st Edn., Vol. IIIA, 1954, p. 34 *et seq.*) has consistently regarded this as having in such cases the dominating influence on orientation and reactivity. The question of whether the —*M* effect (XV) also plays a part has been considered by *E. Baciocchi* and *G. Illuminati* (J. Amer. chem. Soc., 1964, **86**, 2677), who have shown that steric interference with conjugation between the nitro group and the ring only slightly diminishes the influence of the group on the rate of *para*- and of *meta*-bromination. Perhaps here we should best consider that there are potentially electron-releasing as well as electron-withdrawing conjugative effects, and that these approximately cancel. It may be significant in this regard that for electrophilic aromatic substitution the nitro and the trimethylammonium substituents have very similar deactivating power (see above), whereas for nucleophilic aromatic substitution, where the (—*M*) effect of the nitro group might respond to interaction with the reagent, the nitro group appears to be much superior in its activating influence (cf. *J. F. Bunnett* and *R. E. Zahler*, Chem. Reviews, 1951, **49**, 273).

(*vi*) *Groups of dual* (−I, +M) *polar character*

Many groups combine (−*I*) and (+*M*) character. As one crosses the periodic table, from $\cdot NR_2$ to $\cdot OR$ to $\cdot F$, the electron-releasing conjugative power of the substituent diminishes (despite the increasing number of lone pairs) because the increasing nuclear charge progressively increases in its influence on the electrons; and for the same reason the electron-withdrawing inductive effect of the substituent increases.

The combined changes can influence the reactivity greatly; for example, for bromination by molecular bromine, relative rates of substitution *para* to the groups $\cdot NMe_2$, $\cdot OMe$, and $\cdot F$ are approximately in the ratio 10^{18}: 10^{9}: 1 (*P. W. Robertson, de la Mare* and *B. E. Swedlund*, J. chem. Soc., 1953, 782). For the halogenobenzenes, (diagrams XXV–XXVIII; *Ingold, loc. cit.*; *J. D. Roberts et al.*, J. Amer. chem. Soc., 1954, **76**, 4525), we have examples of exceptions to Holleman's rule that *ortho,para*-orientation is associated with activation.

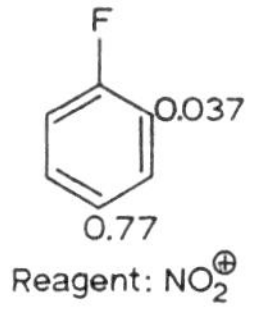

F: 0.037, 0.77 — Reagent: $NO_2^{\oplus}$ (XXV)

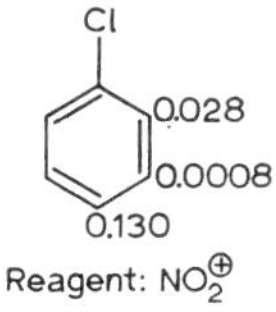

Cl: 0.028, 0.0008, 0.130 — Reagent: $NO_2^{\oplus}$ (XXVI)

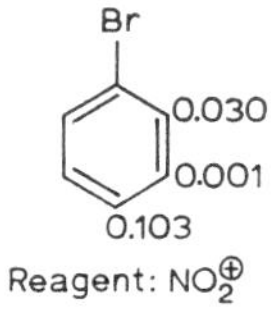

Br: 0.030, 0.001, 0.103 — Reagent: $NO_2^{\oplus}$ (XXVII)

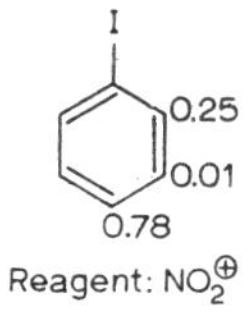

I: 0.25, 0.01, 0.78 — Reagent: $NO_2^{\oplus}$ (XXVIII)

For these compounds, the inductive effects of the halogens are seen to be sufficiently important to make the overall reactivity less than in benzene;

but the associated conjugative effects are, as would be expected, more selective in their influence on the positions in the ring, so that deactivation is associated with *ortho,para*-direction.

Other groups rather balanced in their effect on reactivity, but still clearly *ortho,para*-directing, include some negatively substituted alkyl groups (*e.g.*, $\cdot CH_2Cl$) and the $\cdot CH{:}CH\cdot CO_2H$ and $\cdot CH{:}CH\cdot NO_2$ groups. Some relevant partial rate factors are shown in structures XXIX and XXX (*Ingold* and *F. R. Shaw*, J. chem. Soc., 1949, 575):

CH_2Cl — 0.29 (ortho), 0.14 (meta), 0.95 (para); Reagent: $NO_2^{\oplus}$ (XXIX)

$CH_2 \cdot CO_2Et$ — 4.6 (ortho), 1.2 (meta), 10.4 (para); Reagent: $NO_2^{\oplus}$ (XXX)

Even the phenyl group has $(-I)$ character, as shown by the fact that it deactivates positions *meta* to it (cf. XVII). *A. D. Walsh*'s interpretation (Disc. Faraday Soc., 1947, **2**, 18) that carbon atoms differ in their electronegativity according to their state of hybridisation in the order $sp^3 < sp^2 < sp$ (essentially because the screening effect of the electron pairs on the influence of the nuclear charge becomes progressively less) has been generally accepted.

(*vii*) *Effects of polarisability: the electron-releasing electromeric effect* (*symbol*: +E) *and the inductomeric effect* (*symbol*: $+I_d$)

We have not up to this point taken up the question of whether the electron-releasing conjugative effects, established for groups such as ·OMe and ·CH:CHX, are dominantly mesomeric or dominantly electromeric in character. In considering this matter, it should first be appreciated clearly that Ingold's two classes (*loc. cit.*) of electronic effects, namely permanent (inductive, mesomeric; sometimes called effects of polarisation) and time-variable (inductomeric, electromeric; sometimes called effects of polarisability) cannot be identified with, respectively, effects of the substituent on the energies of ground-state and transition-state. For if, for example, one were to seek to identify the permanent mesomeric effect of a substituent with the conjugative effect of that substituent on the stability of the ground-state, the influence could only be in one direction, that of deactivation for any reaction; since, by their very nature, mesomeric effects stabilise the ground state, and to this extent make it energetically more difficult to reach the transition state. Ingold's way of approaching the problem is to say that there is, for example, a mesomeric effect *on the reactivity* (and hence on both the initial and transition states) in a particular reaction; this effect reflects the permanent conjugative electronic movements in the molecule. His time-variable effects are regarded as specific to the interaction between particular reagents in the transition state, and thus are made manifest differentially when two or more reactions are compared.

The fact that halogen substituents, for example, are *ortho,para*-orienting, although they are known to have a large permanent electron-displacement away from the ring, has been adduced to argue (*Ingold*, in *Rodd*, C.C.C., 1st Edn., Vol. III A, p. 32) that the electron-displacements which are so effective in leading to *ortho,para*-substitution must actually be temporary ($+E$). Evidence pointing in the same direction comes from observations of the difference in activating power when the reagent is changed. Thus for nitration, the fluorine substituent is deactivating; but for molecular bromination or chlorination it is activating. No such marked difference is shown by chlorine and bromine substituents (*de la Mare* and *Robertson*, J. chem. Soc., 1948, 100; *G. Illuminati* and *G. Marino*, J. Amer. chem. Soc., 1956, **78**, 4975). It would be difficult to interpret these structural relationships except in terms of an electromeric effect specific to the reagent.

Evidence for a second polarisability sequence, of an inductomeric character has also been presented (*de la Mare* and *Robertson*, *loc. cit.*).

(*viii*) *Effects of* (+M, +E) *groups on the* meta-*position*

To a first approximation, conjugative electron-release does not reach the *meta*-position; thus the phenyl group, by virtue of its ($-I$) effect, is *meta*-deactivating. There is, however, some evidence for leakage of conjugative effects to the *meta*-position, particularly with reagents for which conjugative effects are important (cf. *de la Mare* and *C. A. Vernon*, J. chem. Soc., 1951, 1764; *de la Mare* and *Ridd*, "Aromatic Substitution – Nitration and Halogenation", Butterworths, London, 1958). It is likely that this also is a manifestation of electromeric polarisability.

(*ix*) *Electronic effects on the* $\frac{1}{2}$o:p *ratio*

It was first proposed by *Ingold* (Ann. Reports, 1926, **23**, 141) that conjugative electron release increases *para*-reactivity more than *ortho*-reactivity. Despite the fact that simple quantum-mechanical treatments (cf. *G. W. Wheland*, J. Amer. chem. Soc., 1942, **64**, 900) do not support this view, it has been generally accepted, partly on the basis of *W. A. Waters*' analogy (J. chem. Soc., 1948, 727) that the quinonoid-like structures involved respectively in *ortho*- and in *para*-substitution should parallel in relative stability the *ortho*- and *para*-quinones. Much of the experimental evidence is of a statistical and general nature, and its unambiguous interpretation is complicated by electrostatic, inductive, and steric effects.

It has been shown, however, (*C. MacLean* and *E. L. Mackor*, J. chem. Phys., 1961, 34, 2208) that the proton magnetic resonance spectra of the positive ions, $ArH_2^{\oplus}$, derived by protonation of the methylbenzenes, are best interpreted on the view that the positive charge is located more on the position *para* to the site of protonation than on the corresponding *ortho*-position.

The reverse behaviour, that conjugative electron-withdrawal should deactivate the *para*- more than the *ortho*-position, has also been considered. The relatively high proportions of substitution *ortho* to such ($-M$) substituents as the nitro and ethoxycarbonyl groups (XXIII, XXIV) have been taken to illustrate this point (*Ingold, loc. cit.*; cf. *Rodd*, C.C.C., 1st Edn. Vol. III A, pp. 44–45). Although contribution from an alternative possibility involving differential activation of the *ortho*-position by way of a cyclic transition state (*Lapworth* and *Robinson*, Mem. Manchester phil. Soc., 1928, **72**, 243) is not completely excluded for every example of this kind*, the fact that changes in $\frac{1}{2}o:p$ ratio parallel changes in the $\frac{1}{2}m:p$ ratio over a wide range of structure and reactivity strongly suggests that the enhanced *ortho*-substitution is usually of electronic origin (*de la Mare* and *Ridd*, *loc. cit.*, p. 81 *et seq.*; cf. *M. Charton*, J. org. Chem, 1969, 34, 278).

Evidence that ($-I$) groups deactivate the *ortho*- more than the *para*-position is substantial. Diagrams XXV–XXVIII illustrate this: despite the opposed influence of steric hindrance, the $\frac{1}{2}o:p$ ratio for nitration falls with increase in the ($-I$) effect from 0.32 for the iodine substituent to 0.05 for the fluorine substituent.

(x) *The* ½o:p *ratio; steric effects and primary steric hindrance*

Since *ortho*- and *para*-substitution are theoretically linked, it might naturally be assumed that the ratio of reactivities in these positions would be determined by electronic factors. But considerations of atomic and molecular size, and of the probable geometry of the transition states for typical aromatic substitutions, suggest that this ratio would in many cases be modified by steric hindrance to electrophilic attack. Such a proposal was made by *Holleman* (*loc. cit.*), who suggested that the change in orientation towards predominant *para*-substitution as one proceeds from nitration through bromination to sulphonation is a manifestation of increasing steric hindrance as the size of the reagent is increased.

* For recent discussions of some of the possible consequences of the intervention of mechanisms involving cyclic transition states in aromatic substitutions, see *R. O. C. Norman* and coworkers (J. chem. Soc. 1961, 3030, 3604, 3610, 3888); *de la Mare, I. C. Hilton* and *S. Varma*, *ibid.*, 1960, 4044).

There is little doubt that this theory is correct in principle. The partial rate factors shown below for nitration and for bromination by positive bromine of toluene and *tert*-butylbenzene (XII; XXXI–XXXIII) illustrate that primary steric hindrance becomes important for alkyl groups larger than the methyl group (*L. M. Stock*, J. org. Chem., 1961, **26**, 4120; *de la Mare* and *J. T. Harvey*, J. chem. Soc., 1956, 36; 1957, 131):

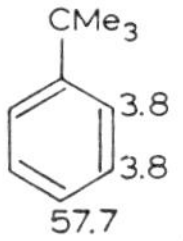

CMe_3: 3.8, 3.8, 57.7 — Reagent: $NO_2^{\oplus}$ (XXXI)

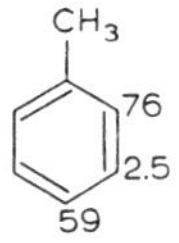

CH_3: 76, 2.5, 59 — Reagent: 'Br$^{\oplus}$' (XXXII)

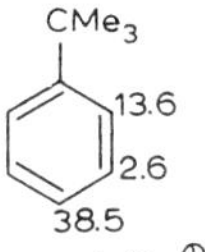

CMe_3: 13.6, 2.6, 38.5 — Reagent: 'Br$^{\oplus}$' (XXXIII)

They help to illustrate also that the detailed composition and geometry of the transition state are important. In nitration, the reagent as it attacks the aromatic ring has a greater effective radius in the direction of a bulky *ortho*-substituent than has the more nearly spherical bromine, which therefore is not so powerfully impeded. This particular comparison can be regarded as satisfactory, since both for the methyl and the *tert*-butyl groups the two reagents give very similar $\frac{1}{2}m:p$ ratios. Comparison of the relative sizes of positive and molecular bromine, or of the nitronium ion and molecular bromine, cannot so easily be made in this way. The partial rate factors for molecular bromination are shown below (XXXIV, XXXV); they are very much larger than those for nitration or for bromination by positive bromine, and this difference must in itself contribute in determining the $\frac{1}{2}o:p$ ratio (*H. C. Brown* and *Stock*, J. Amer. chem. Soc., 1957, **79**, 1421):

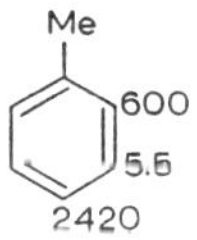

Me: 600, 5.5, 2420 — Reagent: Br_2 (XXXIV)

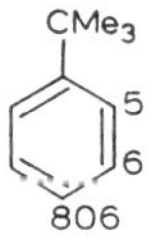

CMe_3: 5, 6, 806 — Reagent: Br_2 (XXXV)

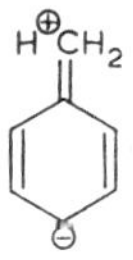

$H^{\oplus}CH_2$ (XXXVI)

The sequences of effective molecular size, $Cl^{\oplus} < Br^{\oplus} < NO_2^{\oplus}$ and $Cl_2 < Br_2$, can be regarded as established, but the exact positions of many other reagents can be estimated only approximately. The close balance between steric effects in initial and transition states is emphasised by the fact that in certain aromatic displacements, particularly where a large group is replaced by a smaller one, steric acceleration is to be expected. One of the first definite examples of steric acceleration in an aromatic replacement was given by *R. A. Benkeser* and *R. A. Krysiak* (*ibid.*, 1954, **76**, 6353; cf. *C. Eaborn* and *R. C. Moore*, J. chem. Soc., 1959, 3640) (cf. p. 73).

(xi) Conjugative effects of alkyl groups; hyperconjugation

The data presented in diagrams XXXI–XXXV bring us now to a more complete account of the electronic effects of alkyl groups. Relative to hydrogen, these substituents are electron-releasing, independently of whether or not they are conjugated with the centre they are influencing. For this reason, they have been ascribed a $(+I)$ effect, greater for the *tert*-butyl group than for the methyl group. Accordingly the *tert*-butyl group is usually superior to the methyl group in activating the *meta*-position for aromatic electrophilic substitution.

In 1935, *J. W. Baker* and *W. S. Nathan* (J. chem. Soc., 1935, 1844) showed that sometimes the opposite order of electron-release, Me > *tert*-Bu, could be observed, especially when the alkyl group could be regarded as conjugated with the reaction centre. They proposed a mode of electron-release which in valence-bond terms can be represented for the ground-state of toluene as in structure XXXVI, and has variously been called the "Baker-Nathan effect", "alkyl conjugation", "no-bond resonance", and "hyperconjugation". Quantum-mechanical calculations based on the molecular orbital approach (*R. S. Mulliken, C. A. Rieke* and *W. G. Brown*, J. Amer. chem. Soc., 1941, **63**, 41) provided substantial theoretical justification for this suggestion (cf. *N. Muller* and *Mulliken, ibid.*, 1958, **80**, 3489; *Mulliken*, Tetrahedron, 1959, **5**, 253; **6**, 68). Cogent experimental support has been provided also by the work of *V. J. Shiner* and coworkers (J. Amer. chem. Soc., 1960, **82**, 2655; 1963, **85**, 2416; 1964, **86**, 945), who showed that the isotope effect on electron-release by C–H groups varies with the geometry of the system in the way expected if the electron release involves overlap of the C–H bonding electrons with an adjacent unsaturated group*.

The concept of hyperconjugation has been extensively applied in the interpretation both of the ground-state and of the transition-state properties of alkyl-substituted unsaturated systems. An excellent example of the Baker-Nathan order of electron release, Me > *tert*-Bu > H, was provided by *E. D. Hughes, Ingold* and *N.A. Taher* (J. chem. Soc., 1940, 949) from the rates of solvolysis of the alkyl-substituted diphenylmethyl chlorides. The first application of the theory to aromatic substitution was provided by *de la Mare* and *Robertson* (*ibid.*, 1943, 279), and independently by *E. Berliner*

* Although the writer does not seek to be accused of breaking the Second Commandment (Exodus 20, 4), he confesses not to subscribe to the agnosticism on this subject referred to in Vol. I A, p. 265.

and *F. J. Bondhus* (J. Amer. chem. Soc., 1946, **68**, 2355); later investigations have shown that *para*-substitution in alkylbenzenes often, though not invariably, shows the Baker-Nathan order. Diagrams XXXII–XXXV illustrate that bromination provides examples. Among the many other cases now established, the heats of formation of the benzenonium ions, $[R \cdot C_6H_5 \cdot H]^{\oplus}$ from the hydrocarbon and fluorosulphuric acid in antimony pentafluoride give a particularly clear-cut illustration (*E. M. Arnett* and *J. W. Larsen*, *ibid.*, 1969, **91**, 1438).

Substituted alkyl groups can, of course also contribute electrons through C–H hyperconjugation, and it has been known for a long time (*Ingold* and *Shaw*, *loc. cit.*; cf. also *J. R. Knowles* and *R. O. C. Norman*, J. chem. Soc., 1961, 2938; *F. L. Riley* and *E. Rothstein*, *ibid.*, 1964, 3860, 3872) that groups of the type $\cdot CH_2X$, where X is sufficiently electron-withdrawing, direct substituents into the *ortho*- and *para*-positions, which are in some cases activated and in others deactivated (cf. diagrams XXIX–XXX). *Mulliken, Riecke* and *Brown*'s theory (*loc.cit.*) is furthermore applicable not only to C–H bonds but also to other bonds similarly situated with respect to an unsaturated system; and in 1948 *Berliner* and *Bondhus* (J. Amer. chem. Soc., 1948, **70**, 854) extended the range over which hyperconjugation should be applied in the field of aromatic substitution by suggesting that the powerful activation of the *para*-position in the bromination of *tert*-butyl benzene (XXXV) cannot plausibly be attributed entirely to the inductive effect, and therefore may be partly of conjugative origin. That both the conjugative and the inductive effects of alkyl groups are closely balanced and subject to modification by other features of the system under study has been recognised for some time (cf. *Berliner*, Tetrahedron, 1959, **5**, 143); and the complicated influences of solvent on the Baker-Nathan order have become manifest in many investigations (cf. *A. Himoe* and *Stock*, J. Amer. chem. Soc., 1969, **91**, 1452).

Hyperconjugation by σ-electrons has been considered also to contribute to the electron-releasing properties of the $\cdot$O–H group (*de la Mare*, Tetrahedron 1969, **5**, 107). The solvent would be expected to play a part in O–H hyperconjugation, and one manifestation of this may be the prominence of the S_E2' path for substitution in phenols and naphthols (p. 79); "no-bond resonance" is taken to its limit when in the transition state the hyperconjugating bond becomes broken. Quantum-mechanical theory(*Mulliken*, *loc. cit.*) indicates that other groups generally will be capable of hyperconjugative electron release; in discussing the electron-withdrawing properties of $\cdot NR_3^{\oplus}$ groups (cf. p. 51) it is difficult to assess this factor, which may enhance the reactivities of the *para*-positions in $Ph \cdot NMe_3^{\oplus}$ and $Ph \cdot NH_3^{\oplus}$.

(c) Quantitative treatment of directive effects

The qualitative electronic theory of valency, some aspects of which are outlined above, stimulated widespread interest in careful quantitative studies of physical properties of organic molecules. Investigation, for example, of dipole moments (*L. E. Sutton,* Proc. roy. Soc. [London], 1931, **A, 133,** 668) and of strengths of acids and bases (*J. F. J. Dippy, H. B. Watson* and *F. R. Williams,* J. chem. Soc., 1935, 346; cf. *H. B. Watson,* "Modern Theories of Organic Chemistry", University Press, Oxford, 1937) revealed correlations between the effects of change of structure on these properties on the one hand, and on aromatic reactivity and orientation on the other, which played an important part in the recognition of the importance of the resulting new approach to organic chemistry. That such correlations are qualitatively successful has led a number of investigators to attempt to express these relationships in quantitative form. Although these attempts have been only in part successful, they have made significant contributions towards clarification of theories of organic reactivity.

(i) Linear free-energy relationships in aromatic substitution

The starting-point for this type of approach has been to assume that substituents affect the free-energy of activation (generally equated with the heat of activation) in a way which is characteristic of the substituent, and is in principle separable from the effect of altering the reaction. It is assumed, therefore, that the effects of substituents on aromatic substitutions can be correlated, either between themselves or with the related influences on other equilibria or reaction rates, by way of plots, which should in principle be linear, of energy-differences.

As far as aromatic substitution is concerned, a treatment of this type was first suggested by *F. E. C. Scheffer* (Proc. k. Akad. Wetenschap. Amsterdam, 1913, **15,** 1109, 1118). Later, *A. E. Bradfield* and *B. Jones* (J. chem. Soc., 1928, 1006, 3073; Trans. Faraday Soc., 1941, **37,** 726) presented evidence that for a series of substituted ethers and anilides the effects of substituents on the rate of chlorination in acetic acid are independent and additive in heat of activation. Later workers have used the same hypothesis in interpreting the relative reactivities of polyalkylbenzenes and their derivatives (cf. *F. E. Condon,* J. Amer. chem. Soc., 1948, **70,** 1963; 1949, **71,** 3544; 1952, **74,** 2528; *H. C. Brown* and *L. M. Stock, loc. cit.*; *E. Baciocchi* and *G. Illuminati,* Gazz., 1962, **92,** 89). Such an *additivity principle* had earlier been considered by *A. F. Holleman* (*loc. cit.*) as an approach to treatment of the combined orientational influence of more than one substituent. It is very useful in providing a method for determining indirectly the partial rate factors for substitution in positions for which a direct measurement is impracticable; as for example for substitution *meta* to powerfully activating substituents (cf. *P. B. D. de la Mare* and *Vernon, loc. cit.*; *de la Mare* and *M. Hassan*, J. chem. Soc., 1958, 1519; *Stock* and *Brown*, J. Amer. chem. Soc., 1960, **82**, 1942). Some of its limitations, especially those of steric origin which have been discussed for example by *Baciocchi* and *Illuminati* (*loc. cit.*) and by *de la Mare* and *Hassan* (*loc. cit.*), are theoretically understood; others are still obscure.

The most generally used treatment of reaction-correlation based on linear free-energy relationships was proposed by *L. P. Hammett* (Trans. Faraday Soc., 1938, **34**, 156). He suggested that the effects of substituents in the reactions of benzyl and related compounds could be correlated through an equation such as [1].

$$\log_{10}(k_{R}/k_{H}) = \sigma_{R}\varrho \qquad [1]$$

In this equation, k_R is the rate- or equilibrium-constant for a compound containing the substituent R; ϱ is a reaction constant independent of the substituent; and σ_R is a substituent constant independent of the reaction. Standard values of σ_R were determined from the effects of substituents on the dissociation constants of substituted benzoic acids, with ϱ taken as 1·00 for this reaction. The values of σ_R for *meta*-substituents could be considered to give an approximation to the relative inductive effects of these groups, but the corresponding values for *para*-substituents were quite clearly composite of inductive and conjugative effects. It was already obvious that there was a general but not wholly satisfactory correlation with results for aromatic electrophilic substitution; a more specialised equation, proposed by *T. Ri* and *H. Eyring* (J. chem. phys., 1940, **8**, 433) to correlate rates of aromatic substitution with dipole moments, carried similar implications and had similar unsatisfactory features.

Equations of this kind carry the implication that the partial rate factors for *meta*- and *para*-substitution should be related by an equation such as [2] (cf. *de la Mare*, J. chem. Soc., 1954, 4450):

$$\log_{10} f_m = \text{const.} \log_{10} f_p \qquad [2]$$

In 1953, *Brown* and *K. L. Nelson* (J. Amer. chem. Soc., 1953, **75**, 6292) proposed a more complicated form of this equation, and called it the "selectivity relationship" [equation 3].

$$\log_{10}(f_p/f_m) = \text{const.} \log_{10} f_p \qquad [3]$$

Reasonable agreement with experiment was found for a number of electrophilic substitutions in toluene. Seeking, therefore, a more satisfactory reaction sequence than the dissociation constants of the benzoic acids to compare with aromatic substitutions, Brown chose the unimolecular solvolyses of substituted phenyldimethylcarbinyl chlorides. These reactions were shown to correlate well with a number of electrophilic aromatic substitutions, and by using the data for the rates of the solvolytic process, a series of so-called "electrophilic substituent constants" was defined by equation [4].

$$\log_{10}(k_R/k_H) = \sigma_R{}^+\varrho \qquad [4]$$

Here ϱ was taken as numerically $-4{\cdot}620$ for the standard reaction, because with this value the σ_R and $\sigma_R{}^+$ constants were put as nearly as possible on the same scale (*Y. Okamoto* and *Brown*, J. org. Chem., 1957, **22**, 485.).

Since that time, *Brown* and his coworkers have made an extensive examination of the applicability of this treatment. The results have been summarised ("A Quantitative Treatment of Directive Effects in Aromatic Substitution," by *Stock* and *Brown*, in "Advances in Physical Organic Chemistry", Ed. *V. Gold*, Academic Press, New York, 1963, p. 35). The general conclusion is that the treatment is partly successful; and that, where it fails, the reaction-path is complicated by effects specific to the electrophile. Some of these special features are discussed below; modified equations which take into account factors of polarisability by assuming that the conjugative contribution to the substituent constant is variable have been proposed by *H. van Bekkum, P. E. Verkade* and *B. M. Wepster* (Rec. Trav. chim., 1959, **78**, 815), and by *Y. Yukawa* and *Y. Tsuno* (Bull. chem. Soc. Japan, 1959, **32**, 971). *Charton* (J. org. Chem., 1969, **34**, 278) has discussed application of such treatments to substitution at the *ortho*-position and recently *Ingold* ("Structure and Mechanism in Organic Chemistry", 2nd Edn., G. Bell, London, 1969) has provided an important general survey.

(*ii*) *Polycyclic aromatic hydrocarbons*

Theories of aromatic substitution which imply linear free-energy relationships have been extended to discussions of the reactions of polycyclic aromatic hydrocarbons with electrophiles. The most notable feature of these processes is that, whereas benzene because of its stabilisation by resonance is rather unreactive, bi- and poly-cyclic hydrocarbons become progressively more activated, and the reactions are often, though not always, orientationally rather selective. Table 1 shows some examples taken mainly from a summary given by *de la Mare* and *Ridd* (*loc. cit.*).

TABLE 1

CHLORINATION OF SOME POLYCYCLIC AROMATIC HYDROCARBONS

Compound		Rel. rate of chlorination ($PhH = 1$; Cl_2, HOAc, 25°)	Main position of substitution
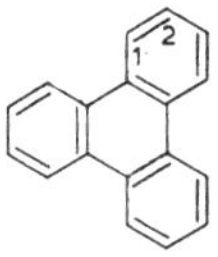	(XXXVII)	2.3×10^4	1,2
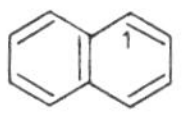	(XXXVIII)	6.6×10^4	1
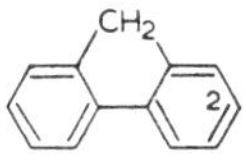	(XXXIX)	1.1×10^5	2
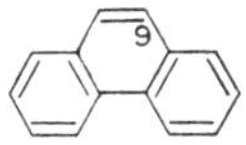	(XL)	3.0×10^5	9
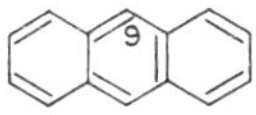	(XLI)	very rapid ($> 10^8$)	9

Addition accompanies substitution in all these cases, and probably also in many other reactions of these compounds with electrophiles. The substitutions, however, are not necessarily preceded by completion of the addition process, as was shown for the bromination of phenanthrene by *C. C. Price* (J. Amer. chem. Soc., 1936, 58, 2101; Chem Reviews, 1941, 29, 37) and has been confirmed for most of the above chlorinations.

Equation [5] is an example of the type of relationship which has been applied extensively in theoretical treatments of these and related data (cf. *M. J. S. Dewar* and coworkers, J. Amer. chem. Soc., 1952, 74, 3357; J. chem. Soc., 1956, 3576, 3581).

$$\log_{10}(k_x/k_o) = \text{const.}(N_x - N_o) \qquad [5]$$

Here, k_x is the rate of substitution at a particular position (x) in a polycyclic compound relative to the rate (k_0) at one position in benzene, and N_x (the reactivity number) is a theoretical parameter, derivable by calculation, and representing the potential reactivity at that position.

Comparison of the theoretical value for the constant term in equation [5] with the values derived from experiments on nitration and chlorination indicate that the experimental value for chlorination is not very different from the theoretical value, but that the experimental value for nitration is much lower. This result accords with an earlier conclusion (*de la Mare* and *P. W. Robertson*, J. chem. Soc., 1948, 100) that conjugative effects are more important in molecular halogenation than in nitration. Accordingly, it was suggested (*Dewar* and *T. Mole*, *ibid.*, 1957, 342) that the transition state for this reaction should be described as in XLII, with only partial bonding between the entering electrophile and the ring, rather than as in the carbonium ionic form (XLIII) discussed* by *P. Pfeiffer* and *R. Wizinger* (Ann. 1928, 461, 132), and adopted as a model for the transition state by many later workers (cf. *G. W. Wheland*, J. Amer. chem. Soc., 1942, 64, 900):

(XLII) (XLIII)

Details of orientation and rate of substitution are not, however, always well predicted by this treatment; and, though a multitude of other theoretically derived parameters have been tested for correlation with rates of substitution in polycyclic aromatic systems (cf. *A. Streitwieser Jr.*, "Molecular Orbital Theory for Organic Chemists", Wiley, New York, 1961) none of them can be claimed to be much superior to Dewar's. Two extra types of structural influence not allowed for in the calculations have been considered to make significant contributions to the reactivities of such systems. The first is steric hindrance, which has been considered (*Dewar et al.*, *loc. cit.*) to explain anomalies in the rate of substitution in the positions analogous to the 1-position in naphthalene and to the 4-position in phenanthrene. Much more work needs to be done to establish the area of validity of this concept; the effect is probably important only for large reagents (cf. *R. Bolton*, *de la Mare* and *L. Main*, J. chem. Soc., B, 1969, 170).

* Pfeiffer and Wizinger's paper preceded the general acceptance of the theory of resonance between valence-bond structures, so they formulated their intermediate without indicating electron-delocalisation in the cyclohexadienyl cation; they recognised, however, not only the stabilising effect of substituents, like ·OMe and $\cdot NH_2$, which can provide an alternative site for the positive charge, but also the necessity of conjugation between such substituents and the point of attachment of the electrophile.

The second is the influence of internal strain on reactivity. Careful comparison of partial rate factors for chlorination of fluorene (XXXIX) with those for its unstrained analogues with larger bridges (XLIV; n = 2, 3, 4), due allowance being made for planarity, led *de la Mare, E. A. Johnson* and *J. S. Lomas* (J. chem. Soc., 1964, 5317) to the conclusion that the strain introduced by the methylene bridge in XXXIX was responsible for a significant increase in reactivity in the 2-position:

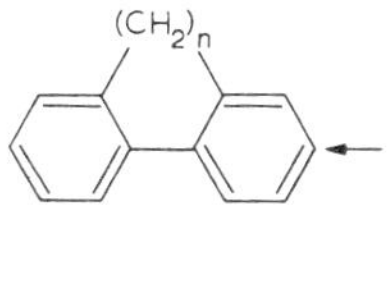

$(CH_2)_n$ $H_2C—CH_2$

(XLIV) 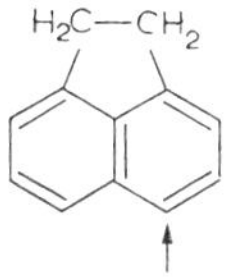 (XLV)

Similarly, *Berliner, D. M. Falcione* and *J. L. Riemenschneider* (J. org. Chem., 1965, 30, 1812) have concluded that acenaphthene (XLV) has for the same reason enhanced reactivity in bromination. It seems probable that strain is a general feature which, when it exists, can affect the reactivity of aromatic systems, enhancing it in some positions and perhaps diminishing it in others (*R. Taylor*, J. chem. Soc., B, 1968, 1559).

(d) Substitution in heterocyclic aromatic systems

Since electrophilic substitution was for many years considered to be the main reaction which characterised aromaticity, it is natural that the electrophilic substitution reactions of heterocyclic systems should have been the subject of much study. Unfortunately, theoretical studies in this field often have not been relevant to the conditions of the experimental observations, and experimental work in its turn has until recently not been subject to mechanistic control sufficient to define clearly what theoretical approach would be appropriate. *J. H. Ridd* (in "Physical Methods in Heterocyclic Chemistry", Vol. 1, Ed. *A. R. Katritsky*, Interscience, New York, 1963, p. 109 *et seq.*) has outlined some of the outstanding areas of difficulty.

Many of the theoretical calculations of reactivity in heterocyclic systems have been concerned with pyridine and the related nitrogen heterocycles. In almost all these treatments, the approach has been to consider the heterocyclic ring as being analogous with the benzene ring in which one –CH= group is considered to be replaced by a more electronegative group. It has been qualitatively irrelevant in these calculations whether or not the nitrogen atom was protonated; the difference between =N–, and $=N^{\oplus}H–$ could be taken into account by a quantitative difference in the assumed electronegativity. By implication, other groups of the type $=N^{\oplus}R-$ could be treated within the same framework, though of course it was envisaged that extreme cases (*e.g.* when R is $O^{\ominus}$) would involve separate consideration.

Calculations based on this approach, with different models for the transition state, have provided a variety of so-called reactivity indices, which could be converted into orientational predictions and into predicted rates of reaction relative to benzene by the assumptions involved in a linear free-energy treatment. Perhaps the most commonly quoted are the charge (π-electron) densities (cf. *R. D. Brown*, Quart. Reviews, 1952, 6, 83); predictions based on these quantities take the initial state as a model for the transition state, an assumption known to be inappropriate for most electrophilic substitutions. Localisation energies, being based on the carbonium ionic transition state (cf. XLIII), might theoretically be thought to be more suitable. These quantities have been used to predict the orientation of substitution with about the same degree of success as is found for the polycyclic aromatic hydrocarbons. But data are now accumulating which establish approximate partial rate factors for substitution in some of the important reference compounds under conditions of known mechanism, largely as the result of work by *Ridd* and coworkers (cf. J. chem. Soc., 1963, 4204; 1965, 1051), *K. Schofield* and coworkers (cf. *ibid.*, B, 1966, 870), and *A. R. Katritzky* and coworkers (cf. *ibid.*, 1963, 3753, 3764; B, 1967, 1204). Some values for nitration are shown in diagrams XLVI–XLIX:

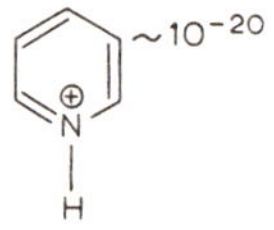

~10^{-20} N$^{\oplus}$ H

Reagent: $NO_2^{\oplus}$

(XLVI)

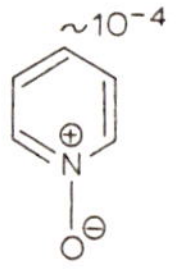

~10^{-4} N$^{\oplus}$ $O^{\ominus}$

Reagent: $NO_2^{\oplus}$

(XLVII)

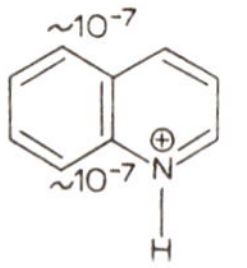

~10^{-7} ~10^{-7} N$^{\oplus}$ H

Reagent: $NO_2^{\oplus}$

(XLVIII)

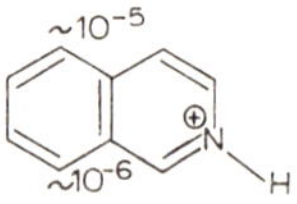

~10^{-5} ~10^{-6} N$^{\oplus}$ H

Reagent: $NO_2^{\oplus}$

(XLIX)

These experimental values are widely different from the predictions of the various theoretical approaches (*Ridd, loc. cit.*), and the subject awaits some major advance which will supplant the existing, quite unsatisfactory, theories, and perhaps at the same time improve the theory of substitution in polycyclic hydrocarbons.

The other type of heterocycle which has been much studied and discussed is that in which the hetero-atom is highly activating for electrophilic substitution because it carries a lone pair of electrons conjugated with the double-bond system. Pyrrole (L), furan (LI), thiophene (LII), and their benzo- and dibenzo-annelated derivatives fall into this class, which may also be held to include imidazole and benzimidazole. Partial rate factors for bromination of some of these systems have been estimated by *P. Linda* and *G. Marino* (Chem. Comm., 1967, 499), and are given in the diagrams (L–LII):

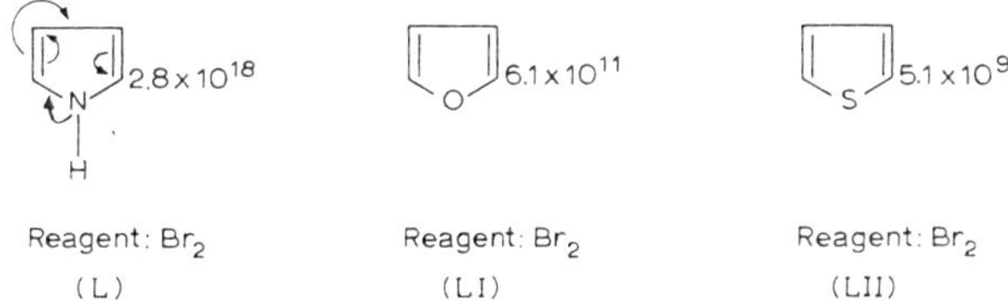

Qualitatively, the results are interpretable satisfactorily by using the theory of mesomeric electron-release developed for derivatives of benzene. The arrows in L indicate how both α- and β-positions can become activated; the observed sequence of electron-release ($+M$): (NH $>$ O $>$ S) is well established in other studies. That the β-positions in such molecules are less activated than corresponding α-positions, but are nevertheless more activated than a single position in benzene, has been shown by *F. B. Deans* and *C. Eaborn* (J. chem. Soc., 1959, 2303) for protodesilylation of thiophene.

Since current theoretical treatments are not sufficiently refined to describe satisfactorily the details of orientation (particularly the $\frac{1}{2}o{:}p$ ratio) determined by exocyclic substituents such as the amino group, it is not surprising that theoretical attempts similarly to predict orientation in activated heterocycles are not uniformly successful (cf. *Ridd, loc. cit.*). In only a few cases of this kind have partial rate factors for aromatic substitution been measured. *Eaborn* and *J. A. Sperry* (J. chem. Soc., 1961, 4921) have compared the partial rate factors for the protodesilylation of diphenyl ether (LIII) and dibenzofuran (LIV), and *M. J. S. Dewar* and *D. S. Urch* (*ibid.*, 1957, 345; 1958, 3079) have made related comparisons for the nitration of diphenylamine and carbazole. In each case, on going from the acyclic to the cyclic structure, there was a reduction in reactivity *para* to the hetero-atom. Part of this reduction is the result of the inductive effect of the aryl group; but there is probably also a further rate-reduction, which has been attributed to the special resonance of the heterocyclic ring. The quantitative contribution from this latter factor seems, however, to be small; the reactivities of the various positions in dibenzofuran and in carbazole are not very different from what would be expected for similarly substituted compounds in which the hetero-atom is not part of the cyclic conjugated system (cf. also *P. B. D. de la Mare, O. M. H. el Dusouqui* and *E. A. Johnson, ibid.*, 1966, 521).

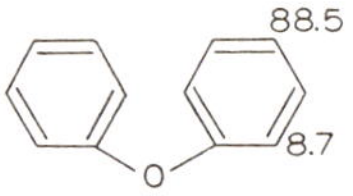

88.5 8.7

Reagent: $H^{\oplus}$, displacing $SiMe_3^{\oplus}$

(LIII)

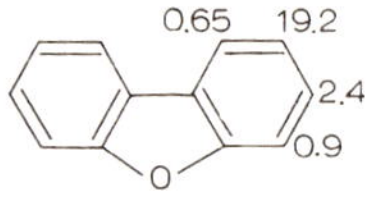

0.65 19.2 2.4 0.9

Reagent: $H^{\oplus}$, displacing $SiMe_3^{\oplus}$

(LIV)

(e) Reagents concerned in electrophilic aromatic substitution

So far, in considering how the nature of the electrophile involved in attack on the aromatic nucleus may affect the relative rates of substitution in the various nuclear positions, we have noted only:

(a) that steric effects (notably those of primary steric hindrance) may influence the rate of substitution adjacent to bulky groups, and thus with large reagents may reduce the $\frac{1}{2}o:p$ ratio;

(b) that reagents differ in the extent to which change in rate responds to change in structure (*i.e.* have different values of ϱ in the Hammett equation or its modifications), and that this has its own intrinsic influence on isomer ratios as well as on rates of reaction.

Physico-chemical studies of various kinds have been made of recent years to define as far as possible the nature of the reagents concerned in the important aromatic substitutions and replacements. Nitration was examined in detail by *C. K. Ingold, E. D. Hughes,* and coworkers in a series of most important investigations, summarised by *Ingold* ("Structure and Mechanism in Organic Chemistry", G. Bell, London, 1953). These investigations made it reasonably certain that the most important reagent concerned in this reaction is the nitronium ion, $NO_2^{\oplus}$, which can be formed in pre-equilibrium or in the rate-determining stage of the reaction.

It was made probable also that covalently combined forms of the nitronium ion (*e.g.* $NO_2\cdot ONO_2$; $NO_2\cdot O\cdot CO\cdot CH_3$) could become effective under rather special conditions (*Ingold, Hughes et al.*, J. chem. Soc., 1950, 2452, 2467; cf. *F. G. Bordwell* and *E. W. Garbisch*, J. Amer. chem. Soc., 1960, **82**, 3588).

Nitration through nitrosation can also become important, particularly for reactive substrates (*Ingold, loc. cit.*), for which the rate of nitration by the nitronium ion can reach a limiting value determined by the encounter rate (*K. Schofield et al.*, J. chem. Soc. B, 1968, 800; 1969, 1).

As far as halogenation is concerned, it was first shown by *E. Shilov* and *N. P. Kaniaev* (C. r. Acad. Sci., U.R.S.S., 1939, **24**, 890) that a positively charged brominating species, either $Br^{\oplus}$ or $BrOH_2^{\oplus}$, can be an effective reagent. Later investigations confirmed this conclusion, extended it to chlorination (*D. H. Derbyshire* and *W. A. Waters*, J. chem. Soc., 1950, 564; 1951, 73; *P. B. D. de la Mare, A. D. Ketley* and *C. A. Vernon, ibid.*, 1954, 1290; *S. J. Branch* and *B. Jones, ibid.*, 1954, 2317), and established also that, apart from minor details, the response of reaction-rate to change in structure is rather similar in these reactions and in nitration (*de la Mare* and *Harvey, loc. cit.*; *de la Mare* and *I. C. Hilton*, J. chem. Soc., 1962, 997).

The more usual modes of chlorination and bromination, however, respond much more strongly to change in structure, and particularly to electron-releasing conjugative effects. Reagents of the type Hal-X, where $X^{\ominus}$ may be almost any nucleophilic anion, can be made effective; and where X is halogen, kinetic measurements show that the whole of the halogen molecule is concerned in the transition state, perhaps in a complex of stoichiometry $ArH \cdot X_2$. It has been argued, again in part from the response of reactivity to change in structure, that such complexes involve covalent attachment of the halogen molecule to the aromatic compound, so that they should be regarded as structurally similar to the well known trihalide ions (*e.g.*, $Br_3^{\ominus}$; *de la Mare*, "The Kekulé Symposium", Butterworths, London, 1958, p. 219). Catalysis of the stage of the reaction which involves the breaking of the Br–Br bond in the complex $ArH \cdot Br_2$ has also been recognised kinetically (*P. W. Robertson et al.*, J. chem. Soc., 1943, 276).

Another reaction in which it is known that the electrophile is often not completely removed from its associated anion in the transition state is nitrosation (*E. J. Blackall, Hughes* and *Ingold, ibid.*, 1952, 28; *Ingold, Hughes* and *J. H. Ridd, ibid.*, 1958, 88). Reagents such as NOCl, $NO \cdot NO_2$, and even the nitrosonium ion, $NO^{\oplus}$, attack readily only highly activated centres, such as the *para*-position in $Ph \cdot NMe_2$; in fact, *N*-nitrosation has been much more extensively studied than *C*-nitrosation.

Sulphonation, though often reversible at high temperatures, can normally be effected under conditions of kinetic control. Deactivated substrates such as *p*-nitrotoluene are sulphonated at rates suitable for measurement in oleum (*W. A. Cowdrey* and *D. S. Davies, ibid.*, 1949, 1871), whereas activated substrates such as toluene are sulphonated quite rapidly in aqueous sulphuric acid (*C. Eaborn* and *R. Taylor, ibid.*, 1960, 1480). The kinetics of sulphonation in aqueous sulphuric acid, when proper allowance is made for protonation of the aromatic substrate and for effects of change in medium, can generally be interpreted in terms of sulphur trioxide or a solvated form of this entity as the kinetically effective electrophile. The earlier work is summarised by *Ingold* (*loc. cit.*); the more recent studies are by *H. Cerfontain* and his co-workers (Rec. Trav. chim., 1960, **79,** 935; 1961, **80,** 11, 296; 1962, **81,** 969; 1963, **82,** 113, 565, 659, 923; 1964, **83,** 226, 493, 1103; 1968, **87,** 24, 873; "Mechanistic Aspects of aromatic sulfonation and desulfonation" Interscience, New York, 1968). An extra proton may be involved in sulphonation in oleum (*J. C. D. Brand, A. W. P. Jarvie* and *W. C. Horning*, J. chem. Soc., 1959, 3844).

Solutions of sulphur trioxide in aprotic solvents have also been used for kinetic investigations (*C. N. Hinshelwood et al., ibid.*, 1939, 1372; 1944, 469, 649). Under these conditions, the rate is proportional to the concentration of aromatic compound and to the square of the concentration of sulphur trioxide. As with molecular bromine, a second molecule of sulphur trioxide is apparently concerned in the transition state; perhaps it acts as a catalyst for removal of a proton, but more probably it is involved in polarising the electrophilic sulphur trioxide.

Attempts have been made to establish the nature of the attacking electrophile in a number of reactions of the Friedel-Crafts type. The results have been surveyed by *G. Olah* ("Friedel-Crafts and Related Reactions", Vols. 1–4, Interscience, N.Y., 1963 ff; cf. also *L.M. Stock* and *H. C. Brown*, "Advances in Physical Organic Chemistry", Ed. *V. Gold*, Academic Press, New York, 1963); equations 6–9 illustrate some of the conditions chosen for study.

$$\text{Ar·H} + \text{Ph·CO·Cl} \xrightarrow{\text{AlCl}_3,\ \text{Ph.NO}_2} \text{Ar·CO·Ph} + \text{HCl} \qquad [6]$$

$$\text{Ar·H} + \text{Me·CO·Cl} \xrightarrow{\text{AlCl}_3,\ \text{C}_2\text{H}_4\text{Cl}_2} \text{Ar·CO·Me} + \text{HCl} \qquad [7]$$

$$\text{Ar·H} + \text{CH}_2\text{O} + \text{HOAc} \xrightarrow{\text{HCl},\ \text{ZnCl}_2} \text{Ar·CH}_2\text{·OAc} + \text{H}_2\text{O} \qquad [8]$$

$$\text{Ar·H} + \text{EtBr} \xrightarrow{\text{GaBr}_3} \text{Ar·Et} + \text{HBr} \qquad [9]$$

Most of the reactions cannot be carried out except in essentially anhydrous, or dehydrating, conditions, and this makes it very difficult to establish the exact mode of combination of the catalyst in the transition state. Probably, polarised complexes, $[\text{R}^{\delta\oplus}\text{--X--MX}_{n-1}^{\delta\ominus}]$, or ion-pairs, $[\text{R}^{\oplus}\ \text{MX}_n^{\ominus}]$, where M is a metal, X is halogen, and R is an alkyl or acyl group, are often implicated. Structural effects, as far as they have been investigated, can be interpreted on the lines already outlined.

Many reactions are known in which groups other than hydrogen are replaced by an electrophile, and some of them have been investigated mechanistically; desilylation and related replacements for example particularly by *Eaborn* and his coworkers (equations 10, 11; J. chem. Soc., 1956, 4858; 1957, 4449; 1959, 3034, 3640; 1960, 179, 1566; 1961, 297, 3715, 4921, 5082); deboronation by *R. A. Benkeser* and coworkers (equations 12, 13; J. Amer. chem. Soc., 1953, **75**, 4528; 1954, **76**, 6353; 1958, **80**, 2279, 2283, 5294); and decarbonylation by *W. M. Schubert* and his co-workers (equation 14; *ibid.*, 1956, **78**, 64; 1958, **80**, 1755; 1959, **81**, 3923). Structural effects,

$$Ar{\cdot}SiMe_3 \xrightarrow{Br_2,\ HOAc} Ar{\cdot}Br + \text{other products} \qquad [10]$$

$$Ar{\cdot}SiMe_3 \xrightarrow{H^{\oplus},\ MeOH,\ H_2O} Ar{\cdot}H + \text{other products} \qquad [11]$$

$$Ar{\cdot}B(OH)_2 \xrightarrow{Br_2,\ HOAc,\ H_2O} ArBr + \text{other products} \qquad [12]$$

$$Ar{\cdot}B(OH)_2 \xrightarrow{H^{\oplus},\ H_2O} ArH + \text{other products} \qquad [13]$$

$$Ar{\cdot}CHO \xrightarrow{H^{\oplus}} ArH + CO \qquad [14]$$

summarised by *Stock* and *Brown* (*loc. cit.*) show fairly good correlations with the more usual aromatic substitutions. A further general survey, which covers most of the investigated aromatic replacements, has been given by *R. O. C. Norman* and *R. Taylor* ("Electrophilic Substitution in Benzenoid Compounds", Elsevier, Amsterdam, 1965).

(f) Stages on the reaction path

(i) π-Complexes as intermediates

The simplest conceptual picture for an electrophilic aromatic substitution is of a one-stage bimolecular process (S_E2), in which the transition state is reached from the starting materials without the intervention of complexes of finite life. Evidence is forthcoming, however, to establish that the reaction path is often more complicated than this. A number of authors have surveyed this matter, and among the important general accounts of the experimental and theoretical problems are the substantial articles by *L. Melander* (Ark. Kemi, 1950, **2**, 211); by *E. Berliner* (in "Progress in Physical Organic Chemistry 2" Eds. *S. G. Cohen, A. Streitwieser Jr.* and *R. W. Taft*, Interscience, New York, 1964, p. 253); and by *H. Zollinger* (in "Advances in Physical Organic Chemistry 2", Ed. *V. Gold*, Academic Press, New York, 1964, p. 163). The earliest stage which has been considered as potentially important is that in which the unsaturation electrons of the ring are considered to act as a unit in the manner of a general base to form "charge-transfer" or π-complexes with electrophilic ions or molecules. Such interactions are recognised, for example, from studies of the spectra of solutions of halogens in aromatic hydrocarbons (cf. *F. Fairbrother,* J. chem. Soc., 1948, 1051; *H. A. Benesi* and *J. H. Hildebrand,* J. Amer. chem. Soc., 1948, **70**, 2832; 1949, **71**, 2703). Infrared (*E. E. Ferguson,* J. chem. Phys., 1956, **25**, 1115; 1957, **26**, 1357) and crystallographic (*O. Hassel* and *C. Roming,* Quart. Reviews, 1962, **16**, 1) studies confirm that in such complexes the symmetry of the aromatic nucleus is not much perturbed.

Such complexes would represent an early stage on the reaction path leading from the aromatic compound and the electrophile to the product of substitution, and the possibility has been considered that they provide a suitable model for the transition state for these reactions (cf. *M. J. S. Dewar,* J. chem. Soc., 1946, 406, 777). Critical examination of this proposal was made by *H. C. Brown* and *J. D. Brady* (J. Amer. chem. Soc., 1952, **74**, 3570). By studying the effects of substituents on the equilibrium constants for the formation of complexes, first between aromatic compounds and hydrogen chloride at low temperatures, and secondly between the same aromatic compounds and hydrogen chloride in the presence of aluminium chloride, they made it probable that there are two types of complex. One of these, they suggest, involves the more general type of interaction envisaged by the term π-complex (LV), and the other, described as a σ-complex, requires specific interaction between the electrophile and a particular carbon atom of the aromatic ring (LVI; cf. XLIII). Since correlations between the effects of substituents on the rate of aromatic halogenation were better with the latter than with the former measure of basicity, it was concluded, in agreement with the main stream of chemical thought as developed in the preceding sections, that the transition states for aromatic substitution approximate more closely in structure to σ-complexes than to π-complexes.

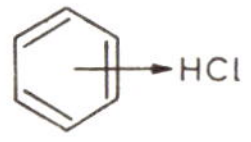

HCl (LV) $AlCl_4^{\ominus}$ H H (LVI) $SO_3^{\ominus}$ OH $^{\ominus}O_3S$ (LVII)

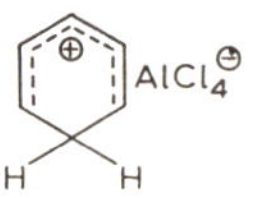

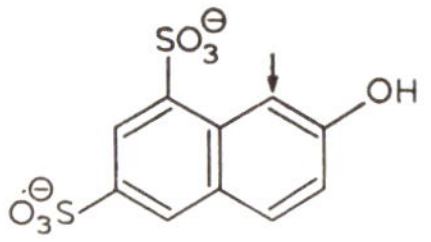

The fact that π-complexes can be shown to exist means, however, that in principle their rates of formation might in some cases be measurable, and that circumstances could be envisaged in which their rate of formation could be rate-limiting for an aromatic substitution. *Zollinger* and coworkers (Helv., 1962, **45**, 2057, 2066, 2077; 1965 **48**, 554) have made a most interesting contribution by examining the proton magnetic resonance spectrum of a complex formed in the reaction of iodine with 2-naphthol-6,8-disulphonic acid (LVII) and have shown that its properties are so different from those of the corresponding intermediate in bromination that the former must be a π-, whereas the latter is clearly a σ-complex. The reason suggested for the difference between the two halogens is the greater steric difficulty of bringing electrophilic iodine close to the bulky $\cdot SO_3^{\ominus}$ group.

(ii) σ-Complexes as intermediates

Having concluded that the π-complex does not provide the most generally suitable model for the transition state for aromatic electrophilic substitution, we have noted also that the carbonium ionic structure (XLIII) gives a preferable approximate description; but that, at the energy-maximum which usually is rate-limiting, the quinonoid structure is not always fully developed, so that structure XLII may be a still better representation of the transition state. It is now natural to ask whether complexes corresponding with the carbonium ionic structure can ever be shown to be discrete intermediates in aromatic substitution. The first definite suggestion, that aromatic substitution may be a two-stage process involving a quinonoid intermediate, was apparently made by *A. Lapworth* (J. chem. Soc., 1901, **79**, 1265). Some of the first experiments towards isolating such complexes were made by *P. Pfeiffer* and *R. Wizinger* (Ann., 1928, **461**, 132; cf. *Wizinger*, Chimia, 1953, **7**, 273), who obtained salts of the type $[Ar_2C{\cdot}CH_2Br]^{\oplus}X^{\ominus}$ as intermediates in the bromination of 1,1-diarylethylenes, and regarded them as analogues of the intermediates concerned in aromatic substitution.

Over the past decade or so, many investigators have contributed towards establishing that positively charged electrophiles can react under suitable conditions with aromatic compounds to give relatively stable, identifiable compounds of the carbonium ionic type. Some of the evidence has been discussed by *Brown* and his coworkers (cf. *Brown* and *Brady*, *loc. cit.*). Studies of solubility (*D. A. McCaulay* and *A. P. Lien*, J. Amer. chem. Soc., 1951, **73**, 2013; Tetrahedron, 1959, **5**, 186); conductivity (*M. Kilpatrick* and *F. E. Luborsky*, J. Amer. chem. Soc., 1953, **75**, 577); of ultraviolet absorption spectra (*Gold* and *F. L. Tye*, J. chem. Soc., 1952, 2172), and of N.M.R. spectra (*C. MacLean*, *J. H. van der Waals* and *E. L. Mackor*, Molecular Phys., 1958, **1**, 247) all support the view that carbonium ions (*e.g.* LVIII, from anthracene) can be formed by protonation of aromatic rings under strongly acid conditions. More recently still, *G. A. Olah* and his coworkers (*Olah* and *S. J. Kuhn*, J. Amer. chem. Soc., 1958, **80**, 6535, 6541) succeeded in isolating σ-complexes from reactions of hydrocarbons with hydrogen fluoride and boron fluoride at low temperatures. The properties and reactions of the isolated products confirmed their nature; it is significant, in connection with their role as analogues of intermediates involved in the conventional aromatic substitution, that when deuterium fluoride was used instead of hydrogen fluoride, the complexes were converted on heating into a mixture of deuterated and protonated hydrocarbon.

The work was extended by the isolation of a series of complexes which can be considered to be the intermediates involved in alkylation, acylation and nitration. In all cases the intermediates were formed in the correct stoichiometric ratio, were coloured, and when dissolved in aprotic solvents had conductivities of the expected magnitude. Structures (LIX–LXI) are typical.

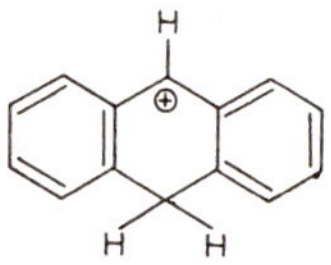

H ⊕ H H

(LVIII)

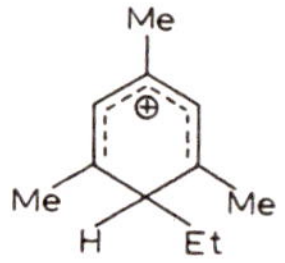

Me ⊕ Me Me H Et

(LIX)

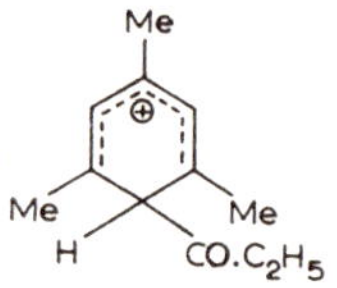

Me ⊕ Me Me H $CO.C_2H_5$

(LX)

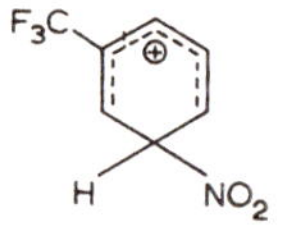

F_3C ⊕ H NO_2

(LXI)

Similarly, *W. von E. Doering et al.* (Tetrahedron, 1958, 4, 178) have made a very thorough study of the analogous complex LXII which is formed as bright yellow needles by the reaction of hexamethylbenzene with methyl chloride and aluminium chloride.

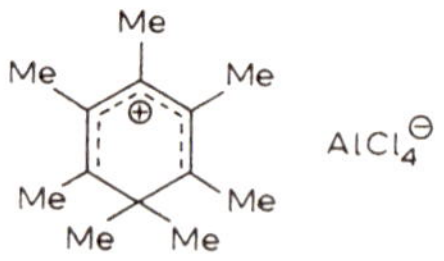

Me Me Me ⊕ Me Me Me Me $AlCl_4^{\ominus}$

(LXII)

More fleeting intermediates, still observable spectroscopically in the course of substitution, have been investigated in special cases; thus *H. Cerfontain* and *A. Telder* (Rec. Trav. chim., 1967, **86**, 371) describe the properties of the nitro-analogue of LVIII, formed in the course of nitration of anthracene in sulpholane.

Much other indirect evidence exists for the intervention of carbonium ionic species in electrophilic substitutions. Studies of primary isotope effects provide much relevant information, as is discussed below. *P. C. Myrhe* and *M. Beug* (J. Amer. chem. Soc., 1966, **88**, 1568, 1569) have also obtained evidence for carbonium-ionic rearrangements accompanying nitrations of hindered systems, where it can be presumed that the course of the reaction is partly determined by the necessity of releasing internal steric strains in the intermediates.

It has been presumed also, from permissive though not absolutely compelling evidence based on studies of kinetic forms, structural effects, and products, that the analogous neutral σ-complexes, $ArH \cdot X_2$, are concerned in the halogenation of aromatic compounds (cf. *P.B.D. de la Mare*, "The Kekulé Symposium", Butterworths, London, 1958).

(iii) Proton-removal, primary isotope effects, and reversibility

We have so far been considering electrophilic aromatic substitution essentially as a process leading to a transition state, and thence to an intermediate, in which the electrophile is partly or wholly covalently attached to the aromatic ring (as in structures XLII or XLIII), but in which the hydrogen to be replaced is still fully bound to the nucleus. Within this framework we have considered the possibilities (a) that the electrophile may have become free from its nucleophilic partner, or may be still attached to it; (b) that the rate-determining step may be the formation of the electrophile, the formation of a π-complex, or the formation of a σ-complex; and (c) that sometimes special forms of catalysis may modify the observed kinetic form for the reaction.

There are circumstances, however, under which a still later stage in the reaction, the removal of the displaced group, can be concerned in the rate-determining stage. *Melander*'s classic paper (*loc. cit.*) formulated the method by which studies of hydrogen isotope effects could be applied to this problem, and showed for the nitration and bromination of some common substrates that hydrogen and tritium are removed from similar positions at substantially the same rate. For these, as for other reactions in which it can be shown that the primary hydrogen-deuterium isotope effect is small, it can be concluded that the breaking of the C–H bond has made little progress in the transition state; for nitration and halogenation, this confirms the kinetic arguments adduced by *Ingold, E. D. Hughes* and their co-workers (*Ingold*, "Structure and Mechanism in Organic Chemistry", G. Bell, London, 1953) and by *P. W. Robertson, de la Mare* and *B. E. Swedlund* (J. chem. Soc.,

1953, 782). For these reactions, isotope effects are observed only in structurally unusual circumstances. In these, steric hindrance at the reaction site can play a part as in the bromination of 1,3,5-tri-*tert*-butylbenzene (*E. Baciocchi et al.* J. Amer. chem. Soc., 1967, **89**, 125). Other factors probably can contribute also, however; as for example in the nitration of 9-deuterioanthracene (*Cerfontain* and *Telder, loc. cit.*). Here the kinetic isotope effect, $k_H/k_D = 2.6$, establishes that the stage of reaction involving proton-loss has become kinetically significant.

Certain other reactions show isotope effects more frequently. Thus Melander noted a small but significant isotope-effect in the sulphonation of simple aromatic hydrocarbons. Since that time, many other systems have been examined; the results have been surveyed by *Zollinger* (*loc. cit.*), by *Berliner* (*loc. cit.*), and by *Melander* ("Isotope Effects on Reaction Rates", Ronald, N.Y., 1960). Large kinetic isotope effects have been established for iodination (*Berliner et al.*, J. Amer. chem. Soc., 1960, **82**, 5435; Chem. and Ind., 1960, 177; *E. Grovenstein* and coworkers, J. Amer. chem. Soc., 1957, **79**, 2972; *E. A. Shilov et al.*, Nature, 1958, **182**, 1300) and for mercuration (*A. J. Kresge* and *J. F. Brennan*, Proc. chem. Soc., 1963, 215). A particularly important contribution has been made by *Zollinger* (*loc. cit.*; cf. Experientia, 1956, **12**, 165), who has studied diazo-coupling. Here the electrophile is a cation of the type $Ar \cdot N_2^{\oplus}$; and by altering the degree of steric hindrance to approach of the cation to the attacked centre, it was found to be possible to alter the kinetic form from that of a reaction in which the loss of the proton is not kinetically significant, to one in which an isotope effect is observed and the reaction is subject to catalysis by bases.

Although the structural factors facilitating electrophilic attack conflict with those influencing proton-loss, the effects of change in structure on reactions exhibiting an isotope effect appear usually to give reasonable linear free-energy correlations with similar effects on reactions which do not show an isotope effect. It seems, therefore, that the earlier stages on the reaction-path usually have the dominant influence on the reactivity; in part, perhaps, because they are generally more responsive to change in structure, and in part because the bond-forming process is often more strongly developed than the bond-breaking process when the transition state is reached. Orientational specificity is also generally maintained under these conditions. It is significant, of course, that a number of aromatic substitutions for which isotope effects have been observed are reactions for which reversibility can easily be realised; as, for example, is the case with iodination and sulphonation. Clearly, however, thermodynamic control of the isomeric product-ratio is not necessarily associated with reversibility or with the observations of isotope-effects.

Hydrogen isotope-exchange in aromatic substitution is necesarily a reversible reaction, and one which must therefore show a kinetic isotope effect, study of which can be complicated by solvent isotope-effects. That this reaction, when carried out in polar solvents under acid catalysis, is subject to the usual orientational rules of aromatic substitution, and is influenced by structure in the characteristic manner, was shown by *Ingold, C. L. Wilson* and their co-workers (*Ingold, C. G. Raisin* and *Wilson*, Nature, 1934, 134, 734; J. chem. Soc., 1936, 915; *A. P. Best* and *Wilson, ibid.*, 1938, 28). Of recent years, interest has focussed on the details of the dependence of the rate on the acidity of the medium (*V. Gold* and *D. P. N. Satchell, ibid.*, 1955, 3609, 3619, 3622; *Gold, R. W. Lambert* and *Satchell, ibid.*, 1960, 2461; *C. Eaborn* and *R. Taylor, ibid.*, 1960, 3301; 1961, 247). For a time, the fact that the rate of reaction follows the acidity of the medium as measured by Hammett's acidity function, h_0, (*L. P. Hammett* and *A. J. Deyrup*, J. Amer. chem. Soc., 1932, 54, 2721), just as does the protonation of a neutral base, was taken to imply that the exchange reaction involves of necessity a complex, $ArH \cdot H^{\oplus}$, in pre-equilibrium with the starting materials. *A. J. Kresge* and *Y. Chiang* (*ibid.*, 1959, 81, 5509; 1961, 83, 2877; 1962, 84, 3976), however, by showing that this reaction is subject to general acid catalysis, have not only established that both charged and uncharged acids can be involved kinetically in electrophilic attack on a specific centre in an aromatic molecule, but have also made it very probable that the dependence of the reactivity on Hammett's acidity-function merely reflects the rate of a slow proton-transfer from the medium to unsaturated carbon, as has also been proposed by *F. A. Long* and coworkers (*J. Schulze* and *Long, ibid.*, 1964, 86, 331; *B. C. Challis* and *Long, ibid.*, 1965, 87, 1196).

(g) Less common mechanisms of electrophilic aromatic substitution

(i) The S_E2' reaction

The above treatment of electrophilic aromatic substitution is an elaboration of the S_E2 (*Ingold, loc. cit.*) process in terms of the reaction sequence:

$$ArH + X{-}Y \rightleftharpoons [ArH,XY] \rightleftharpoons \left[\overset{\oplus}{Ar}\!\!\begin{smallmatrix}H\\ \ominus\\ X{-}Y\end{smallmatrix}\right] \underset{+Y^-}{\overset{-Y^-}{\rightleftharpoons}} \overset{\oplus}{Ar}\!\!\begin{smallmatrix}H\\ X\end{smallmatrix} \rightleftharpoons ArX + H^{\oplus}$$

π-Complex $\quad \sigma$-Complex $\quad$ Carbonium ionic intermediate

An alternative sequence can lead also to the carbonium ionic intermediate and thence to the products, namely:

$$XY \underset{+Y^{\ominus}}{\overset{-Y^{\ominus}}{\rightleftharpoons}} X^{+} \underset{-ArH}{\overset{+ArH}{\rightleftharpoons}} Ar^{\oplus}\begin{matrix} H \\ X \end{matrix} \rightleftharpoons ArX + H^{\oplus}$$

Minor modifications within the above framework, as for example in various proposals involving cyclic transition states (cf. p. 58) allow interpretation of special features whilst leaving the general nature of the reaction path unchanged.

Certain important electrophilic aromatic substitutions, however, undoubtedly occur in quite a different way. One mechanism which has been established unambiguously is the S_E2' process, in which a proton is displaced not from the attacked carbon atom, but from the position conjugated with it. One way of establishing this mechanism is to find conditions in which the intermediate does not immediately rearrange to the normal product. A clear case is the bromination of 2,6-di-*tert*-butylphenol (LXIII). The dienone LXIV, which is formed rapidly, can be isolated; it rearranges under acidic or basic conditions to give the normal product of substitution (LXV) (*V. V. Ershov* and *A. A. Volodkin*, Isvest. Akad. Nauk. U.S.S.R., Otdel Khim. Nauk., 1962, 730).

OH, Bu^t, Bu^t — $\xrightarrow[-HBr]{Br_2}$ — O, Bu^t, Bu^t, H, Br — $\longrightarrow$ — OH, Bu^t, Bu^t, Br

(LXIII) (LXIV) (LXV)

Only preliminary investigations have been made of the mechanism of this reaction sequence, the details of which await further study (cf. *P. B. D. de la Mare* and coworkers, J. chem. Soc., 1964, 5306; B, 1967, 251). Some simpler phenols (*e.g.*, 2,6-dimethylphenol), which react more rapidly with bromine, can be shown also to give a dienone as intermediate, which then decomposes by rearrangement still more rapidly. The chemistry of phenols generally reveals that this type of reaction is not uncommon; products of structure analogous to that of the dienone LXIV have been recognised as formed from phenols which have no activated position available for substitution (*e.g.*, LXVII, from LXVI; cf. *G. M. Coppinger* and *T. W. Campbell*, J. Amer. chem. Soc., 1953, 75, 734), and from naphthols (*e.g.* LXIX, from LXVIII; cf. *K. Fries* and *K. Schimmelschmidt*, Ann., 1930, 484, 245). *Zollinger et al.* (*loc. cit.*) have recently discussed a closely analogous situation in the bromination of naphthols. Furthermore, careful kinetic and spectroscopic investigation of the bromodesulphonation of sodium 3,5-

dibromo-4-hydroxybenzenesulphonate (LXX) showed that a dienone (LXXI) is a metastable intermediate on the reaction path (*L. Cannell*, J. Amer. chem. Soc., 1957, 79, 2927, 2932); and similar evidence has been obtained for the bromodecarboxylations of substituted hydroxybenzoic acids (*E. Grovenstein Jr.* and *U. V. Henderson Jr.*, *ibid.*, 1956, 78, 569). Some demethylations of ethers and deacylations of esters accompanying nitrations and halogenations may be analogous (*H. J. Lewis* and *R. Robinson*, J. chem. Soc., 1934, 1253; *Ingold et al.*, *loc. cit.*; *P. B. D. de la Mare, S. de la Mare*, and *H. Suzuki*, J. chem. Soc. B, 1969, 429; *G. Antinovi, E. Baciocchi* and *G. Illuminati*, *ibid.*, B, 1969, 373; *P. B. D. de la Mare and B. N. B. Hannan*, Chem. Comm., 1970, 156):

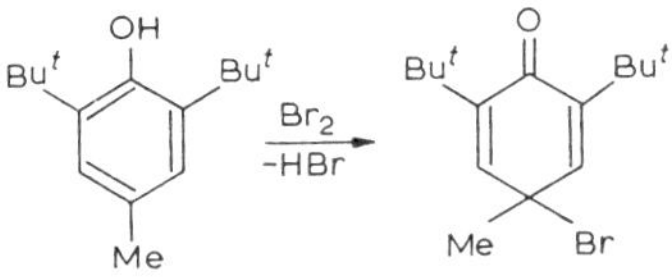

(LXVI) (LXVII)

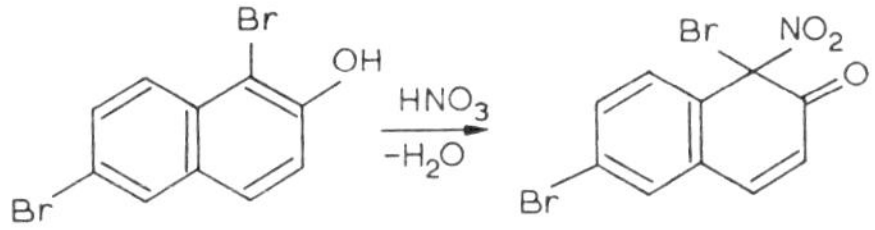

(LXVIII) (LXIX)

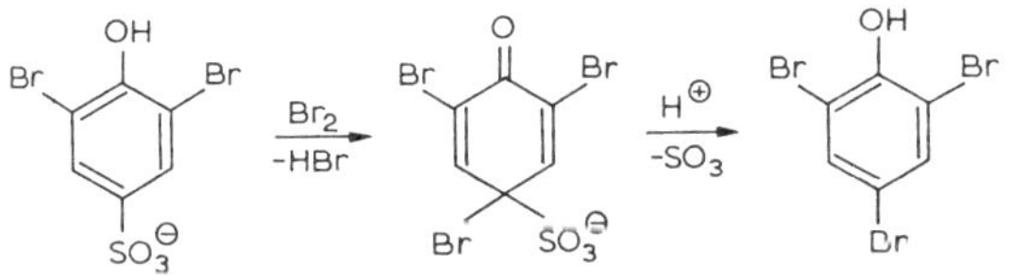

(LXX) (LXXI) (LXXII)

We have already noted (p. 61) that part of the unexpectedly great power of electron release of the ·OH group (specifically, its greater influence than that of the ·OMe group) may be interpreted as an indirect result of H–O hyperconjugation. Whether anything similar can be identified in the reactions of amines or substituted amines is not known; it is possible, however, that part of the reactivity of the phenylammonium ion in the *para*- and *ortho*-positions should be interpreted in this way (*J. H. Ridd et al.*, Proc.

chem. Soc., 1962, 228; 1964, 24). Recent experiments (*Baciocchi et al.*, Gazz., 1962, **92**, 89; Tetrahedron Letters, 1962, **15**, 637; J. Amer. chem. Soc., 1965, **87**, 3953; Progr. phys. org. Chem., 1967, **5**, 1) make it probable that a related sequence is concerned in the so-called "heterolytic side-chain substitution" in hexamethylbenzene and related compounds; the indicated sequence illustrates types of intermediate (LXXIV, LXXV) which may be concerned:

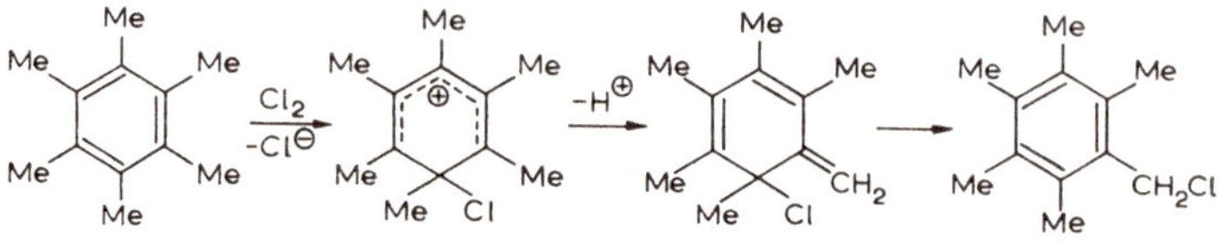

(LXXIII) $\xrightarrow[-Cl^{\ominus}]{Cl_2}$ (LXXIV) $\xrightarrow{-H^{\oplus}}$ (LXXV) $\longrightarrow$ (LXXVI)

Analogous processes have been proposed in the chemistry of alkylnaphthalenes (*P. B. D. de la Mare et al.*, J. chem. Soc. C, 1967, 1586, 1590) and of heterocyclic aromatic compounds (see below).

(ii) Addition-elimination sequences

Electrophilic substitution at an unsaturated centre can in principle occur by way of an addition*-elimination sequence, which for a simple unsaturated compound can be represented:

$$R\cdot CH{:}CH\cdot R' + X\text{–}Y \longrightarrow \begin{cases} R\cdot CH(X)\cdot CH(Y)\cdot R' \longrightarrow \begin{cases} R\cdot CH{:}CYR' + HX \\ R\cdot CX{:}CH\cdot R' + HY \end{cases} \\ R\cdot CH(Y)\cdot CH(X)R' \longrightarrow \begin{cases} R\cdot CY{:}CH\cdot R' + HX \\ R\cdot CH{:}CXR' + HY \end{cases} \end{cases}$$

Kinetic and orientational information can be used to determine whether the reaction involves electrophilic attack, and whether X or Y is the dominant electrophilic partner. The stabilities and the chemical properties of the adducts under the reaction conditions will determine whether or not they will be isolatable as intermediates on the reaction path, and whether or not the ratio of derived isomeric products will be what would be expected for reactions following other reaction paths.

* In the sense in which this term is used here, addition is effected only when the double bond becomes saturated; the stages of an S_E2 reaction leading to a carbonium ionic intermediate (p. 79) are not considered to effect addition.

Addition-elimination as a general mechanism for electrophilic substitution in aromatic and also in aliphatic systems has been under consideration for many years, and many distinguished contributors have expressed opinions concerning its availability. An excellent summary of the relevant history is given by *L. Fieser* (in *Gilman*, 'Organic Chemistry', 2nd Edn., Vol. I, Wiley, New York, 1943, p. 174 *et seq.*). With the coming of resonance theory and its illumination of the theory of aromaticity, however, it has become customary to consider aromatic substititution essentially as a carbonium-ion reaction, in which the thermodynamic stability of the aromatic product of proton-loss, in comparison with that of the non-aromatic or only partly aromatic adduct, determines that the former is the major product.

This view is certainly acceptable for most substitutions involving benzene and its derivatives. In many cases, indeed, a primary addition would hardly be expected to lead exclusively or even principally to mono-substitution, because the olefinic system so produced would be relatively reactive, and so would undergo further addition rather than elimination. Partly or fully saturated materials, or alternatively di- or tri-substituted products, would then be expected, as when benzene is chlorinated in sunlight. It is, however, perhaps surprising that addition-elimination routes have been given so little theoretical consideration over recent years, in view of the many examples in the literature, and the fact that carbonium ions often undergo further reaction to give products which are not the most thermodynamically stable of those available (cf. *Ingold, loc. cit.*).

Additions can accompany substitutions in the reactions of quite simple aromatic hydrocarbons under conditions conducive to heterolytic processes. Thus derivatives of biphenyl including bridged derivatives undergo chlorination in acetic acid to give products of substitution as the main components, accompanied by small but significant proportions of products of addition (*de la Mare, E. A. Johnson* and *J. S. Lomas*, J. chem. Soc., 1964, 5317). The latter substances include tetrachlorides and acetoxytrichlorides; clearly the intermediate carbonium ion can be captured by the solvent as well as by chloride ions, as is common in heterolytic additions (cf. *G. Williams*, Trans. Faraday Soc., 1941, 37, 749). Some of the products are relatively stable under the conditions of the reaction; others are not. Estimation of the proportions of minor components of the reaction mixture can be seriously misleading if this fact is not realised.

Naphthalene and phenanthrene also react with chlorine in acetic acid to give products of addition as well as products of substitution (*de la Mare* and *N. V. Klassen*, Chem. and Ind., 1960, 498; *de la Mare* and *R. Koenigsberger*, J. chem. Soc., 1964, 5327; *de la Mare et al.*, Rec. Trav. chim., 1965, 84, 109):

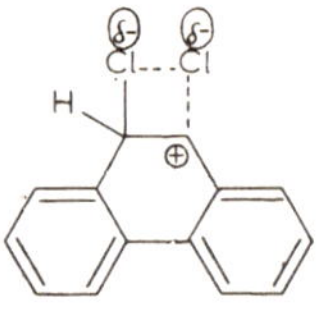

(LXXVII) 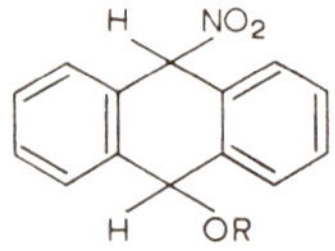 (LXXVIII)

In both cases the reactions are heterolytic; the homolytic (light-catalysed) reactions give adducts of quite different stereochemistry. In both cases, both acetoxychlorides and di- or poly-chlorides are obtained. The stereochemistry of the addition reveals the complexity of the reaction path. It is probable that the *cis*-dichloride of phenanthrene is produced through a transition state LXXVII involving internal capture of chloride ion as it heterolyses from the ArH·Cl_2 complex; whereas the *trans*-adducts are formed at a later stage in the reaction path, perhaps through attack by external nucleophiles on the carbonium ionic intermediate.

It is reasonable, therefore, to regard the adducts and the products of substitution as branching stages on a common reaction path, as was proposed (in a simpler form, since *cis*-addition had not been recognised) by *C.C. Price* (J. Amer. chem. Soc., 1936, **58**, 2101; Chem. Reviews, 1941, **29**, 37) for the bromination of phenanthrene. It is perfectly correct to say that substitution is not necessarily preceded by addition; but it must be recognised that, when addition occurs, this may be reflected in the course of the reaction. Thus the product-ratio may depend on the method used for working up the reaction-mixture; for example, the reaction of phenanthrene with chlorine in acetic acid can appear to give significant amounts of acetoxyphenanthrene, though this is not a primary product (*de la Mare* and *Koenigsberger*, *loc. cit.*). Sometimes, indeed, such procedures can be the bases of preparative methods, as for example for 9-methoxyphenanthrene and thence phenanthr-9-ol (*Fieser*, *R. P. Jacobsen* and *Price*, J. Amer. chem. Soc., 1936, **58**, 2163).

Attack by other electrophiles can also lead to addition accompanying, or preceding, substitution. Thus nitration in acetic anhydride can lead to the formation of adducts from hydrocarbons (*D. J. Blackstock et al.*, Chem. Comm., 1970, 641). Classical examples come from the reaction of anthracene with nitrating agents (*J. Meisenheimer* and *E. Connerade*, Ann., 1904, **330**, 133); various adducts of the type shown in structure LXXVIII (R = Ac, NO, Alkyl, etc.) have been shown to be obtainable, and some of these substances can be decomposed easily to give products of substitution. The reactions of some heterocyclic systems, however, provide more spectacular examples of reaction paths of this kind; the product of direct substitution may be very different from that obtained by addition-elimination. Thus the 5- and 8-nitration of quinoline probably involves the quinolinium cation reaction by the usual (S_E2) mechanism. The same is probably true of 5- and 8-halogenation in sulphuric acid (*de la Mare, M. Kiamud-din* and *Ridd*, J. chem. Soc., 1960, 561). The more usual methods of halogenation of quinoline, however, give quite a different orientation; bromination with molecular bromine, for example, gives 3-bromoquinoline, 3,6-dibromoquinoline, and 3,6,8-tribromoquinoline (cf. *W. La Coste*, Ber., 1881, **14**, 915; *A. Claus* and *F. Collischonn*, *ibid.*, 1886, **19**, 2763). The exact details of the sequence leading to these products is still a matter for discussion (cf. *J. J. Eisch*, Adv. Heterocyclic Chem., 1966, **7**, 1), but it has been proposed by *M. D. Johnson* and *Ridd* (J. chem. Soc., 1962, 291) from the results of model experiments that covalent adducts are the intermediates which determine 3-, 3,6-, and 3,6,8-halogenation (cf. Scheme 1).

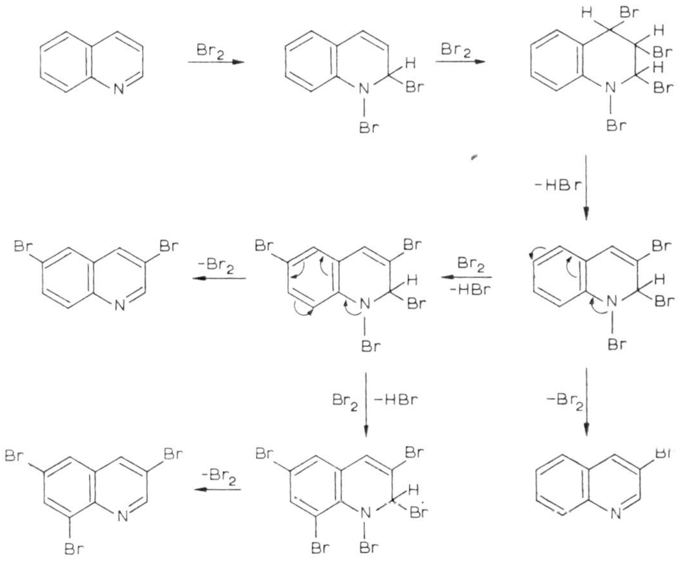

Scheme 1. Possible addition-elimination sequence for 3-, 3,6-, and 3,6,8-bromination of quinoline.

Substitution in the 3-position arises through addition followed by elimination; substitution in the 6-(and analogously in the 8-)position is determined by the conversion of the quinoline nucleus into a derivative of di- or tetrahydroquinoline, in which the lone pair of electrons of nitrogen is now able to conjugate with the remaining aromatic ring and hence to activate the positions *ortho* and *para* to itself.

Addition-elimination sequences thus are potentially of great significance in determining the course of substitution in heterocyclic systems. This is true not only for halogenation, but also for other substitutions; there are, for example, modes of nitration which from their unusual orientational characteristics probably involve such routes (*M. J. S. Dewar* and *P. M. Maitlis*, J. chem. Soc., 1957, 944). The patterns of orientation that have already been noted for quinoline are not the only ones that could in principle be established. Substitution in the 4-position might follow the formation of such an adduct as LXXIX from quinoline and XY if the course of elimination were suitable; and if protonated adducts (*e.g.* LXXX) were formed, their further substitution would probably be in the 7- and 5-positions.

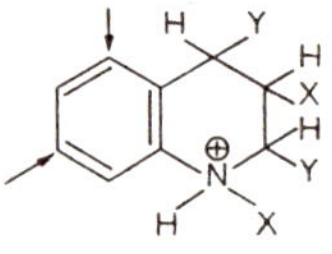

(LXXIX) (LXXX)

Substitution in activated heterocycles (pyrrole, furan, etc.) may also in appropriate cases take place by addition-elimination sequences. Very little has been done towards establishing systematically the paths available; we note here just one example in the indole series, from the work of *S. G. P. Plant* and *M. L. Tomlinson* (*ibid.*, 1933, 955). Although it would be desirable to reinvestigate by modern methods the structures of the intermediates (LXXXI–LXXXIV) discussed by these authors, there is no doubt that the experimental observations establish that complex reactions paths of these kinds are available in this system. A proposed intermediate (LXXXIII) here is of the S_E2' type, a result which illustrates how closely inter-related the various unusual modes of electrophilic aromatic substitution can be.

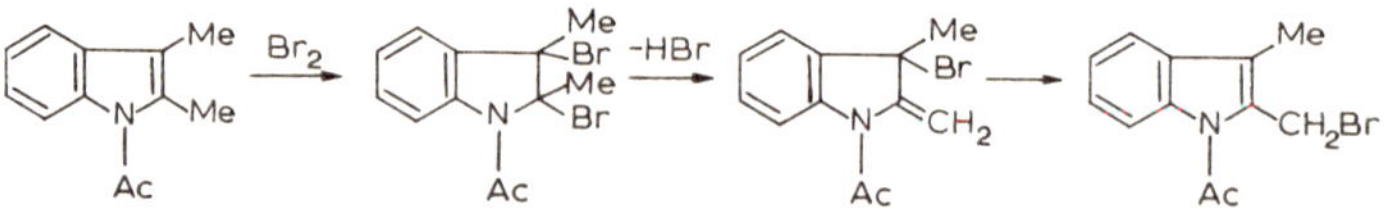

$\xrightarrow{Br_2}$ $\xrightarrow{-HBr}$ $\longrightarrow$

(LXXXI) (LXXXII) (LXXXIII) (LXXXIV)

(iii) "Indirect substitution"

Aromatic compounds having the groups $\cdot OH$, $\cdot NH_2$, or $\cdot NHR$ attached to the nucleus can in principle undergo substitution by yet another distinct mechanism, in which first *N*- or *O*-substitution occurs, and the product then undergoes rearrangement. Under some circumstances, the intervention of paths leading to *N*- or *O*-substituted compounds will be irrelevant to the course of *C*-substitution, because the rearrangement occurs by the reversal of the *N*- or *O*-substitution. This is the case, for example, in the the reaction of acetanilide with chlorine (Scheme 2; R = Ac, X = Y = Cl), where even if the *N*-halogenated compound were formed by path 2, its rearrangement to the product of normal substitution would under most conditions proceed through acetanilide and the halogen (paths 1 + 3).

$C_6H_5\cdot NRX + HY$ —(4)→

(1) ↓ ↑ (2) $X\cdot C_6H_4\cdot NHR + HY$

$C_6H_5\cdot NHR + XY$ —(3)→

Scheme 2. Illustrative paths in systems where "indirect substitution" is possible.

As *Hughes* and *Ingold* have pointed out, however, (Quart. Reviews, 1952, **6**, 34), direct substitution by route 3 is sometimes not easy to distinguish from the indirect route, 2 + 4. There are certain instances in which the distinction can be made by using evidence based primarily on kinetic measurements, as when the rates of the individual stages 1, 2, and 3 can be determined separately and compared with the overall rate of rearrangement. This is the situation for the chlorination of acetanilide (*K. J. P. Orton, F. G. Soper* and *G. Williams*, J. chem. Soc., 1928, 998), at least under the conditions which have been most thoroughly investigated. It also probably represents the situation for some nitrosations and alkylations (*Hughes* and *Ingold, loc. cit.*). These authors observed also, however, that these conclusions do not exclude the possibility that the indirect path might become effective with other reactants or conditions. Evidence suggestive of the possibility that a reaction path of the type 2 + 4 can under some conditions replace path 3, and in doing so can affect the isomeric product-ratio, is available from studies of the nitramine rearrangement (Scheme 2; R = H, X = NO_2; *A. F. Holleman, J. C. Hartogs* and *T. van der Linden*, Ber., 1911, **44**, 704; *Hughes* and *G. T. Jones*, J. chem. Soc., 1950, 2678). Some of the relevant data are shown in Table 2.

The very high proportion of *ortho*-substitution found for the nitration of aniline with nitric acid in acetic anhydride, when put together with

TABLE 2

PRODUCTS IN THE NITRATION OF ANILINE, AND IN THE REARRANGEMENT OF *N*-NITROANILINE (PHENYL NITRAMINE)

Process	Conditions	Isomeric proportions *o*- (%)	*m*- (%)	*p*- (%)
Nitration	$Ph\cdot NH_3^{\oplus}NO_3^{\ominus}$ 95% aq. H_2SO_4, –20°	4	39	56
Nitration	$Ph\cdot NH_3^{\oplus}NO_3^{\ominus}$ 80% aq. HNO_3	5	32	62
Nitration	$Ph\cdot NH_3^{\oplus}NO_3^{\ominus}$ Ac_2O	82	3	15
Rearrangement	$Ph\cdot NH\cdot NO_2$ 74% aq. H_2SO_4, –20°	95	1.5	3.5

the very similar proportions observed for the nitramine rearrangement and the very different isomeric proportions found for nitration in nitric acid or sulphuric acid, makes it very likely that this nitration in acetic anhydride follows the indirect path. That such a marked difference is found in the isomeric proportions when the two routes are compared probably arises because the intramolecular rearrangement of phenylnitramine has a special mechanism which favours *ortho*-substitution. It seems likely that indirect paths leading to substitution are also available for the sulphonation of some amines (cf. *Hughes* and *Ingold, loc. cit.*).

Indirect paths may also be concerned in some substitutions in heterocyclic systems, being invoked because electrophilic attack at the heteroatom gives an initially stable, but rearrangeable, product. The rearrangement first studied by *A. Ladenburg* (Ann., 1888, **247**, 1), in which 1-alkylpyridinium iodides are converted on heating into 2- and 4-alkylpyridines, may be an extreme example, though *N.V. Sidgwick* ("Organic Chemistry of Nitrogen", University Press, Oxford, 1937) treats the reaction as a conventional substitution involving the products of dissociation of the salt.

3. Nucleophilic aromatic substitution*

N. B. CHAPMAN

The term "aromatic substitution" is often identified with electrophilic substitution, in which both electrons of the bond formed between the entering substituent and the aromatic nucleus during the reaction are furnished by the carbon atom at the seat of substitution. This view has arisen because the best-known aromatic substitutions such as nitration, halogenation and sulphonation, are those in which hydrogen is replaced, and are electrophilic in character. The foregoing discussion was concerned with this type of substitution. However, many reactions are known in which one or both of the electrons of the new bond are supplied by the reagent: these are homolytic and nucleophilic substitutions, respectively, and as groups of atoms rather than hydrogen are often involved, such reactions are frequently referred to as replacement or displacement reactions.

Reactions of this type differ from the electrophilic substitutions discussed in Section 2 in two main respects. First, in nucleophilic substitution the displaced group is usually ejected as an anion, and therefore the groups which are most easily replaced in this type of reaction are those which form sufficiently stable anions, such as the halogens; *e.g.*, the reaction of 1-chloro-2,4-dinitrobenzene with alkali to give 2,4-dinitrophenol is familiar. This substance also reacts readily with other nucleophilic reagents

$$C_6H_3Cl(NO_2)_2 + OH^{\ominus} \longrightarrow C_6H_3(OH)(NO_2)_2 + Cl^{\ominus}$$

such as ammonia or hydrazine to give 2,4-dinitroaniline or 2,4-dinitrophenylhydrazine. For the above reason, nucleophilic replacement of hydrogen is

* The reader is referred to reviews or discussions in the following works: *J. Kenner*, J. chem. Soc., 1914, **105**, 2717; *G. M. Bennett*, Ann. Reports, 1929, **26**, 132; *W. Bradley* and *R. Robinson*, J. chem. Soc., 1932, 1254; *F. W. Bergstrom*, Chem. Reviews, 1944, **35**, 77; *H. S. Mosher*, *Elderfield's* "Heterocyclic Compounds", Vol. I, John Wiley and Sons, Inc., New York, 1950; *J. F. Bunnett* and *R. E. Zahler*, Chem. Reviews, 1951, **49**, 273; *Bunnett*, Quart. Reviews, 1958, **12**, 1; *J. Miller*, Reviews of Pure and Applied Chemistry, 1951, **1**, 171; *G. M. Badger*, "The Structure and Reactions of Aromatic Compounds", Cambridge University Press, 1954; *C. K. Ingold*, "Structure and Mechanism in Organic Chemistry", 2nd Edn., G. Bell & Sons Ltd., London, 1969; *W. Hückel*, "Theoretical Principles of Organic Chemistry", Elsevier, Amsterdam, 1955; *J. Sauer* and *R. Huisgen*, Angew. Chem., 1960, **72**, 91; *S. D. Ross* in "Progress in Physical Organic Chemistry" Vol. I, Eds. *S. G. Cohen*, *A. Streitwieser Jr.* and *R. W. Taft Jr.*, Interscience, New York, 1963, p. 31; *Miller*, "Nucleophilic Aromatic Substitution", Elsevier, Amsterdam, 1969; *F. Pietra*, Quart. Reviews, 1969, **23**, 504.

rare. As examples of reactions involving nucleophilic replacement of hydrogen, the reactions of nitrobenzene with potassium carbazole to give *N-p*-nitrophenylcarbazole (*G.* and *M. de Montmollin*, Helv., 1923, 6, 94):

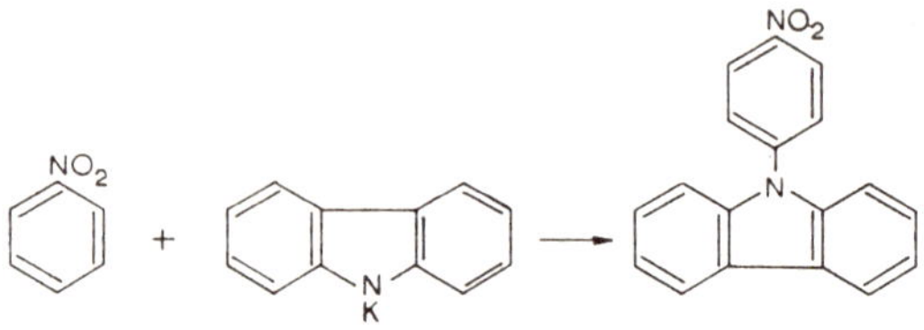

NO_2 + N K → N (NO_2)

of nitrobenzene with potassium diphenylamide to give *N-p*-nitrophenyldiphenylamine (*Bergstrom et al.*, J. org. Chem., 1942, 7, 98):

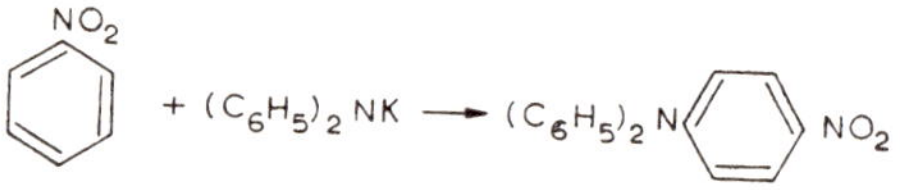

NO_2 + $(C_6H_5)_2NK$ → $(C_6H_5)_2N$ NO_2

and of potassium hydroxide plus an oxidant with mesobenzanthrone to give a mixture of 4- and 6-hydroxymesobenzanthrone (*W. Bradley* and *G. V. Jadhav*, J. chem. Soc., 1937, 1791):

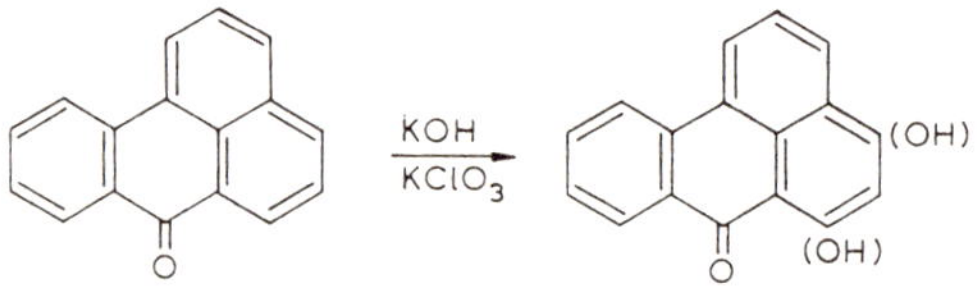

O $\xrightarrow[KClO_3]{KOH}$ (OH) (OH) O

may be cited. Secondly, nucleophilic aromatic substitution generally occurs only with difficulty unless there is present some structural feature whereby the compound is activated towards such substitution. The activating structure may be one or more electron-attracting substituents at the *ortho*- or *para*-positions with respect to the seat of substitution, or it may be a hetero nitrogen atom (*e.g.* in pyridine). However, some well-known nucleophilic substitutions, such as the production of phenols by the fusion of sulphonates with alkali, occur in the absence of any activating structure. In such reactions extremely vigorous conditions are generally necessary in order to bring about substitution.

(a) Mechanism of substitution

Two different major mechanisms, together with minor variants of them, have long been recognised for nucleophilic substitution at a saturated carbon atom. These are the unimolecular and the bimolecular mechanism, designated S_N1 and S_N2, respectively (*E. D. Hughes*, Trans. Faraday Soc., 1941, 37, 603; cf. Vol. IA, p. 270 *et seq.*). In this context a mechanism is said to be bimolecular if two species are involved in *covalency change* in the rate-limiting stage. In the S_N1 mechanism we usually have a two-stage process, in which the first and rate-limiting stage is the heterolysis of the aliphatic compound (RX in the following scheme). The carbonium ion so produced then rapidly combines with the nucleophile, Y, which is often, though not necessarily, an anion:

$$R\text{–}X \underset{v_2}{\overset{v_1}{\rightleftharpoons}} R^{\oplus} + X^{\ominus}$$

$$Y\!: + R^{\oplus} \xrightarrow{v_3} Y - R^{\oplus}$$

(In the limit, $v_3 \gg v_2$ and the observed rate $= v_1$).

in such a case, the product is electrically neutral. Reactions of this type are favoured by electron release in R in the direction of X. A reaction pursuing the S_N2 mechanism usually but not always is a one-stage process, and involves the synchronous approach of Y and recession of X, as follows:

$$Y\!: + RX \rightarrow \underset{\text{Transition state}}{Y^{\delta+}\ldots R \ldots X^{\delta-}} \rightarrow Y - R^{\oplus} + X^{\ominus}$$

In the above schemes the charges indicated apply to the case when both Y and RX are uncharged: the extension to other cases is obvious. It is usually found experimentally that electron-recession from the seat of substitution facilitates S_N2 reactions, but theoretically the electronic demands of this mechanism are equivocal, bond breaking being facilitated by electron-accession to, and bond formation by electron-recession from the seat of substitution. S_N2 reactions are normally of the second order kinetically (first order in each reactant individually). Both S_N1 and S_N2 mechanisms are influenced by the stability as an anion of the expelled group or atom X. Very many examples of the bimolecular mechanism have been observed in nucleophilic aromatic substitution, but there are some associated complexities (see p. 99 *et seq.*). The unimolecular mechanism is but rarely encountered and what was once thought to be the one clear case has recently been shown to be more complicated than was previously believed.

(i) Unimolecular substitution

Until recently it was thought that the uncatalysed thermal decomposition of arenediazonium salts in acidic aqueous solution, even in the presence of strongly nucleophilic anions, was a well-authenticated example of an aromatic S_N1 reaction and the only one, and that it involved a more or less kinetically free aryl carbonium ion. The work of *E. S. Lewis* and his coworkers (see p. 94) has shown that a more subtle mechanistic description is now necessary.

The S_N1 mechanism for the above group of reactions was proposed by *W. A. Waters* (J. chem. Soc., 1942, 266) and by *DeLos F. DeTar* and *S. V. Sagmanli* (J. Amer. chem. Soc., 1950, **72**, 965). The displaced group is $-N_2^{\oplus}$, which is liberated as gaseous nitrogen and usually the main nucleophile is solvent water and the main product a phenol. *E. A. Moelwyn-Hughes* and *P. Johnson* (Trans. Faraday Soc., 1940, **36**, 948) showed with care and precision that the reaction of $C_6H_5N_2^{\oplus}$ in most conditions, was of the first order with respect to diazonium cation, a result compatible with the S_N1 mechanism or with S_N2 solvolysis. Reaction rates usually show little or no dependence on the nature or concentration of the anions present (*Moelwyn-Hughes* and *Johnson, loc. cit.*; *J. C. Cain*, Ber., 1905, **38**, 2511; *H. A. H. Pray*, J. phys. Chem., 1926, **30**, 1417; *M. L. Crossley, R. H. Kienle* and *C. H. Benbrook*, J. Amer. chem. Soc., 1940, **62**, 1400). Some of this work, together with other relevant investigations, has been pertinently criticised by *Ross* (*loc. cit.*). In particular, it is pointed out that the work of *A. N. Nesmeyanov, L. G. Makarova* and *T. P. Tolstaya* (Tetrahedron, 1957, **1**, 145) which was directed to identifying the intermediate aryl cation by capture as an oxonium, brominium, chlorinium or pyridinium salt, provides useful permissive but not compelling evidence in favour of the S_N1 mechanism.

Substituent effects in the aromatic nucleus mainly conform with the electronic requirements of the S_N1 mechanism: electron-releasing *ortho*- and *meta*-substituents facilitate the reaction and *vice-versa*, although there are some unexplained details with *ortho*-substituents. As expected, electron-attracting ($-K$) *para*-substituents retard the reaction, but it is also retarded by electron-releasing ($+K$) *para*-substituents. This anomaly was convincingly explained by *Hughes*, quoted by *Bunnett* and *Zahler* (*loc. cit.*), as follows. Hughes suggested that $+K$ substituents, by increasing the double-bond character of and so strengthening the carbon-nitrogen bond which is broken in the rate-limiting heterolysis, increased the related activation energy, *i.e.*, the initial state for reactions of this type was to be formulated as follows:

$$X-C_6H_4-\overset{\oplus}{N}\equiv N \longleftrightarrow \overset{\oplus}{X}=C_6H_4=\overset{\oplus}{N}=\overset{\ominus}{N}$$

(I)

The rareness of the S_N1 mechanism in substitution at aromatic carbon compared with aliphatic is attributable to the absence of structural features which stabilise the aryl carbonium ion, and to the stronger bonding in aromatic compounds of the common displaced groups. *Taft* (J. Amer. chem.

Soc., 1961, 83, 3350) has suggested an additional factor which may, however, favour the S_N1 mechanism, *viz.* that an intermediate aryl carbonium ion would have the character of a triplet ion-radical and so gain some extra stability, which would also appear, but to a less extent, in the transition state for its formation. The intermediate II is thus formulated as follows:

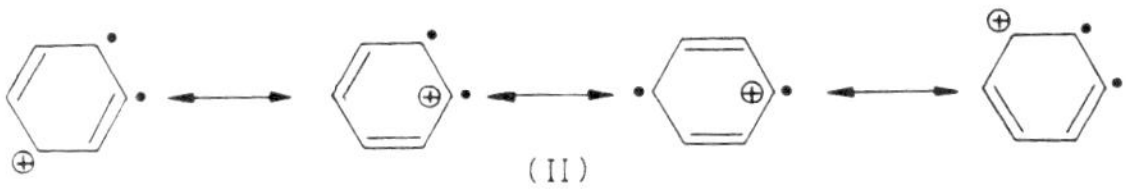

(II)

cf.

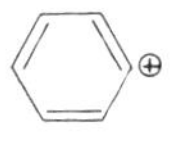

(III)

In structure II a π-electron of the aromatic sextet has entered "with concerted uncoupling" the vacant σ-orbital of III formerly occupied by the C–N bonding pair. The main evidence in favour of this suggestion is that electron-releasing *meta*-substituents facilitate the reaction more strongly than would be expected on the basis of the appropriate Hammett-type σ-values. The substituted ion-radical may be formulated as in IV.

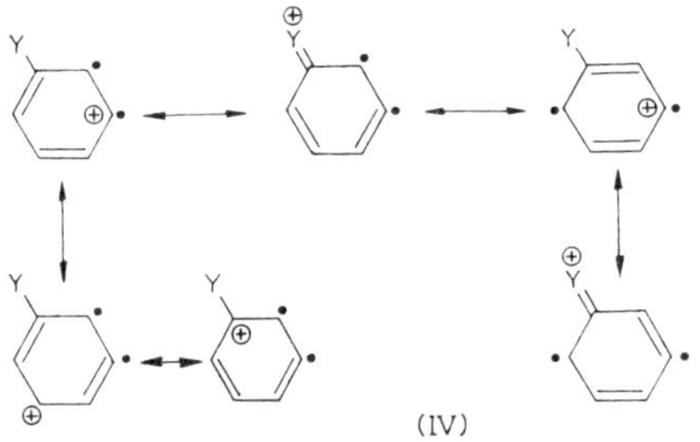

(IV)

Further evidence in favour of this view of the reactive intermediate has been provided by *E. A. Abramovitch* and *J. G. Saha* (Canad. J. Chem., 1965, 43, 3269) from a detailed study of the arylation of aromatic compounds with benzenediazonium tetrafluoroborate.

The simple S_N1 mechanism was questioned by *F. H. Field* and *J. L. Franklin* ("Electron Impact Phenomena", Academic Press, N.Y., 1957, p. 252) who provided mass spectrometric evidence that the ion $C_7H_7^{\oplus}$ is apparently the same whether it is produced from tolyl or from benzyl derivatives, but pointed out that toluenediazonium salts decompose in aqueous solution to cresols without the formation of rearrangement products. The major refinement of view about this mechanism has, however, stemmed from the kinetic investigations of *E. S. Lewis* and his co-workers (*Lewis* and *W. H. Hinds*, J. Amer. chem. Soc., 1952, **74**, 304; *Lewis* and *J. E. Cooper*, *ibid.*, 1962, **84**, 3847; *Lewis* and *J. M. Insole*, *ibid.*, 1964, **86**, 32, 34).

From their study of the kinetics of the reactions of arenediazonium salts with aqueous thiocyanate to give phenols, aryl thiocyanates and isothiocyanates, and derived products, Lewis and Cooper concluded that the reaction involves a highly reactive and therefore unselective intermediate which reacts at similar rates with water or the more powerfully nucleophilic thiocyanate ion, and which may also revert to the diazonium ion; but that their results were inconsistent with the usual S_N1 scheme (p. 91). By using diazonium salts containing ^{15}N, Lewis and Cooper showed that the substitution reactions were accompanied by a slower rearrangement thus:

$$Ar{-}^{15}N^{\oplus}{\equiv}N \rightarrow Ar{-}N^{\oplus}{\equiv}^{15}N$$

That a "nitrogen fixation path" *i.e.*, $Ar^{\oplus} + N_2 \longrightarrow ArN_2^{\oplus}$, was unlikely for this last reaction was shown by the failure of the isoelectronic but more nucleophilic carbon monoxide ($C^{\ominus} \equiv O^{\oplus}$) to react, despite its known power of reacting with relatively unreactive carbonium ions. These observations and detailed analysis of measured rates led them to suggest a scheme involving two intermediates, X and Y, as shown below:

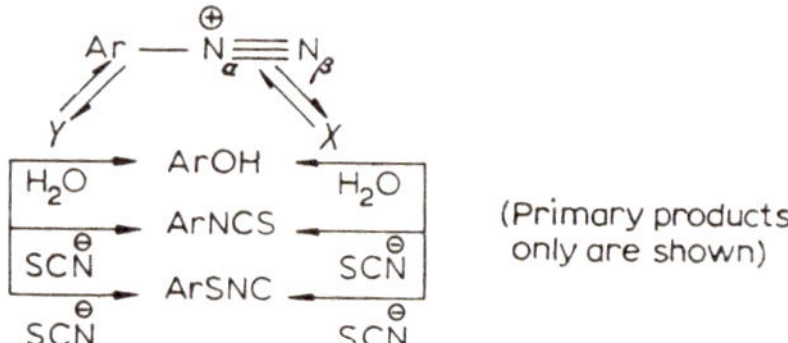

The reactive and unselective intermediate X returns predominantly to $ArN_2^{\oplus}$, and its two nitrogen atoms are not equivalent: it resembles but is not identical with an aryl cation. Lewis and Cooper suggest that X is a vibrationally excited state of the diazonium ion with a weak $C{-}N\alpha$ bond, having B (below) as main structure, and so possessing nearly all the activation energy necessary for decomposition:

(A) (B) (C)

(E) (D)

Y is less reactive and more selective, and the specific rate of return from Y to $ArN_2^{\oplus}$ was almost exactly twice that of isotopic rearrangement. A spirocyclic structure, V, was therefore suggested for Y:

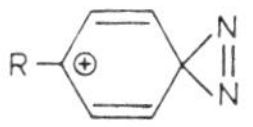

(V)

Y might be formed *via* X but there was no compelling evidence in favour of this. *T. Cohen* and his co-workers (J. org. Chem., 1962, **27**, 3385; 1965, **30**, 3891; J. Amer. chem. Soc., 1964, **86**, 2514; Tetrahedron Letters, 1964, 3721) have provided evidence in favour of an aryl carbonium ion being formed in certain thermal decompositions of diazonium cations in various solvents by demonstrating that the formation of the products involved a 1,5-hydride shift*.

The unimolecular mechanism has been suggested also for some other reactions, notably the alkaline hydrolysis of *o*- and *p*-halogenophenols, and the high-temperature hydrolysis of phenyl halides, for which *W. J. Hale* and *E. S. Britton* (Ind. Eng. Chem., 1928, **20**, 114) found that the rate of reaction is not dependent on the concentration or the nature of the nucleophilic reagent. These examples are, however, less well documented. In addition, the unimolecular mechanism has been proposed for certain nucleophilic aromatic rearrangements.

Finally, there is a recent claim (*P. J. Hutchison* and *R. J. L. Martin*, Austral. J. Chem., 1965, **18**, 699) that *N-tert*-butyl-2,4,6-trinitrobenzamide reacts with hydroxide ion to liberate nitrite ion in a reaction of zeroth order in hydroxide. It is formulated thus:

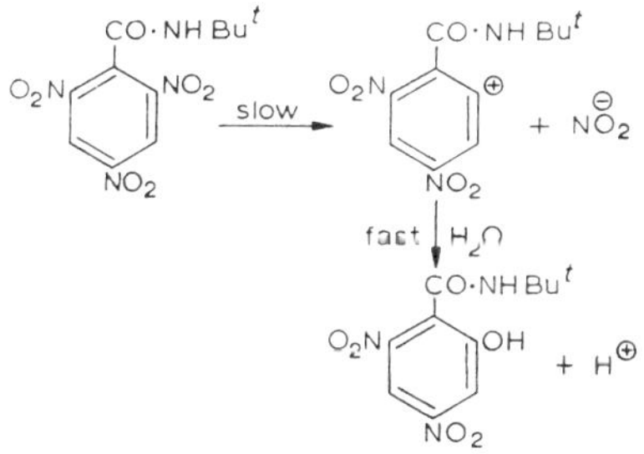

Although the experimental work appears sound, the slenderness of the published evidence for the proposed kinetics and the intrinsic unlikeliness of the first stage of this scheme are such that judgement must be reserved as to its validity.

* Recently Lewis has changed his views on this mechanism (*Lewis et al.*, J. Amer. chem. Soc., 1969, **91**, 419, 426).

To sum up, a really convincing demonstration of a reaction pursuing the simple limiting S_N1 mechanism, in close similarity with aliphatic substitution, is still awaited.

(ii) The elimination-addition mechanism – arynes and hetarynes

The elimination of hydrogen halides from vinyl halides to form alkynes and the nucleophilic addition reactions of the latter are well-known. The possibility of nucleophilic aromatic substitution by an elimination-addition reaction involving an intermediate aryne, *e.g.* benzyne itself (VI), or a hetaryne *e.g.* 2-pyridyne (VII), thus arises:

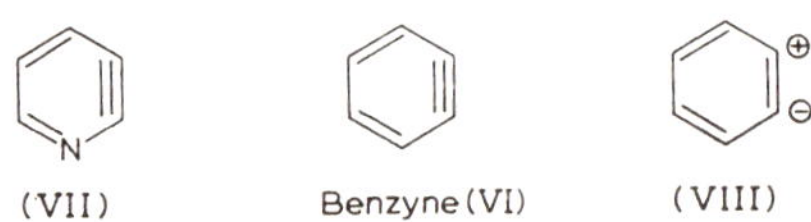

(VII) Benzyne (VI) (VIII)

In their review, *Bunnett* and *Zahler* (Chem. Reviews, 1951, 49, 273) drew attention to a number of cases of comparatively facile substitution involving so-called *ortho cine*-substitution (substitution with rearrangement, with entry of the substituent into the *ortho*-position), clearly a possibility in this mechanism. Strongly basic reagents at low temperatures were involved. *G. Wittig* and his co-workers had previously advanced the idea of a di-dehydrobenzene intermediate, formulated in the dipolar form VIII, in the reactions of halogenobenzenes with phenyl-lithium (*Wittig, G. Pieper* and *G. Fuhrmann*, Ber., 1940, 73, 1193; *Wittig* and *H. Witt, ibid.*, 1941, 74, 1480; *Wittig*, Angew. Chem., internat. Edn., 1965, 4, 731).

Critical evidence was provided by *J. D. Roberts* and his co-workers (*Roberts, H. E. Simmons et al.*, J. Amer. chem. Soc., 1953, 75, 3290; *Roberts, D. A. Semenov et al., ibid.*, 1956, 78, 601, 611; *E. F. Jenny* and *Roberts*, Helv., 1955, 38, 1248), who showed that [1-^{14}C]chlorobenzene reacts with potassamide in liquid ammonia to give almost equal amounts of [1-^{14}C]-aniline and [2-^{14}C]aniline, the small divergence from exact equality being attributed to a [^{12}C-^{14}C]isotope effect:

Cl —(KNH_2, in liquid NH_3)→ * —(NH_3)→ NH_2 / NH_2

Compounds lacking hydrogen *ortho* to the halogen were unreactive under these conditions (*Roberts, Semenov et al., loc. cit.*; *R. A. Benkeser* and *W. E. Buting*, J. Amer. chem. Soc., 1952, 74, 3011). More detailed information about the mechanism and its dependence on the halogen displaced has been obtained by using *ortho*-deuterated halogenobenzenes (*Roberts et al., loc. cit.*) with the following results:

$$NH_2^{\ominus} + C_6H_4(H(D))X \underset{k_{-1}}{\overset{k_1}{\rightleftharpoons}} C_6H_4^{\ominus}X + NH_3(NH_2D)$$

$$\downarrow k_2$$

$$C_6H_5NH_2 \xleftarrow{NH_3} C_6H_4 + X^{\ominus}$$

For X = F, $k_{-1} \gg k_2$ and rapid isotopic exchange with little elimination takes place; for X = Br, $k_{-1} \ll k_2$, *i.e.*, isotopic exchange is slow, but hydrogen (or deuterium) bromide is readily eliminated. For X = Cl, *e.g.* with chlorobenzene, intermediate behaviour is observed. The involvement of 1,2- and 2,3-naphthalyne in suitable nucleophilic substitution reactions of 1- or 2-halogenonaphthalenes has been demonstrated by *R. Huisgen* and his co-workers (*Huisgen* and *H. Rist, Naturwiss.*, 1954, **51**, 358; *Huisgen* and *H. Zirngibl*, Ber., 1958, **91**, 1438; *Sauer, Huisgen* and *A. Hauser, ibid.*, 1958, **91**, 1461) and by *Bunnett* and *J. Brotherton* (J. Amer. chem. Soc., 1956, **78**, 155). The whole subject has been much elaborated quantitatively by *Huisgen* and his co-workers and reviewed by him (in "Theoretical Organic Chemistry", Kekulé Symposium, Chem. Soc., London, 1958, p. 158; *Huisgen* and *Sauer*, Angew. Chem., 1960, **72**, 91). A further review by *H. Heaney* (Chem. Reviews, 1962, **62**, 81) contains much information about substituent effects in this mechanism, with particular reference to isomer distribution in the product. It will be appreciated that an *ortho*-substituted halogenobenzene can form only a 2,3-benzyne, a *para*-substituted compound a 3,4-benzyne, whereas a *meta*-substituted compound may form both:

A 2,3-benzyne   A 3,4-benzyne

The orientation of the benzyne formed in this last case appears to be mainly controlled by the inductive effect of the substituent, likewise the subsequent nucleophilic addition stage.

Much recent work has concerned the detection and isolation of benzyne. It is, however, a highly fugitive species. It may readily be produced by thermolysis of suitable substrates, *e.g.* benzenediazonium-2-carboxylate, and, in quantitative yield, by oxidation (under mild conditions) of 1-amino-

$$\text{(benzene-diazonium-2-carboxylate)} \longrightarrow \text{(benzyne)} + CO_2 + N_2 \longrightarrow \text{(biphenylene)}$$

benzotriazole with lead tetra-acetate under nitrogen (*C. D. Campbell* and *C. W. Rees*, Proc. chem. Soc., 1964, 296; Chem. Comm., 1965, 192). It is detected and estimated by observing its dimerisation products, products of cyclo-additions, and addition product with 2,3,4,5-tetraphenylcyclopentadienone. Arynes may also be prepared usefully by photolysis of *o*-diiodoarenes (*J. A. Kampmeier* and *E. Hoffmeister*, J. Amer. chem. Soc., 1962, **84**, 3787; *N. Kharasch* and *K. Sharma*, Chem. Comm., 1967, 492). A stable nickel complex of benzyne has recently been isolated (*E. W. Gowling, S. F. A. Kettle* and *G. M. Sharples*, *ibid.*, 1968, 21). Physico-chemical work on benzyne includes studies of its absorption spectrum and its behaviour in the mass spectrometer — both useful for detection (*E. Le Goff*, J. Amer. chem. Soc., 1962, **84**, 3786; *Kampmeier* and *Hoffmeister*, *loc. cit.*; *Kharasch* and *Sharma*, *loc. cit.*; *Wittig* and *R. W. Hoffmann*, Ber., 1962, **95**, 2728; *R. S. Berry*, *G. N. Spokes* and *M. Stiles*, J. Amer. chem. Soc., 1960, **82**, 5240; 1962, **84**, 3570; *Berry*, *J. Clardy* and *M. E. Schafer*, *ibid.*, 1964, **86**, 2738). *Campbell* and *Rees* (*loc. cit.*) have suggested that benzyne may be formed in a triplet state.

Simple heterocyclic analogues of the aryne, benzyne, *e.g.* the pyridynes, have now been known for some years (*R. Levine* and *W. W. Leake*, Science, 1955, **121**, 780; *Th. Kauffmann* and *F.-B. Boettcher*, Angew. Chem., 1961, **73**, 65; Ber., 1962, **95**, 949, 1528; *M. J. Pieterse* and *H. J. Den Hertog*, Rec. Trav. chim., 1961, **80**, 1377; *R. J. Martens* and *Den Hertog*, *ibid.*, 1964, **83**, 621). Lack of space precludes further discussion of this topic, which has, however, been well reviewed by *Den Hertog* and *H. C. Van Der Plas* in "Advances in Heterocyclic Chemistry", Vol. 4, *A. R. Katritzky* (Ed.), Academic Press, New York, 1965, pp. 121–144), and by *Kauffmann* (Angew. Chem. intern. Edn., 1965, **4**, 543).

There have now been some fifteen years of vigorous work on arynes, a still growing area of organic chemistry, of which the above account, limited as it is by the space available, can give only a glimpse. A useful brief monograph by *T. C. Gilchrist* and *C. W. Rees* ("Carbenes, Nitrenes and Arynes", Nelson, London, 1969) contains a valuable up to date summary of this area and a more extensive monograph by *R. W. Hoffman* ("Dehydrobenzene and Cycloalkynes", Academic Press, New York, 1968) is also available.

(iii) The S_N2 mechanism: "activated" nucleophilic substitution

The S_N2 mechanism is applicable to very many "activated" nucleophilic aromatic substitution reactions. A large number of kinetic investigations have been carried out, and these reactions have been found to conform to the criteria normally employed for the diagnosis of S_N2 reactions. Second-order kinetics are observed, the velocity of the reactions is proportional to the concentration and dependent on the nature of the nucleophilic reagents, and the reactions are facilitated by electron-attracting substituents and hindered by electron-releasing substituents in the aromatic nucleus.

The detailed nature of the bimolecular mechanism of substitution in "activated" compounds has been the subject of much investigation and discussion during the past two decades. The dominant view, to which there has been some dissent, is that the reaction involves first the formation of an intermediate of some stability, and then its decomposition to yield the products. This mechanism is referred to as the addition-elimination mechanism.

There is ample evidence that complexes of two different kinds, (*a*) charge-transfer complexes, (*b*) so-called Meisenheimer complexes, are formed between nucleophiles and aryl compounds containing the electron-withdrawing groups, usually nitro groups, necessary for "activation". Charge-transfer complexes appear to play a minor but not negligible role in nucleophilic aromatic substitution when aromatic amines, such as aniline, are the nucleophiles (cf. *Ross, loc. cit.*, and *Ross* and *I. Kuntz*, J. Amer. chem., Soc., 1954, **76**, 3000). According to Ross there are two possibilities: (*a*) aniline and chloro-2,4-dinitrobenzene, for example, are in equilibrium with a charge-transfer complex and concurrently react to form products, in which case appropriate kinetic investigation gives the second-order rate coefficient for the reaction of the amine and the chloro compound, as well as a value of the equilibrium constant for charge-transfer complex formation, which agrees well with that obtained by an independent spectroscopic method (for a revealing n.m.r. study of such complexes, see *R. A. Foster* and *C. A. Fyfe*, Trans. Faraday Soc., 1965, **61**, 1626); and (*b*) the two reagents are in equilibrium with the complex which decomposes to products. Although no certain choice can be made between these two possibilities, the writer prefers the former since the geometry of the charge-transfer complexes makes it unlikely that they would readily decompose to the observed products.

The formation of deeply coloured products in the reaction of picryl chloride with an excess of methanolic sodium methoxide, or of 2,4,6-trinitroanisole with the same reagent or with ethanolic sodium ethoxide, has been known since the turn of the century (*e.g., P. Hepp*, Ann., 1882, **215**, 345; *C. A. L. de Bruyn*, Rec. Trav. chim., 1895, **14**, 89; *J. Meisenheimer*, Ann., 1902, **323**, 205). Meisenheimer noted that 2,4,6-trinitrophenetole and methanolic sodium methoxide apparently gave the same product as 2,4,6-

trinitroanisole and ethanolic sodium ethoxide, for with mineral acid, each product gave the same mixture of trinitroanisole and trinitrophenetole. Meisenheimer assigned a salt-like structure IX to the compound. The structures proposed for intermediates in substitutions have naturally all been modelled on IX. *J. B. Ainscough* and *E. F. Caldin* (J. chem. Soc., 1956, 2528) provided evidence that the reaction producing IX is preceded by a very rapid reversible reaction, and whose product, they suggest, is a charge-transfer complex.

MeO OEt
O_2N NO_2
$K^{\oplus}$

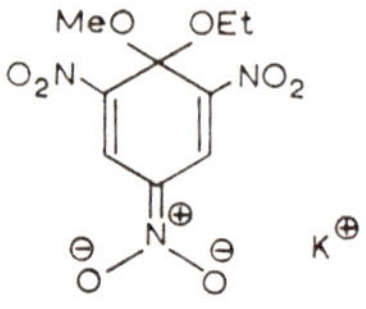

(IX)

Meisenheimer complexes* have been intensively investigated by optical spectroscopy (infrared and visible) (*R. Foster* and *D. Ll. Hammick*, *ibid.*, 1954, 2153; *Foster*, Nature, 1955, **176**, 746; 1959, **183**, 1052; *Foster* and *R. K. Mackie*, Tetrahedron, 1961, **16**, 119; 1962, **18**, 161; J. chem. Soc., 1963, 3796; *R. C. Farmer*, *ibid.*, 1959, 3425, 3430; *L. K. Dyall*, *ibid.*, 1960, 5160; *V. Gold* and *C. H. Rochester*, *ibid.*, 1964, 1687, 1692, 1697, 1704, 1710, 1717, 1722, 1727). Proton magnetic resonance studies have been carried out by *M. R. Crampton* and *Gold* (*ibid.*, 1964, 4293; Proc. chem. Soc., 1964, 298), by *Foster* and his co-workers (Rec. Trav. chim., 1965, **84**, 516; Tetrahedron, 1965, **21**, 3363) and by *K. L. Servis* (J. Amer. chem. Soc., 1965, **87**, 5495; 1967, **89**, 1508). The outcome of this extensive work has been to confirm fully the original structure proposed by Meisenheimer. It should be emphasised, however, that *Caldin* and *G. Long* (Proc. roy. Soc., 1955, **228A**, 263) and *Ainscough* and *Caldin* (J. chem. Soc., 1956, 2546) have shown that 2,4,6-trinitrotoluene may react with ethanolic ethoxide at different concentrations to yield either or both of a brown charge-transfer complex, and a purple salt formed by proton loss from the methyl group (cf. *W. Slough*, Trans. Faraday Soc., 1961, **57**, 366). Moreover, 1,3,5-trinitrobenzene with ethanolic sodium ethoxide shows a rapid reversible reaction which, according to *Caldin* and *Ainscough* (*loc. cit.*) yields an alternative type of charge-transfer complex, but according to *Gold* and *Rochester* (J. chem. Soc., 1964, 1692) an unstable complex of the Meisenheimer type, cf.

* In modern usage "Meisenheimer complex" refers to IX without the potassium ion, and structure IX carries the implication of charge delocalisation involving the *ortho*-nitro groups.

IX. The colour due to this complex fades at a measurable rate, and 3,5-dinitroanisole and nitrite ions are produced in high yield, in a reaction of first order in the complex, which, nevertheless, is not considered by Gold and Rochester to be an intermediate in this reaction. *Gold* and *Rochester* (*loc. cit.*) have also measured the equilibrium constant (7700 l·mole^{-1}) and rate of attainment of equilibrium in complex formation between 2,4,6-trinitroanisole and methoxide ion. Moreover, a further complex, probably involving two methoxide ions per molecule of trinitro compound, is formed at higher methoxide ion concentration. These reactions are also complicated by the influence on them of visible light (see Gold and Rochester's papers for further details), particularly in relation to the displacement of nitro groups.

Finally, by treating 2,4,6-trinitro[methyl-^{14}C]anisole with methanolic methoxide at room temperature, and examining quantitatively the fate of the radioactive group, *J. H. Fendler* (Chem. and Ind., 1965, 764) confirmed the accepted reaction scheme (below) for formation of Meisenheimer complexes:

OMe* | O_2N | NO_2 | NO_2 + $\overset{\ominus}{OMe}$ ⇌ (fast / slow) MeO OMe* ... $N^{\oplus}$ $O^{\ominus}$ $O^{\ominus}$ (X) ⇌ (slow / fast) OMe | O_2N | NO_2 | NO_2 + $\overset{\ominus}{OMe}$*

(X)

T. Abe (Bull. chem. Soc. Japan, 1964, **37**, 508) by quantum mechanical methods, and *J. Miller* (J. Amer. chem. Soc., 1963, **85**, 1628) by a thermochemical method, have each calculated the energy of activation for decomposition to its factors of the Meisenheimer complex X. The values obtained (~16 kcal·mole^{-1}) agree well, but differ significantly from Fendler's experimental value (19·4 kcal·mole^{-1}). The upshot of all these investigations is that the combined formation and decomposition of a Meisenheimer complex provides a clear example of nucleophilic aromatic substitution pursuing the addition-elimination mechanism and provides a probable model for the reaction mechanism in other cases in which such direct and definite evidence of the formation of an intermediate of some stability has not been obtained.

Recent extensive reviews of this topic are those of *E. Buncel, A. R. Norris* and *K. E. Russell* (Quart. Reviews, 1968, **22**, 123), and *M. R. Crampton* in "Advances in Physical Organic Chemistry", Ed. *V. Gold*, Academic Press, London, Vol. 7, 1969, p. 211.

There is important evidence of intermediate formation in a more typical example of nucleophilic aromatic substitution, *viz.* in a displacement reaction of an "activated" chloro-compound, but one has to enter the field of aromatic heterocyclic compounds to find it (*R. P. Mariella, J. J. Callahan* and *A. O. Jibril,* J. org. Chem., 1955, **20**, 1721). 2-Chloro-3-cyano-6-methyl-5-nitropyridine (XI) and the related compounds XII and XIII each reacted with alcoholic alkoxides to give solutions of an intense purple colour, which was

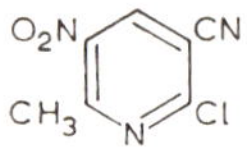

(XI) 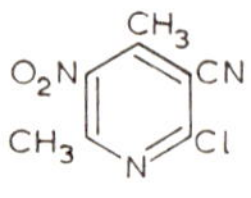 (XII) 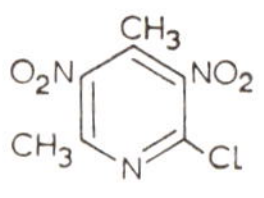 (XIII)

discharged by the addition of water, after which the corresponding 2-alkoxy compound could be isolated. The 2-methoxy compound corresponding to XII also gave the colour with methanolic methoxide. Addition of base to compounds XIV and XV gave no colour, but each reacted readily with methanolic methoxide to give the related 2-methoxy compound. The colour was attributed to the formation of an intermediate of type XVI. The fundamental similarity between XVI and X is clear.

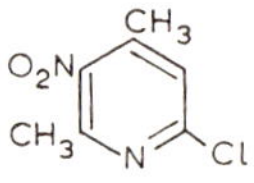

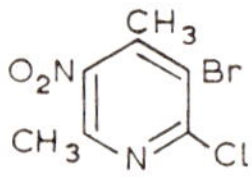

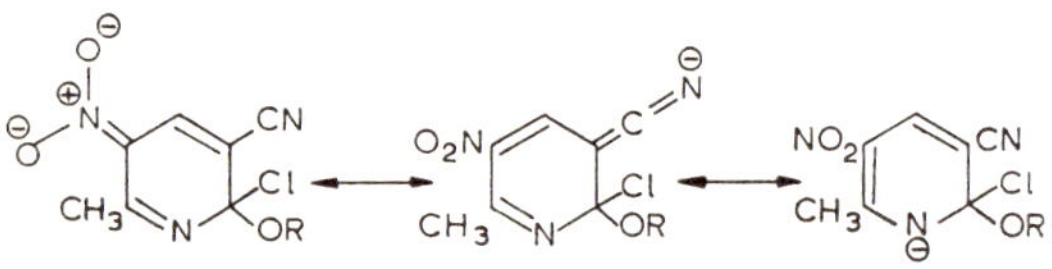

(XIV) (XV)

(XVI)

The addition-elimination mechanism may be formulated in a general way, following *Miller* (*loc. cit.*) as shown below. *Bunnett* and *Zahler* (*loc. cit.*) gave a similar formulation differing somewhat in detail, however.

This mechanism is of the form

$$\mathrm{A} + \mathrm{B} \underset{k_{-1}}{\overset{k_1}{\rightleftharpoons}} \mathrm{C}$$

$$\mathrm{C} \underset{k_{-2}}{\overset{k_2}{\rightleftharpoons}} \mathrm{D} + \mathrm{E}$$

Initial State | Transition State for formation of the complex | Intermediate complex | Transition State for decomposition of the complex | Final State

If C may be regarded as a reactive intermediate, and assuming that $k_{-2} = 0$, application of the stationary-state treatment gives $d[D]/dt = k_2k_1[A][B]/(k_{-1} + k_2)$. There are the usual two important extreme cases, viz., $k_2 \gg k_{-1}$, when $d[D]/dt = k_1[A][B]$, and the observed rate coefficient $k_{obs} = k_1$. If $k_2 \ll k_{-1}$, $k_{obs} = Kk_2$, where K is the equilibrium constant for the (now) rapid reversible formation of the complex. The corresponding energy profiles are shown in Figs. 1 and 2, respectively. This mechanism, of course, conforms well with the second-order kinetics always observed for "activated" substitutions. On the basis of this mechanism *Miller* (*loc. cit.*)

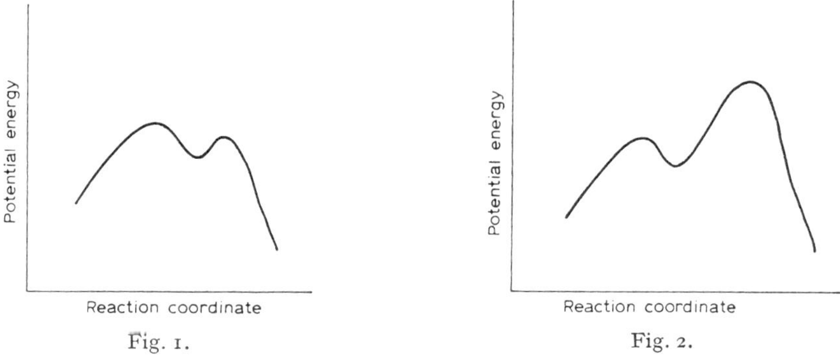

Fig. 1. Fig. 2.

has provided a convincing explanation of the order of ease of replacement of aromatic halogen (F $\gg$ Cl > Br > I) towards reagents whose nucleophilic atom belongs to the first row of the Periodic Table, and of the reversal of this order when the nucleophilic atom of the reagent belongs to a higher row of the Periodic Table. In the first case formation of the intermediate complex is thought to be rate-limiting, in the second case its decomposition. This view is strongly supported by the observed values of the Hammett ϱ parameter for the reactions of azide ion and thiocyanate ion with "activated" aromatic compounds in methanol, *viz.*, 3·77 and 4·13 respectively (*B. A.*

Bolto, Miller and *A. J. Parker*, J. Amer. chem. Soc., 1957, **79**, 93). Further important evidence was provided by *Bunnett* and *K. M. Pruitt* (J. Elisha Mitchell Sci. Soc., 1957, **73**, 297) and *Bunnett* and *J. J. Randall* (J. Amer. chem. Soc., 1958, **80**, 6020) who showed that the reaction between 1-fluoro-2,4-dinitrobenzene and *N*-methylaniline in methanol was subject to general base catalysis (in their case by acetate). The effect was much too large to be a salt effect. The reaction was formulated as follows:

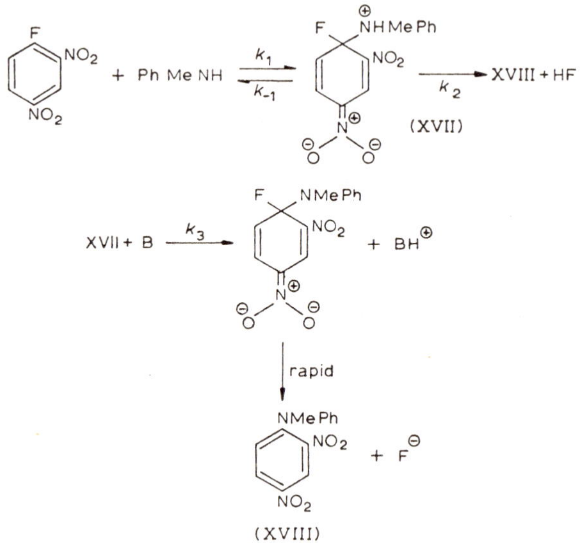

F, NO_2, NO_2 + Ph Me NH $\underset{k_{-1}}{\overset{k_1}{\rightleftharpoons}}$ F, $\overset{\oplus}{N}$H Me Ph, NO_2, $\overset{\oplus}{N}$, $\overset{\ominus}{O}$, $\overset{\ominus}{O}$ (XVII) $\xrightarrow{k_2}$ XVIII + HF

XVII + B $\xrightarrow{k_3}$ F, NMePh, NO_2, $\overset{\oplus}{N}$, $\overset{\ominus}{O}$, $\overset{\ominus}{O}$ + $BH^{\oplus}$

rapid

NMePh, NO_2, NO_2 (XVIII) + $F^{\ominus}$

Application of the steady state approximation gives

$$k_{obs} = \frac{k_1k_2 + k_1k_3[B]}{k_{-1} + k_2 + k_3[B]}$$

$$\text{If } k_{-1} \gg (k_2 + k_3[B])$$

$$k_{obs} = k_1/k_{-1}\{k_2 + k_3[B]\}$$

i.e. the observed rate coefficient should be a linear function of [B], which was indeed observed. If $\{k_2 + k_3[B]\} \gg k_{-1}$, $k_{obs} = k_1$ and base catalysis will not be observed, as was found for the chloro and bromo compounds. Bunnett's conclusions have been criticised by *Ross* (*loc. cit.*), but supported by other work. Recently *K. B. Lam* and *J. Miller* (Chem. Comm., 1966, 642) have demonstrated large electrophilic catalytic effects in the reactions of iodide and thiocyanate ion ("heavy" nucleophiles) with 1-fluoro-2,4-dinitrobenzene, but not in the reaction of azide ion. With the heavy nucleophiles, theoretical analysis indicates that decomposition of the complex is rate-limiting and hence electrophilic catalysis is possible. This investigation provides further evidence in favour of the two-stage mechanism.

(iv) Substituent effects in the S_N2 mechanism

There is by now a tremendous wealth of quantitative information on this subject in the form of rate coefficients, and usually, activation parameters for well-defined reaction series. The most up-to-date and comprehensive account is that given by *Miller* in his monograph "Aromatic Nucleophilic Substitution", Elsevier, Amsterdam, 1969, in which the applicability of quantum chemical calculations and linear free energy relationships, especially the Hammett equation, in rationalising the results is critically assessed. These methods, especially the former, are subject to considerable limitations, so it is necessary to fall back on well-ordered empiricism.

A major classification is based on the position of the substituent, *ortho*, *meta*, or *para*, with respect to the seat of substitution, and a further sub-classification depends on the polar character of the substituent. *Miller* has defined a useful quantity, the substituent rate factor (s.r.f.), *viz.* the relative rate coefficient for the reactions of H·Ar·X and R·Ar·X (X is the displaced group) with a given reagent under fixed reaction conditions. These s.r.f.'s are usually symbolised $f_p^{NO_2}$, for example, as the s.r.f. for the *para*-nitro group. Values of f range from $\sim 10^{16}$ (for R = p-$N_2^{\oplus}$ in the reaction of methanolic methoxide with substituted fluorobenzenes at 0^0) through $\sim 10^5$ for R = p-$SMe_2^{\oplus}$ in a similar reaction of substituted o-chloronitrobenzenes at 50^0, and through $\sim 10^2$ to $\sim 10^4$ for numerous dipolar substituents such as p-CO·NH_2 and p-SO_2·Me, to $\sim 10^{-1}$ to 10^{-2} for electron releasing p-alkyl groups and other electron-releasing groups under similar conditions. The qualitative pattern described by *Bunnett* and *Zahler* (*loc. cit.*) has not been significantly disturbed by these investigations, but has been greatly enriched in detail. Bunnett and Zahler's table is given below, and Miller's monograph should be consulted for the details.

It is noteworthy that halogens are all activating in the order m- > p-halogen, and the values of f_p/f_m lie in the range $\sim 0 \cdot 2$ to $0 \cdot 5$.

The effect of the leaving group on reactivity has been mentioned earlier, particularly the important halogen mobility order, F ≫ Cl > Br > I for "light" nucleophiles, and *vice versa* for "heavy" nucleophiles. Again *Miller*'s monograph should be consulted for detailed information, in view of the wealth of information now available.

TABLE 1

ACTIVATING GROUPS IN ORDER OF DECREASING ACTIVATING POWER

Diazonium salt group ($-N_2^{\oplus}$) Cationic carbon ($-CH_2^{\oplus}$)	Activate halogen exchange at room temperature or below.
Alkylated hetero nitrogen $N^{\oplus}$–R	Activate replacement by strongly nucleophilic reagents at room temperature.
Nitroso group (NO) Nitro group (NO_2)	Activate reaction with strongly nucleophilic reagents at ∼80–100°C.
Hetero nitrogen atom N Methylsulphonyl (CH_3SO_2) Trimethylammonio ($(CH_3)_3N^{\oplus}$) Trifluoromethyl (CF_3) Acyl groups (RCO) Cyano (CN) Carboxyl (CO_2H)	With nitro also present, activate reaction with strongly nucleophilic reagents at room temperature.
Ionized sulphonate ($SO_3^{\ominus}$) Halogens (Cl, Br, I) Carboxylate ($CO_2^{\ominus}$) Phenyl	With nitro also present, activate reaction with strongly nucleophilic reagents at ∼40–60°C.

(*v*) *Influence of the nucleophilic reagent*

The last two decades have seen substantial developments in attempts at quantification of the concept of kinetic nucleophilicity. *Miller* has concentrated on and studied in detail kinetic nucleophilicity in "activated" aromatic S_N2 reactions (*K. C. Ho, Miller* and *K. W. Wong*, J. chem. Soc., B, 1966, 310; *D. C. Hill, Ho* and *Miller, ibid.*, 299; *Miller* and *A. J. Parker*, J. Amer. Chem. Soc., 1961, 83, 117; *Miller, ibid.*, 1963, 85, 1628; *idem*, "Organic Reaction Mechanisms", Chemical Society Special Publication No. 19, 1965, 193). The theoretical treatment depends greatly on a knowledge of the reaction mechanism and the formulation of a satisfactory hypothesis about the relevant energy profile. For this purpose reagents are divided into two classes – anionic and neutral. Moreover kinetic nucleophilicity depends also on the substrate and reaction conditions. This is an intricate and involved subject, detailed discussion of which is inappropriate here. Suffice it to say that with suitable assumptions the theoretical treatment and experimental results agree reasonably well. A typical (qualitative) kinetic nucleophilicity series is as follows: $SPh^{\ominus} > SMe^{\ominus} > OMe^{\ominus} > N_3^{\ominus}$ ($> SCN^{\ominus} > I^{\ominus} > Br^{\ominus} > Cl^{\ominus} > F^{\ominus}$). [That part of the series in parentheses was obtained only by calculation; the other part by calculation and experiment, which agree]. The nucleophilicity series must necessarily refer to a standard substrate *viz.*, *p*-iodonitrobenzene in methanol at 25°. *Miller*'s monograph (*loc. cit.*) gives much detailed information for both anionic and neutral reagents – on the whole, for the latter the qualitative kinetic nucleophilicity order is broadly what one would expect from basicities, tempered by steric considerations. For lack of space no attempt is made to deal here systematically with salt, solvent and steric effects and with various other topics. The interested reader is referred to *Miller*'s monograph.

(b) Nucleophilic rearrangements

(*i*) *The Smiles rearrangement*

This subject has been reviewed by *H. B. Watson* (Ann. Reports, 1939, 36, 197), by *Bunnett* and *Zahler* (*loc. cit.*), by *Bunnett* (Quart. Reviews, 1958, **12**, 1), and by *H. J. Shine*, in "Aromatic Rearrangements", Elsevier Publishing Co., Amsterdam, 1967, pp. 307–316.

Smiles rearrangements (so called after Samuel Smiles, who largely developed the field), can be represented generally by the following scheme:

(XIX) (XX)

In suitably activated systems, rearrangement has been found with X = O, S, SO, SO_2, CO·O and SO_2·O, and with YH = NHR, CO·NH_2, SO_2·NH_2, OH, NH_2, SH and SO_2H, although it does not follow that a compound containing any given combination of these will undergo rearrangement. Many rearrangements of this type are recorded in the literature, particularly in a number of papers by Smiles and his collaborators. The following two reactions are typical and proceed under mild conditions in the presence of aqueous alkali:

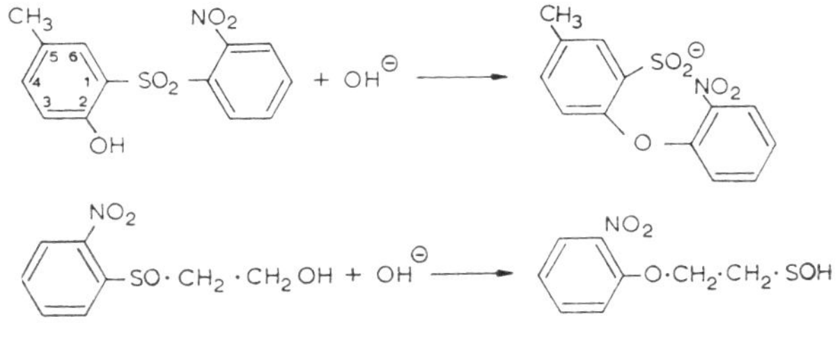

A. A. Levy, H. C. Raines and *S. Smiles* (J. chem. Soc., 1931, 3264) established the mechanism of the rearrangement as, first, conversion of YH to $Y^{\ominus}$ by the strong base present, followed by nucleophilic replacement of X by Y at carbon atom C (formulae XIX and XX). This mechanism was confirmed in later papers by *Smiles*, who showed that the rate of rearrangement was influenced by the groups present in the aromatic nucleus, the replaceability of X, the nucleophilic activity of $Y^{\ominus}$, and the acidity of $-YH$. It had also been known for some time from the work of *C. S. McClement* and *Smiles* (*ibid.*, 1937, 1016) that the rearrangement of 6-methyl substituted derivatives of XXI is much more rapid than that of XXI itself, and these authors

(XXI)

attributed this to the polar effect of the methyl group. *Bunnett* and *Zahler* (*loc. cit.*) first proposed that the acceleration was due to a steric effect and *Bunnett* and *T. Okamoto* (J. Amer. chem. Soc., 1956, 78, 5363) have provided powerful evidence in favour of this by a kinetic study the results of which are summarized in Table 2. They were able to show that substituents in the 6-position of opposite polar character to the methyl group, but of suitable size, also exercised the powerful accelerative effect observed with the methyl group. The effect is very large indeed, a rate ratio at 0° for the 6-chloro

TABLE 2

MEAN RATE COEFFICIENTS

Substituents	Mean rate coefficient, min^{-1}		ΔE, cal. $mole^{-1}$	$\Delta S^{\ddagger}$, e.u.
	At 0·0°	At 46·0°		
$5\text{-}CH_3$	—	$1{\cdot}94 \times 10^{-2}$	23,800	−2·0
$4,5\text{-Di-}CH_3$	$(3{\cdot}6 \times 10^{-5})$*	$1{\cdot}66 \times 10^{-2}$	23,200	−4·2
$5,6\text{-Di-}CH_3$	>3	—	—	—
$4\text{-Cl-}5\text{-}CH_3$	$(1{\cdot}8 \times 10^{-6})$*	$1{\cdot}44 \times 10^{-3}$	25,300	−2·6
$6\text{-Cl-}5\text{-}CH_3$	$9{\cdot}2 \times 10^{-1}$	—	—	—
$4\text{-Br-}5\text{-}CH_3$	—	$1{\cdot}23 \times 10^{-3}$	—	—
$6\text{-Br-}5\text{-}CH_3$	2·1	—	—	—
$3\text{-Cl-}4,5\text{-di-}CH_3$	$(7{\cdot}4 \times 10^{-7})$*	$1{\cdot}0 \times 10^{-3}$	27,200	2·6
$3\text{-Cl-}5,6\text{-di }CH_3$	$1{\cdot}29 \times 10^{-1}$	—	20,200	1·5

* Extrapolated from measurements at higher temperatures.

compound of $\sim 5 \times 10^5$ being estimated and for the 6-methyl group a steric acceleration of $\sim 2 \times 10^5$ is observed. In the only case where relative heats and entropies of activation were determined, the accelerative effect was located mainly in a decrease in ΔE of ~ 7 kcal/mole. *Bunnett* and *Okamoto* (*loc. cit.*) ascribe this not to steric effects in the transition state, in which the 6-substituent is in a relatively open situation, but to an increase in the enthalpy of the initial state, *i.e.* the unrearranged sulphone anion. A study of molecular models shows that the net effect of a bulky 6-substituent is to favour just those rotational conformations of the unrearranged compound which are the necessary pre-requisite of rearrangement, *i.e.* in which the ionized hydroxyl group is brought as near as possible to the 1-position in the second benzene ring. When a sufficiently large 6-substituent is present the "rearranging" conformations become the most favourable as far as compressions with the 6-substituent are concerned, whereas in the absence of a 6-substituent the "rearranging" conformations are energetically less favoured.

A rearrangement related to the Smiles rearrangement has been discovered by *W. E. Truce* and his coworkers (J. Amer. chem. Soc., 1958, **80**, 3625; 1959, **81**, 481, 484). Examples are given below (R = H or CH_3):

CH_3 SO_2 R CH_3 CH_3 —BuLi in Et_2O at b.p.→ CH_3 SO_2 R CH_3 $CH_2^{\ominus}$ → CH_3 $SO_2^{\ominus}$ CH_3 CH_2 R

(*ii*) *Arylhydroxylamine rearrangements*

This subject has been reviewed by *E. D. Hughes* and *C. K. Ingold* (Quart. Reviews, 1952, 6, 45) who arrived at different conclusions from *M. J. S. Dewar* ("Electronic Theory of Organic Chemistry", Oxford Univ. Press, 1949, p. 225). An up to-date and comprehensive review is provided by *H. J. Shine* in "Aromatic Rearrangements", Elsevier Publishing Co., Amsterdam, 1967, pp. 182–190.

E. Bamberger (Ber., 1894, **27**, 1347, 1548, and subsequent papers) investigated the acid-catalyzed rearrangement of arylhydroxylamines, such as the conversion of phenylhydroxylamine into *p*-aminophenol by treatment with dilute sulphuric acid. *Hughes* and *Ingold* (*loc. cit.*) consider the mechanism of this intermolecular rearrangement to be as follows, for replacement at the *para*-position:

$$H\text{-}C_6H_4\text{-}NH{\cdot}OH + H^{\oplus} \xrightarrow{(1)} H\text{-}C_6H_4\text{-}NH{\cdot}\overset{\oplus}{O}H_2$$

$$\xrightarrow{(2)} \overset{\oplus}{C}_6H_5{=}NH \xrightarrow[(3)]{HY:} (\overset{\oplus}{H}Y)(H)C_6H_4{=}NH \longrightarrow Y\text{-}C_6H_4\text{-}NH_2 + H^{\oplus}$$

The extension to the case of *ortho*-replacement is obvious. If HY: is a water molecule, the isomeric aminophenol is obtained, although many products have been isolated from time to time from reactions carried out in the presence of other nucleophilic reagents HY:, such as *o*- and *p*-anisidines and -phenetidines when methanol and ethanol, respectively, are used to dilute the acid. In principle, a bimolecular mechanism analogous to the unimolecular one formulated above is also feasible. In such a reaction the approach of HY: (stage 3) and the recession of H_2O (stage 2) become synchronous.

(iii) The von Richter reaction

This reaction was discovered by *V. von Richter* (Ber., 1871, **4**, 21, and subsequent papers). In it substitution by a carboxyl group is effected at a position *ortho* to an ejected nitro group. The reaction is carried out by heating the nitro compound with ethanolic potassium cyanide. Thus, for example, *p*-bromonitrobenzene gives *m*-bromobenzoic acid, whilst *m*-bromonitrobenzene gives a mixture of *o*- and *p*-bromobenzoic acid.

The reaction was little studied for several decades until *Bunnett* and his co-workers carried out a series of studies which resulted in the definition of its salient features and a proposed reaction mechanism (*Bunnett, J. F. Cormack* and *F. C. McKay*, J. org. Chem., 1950, **15**, 481; *Bunnett et al.*, J. Amer. chem. Soc., 1954, **76**, 5755; *Bunnett* and *M. M. Rauhut*, J. org. Chem., 1956, **21**, 934, 939, 944). Further work by *M. Rosenblum* (J. Amer. chem. Soc., 1960, **82**, 3796) and by *E. F. Ullman* and *E. A. Bartkus* (Chem. and Ind., 1962, 93) has led to a modification of the mechanism and support for this modification. The essential features of the reaction are: (*a*) the carboxyl group, which occupies the position *ortho* to the nitro group lost, is *not* formed from the related cyano or carboxamide group, (*b*) hydrogen from the hydroxylic solvent is incorporated in the product, (*c*) substituents *ortho* to the nitro group inhibit but do not necessarily abolish reaction, (*d*) activating substituents (not too strongly so) should be present, (*e*) gaseous nitrogen is evolved in a yield corresponding to that of the carboxylic acid, one of the nitrogen atoms of N_2 coming from the NO_2-group (as shown by ^{15}N-labelling) and the other from $CN^{\ominus}$ by a path not involving ammonia, (*f*) one oxygen atom of the carboxyl group comes from the solvent and one from the NO_2-group, as shown by ^{18}O-labelling of the solvent. The mechanism based on these facts and supported by the work of *Ullman* and *Bartkus* is as follows:

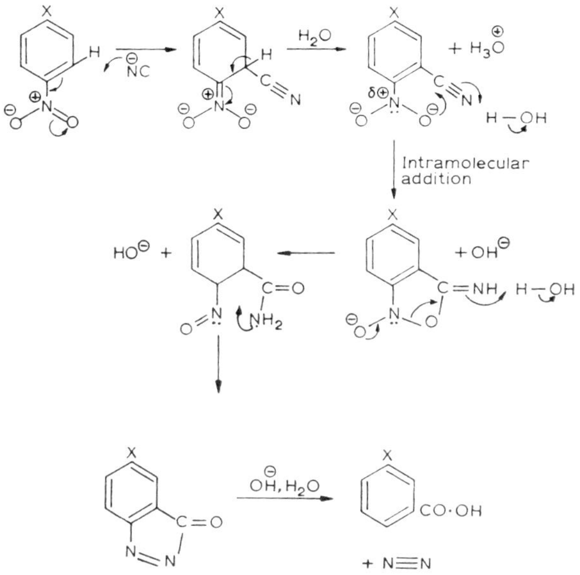

X H NC H H_2O $+ H_3O^{\oplus}$ C≡N H–OH Intramolecular addition $HO^{\ominus}$ + C=O NH₂ + $OH^{\ominus}$ C=NH H–OH $\overset{\ominus}{O}H, H_2O$ CO·OH + N≡N

This subject is excellently reviewed by *H. J. Shine* (*loc. cit.*, pp. 326–335).

4. Homolytic aromatic substitution

D. H. HEY and G. H. WILLIAMS

The suggestion that certain aromatic substitution reactions can take place by means of electrically neutral radicals was put forward by *D. H. Hey* (J. chem. Soc., 1934, 1966, 1977), and later developed by *Hey* and *W. A. Waters* (Chem. Reviews, 1937, **21**, 169). Such reactions are classified as homolytic substitution reactions in order to distinguish them from the heterolytic substitution reactions brought about by electrophilic and nucleophilic reagents. The subject has been reviewed by *D. R. Augood* and *G. H. Williams* (*ibid.*, 1957, **57**, 123), by *O. C. Dermer* and *M. T. Edmison* (*ibid.*, 1957, **57**, 77), by *Williams* ("Homolytic Aromatic Substitution", Pergamon Press, Oxford, 1960), and by *Hey* ("Advances in Free Radical Chemistry", Logos Press – Academic Press, London, 1967, **2**, 47).

It has already been pointed out (Vol. IA, p. 323) that with free radicals of short life, aromatic compounds and heterocyclic compounds of aromatic type give rise to nuclear substitution rather than hydrogen abstraction:

$$R\cdot + ArH \rightarrow RAr + [H\cdot]$$

The hydrogen atom thus liberated may have no free existence but usually enters into combination with other radicals or molecules present in the system. The predominance of this type of reaction with aromatic compounds leads to the existence of a wide range of substitution reactions, some of which are of considerable preparative value. In addition, it also provides a vehicle for a qualitative and quantitative study of the substitution processes with reference to both the directing influence and the activating or deactivating influence of substituent atoms or groups. It was indeed largely from qualitative observations on these phenomena that the existence of such substitution reactions was first recognised. The substitution reactions of this type which have received most attention are (a) alkylation, (b) amination, (c) hydroxylation, (d) thienylation and (e) arylation.

(a) Alkylation

L. F. Fieser and his collaborators have shown that methylation of quinones and aromatic nitro compounds can be readily effected with acetyl peroxide (J. Amer. chem. Soc., 1942, **64**, 2060; 1947, **69**, 2338). When 2-methylnaphthaquinone is heated with acetyl peroxide in acetic acid solution a methyl group is introduced into the 3-position of the naphthalene nucleus to give 2,3-dimethylnaphthaquinone, and similar treatment of 2,4,6-

trinitrotoluene gives 2,4,6-trinitro-*m*-xylene:

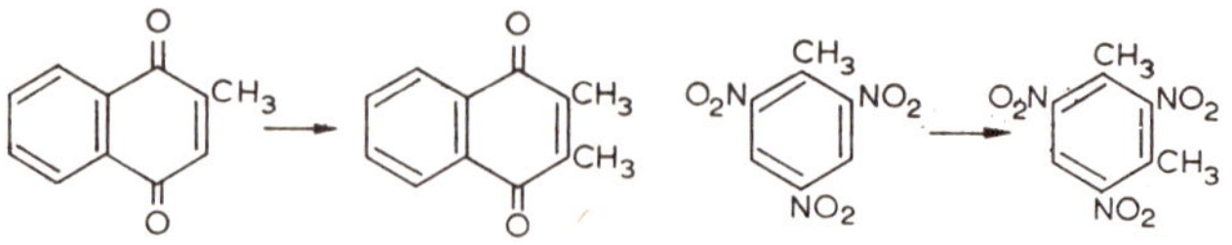

The latter reaction can also be effected with lead tetra-acetate and by the electrolysis of sodium acetate, both of these processes providing alternative sources of the free methyl radical (*Fieser et al., ibid.*, 1942, 64, 2043, 2052). The production of free methyl radicals from acetyl peroxide may be represented thus:

$$CH_3{\cdot}CO{\cdot}O{\cdot}O{\cdot}CO{\cdot}CH_3 \rightarrow 2\ CH_3{\cdot}CO{\cdot}O{\cdot} \rightarrow 2CO_2 + 2\ CH_3{\cdot}$$

In similar manner *m*-dinitrobenzene is converted into 2,4-dinitrotoluene and 2,4-dinitro-*m*-xylene. Nitrobenzene may also be methylated, but the yields are low. Closely related methylation reactions can be effected with phenyliodosyl acetate (*R. B. Sandin* and *W. B. McCormack, ibid.*, 1945, 67, 2051). Similar alkylation reactions can be brought about by means of the peroxides of long chain aliphatic acids and dibasic acids (*Fieser* and *R. B. Turner, ibid.*, 1947, 69, 2338; *Fieser* and *E. M. Chamberlin, ibid.*, 1948, 70, 71).

The methylation of benzene and of toluene has been reported by *E. L. Eliel, K. Rabindran* and *S. H. Wilen* (J. org. Chem., 1957, **22**, 859), who used high concentrations of acetyl peroxide. Pyridine has been alkylated in good yields with a number of diacyl peroxides and by the electrolysis of solutions of aliphatic acids (*S. Goldschmidt* and *H. Minsinger*, Ber., 1954, 87, 956). The methylation of aromatic compounds has also been effected with methyl radicals derived from *tert*-butyl peroxide (*B. R. Cowley, R. O. C. Norman* and *Waters*, J. chem. Soc., 1959, 1799) and, in good yield, by the photolysis of methylmercuric iodide (*G. E. Corbett* and *Williams, ibid.*, 1964, 3437).

M. Szwarc and his co-workers (J. Amer. chem. Soc., 1955, 77, 1949, and subsequent papers) have reported determinations of the methyl affinities of aromatic and heterocyclic compounds in reactions with methyl radicals derived from acetyl peroxide. The relative reactivities of a number of polycyclic aromatic hydrocarbons towards trichloromethyl radicals have been measured by *E. C. Kooyman* and *E. Farenhorst* (Trans. Faraday Soc., 1953, 49, 58).

A mechanism for the alkylation of benzene by alkyl radicals derived from the photolysis of alkylmercuric iodides, based on addition followed by dehydrogenation, has been put forward by *Corbett* and *Williams* (*loc. cit.*). The existence of σ-complexes was inferred from the formation of residues containing their binuclear dimerisation products, while the formation of nearly theoretical yields of alkanes indicated that alkyl radicals were responsible for the dehydrogenation, thus:

$$Me\cdot + C_6H_6 \longrightarrow \text{(Me, H σ-complex)} \xrightarrow{Me\cdot} C_6H_5{\cdot}Me + MeH$$

$$2\,\text{(Me, H σ-complex)} \longrightarrow \text{(Me, H, H, H, H, Me dimer)} \quad \text{(and isomers)}$$

In an extension of this study, the same authors (J. chem. Soc, B, 1966, 877) showed that alkylbenzenes undergo hydrogen-abstraction from the side-chain as well as nuclear methylation, and reported the analogous ethylation and *n*-butylation reactions. The formation of iodine atoms by breakdown of HgI· fragments was considered possible, and the following alternative mechanism for the dehydrogenation of σ-complexes was suggested:

$$RArH\cdot + I\cdot \rightarrow RAr + HI$$

$$RHgI + HI \rightarrow RH + HgI_2$$

Mono-substituted benzenes undergo cyclohexylation with cyclohexyl radicals formed from the decomposition of *tert*-butyl peroxide in the presence of cyclohexane (*J. R. Shelton* and *C. W. Uzelmeier*, J. Amer. chem. Soc., 1966, **88**, 5222). The partial rate factors conform to a Hammett ρ-factor of +1·1 for the cyclohexyl radical. Comparison of partial rate factors for the appropriate reactions indicates that the nucleophilicity of the radicals increases in the order phenyl < methyl < cyclohexyl. This conclusion (for methyl and phenyl radicals) is supported by a comparison of the ratios of isomeric products formed in the methylation and phenylation of 3- and 4-picoline (*R. A. Abramovitch* and *K. Kenaschuk*, Canad. J. Chem., 1967, 45, 509).

(*b*) *Amination*

When the bromination of cyclohexene with *N*-bromosuccinimide is carried out in benzene solution the product contains some *N*-phenylsuccinimide in addition to 3-bromocyclohexene. The free succinimido-radical involved in the allylic bromination of cyclohexene thus appears to be able to attack benzene, resulting in the direct introduction of the substituted amino group into the aromatic nucleus (*D. R. Howton*, J. Amer. chem. Soc., 1947, **69**, 2060). The direct phthalimidation of aromatic compounds by the action of heat on *N*-chlorosulphonylphthalimide has been reported by *R. A. Lidgett, E. R. Lynch* and *E. B. McCall* (J. chem. Soc., 1965, 3754).

Although dimethylamino-radicals formed by the thermal decomposition of tetramethyl-2-tetrazene do not effect substitution in aromatic nuclei (*R. E. Jacobson, K. M. Johnston* and *Williams*, Chem. and Ind., 1967, 157), substitution products are obtained by treatment of several aromatic compounds with mono- and di-alkyl-*N*-halogenoamines in the presence of ferrous, cuprous or titanous salts:

$$R_2NCl + Fe^{2\oplus} \rightarrow R_2N\cdot + Fe^{3\oplus} + Cl^{\ominus}$$

(*F. Minisci* and *R. Galli*, Tetrahedron Letters, 1965, 433; *Minisci, Galli* and *G. Pollina*, Chim. Ind. [Milan], 1965, **47**, 736; *Minisci, Galli* and *R. Bernardi*, Tetrahedron Letters, 1966, 699; *Minisci, Galli* and *M. Cecere*, Chim. Ind. [Milan], 1966, **48**, 725; *Minisci, V. Trabucchi* and *Galli, ibid.*, p. 716; *Minisci et al., ibid.*, p. 1147; *idem, ibid.*, p. 484; *Minisci, Galli* and *Bernardi*, Chem. Comm., 1967, 903; *Minisci* and *Cecere*, Chim. Ind. [Milan], 1967, **49**, 1333; *Minisci et al., ibid.*, 1969, **51**, 280; *H. Bock* and *K.-L. Kompa*, Angew. Chem. internat. Edn., 1965, **4**, 783; *idem*, Ber., 1966, **99**, 1347, 1357; *Montecatini Edison S.p.A.*, Netherlands Patent Appl., 6,614,947/1967). The *ortho-para* orientation exhibited by anisole in this reaction would indicate that the radicals are appreciably electrophilic, although chlorobenzene is substituted mainly (60%) in the *meta*-position.

Hydroxylamine and hydroxylaminesulphonic acid have also been used as sources of amino-radicals for the direct amination of benzene derivatives and naphthalene, sometimes in good yield. Ferrous and titanous salts have been used as initiators (*Z. Yoshida, T. Matsumoto* and *R. Oda*, Kogyo Kagaku Zasshi, 1962, **65**, 46; *idem, ibid.*, 1964, **67**, 64; *Minisci, Galli* and *Cecere*, Tetrahedron Letters, 1965, 4663; *Minisci* and *Galli, ibid.*, p. 1679; *Minisci et al.*, Chim. Ind. [Milan], 1966, **48**, 264). In acid solution amino-radicals are thought to be formed from hydroxylamine as follows:

$$NH_2OH\cdot HCl + M^{n\oplus} \rightarrow \cdot NH_2 + M^{(n+1)\oplus} + Cl^{\ominus} + H_2O$$

where M is Fe or Ti. Hydroxylaminesulphonic acid with ferrous salts gives protonated amino-radicals which lead to amination of aromatic substrates as follows:

$$\overset{\oplus}{N}H_3\,\overset{\ominus}{O}SO_3 + Fe^{2\oplus} \rightarrow \overset{\oplus}{N}H_3\cdot + SO_4^{2\ominus} + Fe^{3\oplus}$$

$$ArH + \overset{\oplus}{N}H_3\cdot \rightarrow [HArNH_3]^{\oplus}\cdot$$

$$[HArNH_3]^{\oplus}\cdot + Fe^{3\oplus} \rightarrow ArNH_3^{\oplus} + H^{\oplus} + Fe^{2\oplus}$$

Measurements of isomer ratios with benzene derivatives indicate that, as would be expected, the aminium radical cation from hydroxylaminesulphonic acid is substantially more electrophilic than the amino-radical from hydroxylamine. Yields in general also tend to be considerably higher in the hydroxylaminesulphonic acid reaction.

(c) Hydroxylation

Aromatic hydroxylation has been effected by means of hydroxyl radicals derived from aqueous ferrous sulphate in the presence of hydrogen peroxide (Fenton's reagent) and by the action of ionising radiations on aqueous solutions (*J. Weiss et al.*, J. chem. Soc., 1949, 2074, 3245; 1950, 2704; 1951, 3265, 3275; *Norman* and *G. K. Radda*, Proc. chem. Soc., 1962, 138; *J. R. Lindsay Smith* and *Norman*, J. chem. Soc., 1963, 2897). The products of a number of such hydroxylation reactions with mono-substituted benzene derivatives have been estimated quantitatively in order to determine the influence of the substituent atom or group on the ratio of the isomers formed. Some of the results obtained are given in Table 1.

Kinetic measurements on the hydroxylation of substituted benzenes and benzoate ions give a Hammet ρ-factor of $-0{\cdot}41$ for hydroxylation (*M. Anbar, D. Myerstein* and *P. Neta*, J. phys. Chem., 1966, 70, 2660).

The mechanism of aromatic hydroxylation has been discussed by *W. T. Dixon* and *Norman* (J. chem. Soc., 1964, 4857), who obtained evidence for the intermediate formation of substituted cyclohexadienyl radicals (σ-complexes) by the use of electron spin resonance spectroscopy. The hydroxylating agent was hydrogen peroxide in the presence of acidified titanous chloride. The spectra were consistent with the existence of the resonance-stabilised radical:

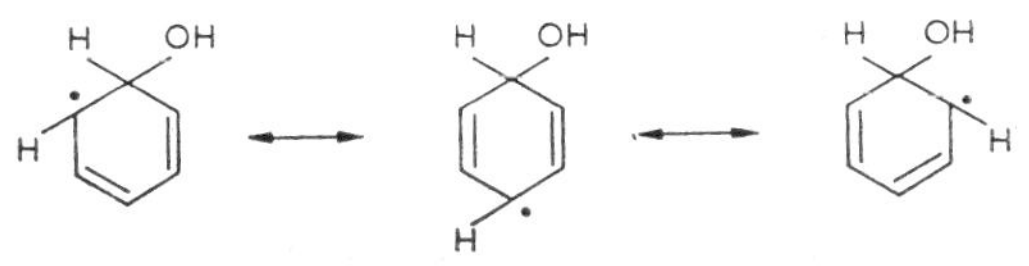

TABLE 1

HYDROXYLATION BY HYDROXYL RADICALS

Source of radical	Aromatic compound	Yield of isomerides, %		
		ortho	*meta*	*para*
$FeSO_4$-H_2O_2	$PhNO_2$	24	30	46
$H_2O \rightarrow H\cdot + \cdot OH$ (pH = 6)	$PhNO_2$	35·5	29	35·5
$FeSO_4$-H_2O_2	PhCl	42	29	29
$H_2O \rightarrow H\cdot + \cdot OH$ (pH = 6)	PhCl	15–20	20–25	50–60
$FeSO_4$-H_2O_2	PhF	37	18	45
$FeSO_4$-H_2O_2	PhOMe	84	—	16

In acid solution, the σ-complexes formed with some benzene derivatives undergo acid-catalysed heterolytic bond-cleavages to give radicals formed by loss of the hydroxyl group from the nucleus, and a group (H, CO_2H or CH_2OH) from the side-chain (*Norman* and *R. J. Pritchett, ibid.*, 1967, 926).

This interpretation of the mechanism of hydroxylation is supported by measurements of *G*-values for the γ-radiolysis of *p*-nitrophenol in aqueous solution (*D. Graesslin et al.*, Z. phys. Chem., 1966, 51, 84). 1,2-Dihydroxy-4-nitrobenzene is formed *via* a σ-complex which is dehydrogenated by excess *p*-nitrophenol, or its radical-ion, or the conjugate acid thereof, leading ultimately to the formation of *p*-aminophenol together with the hydroxylation product.

Aromatic hydroxylation is known to take place *in vivo* and is of interest in relation to the metabolism of drugs. A review of the available data suggests some correlation with the results of hydroxylation by free hydroxyl radicals.

The closely related reactions of acetoxylation and benzoyloxylation are also well established. Benzoyloxylation usually accompanies phenylation as a side reaction when benzoyl peroxide is decomposed in aromatic solvents. Homolytic benzoyloxylation is most frequently encountered with polycyclic systems which are more reactive than monocyclic aromatic systems and can therefore compete more successfully with decarboxylation. *I. M. Roitt* and *Waters* (J. chem. Soc., 1952, 2695) studied the reactions of benzoyl peroxide with a series of polycyclic hydrocarbons, and similar reactions have been reported with naphthalene (*R. L. Dannley* and *M. J. Gippin*, J. Amer. chem. Soc., 1952, 74, 332; *D. I. Davies, Hey* and *Williams*, J. chem. Soc. 1958, 1878). Acetoxylation of naphthalene has been effected with acetyloxyradicals derived from lead tetra-acetate (*Fieser, R. Clapp* and *W. Daudt*, J. Amer. chem. Soc., 1942, **64**, 2052).

(d) Thienylation

2- and 3-Thienyl radicals formed by photolysis of 2- and 3-iodothiophene replace hydrogen in aromatic nuclei (*L. Benati* and *M. Tiecco*, Boll. sci. Fac. Chim. ind. Bologna, 1966, **24**, 255; *G. Martelle, P. Spagnolo* and *M. Tiecco*, J. chem. Soc., B, 1968, 901). Partial rate factors (*q.v.*, p. 122) for these reactions indicate a homolytic mechanism.

(e) Arylation

Substitution reactions of aromatic nuclei with aryl radicals lead to the formation of biaryl derivatives, frequently in good yield, and many such reactions have been developed into useful preparative methods (see *W. E. Bachmann* and *R. A. Hoffman*, Organic Reactions, 1944, Vol. II, Ch. 6.) Reactions of this type have also been used in quantitative investigations on the influence on the course of such reactions of substituent atoms or groups. The earlier recognition that these influences were different in both kind and magnitude from those encountered in electrophilic substitution processes such as nitration, led to the hypothesis of homolytic substitution in which the course of the reaction is not controlled by electrostatic forces. The vast majority of the substitution reactions discussed below involve replacement of hydrogen by an aryl group, but similar reactions have been reported with hexafluorobenzene involving replacement of a fluorine atom (*P. A. Claret, Williams* and *J. Coulson*, J. chem. Soc. B, 1968, 341; *Williams*, Intrascience Chem. Reports, 1969, **3**, 229). The following reactions give rise to homolytic arylation.

(1) *The Gomberg-Bachmann reaction* utilises the action of alkali metal diazoates in aqueous solution with an aromatic compound PhX (*M. Gomberg* and *W. E. Bachmann*, J. Amer. chem. Soc., 1924, **46**, 2339; *Gomberg* and *J. C. Pernert*, *ibid.*, 1926, **48**, 1372):

$$\mathrm{Ar{\cdot}N{:}N{\cdot}ONa + PhX \rightarrow Ar{\cdot}C_6H_4{\cdot}X + N_2 + NaOH}$$

This reaction, which takes place in a two-phase system, is capable of several modifications (*W. S. M. Grieve* and *Hey*, J. chem. Soc., 1938, 108; *J. Elks, J. W. Haworth* and *Hey*, *ibid.*, 1940, 1284; *M. R. Pettit* and *J. C. Tatlow*, *ibid.*, 1954, 1941; *R. A. Abramovitch* and *J. G. Saha*, Tetrahedron, 1965, **21**, 3297).

(2) *The nitrosoacylarylamine reaction,* first reported by *E. Bamberger* (Ber., 1897, **30**, 366), takes place in homogeneous solution. The reaction involves the preliminary rearrangement of the nitroso-compound into the acyl derivative of the diazoic acid, which then undergoes reaction with the aromatic compound PhX (*R. Huisgen* and *G. Horeld,* Ann., 1948, **562**, 137; *Hey et al.*, J. chem. Soc., 1952, 4657):

$$\underset{\displaystyle\text{NO}}{\text{Ar}\cdot\underset{|}{\text{N}}}\cdot\text{CO}\cdot\text{CH}_3 \xrightarrow{\text{slow}} \text{Ar}\cdot\text{N}:\text{N}\cdot\text{O}\cdot\text{CO}\cdot\text{CH}_3$$

$$\text{Ar}\cdot\text{N}:\text{N}\cdot\text{O}\cdot\text{CO}\cdot\text{CH}_3 + \text{PhX} \rightarrow \text{Ar}\cdot\text{C}_6\text{H}_4\cdot\text{X} + \text{N}_2 + \text{CH}_3\cdot\text{CO}_2\text{H}$$

A closely related reaction involving an *N*-aryl-*N*-nitrosophosphoramidate which rearranges to a diazonium phosphate has been reported (*P. J. Bunyan* and *J. I. G. Cadogan,* J. chem. Soc., 1962, 1304).

(3) *The triazene reaction.* Dry hydrogen chloride is passed into a boiling solution of a 1-aryl-3,3-dimethyltriazene in the aromatic compound PhX (*I. G. Farbenind.*, B.P. 513846/1938; *Elks* and *Hey,* J. chem. Soc., 1943, 441):

$$\text{Ar}\cdot\text{N}:\text{N}\cdot\text{NMe}_2 + 2\,\text{HCl} \rightarrow \text{Ar}\cdot\text{N}:\text{N}\cdot\text{Cl} + \text{Me}_2\text{NH}\cdot\text{HCl}$$

$$\text{Ar}\cdot\text{N}:\text{N}\cdot\text{Cl} + \text{PhX} \rightarrow \text{Ar}\cdot\text{C}_6\text{H}_4\cdot\text{X} + \text{N}_2 + \text{HCl}$$

(4) *The reaction of primary aromatic amines with pentyl nitrite* (*Shu Huang,* Acta Chim. Sinica, 1959, **33**, 171; *J. I. G. Cadogan,* J. chem. Soc., 1962, 4257). This reaction takes place in an excess of benzene at 60°–80°:

$$\text{Ar}\cdot\text{NH}_2 + \text{C}_5\text{H}_{11}\cdot\text{ONO} + \text{PhX} \rightarrow \text{Ar}\cdot\text{C}_6\text{H}_4\cdot\text{X} + \text{N}_2 + \text{C}_5\text{H}_{11}\text{OH} + \text{H}_2\text{O}$$

(5) *The areneazotriphenylmethane reaction* (*Hey, ibid.*, 1934, 1966; *H. Wieland et al.*, Ann., 1934, **514**, 145; *Huisgen* and *R. Grashey, ibid.*, 1957, **607**, 46). This reaction is usually carried out at about 80° in an excess of the aromatic compound PhX:

$$\text{Ar}\cdot\text{N}:\text{N}\cdot\text{CPh}_3 + \text{PhX} \rightarrow \text{Ar}\cdot\text{C}_6\text{H}_4\cdot\text{X} + \text{N}_2 + \text{Ph}_3\text{CH}$$

(6) *The acyl peroxide reaction* (*H. Gelissen* and *P. H. Hermans,* Ber., 1925, **58**, 285, 476; *Hey,* J. chem. Soc., 1934, 1966; *Wieland et al.*, Ann., 1934, **513**, 93). This reaction is usually carried out at about 80°–100° in an excess of the aromatic compound PhX.

$$\text{Ar}\cdot\text{CO}\cdot\text{O}\cdot\text{O}\cdot\text{CO}\cdot\text{Ar} + \text{PhX} \rightarrow \text{Ar}\cdot\text{C}_6\text{H}_4\cdot\text{X} + \text{Ar}\cdot\text{CO}_2\text{H} + \text{CO}_2$$

(7) *Reactions with lead tetrabenzoate and phenyliodosyl benzoate.* These compounds decompose at 125°–130° in aromatic solvents to give phenyl and benzoyloxy-radicals (*Hey, C. J. M. Stirling* and *Williams,* J. chem. Soc., 1954, 2747 and 1956, 1475; *B. M. Lynch* and *K. H. Pausacker,* Australian J. Chem., 1957, **10,** 329):

$$Pb\cdot(O\cdot CO\cdot C_6H_5)_4 \rightarrow Pb(O\cdot CO\cdot C_6H_5)_2 + 2\ C_6H_5\cdot CO\cdot O\cdot$$

$$C_6H_5I(O\cdot CO\cdot C_6H_5)_2 \rightarrow C_6H_5I + 2\ C_6H_5\cdot CO\cdot O\cdot$$

$$C_6H_5\cdot CO\cdot O\cdot \rightarrow C_6H_5\cdot + CO_2$$

The phenyl radicals thus liberated give rise to the phenylation of the aromatic solvent. A closely related reaction which involves the thermal decomposition of silver halide dibenzoates has been reported (*D. Bryce-Smith* and *P. Clarke,* J. chem. Soc., 1956, 2264).

(8) *Photolysis of organo-metallic compounds.* Diphenylmercury, tetraphenyl-lead (*J. M. Blair, Bryce-Smith* and *B. W. Pengilly, ibid.,* 1959, 3174), and triphenylbismuth (*Hey, D. A. Shingleton* and *Williams, ibid.,* 1963, 5612) have been used as sources of phenyl radicals in homolytic substitution reactions.

(9) *Photolysis of aromatic iodo-compounds.* The photolysis of iodobenzene as a source of phenyl radicals has been reported by *Blair* and *Bryce-Smith, ibid.,* 1960, 1778. The use of this reaction for the synthesis of biaryls in good yield has been extensively developed by *N. Kharasch* and his co-workers [*e.g., W. Wolf* and *Kharasch,* J. org. Chem., 1965, **30,** 2493; *R. K. Sharma* and *Kharasch,* Angew. Chem. (internat. Edn.), 1968, **7,** 36].

(10) *Pyrolysis of aromatic sulphonyl halides.* The thermal homolysis of benzenesulphonyl chloride takes place at about 255° and the phenyl radicals thus liberated give rise to phenylation of the substrate (*P. J. Bain et al.,* Proc. chem. Soc., 1962, 186; *G. M. Badger* and *C. P. Whittle,* Australian J. Chem., 1963, **16,** 440). This method is particularly useful with substrates which are solid at the temperatures used in other methods.

(11) *Radiolysis of aromatic halides.* The action of γ-radiation on bromobenzene gives rise to phenyl radicals (*A. F. Everard et al.,* J. chem. Soc., 1962, 905).

(12) *Reaction of diazonium salts in aqueous acetone in presence of cupric chloride* (The Meerwein Reaction). *S. C. Dickerman* and *K. Weiss* (J. org. Chem., 1957, **22,** 1070) have shown that this reaction can be used to effect homolytic arylation in a homogeneous medium.

(13) *Pyrolysis of diazoaminobenzenes.* The decomposition of diazoaminobenzene at 150–160° gives phenyl and anilino-radicals and the former

react with the aromatic solvent (*R. L. Hardie* and *R. H. Thomson*, J. chem. Soc., 1958, 1286).

(14) *Oxidation of phenylhydrazine with silver oxide.* This reaction when carried out in an aromatic solvent gives phenyl radicals which react with the solvent (*idem, ibid.*, 1957, 2512).

(15) *Pyrolysis of nitrobenzene at 600°.* This reaction generates phenyl radicals, which can then be used for the phenylation of aromatic compounds (*E. K. Fields* and *S. Meyerson*, J. Amer. chem. Soc., 1967, **89**, 724, 3224; J. org. Chem., 1967, **32**, 3114; *ibid.*, 1968, **33**, 2315).

(16) *Electrochemical reduction of diazonium tetrafluoroborates in aprotic solvents.* This reaction, which may be represented as

$$ArN^{\oplus}_2 + e \rightarrow ArN_2\cdot \rightarrow Ar\cdot + N_2$$

has been used for the phenylation of benzene, toluene, anisole, benzonitrile, nitrobenzene, bromobenzene and naphthalene (*F. F. Gadallah* and *R. M. Elofson*, J. org. Chem., 1969, **34**, 3335).

In all of the above reactions the orientation of the product $Ar\cdot C_6H_4\cdot X$ does not conform to the accepted pattern for either electrophilic or nucleophilic substitution, and the outstanding feature is the absence of polar effects of the substituent X. In parallel reactions with pyridine in place of PhX the product consists of a mixture of all three isomeric arylpyridines (*e.g. J. W. Haworth, I. M. Heilbron* and *Hey*, J. chem. Soc., 1940, 349; *Hey, Stirling* and *Williams, ibid.*, 1955, 3963; *P. J. Bunyan* and *Cadogan, ibid.*, 1962, 1304).

The application of quantitative methods to these reactions has shown the characteristic features of homolytic substitution in more precise detail. In investigations to determine the proportions of the isomeric substituted biphenyl derivatives formed in these reactions coupled with competitive experiments to evaluate the activating (or deactivating) influence of the substituent atom or group with reference to hydrogen (*i.e.* comparison with benzene), it has been possible to determine the *partial rate factors* for a large number of arylation reactions. These factors give a quantitative expression of the velocity of substitution at each nuclear position relative to that at one carbon atom in benzene. The results obtained in this manner for phenylation with benzoyl peroxide are summarised in Table 2 (see *Hey, S. Orman* and *Williams*, J. chem. Soc., 1961, 565; *R. L. Dannley* and *E. C. Gregg*, J. Amer. chem. Soc., 1954, **76**, 2997; *Williams*, Chem. and Ind., 1961, 1286; *Hey*, "Advances in Free Radical Chemistry", Logos Press-Academic Press, London 1967, **2**, 47).

TABLE 2

PARTIAL RATE FACTORS FOR HOMOLYTIC PHENYLATION WITH BENZOYL PEROXIDE AT 80°

	Rate ratio $C_6H_6 = 1$	Composition %			Partial Rate Factors		
		o	*m*	*p*	F_o	F_m	F_p
$PhNO_2$	2·94	62·5	9·8	27·7	5·5	0·86	4·9
PhF	1·03	54·1	30·7	15·2	1·7	0·95	0·86
PhCl	1·06	50·1	31·6	18·3	1·6	1·0	1·2
PhBr	1·29	49·3	33·3	17·4	1·9	1·3	1·3
PhI	1·32	51·7	31·6	16·7	2·0	1·3	1·3
PhMe	1·23	66·5	19·3	14·2	2.5	0·71	1·0
PhEt	0·90	53	28	19	1·4	0·76	1·0
$PhPr^i$	0·64	31	42	27	0·60	0·81	1·0
$PhBu^t$	0·64	24	49	27	0·46	0·94	1·0
PhCN	3·7	60	10	30	6·5	1·1	6·5
$PhCO_2Me$	1·77	57·0	17·5	25·5	3·0	0·93	2·7
Ph_2	2·94	48·5	23·0	28·5	2·1	1·0	2·5

For purposes of comparison some partial rate factors for nitration are given below:

Partial Rate Factors for Nitration of C_6H_5X ($C_6H_6 = 1$)

X	F_o	F_m	F_p
Cl	0·029	0·0009	0·137
Br	0·033	0·0011	0·112
NO_2	$\sim 0{\cdot}24 \times 10^{-6}$	$\sim 2{\cdot}75 \times 10^{-6}$	$\sim 0{\cdot}03 \times 10^{-6}$

Relative rate and partial rate factors for the arylation of a series of substituted benzenes by the nitrosoacetanilide reaction at 20° have been investigated by *O. Simamura* and his co-workers (see "Advances in Free Radical Chemistry", 1967, **2**, 76).

In the phenylation reaction all of the substituted benzene derivatives so far studied, with the exception of the alkylbenzenes referred to below, are more reactive than benzene itself, and pyridine and benzene react at approximately the same rates. The influence of the substituent atom or group is, however, much smaller in magnitude than that encountered in electrophilic substitution and the orientation shows no obvious relationship to the polar character of the substituent group or atom. Most of the phenylation reactions are characterised by an unusually high percentage of substitution at the *ortho* position. Further, these reactions, while insensitive to the polar character of nuclear substituents in the aromatic compound undergoing substitution, are nevertheless sensitive, as in electrophilic substitution, to steric influences. *tert*-Butylbenzene and isopropylbenzene provide examples of overall deactivation, which is attributed to a marked steric effect on the *ortho* positions. On the other hand, the presence of a substituent atom or group in the phenyl radical can give rise to a polarised radical, which shows some measure of electrophilic or nucleophilic character (*R. L. Dannley* and *M. Sternfield,* J. Amer. chem. Soc., 1954, **76**, 4543; *Cadogan, Hey* and *Williams,* J. chem. Soc., 1955, 1425). The results of a large number of reactions carried out with phenyl radicals having a substituent atom or group in the *ortho, meta,* or *para* position give support to the existence of polarisation effects in the radical (see for example, *Chang Shih, Hey* and *Williams, ibid.*, 1958, 1885 and 4403; *Hey, H. N. Moulden* and *Williams, ibid.*, 1960, 3769; *J. K. Hambling, Hey* and *Williams, ibid.*, 1960, 3782 and 1962, 486; *Hey, Orman* and *Williams, ibid.*, 1965, 101). Quantitative reactions carried out with the *p*-chlorophenyl and *p*-nitrophenyl radicals derived from the appropriate *p*-substituted nitrosoacetanilide have also confirmed the existence of polarisation effects in these radicals (*R. Ito et al.*, Tetrahedron, 1965, **21**, 955).

The mechanism of the homolytic arylation of aromatic compounds has been studied extensively. The conclusion that aryl radicals (R·) react with aromatic substrates (ArH) by an *addition* rather than an *abstraction* mechanism is evidenced by repeated observations that the main binuclear products are of the type RAr, compounds RR being formed only in low yield and compounds ArAr not at all. The absence of a kinetic isotope effect in the unreacted solvent in reactions of benzoyl peroxide with deuterated or tritiated aromatic substrates (*R. J. Convery* and *C. C. Price,* J. Amer. chem. Soc., 1958, **80**, 4101; *Chang Shih, Hey* and *Williams,* J. chem Soc., 1959, 1871) has established that the removal of the hydrogen atom is not kinetically significant in this process, which must therefore proceed through the formation of a σ-complex (a substituted arylcyclohexadienyl radical), which on oxidation gives the biaryl, as was first postulated by *C. Walling* ("Free Radicals in Solution", Wiley, New York 1957, p. 483) and by *Lynch* and *K. H. Pausacker* (Austral. J. Chem., 1957, **10**, 40), thus:

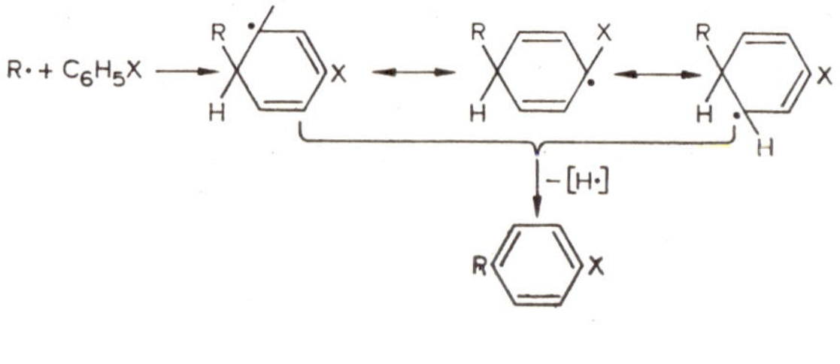

$R\cdot + C_6H_5X \longrightarrow$ [σ-complex resonance structures] $\xrightarrow{-[H\cdot]}$ RC_6H_4X

This σ-complex, being a resonance-stabilised radical, is particularly prone to further reaction by dimerisation and disproportionation, and products of these reactions have been observed. Thus Pausacker (*ibid.*, 1957, **10**, 49) obtained 4,4‴-dichloro-*p*-quaterphenyl which could not have been formed by successive arylation, from the reaction of *p*-chlorobenzoyl peroxide with benzene. Similarly, *G. A. Razuvaev, B. G. Zateev* and *G. G. Petukhov* (Proc. Acad. Sci. U.S.S.R., 1960, **130**, 336) have shown that the reaction of benzoyl peroxide with [1-^{14}C]benzene gives a little *p*-terphenyl labelled in only one nucleus, but a much higher yield of *pp'*-quaterphenyl labelled in two nuclei. *D. F. DeTar* and *R. J. Long* (J. Amer. chem. Soc., 1958, **80**, 4742) have isolated an isomer of tetrahydro-*p*-quaterphenyl (II), as well as 1,4-dihydrobiphenyl (III), which is formed together with biphenyl by disproportionation of the σ-complex (I), from the reaction of benzoyl peroxide with benzene:

All three isomers (2,2'-, 2,4'-, and 4,4'-) of quaterphenyl were obtained from this reaction by *Hey, M. J. Perkins* and *Williams*, (J. chem. Soc., 1963, 5604) by oxidation of the hydroaromatic products with *o*-chloranil. Thus at least some of the biaryl must arise directly, or by subsequent atmospheric oxidation of dihydrobiaryls, from the disproportionation reaction, while dimerisation of σ-complexes is the main source of the high-boiling residue which is always formed in this reaction.

These processes do not, however, give a complete mechanistic picture of the benzoyl peroxide reaction, since the early kinetic studies of *K. Nozaki* and *P. D. Bartlett* (J. Amer. chem. Soc., 1946, **68**, 1686; 1947, **69**, 2299) revealed the existence of an induced process leading to a reaction of order 1·5 in the peroxide, which accompanies the first-order primary homolysis into benzoyloxy-radicals. More recent kinetic studies of the reaction of benzoyl peroxide with alkylbenzenes by *W. R. Foster* and *Williams* (J. chem. Soc., 1963, 2862) and of the corresponding reaction with benzene and halogenobenzenes by *G. B. Gill* and *Williams* (*ibid.*, 1965, 995, 7127); *P. Lewis* and *Williams* (*ibid.*, [B] 1969, 120) have revealed that the radical which brings about the main part of the induced decomposition of the peroxide is the σ-complex:

$$[Ph \cdot C_6H_6]\cdot + Ph \cdot CO \cdot O \cdot O \cdot CO \cdot Ph \rightarrow Ph_2 + PhCO_2H + PhCO \cdot O\cdot$$

Thus some σ-complexes are oxidised to biaryls by molecular benzoyl peroxide. This reaction can give rise to 1·5-order kinetics provided chains are terminated by reactions between like radicals, such as dimerisation or disproportionation of σ-complexes, but not by reaction between unlike radicals such as a σ-complex and a benzoyloxy-radical. In a study of the way in which the yields of the reaction products with benzene vary over a wide range of initial peroxide concentration, it has been demonstrated that at infinite dilution the yield of benzoic acid would be virtually zero and the yields of biphenyl and dihydrobiphenyl would be equal *i.e.*, exclusive formation by the disproportionation reaction (*Hey, Perkins* and *Williams, ibid.*, 1964, 3412). Reactions with bromobenzene, however, show a different kinetic picture in which the reaction between unlike radicals (σ-complex and the benzoyloxy-radical) constitutes the main termination reaction, which results in very high yields of biaryls and benzoic acid.

The efficiency of the conversion of σ-complexes to biaryls in the aroyl peroxide reactions can be improved by the presence of certain additives, which will oxidise the σ-complexes before they can dimerise or participate in any other reactions. Catalytic quantities of nitrobenzene (*idem*, Chem. and Ind., 1963, 83), cupric salts (*Hey, K. S. Y. Liang* and *Perkins*, Tetrahedron Letters, 1967, 1477), or ferric benzoate (*B. N. Dailly* and *Williams*, unpublished observations) as well as molecular oxygen (*M. Eberhardt* and *E. L. Eliel*, J. org. Chem., 1962, **27**, 2289; *R. T. Morrison et al.*, J. Amer. chem. Soc., 1962, **84**, 4152), are capable of modifying the reaction in this way. In spite of the marked increase in yield, however, the ratios of the isomeric biphenyls formed in these reactions remained unaltered. These observations establish the conclusion that the σ-complexes formed in arylation reactions are not selectively removed by dimerisation, and that the partial rate factors summarised earlier may be taken to be valid measures of the relative ease of addition of aryl radicals to the various sites in aromatic nuclei.

The increase in the yield of biaryls which results from the addition of a nitro compound or oxygen to the reaction with peroxides increases significantly the preparative value of these processes. By means of a kinetic and product study of the reactions of benzoyl peroxide with benzene containing 1 per cent of nitrobenzene it has been shown that there is almost complete suppression of the formation of high boiling products and of dihydrobiphenyls and that the nitrobenzene is not consumed (*G. B. Gill* and *Williams*, J. chem. Soc., 1966, 880). Some of the reduction products of nitrobenzene (*e.g.*, nitrosobenzene, phenylhydroxylamine) are found to be more effective than nitrobenzene itself (*G. R. Chalfont et al.*, Chem. Comm., 1967, 367). An explanation of these results is suggested by the detection of a stable diaryl nitroxide radical formed from the various nitrogenous additives (*e.g.* Ar· + ArN:O $\longrightarrow$ $Ar_2NO\cdot$). The nitroxide, present in relatively high stationary-state concentration, intercepts the σ-complex to give biaryl and a hydroxylamine. The hydroxylamine is reoxidised to nitroxide by molecular aroyl peroxide:

$$Ar_2NO\cdot + (I) \rightarrow Ph_2 + Ar_2N\cdot OH$$

$$Ar_2N\cdot OH + PhCO\cdot O\cdot O\cdot COPh \rightarrow Ar_2NO\cdot + PhCO_2H + PhCO_2\cdot$$

$$PhCO_2\cdot \rightarrow Ph\cdot + CO_2$$

The mechanisms of arylation reactions with radical sources other than the aroyl peroxides have been less extensively investigated, although it is generally agreed that aryl radicals are responsible for the initial attack upon the aromatic nucleus in all of them. This follows in the main from common patterns of isomer-distribution. A suggestion by *Eliel et al.*, (Tetrahedron Letters, 1962, 749) that in the areneazotriarylmethane reaction, the σ-complexes were not free, but rather were formed and dehydrogenated within a solvent cage, has been shown to be unfounded (*Hey, Perkins* and *Williams, ibid.*, 1963, 445; J. chem. Soc., 1965, 110; *J. F. Garst* and *R. S. Cole*, Tetrahedron Letters, 1963, 679; and *G. A. Russell* and *R. F. Bridger, ibid.*, p. 737). It therefore seems likely that, in this reaction, σ-complexes are formed by addition of aryl radicals to the aromatic nucleus, and subsequently dehydrogenated by triarylmethyl radicals, *e.g.*:

$$PhArH\cdot + \cdot CPh_3 \rightarrow PhAr + CHPh_3$$

The kinetics of the decomposition of areneazotriarylmethanes in aromatic solvents has received considerable attention (*e.g.*, *S. G. Cohen* and *C. H. Wang*, *J.* Amer. chem. Soc., 1953, **75**, 5504; *Huisgen* and *H. Nakaten*, Ann., 1954, **586**, 70; *G. L. Davies, Hey* and *Williams*, J. chem. Soc., 1956, 4397).

The nitrosoacylarylamine reaction has long been recognised to be anomalous in respect of the very small quantities of carbon dioxide which are formed, since appreciable yields of this product would be expected from the known instability of the acetyloxy-radical (*Huisgen* and *G. Horeld*, Ann., 1949, **562**, 137). *C. Rüchardt* and *B. Freudenberg* (Tetrahedron Letters,

1964, 3623) have shown that free acyloxy-radicals are not formed during these reactions and have suggested the formation of a diazo-anhydride as the intermediate, which undergoes homolysis to give aryl and diazoate radicals, and nitrogen. This sequence of reactions is represented as follows:

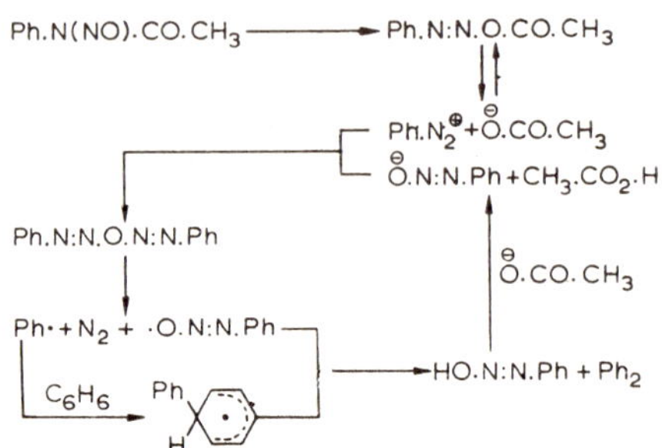

A similar diazo-anhydride intermediate also has been shown by *Rüchardt* and *E. Merz* (Tetrahedron Letters, 1964, 2431) to be involved in the decomposition of diazoic acids (the Gomberg-Bachmann reaction), which does not, as previously believed, give rise to the formation of hydroxyl radicals. The reaction was shown to be of the second order with respect to the diazonium salt, and this and other observations were rationalised in the following scheme:

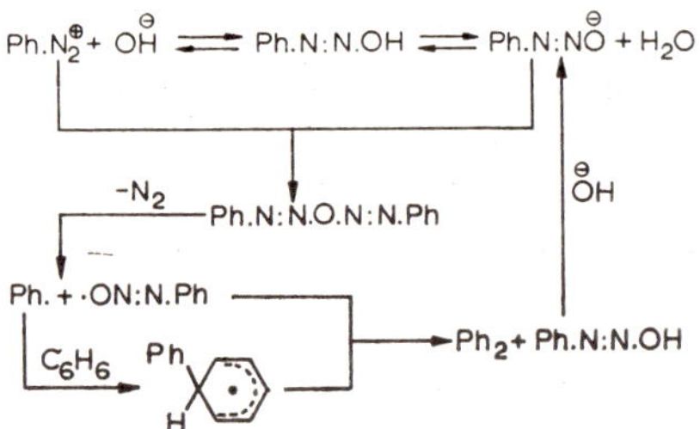

This mechanism for the Gomberg reaction is supported by the work of *Eliel, J. G. Saha* and *S. Meyerson* (J. org. Chem., 1965, 30, 2451) on reactions with benzene-*d*, the results of which are consistent with the intermediate formation of a relatively stable radical, which is capable of scavenging phenylcyclohexadienyl radicals before they can reach a sufficient concentration for dimerisation or disproportionation to become important.

The detection by electron spin resonance spectroscopy of a relatively stable radical considered to be Ph·N:N·O· from the nitrosoacetanilide reaction was held to support these mechanisms (*G. Binsch* and *Rüchardt*, J. Amer. chem. Soc., 1966, 88, 173 and *Binsch, Merz* and *Rüchardt*, Ber., 1967, 100, 247). However, more recent evidence has been presented (*Chalfont* and *Perkins*, J. Amer. chem. Soc., 1967, 89, 3054), which favours an alternative structure, IV, for this radical, which could arise as shown,

```
Ph· +         N=O            Ph-N-O·
              |       --->     |
         Ph-N-COCH3          Ph-N-COCH3
                               (IV)
```

and a revised mechanism incorporating IV closely parallels that indicated above for the modified peroxide reactions in the presence of nitrosobenzene:

```
(IV) + (I) -> Ph2 + Ph-N-OH
                      |
                    Ph-N-COCH3

Ph-N-OH + PhN:N·OCOCH3 -> Ph-N-O· + CH3·CO2H + Ph· + N2
  |                          |
Ph-N-COCH3                 Ph-N-COCH3
```

5. Formation and fission of the benzene nucleus

NEIL CAMPBELL

(a) Formation of the benzene nucleus

"The benzene nucleus has been built up from aliphatic compounds in a great variety of ways, very few of which, however, can be termed rational syntheses, if by that is meant that some definite structure is indicated" (*C. K. Ingold*, J. chem. Soc., 1922, **121**, 1143). The truth of this statement still obtains, but although few of these methods are of value in preparative or structural work many provide an interesting connecting link between the aromatic series on one hand and the aliphatic, alicyclic, and heterocyclic series on the other.

The reactions which lead to the formation of the benzene nucleus may conveniently be classified as follows:

(i) The pyrolysis, polymerisation and condensation of aliphatic compounds.

(ii) The dehydration, dehydrohalogenation, and dehydrogenation of fully or partially reduced six-membered ring compounds.

(iii) The isomerisation of alicyclic compounds.

(iv) The conversion of 3-, 5-, 7-, and 8-membered ring compounds and of heterocyclic compounds into aromatic compounds.

(i) Pyrolysis, condensation and polymerisation reactions

The pyrolysis of aliphatic compounds to yield benzene derivatives has been extensively studied in the hundred years which have elapsed since the pioneer investigations of Berthelot. Methane when passed through a porcelain tube heated to dull redness yields traces of benzene and naphthalene. The lower paraffins with a chromium catalyst at 450–550° yield aromatics and with a Cr_2O_3–Al_2O_3 catalyst *n*-heptane gives a 75 % yield of toluene (*R. C. Archibald* and *B. S. Greensfelder*, Ind. Eng. Chem., 1945, **37**, 356). Unsaturated hydrocarbons are probably intermediates in such reactions and the conversion of propane into benzene may be formulated as follows:

$$C_3H_8 = CH_2{:}CH_2 + CH_4$$

$$2\,CH_2{:}CH_2 = CH_2{:}CH{\cdot}CH{:}CH_2 + H_2$$

$$CH_2{=}CH{-}CH{=}CH_2 + CH_2{=}CH_2 = C_6H_6 \text{ (benzene)} + 2H_2$$

The final stage has been investigated at 550° (*J. Shabtai* and *E. Gil-Av*, Tetrahedron, 1962, **18**, 87). Unsaturated intermediates are also obtained when paraffins capable of forming six-membered rings are aromatised (*H. Hoog, J. Verheus* and *F. J. Zuiderweg,* Trans. Farad. Soc., 1939, **35**, 993, 1007):

$$CH_3CH_2CH_2CH_2CH_2CH_3 \xrightarrow{-H_2} CH_3CH_2CH_2CH_2CH{=}CH_2 \xrightarrow{-3H_2}$$

The *dehydrocyclisation* of alkanes over chromia catalysts has long been known. Moderately high temperatures (300° and above) are required as well as catalysts such as chromium oxide, which activate the C–H bonds, but not the C–C linkage (*H. Pines* and *C. T. Goetschel,* J. org. Chem., 1965, **30**, 3530). The mechanism is complex.

Of interest is the conversion of carbon monoxide and heated potassium into the potassium salt of hexahydroxybenzene (*R. Nietzki* and *Th. Benckiser,* Ber., 1885, **18**, 1833), an example of single carbon atoms uniting to form the stable six-membered benzene nucleus.

Acetylene at 500° affords benzene (Berthelot), but more effective syntheses can be carried out at much lower temperatures when suitable catalysts are used. An extreme example is the conversion of acetylene to benzene in avocado pears (*E. F. Jansen* and *J. M. Wallace,* J. biol. Chem., 1965, **240**, 1042). The polymerisation of acetylene derivatives (but not acetylene itself) is catalysed by palladium, dimethyl acetylenedicarboxylate, for instance, in boiling benzene with palladium – charcoal giving a 93 % yield of hexamethyl mellitate (*D. Bryce-Smith,* Proc. chem. Soc., 1964, 239). Catalysts in the form of complex compounds bring about polymerisation under very mild conditions. Thus while the pyrolysis of acetylene without a catalyst in an unpacked silica tube at 700° yields benzene (28 %), toluene (3 %), etc. (*G. M. Badger et al.,* J. chem. Soc., 1960, 2825), acetylene at 60–70°/15 at. in the presence of $Ni(CO)_2[Ph_3P]_2$ gives an 88 % yield of benzene and 12 % of styrene (*W. Reppe* and *W. J. Schweckendiek,* Ann., 1948, **560**, 104), while tolane, diphenylacetylene, yields hexaphenylbenzene with $(PhCN)_2 \cdot PdCl_2$ as catalyst (*A. T. Blomquist* and *P. M. Maitlis,* J. Amer. chem. Soc., 1962, **84**, 2329) or a nickel catalyst (*G. N. Schrauzer,* Ber., 1961, **94**, 1403; *H. H. Zeiss* and *M. Tsutsui,* J. Amer. chem. Soc., 1960, **82**, 6255). 2-Butyne at 20° for a short period with triphenylchromium tri-tetrahydrofuranate gives hexamethylbenzene and some 1,2,3,4-tetramethylnaphthalene (*Zeiss* and *W. Herwig, ibid.,* 1958, **80**, 2913). The products obtained sometimes depend on conditions and catalyst, phenylacetylene, for example, with a nickel catalyst yielding at 70°, 1,2,4-triphenylbenzene (*J. D. Rose* and *F. S. Statham,* J. chem. Soc., 1950, 69), while with methylamine at 250°

the product is 1,3,5-triphenylbenzene (*K. Krasouskii* and *A. Kipryanov*, J. Russ. phys. Chem. Soc., 1925, **56**, 1. C.A., 1925, **19**, 2817).

Di- and tri-enes (and -ynes) frequently yield aromatic products. Diacetylenes of the formula CH⫶C·$(CH_2)_n$·C⫶CH, where n may be 4, 5, 6, 10 and 17, give *o*-dialkylbenzenes (*R. A. Raphael et al.*, J. chem. Soc., 1964, 2597). Thus hepta-1,6-diyne, CH⫶C·$(CH_2)_3$·C⫶CH, with a basic catalyst yields toluene. Other instances include the photolysis of *cis,trans*-1,3,5-hexatriene to give *inter alia* benzene and hydrogen (*R. Srinivasan*, J. Amer. chem. Soc., 1961, **83**, 2806) and the conversion of 2,5-dienoic acids into phenols by means of basic catalysts such as sodium acetate in acetic anhydride and acetic acid. Thus hepta-2,5-dienoic acid gives a 65 % yield of *o*-cresol (*G. P. Chiusoli* and *G. Agnès*, Proc. chem. Soc., 1963, 310):

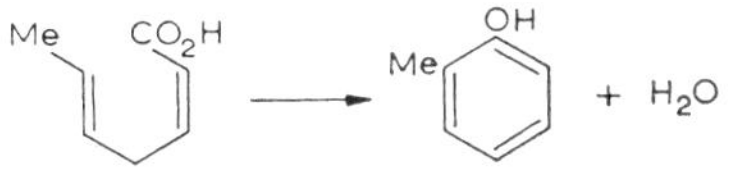

The polymerisation (*bicyclotrimerisation*) of *tert*-butylfluoracetylene has led to spectacular results for it yields not only tri-*tert*-butyltrifluorobenzene (I, R = *tert*-butyl, but also the valence-bond isomers, II, III, and IV, all stable compounds (*H. G. Viehe et al.*, Angew. Chem., internat. Edn., 1964, **3**, 755; 1965, **4**, 746). The prismane formula II is reminiscent of Ladenburg's prism formulae for benzene (1869) and formula III for the so-called Dewar

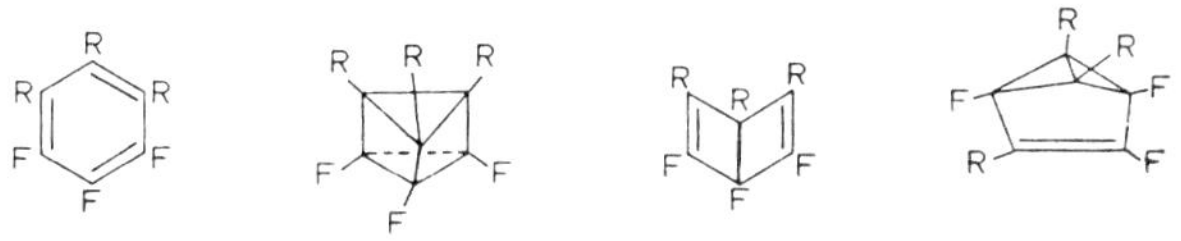

(I) (II) (III) (IV)

structure (see *Wilson Baker*, Chemistry in Britain, 1965, 191), while the benzvalene structure IV is quite remarkable. The unsubstituted Dewar form of benzene has also been obtained (*E. E. van Tamelen* and *S. P. Pappas*, J. Amer. chem. Soc., 1963, **85**, 3297; Angew. Chem., internat. Edn., 1965, **4**, 738) and its structure established by its quantitative isomerisation to benzene and hydrogenation to bicyclo[2.2.0]hexane (see also *van Tamelen* and *D. Carty*, J. Amer. chem. Soc., 1967, **89**, 3922).

Valence-bond isomers of other benzene derivatives have also been prepared (see, *e.g.*, *R. Criegee* and *F. Zanker*, Angew. Chem., internat. Edn., 1964, **3**, 695; *E. M. Arnett* and *J. M. Bollinger*, Tetrahedron Letters, 1964, 3803; *K. F. Wilzbach* and *L. Kaplan*, J. Amer. chem. Soc., 1965, **87**, 4004; *W. Schäfer* and *H. Hellmann*, Angew. Chem. internat. Edn., 1967, **6**, 518). Kilogram quantities of the Dewar form of hexamethylbenzene are obtained by treatment of but-2-yne with aluminium trichloride in benzene at 20–30° (*Schäfer et al., ibid.*, 1966, **5**, 669).

Unsaturated aliphatic compounds are converted into aromatic products by means of the Diels-Alder reaction, which is discussed elsewhere (see p. 169). The initial products are generally alicyclic compounds, aromatisation of which is accomplished by the loss of small fragments such as ethylene, carbon monoxide or by dehydrogenation. Dehydrogenation is particularly easily effected from maleic anhydride products, the attached carboxyl groups apparently facilitating dehydrogenation. Such adducts are dehydrogenated by the mild oxidising agent, alkaline potassium ferrocyanide, dehydrogenation being accompanied by decarboxylation (*N. Campbell* and *R. S. Gow*, J. chem. Soc., 1949, 1555; *L. F. Fieser* and *M. J. Haddadin*, J. Amer. chem. Soc., 1964, **86**, 2392).

Aromatic compounds are frequently obtained by the cyclic condensations of aldehydes and ketones, one of the earliest examples of which is the self-condensation of acetone to give mesitylene (*R. Kane*, J. pr. Chem., 1838, **15**, 131; *R. Adams* and *R. W. Hufferd*, Org. Synth., 1922, **2**, 41):

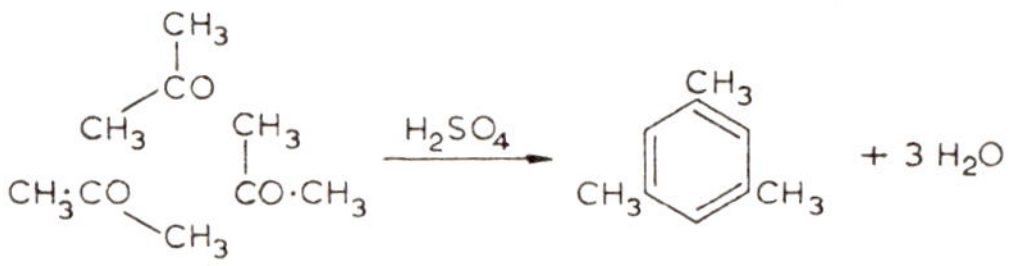

CH_3COCH_3, CH_3COCH_3, CH_3COCH_3 $\xrightarrow{H_2SO_4}$ mesitylene (1,3,5-trimethylbenzene) $+ 3\,H_2O$

β-Polyketones undergo self-condensation in the presence of alkali to give aromatic compounds, heptane-2,4,6-trione, for instance, yielding orcinol (*J. N. Collie*, J. chem. Soc., 1893, **63**, 122, 329) or the benzene derivative 2,6-diacetyl-3-acetonyl-5-methylphenol (*A. J. Birch, D. W. Cameron* and *R. W. Rickards, ibid.*, 1960, 4395):

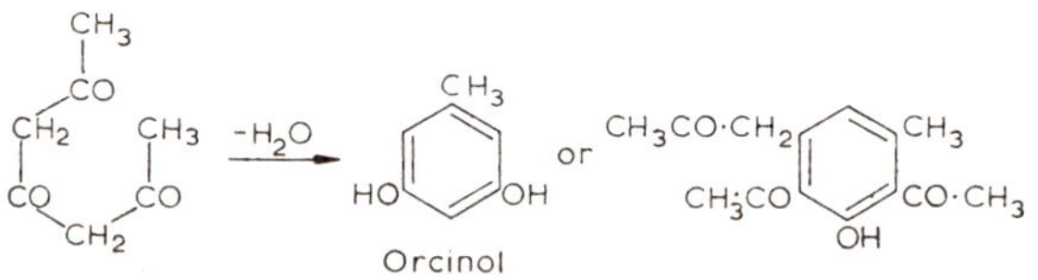

$CH_3CO{\cdot}CH_2{\cdot}CO{\cdot}CH_2{\cdot}CO{\cdot}CH_3$ $\xrightarrow{-H_2O}$ Orcinol (5-methylbenzene-1,3-diol) or 2,6-diacetyl-3-acetonyl-5-methylphenol ($CH_3CO{\cdot}CH_2$, CH_3, $CO{\cdot}CH_3$, OH, CH_3CO)

Orcinol

Such condensations frequently play a significant role in the biosynthesis of aromatic compounds (*R. Robinson*, "Structural Relationships of Natural Products", Oxford, University Press, 1955). *J. N. Collie* many years ago noted "how easily the acetyl group condenses with itself" (J. chem. Soc., 1907, **91**, 1806) – a fact substantiated by the work of Rittenberg in his studies on the biosynthesis of fatty acids – and suggested that in the plant world phenolic substances originate from acetate units *via* polyacetyl intermediates by polymerisation or condensation. This hypothesis was revived by *A. J. Birch* who verified the correctness of the theory by means of labelled acetate (*Birch* and *F. W. Donovan*, Austral. J. Chem., 1953, **6**, 361; *Birch* Proc. chem. Soc., 1962, 3). For example, four acetate units may be pictured as combining to form the triketonic acid, 3,5,7-trioxo-octanoic acid, which after reduction to 5-hydroxy-3,7-dioxo-octanoic acid, undergoes ring-closure to give 2-hydroxy-6-methylbenzoic acid (6-methylsalicylic acid) (V). By using acetate containing the carboxyl group labelled with ^{14}C in the presence of the mould metabolite *Penicillium griseofulvum* Birch synthesised 6-methyl-[2,4,6-*ar-carboxy*-^{14}C]salicylic acid (Va); he then degraded the acid and by measuring the radioactivity of the fragments proved that the

$$4\ CH_3{\cdot}CO_2H \rightarrow CH_3{\cdot}CO{\cdot}CH_2{\cdot}CO{\cdot}CH_2{\cdot}CO{\cdot}CH_2{\cdot}CO_2H$$

$$\downarrow$$

$$CH_3{\cdot}CO{\cdot}CH_2{\cdot}CHOH{\cdot}CH_2{\cdot}CO{\cdot}CH_2{\cdot}CO_2H$$

$\longrightarrow$

6-Methylsalicylic acid
(V)

(Va)

synthesis had been effected by the head-to-tail combination of acetate units (*Birch, R. A. Massy-Westropp* and *C. J. Moye*, Austral. J. Chem., 1955, **8**, 539; *S. W. Tanenbaum* and *E. W. Bassett*, J. Biochem., 1959, **234**, 1860). The cyclisation of 7-aryl-3,5,7-trioxoheptanoic acids to phenolic acids has been accomplished in the laboratory (*T. M. Harris* and *R. L. Carney*, J. Amer. chem. Soc., 1967, **89**, 6734; Chem. Comm., 1968, 1254).

It has been shown conclusively by *Bassett* and *Tanenbaum* (Biochim. Biophys. Acta, 1960, **40**, 535) that the acetate participates in biosynthesis in the form of acetyl-coenzyme A, $CH_3{\cdot}COSCoA$ (HSCoA = coenzyme A).

Further confirmation of the correctness of the head-to-tail condensation mechanism has been provided by *B. D. Davies* using the mutants of micro-organisms (for a review, see *R. Bentley*, Ann. Review Biochemistry, 1962, **31**, 589). It transpires that in some instances as suggested by *F. Lynen* (J. Cellular Comp. Physiol., 1959, **54**, suppl. 1,53) both acetate and malonate (derived from acetate) units are required. Orsellinic acid, for example, is formed by the condensation of one molecule of acetyl-coenzyme A and three of malonyl-coenzyme A to form an intermediate polyketone which on decarboxylation forms orsellinic acid (*K. Mosbach*, Acta Chem. Scand., 1960, **14**, 457; *R. Bentley* and *J. G. Keil*, Proc. chem. Soc., 1961, 111):

1 Acetyl-SCoA + 3 Malonyl-SCoA → $CO_2H \cdot CH(CO \cdot Me) \cdot CO \cdot CH(CO_2H) \cdot CO \cdot CH(COSCoA) \cdot CO_2H$ → Orsellinic acid

For other examples of malonate participation in biosynthesis see *J. D. Bu'-Lock* and *H. M. Smalley, ibid.*, 1961, 209; *Birch, A. Cassera* and *Rickards*, Chem. and Ind., 1961, 792.

The acetate hypothesis not only supplements the well known isoprene rules, but is also very useful in predicting the structure of naturally-occurring compounds. There is, however, another route by which aromatic compounds can be synthesised. In this pathway, discovered by *Davis* (1955), carbohydrate material is converted into aromatic products through the intermediates, 5-dihydroshikimic acid and shikimic acid whose significance in plant synthesis had earlier been mooted by *H. O. L. Fischer* and *G. Dangschat*, (Helv., 1935, **18**, 1206). Here again the route has been investigated by the use of micro-organism mutants. The conversion of carbohydrate material into aromatic products is exemplified by the interaction of enolic pyruvate phosphate and D-erythrose 4-phosphate to give 3-deoxy-D-arabinoheptulosonic acid (VI). This acid undergoes aldol cyclisation to yield 5-dehydroquinic acid and hence shikimic acid (*Srinivasan, M. Katagiri* and *D. B. Sprinson*, J. biol. Chem., 1959, **234**, 713):

$CO_2H \cdot CO \cdot CH_2 \cdot (CHOH)_3 \cdot CH_2OH$ (VI) → Dehydroquinic acid → Shikimic acid → Chorismic acid → Prehenic acid

Shikimic acid 5-phosphate yields chorismic acid, 3-enolpyruvylshikimic acid (*M. I. Gibson* and *F. Gibson*, Biochem. J., 1964, **90**, 248), which can be converted into prehenic acid. These three acids are important in the biosynthesis of aromatic compounds. Chorismic acid, for example, is converted enzymatically into anthranilic acid, prehenic acid and phenylpyruvic acid (*F. Gibson*, Biochem. J., 1964, **90**, 256); prehenic acid gives phenylalanine, tyrosine, and phenylpyruvic acid (*H. Plieninger*, Angew. Chem., internat. Edn., 1962, **1**, 367; *M. Sprecher et al.*, Biochemistry, 1965, **4**, 2855); while shikimic acid gives *o*- and *p*-aminobenzoic acids and is clearly related to gallic acid.

With acid and alkali prehenic acid gives phenylpyruvic acid and *p*-hydroxyphenylpyruvic acid, respectively.

Useful reviews of the shikimic acid pathway are to be found [see *B. A. Bohm*, Chem. Reviews, 1965, **65**, 435; *J. B. Hendrickson*, "The Molecules of Nature" (W. A. Benjamin, Inc., New York, 1965); *Bu'Lock*, "The Biosynthesis of Natural Products" (McGraw-Hill Publishing Co., Ltd., New York, 1965].

(ii) Aromatisation of alicyclic compounds by dehydrogenation

Alicyclic compounds can be converted into aromatic compounds by a variety of methods including dehydrogenation, dehydrohalogenation and dehydration.

Dehydrogenation, especially of polycyclic compounds, has been extensively used in the preparation and structural determination of aromatic compounds (*E. Clar*, "Polycyclic Hydrocarbons", Vol. 1, p. 169, Academic Press, 1964). The simplest example is the conversion of cyclohexane into benzene and this and similar dehydrogenations have been effected by a variety of reagents.

Sulphur. The use of sulphur for dehydrogenation is associated with the name of *A. Vesterberg* (Ber., 1903, **36**, 4200) although it had been used by previous workers for the same purpose. The yields are frequently good, but the products may be difficult to purify. The use of a solvent such as quinoline is often advantageous (*see, e. g., C. L. Hewett*, J. chem. Soc., 1938, 1286).

Selenium. Dehydrogenation with selenium instead of sulphur proceeds more smoothly and gives better yields than sulphur (*O. Diels, W. Gadke* and *P. Körding*, Ann., 1927, **459**, 1). The temperature used is 320–340°. At higher temperatures dehydrogenation may be accompanied by rearrangement accompanied by removal of small molecules such as methane. Thus 2-*n*-butyl-1,1,3-trimethylcyclohexane with selenium at 390° gives a mixture of *m*-xylene and 2-*n*-butyl-*m*-xylene (*L. Ruzicka* and *C. F. Seidel*, Helv., 1936, **19**, 424):

Me Me $(CH_2)_3Me$ Me ⟶ Me Me + Me $(CH_2)_3Me$ Me

Catalytic metals. The hydrogenation catalysts platinum, palladium, and nickel are also efficient dehydrogenating catalysts (*N. D. Zelinski*). With platinum or palladium at temperatures between 170° and 300° cyclohexane yields benzene and with platinised charcoal at 240–300°, pinane yields *o*-

and *p*-cymene, 1,2,4- and 1,2,3-trimethylbenzene (*H. Pines, R. C. Olberg* and *V. N. Ipatieff*, J. Amer. chem. Soc., 1948, 70, 533).

Since hydrogenating and dehydrogenating catalysts have the power to transfer hydrogen from one organic compound to another disproportionation is frequently encountered when platinum or palladium is used for dehydrogenation. Thus cyclohexa-1,3-diene when heated with platinum gives a mixture of benzene and cyclohexane (*Zelinski* and *G. Pawlow*, Ber., 1924, 57, 1066; 1933, 66, 1420). Aromatisation of the terpenes is often accompanied by disproportionation (*R. P. Linstead et al.*, J. chem. Soc., 1940, 1139). Limonene (3 moles) with platinised charcoal at 140° gives *p*-cymene (2 moles) and *p*-menthane (1 mole).

$$3\,C_{10}H_{16} = 2\,C_{10}H_{14} + C_{10}H_{20}$$

Limonene *p*-Cymene *p*-Menthane

At higher temperatures hydrogen is evolved and high yields of aromatic products are obtained, *e.g.* limonene at 305° giving *p*-cymene in 80 % yield.

Bromine. Bromination followed by dehydrobromination is a useful method for aromatising alicyclic compounds and can often be effected under mild conditions thereby minimising the chance of decomposition and rearrangement. Bromosuccinimide is very effective (*R. A. Barnes*, J. Amer. chem. Soc., 1948, 70, 145; *R. Filler*, Chem. Reviews, 1963, 63, 39), the alicyclic compounds being first brominated by this reagent in carbon tetrachloride solution and the products then dehydrobrominated by treatment with potassium acetate. Tetralin thus affords naphthalene (74 %) and cyclohexene gives a 58 % yield of a mixture of *m*- and *p*-dibromobenzenes. Dehydrohalogenation may sometimes be effected by basic ion-exchangers, α- and β-benzene hexachloride, for example, when boiled with Amberlite IRA-400 gives an almost quantitative yield of 1,2,4-trichlorobenzene (*A. Galat*, J. Amer. chem. Soc., 1952, 74, 3890):

Bromine or chlorine can be used to convert cyclic ketones into phenols. Cyclohexanone with chlorine gives a tetrachloro compound, which when heated at 260° gives 2,6-dichlorophenol (*R. Riemschneider*, Monatsh., 1954, 85, 417); this same compound is obtained likewise by heating 2,3,5,6-tetrachlorocyclohexanone (*O. Hassel* and *K. Lunde*, Acta Chem. Scand., 1950, 4, 200):

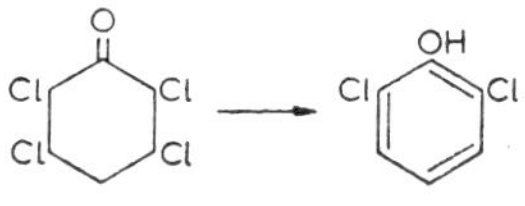

Dehydrohalogenation is frequently effected by bases such as aniline, pyridine or quinoline. Phenol is thus obtained from cyclohex-2-en-1-one by bromination and subsequent dehydrobromination with aniline (*A. Kötz* and *C. Götz*, Ann., 1907, **358**, 183).

Aromatisation may be accompanied by rearrangement. Isophorone (3,5,5-trimethylcyclohex-2-enone) (Vol. IC, p. 63; Vol. IIB, p. 117) when heated with bromine and boiled in trichlorobenzene gives a 44–47 % yield of 2,3,5- and 3,4,5-trimethylphenols in the ratio 1:3 (*F. M. Beringer* and *E. J. Geering*, J. Amer. chem. Soc., 1953, **75**, 2633).

Quinones. Chloranil, tetrachloro-1,4-benzoquinone, is a useful dehydrogenating agent with which in dry, boiling toluene or xylene many alicyclic-aromatic dehydrogenations can readily be effected, phenylcyclohexene, for instance giving biphenyl (52 % yield) (*R. T. Arnold* and *C. J. Collins*, *ibid.*, 1939, **61**, 1407; *idem* and *W. Zenk*, *ibid.*, 1940, **62**, 983).

Chloranil is cheap and readily available, but more reactive quinones of high potential such as tetrachloro-1,2-benzoquinone, 3,3′,5,5′-tetrachloro-4,4′-biphenoquinone, and 2,3-dichloro-5,6-dicyano-1,4-benzoquinone are often more effective (*D. Walker* and *J. D. Hiebert*, Chem. Reviews, 1967, 67, 153). The last-named quinone, for example converts tetralin into naphthalene in nearly quantitative yield at room temperature (5 days) or in boiling benzene solution (2 hours).

On the other hand, dehydro-β-ionone (VII) is rapidly dehydrogenated with tetrachloro-1,2-benzoquinone to the trimethylphenyl ketone, VIII, dehydrogenation being accompanied by rearrangement (*E. A. Braude et al.*, J. chem. Soc., 1960, 3123):

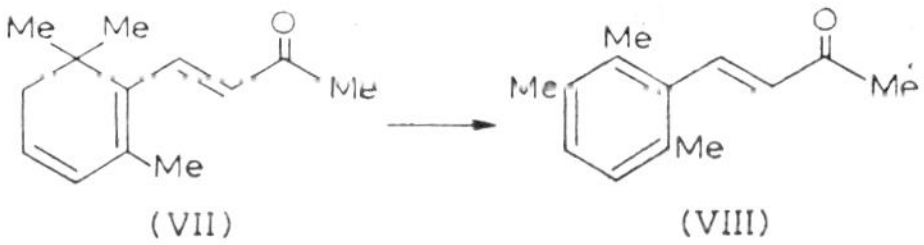

(VII) (VIII)

Carbanions. Aromatisation and disproportionation can be effected by certain carbanions. Limonene, for example, yields *p*-cymene (95 %) when treated with potassium *tert*-butoxide (*Pines* and *L. Schaap*. J. Amer. chem. Soc., 1957, **79**, 2956) or in 91 % yield with lithium–ethylenediamine (*L. Reggel, S. Friedman* and *I. Wender*, J. org. Chem., 1958, **23**, 1137). Other terpenes such as terpinolene and α-phellandrene react similarly. Disproportionation is exemplified by the conversion of cyclohexa-1,3-diene in dimethyl sulphoxide by potassium *tert*-butoxide quantitatively into benzene and cyclohexene (*J. E. Hofmann, P. A. Argabright* and *A. Schriesheim,* Tetrahedron Letters, 1964, 1005):

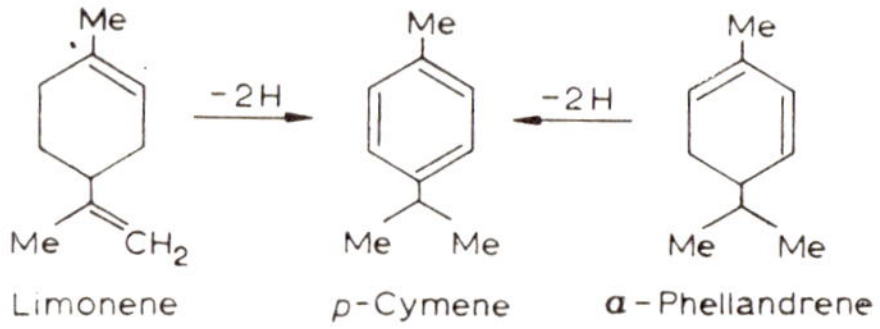

Me -2H Me -2H Me

Me CH2 Me Me Me Me

Limonene *p*-Cymene α-Phellandrene

(*iii*) *Aromatisation by isomeration of alicyclic compounds*

Certain poly-functional compounds show well defined tautomeric characteristics and behave both as aromatic and as alicyclic compounds (*R. H. Thomson,* Quart. Reviews, 1956, **10**, 1). The classic example is phloroglucinol, which behaves as a trihydric phenol and as a tri-oxocyclohexane. In general, the aromatic enol forms are more stable than the alicyclic oxo forms, but many oxo forms can be isolated. Cyclohex-2-ene-1,4-dione, for example, is a crystalline compound stable at 10^0, but rapidly tautomerises in aqueous or ethanolic solution to hydroquinone (*E. W. Garbisch,* J. Amer. chem. Soc., 1965, **87**, 4971), while the "benzoquinone dihalides" (*e.g.*, IX) with acid likewise yield substituted hydroquinones (*e.g.*, X) (*R. K. Norris* and *S. Sternhell,* Chem. Comm., 1965, 608). The natural product gliorosein (XI) with alkali gives 2,3-dimethoxy-5,6-dimethylhydroquinone (XII) (*E. B. Vischer,* J. chem. Soc., 1953, 815):

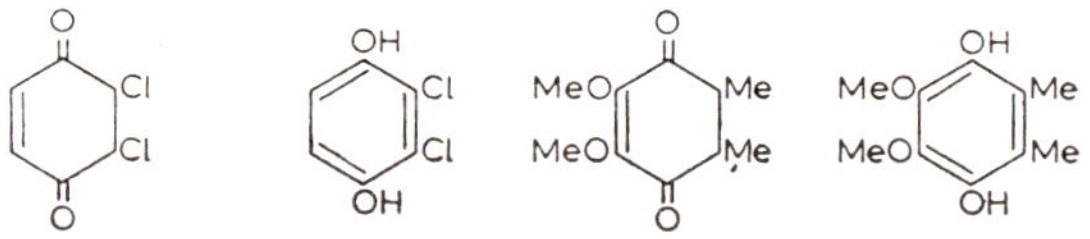

(IX) (X) (XI) (XII)

In the naphthalene series stable diketones such as XIII can be isolated and enolise readily to the corresponding naphthol derivatives (*e.g.*, XIV) (*Thomson, ibid.*, 1950, 1737). Other instances of the conversion of cyclic unsaturated diketones into aromatic derivatives by enolisation are found in the conversion of 6-chloro-5,8,4a,8a-tetrahydro-1,4-naphthaquinone (XV) into the hydroquinone (XVI) when treated with hydrobromic acid and acetic acid (*C. A. Grob* and *W. Jundt,* Helv., 1952, **35**, 2111), and the isomerisation of a decahydro-1,4-dioxophenanthrene into the hexahydro-1,4-dihydroxyphenanthrene by means of acid or alkali (*P. A. Robins* and *J. Walker,* J. chem. Soc., 1952, 642). Many other examples of this type of

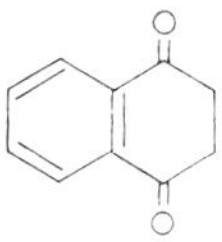

(XIII) 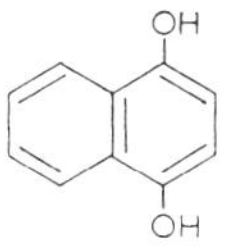 (XIV) 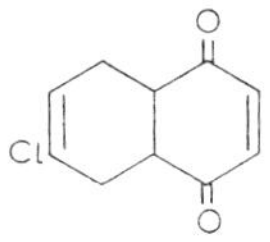 (XV) 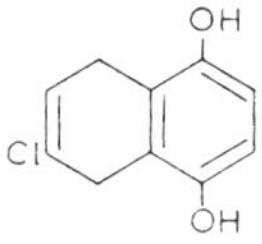 (XVI)

aromatisation are to be found in the literature (see *e.g.*, *idem, ibid.*, 1956, 3260).

Aromatisation by the migration of a hydrogen atom from the nucleus of an alicyclic hydrocarbon to an *exo*-double bond is exemplified by the interaction of the tri-ene, penta-1,2,4-triene, with dimethyl acetylenedicarboxylate to give dimethyl 3-methylphthalate (XVIII) the intermediate triene XVII undergoing aromatisation by the migration of the double bonds (*E. R. H. Jones, H. H. Lee* and *M. C. Whiting, ibid.*, 1960, 341):

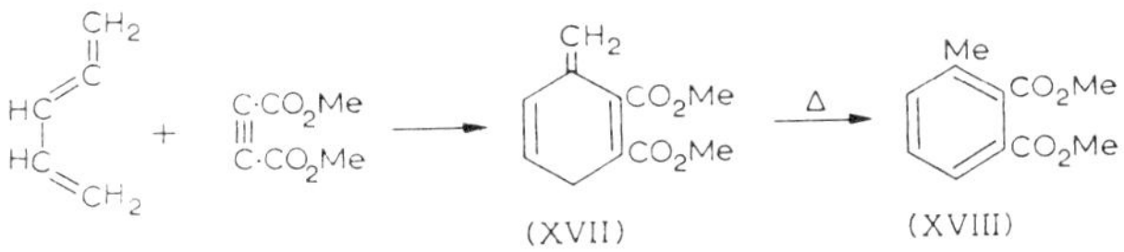

(XVII) (XVIII)

Other examples are the conversion of 1,2-dimethylenecyclohex-4-ene into *o*-xylene (97 % yield) when heated with palladised charcoal (*W. J. Bailey* and *J. Rosenberg*, J. Amer. chem. Soc., 1955, **77**, 73); 2,6-dibenzylidenecyclohexanone into 2,6-dibenzylphenol in the same way (*E. C. Horning*, J. org. Chem., 1945, **10**, 263); and the cyclic sulphone XIX by means of alkali into the aromatic sulphone XX (*H. Kloosterziel* and *H. J. Backer*, Rec. Trav. chim., 1952, **71**, 1235):

(XIX) (XX)

Aromatisation effected by isomerisation is most commonly encountered in the conversion of cyclic ketones into phenols by enolisation or related intramolecular hydrogen transfer. The stable ketone carvone, for instance, is converted into carvacrol (25 % yield) when heated with clay chips at 450° (*W. Treibs*, Ber., 1928, **61**, 683), by heating with palladised charcoal (*Horning, loc. cit.*), or, in 87 % yield, by heating with hydrochloric acid at 125° (*A. Müller*, J. pr. Chem., 1916, [ii], **93**, 10):

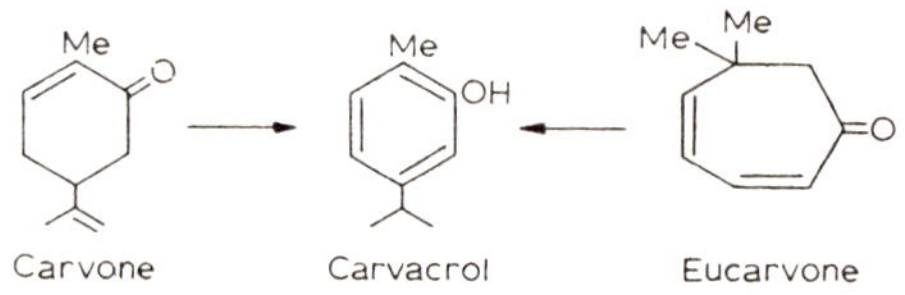

Carvone Carvacrol Eucarvone

Carvacrol is also obtained by the isomerisation of the seven-membered ring compound eucarvone (*T. M. M. Dormaar*, Rec. Trav. chim., 1904, **23**, 394).

Ring-fission of bridged compounds can yield hydrocarbons and phenols. Camphor when dehydrated with phosphorus pentoxide gives *p*-cymene and with mild oxidising agents such as iodine yields carvacrol. There is evidence that in both reactions carvenone is first formed, since this ketone is not only obtained along with 3,4-dimethylacetophenone when camphor is treated with sulphuric acid, but also undergoes the changes just mentioned:

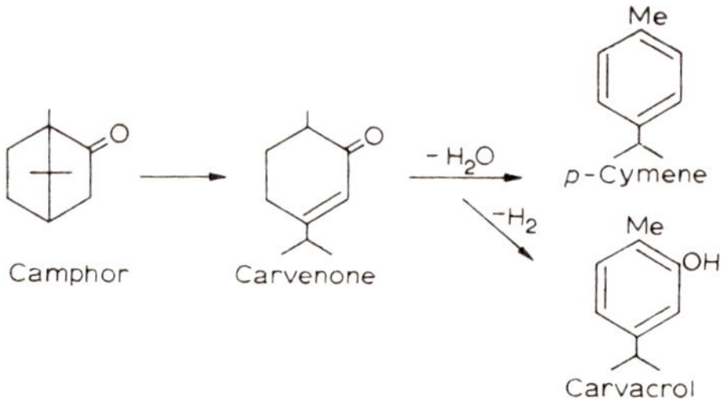

Camphor Carvenone $-H_2O$ *p*-Cymene $-H_2$ Carvacrol

Umbellulone when exposed to ultraviolet light smoothly undergoes ring-enlargement to give a quantitative yield of thymol (*J. W. Wheeler Jr.* and *R. H. Eastman*, J. Amer. chem. Soc., 1959, 81, 236), while on thermal decomposition it yields both thymol and *symm*-thymol:

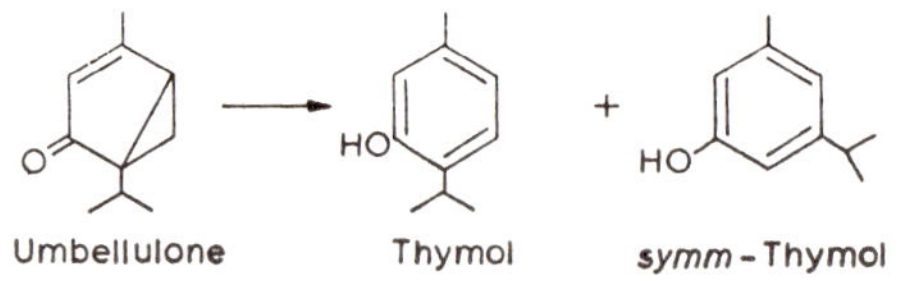

Umbellulone Thymol *symm*-Thymol

Ascaridole with oxygen in the bridge yields *p*-cymene when deoxygenated with triethyl phosphite (*T. Kametani* and *K. Osagawara*, Chem. and Ind., 1968, 1772).

Cyclic dienones and quinols frequently undergo rearrangement to yield aromatic products, particularly phenols. Thus 4,4-diphenylcyclohexadienone when exposed to ultraviolet light gives 2,3- and 3,4-diphenylphenol along with an unsaturated acid (*H. E. Zimmerman*, "Advances in Photochemistry",

Interscience Publishers, New York, Vol. 1, p. 187, 1963; *D. I. Schuster* and *V. Y. Abraitys*, Chem. Comm., 1969, 419):

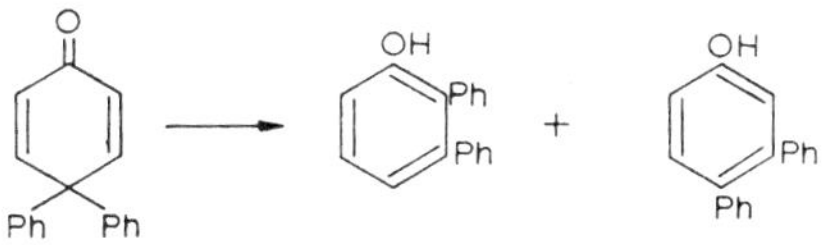

Quinols likewise undergo rearrangement, *p*-toluquinol either by enzymes or trifluoracetic anhydride yielding toluhydroquinone and cresorcinol:

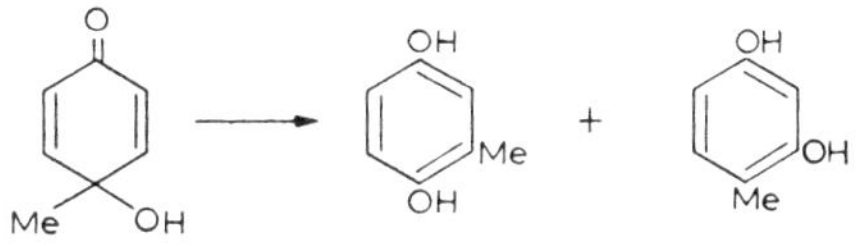

p-Toluquinol Toluhydroquinone Cresorcinol

Toluhydroquinone is also obtained by the action of dilute sulphuric acid on the quinol, while Thiele acetylation of the quinol (acetic anhydride and sulphuric acid) yields the diacetate of cresorcinol (*J. S. Davies, C. H. Hassall* and *J. A. Schofield*, J. chem. Soc., 1964, 3126). The migration of the methyl group is related to the rearrangement of the cross conjugated dienone XXI, obtained from *p*-benzoquinone and diphenylketene, which in sunlight yields the phenolic product XXII (*H. Staudinger* and *St. Bereza*, Ann., 1911, 380, 243):

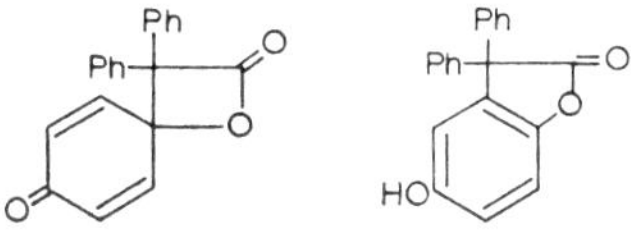

(XXI) (XXII)

p-Dienones may play a part as intermediates in biochemical changes (see, *e.g.*, *Davies, Hassall* and *Schofield, loc. cit.*).

ortho-Type cyclohexadienones (*o*-quinols) likewise give high yields of aromatic products when exposed to ultraviolet light, *e.g.* XXIII → XXIV (*D. H. R. Barton* and *G. Quinkert*, J. chem. Soc., 1960, 1):

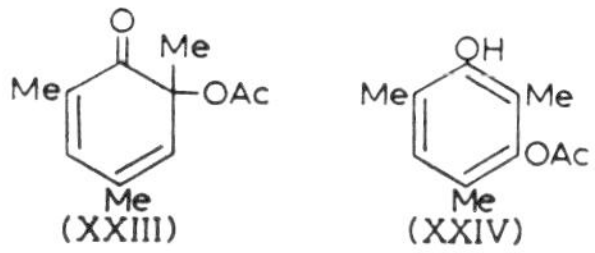

(XXIII) (XXIV)

Similar rearrangements are effected chemically by reagents such as boron trifluoride or the Thiele reagent (*J. D. Loudon*, "Progress in org. Chem., Butterworths, London, 1961, Vol. 5, p. 46). *o*-Quinols are considered to be intermediates in the Reimer-Tiemann reaction in which phenols are converted into *o*- and *p*-hydroxybenzaldehydes by chloroform and alkali (*H. Wynberg*, J. Amer. chem. Soc., 1954, **76**, 4998).

Related to the *dienone-phenol* rearrangement discussed above is the *dienol-benzene* rearrangement (see *e.g.*, *H. Plieninger et al.*, Ber., 1965, **98**, 1765) in which cyclohexadienols with acid undergo aromatisation to give dialkylbenzenes, *e.g.*:

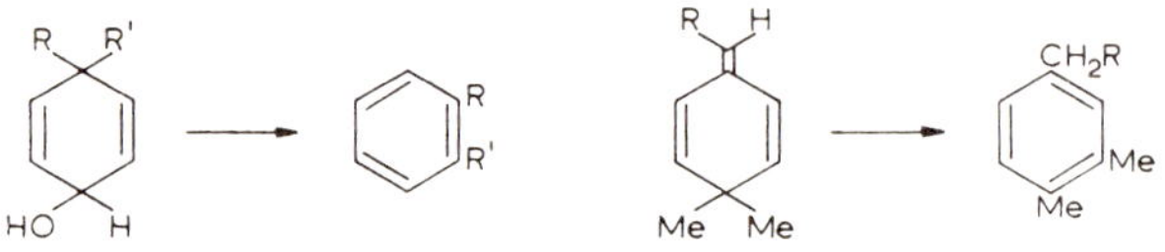

(XXV) (XXVI)

An example of this type of rearrangement is found in the spontaneous aromatisation of prehenic acid (p. 136). Carbonium ions are probably involved and the rearrangement is reminiscent of the conversion of semibenzenes (*e.g.* XXV) into trialkylbenzenes (*e.g.*, XXVI) by acids.

(iv) Aromatisation of 5-, 7-, *and* 8-*membered ring compounds and of heterocyclic compounds*

Carbocyclic compounds other than those containing six-membered rings can be aromatised by ring-enlargement or ring-contraction. Examples of the conversion of five-membered rings are found in the formation of benzene from methylcyclopentane (Vol. II A, p. 113) and the interaction of sodium cyclopentadienide with chloroform to give a 23% yield of chlorobenzene, the reactive agent being dichlorocarbene (*A. P. ter Borg* and *A. F. Bickel*, Rec. Trav. chim., 1961, **80**, 1217):

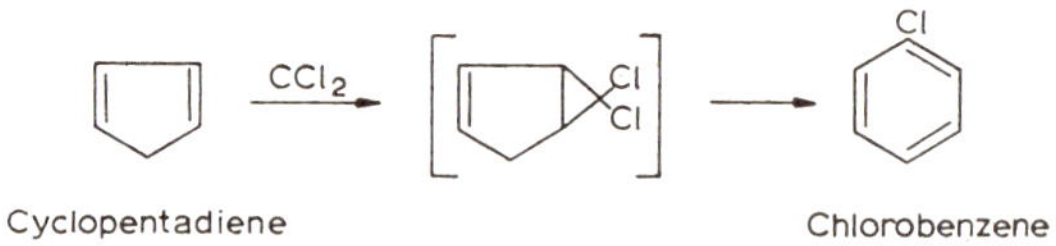

Cyclopentadiene Chlorobenzene

Seven-membered rings often yield aromatic compounds and the isomerisation of eucarvone has already been mentioned (p. 142). Cycloheptane when dehydrogenated with selenium at 440° gives toluene (*L. Ruzicka* and *C. F. Seidel*, Helv., 1936, **19**, 424) and cycloheptatriene undergoes photoisomerisation when irradiated in the vapour phase with ultraviolet light to give toluene and bicyclo[3.2.0]hepta-2,6-diene (*R. Srinivasan*, J. Amer. chem. Soc., 1962, **84**, 3432):

Me

Cycloheptatriene → Toluene + Bicyclo[3.2.0]hepta-2,6-diene

Cyclo-octane with selenium at 390° gives *p*-xylene (*Ruzicka* and *Seidel*, *loc. cit.*) and with chromium oxide at 425–455° styrene (*S. Goldwasser* and *H. S. Taylor*, J. Amer. chem. Soc., 1939, **61**, 1260), but the most interesting and important aromatisations of eight-membered rings are those of cyclo-octatetraene which is readily converted into mono- and disubstituted benzene derivatives (Vol. IIB, p. 392 *et seq.*) and with ultraviolet light yields benzene and acetylene (*H. Yamazaki* and *S. Shida*, J. chem. Phys., 1956, **24**, 1278).

Tropylium salts and related compounds readily yield aromatic compounds (see Vol. IIB, p. 359 *et seq.*). Tropylium salts, for example, with chromic oxide and acetic acid yield benzaldehyde (*W. E. Doering* and *L. H. Knox*, J. Amer. chem. Soc., 1957, **79**, 352), while with aqueous hydrogen peroxide they yield benzene as well as carbon monoxide or formic acid (*M. E. Volpin*, *D. N. Kursanov*, and *V. G. Dulova*, Tetrahedron, 1960, **8**, 33):

$$C_7H_7Br + H_2O_2 = C_6H_6 + CO + H_2O + HBr$$

Tropone with ultraviolet light gives a small yield of benzene (*O. L. Chapman*, "Advances in Photochemistry", Interscience, New York, 1963, Vol. 1, p. 326), but the most characteristic aromatisation of the series is the benzilic acid type of rearrangement which tropolones undergo with alkali to give benzenoid products, frequently under mild conditions. Thus the nitro-product of hinokitiol when warmed with 50% acetic acid yields 4,6-dinitro-*m*-cumic acid, while dinitro-*β*-thujaplicin gives the same product merely by warming with water (for other examples, see *P. Pauson*, Chem. Reviews, 1955, **55**, 9; *T. Nozoe*, "Progress in the Chemistry of Natural Products", Springer Verlag, Berlin, 1956, Vol. 13, p. 274).

Heterocyclic compounds, especially those which undergo hydrolytic ring-fission to ketonic products, can be converted into aromatic compounds. Dehydracetic acid for instance, yields on hydrolysis with sodium hydroxide 1-aceto-1-acetylacetoacetic acid (XXVII), which by intramolecular condensation and decarboxylation gives orcinol (*Collie* and *W. S. Myers*, J. chem. Soc., 1893, **63**, 122):

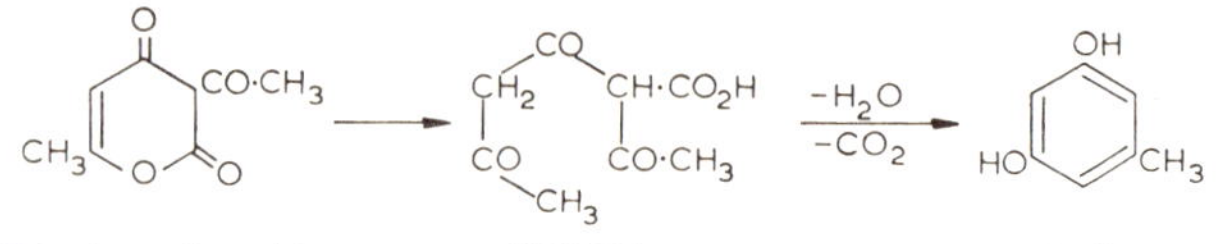

Dehydracetic acid (XXVII) Orcinol

For the significance of such reactions in biogenesis see p. 134.

α-Pyrones and Grignard reagents under suitable conditions yield a variety of aromatic products including hydrocarbons (*R. Gompper* and *O. Christmann* Ber., 1961, 94, 1795; *G. Köbrich* and *D. Wunder*, Ann., 1962, 654, 131), 4,6-dimethyl-α-pyrone and phenylmagnesium bromide, for example, giving 3,5-diphenyltoluene:

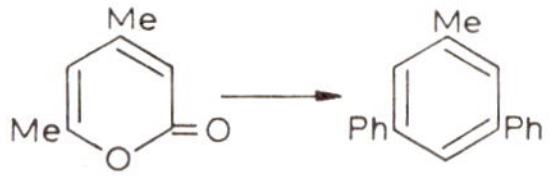

Aromatic hydrocarbons can also be obtained from pyrylium salts and Wittig reagents (triphenylphosphinemethylene derivatives) in which the cyclic oxygen atom is exchanged for a methine grouping (*G. Märkl*, Angew. Chem., internat. Edn., 1962, 1, 511). Thus 1,3,5-triphenylbenzene is obtained from 2,4,6-triphenylpyrylium borofluoride:

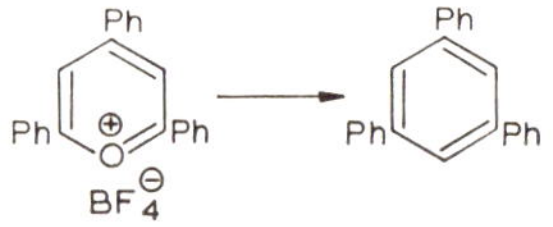

Pyrylium salts are also a fruitful source of aromatic compounds including phenols, amines, and nitro compounds (*K. Dimroth*, Zeit. angew. Chem., 1960, 72, 331). Thus alkylated phenols result when α-methylpyrylium salts are treated with alkali, the intermediate pseudo-base undergoing ring-closure. For the synthesis of aromatic amines from pyrylium salts see *O. Diels* and *K. Alder*, Ber., 1927, 60, 716, and for nitro-compounds, *Dimroth et al.*, Ber., 1957, 90, 1634, 1668; *Dimroth* and *K. H. Wolf*, "Preparative Organic Chemistry", Academic Press, Inc., N.Y., 1964, Vol. 3, p. 357.

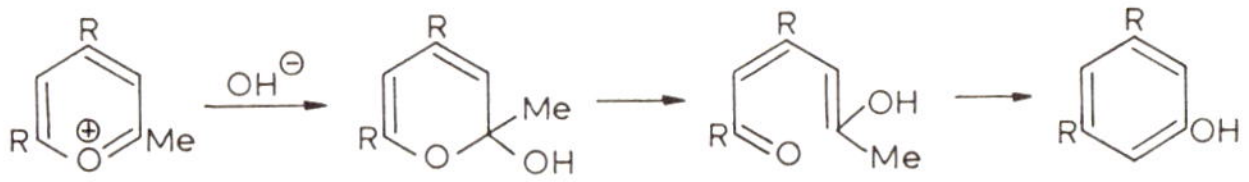

Isolated instances are known of conversion of 3- and 4-membered ring compounds into aromatic hydrocarbons. Biscyclopropenyls are isomerised by light or heat to substituted benzenes, bistriphenylcyclopropenyl (XXVIII)

by light or heat giving hexaphenylbenzene (*R. Breslow et al.*, J. Amer. chem. Soc., 1965, 87, 5139):

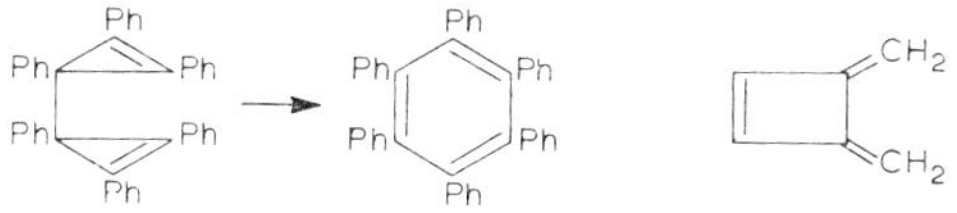

(XXVIII) Hexaphenylbenzene (XXIX)

and 3,4-dimethylenecyclobutene (XXIX) undergoes thermal rearrangement at 420° to give a 50 % yield of benzene (*M. L. Hefferman* and *A. J. Jones*, Chem. Comm., 1966, 120).

(b) *Rupture of the benzene ring*

The benzene ring is noted for its stability, but it can be broken down by ultrasonic waves into acetylene (*D. L. Currell* and *L. Zechmeister*, J. Amer. chem. Soc., 1958, 80, 205). Irradiation of liquid benzene yields fulvene and benzvalene (*H. J. F. Angus, J. McDonald Blair* and *D. Bryce-Smith*, J. chem. Soc., 1960, 2003; *K. E. Wilzbach, J. S. Ritscher* and *L. Kaplan*, J. Amer. chem. Soc., 1967, 89, 1031), while the photolysis of benzene vapour gives fulvene and *cis*-hexa-1,3-dien-5-yne (*Kaplan* and *Wilzbach*, *ibid.*, 1967, 89, 1030; 1968, 90, 5646). In such transformations biradicals like XXX may be the precursors of fulvene and benzvalene (*Bryce-Smith* and *H. C. Longuet-Higgins*, Chem. Comm., 1966, 593):

Benzene (XXX) Fulvene (XXXI)

Photolysis in rigid glasses or viscous solids can yield substituted hexatrienes XXXI by a four-centre interaction between benzene and solvent (RH) (*E. J. Anderton, H. T. J. Chilton* and *G. Porter*, Proc. chem. Soc., 1960, 352).

Disruption of the benzene ring is generally effected by oxidation, some examples of which are given below.

(*i*) *Oxidation*

Benzene is not readily attacked by the usual oxidising agents, potassium permanganate, for example, oxidising it very slowly to formic and oxalic acids. This resistance of the ring is also shown by the behaviour of the benzene homologues, the side-chains of which are readily oxidised by potassium permanganate or dilute nitric acid to benzene carboxylic acids, further oxidation being extremely difficult. Nevertheless by the use of catalysts, high temperatures, or certain agents such as ozone the degradative oxidation of the benzene nucleus can be effected. The catalytic oxidation of benzene vapour to maleic anhydride is a commercial process (Vol. I D, p. 340), while oxidation by the Milas reagent (hydrogen peroxide, *tert*-butyl hydroperoxide and osmium tetroxide) gives *cis-cis*-muconic acid along with allomucic acid, mesotartaric acid, and oxalic acid (*J. W. Cook* and *R. Schoental*, J. chem. Soc., 1950, 47):

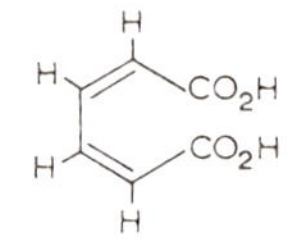

cis-cis-Muconic acid

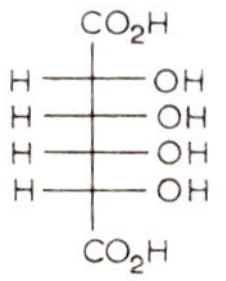

Allomucic acid

Metabolic oxidation of benzene to *trans-trans*-muconic acid occurs in rabbits dosed with benzene (*K. Bernhard* and *E. Gressly*, Helv., 1941, **24**, 83).

It is well known that the presence of amino and hydroxyl groups greatly increases the ease with which the benzene nucleus is broken. This is illustrated by the oxidation of catechol to carbon dioxide merely by bubbling air through an alkaline solution of the phenol (*G. R. Clemo* and *F. K. Duxbury*, Chem. and Ind., 1953, 195). Catechol is oxidised by nickel dioxide to *cis-cis*-muconic acid (*K. Nakagawa* and *H. Onoue*, Tetrahedron Letters, 1965, 1433) and *o*-phenylenediamine with the same reagent or lead tetra-acetate yields 1,4-dicyano-1,3-butadiene (*cis-cis*-mucononitrile) (XXXIII), possibly *via* the nitrene XXXII (*idem*, Chem. Comm., 1965, 386):

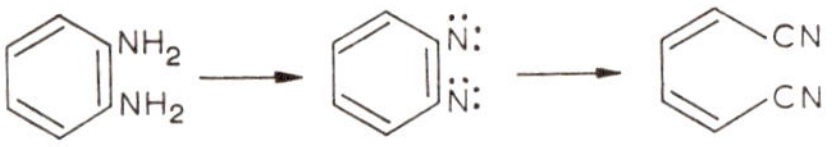

(XXXII) (XXXIII)

Related to this reaction is the isolation of the same dinitrile (79 % yield) when *o*-diazobenzene is thermally decomposed in boiling decalin (*J. H. Hall*, J. Amer. chem. Soc., 1965, **87**, 1147).

A novel breakdown is the conversion of salicylatobisethylenediamine-cobalt by nitric acid into the oxalic acid derivative (90 % yield) (*K. Garbett* and *R. D. Gillard*, Chem. Comm., 1967, 694).

Mild oxidation of polyhydric phenols can be effected with oxygen, 4,6-di-*tert*-butylpyrogallol (XXXIV), for example, yielding an intermediate *ortho*-quinone XXXV, which undergoes ring-fission to a dibasic acid XXXVI (*T. W. Campbell*, J. Amer. chem. Soc., 1951, **73**, 4190; *H. Schulze* and *W. Flaig*, Ann., 1952, **575**, 231; see also *J. E. Baldwin, H. H. Basson* and *H. Krauss*, Chem. Comm., 1968, 984).

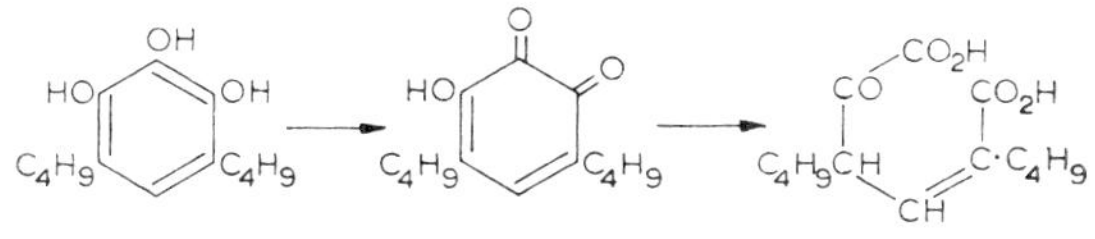

(XXXIV) (XXXV) (XXXVI)

Phenols may be disrupted by oxidation to dienones, which undergo fission when irradiated. The quinol acetate XXXVII, prepared by the action of

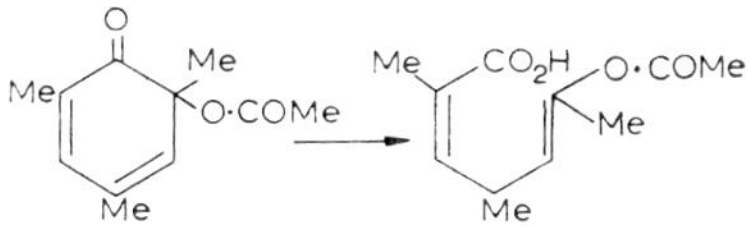

(XXXVII) (XXXVIII)

ead tetra-acetate on 2,4,6-trimethylphenol (*F. Wessely* and *F. Sinwel*, Monatsh., 1950, **81**, 1055) yields the dienoic acid XXXVIII when exposed to ultraviolet light (*D. H. R. Barton* and *G. Quinkert*, J. chem. Soc., 1960, 1).

(ii) Oxidative breakdown of phenols by micro-organisms

It is well known that muconic acid can be obtained by the oxidation of phenol or catechol by peracetic acid (*J. Boeseken* and *C. L. M. Kerkhoven*, Rec. Trav. chim., 1932, **51**, 964; *J. A. Elvidge et al.*, J. chem. Soc., 1950, 2235) or by the action of pyrocatechase on catechol (*O. Hayaishi* and *K. Hashimoto*, J. Biochem. Japan, 1950, **37**, 231). More recent investigations show that micro-organisms form enzymes which catalyse the ring-fission of phenolic substances (for a review see *D. W. Ribbons*, Ann. Reports, 1965, **62**, 445). In such reactions the micro-organisms use aromatic substances as the sole source of the organic carbon necessary for their growth and the disrupted products are then utilised in the various metabolic processes in the cell. One product is generally carbon dioxide, but methane is sometimes encountered (see, *e.g.*, *G. E. Symons* and *A. M. Buswell*, J. Amer. chem. Soc., 1953, **55**, 2028).

In the oxidative fission of phenols two types of cleavage can be postulated (*S. Dagley et al.*, Nature, 1964, **202**, 775): (1) oxidative cleavage of bonds linking two carbon atoms each bearing a hydroxyl group and (2) cleavage of a bond between a hydroxyl-bearing carbon atom and non-hydroxylated carbon atom.

The first process is illustrated by the enzymic breakdown to *cis-cis*-muconic acid and other products including 3-oxoadipic acid (*W. C. Evans et al.*, *ibid.*, 1951, **168**, 772). The initial stages involve oxidation of phenol to catechol and thence to muconic acid which by means of lactonising and lactone-splitting enzymes yields 3-oxo-adipic acid (see, *e.g.* *W. R. Sistrom* and *R. Y. Stanier*, J. biol. Chem., 1954, **210**, 821):

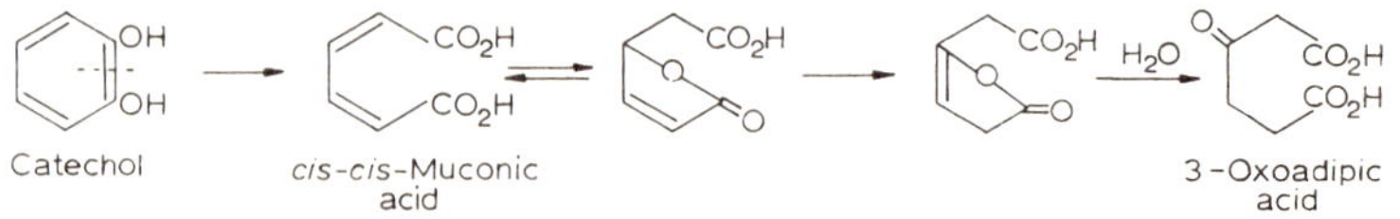

Catechol — *cis-cis*-Muconic acid — 3-Oxoadipic acid

Protocatechuic acid likewise undergoes fission by means of protocatechuic acid oxidase to *cis-cis-β*-carboxymuconic acid and if the enzyme is contaminated with a decarboxylase the product is 3-oxoadipic acid (*S. R. Gross, R. D. Gafford* and *E. L. Tatum, ibid.*, 1956, **219**, 781). When protocatechuic acid 4,5-oxygenase is used one of the products is 2,4-lutidinic acid, the inter-

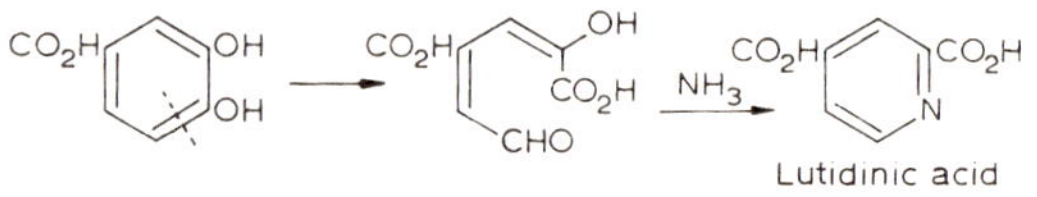

Lutidinic acid

mediate α-hydroxy-carboxymuconic semialdehyde with ammonia (from the ammonium sulphate used in the enzyme preparation) by a non-enzymic reaction yielding the heterocyclic acid (*S. Trippett, S. Dagley* and *D. A. Stopher*, Biochem. J., 1960, **76**, 9P). This illustrates the second type of fission, which is also exemplified by homogentisic acid yielding maleylacetoacetic acid by the action of homogentisicase (*W. E. Knox* and *S. W. Edwards*, J. biol. Chem., 1955, **216**, 479):

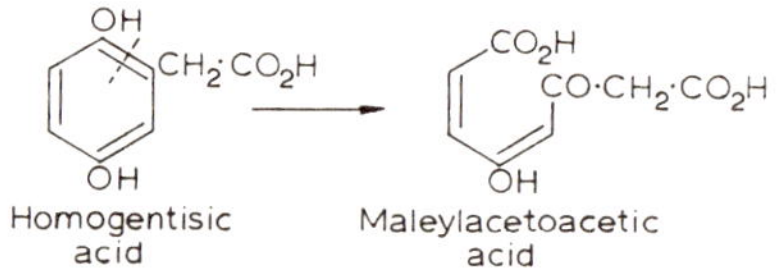

Homogentisic acid — Maleylacetoacetic acid

Gentisic acid likewise with gentisicase gives maleylpyruvic acid (*L. Lack*, Biochim. Biophys. Acta, 1959, **34**, 117; *S. Sugujama et al.*, Bull. agric. chem. Soc. Japan, 1960, **24**, 255), while 3,4-dihydroxyphenylacetic acid in *Pseudomonas ovalis* yields δ-carboxymethyl-α-hydroxymuconic semialdehyde (*K. Adachi et al.*, Biochim. Biophys. Acta, 1964, **93**, 483).

(iii) Ozonolysis

Ozonolysis of benzene yields glyoxal and *G. Reddelien* (J. pr. Chem., 1915, [ii], **91**, 225) suggested that ozonolysis of benzene derivatives might profitably be applied to the "benzene problem". This was carried into effect by later workers, particularly *P. W. Haaijman* and *L. P. Wibaut* (Rev. Trav. chim., 1941, **60**, 842) who ozonised *o*-xylene, decomposed the ozonides, isolated glyoxal, methylglyoxal, and diacetyl (dimethylglyoxal), and estimated these products quantitatively. For a discussion on the interpretation of these results see *P. S. Bailey* (Chem. Reviews, 1958, **58**, 958).

Me, Me → O_3 → 2 Me·CO·CHO, CHO·CHO

Me, Me → O_3 → Me·CO·CO·Me, 2CHO·CHO

The ease of ozonolysis of benzene homologues increases with the number of methyl groups in the ring, the order of increasing activity being: toluene < *m*-xylene < mesitylene < hexamethylbenzene (*J. van Dijk,* Rec. Trav. chim., 1948, **67**, 945).

(iv) Disruption of chlorinated compounds

Fission of phenols can be effected by chlorination followed by treatment of the products with alkali. These products contain the grouping CCl_2·CO; and C–C rupture occurs at this group. Hydroquinone, for example, with chlorine yields chloranil and then hexachlorocyclohex-2-ene-1,4-dione

Chloranil → (XXXIX) → (XL) → Dichloromaleic acid + Trichloroethene

(XXXIX). The latter substance with ethanolic potassium hydroxide gives the pentachloroketonic acid XL. Both the dione and the acid are disrupted by aqueous sodium hydroxide to dichloromaleic acid and trichloroethene (*Th. Zincke* and *O. Fuchs,* Ann., 1892, **267**, 1). Similar ring-fissions are observed in the degradations of catechol to ethylidenepropionic acid (*Zincke* and *Fr. Küster,* Ber., 1890, **23**, 2200) and of resorcinol to 2,2,4,6,6-pentachloro-5-oxohex-3-enoic acid, $Cl_2CH \cdot CO \cdot CCl{:}CH \cdot CCl_2 \cdot CO_2H$, and other products (*C. J. Moye,* Chem. Comm., 1967, 196).

(v) Reduction in alkaline solution

m-Hydroxybenzoic acids are reduced by sodium and boiling amyl alcohol to cyclohexane compounds, but *o*-hydroxybenzoic acids are reduced with ring fission to substituted pimelic acids (*A. Einhorn* and *J. S. Lumsden*, Ann., 1895, **286**, 257). Salicylic acid yields pimelic acid in excellent yield, and *β*- and *γ*-isopropylpimelic acids are obtained in this way from 2-hydroxy-4-isopropylbenzoic acid and 2-hydroxy-5-isopropylbenzoic acid, respectively (*A. B. Anderson* and *J. Gripenberg*, Acta Chem. Scand., 1948, **2**, 644; *H. Erdtman* and *Gripenberg*, *ibid.*, p. 625).

(*vi*) *Miscellaneous*

Certain nitrophenols undergo ring-fission when treated with sulphuric acid (G. Schultz). Thus 3-nitro-*p*-cresol yields 2-methylmuconic acid in the form of its lactone. The reaction takes the following course (*H. Pauly* and *G. Will*. Ann., 1918, **416**, 1):

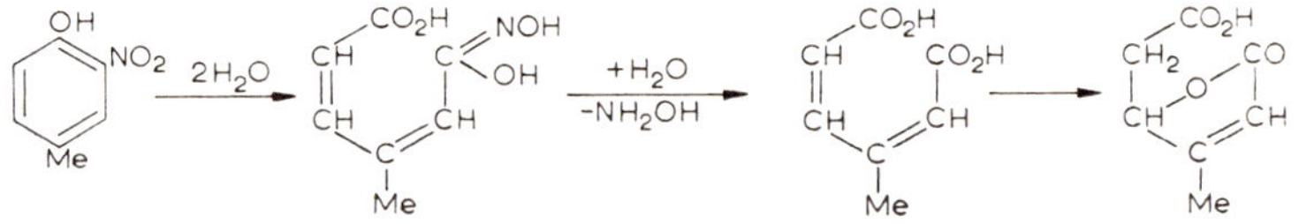

$\xrightarrow{2H_2O}$ $\xrightarrow[-NH_2OH]{+H_2O}$ $\longrightarrow$

3-Nitro-*p*-cresol

2-Methylmuconic lactone

Similarly 4-bromo-2-nitrophenol with concentrated sulphuric acid at 110° gives a good yield of 2-bromomucolactone (*I. J. Rinkes*, Rec. Trav. chim., 1943, **62**, 12). For this fission of other *o*-nitrophenols see *O. Neunhoffer* and *H. Kölbel*, Ber., 1935, **68**, 255, 1774.

Chapter 2

Mononuclear Hydrocarbons: Benzene and its Homologues

J. D. DOWNER

1. Introduction

The homologous series known as the alkylbenzenes is one of the largest and most important of the hydrocarbon classes. Benzene, C_6H_6, is the parent member, and its structure and aromatic character are discussed in Chapter 1. The alkylbenzenes can be derived formally from benzene by substituting hydrogen atoms in the benzene nucleus successively by alkyl groups. Mono- up to hexa-alkyl derivatives of benzene are obtained according to the number of hydrogen atoms replaced.

Apart from the *isomerism* occurring within the alkyl substituent(s) (Vol. I A, p. 357), di- and poly-alkylbenzenes can also exhibit isomerism arising from the relative position of the alkyl substituents in the ring (p. 5). Formulae have been derived for calculating the number of possible isomers for mono-, di-, and tri-alkylbenzenes (*B. F. Pishnamazzade,* Azerb. Khim. Zhur., 1962, No. 2, 41; No. 3, 27; No. 5, 77).

2. Natural occurrence of mononuclear hydrocarbons

The only natural occurrence of benzene and its homologues in significant amounts is to be found in *crude oil* (petroleum) which contains three broad classes of hydrocarbons, *viz.* paraffins, naphthenes (cycloparaffins), and aromatics, although mixed-type hydrocarbons are also present (*K. Van Nes* and *H. A. Van Westen,* "Mineral Oils", Elsevier Publishing Company, Amsterdam, 1951). The primary liquid products, gasoline (petrol), kerosine and gas oil, obtained in refining a crude oil by distillation (see Vol. I A, p. 392), all contain mononuclear hydrocarbons. As the number of carbon atoms increases, however, more of the benzene derivatives are of the aromatic-cycloparaffin sub-class. A representative petroleum (Ponca Crude) has been investigated continuously since 1928 under the American Petroleum Institute Research Project 6 in order to extend the knowledge of the hydrocarbon components of petroleum (see *F. D. Rossini*, J. chem. Educ., 1960, **37**, 554; *B. J. Mair*, Oil Gas J., 1964, **62**, No. 37, 130). This project has also examined the composition of the gasoline fraction from a variety of crude oils. The relative proportions of each broad hydrocarbon class varies widely in different crude oils but it would seem that they all contain substantially the same hydrocarbon compounds. Also the relative amounts of individual compounds within a class are of similar order for different crude oils. Table 1 shows how the relative amounts of benzene, toluene and the four C_8 alkylbenzenes have nearly the same values in gasolines from seven different crude oils. Similarly, the ratio of *tert*-butylbenzene to 1,2,4-trimethylbenzene (pseudocumene) appears to be about 1 to 50 in all petroleum. At least 48 mononuclear hydrocarbons have been isolated from Ponca crude including all the 32 possible alkylbenzenes boiling up to 190° (see Table 2, p. 179). In any given group of isomeric alkylbenzenes the most abundant are those which are thermodynamically more stable or which have more substituents.

TABLE 1

THE RELATIVE AMOUNTS OF INDIVIDUAL MONONUCLEAR HYDROCARBONS IN SEVEN DIFFERENT GASOLINES (PETROLS) BOILING UP TO 180°

	A	B	C	D	E	F	G
Benzene	0·09	0·04	0·04	0·13	0·10	0·05	0·06
Toluene	0·30	0·29	0·31	0·37	0·20	0·30	0·38
C_8 alkylbenzenes	0·61	0·67	0·65	0·50	0·70	0·65	0·56
Percentage of these aromatics in the Gasoline	5·4	5·9	4·9	4·8	1·3	4·3	19·5

A. Ponca, Oklahoma
B. East Texas
C. Bradford, Pennsylvania
D. Greendale-Kawkawlin, Michigan
E. Winkler, Texas
F. Midway, California
G. Conroe, Texas

Practically all of the pure aromatics which have been isolated from petroleum so far are contained in the straight run gasoline fraction. The aromatic content of straight run gasolines varies from >1 to <20 % according to the crude source (see Table 1) and it is significant that these aromatics have high octane ratings. Individual hydrocarbons have been isolated from the kerosine and gas oil fractions in exceptional cases but it is usually possible to separate only groups of closely related components. *Mair* and *J. M. Barnewall* (J. Chem. Eng. Data, 1964, **9**, 282) have investigated the composition of mononuclear aromatic hydrocarbons in the 230 to 305° fraction from Ponca crude, by spectroscopic methods. They found that the C_{13-15} alkylbenzenes in this fraction were principally di- and tri-substituted, and the average disubstituted alkylbenzene has one methyl group and one long chain alkyl group attached to the benzene ring. Also, in most of the molecules, the long chain alkyl group contained a branch methyl, and most of the trisubstituted alkylbenzenes contained one methyl group and two intermediate length alkyl side-chains.

The naturally-occurring mononuclear aromatic hydrocarbons represent a very small volume of the crude (v. 1·7 % would be typical) but the total amount is enormous in view of the World proved and probable reserves of crude oil which for 1968 have been estimated at 509,877,000,000 barrels (World Petroleum Report, 1969) and continue to grow at a faster rate than they are depleted. There are more than enough aromatic compounds present in the yearly crude production to meet all demands but the complexity of petroleum makes isolation of these aromatics uneconomical. However, aromatic hydrocarbons are now "synthesised" from petroleum on an ever increasing scale and the processess used are described later in this chapter (see p. 222).

In studies connected with the genesis of crude oil it has been shown that low molecular weight aromatic hydrocarbons are present in aquatic sediments and both recent and ancient carbonaceous sediments (*J. G. Erdman et al.*, Amer. chem. Soc., Div. Pet. Chem., Preprints, 1958, **3**, No. 4, C39).

Mononuclear aromatic hydrocarbons do not occur naturally associated with coal in significant amounts, but benzene, toluene, and the C_8 aromatic hydrocarbons are evolved in small amounts from coal during storage (*R. P. W. Scott* and *G. W. Girling*, Chem. and Ind., 1961, 1570). The basic unit in coal, the vitrinite molecule, appears to be a polycyclic aromatic nucleus containing on the average about 2 to 4 fused rings in low rank coals and up to 30 in the anthracites. This nucleus is surrounded mainly by short aliphatic side-chains or hydrogen saturated structures, and oxygen may be present in the form of phenolic hydroxyl and quinone groups as well as smaller amounts of sulphur and nitrogen. The vitrinite molecules are linked to each other, in amounts varying with the type of coal, forming relatively flat molecules which are stacked more or less regularly in the coal (*W. Idris Jones*, R.I.C. Lectures, Monographs and Reports, 1956, No. 3). The tar and gas resulting from the carbonization and hydrogenation of coal are important sources of benzene and its homologues (p. 221).

p-Cymene appears to be the only benzenoid hydrocarbon which occurs free in vegetable products presumably because of its direct relationship with α-pinene. It is widely distributed in the essential oils of the gymnosperms and angiosperms, particularly the *umbelliferae* and *labiatae*. It was first isolated from Roman oil of cumin by *C. Gerhardt* and *A. Cahours* (Ann. Chim., 1841, [iii], **1**, 63, 102).

3. Laboratory methods of preparing benzene and its homologues

Many of the individual hydrocarbons prepared under the American Petroleum Institute Research Project 45 were made by routes involving Grignard reactions and these have been reviewed by *C. E. Boord* and co-workers (Ind. Eng. Chem., 1949, **41**, 609). Alkylation of benzene is a most convenient way of making the lower molecular weight mono- and certain poly-alkylbenzenes. A general method for introducing methyl groups into an alkylbenzene is by chloromethylation followed by replacement of the chlorine by hydrogen. The Wurtz-Grignard synthesis of alkylbenzenes using an alkyl sulphate in place of the alkyl halide is reported to be superior to the Fittig synthesis as a route to *o*- and *p*-dialkylbenzenes (*J. V. Karabinos et al.*, J. Amer. chem. Soc., 1946, **68**, 2107). Catalytic trimerization of substituted acetylenes offers the widest scope for making individual hydrocarbons. Good examples of synthetic methods other than alkylation, used in preparing monoalkylbenzenes, are described by *E. R. Lynch* and *E. B. McCall* (J. chem. Soc., 1960, 1254).

These main general methods are discussed in more detail in sections (*a*) – (*k*) below.

Most of the isomeric *o*-, *m*-, and *p*-dialkylbenzenes boil too closely to be separated easily by fractional distillation, and in many cases at least two of the positional isomers boil at virtually the same temperature. Methods other than distillation which have been used to separate certain of the dialkylbenzenes, chiefly the C_8 aromatic hydrocarbons, include crystallization (*C. J. Egan* and *R. V. Luthy*, Ind. Eng. Chem., 1955, **47**, 250), combined crystallization and distillation (*J. A. Weedman*, U.S.P., 3,067,270/1962), sulphonation-desulphonation (*N. F. Yates*, U.S.P., 2,585,525/1952), selective alkylation-dealkylation (*B. B. Corson et al.*, Ind. Eng. Chem., 1956, **48**, 1180; *S. D. Mekhtiev et al.*, Neftekhimiya, 1961, **1**, 54), and differential adsorption-desorption on solids (*I. A. Kuzin* and *T. G. Plachenov*, J. appl. Chem. [U.S.S.R.], 1952, **25**, 224).

Methods used for preparing benzene and alkylbenzenes with ^{14}C-labelled rings have been reviewed by *P. I. Petrovich* (Khim. Prom., 1959, No. 4, 42). In this connection see also *J. R. Catch*, "Carbon-14 Compounds", Butterworths, London, 1961.

(a) Nuclear alkylation of aromatic hydrocarbons

Alkylbenzenes with one or more alkyl substituents, which may or may not be similar, are prepared by *alkylation* of the appropriate arene with a variety of reagents. Alkylation can be accomplished thermally without a catalyst but the presence of proton-donating catalysts greatly facilitates the reaction which in its general form may be expressed thus:

$$\text{ArH} + \text{RX} \rightarrow \text{ArR} + \text{HX}$$

Alkyl halides, olefins, paraffins, alcohols, ethers, and esters have all been used as alkylating agents ("Friedel-Crafts and Related Reactions", Ed. *G. A. Olah*, Interscience Publishers, London, 1964, Vol. 2, Part I). In liquid phase alkylations suitable catalysts are the Lewis acids, *e.g.*, boron trifluoride, aluminium chloride and other metal halides, and the Brønsted acids, *e.g.*, sulphuric acid, phosphoric acid, and hydrogen fluoride. Anhydrous aluminium chloride is the catalyst most frequently used (*C. A. Thomas*, "Anhydrous Aluminium chloride in Organic Chemistry", Reinhold Publishing Corp., New York, 1941). Crystalline aluminosilicates are versatile catalysts for liquid phase alkylation of aromatic hydrocarbons (*P. B. Venuto et al.*, J. Catalysis, 1966, **5**, 81). The Lewis acids require a proton-donating compound as co-catalyst. Excess aromatic substrate often suffices as medium for carrying out the reaction but inert solvents, such as carbon disulphide, nitroalkanes (*L. Schmerling*, Ind. Eng. Chem., 1948, **40**, 2072), and nitrobenzene are frequently used. In vapour phase alkylations acidic solid state catalysts may be used, *e.g.*, mixed acidic metal oxides (silica-alumina) and Brønsted or Lewis acids, supported on metal oxides or clays.

The reaction is considered to proceed by an electrophilic mechanism (*C. C. Price*, Org. Reactions, 1946, **3**, 1; *H. C. Brown* and *M. Grayson*, J. Amer. chem. Soc., 1953, **75**, 6285; *Schmerling*, Ind. Eng. Chem., 1953, **45**, 1447; *Brown et al.*, J. Amer. chem. Soc., 1956, **78**, 2185; *R. M. Roberts* and *D. Shiengthong*, *ibid.*, 1960, **82**, 732; *Olah et al.*, *ibid.*, 1964, **86**, 1046; *R. Nakane* and *A. Natsubori*, *ibid.*, 1966, **88**, 3011). More than one nuclear hydrogen can be replaced and by using an excess of the alkylating agent hexa-alkylbenzenes have been made. In general, however, the use of this method for preparing poly-alkylbenzenes is limited by the uncertainty of the orientation of the groups, and the reaction is of most use for the preparation of mono-alkylbenzenes. A further disadvantage is that the alkyl groups from propyl upwards may suffer isomerization under the conditions of the reaction. The same catalyst which promotes the alkylation reaction also facilitates isomerization, trans-alkylation, dealkylation, polymerization, cracking, etc., so that it is not surprising that the composition of the alkylation product varies widely depending on the reaction conditions. Indeed it is seldom possible to prepare pure alkylbenzenes containing more than two carbon atoms in the side chain by the Friedel-Crafts reaction (*V. G. Plyusnin et al.*, Zhur. fiz. Khim., 1960, **34**, 726; *R. H. Allen*, and *L. D. Yats*, J. Amer. chem. Soc., 1961, **83**, 2799; *S. H. Sharman*, *ibid.*, 1962, **84**, 2945; 2951; *K. L. Marsi* and *S. H. Wilen*, J. chem. Educ., 1963, **40**, 214; *G. Geiseler et al.*, Ber., 1965, **98**, 1695; *B. V. Ioffe* and *B. V. Stolyarov*, Doklady Akad. Nauk SSSR., 1965, **161**, 1339; *P. G. Nield*, J. chem. Soc., 1964, 2278; C 1966, 712). The alkylbenzenes are usually alkylated more readily and under milder conditions than benzene itself (*F. E. Condon*, J. Amer. chem. Soc., 1948, **70**, 2265). The alkyl groups already in the nucleus do not exert a strong directive influence on the orientation (*A. W. Francis*, Chem. Reviews, 1948, **43**, 257; *Olah et al.*, J. Amer. chem. Soc., 1964, **86**, 1046). However, when alkylating a di- or poly-alkylbenzene the reaction may be influenced sterically. Thus a *tert*-butyl blocking group has been used in the synthesis of 1,2-dialkyl- and 1,2,3-trialkyl-benzenes (*M. J. Schlatter*, Amer. chem. Soc., Div. Pet. Chem. Symposium, 1955, No. 35-S, 79). Steric hindrance of the alkylation reaction can also be used to recover certain poly-alkylbenzenes. For example, 1,3,5-trimethylbenzene in a mixture of C_9 aromatic isomers remains unchanged when treated in the presence of aluminium chloride, with a C_4-C_6 tertiary olefin, and can be separated from the alkylated isomers by distillation (*W. G. De Pierri Jr. et al.*, U.S.P., 3,052,741/1962).

The alkylation of aromatic hydrocarbons with alkyl halides was discovered by *C. Friedel* and *J. M. Crafts* (Compt. rend., 1877, **84**, 1392; 1450). It is easy to produce alkylbenzenes by alkylation with alkyl halides and much information is available on their use as alkylating agents. In general, reaction occurs most readily with tertiary halides and least readily with primary halides. The order of activity with respect to the halide, using aluminium chloride as catalyst, is fluorine > chlorine > bromine > iodine (*N. O. Calloway*, J. Amer. chem. Soc., 1937, **59**, 1474). Nickel has been used as an alkylation catalyst with *tert*-butyl chloride (*W. A. Bonner* and *R. A. Grimm*, J. org. Chem., 1966, **31**, 4304).

Olefins combine additively with aromatic hydrocarbons in the presence of Friedel-Crafts catalyst to give alkylbenzenes, Aluminium chloride brings about the combination of ethylene with benzene yielding mono- up to hexa-ethyl-benzenes (*M. Balsohn*, Bull. Soc. chim. France, 1879, [II], **31**, 539; *E. E. Reid et al.*, J. Amer. chem. Soc., 1922, **44**, 206; 1927, **49**, 3142; 1938, **60**, 2606; J. org. Chem., 1944, **9**, 13; *Plyusnin* and *T. I. Sukhorosova*, Tr. Inst. Khim., Akad. Nauk SSSR., Ural. Filial, 1960, No. 4, 21; *T. Tecza et al.*, Brennstoff-

Chem., 1966, **47**, 97). Ethylation of arenes with strong acid catalysts produces high yields of single isomers (*D. A. McCaulay* and *A. P. Lien*, J. Amer. chem. Soc., 1955, **77**, 1803). The homologues of ethylene react more readily, propylene giving isopropyl derivatives, and isobutylene, *tert*-butyl derivatives (*W. M. Potts* and *L. L. Carpenter*, *ibid.*, 1939, **61**, 663). By proper choice of catalyst, alkylation of aromatic hydrocarbons with di-isobutylene can be made to produce either octyl derivatives or *tert*-butyl derivatives by *depolyalkylation* (*V. N. Ipatieff* and *H. Pines*, *ibid.*, 1936, **58**, 1056; *R. J. Lee et al.*, Ind. Eng. Chem., 1958, **50**, 1001; *R. A. Sanford et al.*, *ibid.*, 1959, **51**, 1455). Concentrated sulphuric acid is a useful reagent for the introduction of the higher alkyl groups into the nucleus (*Ipatieff et al.*, J. Amer. chem. Soc., 1936, **58**, 919; J. org. Chem., 1940, **5**, 253; *A. Newton*, J. Amer. chem. Soc., 1943, **65**, 320; *Yu G. Mamedeliev et al.*, Uchenye Zapiski Azerbaidzhan Gosudarst. Univ. im. S. M. Kirova, Ser. Fiz.-Mat. i Khim. Nauk, 1956, No. 7, 23; *A. C. Olson*, Ind. Eng. Chem., 1960, **52**, 833; *E. S. Pokrovskaya* and *N. A. Shimanko*, Neftekhimiya, 1962, **2**, 657). Hydrogen fluoride (*J. H. Simons* and *S. Archer*, J. Amer. chem. Soc., 1938, **60**, 2952), and crystalline aluminosilicates of rare earth elements (*Kh. M. Minachev* and *Ya. I. Isakov*, Doklady Akad. Nauk SSSR., 1966, **170**, 99) have also been used in this reaction.

The alkylation of aromatic hydrocarbons with paraffins was pioneered by *Ipatieff* and coworkers (J. Amer. chem. Soc., 1935, **57**, 2415; 1936, **58**, 918; J. org. Chem., 1938, **3**, 137; 448). By treatment of the paraffins under suitable conditions with Friedel-Crafts type or silica-alumina catalysts, carbonium ions are generated which then alkylate the aromatic hydrocarbon. Isomerization and cracking of the paraffin also occur and the reaction, *destructive alkylation*, normally produces a broad spectrum of alkylbenzenes (*A. A. Buniyat-Zade et al.*, Uchenye Zapiski Azerbaidzhan-Gosudarst. Univ. im. S. M. Kirova, Ser. Fiz.-Mat. i Khim. Nauk, 1960, No. 1, 91; *I. M. Tolchinskii et al.*, Izvest. Akad. Nauk, SSSR., Otdel khim. Nauk, 1955, 512). The reaction is facilitated by the presence of hydrogen acceptors (*Condon* and *M. P. Matuszak*, J. Amer. chem. Soc., 1948, **70**, 2539; *J. T. Kelly* and *Lee*, Ind. Eng. Chem., 1955, **47**, 757; *A. Rieche* and *H. Seeboth*, Angew. Chem., 1963, **75**, 861). Thus *tert*-butylarenes have been made in good yield by this reaction, *dehydroalkylation*, using isobutane:

$$Me_3CH + C_6H_5Me + CH_3{\cdot}CHMe{\cdot}CH{=}CHMe \xrightarrow[\text{acid catalyst}]{10^\circ} p\text{-}MeC_6H_4Bu^t + CH_3{\cdot}CHMe{\cdot}CH_2{\cdot}CH_2{\cdot}CH_3$$

Alcohols and ethers have also been used successfully to prepare alkylbenzenes (*Olah*, *op. cit.*, p. 477; *R. L. Burwell* and *L. M. Elkin*, J. Amer. chem. Soc., 1951, **73**, 502; *A. I. Nogaideli* and *N. N. Skhirtladze*, Zhur. obshchei Khim., 1963, **33**, 1414). A series of primary, secondary and tertiary alcohols have been used to alkylate benzene (*R. C. Huston et al.*, J. org. Chem., 1941, **6**, 252; J. Amer. chem. Soc., 1942, **64**, 1576; *Yu. K. Yur'ev* and *M. N. Savosina*, Zhur. obshchei Khim., 1959, **29**, 432; *I. Romadane et al.*, *ibid.*, 1959, **29**, 103; 1960, **30**, 420; *G. Kh. Khakimov* and *I. P. Tsukervanik*, *ibid.*, 1963, **33**, 493).

With organic esters, aralkyl ketones as well as alkylbenzenes are usually formed the proportion depending on the reaction conditions (*E. Bowden*, J. Amer. chem. Soc., 1938, **60**, 645; *D. N. Kursanov* and *R. R. Zelvin*, J. gen. Chem. [U.S.S.R.], 1939, **9**, 2173; *Simons et al.*, J. Amer. chem. Soc., 1939, **61**, 1821; *J. F. Norris* and *P. Arthur*, *ibid.*, 1940, **62**, 874):

$$RCO_2R' + 2\ ArH \rightarrow RCOAr + R'Ar + H_2O$$

Many alkyl esters of inorganic acids are known which react with aromatic substrates in the presence of Lewis acids and yield alkylates as the major product (*F. A. Drahowzal*, in *Olah*, *op. cit.*, p. 641). Thus 90% yields of *tert*-butylxylene result from tri-isobutyl borate and benzene (*A. Kaufmann*, G.P., 555,403/1930; *V. K. Kushov* and *B. M. Sheiman*, Doklady. Akad. Nauk, SSSR., 1956, **106**, 479):

$$(RO)_3B + 3\ ArH + AlCl_3 \rightarrow 3\ RAr + AlBO_3 + 3\ HCl$$

The reaction has also been applied successfully to dialkyl sulphates R_2SO_4, dialkyl sulphites R_2SO_3, alkyl phosphates R_3PO_4, alkyl orthosilicates $(RO)_4Si$, alkyl hypochlorites ROCl, alkyl chlorosulphonates $ROSO_2Cl$, alkyl chlorosulphites ROSOCl, alkyl arenesulphonates $ArSO_3R$, and alkyl perchlorates $RClO_4$.

Alkylation of the aromatic ring has been detected in the diazotization of isopropylamine with butyl nitrite and an aliphatic carboxylic acid in an excess of a mononuclear aromatic hydrocarbon (*D. E. Pearson et al.*, J. Amer. chem. Soc., 1964, **86**, 5054). Alkylation of benzene and toluene by diazotization of amines with nitrosonium salts in nitromethane gives relative reactivities in good agreement with data reported for the more conventional Friedel-Crafts type alkylations (*Olah et al.*, *ibid.*, 1965, **87**, 5785).

Alkylation of aromatic hydrocarbons is an important petrochemical process for the commercial production of alkylbenzenes (p. 225).

(b) Dealkylation, isomerization, and disproportionation of aromatic hydrocarbons

Friedel-Crafts type catalysts are not only employed for the introduction of alkyl groups, they can also be used to remove (dealkylation), rearrange (isomerization), and transfer (transalkylation or disproportionation) the alkyl group(s) in an alkylbenzene (*R. M. Roberts et al.*, J. Amer. chem. Soc., 1955, **77**, 1764; *D. A. McCaulay*, in *Olah*, *op. cit.*, Vol. 2, Part II, p. 1049). Use is made of these reactions to prepare and purify alkylbenzenes. Processes for the interconversion of the isomeric C_8 aromatic hydrocarbons based on such reactions, are especially important because there is a wide variation in demand for these chemical intermediates.

(*i*) *Dealkylation*

It has long been known that alkylbenzenes can be *dealkylated* by the action of aluminium chloride (*O. Jacobsen*, Ber., 1885, **18**, 338; *R. Anschutz* and *H. Immendorf*, *ibid.*, p. 657; *Friedel* and *Crafts*, Compt. rend., 1885, **100**, 692). In general Friedel-Crafts type catalysts employed in alkylation can be used for dealkylation but the reaction is usually accompanied by much disproportionation unless certain precautions are taken. For example, the vapour phase demethylation of *o*-xylene to toluene using a silica-alumina catalyst is much improved by the presence of benzene (*A. V. Topchiev et al.*, Invest. Akad. Nauk SSSR., Otdel. khim. Nauk, 1956, 1390). The order of susceptibility of the alkyl side chain to dealkylation is tertiary > secondary > primary > methyl (*Roberts et al.*, J. Amer. chem. Soc., 1963, **85**, 3454). Polyalkylbenzenes are dealkylated in varying degrees according the reaction conditions. Hexamethylbenzene heated at 200° with aluminium chloride gives methyl chloride and a mixture of hydrocarbons: pentamethylbenzene, durene, isodurene, trimethylbenzenes, xylenes, and small amounts of benzene and toluene. If a stream of dry hydrogen chloride is passed through the mixture the dealkylation is easier and more complete (*Jacobson*, *loc. cit.*). Mixed polyethylbenzenes (b.r. 175–350°) at 100° with aluminium chloride yield ethylbenzene (*M. M. Movsumzade* and *L. S. Dedusenko*, Azerb. Khim. Zhur., 1961, No. 4, 53). Fragmentation dealkylation reactions are known in which lower alkylbenzenes are produced together with alkanes with fewer carbon atoms than the original alkyl side chains (*Roberts et al.*, *loc. cit.*).

(*ii*) *Hydrodealkylation*

Dealkylation proceeds without a catalyst at high temperature, but the presence of a substantial molar excess of hydrogen is necessary to minimize the formation of polycyclic products. Thus toluene gives up to 85 % benzene at 700–790° and 40 atmospheres hydrogen (*A. Z. Dorogochinskii et al.*, Neftekhimiya, 1961, **1**, 46; 501; *M. J. Fowle* and *P. M. Pitts*, Chem. Eng. Progr., 1962, **58**, No. 4, 37; *C. C. Zimmerman* and *R. York*, Ind. Eng. Chem., Process Design Develop., 1964, **3**, No. 3, 254). The loss of the first methyl group from xylene proceeds readily at 460–560° (*M. C. Gonikberg et al.*, Izvest. Akad. Nauk, SSSR., Otdel. khim. Nauk, 1961, 1711). Mesitylene dealkylates stepwise to benzene at 527° (*S. E. Shull* and *A. N. Hixson*, Ind. Eng. Chem., Process Design Develop., 1966, **5**, No. 2, 146). Dealkylation with hydrogen, or *hydrodealkylation* as it is called, proceeds cleanly in the presence of a hydrogenation catalyst with very little undesirable by-products. Many commercial processes for making alkylbenzenes are based on this method of preparation (p. 224). Toluene is readily hydrodealkylated catalytically to benzene at 550–580° (*G. J. F. Stijntjes et al.*, Erdöl u. Kohle, 1961, **14**, 1011; *A. S. Sultanov et al.*, Neftekhimiya, 1965, **5**, 187), *tert*-butylbenzene to benzene at 350° (*Pines* and *W. S. Postl*, J. Amer. chem. Soc., 1957, **79**, 1769), *n*-amylbenzene to benzene, toluene and ethylbenzene at 490° (*Kh. Dimitrov* and *R. Pelova*, Kinetika i Kataliz, 1966, **7**, 116), xylenes to toluene at 520° (*A. A. Krichko* and *L. S. Sovetova*, Izvest. Akad. Nauk SSSR., Otdel. khim. Nauk, 1961, 1704) and trimethylbenzenes to benzene, and/or toluene, and xylenes at 520–600° (*idem*, Tr. Inst. Goryuch. Iskop., Akad. Nauk SSSR., 1962, **17**, 246; Kinetika i Kataliz, 1962, **3**, 399; *A. D. Sulimov et al.*, C.A. 1964, **60**, 2805b). The hydrodealkylation of C_9 alkylbenzenes with various catalysts has been studied by *V. Berti* and coworkers (Riv. Combust., 1966, **20**, [1], 3). The hydrogen can be replaced effectively by steam in the catalytic hydrodealkylation reaction (*A. A. Balandin et al.*, Vestnik Moskov. Univ., Ser. Mat., Mekh., Astron., Fiz. i Khim., 1957, **12**, No. 1, 101; *G. N. Maslyanskii et al.*, Neftekhimiya, 1962, **2**, 709; 1964, **4**, 421; 426; Petroleum Chem., 1966, **5**, 112), or both steam and hydrogen can be used together (*T. F. Doumani*, Ind. Eng. Chem., 1958, **50**, 1677; *F. H. Senbold*, U.S.P., 2,960,545/1960). For a review of both catalytic and thermal hydrodealkylation see *L. C. Doelp et al.*, Ind. Eng. Chem., Process Design Develop., 1965, **4**, No. 1, 92.

(*iii*) *Isomerization*

In the presence of Friedel-Crafts catalysts the alkyl group(s) of alkylbenzenes can undergo nuclear *isomerization* by intramolecular transfer, and/or side chain isomerization by internal rearrangement of the alkyl substituent. In the reaction of xylenes with metal halides nuclear isomerization proceeds more readily than disproportionation to give an equilibrium mixture (16 % *o*; 60 % *m*; 24 % *p* at 27°) of the three isomeric xylenes (*N. I. Shuikin et al.*, Izvest. Akad. Nauk SSSR., Otdel. khim. Nauk, 1955, 181; *R. H. Allen* and *L. D. Yats*, J. Amer. chem. Soc., 1959, **81**, 5289). A similar transformation is also effected with triethyl phosphate (*S. Tsutsumi* and *C. Matsumoto*, Technol. Reports Osaka Univ., 1960, **10**, 505). The isomerization of xylenes at 500° in the presence of steam over a silica-alumina cracking catalyst has been studied by *T. Amemiya* and coworkers (Bull. Japan Petrol. Inst., 1961, **3**, 14), and without steam by *K. L. Hauson* and *A. J. Engel* (Amer. Inst. chem. Engineers *J.*, 1967, **13**, 260). The aluminium chloride-catalyzed isomerization of diethyl-, diisopropyl-benzenes and *tert*-butyltoluene is described by *Olah* and coworkers (J. org. Chem., 1964, **29**, 2310; 2313; 2315). It has been shown that the basicity of alkylbenzenes and the stability of the complexes formed between the Friedel-Crafts catalyst and these aromatics, increases with the number of alkyl groups on the ring and is highest with the *meta* derivatives (*D. A. McCaulay* and *A. P. Lien*, J. Amer. chem. Soc., 1951, **73**, 2013; 1952, **74**, 6246; 1955, **77**, 1803). The stability of the complexes is also a function of the acid strength of the catalyst (*McCaulay*, in *Olah*, *op. cit.*, Vol. 2, Part II, p. 1049). Thus *p*-xylene and *o*-xylene can each be isomerized to *m*-xylene in the presence of Friedel-Crafts catalysts, without disproportionation, and the isomerization can be shifted beyond thermodynamic equilibrium by an excess of the catalyst due to selective complex formation. The isomerization of xylenes to give an equilibrium mixture is important in the preparation of *p*-xylene, a raw material for producing terephthalic acid. The preference for *meta* orientation in greater than equilibrium amounts in the presence of excess catalyst is of practical importance in the preparation of certain alkylbenzenes, *e.g.*, mesitylene.

Dual function catalysts, or catforming catalysts, which promote both the carbonium ion type reaction and hydrogenation-dehydrogenation reactions, have been applied with advantage to the interconversion of C_8 and C_9 isomeric alkylbenzenes (*P. M. Pitts et al.*, Ind. Eng. Chem., 1955, **47**, 770; *I. I. Eru et al.*, Koks i Khim., 1960, No. 3, 51).

Alkylbenzenes with alkyl groups having three or more carbon atoms are rearranged in their side chains by Friedel-Crafts catalysts (*G. L. Hervert*, U.S.P., 2,830,103/1958; *C. D. Nenitzescu et al.*, Ber., 1959, **92**, 10; *Roberts et al.*, J. Amer. chem., 1959, **81**, 640; 1963, **85**, 1168; J. org. Chem., 1963, **28**, 1225; 1229; 1964, **29**, 1511). Side chain isomerization has also been achieved by free radical rearrangement of alkylbenzenes using free radical initiators and halogen- or sulphur-containing promotors (*L. H. Slaugh* and *J. H. Raley*, J. Amer. chem. Soc., 1960, **82**, 1259; 1962, **84**, 2640).

The Jacobsen reaction, in which tetra-alkylbenzenes are isomerized (and penta-alkylbenzenes disproportionated) by sulphuric acid to 1,2,3,4-tetra-alkylbenzene, is a special case of conversion to a single isomeric product, the equilibrium being shifted towards the formation of the most stable sulphonic acid (*p.* 210). Decomposition is appreciable when the side chain is long (*H. Suzuki* and *R. Goto*, Nippon Kagaku Zasshi, 1963, **84**, 435).

(*iv*) *Disproportionation*

Intermolecular alkyl group transfer, or *disproportionation*, is a predominant reaction of alkylbenzenes in the presence of acid type catalysts (*D. V. Nightingale*, Chem. Reviews, 1939, **25**, 329). In general isomerization, alkylation, and dealkylation all play a part in a disproportionation reaction which proceeds stepwise to give a mixture of hydrocarbons, some with fewer and some with more alkyl groups than the original alkylbenzene(s). A large number of species can be involved in the reaction which is usually slower than isomerization. Such interconversions are used to prepare certain alkylbenzenes and are of potential importance in the petrochemical industry for their commercial production. The monoalkylbenzenes disproportionate in the presence of Friedel-Crafts catalysts to give benzene and a mixture of polyalkylbenzenes which may include all possible products up to the hexa-alkylbenzene depending on the degree of equilibrium attained. Toluene in the presence of aluminium bromide – hydrogen bromide or boron trifluoride – hydrogen fluoride readily disproportionates to benzene, xylenes, and varying amounts of C_9 and C_{10} aromatics depending on the temperature and time of reaction (*A. Schriesheim*, J. org. Chem., 1961, **26**, 3550; *McCaulay et al.*, U.S.P., 3,006,977/1960; 3,009,004/1961), *m*-xylene in the presence of aluminosilicate catalysts at 450° gives mainly toluene and trimethylbenzenes (*G. M. Mamedaliev et al.*, Khim. i Tekhnol. Topliv i Masel, 1966, **11**, No. 6, 10), ethylbenzene heated with boron trifluoride – hydrogen fluoride at 11–80° or with aluminium silicate at 150–300° gives benzene, diethylbenzenes, and 1,3,5-triethylbenzene (*B. G. Gavrilov* and *M. M. Visnevskaya*, Uchenye Zapiski Leningrad, Gosudarst. Univ. im. A. A. Zhdanova No. 211, Ser. Khim. Nauk, 1957, No. 15, 163; *McCaulay* and *Lien*, J. Amer. chem. Soc.,

1957, **79**, 5953), *n*-butylbenzene in the presence of aluminium chloride forms benzene and di-*n*-butylbenzene (*R. E. Kinney* and *L. A. Hamilton*, *ibid.*, 1954, **76**, 786; *Roberts et al.*, *ibid.*, 1958, **80**, 2507), while octyl- and octadecyl-benzenes with aluminium chloride give isomeric dioctyl- and di-(octadecyl)-benzenes, respectively (*V. Vîntu* and *M. Popescu*, C.A. 1960, **54**, 24466b). Di- and poly-alkylbenzenes in the presence of acid type catalysts disproportionate in a similar manner to give a broad spectrum of alkylbenzenes (*V. V. Tishchenko* and *S. G. Chepurina*, Uchenye Zapiski Leningrad, Gosudarst, Univ. Im. A. A. Zhdanova No. 211, Ser. Khim. Nauk, 1957, No. 15, 155; *I. O. Delone et al.*, Neftekhimiya, 1962, **2**, 189). A higher proportion of a desirable component in the product may be obtained either by choice of reaction conditions and type of catalyst, or by reaction with more than one alkylbenzene. Undesirable products may be repressed by addition of the proper concentration of certain alkylbenzenes and/or benzene to the feedstock; for example, *tert*-butylbenzene is formed in 85% yield from benzene and *p*-di-*tert*-butylbenzene in the presence of ferric chloride (*V. N. Ipatieff* and *B. B. Corson*, J. Amer. chem. Soc., 1937, **59**, 1417), ethylbenzene and xylenes interact in the presence of boron trifluoride – hydrogen fluoride to give benzene and 1,3,5-dimethylethylbenzene (*McCaulay et al.*, *ibid.*, 1957, **79**, 5808; *M. C. Hoff*, *ibid.*, 1958, **80**, 6046), trimethylbenzenes and toluene with aluminium chloride or silica-alumina catalysts give mainly xylenes (*A. V. Topchiev et al.*, Doklady Akad. Nauk, SSSR., 1957, **112**, 1071; 1957, **117**, 1007; *C. Holszky et al.*, Acad. Rep. Populare Romine Studii Cercetari Chim., 1962, **10**, 385) and mixed polyisopropylbenzenes, cumene, and benzene interact with aluminium silicate or silico-tungstic acid to yield cumene (*A. de Keizer*, Dutch Pat., 85204/1957; *J. M. Oelderik* and *H. I. Waterman*, Brennstoff-Chem., 1959, **40**, 13). A general method for preparing trialkylbenzenes which involves treating a polyalkylbenzene of 4–6 alkyl groups with an alkyl acceptor, *e.g.*, benzene or toluene, in the presence of HF/BF_3, is described by *Lien* and *McCauley* (U.S.P., 2,795,630/1957). Data on the thermodynamic properties of the methylbenzenes up to hexamethylbenzene have been calculated and can be applied in designing a reaction to increase the yield of a desirable component (*C. J. Egan*, J. chem. Eng. Data, 1960, **5**, 298; *S. H. Hastings* and *D. E. Nicholson*, *ibid.*, 1961, **6**, 1).

(c) Side-chain alkylation of aromatic hydrocarbons

Whilst the acid-catalyzed reaction of olefins with aromatic hydrocarbons results in nuclear alkylation, base-catalyzed reaction results in side-chain alkylation as long as benzylic hydrogens are available, and is a useful method for enlarging the alkyl group of an alkylbenzene (*H. Pines*, Advances in Catalysis, 1960, **12**, 117). Ethylene, propylene, and to a lesser extent butylenes and higher olefins react with the benzylic hydrogens of alkylbenzenes at 100–300° in the presence of basic catalysts such as alkali metal hydrides (*Pines* and *V. Mark*, J. Amer. chem. Soc., 1956, **78**, 4316; *S. Voltz*,

J. org. Chem., 1957, **22**, 48), alkali metals (*R. M. Schramm* and *G. E. Langlois*, J. Amer. chem. Soc., 1960, **82**, 4912), organo sodium compounds (*Pines et al.*, *ibid.*, 1955, **77**, 554; 1957, **79**, 4967; 1958, **80**, 3076; *R. D. Closson et al.*, J. org. Chem., 1957, **22**, 646) and potassium graphite (*H. Podall* and *W. E. Foster*, *ibid.*, 1958, **23**, 401) to yield the corresponding alkylbenzene. Disubstitution on the α-carbon atom may also occur if there are two benzylic hydrogens:

$$Ar\cdot\overset{R}{\underset{R^1}{CH}} + CH_2{=}\overset{R^2}{\underset{R^3}{C}} \rightarrow Ar\cdot\overset{R}{\underset{R^1}{C}}-\overset{R^2}{\underset{R^3}{C}}\cdot CH_3$$

The acidity of the benzylic hydrogen decreases with increasing alkyl substitution and metalation of the ring becomes competitive (*H. Hart* and *R. E. Crocker*, J. Amer. chem. Soc., 1960, **82**, 418). In the reaction of toluene with sodium alkyls both nuclear and side chain alkylation occurs (*Vîntu et al.*, C.A., 1964, **61**, 595d).

A benzylic carbanion is probably formed by reaction of the aromatic hydrocarbon with the catalyst which adds to the olefin, and the resultant carbanion is then involved in a transmetalation reaction with more of the aromatic, thus perpetuating a chain reaction (*Hart*, J. Amer. chem. Soc., 1956, **78**, 2619):

$$Ar\overset{R}{\underset{R^1}{C}}Na^{\oplus} + CH_2{=}CH_2 \longrightarrow Ar\overset{R}{\underset{R^1}{C}}\cdot CH_2\cdot C^{\ominus}H_2Na^{\oplus} \xrightarrow{ArCHRR^1} Ar\overset{R}{\underset{R^1}{C}}\cdot CH_2\cdot CH_3 + Ar\overset{R}{\underset{R^1}{C}}Na^{\oplus}$$

Sodium catalysts are more selective than potassium catalysts, the latter catalyzed alkylations yielding indane as well as the expected alkylbenzene. This is brought about by an intramolecular alkylation of the aromatic ring by a carbanion. Lithium compounds are not very effective catalysts.

Long chain phenylalkanes can be made from aromatic hydrocarbons by the telomerization reaction of ethylene with aluminium triaralkyl compounds (*Esso Research & Engineering Co.*, B.P., 876,536/1959). The primary telomerization products obtained from benzene and ethylene using an organolithium compound as catalyst, were straight chain, even-numbered 1-phenyl-*n*-alkanes. From toluene the product consisted of essentially odd-numbered 1-phenyl-*n*-alkanes (*G. E. Eberhardt* and *W. A. Butte*, J. org. Chem., 1964, **29**, 2928).

Alkylbenzenes are also alkylated by olefins at high temperature and pressure by a free radical mechanism (*Pines et al.*, J. Amer. chem. Soc., 1957, **79**, 4958;

1959, **81**, 3629; Chem. Age. Ind., 1961, **12**, 91). The reaction is enhanced by the presence of free radicals but the alkylation is usually accompanied by extensive cracking, isomerization and polymerization of the olefin, rearrangement of the alkylation products, and in some cases nuclear alkylation.

Alkylbenzenes are *alkenylated* in the side-chain alkali-metal catalyzed reaction with isoprene (*F. Hoffman* and *A. Michael*, U.S.P., 2,448,641/1928; *Pines* and *N.C. Sih*, J. org. Chem., 1965, **30**, 280), and the alkenyl compound hydrogenated to the alkylbenzene.

(d) Dehydrogenation of saturated and partially saturated acidic and alicyclic hydrocarbons

Alicyclic hydrocarbons can be dehydrogenated at high temperatures to give the corresponding aromatic hydrocarbon. Thus cyclohexane has been dehydrogenated homogeneously to benzene, and methylcyclohexane to toluene, by heating (500–800°) in the presence of hydrogen (*P. J. Owen et al.*, Ind. Eng. Chem., 1961, **53**, 10; *A. C. Reeve* J. appl. Chem., 1963, **13**, 403). Quantitative conversion of cyclohexanes to aromatic compounds has been achieved in the presence of activated carbon at 600° (*N. I. Shuĭkin* and *T. I. Naryshkina*, Doklady Akad. Nauk, SSSR., 1961, **136**, 849). This is a principal reaction in the thermal reforming of petroleum naphthas into high octane petrols. Moderate yields of benzene are obtained if cyclohexene is reacted with hexabromoethane in quinoline at 150° (*N. A. Domnin* and *V. A. Cherkasova*, Zhur. obshch. Khim., 1958, **28**, 2334). The isomerization of 1,2,4-trivinylcyclohexane to 1,2,4-triethylbenzene in 72 % yield is a unique example of the dehydrogenation of the cyclohexane ring (*Chem. werke Hüls*, B.P., 930,054/1963).

(i) Pyrolysis of coal and other hydrocarbons

The destructive distillation of coal at a relatively low temperature, about 600°, yields a condensate containing no aromatic hydrocarbons and is in complete contrast to high temperature distillation (p. 221). This suggests that the aromatic hydrocarbons are derived from aliphatic and alicyclic hydrocarbons under conditions of high temperature and exposure to the surface action of hot coke and the walls of the retorts. This view is supported by the fact that aromatic hydrocarbons are formed when aliphatic hydrocarbons are heated to high temperatures. Indeed benzene was first isolated from the condensate of a compressed gas from the pyrolysis of fish oil (*M. Faraday*, Phil. Trans., 1825, p. 440). The pyrolysis of methane and natural gases rich in methane is known to give benzene and its homologues (*W. H. Cadman*, Ind. Eng. Chem., 1934, **26**, 315; *A. E. Dunstan et al.*, *ibid.*, 1934, **26**, 307; *P. K. Frohlich* and *P. J. Wiezevich*, *ibid.*, 1935, **27**, 1055). Yields of benzene up to 44 % have been obtained from propane and butane at temperatures above 900° (*K. H. Schmidt*, Brennstoff-Chem., 1956,

37, 175). However, the pyrolytic formation of arenes from aliphatic hydrocarbons is not a simple dehydrogenation-cyclization reaction. Degradation of the hydrocarbon to smaller unsaturated units followed by resynthesis is the main course of the reaction (*G. M. Badger et al.*, Austral. J. Chem., 1962, **15**, 605; *Owen et al.*, *loc. cit.*; *Reeve* and *R. Long*, J. appl. Chem., 1963, **13**, 176).

Pyrolysis of the lower olefins (*R. V. Wheeler* and *W. Wood*, J. chem. Soc., 1930, 1819), butadiene (*H. Staudinger et al.*, Ber., 1913, **46**, 2466; *E. Gil-Av et al.*, J. chem. Eng. Data, 1960, **5**, 98), mixtures of butadiene with ethylene, propylene, 1- or 2-butene, or trimethylethylene (*E. M. Tarasenkova*, C.A., 1959, **53**, 4710d), and acetylene (*M. Berthelot*, Ann. Chim. Phys., 1866, [iv], **9**, 445; *N. D. Zelinskii*, Ber., 1924, **57**, 264; cf. *P. Sabatier* and *J. B. Senderens*, Compt. rend., 1900, **130**, 250; *P. Pascal* and *C. Coupard*, *ibid.*, 1942, **214**,, 757), leads to aromatization with the formation of mononuclear aromatics.

A survey of the pyrolytic formation of arenes from hydrocarbons, and the various hypotheses proposed to explain their formation, has been made by *C. D. Hurd* and coworkers (J. Amer. chem. Soc., 1962, **84**, 4509). In general, non-catalytic pyrolysis of aliphatic hydrocarbons as a preparative route to the alkylbenzenes is not attractive. The degradation reaction may be considerably depressed if the thermal reaction is carried out in the presence of one or more molecules of iodine at lower temperatures (*R. D. Mullineaux* and *J. H. Raley*, *ibid.*, 1963, **85**, 3178; *V. P. Musienko* and *A. P. Rezuik*, C.A., 1968, **69**, 18719a). Thus 75 % of *n*-hexane is converted to benzene at 500° while cyclohexane is dehydrogenated almost exclusively to benzene. An analogous reaction is the formation of alkylbenzenes from paraffins and sulphur dioxide at 500° (*V. J. Frilette* and *G. W. Munns*, Chim. Ind. [Paris], 1962, **88**, 487).

(*ii*) *Dehydrocyclization*

Hydrogenation catalysts, *e.g.*, chromia, molybdena, oxides of rare-earth elements, nickel, platinum, palladium, etc., enhance considerably the dehydrogenation reaction, and have been widely used as catalysts for the preparation of alkylbenzenes from alkyl-cyclohexanes and -cyclohexenes (*Zelinskii*, Ber., 1912, **45**, 3678; 1923, **56**, 787; *T. W. Reynolds et al.*, Ind. Eng. Chem., 1948, **40**, 1751; *B. B. Elsner* and *H. E. Strauss*, J. chem. Soc., 1957, 583; *Pines* and *C-T. Chen*, J. Amer. chem. Soc., 1960, **82**, 3562; *Kh. M. Minachev et al.*, Neftekhimiya, 1961, **1**, 489). Dehydrogenation begins above 250° and takes place even in the presence of excess hydrogen; below 200° reduction of the aromatic hydrocarbon occurs. The dehydrogenation reaction increases with increase in temperature and decrease in hydrogen partial pressure (*B. S. Greensfelder et al.*, Chem. Engr. Progr., 1947, **43**, 561; *A. A. Draeger et al.*, Petroleum Refiner, 1951, **30**, No. 8, 71). Benzene is an effective hydrogen acceptor for liquid phase catalytic dehydrogenation of alicyclic hydrocarbons (*H. Adkins et al.*, J. Amer. chem. Soc., 1941, **63**, 1320; 1948, **70**, 381). Aromatic hydrocarbons can also be made by the dehydrogenation of paraffinic and olefinic hydrocarbons containing six or more carbon atoms if special catalysts are employed. The reaction involves ring closure and is known as *dehydrocyclization* (*C. Hansch*, Chem. Reviews, 1953, **53**, 353).

It was first discovered by *B. L. Moldavsky* and *H. D. Kamuscher* (Compt. rend. Acad. Sci. URSS., 1936, **1**, 355), who obtained a considerable quantity of toluene by passing *n*-heptane vapours over a chromium oxide catalyst; negligible cracking occurred. In the same year *B. A. Kazansky* and *A. F. Plate* (Ber., 1936, **69**, 1862) described the aromatization of alkanes using platinum catalysts. A number of U.S. patents describe the catalytic cyclization of *n*-paraffins to the corresponding aromatics using oxides of the metals belonging to the 5th and 6th sub-groups of the periodic table (*A. V. Grosse* and *J. C. Morrell*, U.S.P., 2,124,566–7/1938; 2,124,583–6/1938) and many papers and patents on the dehydrocyclization reaction have appeared since. Reactivity of *n*-paraffins increases from C_6 to C_9 and then levels out (*H. Suzumura et al.*, C.A., 1966, **65**, 15128c). Long chain aliphatic compounds dehydrocyclize to a mixture of aromatics. Thus dodecene gives a mixture of pseudocumene, mesitylene, propylbenzene, benzene, toluene, and xylene (*A. Schmetterling* and *W. Dimmling*, U.S.P., 2,941,016/1960). Olefinic hydrocarbons containing less than six carbon atoms have been aromatized by catalytic dehydrocyclopolymerization (*A. D. Petrov et al.*, Izvest. Akad. Nauk SSSR., Ser. Khim., 1964, 1866).

(iii) Dehydroisomerization

In many reactions isomerization plays an integral part in dehydrogenation reactions and the term *dehydroisomerization* has been introduced. *n*-Octane isomerizes before dehydrocyclization at 528° with an alumina-chromia-potassium catalyst (*M. I. Rozengart et al.*, *ibid.*, 1966, 1323). Hydrocarbons having a straight chain of less than six carbon atoms, *e.g.*, 2,2,3-trimethylpentane (I), rearrange before cyclization (*E. F. G. Herington* and *E. K. Rideal*, Proc. roy. Soc., 1945, **A184**, 434; 447). Similarly alicyclic rings containing five or more than six carbon atoms, *e.g.*, ethylcyclopentane (II), undergo ring expansion or contraction (*A. I. M. Keulemans* and *H. H. Voge*, J. phys. Chem., 1959, **63**, 476; *O. E. Morozova et al.*, Neftekhimiya, 1961, **1**, 760; *D. Y. Waddan*, B.P., 982,755/1965):

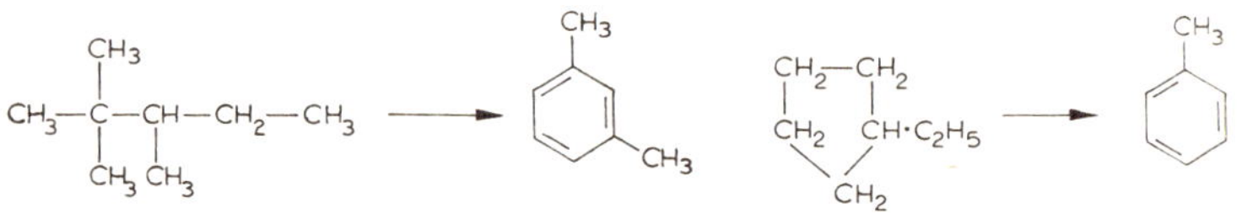

(I) (II)

For the combined dehydrogenation-dehydrocyclization-dehydroisomerization reactions to occur it is necessary for the catalyst to have both an acidity and hydrogenation-dehydrogenation function (*H. Steiner*, "Catalysis", Ed. *P. H. Emmett*, Reinhold Publishing Corpn., 1956, Vol. IV, p. 529; *F. G. Ciapetta et al.*, *ibid.*, 1958, Vol. VI, p. 495; *Pines* and *S. M. Csicsery*, J. Amer. chem. Soc., 1962, **84**, 292). Alkali metals act as promotors causing an increase in aromatization activity (*Z. A. Davydova et al.*, Azerb. Khim. Zhur., 1964, No. 6, 29; Kinetika i Kataliz, 1965, **6**, 1046; 1966, **7**, 566; Neftekhimiya, 1966, **6**, 35). The acidity function may reside in the same dehydrogenation component of the catalyst, *e.g.*, metal oxides of the Group VI metals, or in a second component, *e.g.*, in the silica-alumina component of platinum-silica-alumina type catalyst. Such dual purpose catalysts form the basis of the commercially important processes developed in the petroleum industry since World War II for the catalytic reforming of virgin and cracked naphtha stocks to produce high octane petrols and pure aromatics (p. 223). For a good review on the mechanism of the aromatization of alkanes over chromia-alumina catalysts see *Pines* and *C. T. Goetschel*, J. org. Chem., 1965, **30**, 3530.

Aliphatic alcohols have been succesfully dehydrocyclized to benzene and alkylbenzenes by *V. I. Komarewsky* and coworkers (*J. Amer. chem. Soc.*, 1939, **61**, 2525) using a chromia-alumina catalyst at 450–475^0.

Oxidation of tropylium salts with hydrogen peroxide gives rise to benzene or substituted benzenes with the same number of carbon atoms as in the parent compound (*M. E. Vol'pin et al.*, Doklady Akad. Nauk SSSR., 1959, **126**, 780; Tetrahedron, 1960, **8**, 33).

(e) Cyclotrimerization of alkynes

It is now more than a century ago since *M. Bertholet* produced benzene and other aromatic hydrocarbons by passing acetylene through a hot tube (Compt. rend., 1866, **62**, 905). The pyrolysis of acetylene has since received considerable study. Polymerization commences above 400^0 and is associated with a number of complicated side reactions resulting in a product which is a complex mixture of aromatic hydrocarbons, paraffins, olefins, hydrogen, and carbon. The maximum yield of liquid products appears to be obtained at 650–700^0; benzene is the major constituent but the alkylbenzenes, toluene, *o*-, *m*-, *p*-xylenes, ethylbenzene, pseudocumene, and mesitylene have also been identified (*J. A. Nieuwland* and *R. R. Vogt*, "The Chemistry of Acetylene", Reinhold Publishing Corpn., 1945; *G. M. Badger et al.*, J. chem. Soc., 1960, 2825). Benzene containing deuterium or tritium has been made by this method (*A. Rupprecht*, Acta Chem. Scand., 1962, **16**, 2189).

The pyrogenic cyclotrimerization of alkynes is not attractive as a method for preparing alkylbenzenes because of its complex nature but several catalysts have been reported recently which bring about the cyclotrimerization of alkynes in solution at much lower temperatures and in very good yields. *B. Franzus et al.* (J. Amer. chem. Soc., 1959, **81**, 1514), *E. F. Lutz* (*ibid.*, 1961, **83**, 2551), *K. Ziegler* (G.P., 1,233,375/1967) and *V.O. Reikhsfel'd et al.* ("Cyclotrimerization of Acetylenes, IX", Zhur. org. Khim., 1967, **3**, 270, and preceding parts) have shown that benzene and alkylbenzenes are produced from alkynes in excellent yields at temperatures between ambient and 140^0 in the presence of Ziegler type catalysts. Co-cyclotrimerization is also possible. The reaction, which may be carried out in an inert solvent can be expressed in the general form:

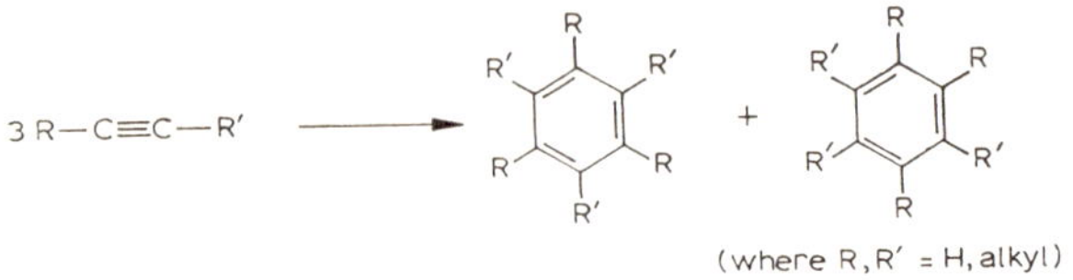

$$3\,R{-}C{\equiv}C{-}R' \longrightarrow$$

(where R, R' = H, alkyl)

This technique offers a convenient method for synthesising substituted benzenes not readily available by other methods. The novel cage compounds (IV) are made by internal cyclotrimerization of the acetylenes (III) in the presence of Ziegler catalysts (*A. J. Hubert* and *M. Hubert*, Tetrahedron Letters, 1966, 5779):

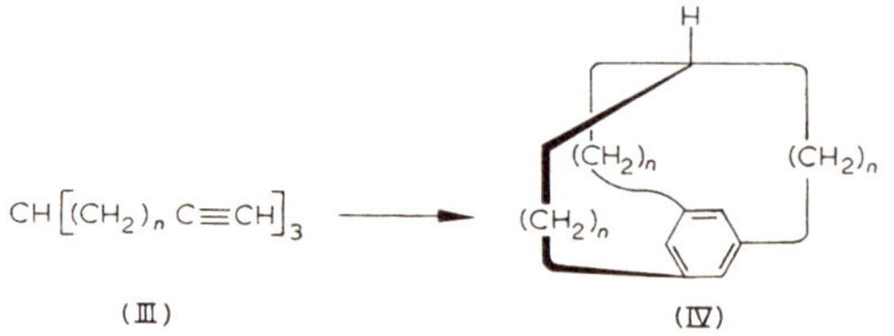

$$CH[(CH_2)_n C{\equiv}CH]_3 \longrightarrow$$

(III) (IV)

The cyclotrimerization reaction is also catalyzed by: alumina promotors containing vanadium pentoxide or molybdenum trioxide-cobaltous oxide mixtures (*J. E. Noakes*, U.S.P., 3,365,510/1968), chromium trioxide on silica-alumina (*W. C. Lanning* and *A. Clark*, U.S.P., 2,819,325/1958), diborane-treated silica-alumina (*I. Shapiro* and *H. G. Weiss*, U.S.P., 2,867,675/1959), metal carbonyl (*W. Hübel* and *C. Hoogzand*, Ber., 1960, **93**, 103; *K. W. Hübel* G.P., 1,142,867/1963), bisacrylonitrilenickel and its triphenylphosphine adduct (*G. N. Schrauzer*, Ber., 1961, **94**, 1403), dimesitylcobalt (*M. Tsutsui* and *H. Zeiss*, J. Amer. chem. Soc., 1961, **83**, 825; *P. Mauret* and *A. Gaset*, Compt. rend., 1967, **264**, 983), tungsten hexachloride and oxychloride (*Ciba*, Belg. P., 609,668/1962; C.A., 1962, **57**, 13672d), and mixed metal halides (*H. Mueller* and *H. Friederich*, B.P., 890,542/1962). [1-^{14}C]Benzene has been prepared from acetylene using triphenylphosphinenickelcarbonyl to catalyze the trimerization (*L. Pichat* and *C. Baret*, Tetrahedron, 1957, **1**, 269) and is an attractive alternative to the 6-stage process of *H. S. Turner* and *R. J. Warne* (J. chem. Soc., 1953, 789).

o-Di-*tert*-butyl-, 1,2,4-tri-*tert*-butyl-, and 1,2,4,5-tetra-*tert*-butyl-benzenes have all been made by trimerization using dicobaltoctacarbonyl complexes and are of special interest because of their highly strained structures (*Hoogzand et al.*, Angew. Chem., 1961, **73**, 680; Ber., 1961, **94**, 2817; Tetrahedron Letters, 1961, 637; *E. M. Arnett et al.*, Chem., and Ind. 1961, 2008; Tetrahedron Letters, 1961, 658; J. Amer. chem. Soc., 1967, **89**, 5389).

Dialkynes of the type $HC{\vdots}C\cdot(CH_2)_n\cdot C{\vdots}CH$ and $Me(CH_2)_n\cdot(C{\vdots}C)_2\cdot(CH_2)_nMe$, readily undergo base catalyzed aromatization to give mixtures consisting mainly of the corresponding *o*-disubstituted and monosubstituted benzenes. Thus toluene is obtained in 65% yield by refluxing 1,6-heptadiyne with 10% potassium isopropoxide in diethylene glycol dimethyl ether (*R. A. Raphael et al.*, J. chem.

Soc., 1964, 2597). Aromatic hydrocarbons with a specific structure have also been prepared from substituted divinylacetylenes by contact with chromic oxide on alumina at 300–500° (*I. L. Kotlyarevskii et al.*, Zhur. priklad. Khim., 1957, **30**, 321; 1356; 1719; Tr. Vost.-Sibirsk, Filiala Akad. Nauk SSSR., Ser. Khim., 1961, **38**, 142).

Benzene is among the products from the photosensitized polymerization of acetylene (*S. Shida et al.*, J. chem. Phys., 1958, **28**, 131).

(f) Fittig synthesis

The interaction of an aryl halide, alkyl halide and alkali metal, an extension of the Wurtz coupling reaction, is known as the Fittig reaction (*R. Fittig* and *B. Tollens*, Ann., 1864, **131**, 303). It has been used extensively for preparing alkylbenzenes (*R. R. Read et al.*, J. Amer. chem. Soc., 1926 **48**, 1606; *W. Griess*, Fette Seifen Anstrichmittel, 1955, **57**, 236; Org. Synth., 1955, Coll. Vol. III, p. 157; *H. L. Holsopple*, C. A., 1964, **60**, 4029e) and may be represented in the general form:

$$Ar(CH_2)_nX + RX \xrightarrow{2\,Na} Ar(CH_2)_nR + 2\,NaX$$

(where n may be 0)

Wurtz (Ar–Ar, R–R) and disproportionation (alkane, alkene) products may be formed as by-products in the reaction but yields of alkylbenzenes are usually good if conditions are carefully controlled. By-products can usually be separated by distillation.

It is probable that the reaction depends on the initial formation of an organosodium compound which can react with a halide to give the alkylbenzene (*A. A. Morton et al.*, J. Amer. chem. Soc., 1936, **58**, 1967; 1938, **60**, 1429; 1940, **62**, 120). Alkylbenzenes with ^{14}C in the side chain have been prepared by this method (*A. A. Korotkov* and *S. P. Mitsengendler*, Izvest. Akad. Nauk SSSR., Otdel. khim. Nauk, 1956, 1507; *N. P. Mel'nikova* and *I. A. Shakhzadova*, Khim. i Tekhnol. Topliv i Masel, 1959, **4**, No. 1, 40)

Sodium is the metal used most for the coupling reaction which is usually carried out in the liquid phase, with or without inert diluent. Highest yields are usually obtained where alkyl and aryl bromides are used (*H. A. Fahim* and *A. Mustafa*, J. chem. Soc., 1949, 519).

Although it appears to be generally assumed that the alkyl group occupies the place vacated by the halogen, it is now known that this method cannot always be relied on to give pure products (*cf. S. F. Birch et al.*, J. Amer. chem. Soc., 1949, **71**, 1362). This may be due partly to metallation of the alkyl group of the alkylbenzene by the sodium alkyl which is formed in the reaction. Indeed higher monoalkylbenzenes can be prepared by reaction between an alkyl chloride, sodium and toluene. A suspension of finely dispersed sodium in toluene gives *n*-butylbenzene on adding *n*-propyl chloride (*Morton et al.*, *ibid.*, 1936, **58**, 2599; 1941, **63**, 327; 1942, **64**, 2239). The reaction apparently occurs in the following stages:

$$PhCH_3 + Na{\cdot}C_3H_7 \rightarrow PhCH_2{\cdot}Na + C_3H_8$$

$$PhCH_2{\cdot}Na + C_3H_7Cl \rightarrow PhCH_2{\cdot}C_3H_7 + NaCl$$

(g) *Synthesis using organo-metallic compounds*

In the Fittig synthesis organo-metallic compounds are involved as intermediates but there are several synthetic routes to arylbenzenes which involve the direct use of organo-metallic substances.

(i) *Wurtz-Grignard reaction*

This reaction includes concurrent reactions of the general type:

$$RMgX + R'X \rightarrow R{\cdot}R' + MgX_2$$

where X is halogen and R is aryl (also aralkyl) or alkyl when R′ is alkyl or aryl (also aralkyl), respectively. The reaction is usually conducted in ether but non-ethereal media have been employed, *e.g.* heptane (*A. D. Petrov* and *E. P. Zakharov*, Zhur. obshchei Khim., 1957, **27**, 2990). It is not an entirely satisfactory method for preparing alkylbenzenes. As in the Fittig reaction, disproportionation and Wurtz type by-products are also produced (*M. S. Malinovskiĭ* and *A. A. Yavorovskii*, *ibid.*, 1955, **25**, 2209). However, it is sometimes useful for the synthesis of certain alkylbenzenes. Thus highly substituted benzenes can be made from polyalkylchloromethylbenzenes by this route and alkyl groups in the *ortho* position to the chloromethyl radical do not hinder the coupling (*R. C. Fuson et al.*, J. Amer. chem. Soc., 1941, **63**, 2652; 1946, **68**, 533). The reaction of benzylmagnesium halides with alkyl halides is relatively more satisfactory and many alkylbenzenes have been made using this method (*E. Spath*, Monatsh., 1913, **34**, 1965; *G. Vavon* and *P. Mottez*, Compt. rend., 1944, **218**, 557; *Petrov et al.*, Zhur. obshchei Khim., 1959, **29**, 49; Izvest. Vysshikh Ucheb. Zavedenii, Khim. i Khim. Tekhnol., 1959, **2**, 384). The Wurtz-Grignard reaction has also been used to make *n*-dialkylbenzenes from a number of alkyl bromines and the bis-Grignard reagent, *p*-$C_6H_4(MgBr)_2$ (*Petrov et al.*, Izvest. Akad. Nauk SSSR., Otdel. khim. Nauk, 1960, 717).

In general, the use of alkyl sulphates and alkyl arenesulphonates instead of alkyl halides gives improved yields in Wurtz-Grignard reactions:

$$ArMgX + R_2SO_4 \rightarrow ArR + RX + MgSO_4$$

$$ArMgX + Ar'SO_3R \rightarrow ArR + Ar'SO_3MgX$$

where Ar may also be aralkyl. This reaction has been used extensively for the preparation of some trimethyl- and ethylpolymethyl-benzenes (*L. I. Smith* and *A. P. Lund,* J. Amer. chem. Soc., 1930, **52,** 4144; *R. W. Maxwell* and *R. Adams, ibid.,* 1930, **52,** 2959; *Birch et al., ibid.,* 1949, **71,** 1362). Isodurene (*Smith,* Org. Synth., 1943, Coll. Vol. II, p. 360; *Smith* and *F. H. MacDougall,* J. Amer. chem. Soc., 1929, **51,** 3001) and other alkylbenzenes (*H. Gilman et al., ibid.,* 1922, **44,** 2621; 2969; 1925, **47,** 518; *C. E. Boord et al.,* Ind. Eng. Chem., 1949, **41,** 609) have also been made by this method.

In the reaction of benzylmagnesium halides with alkyl sulphates a small amount of isomerization may occur (*J. G. Burtle* and *R. L. Shriner,* J. Amer. chem. Soc., 1947, **69,** 2059; *E. Berliner* and *F. Berliner, ibid.,* 1949, **71,** 1196; cf. *Gilman et al., ibid.,* 1932, **54,** 345; Org. Synth., 1941, Coll. Vol. I, p. 471).

Alkyl esters of *p*-toluenesulphonic acid are usually employed in the Grignard-alkyl arenesulphonate reaction and some of the arenesulphonic ester is simultaneously converted to an alkyl halide during the reaction:

$$ArSO_3R + ArSO_3MgX \rightarrow RX + (ArSO_3)_2Mg$$

the yield of hydrocarbon is therefore increased if two moles of ester are used for every mole of Grignard reagent (*Gilman* and *J. Robinson, ibid.,* 1943, Coll. Vol. II, p. 47).

(*ii*) *Reaction of organo-metallic compounds with halides*

This method is similar to the Wurtz-Grignard synthesis and is useful for making monoalkylbenzenes:

$$ArX + RM \rightarrow ArR + MX$$

Alkali metal alkyls are normally used and the reaction often proceeds readily in an inert solvent at ambient temperatures. The metal can be on the aromatic moiety; thus butylbenzene has been made from benzylsodium and *n*-propyl bromide (*D. Bryce-Smith* and *E. E. Turner,* J. chem. Soc., 1950, 1975) and 2-methyl-2-phenylalkanes are conveniently made from α,α-dimethylbenzylpotassium and the appropriate alkyl bromide (*K. Ziegler,* Ann., 1940, **542,** 90). Possible combinations of organo-sodium compounds and organic chlorides in the synthesis of *n*-amylbenzene, have been studied (*Morton* and *F. F. Fallwell,* J. Amer. chem. Soc., 1938, **60,** 1429). As found in the Wurtz-Grignard reaction highest yields were obtained using the benzylmetal reagent instead of the phenyl one.

n-Propylbenzene has been prepared by the action of diethylzinc on benzyl chloride (*R. Schiff*, Ber., 1877, **10**, 294).

(*iii*) *Hydrolysis of organo-metallic compounds*

Alkylbenzenes may be prepared from halogen derivatives by formation of the Grignard reagent followed by hydrolysis (*C. G. Le Fevre et al.*, J. chem. Soc., 1935, 480):

$$\text{RMgX} \xrightarrow[\text{H}^{\oplus}]{\text{H}_2\text{O}} \text{RH (R = aryl, aralkyl)}$$

The reaction is of general application, gives excellent yields and is useful for introducing further methyl groups into alkylbenzenes *via* chloromethylation (p. 217).

(*h*) *Reduction of alkenylbenzenes*

This is a useful method for making pure alkylbenzenes which cannot be made by direct alkylation; the alkenylbenzenes are readily prepared by the Grignard reaction followed by dehydration of the resulting alcohol:

$$\text{ArCOMe} \xrightarrow{\text{RMgX}} \begin{matrix}\text{Ar}\\ \text{R}\end{matrix}\!\!>\!\text{C}\!<\!\!\begin{matrix}\text{Me}\\ \text{OH}\end{matrix} \longrightarrow \begin{matrix}\text{Ar}\\ \text{R}\end{matrix}\!\!>\!\text{C}=\text{CH}_2 \rightarrow \text{ArCHR}\cdot\text{CH}_3$$

Alkenylbenzenes of the general type, $\text{Ar}\cdot(\text{CH}_2)_n\cdot\text{CR}'{:}\text{CR}''\text{R}'''$ (R′, R″, R‴ = H or alkyl, n = 0, 1, 2) may be hydrogenated to the corresponding alkylbenzenes by using a not too active catalyst. Thus 2-phenyltridecane has been made from methyl iodide and phenyl undecyl ketone *via* the Grignard reagent, carbinol and olefin (*F. W. Gray et al.*, J. org. Chem., 1955, **20**, 511). Several isomeric pentylbenzenes have been prepared in this way (*B. V. Ioffe* and *V. Stolyarov*, Neftekhimiya, 1964, **4**, 361). Alternatively, the intermediate Grignard reagent may be prepared from the aryl halide and the aliphatic ketone as described for *p-sec*-butyltoluene (*Pines et al.*, J. Amer. chem. Soc., 1950, **72**, 1563). Reduction of alkenylbenzenes may also be achieved by using sodium and alcohol (*A. Klages*, Ber., 1904, **37**, 1721).

(*i*) *Reduction of aryl halides*

Nuclear and side-chain halogenated benzene derivatives can be reduced by most reducing agents which produce nascent hydrogen. Nuclear halogens are removed by nickel-aluminium alloy in aqueous alkali (*E. Schwenk et al.*, J. org. Chem., 1944, **9**, 1), nickel catalyst in water (*K. Wantanabe*, Bull. chem. Soc. Japan, 1959, **32**, 1280) and magnesium in methanol (*L. Zechmeister* and *P. Pom*, Ann., 1929, 468, 127). Catalytic hydrogenation (supported platinum or palladium) has been used to reduce side-chain halogens (*W. D. Schaeffer* and *F. H. Senbold*, U.S.P., 2,977,395/1961; *B. W. Howk*, U.S.P., 3,110,742/1963). Zinc with aqueous sodium hydroxide solution has also been used (*R. R. Aitken et al.*, J. chem. Soc., 1950, 331). This is a convenient method for introducing the methyl group into alkylbenzenes, the intermediate halogenated compound being made by chloromethylation (p. 217).

(j) Reduction of alkyl aryl ketones, aromatic aldehydes, and other oxygen-containing compounds

The carbonyl group of aldehydes and ketones can be reduced to a methylene group by a variety of methods, the most important being the Clemmensen reduction (*E. Clemmensen*, Ber., 1913, **46**, 1838; 1914, **47**, 681), the Wolff-Kishner reduction (*N. Kishner*, J. Russ. phys. Chem. Soc., 1911, **43**, 582; *L. Wolff*, Ann., 1912, **394**, 86), and catalytic hydrogenation, In general terms the reaction is represented by the following equation:

$$\mathrm{ArCOR} \rightarrow \mathrm{ArCH_2R} \quad (\mathrm{R} = \mathrm{H\ or\ alkyl})$$

The *Clemmensen reduction method* with zinc amalgam and hydrochloric acid has been used extensively for preparing alkylbenzenes and the yields in general are good (*E. J. Martin*, Org. Reactions, 1942, **1**, 155; *H. A. Fahim* and *A. Mustafa* J. chem. Soc., 1949, 519). It provides a convenient route to the poly- and higher *n*-alkylbenzenes not readily obtainable in high purity by Friedel-Crafts reactions (cf. p. 158).

The *Wolff-Kishner method* for the reduction of the carbonyl group consists in the decomposition of the hydrazones or semicarbazones by sodium alkoxides or even caustic potash (*D. Todd*, Org. Reactions, 1948, **4**, 378):

$$\mathrm{ArRC{=}NNH_2} \xrightarrow{\mathrm{KOH}} \mathrm{ArCH_2R} + \mathrm{N_2} \quad (\mathrm{R} = \mathrm{H\ or\ alkyl})$$

The conditions described by Wolff for this reduction require that the decomposition should be carried out in ethanol under pressure at about 160°. Improvements in the technique using glycol or high boiling alcohols as solvents, allow the reduction to be brought about in open vessels (*M. D. Soffer et al.*, J. Amer. chem.

Soc., 1945, **67**, 1435; *F. C. Whitmore et al., ibid.*, p. 2061; *Huang-Minlon, ibid.*, 1946, **68**, 2487; *F. W. Gray et al.*, J. org. Chem., 1955, **20**, 511).

Alkyl aryl ketones are conveniently *hydrogenated* to the corresponding alkylbenzene in excellent yield using copper-alumina and copper-chromia catalysts at 180–250° (*D. Nightingale* and *H. D. Radford, ibid.*, 1949, **14**, 1089) or palladium-charcoal catalyst in the cold (*N. D. Zelinskii et al.*, Ber., 1933, **66B**, 872; 1934, **67B**, 300; *W. H. Hurtung* and *F. S. Crossley*, J. Amer. chem. Soc., 1934, **56**, 158; *G. Shen et al.*, J. Inst. Petroleum, 1940, **26**, 514). A general method for hydrogenating aromatic alcohols, ketones, aldehydes, esters, and acid anhydrides, to benzene or the corresponding alkylbenzene, uses tungsten sulphide or ammonium tungstate as catalyst at 320–340° (*S. Landa* and *J. Mostecký*, Chem. Listy, 1955, **49**, 67; 1956, **50**, 565; Freiberg Forsch., 1955, **A36**, 58).

Phenols have been reduced to aromatic hydrocarbons by slowly heating with phosphorus sulphide to 300° (*W. N. Moulton* and *C. C. Wade*, J. org. Chem., 1961, **26**, 2528). Hydroxyl groups alpha to the aromatic nucleus are reduced by sodium in liquid ammonia (*A. J. Birch*, J. chem. Soc., 1945, 809). Thiophenol yields mainly benzene when allowed to come into contact with reduced iron oxide at 260° (*A. N. Bashkirov* and *N. L. Barabanov*, Doklady Akad. Nauk SSSR., 1955, **104**, 854).

(*k*) *Decarboxylation of carboxylic acids*

It is well over a century ago now that benzene was obtained by distillation of benzoic acid with lime (*E. Mitscherlich*, Ann., 1834, **9**, 43) and in a similar manner from phthalic anhydride (*C. Marignac, ibid.*, 1842, **42**, 217). Several alkylbenzenes may be prepared from the appropriate carboxylic acid by decarboxylation.

Isopropylbenzene can be obtained by distilling cumic acid with soda lime, and *m*-xylene obtained similarly from mesitylenic acid; this presents a means of partial dealkylation since both these acids have been prepared by oxidation of the corresponding polyalkylbenzene. Heating the acid with copper powder or copper oxide in quinoline is an effective method of decarboxylation and has been used for preparing benzene containing ^{14}C (*D. M. Hughes* and *J. C. Reid*, J. org. Chem., 1949, **14**, 516; *V. I. Savushkina et al.*, Doklady Akad. Nauk SSSR., 1955, **102**, 1139).

Esters of benzoic acid are decomposed at 380–420° in the presence of a chromium catalyst to yield benzene, carbon dioxide, and olefin (*D. N. Dolgov et al.*, Zhur. obshchei Khim., 1955, **25**, 1555). Tetrahydrobenzoic acid and its alkyl homologues yield benzene and alkylbenzenes when heated to 500° (*V. R. Skvarchenko et al., ibid.*, 1963, **33**, 1069).

(l) Additional methods of preparation

The removal of the diazonium group from the aromatic nucleus is of general application, the reduction being effected readily by aqueous hypophosphorous acid at 0° (*N. Kornblum,* Org. Reactions, 1944, **2**, 262).

The sulphonic acid group is readily removed from an arenesulphonic acid to give the alkylbenzene, by heating with aqueous sulphuric acid. This method is of particular value in the isolation and purification of alkylbenzenes.

A complex mixture of alkylbenzenes is produced when synthesis gas ($CO+H_2$) is contacted with an alumina catalyst at 480° and 450 psig. (*H. S. Seelig* and *H. I. Week,* U.S.P., 2,727,055/1955). Alkylbenzenes are also made when organic oxides and epoxides are contacted with alumina-silica at 150–480° (*W. J. Mattox,* U.S.P., 2,991,319/1961).

1,3,5-Trisubstituted benzenes have been prepared by the reaction of 2,4,6-trimethylpyrylium perchlorate with alkylmagnesium halide (*R. Gompper* and *O. Christmann,* Ber., 1961, **94**, 1795).

A convenient method of preparing mesitylene is by cyclodehydration of acetone using sulphuric acid (*R. Kane,* J. pr. Chem., 1838, [i], **15**, 129; *R. Adam* and *R. W. Hufford,* Org. Synth., 1956, Coll. Vol. I, p. 341), hydrogen chloride (*V. N. Ipatieff et al.,* Ber., 1930, **63**, 3072), silica gel (*J. A. Mitchell* and *E. E. Reid,* J. Amer. chem. Soc., 1931, **53**, 330), chromic oxide-boric anhydride on alumina (*S. M. Kovach,* U.S.P., 3,201,485/1965), or molybdenum trioxide on alumina (*Y.-T. Hwang et al.,* U.S.P., 3,301,912/1967):

$$3\ CH_3COCH_3 \rightarrow 1,3,5\text{-}C_6H_3(Me)_3 + 3\ H_2O$$

Durene, isodurene, trimethylisobutylbenzene, and pentamethylbenzene are by-products in the reaction (*S. Landa* and *V. Sesulka,* Chem. Listy, 1957, **51**, 1159). The cyclodehydration reaction has been applied successfully to other aliphatic carbonyl compounds, in particular unsaturated ketones and aldehydes, using either polyphosphoric acid, alumina, aluminium silicates, or mixed oxides of aluminium or chromium as dehydrating catalyst; a mixture of alkylbenzenes is obtained (*V.E.B. Farben. Wolfen,* Neth. Pat. Appl., 6,409,506/1966; 6,501,480/1966; *G. Descotes et al.,* Bull. Soc. chim. Fr., 1968 [1], 382).

Diarylmethanes, produced by reaction of alkylbezenes with formaldehyde are converted to methyl substituted aromatic hydrocarbons by catalytic-hydrocracking (*L. C. Fetterly,* U.S.P., 2,819,322/1958; U.S.P., 2,897,245/1959;

J. G. Hendrickson, U.S.P., 2,920,117/1960) according to the scheme:

$$2\ ArH + CH_2O \rightarrow ArCH_2Ar \xrightarrow{H_2} ArMe + ArH$$

p-Xylene is obtained in moderate yield when 2,2,5,5-tetramethyl-di- and -tetra-hydrofurans are heated at 500° in the presence of a mixed metal oxide catalyst (*A. S. Zanina et al.*, Zhur. priklad. Khim., 1963, 36, 203).

4. Physical properties of the mononuclear benzene hydrocarbons

The lower members of the series are mostly liquids under normal conditions and possess a characteristic ethereal odour. The proportion of paraffinic side chains increases as the series is ascended and the characteristics of the alkanes predominate in many of the physical properties of the higher alkylbenzenes. The commoner physical constants of benzene and a number of its homologues are collected in Table 2.

(a) Boiling points

The boiling points of the mono-*n*-alkylbenzenes increase normally with molecular weight. In general the increase for every additional CH_2 group decreases progressively. The branched-chain alkylbenzenes boil at a lower temperature than the corresponding isomeric *n*-alkylbenzenes and in this respect are analogous to the alkanes. For compounds of equal molecular weight the boiling point also increases with increase in the number of side chains, the total number of paraffin carbon atoms attached to the benzene nucleus remaining constant. The sub-series of methylbenzenes shows a fairly constant increase in boiling point for every additional CH_2 group. In general the alkylbenzenes boil above the corresponding cyclohexane compounds.

Vapour pressure data have been reported for aromatic hydrocarbons in the range 10 to 1500 mm. (*C. B. Willingham et al.*, J. Res. nat. Bur. Stand., 1945, **35**, 219; *F. D. Rossini et al.*, "Selected Values of Physical and Thermodynamic Properties of Hydrocarbons and related Compounds", Carnegie Press, Pittsburgh, 1953), and for benzene has been extended to the critical point (*E. J. Gonowski et al.*, Ind. Eng. Chem., 1947, **39**, 1348). The critical temperatures and pressures for a number of alkylbenzenes have been calculated (*K. A. Kobe* and *R. E. Lynn*, Chem. Reviews, 1953, **52**, 117; *D. Ambrose et al.*, Trans Faraday Soc., 1957, **53**, 771; 1960, **56**, 1452; *D. L. Bond* and *G. Thodos*, J. chem. Eng. Data, 1960, **5**, 289).

TABLE 2

PHYSICAL CONSTANTS OF BENZENE AND ITS HOMOLOGUES
(Source – API Project 44)

	M.p.°	B.p.°/ 760 mm	d_4^{20}	n_D^{20}
Benzene	+ 5·533	80·100	0·87901	1·50112
Toluene	–94·991	110·625	0·86696	1·49693
Ethylbenzene	–94·975	136·186	0·86702	1·49588
1,2-Dimethylbenzene (*o*-Xylene)	–25·182	144·411	0·88020	1·50545
1,3-Dimethylbenzene (*m*-Xylene)	–47·872	139·103	0·86417	1·49722
1,4-Dimethylbenzene (*p*-Xylene)	+13·263	138·351	0·86105	1·49581
n-Propylbenzene	–99·50	159·217	0·86204	1·49202
Isopropylbenzene (cumene)	–96·035	152·392	0·86179	1·49146
1-Ethyl-2-methylbenzene	–80·833	165·150	0·88069	1·50456
1-Ethyl-3-methylbenzene	–95·55	161·305	0·86452	1·49661
1-Ethyl-4-methylbenzene	–62·350	161·899	0·86118	1·49500
1,2,3-Trimethylbenzene (hemimellitene)	–25·375	176·084	0·89438	1·51393
1,2,4-Trimethylbenzene (pseudocumene)	–43·80	169·351	0·87582	1·50484
1,3,5-Trimethylbenzene (mesitylene)	–44·720	164·716	0·86518	1·49937
n-Butylbenzene	–87·970	183·270	0·86013	1·48979
Isobutylbenzene	–51·48	172·759	0·85321	1·48646
sec-Butylbenzene	–75·470	173·305	0·86207	1·49020
tert-Butylbenzene	–57·850	169·119	0·86650	1·49266
1-Methyl-2-*n*-propylbenzene	–60·2	184·80	0·8744	1·4998
1-Methyl-3-*n*-propylbenzene	—	181·80	0·8610	1·4936
1-Methyl-4-*n*-propylbenzene	–63·6	183·30	0·8584	1·4919
1-Isopropyl-2-methylbenzene (*o*-Cymene)	–71·540	178·15	0·8766	1·5006
1-Isopropyl-3-methylbenzene (*m*-Cymene)	–63·745	175·14	0·8610	1·4930
1-Isopropyl-4-methylbenzene (*p*-Cymene)	–67·935	177·10	0·8573	1·4909
1,2-Diethylbenzene	–31·240	183·423	0·87996	1·50346
1,3-Diethylbenzene	–83·920	181·102	0·86394	1·49552
1,4-Diethylbenzene	–42·850	183·752	0·86196	1·49483
3-Ethyl-1,2-dimethylbenzene	–49·5	193·91	0·8921	1·5117
4-Ethyl-1,2-dimethylbenzene	–67·0	189·75	0·8745	1·5031
2-Ethyl-1,3-dimethylbenzene	–16·28	190·01	0·8904	1·5107

(continued)

Table 2 (*continued*)

	M.p.°	B.p.°/ 760 mm	d_4^{20}	n_D^{20}
4-Ethyl-1,3-dimethylbenzene	−62·90	181·41	0·8763	1·5038
5-Ethyl-1,3-dimethylbenzene	−84·325	183·75	0·8648	1·4981
2-Ethyl-1,4-dimethylbenzene	−53·68	186·91	0·8772	1·5043
1,2,3,4-Tetramethylbenzene (prehnitene)	− 6·25	205·04	0·9052	1·5203
1,2,3,5-Tetramethylbenzene (isodurene)	−23·685	198·00	0·8903	1·5130
1,2,4,5-Tetramethylbenzene (durene)	+79·240	196·80	0·8875*	1·5116
n-Pentylbenzene	−75	205·4	0·8585	1·4878
n-Hexylbenzene	−61	226·1	0·8575	1·4864
n-Heptylbenzene	−48	245·5	0·8567	1·4854
n-Octylbenzene	−36	264·5	0·8562	1·4845
n-Nonylbenzene	−24	282·0	0·8558	1·4838
n-Decylbenzene	−14·38	300	0·85553	1·48319
n-Dodecylbenzene	3	331	0·8551	1·4824
n-Eicosylbenzene, $C_{26}H_{46}$	44	429	0·8545*	1·4805*
n-Triacontylbenzene, $C_{36}H_{66}$	70	512	0·8543*	1·4795*
n-Hexatricontylbenzene, $C_{42}H_{78}$	80	549	0·8542*	1·4791*

* Undercooled liquid

(b) Melting points

Certain members, *e.g.*, benzene, *p*-xylene, durene and hexamethylbenzene (m.p. 165·3°), possessing high symmetry, have abnormally high melting points. The melting points of the mono-*n*-alkylbenzenes do not show the alternation exhibited by the *n*-alkanes, the melting point approaching more and more closely that of the chain itself as the length of the side chain increases. This is illustrated by the melting points of the following *n*-alkanes and *n*-alkylbenzenes:

	M.p. ° R = Ph, *n*-alkylbenzene	M.p. ° R = H, *n*-alkane
RC_5H_{11}	−75	−129·7
$RC_{10}H_{21}$	−14·38	− 29·7
$RC_{20}H_{41}$	+44	+ 36·4
$RC_{30}H_{61}$	+70	+ 66·0

Branching of the side chain decreases the melting point of the alkylbenzenes.

(c) Crystal structures and bond lengths

The benzene molecule is a regular, planar hexagon with a carbon-carbon distance of 1·392 Å and the closest carbon-carbon distance between molecules 3·8 Å. The molecules pack together like sheets of 6-toothed gear wheels (*E. G. Cox et al.*, Proc. roy. Soc., 1932, **A135,** 491; 1958, **A247,** 1). For durene the aromatic carbon-carbon distance is 1·41 Å and the aromatic carbon-aliphatic carbon distance 1·56 Å, with the closest distance between molecules 3·71 Å (*J. M. Robertson, ibid.*, 1933, **A141,** 594; **A142,** 659).

(d) Viscosity

The viscosities of the mono-*n*-alkylbenzenes increase logarithmically with increase in molecular weight, the least temperature effect on viscosity occurring with the long chain alkylbenzenes. Viscosity also increases with increase in the number of side chains when the total number of paraffinic carbon atoms attached to the benzene nucleus remains constant (*L. A. Mikeska*, Ind. Eng. Chem., 1936, **28,** 970; *H. Koelbel et al.*, Brennstoff-chem., 1949, **30,** 362; *A. M. Kuliev* and *M. G. Sardarly*, Uchenye Zapiski Azerbaidzhan Gosudarst, Univ. im, S. M. Kirova, Ser. Fiz. – Mat i Khim. Nauk, 1955, No. 6, 33).

(e) Physical properties used for the analysis of the mononuclear benzene hydrocarbons

Density and refractive index values for the alkylbenzenes and aromatic hydrocarbons in general, are relatively high compared with those for other classes of hydrocarbons and use can be made of this in determining the aromatic content of petroleum distillates without prior separation (*R. E. Thorpe* and *R. G. Larsen*, Ind. Eng. Chem., 1942, **34,** 853). Dispersion data and refractivity intercept values may be used to analyse benzene, toluene and xylene mixtures (*K. M. Sumer* and *A. R. Thompson*, J. chem. Eng. Data, 1968, **13,** 30).

The analysis of alkylbenzenes in complex mixtures of hydrocarbons is usually preceded by the separation of the aromatic hydrocarbons using a combination of fractional distillation, extraction, and/or chromatography. Aromatic hydrocarbons in general, are selectively extracted from other hydrocarbon classes by solvents such as liquid sulphur dioxide, furfural, nitrobenzene, phenol, cresol, "Chlorex", glycols and sulpholane. Alternatively, they may be isolated by sulphonation. Adsorption on silica gel is an efficient method for separating aromatic hydrocarbons. Further chromatographic separation of the aromatic hydrocarbons into mono-, di-, and poly-nuclear aromatics is possible. A most effective technique for separating and identifying the lower molecular weight alkylbenzenes is a combination of gas liquid chromatography and mass spectrometry (see Vol. IA, p. 369; also *D. M. Ghawrey* and *J. E. Paulson*, Anal. Chem., 1962, **34,** 538; *M. E. Fitzgerald et al., ibid.*, 1962, **34,** 1276). Aromatic hydrocarbons can be separated according to size and shape by the use of molecular sieve adsorbents (*B. J. Mair* and *M. Sharnacengar, ibid.*, 1958, **30**, 276).

The alkylbenzenes show characteristic ultra-violet absorption due to the aromatic ring. Benzene itself has a series of seven peaks ranging from 234·5–269; the most intense being at 255 mμ. Analysis of petroleum fractions for C_6, C_7, and C_8 aromatic hydrocarbons by ultra-violet absorption spectroscopy is a precise and routine procedure. Infra-red spectroscopy is also useful for identifying the alkylbenzenes. Spectral patterns in the 7–10 and the 12–14 micron ranges are characteristic of the positions of the substituent alkyl groups (*L. J. Bellamy*, "Infra-red spectra of Complex Molecules", Methuen, London, 1958; *J. C. Hawkes* and *A. J. Neale*, Spectrochim. Acta, 1960, **16**, 633). The Raman spectroscopic method is especially suited to the analysis of the alkylbenzenes (see *H. M. Tenney* "The Chemistry of Petroleum Hydrocarbons", Ed. *B. T. Brooks et al.*, Reinhold Publishing Corp., 1954, Vol. I, p. 375). Nuclear magnetic resonance spectroscopy is limited in application to structural studies of the alkylbenzenes (*K. Itoh et al.*, Kogyo Kagaku Zasshi, 1967, **70**, 914). It has been used successfully in the analysis of the *tert*-butylbenzenes (*E. M. Arnett et al.*, Tetrahedron Letters, 1961, 658; *C. Hoogzand* and *W. Huebel, ibid.*, 1961, 637). An excellent example of the use of mass and n.m.r. spectroscopy in elucidating the structure of components in a narrow fraction of C_{24}–C_{27} alkylbenzenes from petroleum is given by *Mair* (Amer. chem. Soc. Div. Pet. Chem., Preprints, 1966, **11**, No. 3, 105).

For further data concerning physical properties see: *F. D. Rossini et al.*, "Selected Values of Physical and Thermodynamic Properties of Hydrocarbons and Related Compounds", Carnegie Press, Pittsburgh, 1953; *G. Egloff*, "Physical Constants of Hydrocarbons", Reinhold Publishing Corpn., Vol. III, 1946; Faraday's Encyclopaedia of Hydrocarbon Compounds, Chemindex Ltd., Manchester; "Physical Chemistry of the Hydrocarbons", Ed. *A. Farkas*, Academic Press, New York, Vol. I, 1950; *Rossini et al.*, "Hydrocarbons from Petroleum", Reinhold Publishing Corpn., 1953; "Manual of Hydrocarbon Analysis", ASTM Special Technical Publication No. 332, 1963; "Hydrocarbon Analysis", ASTM Special Technical Publication No. 389, 1965.

5. Chemical properties of the mononuclear benzene hydrocarbons

The resonance structures of the completely conjugated six-membered ring system have already been discussed (p. 6 *ff.*). The high stability of the benzene nucleus is evident in the thermal behaviour of the benzenoid hydrocarbons. The main characteristic of these hydrocarbons on pyrolysis is the multiple condensation of the nuclei to larger molecules rather than rupture of the rings. Similarly, the benzene nucleus is remarkably stable to oxidizing agents and opening of the ring requires drastic reagents or conditions. Various isomers of benzene have been made by photolysis.

Benzene and its homologues do not show the markedly unsaturated character exhibited by the olefinic classes of hydrocarbons. They do, however, reveal their unsaturation under favourable conditions and form addition compounds. *Addition reactions* involving the benzene nucleus are limited to: the formation of molecular complexes with a wide variety of electrophilic agents in which the benzenoid hydrocarbons act as electron donors, clathrate substances where the arene is contained in the crystal cage of the host, and reactants which can overcome the resonance energy of the nucleus, *e.g.*, hydrogen in the presence of a catalyst, halogens in the presence of light, ozone, and carbenes. These last named reactions are in general radical reactions and often occur in the vapour phase at high temperatures or under irradiation. The mechanism is discussed elsewhere (see p. 113). Photo-addition reactions of benzenoid hydrocarbons have been extensively studied recently.

There are two main types of hydrogen atoms in the alkylbenzenes, those attached to the nucleus and those on the side chain; *substitution reactions* occur with both types but in many cases selectivity is possible. The most characteristic reaction of benzenoid hydrocarbons is the displacement of nuclear hydrogen atoms by the more active electrophilic agents, especially if these are activated by the presence of acid catalysts, the aromatic properties being preserved. The mechanism of electrophilic substitution, by far the most important of the substitution reactions of aromatic hydrocarbons, is discussed elsewhere (see Vol. I A, p. 284; this vol. p. 45 *et seq.*). The main reactions in this class are nitration, halogenation, sulphonation and Friedel-Crafts reactions. There seems to be no example of substitution by nucleophilic reactions with an alkylbenzene but this is not surprising because it is unusual for hydrogen to be replaced without the presence of an activating group, *e.g.*, nitro, in the nucleus. However several non-ionic substitutions with free radicals have been demonstrated (*Nelson* and *Brown*, *loc. cit.*, p. 549).

Physiologically benzene and the alkylbenzenes are strong irritants and poisons having a narcotic action in the central nervous system (*N. I. Sax*, "Dangerous Properties of Industrial Materials", 2nd Edn., Reinhold Publishing Co., 1963; *H. W. Gerarde*, "Toxicology and Biochemistry of Aromatic Hydrocarbons", Elsevier Publishing Co., Amsterdam, 1960).

(a) Pyrolysis

The benzene nucleus possesses remarkable thermal stability due to strong carbon-to-carbon bonds. There are marked differences in the thermal behaviour of benzene and that of the alkylbenzenes (see *C. R. Kinney*, "The Chemistry of Aromatic Hydrocarbons", Ed. *B. T. Brooks et al.*, Reinhold Publishing Corp., 1954, Vol. II, p. 113; and *G. M. Badger*, Progr. Phys. Org.

Chem., 1965, 3, 1). Pyrolytic decomposition of benzene involves only the dissociation of hydrogen accompanied by condensation of the radicals formed. Biphenyl is reported to be formed in small amounts at temperatures as low as 300⁰; the reaction being reversible:

$$2\ C_6H_6 \rightleftharpoons C_6H_5\text{–}C_6H_5 + 2\ H$$

Pyrolysis of benzene at 700⁰ yields biphenyl as the major product together with significant amounts of *m*- and *p*-terphenyls and triphenylene (9,10-benzphenanthrene) (*Badger* and *J. Novotny*, J. chem. Soc., 1961, 3400). Above 750⁰ further condensations become excessive with the formation of polyphenyls, resinous material and coke. Contact with various surfaces affects the dehydro-condensation reaction (*A. P. Rudenko*, Vestnik. Moskov. Univ., Ser. II, Khim., 1960, **15**, No. 5, 69; No. 6, 62; 1961, **16**, No. 2, 59; No. 3, 74). At 1100⁰ carbonisation of benzene is very extensive but is depressed by the presence of hydrogen. In the range 800–1100⁰ small quantities of methane and traces of acetylene appear, probably being formed by recombination. The primary decomposition products for short contact times at 1200⁰ are acetylene, diacetylene and hydrogen, suggesting rupture of the benzene ring; at long residence time the main product is carbon (*Kinney* and *R. S. Slysh*, J. phys. Chem., 1961, **65**, 1044). At the temperature of the electric arc benzene yields gaseous mixtures containing hydrogen, acetylene, methane, ethane and related products. To dissociate benzene completely and produce a crystalline graphite, temperatures of 2500 to 3000⁰ are necessary. The technical production of biphenyl consists in passing benzene and superheated steam at 800–900⁰ through a reaction vessel coated with iron oxide. Other conditions are described by *J. P. Wibaut* and coworkers (Rec. Trav. chim., 1934, **53**, 584) and *Y. Hosaka* and *M. Takehisa* (J. chem. Soc. Japan, Ind. Chem. Sect., 1955, **58**, 1001).

The thermal decomposition of the alkylbenzenes involves a number of reactions depending mainly on the temperature and pressure. These reactions, which are discussed in detail by *Kinney* (*loc. cit.*), include dealkylation of the nucleus, cracking of the aliphatic side chain, isomerization, disproportionation, dehydrogenation and condensation, the benzene nucleus usually remaining intact. Relatively high temperatures ($>$ 540⁰) are required before dealkylation becomes significant (see p. 161), the methyl groups being the most stable. Primary and secondary alkyl groups undergo extensive cracking in the region 600–650⁰ but tertiary alkylbenzenes are usually dealkylated. The mode of cracking of the aliphatic side chain depends largely on the number of hydrogens on the α-carbon atom and normally leads to the formation of toluene and xylenes (*A. F. Dobryanskii et al.*, C.A., 1937, **31**, 5334[2]). Thermal isomerization and disproportionation reactions, leading to the formation of more stable configurations, are usually limited to the methylbenzenes since dehydrogenation and cracking are predominant reactions in the longer chain alkylbenzenes. Dehydrogenation of the side chain to produce alkenylbenzenes is an important thermal reaction of certain alkylbenzenes. Thermal dehydrogenation occurs above 650⁰ but at much lower temperatures in contact with suitable catalysts. Thus isopropylbenzene gives an 84 % yield of α-methylstyrene when pyrolyzed in the presence of a mixed metal oxide catalyst at 600°. Similarly, ethylbenzene, *m*-ethyltoluene, and *m*-isopropyltoluene yield styrene, *m*-vinyltoluene and *m*-isopropenyltoluene respectively (*J. E. Nichels*, Ind. Eng. Chem., 1949, **41**, 563; *Yu G. Mamedaliev, et al.*, Azerb. Khim. Zhur., 1962, No. 3, 11; *S. Carra* and *L. Forni*, Ind. Eng. Chem., Process Design Develop., 1965, **4**, No. 3, 281). Dehydrogenation proceeds more rapidly with branched chain alkylbenzenes and is increased by methyl substitution in the nucleus (*O. K. Bogdanova et al.*, Izvest. Akad. Nauk SSSR., Otdel. khim. Nauk, 1963, 611; Neftekhimiya, 1962, **2**, 467). The conversion of ethylbenzene to styrene is an important commercial process; the ethylbenzene being made by alkylation of benzene with ethylene (*W. L. Faith et al.*, "Industrial Chemicals", Chapman and Hall Ltd., London, 1957, p. 730; *Bagdanova et al.*, Neftekhimiya, 1961, **1**, 195; *Dow Chemical Co.*, B.P., 892,779/1962). Dehydrogenation of alkylbenzenes to alkenylbenzenes is also achieved by catalytic oxidative dehydrogenation at 300–700° in the presence of sulphur dioxide (*C. A. Adams*, U.S.P., 3,299,155/1967). Dehydrogenation can also lead to condensation reactions of which there are two types, viz. (1) intramolecular and (2) intermolecular, involving either alkyl groups or only the aromatic nucleus. An interesting example of intramolecular condensation is the pyrolysis of *p*-di-*n*-butylbenzene to yield phenanthrene with no trace of anthracene:

CH_2 — CH_2 — CH_2 — CH_3 / H_3C — CH_2 — CH_2 — CH_2 (on benzene ring) ⟶ (phenanthrene) $+ 6\ H_2$

Intermolecular condensation involving the alkyl group leads to polycyclic aromatics, but for only the nucleus to be involved higher temperatures are usually necessary. Thus toluene, under similar conditions required for condensation of benzene *i.e.* >700°, gives dibenzyl and ditolyls; at higher temperatures stilbene is formed. *p*-Xylene similarly gives 4,4-dimethylbibenzyl (*H. Meyer* and *A. Hofmann*, Monatsh., 1916, **37**, 681; see also *L. A. Errede* and *J. P. Cassidy*, J. Amer. chem. Soc., 1960, **82**, 3653; *M. Takahashi*, Bull. Chem. Soc. Japan, 1960, **33**, 801). In contrast the only products from the pyrolysis of toluene with a hydrogen carrier gas, are methane and benzene (*J. G. Burr et al.*, J. Amer. chem. Soc., 1964, **86**, 3846; 5065). The formation of residual stocks in petroleum refining is largely the result of such condensation reactions.

Of special interest is the rapid pyrolysis of *p*-xylene at 850–900° and low pressure giving the double radical, $\cdot CH_2C_6H_4CH_2\cdot$, which can be condensed in an inert solvent. This diradical, commonly known as *p*-xylylene exists in the quinodimethane form and is more correctly named as dimethylenecyclohexadiene. It has been shown that *p*-methylbenzyl radicals are also formed (*C. J. Brown* and *A. C. Farthing* Nature, 1949, **164**, 915; *M. Szwarc*, J. chem. Phys., 1948,

16, 128; 609; *L. A. Errede* and *B. F. Landrum*, J. Amer. chem. Soc., 1957, **79**, 4952):

$$\text{Me}\text{-}C_6H_4\text{-}\text{Me} \xrightarrow{\Delta} CH_3\text{-}C_6H_4\text{-}\dot{C}H_2 \xrightarrow[\text{fast}]{\text{very}} CH_2{=}C_6H_4{=}CH_2$$

Thermoplastics capable of forming continuous films thinner than one thousandth of an inch have been made from *p*-xylylene.

(b) Photolysis of the benzene ring

Fulvene (I) is obtained by the photolysis of liquid benzene; benzvalene (II), Dewar benzene (III), and prismane (IV) are also formed but are continuously destroyed. The diverse photochemical addition reactions of benze-

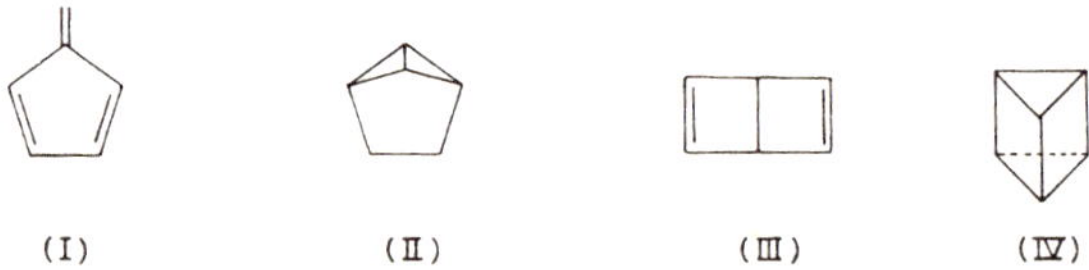

noid hydrocarbons (p. 199) are in part related to the intermediate radicals formed in these transformations (*D. Bryce-Smith et al.*, J. chem. Soc., 1960, 2003; Chem. Comm., 1966, 593; *H. R. Ward et al.*, J. Amer. chem. Soc., 1967, **89**, 162; 1968, **90**, 1085; *L. Kaplan et al., ibid.*, 1967, **89**, 1030; 1031; 1968, **90**, 1116; *E. E. Van Tamelen* and *D. Carty, ibid.*, 1967, **89**, 3922). Similar transformations have been observed with toluene and isopropylbenzene (*Bryce-Smith et al., loc. cit.*). Photolysis of gaseous benzene at 1470Å yields alkenes and alkynes but no cyclic isomers (*W. M. Jackson et al.*, J. phys. Chem., 1967, **71**, 3346); fulvene is formed at 1849 Å (*Kaplan et al.*, J. Amer. chem. Soc., 1968, **90**, 5646). Polyenes are formed by irradiation of benzene in a rigid medium at 77°K (*E. Migirdicyan*, J. Chim. phys., 1966, **63**, 520). For a review of the valence isomers of benzene see *J. H. F. K. Barnick*, Chem. Tech. (Amsterdam), 1969, **24**, 96.

(c) Oxidation

(*i*) *By air or oxygen*

Benzene and its homologues can be oxidized directly with air or oxygen. Reaction occurs well below their ignition point and only the slower type of oxidation is discussed here, complete combustion presumably involving many of the intermediate formulations found in the slow reactions. Both nuclear and side-chain oxidation is possible although the benzene nucleus is more resistant to attack than the paraffinic moiety. The oxidation takes place by a free radical mechanism according to the basic scheme first proposed by *J. L. Bolland* (Proc. roy. Soc., 1946, **A186,** 218).

The *direct oxidation of benzene* itself proceeds *via* successive hydroxylation to the dihydroxybenzenes the ring ultimately breaking and the fission products degrading to C_2 hydrocarbons, formaldehyde, carbon oxides, hydrogen and water (*R. G. W. Norrish* and *G. W. Taylor, ibid.*, 1956, **A234,** 160; *J. Drillat,* Compt. rend., 1961, **252,** 1155; *Drillat* and *M. Odile,* Bull. Soc. chim. Fr., 1962, 519). Some control of the oxidation products formed is possible depending on the reaction conditions. At elevated pressure, *e.g.* 50 atm., oxidation is significant even at temperatures as low as 280° and the main product is phenol. Conversions into phenol of up to 50 % may be obtained and even higher conversions if water and cyclohexane are present (*D. M. Newitt* and *J. H. Burgoyne,* Proc. roy. Soc., 1936, **A153,** 448; *I. I. Ioffe et al.*, Zhur. fiz. Khim., 1954, **28,** 1395; 1959, **33,** 863; *Y. Hosaka* and *Y. Urano,* Tokyo Shikensho Hokohu, 1962, **57,** 295). Phenol itself is not very stable towards oxidation, continued oxidation resulting in the formation of quinol and quinone and rupture of the ring. However, there are a few patents on the production of phenol by vapour-phase oxidation of benzene and some phenol is made commercially by this route (*W. W. Moyer* and *W. C. Klingelhoefer,* U.S.P., 2,367,731/1945; *Oxidation Process Corpn.*, Belg. P., 619,934–5/1962). At pressures below atmospheric, oxidation occurs only above 450° and besides phenol, maleic acid, biphenyl, and *p*-dihydroxybenzene are formed together with fragmentation products (*G. J. Minkoff* and *C. F. H. Tipper,* "Chemistry of Combustion reactions", Butterworth, London, 1962). The most probable mechanism involves phenyl and phenylperoxy radicals:

$$
\begin{aligned}
C_6H_6 + O_2 &\rightarrow C_6H_5\cdot + HO_2\cdot \\
C_6H_5\cdot + O_2 &\rightarrow C_6H_5O_2\cdot \\
C_6H_5O_2\cdot + C_6H_6 &\rightarrow C_6H_5O\cdot + C_6H_5OH \\
C_6H_5O\cdot + C_6H_6 &\rightarrow C_6H_5OH + C_6H_5\cdot
\end{aligned}
$$

the main terminating step being $2C_6H_5\cdot \rightarrow (C_6H_5)_2$. The phenol undergoes a similar oxidation to give dihydroxybenzene which then adds oxygen to form maleic acid:

$$
p\text{-}C_6H_4(OH)_2 + O_2 \longrightarrow \text{[HO-C}_6\text{H}_4\text{(-O-O-)-OH]} \longrightarrow \begin{matrix} CH\cdot CO_2H \\ \| \\ CH\cdot CO_2H \end{matrix} + C_2H_2
$$

Various metal oxides from Groups Vb and VIb catalyze the vapour-phase oxidation of benzene to maleic acid (*S. K. Bhattacharyya* and *N. Venkataraman*, J. appl. Chem., 1958, **8**, 728; *J. K. Dixon* and *J. E. Longfield*, "Catalysis", Ed. *P. H. Emmett*, Reinhold Publishing Corpn., 1950, Vol. VII, p. 183; *B. Dmuchovsky et al.*, J. Catalysis, 1965, **4**, 291). Benzene is the preferred raw material for the manufacture of maleic acid the production of which has risen considerably in recent years. The main commercial process for the production of maleic acid involves the vapour-phase oxidation of benzene using vanadium pentoxide as catalyst in a manner similar to the preparation of phthalic anhydride by oxidation of naphthalene (*C. R. Downs et al.*, Ind. Eng. Chem., 1920, **12**, 228; 1923, **15**, 965; J. Soc. chem. Ind., 1926, **45**, 188T; *W. L. Faith et al.*, "Industrial Chemicals", Chapman and Hall Ltd., London, 1957, p. 500; *B. S. Trehan et al.*, J. Sci. Ind. Res., India, 1959, **18B**, 147; *J. A. Dreibelbis*, U.S.P., 3,005,831/1961).

In the *liquid-phase oxidation of alkylbenzenes* the alkyl group is mainly attacked and at the carbon atom adjacent to the nucleus, the rate and extent of oxidation depending largely on the experimental conditions. Ethylbenzene and its linear homologues change slowly on exposure to air the rate accelerating by rise in temperature, exposure to ultra-violet light, or the presence of catalyst. In general, the kinetics of the oxidation resemble those of the paraffins (see Vol. I A, p. 376). Increase in pressure and reduction of temperature reduce ring fission and increase primary products from the oxidation of the side chain, *e.g.* benzaldehyde, acetophenone and 2-phenylethanol from ethylbenzene (*H. N. Stephens*, J. Amer. chem. Soc., 1928, **50**, 2523; *R. G. Larsen et al.*, Ind. Eng. Chem., 1952, **34**, 183; *P. George et al.*, Nature, 1942, **149**, 601). The rate of oxidation of the longer side chain *n*-alkylbenzenes is greater than that of the shorter side chain alkylbenzenes (*A. N. Bashkirov* and *E. S. Alent'eva*, Neftekhimiya, 1964, **4**, 593). The reaction has been used as a method of preparing ketones directly from alkylbenzenes (*D. T. Mowry*, J. Amer. chem. Soc., 1945, **67**, 1050). The initial product is apparently a hydroperoxide which probably breaks down as shown (*Minkoff* and *Tipper*, *loc. cit.*):

$$C_6H_5\cdot CH(O\cdot O\cdot)\cdot CH_3 \longrightarrow C_6H_5\cdot CO\cdot CH_3 + \cdot OH \text{ or } C_6H_5\cdot CHO + CH_3O\cdot$$

$$C_6H_5\cdot CH(O\cdot O\cdot)\cdot CH_3 \xrightarrow{PhEt} C_6H_5\cdot CH(O\cdot OH)\cdot CH_3 \rightarrow \cdot OH + C_6H_5\cdot CH(O\cdot)\cdot CH_3$$

$$\xrightarrow{PhEt} C_6H_5\cdot CH(OH)\cdot CH_3$$

Branched chain alkylbenzenes form similar products. An alkylbenzene with a tertiary α-carbon atom is oxidized faster than a methylbenzene and the derived hydroperoxide is more stable (*Stephens* and *F. L. Roduta*, J. Amer. chem. Soc., 1935, **57**, 2380; *V. V. Federova*, Okislenie Uglevodorodov v Zhidkoĭ Faze, Akad. Nauk SSSR., Inst. Khim. Fiz., Sbornik Stateĭ, 1959, 197). The first stage in the oxidation of *cumene* is the production of cumene hydroperoxide, and its formation and decomposition have received much study (*H. S. Blanchard*, J. Amer. chem. Soc., 1959, **81**, 4548; *T. G. Traylor* and *P. D. Bartlett*, Tetrahedron Letters, 1960, No. 24, 30; *H. Hock* and *H. Kropf*, J. pr. Chem., 1961, **13**, 285; *S. D. Kaz'min*, Zhur. priklad. Khim., 1962, **35**, 422; *V. A. Shushunov et al.*, *ibid.*, 1962, **35**, 832). The cumene hydroperoxidation process, the most attractive route to phenol at present, involves the liquid phase air oxidation of cumene to yield acetone as by-product (*H. I. Enos* and *J. R. Nixon*, Amer. chem. Soc., Div. Pet. Chem., Symposium on petrochemicals in the post-war years, 1953, September 6–11th; *Hock* and *Kropf*, Angew. Chem., 1957, **69**, 313):

$$C_6H_5CHMe_2 + O_2 \xrightarrow[\text{basic cat.}]{100\text{-}130^\circ} C_6H_5CMe_2{\cdot}O{\cdot}OH \xrightarrow{\text{dil. } H_2SO_4} C_6H_5OH + Me_2CO$$

The liquid-phase oxidation of the branched chain *butylbenzenes* gives hydroperoxides, benzoic acid, and other intermediate oxidation products (*M. S. Eventova* and *G. N. Prytkova*, Vestnik Moskov. Univ., Ser. II, Khim., 1960, **15**, No. 5, 59). A nuclear methyl substituent is apparently more readily attacked than a branched-chain alkyl group in the same ring. Thus *tert*-octylbenzoic acid is obtained by liquid-phase air oxidation of *tert*-octyltoluene in the presence of cobalt oleate (*S. D. Mektiev et al.*, Azerb. Khim. Zhur., 1956, No. 2, 51).

In the series of methylbenzenes the methyl group is oxidized similarly to aldehyde although here the further oxidation to a carboxylic acid goes readily and can be brought about in a suitable solvent by oxidation with air at 140–160° with cobalt, manganese, and lead acetates as catalysts (*G. Ciamician* and *P. Silber*, Ber., 1912, **45**, 38; *Stephens*, J. Amer. chem. Soc., 1926, **48**, 1824; *W. S. Emerson et al.*, *ibid.*, 1949, **71**, 1742). In the commercial liquid-phase *oxidation of toluene* at 140–155° and 6–8 atmospheres pressure in the presence of cobalt salts of carboxylic acids, up to 90 % yields of 99 % pure benzoic acid are obtained (*G. Messina*, Hydrocarb. Process, Petrol. Refiner, 1964, **43**, No. 11, 191). This is also the first stage in the conversion of toluene to phenol by a two-step catalytic oxidation process (*H. B. Richman*, Chem. Eng., 1964, **71**, No. 20, 106; *W. W. Kaeding et al.*, Ind. Eng. Chem., Process Design Develop., 1965, **4**, No. 1, 97). Stable intermediates in the liquid-phase oxidation of toluene to benzoic acid are benzyl alcohol and benzaldehyde, and the principle by-products are benzyl benzoate, biphenyl, and *o*-, *m*-, and *p*-phenyltoluene (*Kaeding*, Hydrocarb. Process. Petrol. Refiner, 1964, **43**, No. 11, 173).

The *xylenes* give mainly the toluic acids when subjected to liquid-phase oxidation in the presence of catalysts although a single stage oxidation to the dicarboxylic acid has been achieved using special catalyst systems (*Y. Kusunoki et al.*, Technol. Reports Osaka Univ., 1960, **10**, 241; *K. E. Khchenyan et al.*, Khim. Prom., 1961, 327). A methyl group in the *ortho* position is more resistant to oxidation. Thus *o*-xylene gives *o*-toluic acid as the main product under conditions where iso- and tere-phthalic acids are readily formed from *m*- and *p*-xylenes. Large quantities of xylenes are produced by the catalytic reforming of napththas (p. 223) and their liquid-phase oxidation is an important route to the phthalic acids. In the commercial processes a mixed xylene stream is oxidized by air at 125–275^0 and pressures up to 40 atmospheres in the presence of a heavy metal oxidation catalyst and a bromine compound, and the resulting solid acids, phthalic, isophthalic, terephthalic and benzoic separated. The individual xylenes can also be oxidized individually to the corresponding phthalic acids (*W. G. Toland* and *E. L. Nimer*, Proc. 4th World Petrol. Congress, 1955, **4**, 39; *D. E. Burney et al.*, Proc. 5th World Petrol. Congress, 1959, **4**, 197; *E. M. Amir*, U.S.P., 3,004,066/1958; *G. H. Whitfield*, B.P., 877,677/1959; *W. F. Brill*, Ind. Eng. Chem., 1960, **52**, 837; *R. Landau* and *R. W. Simon*, Chem. and Ind., 1962, 70; *M. Shigeyasu*, Kogyo Kagaku Zasshi, 1967, **70**, 1150; 1155; 1259).

Liquid-phase oxidation of *p*-diisopropylbenzene at 85^0 in the presence of an alkali peroxy catalyst gives australol (*p*-isopropylphenol) and hydroquinone *via* the mono- and di-peroxides, respectively. One of the hydrogen atoms on the α-carbon atoms of *p*-diisopropylbenzene (and similar dialkylbenzenes) is readily abstracted by oxidation with air at 160–170^0 in the presence of potassium carbonate and aluminium to give 2,3-dimethyl-2,3-bis(*p*-isopropylphenyl)butane *via* an intermediate free radical (*S. Tsutsumi et al.*, J. Fuel Soc., Japan, 1956, **35**, 32; Kogyo Kagaku Zasshi, 1958, **61**, 862). Liquid-phase oxidation of *p*-cymene gives a variety of products, including terephthalic, *p*-toluic and cumic acids, depending on experimental conditions (*Y. Odaira et al.*, *ibid.*, 1956, **59**, 722; *J. Yamamoto* and *K. Hata*, *ibid.*, 1958, **61**, 1217; *H. Boardman*, J. Amer. chem. Soc., 1962, **84**, 1376). For the catalytic oxidation of *para*-substituted isopropylbenzenes to benzoic acids see *R. van Helden* and coworkers (Rec. Trav. chim., 1961, **80**, 1237; 1257).

Of the trimethylbenzenes, *pseudocumene* has received most attention. Mono- and di-basic acids are obtained readily by air oxidation of pseudocumene at 150–160^0 in the presence of a cobalt catalyst, which may be further oxidized to pure trimellitic acid by oxidation with 20% nitric acid at 150–160^0 and 30–35 atm. (*N. Ya. Kachurina et al.*, Neftekhimiya, 1965, **5**, 880). Catalytic oxidation of pseudocumene at 205^0 and 340 psig. in the presence of acetic acid yields trimellitic acid directly (*D. H. Meyer*, U.S.P., 3,261,846/1966). Trimellitic acid is produced commercially by the liquid-phase oxidation of pseudocumene (*H. P. Liao* and *P. H. Towle*, U.S.P., 2,971,011/1961; *Meyer*, U.S.P., 3,161,658/1964). Autoxidation of mesitylene at 100^0 in the presence of cobalt bromide and a peroxide promotor yields 5-methylisophthalic acid and trimesic acid (*Y. Ogata et al.*, Kogyo Kagaku Zasshi, 1967, **70**, 1687). Phthalans and phthalides have been obtained besides carbonyl compounds and acids, by the liquid oxidation of polyalkylbenzenes containing an isopropyl group, in the presence of metal salt catalysts (*R. N. Volkov* and *S. V. Zavgorodnii*, Zhur. obshchei, Khim.,

1961, **31**, 3090; *E. G. E. Hawkins* and *W. F. Maddams*, J. chem. Soc., 1963, 3451):

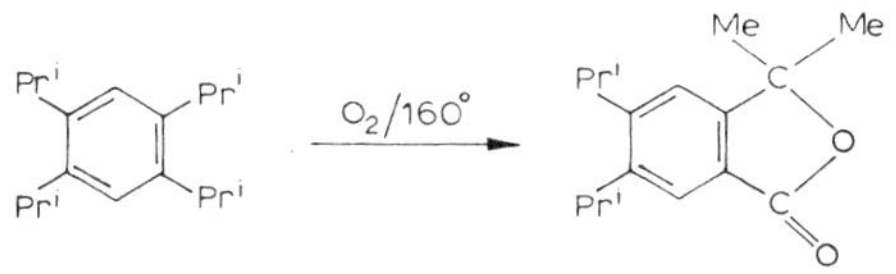

Liquid *durene* has been oxidized by air to durylic acid (*H. F. Lederle* and *R. H. Elkins*, U.S.P., 2,892,868/1959). The autoxidation of *hexaethylbenzene* at 170° in the presence of a metal naphthenate gives a complex mixture of products containing tetraethylphthalic anhydride and tetraethylhydroquinone (*Hawkins* and *Maddams*, J. chem. Soc., 1963, 3456).

In the *vapour-phase oxidation of alkylbenzenes* nuclear oxidation becomes significant (*D. M. Newitt* and *J. H. Burgoyne*, Proc. roy. Soc., 1936, **A153**, 448; *Burgoyne, ibid.*, 1940, **A175**, 539; *F. J. Wright*, J. phys. Chem., 1960, **64**, 1944; *H. Pichler* and *F. Obenaus*, Brennstoff-Chem., 1965, **46**, 258).

Thus the main products obtained by the vapour-phase oxidation of *toluene* are benzaldehyde, benzoic acid, and maleic acid. Nuclear oxidation can be minimized by "cautious" oxidation using excess oxygen at 500° (*J. Jullien* and *L. Minkevitch-Bonazzola*, Bull. Soc. chim. Fr., 1967, 1801–1813). Toluene in benzoic acid is oxidized continuously, essentially to benzoic acid, with oxygen over a heavy metal catalyst at 300–450° under pressure (*J. A. Hundley*, U.S.P., 3,187,038/1965). The oxidation of benzaldehyde can be controlled by the use of catalysts such as the oxides of molybdenum or vanadium on an inert carrier at temperatures up to 550°. More active catalysts favour the oxidation of side chain to carboxyl (*E. B. Maxted*, J. Soc. chem. Ind., 1928, **47**, 101T; *A. P. Kreshkov*, Org. chem. Ind., USSR., 1938, **5**, 551; *A. V. Solomin et al.*, Tr. Inst. Khim. Nauk, Akad. Nauk Kazakh. SSR., 1958, **2**, 182; *M. R. Fenske et al.*, J. Chem. Eng. Data, 1961, **6**, 623; *C. Kroeger* and *G. Bigorajski*, Erdöl u. Kohle, 1962, **15**, 7; 109). Phthalic anhydride is also formed from monoalkylbenzenes containing at least three carbon atoms in a straight chain (*O. Hibino* and *S. Morita*, Nippon Kagaku Zasshi, 1957, **78**, 1764; *Morita*, Bull. chem. Soc. Japan, 1960, **33**, 309). Phthalic and maleic anhydride are the chief products from the vapour-phase oxidation of *o-xylene* over a mixed metal oxide catalyst (*S. K. Bhattacharyya* and *I. B. Gulati*, Ind. Eng. Chem., 1958, **50**, 1719; *D. A. Dowden* and *A. M. V. Caldwell*, U.S.P., 3,012,043/1958). The catalytic vapour-phase air oxidation of *o*-xylene is a commercial route to phthalic anhydride; maleic anhydride is a valuable by-product. With the *m*- and *p*-isomers, products from ring fission are prominent. Air oxidation of *durene* over vanadium pentoxide at 430–450° yields pyromellitic dianhydride (*V. P. Borshchenko et al.*, Neftikhimiya, 1966, **6**, 450). Pyromellitic dianhydride is made commercially by a continuous oxidation process and is used in the manufacture of thermoplastics.

It will be apparent that the aerial oxidation of mononuclear aromatic hydrocarbons has considerable industrial potential. Already intermediates such as the benzenedicarboxylic acids, trimellitic acid, benzoic acid, and phenol, important in the manufacture of synthetic polymers, are being made by such oxidation processes. But these are based on only a few of the benzene homologues potentially available from petroleum. The more interesting commercial advances in the air oxidation of mononuclear aromatic hydrocarbons have been reviewed by *H. L. Riley* (Chem. and Ind., 1966, 979).

(ii) Oxidation in presence of ammonia

Aromatic nitriles may be made by carrying out the vapour-phase oxidation of the hydrocarbons in the presence of ammonia and a catalyst. The reaction is of general application and is a much more attractive process for the manufacture of aromatic nitriles than the classical processes, *e.g.*, diazotisation of aromatic amines, or dehydration of arylamides (*A. D. Kagarlitskii et al.*, Zhur. priklad. Khim., 1963, **36**, 1948; *S. Sakuyama*, Chem. Eng. Progr., 1964, **60**, (9), 48; *S. D. Mekhtiev et al.*, Azerb. Khim. Zhur., 1966, No. 5, 73); for example, oxidative aminolysis of toluene at 400^0 over vanadium pentoxide gives benzonitrile in 87 % yield (*Mekhtiev et al., ibid.*, 1965, No. 2, 18).

(iii) Oxidation with other reagents

Many other reagents and methods can be used to oxidize benzene and its homologues and these are described below. Except for peroxides and electrochemical oxidation, the benzene nucleus is relatively resistant to such reagents. Oxidation of the alkyl chain is comparatively easy and in general the attack is on the α-carbon atom; carbonyl and carboxyl groups being produced.

(1) *Peroxides.* The benzene nucleus is attacked by peroxy compounds. Thus, refluxing *benzene* with benzoyl peroxide under nitrogen yields polyphenyls resulting from phenyl radicals originating from the benzene (*B. M. Lynch* and *K. H. Pausacker*, Austral. J. chem., 1957, **10**, 40; *G. A. Razuvaev et al.*, Doklady Akad. Nauk SSSR., 1960, **130**, 336). *Toluene* reacts with benzoyl peroxide–cupric chloride in acetonitrile to form the three isomeric tolyl benzoates (*P. Kovacic* and *M. E. Kurz*, Tetrahedron Letters, 1966, 2689). There are numerous examples in the literature of aromatic "oxygenation", *i.e.*, the introduction of RO into the nucleus (where R = H, alkyl, acyl, etc.). Oxygenation of the benzene nucleus may be effected with peroxides in the presence of iron salts (Fenton's reagent), presumably by a free radical mechanism (*J. R. L. Smith* and *R. O. C. Norman*, J. chem. Soc., 1963, 2897) or with peroxides in the presence of a Lewis catalyst by an electrophilic mechanism (*J. D. McClure* and *P. H. Williams*, J. org. Chem., 1962, **27**, 24; *D. Z. Denney et al.*, J. Amer. chem. Soc., 1964, **86**, 46), the

latter reaction being more selective. Thus *m-xylene* is oxidized by trifluoroperoxyacetic acid to *m*-xyloquinone, 2,6- and 2,4-xylenols, and *m*-xyloquinol in only low conversion (*R. D. Chambers et al.*, J. chem. Soc., 1959, 1804) but alkyl tolyl carbonates have been obtained in high yields from *toluene* by direct oxygenation with dialkylperoxydicarbonates in the presence of metal halides, and provides a useful method for the direct oxidation of aromatic hydrocarbons (*P. Kovacic et al.*, J. Amer. chem. Soc., 1965, **87**, 1566; 1966, **88**, 2068; 1967, **89**, 4960; Chem. Comm., 1966, 321; J. org. Chem., 1968, **33**, 1950).

o-Nitrosophenol is formed at room temperature by the action of aqueous hydrogen peroxide, hydroxylamine hydrochloride and copper sulphate on *benzene*. Other aromatic hydrocarbons behave similarly (*J. O. Konecny*, J. Amer. chem. Soc., 1955, **77**, 5748). Phenol has also been produced by irradiation oxidation and glow discharge of benzene in water containing metal ions at elevated temperatures (*H. Schueler et al.*, Naturwissenschaften, 1961, **48**, 426; *E. V. Barelko et al.*, Doklady Akad. Nauk, SSSR., 1957, **116**, 74; 1961, **136**, 143; *N. Suzuki* and *H. Hotta*, Bull. chem. Soc. Japan, 1964, **37**, 244; *A. Dauno*, Hydrocarb. Process, Petrol. Refiner, 1964, **43**, No. 4, 131). Benzene is oxidized to *p*-benzoquinone by silver peroxide in the presence of nitric acid, and more satisfactorily by electrochemical oxidation using diluted sulphuric acid, a lead–lead dioxide anode and a cooled lead cathode (*F. Fichter* and *R. Stocker*, Ber., 1914, **47**, 2003).

(2) *Electrolytic oxidation.* Although electrolytic oxidation of benzene yields *p*-benzoquinone as the main product, toluene and xylenes are attacked preferentially in the side chain with the formation of appreciable amounts of aldehydes (*A. Renard*, Compt. rend., 1881, **92**, 965; *K. Puls*, Chem.-Ztg., 1901, **25**, 263; *A. Merzbacher* and *E. F. Smith*, J. Amer. chem. Soc., 1900, **22**, 723; *Fichter*, Z. Electrochem., 1913, **19**, 781; *Fichter et al.*, Helv., 1925, **8**, 74; 1927, **10**, 40). This method in general is not satisfactory and the product usually contains other substances such as quinones and further oxidation products. Using platinium electrodes, *F. Law* and *F. M. Perkin* (Trans. Faraday Soc., 1905, **1**, 31; 251) obtained yields of aldehyde up to 35 %. It is noteworthy that in electrolytic oxidation a diaphragm may have a striking effect on the product. *p*-Xylene oxidized in aqueous sulphuric acid suspensions, gives *p*-tolualdehyde in not very good yield if a diaphragm is used; without a diaphragm *p*-hydrotoluoin ($\alpha\beta$-dihydroxy $\alpha\beta$ di-*p*-tolylethane) is formed in 70 % yield (*Fichter* and *M. Rinderspacher*, Helv., 1926, **9**, 1097).

(3) *Nitric acid.* The alkyl side chain can be oxidized by heating with dilute nitric acid (*R. Fittig* and *W. Bieber*, Ann., 1870, **156**, 242). At temperatures below 100° intermediate oxidation products can be isolated. Thus *p*-methylacetophenone has been made by oxidation of *p*-cymene with 21 % nitric acid at 85° (*J. Kulesza* and *W. Madalinski*, Przemysl Chem., 1955, **11**, 141). At higher temperatures (140–225°) under pressure, smooth oxidation of the alkyl group to carboxyl is achieved with 15 to 50 % nitric acid, up to 82 % yields of benzenedicarboxylic acids being obtained from a wide range of dialkylbenzenes (*I. N. Nazarov et al.*, Doklady Akad. Nauk SSSR., 1954, **99**, 1003; *E. Zaugg* and *R. T. Rapalla*, Org. Synth., 1955, Coll. Vol. III, p. 820; *W. F. Tuley* and *C. S. Marvel*, *ibid.*,

p. 822; *I. V. Butina* and *V. G. Plyusnin,* Tr. Inst. Khim., Akad. Nauk SSSR., Ural. Filial, 1960, No 4, 73; *H. N. Friedlander* and *R. H. Baldwin,* U.S.P., 2,970,169/1961). Similar yields of terephthalic acid have also been obtained from *p*-di-isopropylbenzene and 20 % nitric acid in the presence of a vanadic anhydride catalyst by heating to only 105° (*M. M. Movsimzade* and *A. P. Petrov,* Izvest. Vysshikh Ucheb. Zavendenii, Neft, i Gas, 1961, No. 10, 65). *L. N. Ferguson* and *A. I. Wims* (J. org. Chem., 1960, **25**, 668) have made a study of the selective oxidation of dialkylbenzenes with dilute nitric acid. The conversion of durene to pyromellitic dianhydride has been achieved by liquid-phase oxidation using nitric acid as oxidant (*Anon,* Chem. Eng. News, Oct. 5, 1964, p. 52). Dialkylbenzenes can be oxidized to the corresponding dicarboxybenzenes by treatment with nitric oxide in non-aqueous media at 150° in the presence of a cobalt salt catalyst (*W. A. O'Neill,* B.P., 823,437/1959).

(4) *Manganese dioxide and acid.* Aldehydes may be obtained from the methylbenzenes by use of manganese dioxide and dilute sulphuric acid at 40–50° (*H. Fournier,* Compt. rend., 1901, **133**, 634; *Law* and *Perkin,* J. chem. Soc., 1907, **91**, 258); terephthalic acid has been obtained in good yield by oxidation of *p*-xylene and *p*-cymene with manganese dioxide and sulphuric acid at higher temperatures (*E. Asanagi* and *K. Mizutani,* Kogyo Kagaku Zasshi, 1956, **59**, 690). Aqueous manganic sulphate is also effective in oxidizing the methyl group to aldehyde (*M. S. Venkatachalapathy et al.,* Bull. Acad. polon. Sci., Ser. Sci. chim., geol., geog., 1959, **7**, 629; 1960, **8**, 361; J. Electrochem. Soc., 1963, **110**, 202). Ceric ammonium nitrate in 50 % aqueous acetic acid oxidizes toluene to benzaldehyde (*W. S. Trahanovisky* and *L. B. Young,* J. org. Chem., 1966, **31**, 2033).

(5) *Persulphates. etc.* Persulphates in dilute sulphuric acid oxidize the alkyl side chain to both aldehyde and acid and also effect a union of the two alkyl groups. Thus toluene gives about 15 % of bibenzyl in addition to benzaldehyde and benzoic acid when refluxed with persulphate; ethylbenzene gives 1,2-dimethyl-1,2-diphenylethane and phenylacetaldehyde (*R. Wolffenstein et al.,* Ber., 1899, **32**, 432; 1901, **34**, 2423; 1904, **37**, 3215; 3221; *Law* and *Perkin, loc. cit.*; *P. C. Austin,* J. chem. Soc., 1911, **99**, 262). When ethylbenzene is heated with potassium persulphate in water at 100° without sulphuric acid, acetophenone is mainly formed and a small amount of 2,3-diphenylbutane (*B. D. Kruzhalov* and *P. G. Sergeev,* Tr. Nauchn.-Issled. Inst. Sintetich. Spirtov i Organ. Produktov 1960, 283). Benzoic acid is obtained in 50 % yield by refluxing toluene with aqueous peroxymonosulphate (*R. J. Kennedy* and *A. M. Stock,* J. org. Chem., 1960, **25**, 1901). Toluene and ethylbenzene are oxidzed by cobaltic perchlorate in aqueous acetonitrile yielding side-chain oxidation products, namely, carbonyl derivatives (*T. A. Cooper et al.,* J. chem. Soc., [B], 1966, 793; 1967, 687).

(6) *Permanganates and dichromates.* The complete oxidation of the alkyl side chain to carbonyl can be accomplished most conveniently in the laboratory by the use of acid or alkaline permanganate, the side chain being broken between the α and β carbon atoms (*C. F. Culles* and *J. W. Ladbury, ibid.,* 1955, 555; *D. I. Dimitrov et al.,* Godishnik Khim-Tekhnol. Inst., 1957, **4**, No. 2, 95; 1959,

6, No. 2, 1; 1962, **9**, No. 2, 133; *D. Rakhimov* and *G. Khodzhaev*, Uzbeksk. Khim. Zhur., 1964, **8**, No. 3, 34), or by chromic acid in aqueous sulphuric acid (*S. G. Brandenberger et al.*, J. Amer. chem. Soc., 1961, **83**, 2146; *P. A. Argabright* and *A. H. Petersen*, U.S.P., 2,328,461/1967). This is a characteristic reaction of alkylbenzenes and has been used as a method for determining the orientation of the alkyl group. When two or more alkyl groups are present it is often possible to oxidize one alkyl preferentially; an example is the oxidation of *p*-cymene to *p*-isopropylbenzoic acid. The regulated oxidation of mesitylene gives successively mesitylenic, uvitic, and trimesic acid. Oxidation of alkylbenzenes by autoclaving with aqueous sodium bichromate at 275^0 gives similar results to the conventional acid chromate oxidation (*R. H. Reitsema* and *N. L. Allphin*, J. org. Chem., 1962, **27**, 27).

(7) *Étard's reagent, chromyl chloride.* Chromyl chloride combines with aromatic hydrocarbons in an anhydrous solvent. Treatment of the complex so formed with water under reducing conditions, results in the oxidation of the alkyl group to a carbonyl group. Thus *p*-xylene gives *p*-tolualdehyde and ethylbenzene yields acetophenone (*A. Étard*, Compt. rend., 1877, **84**, 127; Ann. chim. Phys., 1881 [iv], **22**, 218; *O. H. Wheeler*, Canad. J. Chem., 1958,**36**, 667; 1964, **42**, 706; *C. N. Rentea et al.*, Tetrahedron, 1966, **22**, 2037; 3501; Rev. Roum. Chim., 1967, **12**, 1495; 1503).

(8) *Lead tetra-acetate.* The alkyl side chain is oxidized mainly to aldehyde or ketone by treatment of an alkylbenzene with lead tetra-acetate at 120^0 (*E. Detilleux* and *J. Jadot*, Bull. Soc. roy. Sci. Liège, 1955, **24**, 366).

(9) *Sulphur compounds.* Excellent conversion of methyl group to carboxyl is obtained by oxidation of the methylbenzenes with sulphur and water at 200–400^0 under autogenous pressure (*W. G. Toland et al.*, J. Amer. chem. Soc., 1958, **80**, 5423; J. org. Chem., 1961, **26**, 2929). The oxidation of arylmethyl compounds to carboxylic acids is also brought about by heating with an aqueous polysulphide solution (*W. A. Pryor*, J. Amer. chem. Soc., 1960, **82**, 2715), or an inorganic water-soluble sulphate and sulphide at 200–300^0 under pressure (*Toland*, G.P., 1,109,683/1954). Alkylbenzenes are oxidized in the vapour-phase by sulphur dioxide to the corresponding carboxylic acids (*A. J. Shipman*, Adv. Chem. Ser., 1965, **51**, 52). Toluene is converted to benzaldehyde in good yield by rapid reaction with sulphur dioxide at 310–410^0 in the presence of a vanadium pentoxide–alumina catalyst (*T. H. Strickland* and *A. Bell*, U.S.P., 2,928,879/1960; Ind. Eng. Chem., 1960, **52**, 7).

(d) Additive reactions

(i) Hydrogenation

(1) *Vapour and liquid phase hydrogenation.* Benzene and its homologues can be hydrogenated in the vapour or liquid phase using a wide variety of catalysts, to give hexahydro derivatives, the cyclohexanes, in high yield

(*H. A. Smith*, "Catalysis", Ed. *P. H. Emmett*, Reinhold Publishing Corpn., 1957, Vol. V, p. 175; *J. Voelter*, J. catalysis, 1964, **3**, 297).

The reduction can be effected at room temperature using platinum or palladium catalysts, or at elevated temperatures with nickel, copper, cobalt, etc. type catalysts. In general the rate of hydrogenation is little affected by the length or branching of the side chain but decreases with increasing nuclear substitution, and increases with symmetry of substitution. The hydrogenation of benzene to cyclohexane is an important stage in the manufacture of cyclohexane the basic intermediate for making polyamide (nylon) fibres (*H. J. Boonstra* and *P. Zwietering*, Chem. and Ind., 1966, 2039). Destructive hydrogenation of benzene and alkylbenzenes occurs at high temperatures ($>400^0$) using a tungsten sulphide catalyst, (poly)alkylcyclopentanes and ultimately paraffinic hydrocarbons being formed. An increase in temperature increases the amount of isomerization products (*N. I. Shuykin* and *N. G. Berdnikova*, Research [London], 1956, **9**, 132; *I. V. Kalechits et al.*, Tr. Vost.-Sibirsk. Filiala, Akad. Nauk SSSR., Ser. Khim., 1961, **38**, 5; 15; 19).

(2) *Reduction with metals, etc.* Partial catalytic hydrogenation is not possible since the cyclohexadienes and cyclohexenes are hydrogenated much more readily than the aromatics. However, benzene and alkylbenzenes have been reduced to their *dihydro derivatives* in good yields by reaction with sodium in liquid ammonia containing methanol or ethanol (*A. J. Birch*, Quart. Reviews, 1950, **4**, 69; 1958, **12**, 17; *W. Hückel* and *U. Wörffel*, Ber., 1955, **88**, 338), or exclusively to the corresponding *cyclohexene* by sodium or potassium in liquid ammonia without addition of alcohol (*L. H. Slaugh* and *J. H. Riley*, J. org. Chem., 1967, **32**, 369).

Cyclohexadienes and cyclohexenes are also obtained by reduction of benzene and alkylbenzenes with calcium hexamine, $Ca(NH_3)_6$ (*A. V. Dumanskii* and *A. V. Zvyereva*, J. Russ. phys. Chem. Soc., 1916, **48**, 994; *H. Boer* and *P. M. Duinker*, Rec. Trav. chim., 1958, **77**, 346). The benzene nucleus can be reduced to mono-olefins and cycloparaffins by lithium in low molecular weight aliphatic amines and diamines (*R. A. Benkeser et al.*, J. Amer. chem. Soc., 1954, **76**, 631; 1955, **77**, 3230; Tetrahedron Letters, 1960, No 16, 1; J. org. Chem., 1964, **29**, 1313; *L. Reggel et al.*, *ibid.*, 1957, **22**, 891). The reaction may be controlled electrolytically (*Benkeser et al.*, J. Amer. chem. Soc., 1964, **86**, 5272). Potassium and rubidium in neat methylamine at 60–190^0 readily reduce benzene and alkylbenzenes to cyclic mono-olefins, but sodium alone is ineffective (*Slaugh* and *Riley*, J. org. Chem., 1967, **32**, 2861).

(ii) Addition of halogens

Benzene combines additively with *chlorine* or *bromine* in the presence of sunlight or ultra-violet irradiation but in the absence of a catalyst, to give hexahalides of the general formula $C_6H_6X_6$.

Well over a century ago Faraday obtained a hexachlorocyclohexane by bubbling chlorine into benzene in the presence of sunlight. It is considered that the chlorine undergoes homolytic fission and the chlorine atoms so produced then add to the nucleus. The known benzene hexachlorides (1,2,3,4,5,6-hexachlorocyclohexanes) and their stereochemical configurations are discussed in Volume II B (pp. 36–37). Benzene hexachloride, or Gammexane as it is commonly called, is an important insecticide and is made commercially be reacting chlorine and benzene at 20–60^0 in the presence of actinic light. Only the γ-isomer (Lindane) is toxic to insects (for more recent references on the photochlorination of benzene see *A. H. Maude et al.*, U.S.P., 3,006,833/1961; *G. M. Strongin*, Tr. po Khim. i Khim. Tekhnol., 1960, **3**, 398; *T. N. Pliev*, Uchenye Zapiski Yakutsk Univ., 1958, No. 5, 21).

Hypochlorous acid is stated to form an additive compound with benzene in the presence of mercuric oxide (*L. Carius*, Ann., 1865, **136**, 323), but in its absence the reaction leads to α- and β-benzenehexachloride. In the dark, substitution occurs with the formation of chlorobenzene. The homologues of benzene give only mono-substitution products (*F. W. Klingstedt et al.*, C.A., 1929, **23**,, 1399). Additive chlorination of benzene occurs in the dark with a heterogeneous benzene–water mixture in the presence of chlorine and hypochlorous acid or substances capable of forming hypochlorous acid with chlorine (*K. Schwabe* and *G. Gebhardt*, Z. physik. Chem. [Frankfurt], 1960, **24**, 87).

Addition of *fluorine* to benzene goes very readily without activation, (in fact the greater reactivity of the fluorine is liable to destroy the nucleus to form cyclic or straight-chain aliphatic fluorine compounds) to give perfluorocyclohexanes (*R. Stephen et al.*, J. chem. Soc., 1959, 148, 159). Fully halogenated and chlorofluorohydro-cyclohexanes are obtained by the vapour-phase halogenation of benzene by chlorine trifluoride (*R. E. Banks et al.*, Ind. Eng. Chem., Process Design Develop., 1962, 1, No. 4, 262). Polyfluorocyclohexanes are produced when benzene is passed over manganic fluoride at 300 350^0 (*E. J. P Fear* and *J. Thrower*, J. appl. Chem., 1955, **5**, 353). These compounds are discussed more fully in Volume II B, pp. 37 *et seq.*

The homologues of benzene also give additive compounds with the halogens in the presence of ultra-violet irradiation but under these conditions the main products are those in which the halogen is substituted in the side chain (*M. S. Kharasch* and *M. G. Berkman*, J. org. Chem., 1941, **6**, 810). Both substitution and addition of chlorine and fluorine occur when chlorine trifluoride reacts with benzene homologues (*J. F. Ellis* and *W. K. R. Musgrave*, J. chem. Soc., 1953, 1063). It is interesting to note that trifluoromethyl radicals add on to the benzene nucleus and hydrogen abstraction occurs only to a minor extent (*S. W. Charles* and *E. Whittle*, Trans Faraday Soc., 1960, **56**, 794; *J. L. Holmes* and *K. O. Kutschko, ibid.*, 1962, **58**, 333).

(iii) Ozone

Benzene reacts with ozonized oxygen at 5–10° to give a triozonide, $C_6H_6O_9$, which is decomposed by water to give glyoxal. Alkylbenzenes also give ozonides (*C. Harries* and *V. Weiss*, Ber., 1904, **37**, 3431; Ann., 1905, **343**, 371; *E. Bernatek et al.*, Acta Chem. Scand., 1967, **21**, 1229). This reaction has been used in the analysis of mono-aromatics in oil fractions (*Boer*, J. Inst. Petroleum, 1960, **46**, 234). The ring fission of benzenoid hydrocarbons by ozone has been discussed elsewhere (see p. 150).

(iv) Carbenes

Carbenes, R^1R^2C:, add to benzene to produce derivatives of norcaradiene and their transformation products, cycloheptatrienes.

Thus methylene (*e.g.* from the photolysis of ketene or diazomethane) reacts with benzene to give cycloheptatriene although some toluene is formed (*E. Mueller* and *H. Fricke*, Ann., 1963, **661**, 33; *T. Terao* and *S. Shida*, Bull. chem. Soc. Japan, 1964, **37**, 687); cycloheptatriene is also formed when benzene is reacted with synthesis gas in the presence of a cobalt catalyst at 190° due presumably to the formation of methylene radicals by partial reduction of the carbon monoxide (*Ya. Y. Eidus et al.*, Izvest Akad. Nauk SSSR., Otdel. khim. Nauk, 1963, 548). Addition of methyl-lithium to refluxing benzene in dichloromethane gives 7-methyl-1,3,5-cycloheptatriene (*G. L. Closs* and *L. E. Closs*, Tetrahedron Letters, 1960, No. 10, 38), and when benzene is heated with diazoacetic ester nitrogen is evolved to give esters of norcaradienecarboxylic acid (V) and cycloheptatrienecarboxylic ester (VI) (*E. Buchner*, Ber., 1885, **18**, 2377; 1901, **34**, 982; 1903, **36**, 3509; 1920, **53**, 865; Ann., 1907, **358**, 1; 1910, **377**, 259). With durene and prehnitene an acetic ester residue is also introduced into one of the methyl groups (*L. I. Smith et al.*, J. Amer. chem. Soc., 1934, **56**, 2167; 1938, **60**, 648).

CO_2Et (V) CO_2Et (VI)

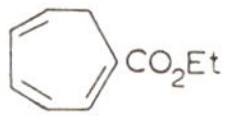

Carbenes have been used in the preparation of benzene derivatives by insertion into side chains; isopropylbenzene and dichlorocarbene giving β,β-dichloro-*tert*-butylbenzene (*E. K. Fields, ibid.*, 1962, **84**, 1744; see also Vol. IA, p. 391).

(v) Photochemical addition reactions

Various compounds react photochemically with benzene and alkylbenzenes, with 1,2, 1,3 and/or 1,4 addition, depending to a large extent on the photochemical transformations of the benzene ring (see p. 186). The reaction of alkenes, alkynes, dienes, etc. with aromatic hydrocarbons in the presence of Friedel-Crafts type catalysts is mentioned on pages 158 and 216; reactions with these compounds also occur in the absence of such catalysts, but under UV irradiation, producing some very novel compounds.

Thus, tricyclo-octenes (VII) are formed by 1,3-cycloaddition of alkenes and cycloalkenes to benzene at room temperature and irradiation at 2537Å (*K. E. Wilzback* and *L. Kaplan, ibid.*, 1966, **88**, 2066; *D. Bryce-Smith, A. Gilbert* and *B. H. Orger*, Chem. Comm., 1966, 512).

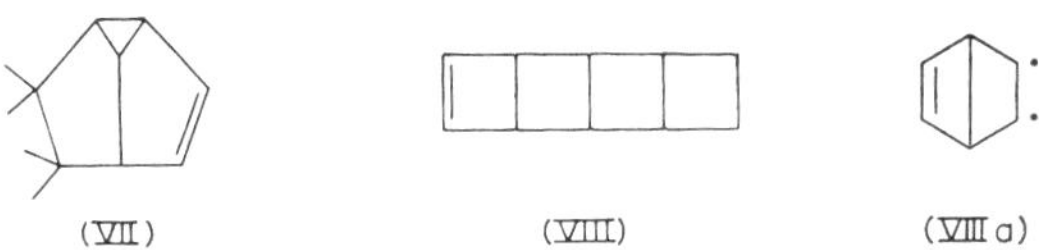
(VII) (VIII) (VIIIa)

The reaction with cycloalkenes can also proceed by a different mechanism to produce tetracyclic products, presumably *via* a diradical (VIIIa) related to Dewar benzene (*R. Srinivasan* and *K. A. Hill*, J. Amer. chem. Soc., 1965, **87**, 4653; *Bryce-Smith, Gilbert* and *Orger, loc. cit.*). For example, *tetracyclo*[$6.2.0^{2,7}.0^{3,6}$]-*dec-9-ene* (VIII), b.p. $24^0/1\cdot5$ mm, is formed by photolysis of a benzene solution of cyclobutene at 25^0 and 2537 Å. 1,2-Addition occurs in the photolysis of a solution of acrylonitrile in benzene at 0^0 leading to the 1:1 adduct; further reaction to give the 2:1 adduct by 1,4-addition of a second molecule of acrylonitrile occurs under reflux (*B. E. Job* and *J. D. Littlehailes*, J. chem. Soc., C, 1968, 886):

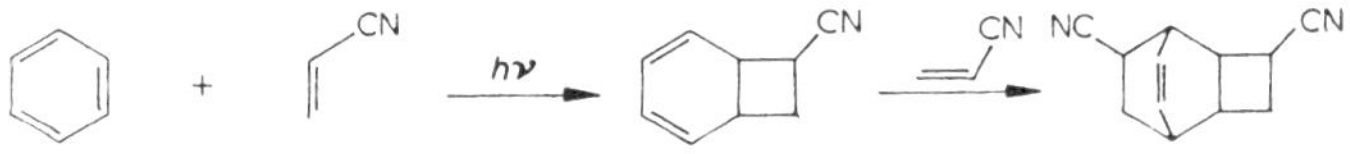

CN + hν → CN → (CN) NC CN

The photochemical reaction of dienes with benzene and alkylbenzenes (*G. Koltzenburg* and *K. Kraft*, Angew. Chem. internat. Edn., 1965, **4**, 981; Tetrahedron Letters, 1966, 389) leads to various adducts of which the 2:2 adducts (IX) and (X) (stereoisomers) formed by 1,4-addition of buta-1,3-diene and isoprene, respectively, to benzene, and the 1:1 adduct (stereoisomers) of probable structure XI, from benzene and isoprene, have been isolated:

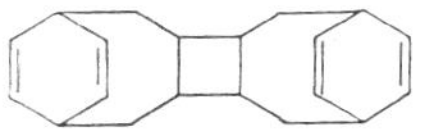

CH_3 H

(IX) 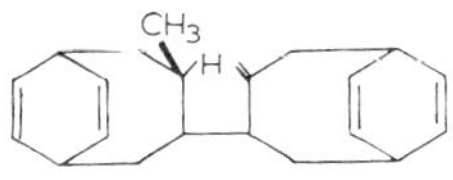 (X) 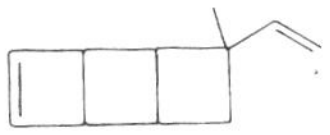 (XI)

Maleic anhydride, maleimide, and *N*-substituted maleimides react with benzene and alkylbenzenes in the presence of UV irradiation to give a 2:1 adduct *via* the 1:1 adduct by successive 1,2 and 1,4-additions (*Bryce-Smith et al.*, J. chem. Soc., 1962, 2675; Tetrahedron Letters, 1966, 1895; *G. B. Vermont et al.*, J. Amer. chem. Soc., 1965, **87**, 4024; J. S. Bradshaw, Tetrahedron Letters, 1966, 2039; *W. M. Hardham* and *G. S. Hammond*, J. Amer. chem. Soc., 1967, **89**, 3200):

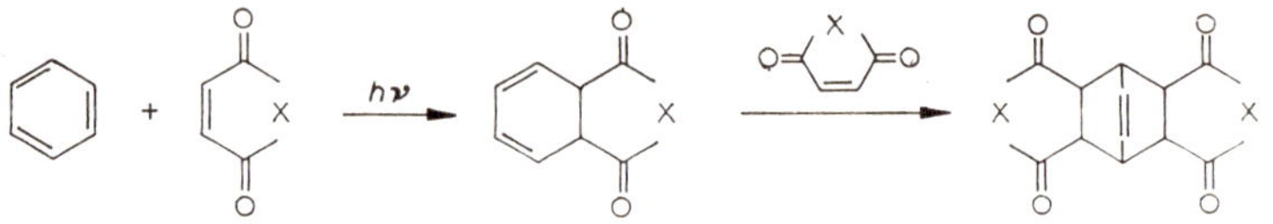

(X = O,NH,NR)

The 1,2-photoaddition of acetylene compounds to benzene is an interesting extension of this type of addition; the ring of the initial adduct opening to form the cyclo-octatetraene derivative (*E. Grovenstein* and *D. V. Rao*, Tetrahedron Letters, 1961, 148; *Bryce-Smith* and *J. E. Lodge*, J. chem. Soc., 1963, 695):

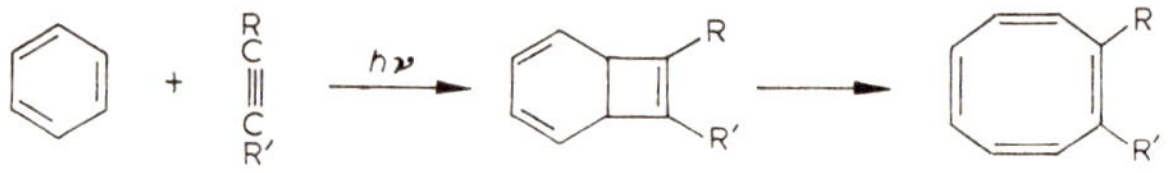

Alkoxy and acyl [3·1·0] bicyclohexenes (XII) are formed by 1,3-photoaddition of alcohols and carboxylic acids, respectively, to benzene and alkylbenzenes, presumably *via* a biradical related to benzvalene (see p. 186) (*Kaplan et al.*, J. Amer. chem. Soc., 1966, **88**, 2881; *E. Farenhorst* and *A. F. Bickel*, Tetrahedron Letters, 1966, 5911). 1,4-Photoaddition of pyrrole to benzene occurs at 25° to give the 1:1 *adduct* (XIII), m.p. 15° (*Bryce-Smith* and *Gilbert*, Chem. Comm., 1967, 263).

(XII) (XIII)

(*vi*) *Molecular complexes*

Benzene and the alkylbenzenes form numerous molecular complexes by sharing their π electrons, as electron donors, with a second component, an electron acceptor, which is additively combined. A 1:1 donor–acceptor ratio is usual. Such complexes play an important role in nuclear electrophilic substitution reactions of the benzenoid hydrocarbons (*K. Le Roi Nelson* and *H. C. Brown*, "The Chemistry of Petroleum Hydrocarbons", Ed. *B. T. Brooks et al.*, Reinhold Publishing Corp., 1955, Vol. III, p. 483). Many of the molecular complexes dissociate so readily into their components that their existence has been recognized only by means of physical measurements (*L. J. Andrews*, Chem. Reviews, 1954, **54**, 713; *A. F. Finch et al.*, Trans. Faraday Soc., 1965, **61**, 2623; *R. F. Weimer* and *J. M. Prausintz*, Spectrochim. Acta, 1966, **22**, 77; *R. J. W. LeFèvre et al.*, J. chem. Soc., B, 1968, 148). The relative stabilities of these complexes is related to the electron-donor capacities or basicities of the benzenoid hydrocarbons. Thus in general stability increases with increasing methyl substitution.

Many classes of electron acceptors are known. The most important are the *Friedel-Crafts acids* (*D. A. McCauley*, "Friedel-Crafts and Related Reactions" Ed. *G. A. Olah*, Interscience Publishers, London, 1964, Vol. 2, Part II, p. 1049). Their formation may be represented by:

$$C_6H_6 + HX + MX_3 \rightleftharpoons [C_6H_7]^{+} + MX_4^{\ominus}$$

and different stoichiometric ratios are found for various hydrocarbons. Thus crystalline complexes of the system mesitylene/hydrogen bromide/aluminium tribromide have been made in the molecular ratios 3:1:2 and 1:1:2. Such complexes undoubtedly exist as intermediates in numerous substitution reactions of the aromatic hydrocarbons, *e.g.*, alkylation. So called proton addition complexes have been made from alkylbenzenes and aluminium chloride, aluminium bromide and gallium halides with hydrogen chloride or bromide. Sigma complexes are apparently formed in the absence of protons [*H-H. Perkampus* and *E. Baumgarten*, Ber. Bunsenges. Physik. Chem., 1964, **68**, (1), 70; Z. physik. Chem. (Frankfurt), 1964, **40**, (3/4), 144; Angew. Chem. intern. Edn., 1964, **3**, 776; *V. A. Koptyug et al.*, Zhur. obshchei Khim., 1964, **34**, 3999; 1965, **35**, 864; 1111]. The crystalline complex Al_2Br_6,C_6H_6 is triclinic with both the Al_2Br_6 and C_6H_6 components centrosymmetric (*D. D. Eley et al.*, J. chem. Soc., 1961, 3867). For a review of the literature on solid complexes of aluminium bromide with aromatic hydrocarbons see *H. C. Brown et al.*, J. Amer. chem. Soc., 1966, **88**, 903; 4128.

The *halogens* form complexes with benzene and the alkylbenzenes, stability increasing in the order $Cl < Br < I < ICl$. The crystalline structures of the benzene–chlorine (1:1) and benzene–bromine (1:1) complexes are monoclinic with chains of alternating benzene and halogen molecules (*O. Hassel* and *K. O. Stroemme*, Acta Chem. Scand., 1958, **12**, 1146; 1959, **13**, 1781). The structures of the alkylbenzene–iodine complexes have been studied by spectroscopy (*R. L. Strong et al.*, J. Amer. chem. Soc., 1960, **82**, 5053; *A. Junghaehnel* and *W. Paetz*, Z. chem., 1964, **4**, No. 3, 110; *W. K. Duerksen* and *M. Tamres*, J.

Amer. chem. Soc., 1968, **90**, 1378), and by measurement of the polar characteristics of the complexes (*J. Gerbier*, Compt. rend., 1965, **261**, 5037).

Hydrogen halides and *halogenomethanes* also complex with benzene and its homologues (*Brown* and *J. J. Melchiore*, J. Amer. chem. Soc., 1965, **87**, 5269). A molecular *complex*, m.p. 34°, has been prepared from mesitylene and hexafluorobenzene (*C. R. Patrick* and *G. S. Prosser*, Nature, 1960, **187**, 1021).

There are many examples of complex formation between *silver* or other *metallic ions* with aromatic hydrocarbons. Silver perchlorate is the classical example forming a solid complex with benzene consisting of chains of alternating benzene molecules and silver ions (*H. G. Smith* and *R. E. Rundle*, J. Amer. chem. Soc., 1958, **80**, 5075). Complexes of benzene and its homologues with mercury, copper, antimony or aluminium salts are also known (*H. Meerwein et al.*, Arch. Pharm., 1958, **291**, 541; *L. W. Daasch*, Spectrochim. Acta, 1959, **15**, 726; *R. Baur et al.*, Helv., 1962, **45**, 775; *R. W. Turner* and *E. L. Amma*, J. Amer. chem. Soc., 1963, **85**, 4046).

The *bisarene metal complexes* constitute a further class of additive complexes formed by the mononuclear benzene hydrocarbons. Iron (II) complexes have been made with benzene, toluene, xylenes, mesitylene, durene, and hexamethylbenzene, the stability increasing in the order given (*M. Tsutsui* and *H. H. Zeiss*, Naturwissenschaften, 1957, **44**, 420). Bisarene complexes of benzene and the alkylbenzenes with molybdenum, chromium, vanadium and tungsten have also been prepared (*E. O. Fischer et al.*, Ber., 1956, **89**, 1805; 1961 **94**, 2204; Naturwissenschaften, 1961, **48**, 452). Arenetricarbonylmanganese salts are readily formed from dichloropentacarbonylmanganese and the appropriate alkylbenzene at 60–80° in the presence of aluminium chloride (*G. Wilkinson et al.*, J. chem. Soc., 1961, 3807).

Benzene and its homologues form *additive compounds with nitro compounds* such as picric acid, styphnic acid, 1,3,5-trinitrobenzene. Benzene picrate separates from a warm solution of picric acid in benzene (*J. Fritzsche*, J. pr. Chem., 1858, [i], **73**, 282); it readily loses benzene on exposure to the air. The picrates of the tetra- and penta-methylbenzenes are more stable, as are the additive compounds with trinitrobenzene (*O. Jacobson*, Ber. 1887, **20**, 898; *A. Töhl, ibid.*, 1888, **21**, 905; *P. Pfeiffer et al.*, Ann., 1916, **412**, 253). The position or nature of the attachment in these complexes in not known except that it is weak in character (*S. C. Abrahams*, J. Amer. chem. Soc., 1952, **74**, 2692; *H. M. Powell et al.*, J. chem. Soc., 1943, 153; 435).

Molecular complexes are formed from benzene and its homologues with *dienophiles*, and such complexes probably precede adduct formation in Diels-Alder reactions (*Andrews* and *R. M. Keefer*, J. Amer. chem. Soc., 1953, **75**, 3776; *W. G. Barb*. Trans. Faraday Soc., 1953, **49**, 143) and certain photoaddition reactions of the benzene ring (*Bryce-Smith et al., loc. cit.*; *Vermont et al., loc cit.*; *Hardmann* and *Hammond, loc. cit.*), (see p. 199). A 1 : 1 molecular complex forms between *acrylonitrile* and benzene (*B. A. Arbusov et al.*, Izvest. Akad. Nauk SSSR., Ser. Fiz., 1963, **27**, 82). The intensely coloured solutions formed when tetracyanoethylene is dissolved in benzenoid hydrocarbons is attributed to the formation of π complexes (*R. E. Merriefield* and *W. D. Phillips*, J. Amer. chem. Soc., 1958, **80**, 2778).

Additive complexes of benzenoid hydrocarbons are also formed with *quinones* (*W. H. Hunter* and *E. H. Northey*, J. phys. Chem., 1933, **37**, 875; *L. Michaelis* and *S. Granick*, J. Amer. chem. Soc., 1944, 60, 1023; *M. Chowdhury*, Trans. Faraday Soc., 1961, **57**, 1482; *E. A. Halevi* and *M. Nussim*, J. chem. Soc., 1963, 876), and *sulphur dioxide* (*Andrews* and *Keefer*, J. Amer. chem. Soc., 1951, **73**, 4169; *D. Booth et al.*, Trans. Faraday Soc., 1959, **55**, 1293).

(vii) Clathrate compounds

Addition of benzene to a solution of *nickel cyanide* in aqueous ammonia containing acetic acid brings about the precipitation of "benzene clathrate", $[Ni(CN)_2NH_3C_6H_6]$ in which the benzene is retained in the closed cavities of the crystalline nickel cyanide–ammonia complex (*K. A. Hofmann* and *F. Z. Kuspert*, Z. anorg. Chem., 1897, **15**, 204). Clathrates similar to the Hofmann type are formed from the double cyanides, $MNi(CN)_4$, where M is bivalent copper, cadmium or zinc (*R. Baur* and *G. Schwarzenbach*, Helv., 1960, **43**, 842), from nickel cyanide–alkylamine complexes, $Ni(CN)_2H_2NR$ (*G. Gawalek et al.*, Chem. Tech. [Berlin], 1964, **16**, 409) and from certain complexes of the type dithiocyanatotetrakis(α-arylalkylamine)nickel (*P. de Radzitzky et al.*, Bull. Soc. chim. Belges, 1965, **74**, 381). The latter complexes also clathrate with alkylbenzenes; the stability of the clathrate decreases in the order benzene $>$ toluene $>$ xylenes, and increases with increasing length of the alkylamine.

Benzene and alkylbenzenes form clathrates with *Werner complexes*, $MA_2(BN)_4$ in which M is a metal, *e.g.*, Mn, Ni, Co, Fe, Cu, Zn; A is an inorganic anion, *e.g.*, SCN, CNO, Cl, NO_3, NO_2, or formate, and BN is a basic nitrogen compound, *e.g.*, substituted pyridine, isoquinoline, nicotinimide or amine. These clathrates are formed rapidly by stirring the complex into the hydrocarbon. There is a marked selectivity for *p*-dialkylbenzenes. The benzenoid hydrocarbon can be recovered by treatment with aqueous acid, distillation, or extraction (*W. D. Schaeffer et al.*, J. Amer. chem. Soc., 1957, **79**, 5870)

Clathration has been used to purify and separate benzene and isomeric alkylbenzenes (*de Radzitsky et al.*, Rev. Inst. francs. Pétrole, 1961, **16**, 886; *R. N. Fleck* and *C. G. Wright*, U.S.P., 3,013,091/1961; *M. Grimberg*, Chem.-Ztg., 1964, **88**, [4], 110; *P. W. Sherwood*, Brit. Chem. Eng., 1965, **10**, 382). A demonstration plant has been built for separating *m*-xylene from a mixture of *m*- and *p*-xylenes using a Werner complex (Hydrocarb. Process. Petrol. Refiner, 1965, **44**, No. 11, 193).

(e) Substitution reactions

(i) Halogenation (*cf. p.* 71)

In general halogenation of benzene and its homologues in non-polar media at elevated temperatures proceeds by a free radical mechanism and is a relatively slow reaction unless catalyzed by light or peroxides. It leads to additive halogenation of the nucleus (p. 196) or substitution in the side chain. On the other hand halogenation at ambient temperatures in the presence of Friedel-Crafts type catalysts and/or in polar media, is a faster reaction, electrophilic in nature and leads to nuclear substitution.

Nuclear substitution products are formed in *liquid-phase halogenation* without catalysts but the reaction is relatively slow; rates and product distribution being highly dependent on the presence and type of solvent (*P. W. Robertson et al.*, J. chem. Soc., 1953, 782; *L. M. Stock et al.*, J. Amer. chem. Soc., 1957, **79**, 1421; 5175; 1959, **81**, 5615; 1961, **83**, 4605; *P. B. D. de la Mare* and *J. H. Ridd*, "Aromatic Substitution", Academic Press Inc., New York, N.Y., 1959; *R.J. Dolinski* and *R.M. Nowak*, J. org. Chem., 1967, **32**, 2936). One, two, three, or more halogens may be introduced and the reaction is highly selective giving *ortho–para* orientation in the nucleus. The reaction is catalyzed by metal halide catalysts *e.g.* aluminium chloride or ferric chloride, and is considered to proceed by a mechanism similar to that for the Friedel-Crafts reaction. Such catalysts suppress completely the additive reaction and also side-chain halogenation (*L. J. Andrew* and *R. M. Keefer*, J. Amer. chem. Soc., 1956, **78**, 4549; *W. Strubell*, Chem. Tech. [Berlin], 1957, **9**, 597; *E. P. Di Bella*, U.S.P., 3,000,975/1959; *G. A. Olah et al.*, J. Amer. chem. Soc., 1964, **86**, 1039; 1055). The general procedure is to add the halogen to the hydrocarbon and catalyst, with or without solvent, until the desired gain in weight has been reached.

In the commercial production of the mono- and the *ortho*- and *para*-di-chlorobenzenes the chlorination of benzene is carried out at 40–60° in the presence of about 1 % of iron turnings; the extent of chlorination being controlled by density. Practically no *m*-dichlorobenzene is formed and the principal component in the residue is 1,2,4-trichlorobenzene (*H. F. Weigandt* and *P. R. Lantos*, Ind. Eng. Chem., 1951, **43**, 2167; *W. L. Faith et al.*, "Industrial Chemicals", Chapman and Hall Ltd., London, 1957, p. 265). Chlorination of toluene in the presence of ferric chloride and sulphur dichloride yields a mixture of *o*- and *p*-chlorotoluenes (Neth. Pat. Appl., 6,511,484/1966). Nuclear substitution is also effected by chlorination of benzene and its homologues at room temperature under the influence of gamma rays but becomes inhibited at elevated temperatures when side-chain chlorination predominates (*J. Imamura* and *N. Ota*, Kogyo Kagaku Zasshi, 1960, **63**, 289).

Chlorine reacts with benzene above 400° in the *vapour-phase* in the absence of catalyst to form a different series of polychlorobenzenes than that obtained by the liquid-phase reactions. *m*-Dichlorobenzene is the principal dichloride formed which chlorinates further to the symmetrical 1,3,5-trichlorobenzene (*J. Mason et al.*, J. chem. Soc., 1931, 3150). The first stage of the Raschig phenol process involves the vapour-phase oxychlorination of benzene with hydrogen chloride and air in the presence of a metal chloride catalyst at 230–250° to give the intermediate monochlorobenzene which is then hydrolyzed to phenol (*J. Gordon*, Petroleum Refiner, 1961, **40** No. 6, 193; *W. H. Prahl*, Chem. Engrs., London, 1964, No. **181**, CE 199; CE 220).

Nuclear substitution of benzene and its homologues by chlorine may be effected by a number of reagents capable of liberating "positive" chlorine and includes hypohalous acids especially in the presence of strong acids (*W. J. Wilson* and *F. G. Soper*, J. chem. Soc., 1949, 3376; *D. H. Derbyshire* and *W. A. Waters*, *ibid.*, 1950, 564; 573; *de la Mare* and *J. T. Harvey*, *ibid.*, 1956, 36; 1957, 131), certain Lewis acid metal halides (*P. Kovacic et al.*, J. Amer. chem. Soc., 1954, **76**, 5491; 1960, **82**, 1917; 5740; J. org. Chem., 1961, **26**, 762; 1963, **28**, 972):

$$ArH + 2\ FeCl_3 \rightarrow ArCl + 2\ FeCl_2 + HCl$$

and nitryl chloride (*F. P. Gintz et al.*, J. chem. Soc., 1958, 445).

Iodine does not react directly with the benzenoid hydrocarbons and substitution of nuclear hydrogen by iodine is brought about by the action of elementary iodine on the hydrocarbon in the presence of an oxidizing agent such as iodic acid (*A. Kekulé*, Ann., 1866, **137**, 161; *H. O. Worth et al.*, Makromol. Chem., 1964, **80**, 120), persulphate (*K. Elbs* and *A. Jaroslawzew*, J. pr. Chem., 1913, [ii], **88**, 92), nitric acid (*A. Edinger* and *P. Goldberg*, Ber., 1900, **33**, 2875; *R. L. Datta* and *N. R. Chatterjee*, J. Amer. chem. Soc., 1917, **39**, 437; *F. B. Dains* and *R. Q. Brewster*, Org. Synth., 1947, Coll. Vol. I, p.323; *B. V. Tronov* and *A. N. Novikov*, Izvest. Vysshikh Ucheb, Zavedenii Khim. i Khim. Tekhnol., 1960, **3**, 872) or peroxyacetic acid (*Y. Ogata* and *K. Nakajima*, Tetrahedron, 1964, **20**, 43). Alternatively mercuric oxide can be used to remove the hydriodic acid which is formed in the reaction. Substitution iodination of alkylbenzenes can also be achieved by reaction with interhalogen compounds, viz. ICl and IBr, in aqueous or halogenated media (*E. Schulek* and *K. Burger*, C.A., 1962, **56**, 2871e; *Andrews* and *Keefer*, J. Amer. chem. Soc., 1957, **79**, 1412).

Unlike the other halogens the greater reactivity of the fluorine molecule tends to destroy the aromatic system. The action of fluorine on benzene is discussed in Volume II B, pp. 37 *et seq.* Nuclear substitution of benzene and its homologues has been achieved using chlorine trifluoride and a Lewis acid or metal fluoride as catalyst (*J. F. Ellis* and *W. K. R. Musgrave*, J. chem. Soc., 1950, 3608; 1953, 1963).

Substitution in the side chain of the alkylbenzenes can be brought about with almost complete exclusion of nuclear substitution by the action of chlorine or bromine on the hydrocarbon at elevated temperatures preferably during exposure to ultra-violet light, sunlight or in the presence of free radical initiators. Thus the methyl groups in xylene are completely chlorinated by reaction at 90° in the presence of benzoyl peroxide (*A. E. Kretov* and *A. D. Syrovatko*, Zhur. obshchei

Khim., 1960, **30**, 3019; *S. D. Mekhtiev et al.*, Azerb. khim. Zhur., 1961, No. 2, 17; *G. Drechsler*, Z. Chem., 1965, **5**, No. 5, 180; *I. P. Azad* and *A. Guillemonat*, C. R. Acad. Sci., Paris, Ser. C, 1967, **264**, 720). It is probable that the complex formed between the halogen and the hydrocarbon (p. 201), preparatory to further reaction by addition, dissociates at the higher temperature; the free halogen atom then attacking the side chain. The halogen is preferentially attached to the α carbon atom. Thus photochemical chlorination or bromination of hexaethylbenzene and hexapropylbenzene without a catalyst gives the corresponding hexakis-(α-halogenoalkyl)benzene, but hexachlorobenzene is obtained if ferric chloride is present (*H. Hopff et al.*, Helv. 1961, **44**, 19; 1965, **48**, 1289; Chimia [Aaran], 1964, **18**, No. 4, 140; *G. Illuminati et al.*, J. Amer. chem. Soc., 1965, **87**, 3953). Dehydrochlorination is liable to occur if the temperature is too high but this can be inhibited by the presence of peroxide. Chlorination of the alkylbenzenes in the gas phase gives ω-chloroalkylbenzenes (*J. W. Engelsma* and *E. C. Kooyman*, Rec. Trav. chim., 1961, **80**, 537). The direct fluorination of the side chain of alkylbenzenes has been little studied. Some benzyl fluoride is formed by the reaction of toluene with chlorine trifluoride (*Ellis* and *Musgrave*, *loc. cit.*). *tert*-Butyl hypochlorite (*C. Walling* and *B. B. Jacknow*, J. Amer. chem. Soc., 1960, **82**, 6108) and sulphuryl chloride (*D. D. Reynolds* and *C. F. H. Allen*, U.S.P., 2,811,486/1957; *E. W. Lane*, U.S.P., 2,926,202/1960) in the presence of initiators give good yields of side-chain chlorinated derivatives.

(*ii*) *Nitration* (*cf. p.* 70)

Direct replacement of hydrogen in the aromatic nucleus by nitro groups is achieved by a wide variety of reagents and conditions. The most common nitrating agent is "mixed acids", a mixture of nitric and sulphuric acids. Sulphuric acid is a very active catalyst for the nitration of alkylbenzenes and maximum activity is obtained with 90 % sulphuric acid (*F. H. Westheimer* and *M. S. Kharasch*, J. Amer. chem. Soc., 1946, **68**, 1871). The introduction of one nitro group is achieved by the use of somewhat more than one molecular proportion of nitric acid in a moderate excess of concentrated sulphuric acid. Two or three nitro groups are introduced by using a greater proportion of nitrating mixture. There appears to be no recorded example of the direct introduction of more than three groups by nitration.

In the commercial nitration of benzene to nitrobenzene a mixed acid composition of approximately 60 to 53 % sulphuric acid, 30 to 39 % nitric acid and 8 % water, is added below the surface of benzene at 50–55°; the ratio of mixed acids to benzene being about 2·5 to 1 by weight (*Faith et al.*, *op. cit.*, p. 542).

The velocity of nitration is increased by the presence of alkyl groups in the nucleus. This is shown in a qualitative manner by a comparison of the nitration of benzene, *o*-xylene, and mesitylene. Thus mesitylene gives dinitromesitylene by the action of fuming nitric acid at 0°, *o*-xylene gives the dinitro compound at 25°, whilst benzene requires refluxing with the acid to give *m*-dinitrobenzene.

In the mono-nitration of the shorter alkylbenzenes the *ortho–para* ratio may be affected by steric factors, *e.g.* when the *tert*-butyl group is present, although larger groups having an α-tertiary carbon atom are apparently rearranged first so that the tertiary carbon atom becomes secondary. In the mono-nitration of longer chain alkylbenzenes the *para* isomer is formed almost exclusively and the introduction of a second nitro group into the nucleus requires more drastic conditions. If fuming mixed acids are used dinitration occurs rapidly at low temperatures and in good yield without oxidation of the side chain (*G. A. Bonetti,* Amer. chem. Soc. Div. Pet. Chem., Preprints, 1961, **6,** No. 3B, 107; U.S.P., 2,934,571/1960).

The nitration of tetra- and penta-methylbenzenes is abnormal. Durene gives a dinitro-compound only and attempts to obtain a mononitrodurene by direct nitration failed (*J. C. Cain,* Ber., 1895, **28,** 967; *L. Rugheimer* and *M. Hankel, ibid.*, 1896, **29,** 2163; *R. Willstätter* and *H. Kubli, ibid.*, 1909, **42,** 4151). Pentamethylbenzene loses a methyl group when nitrated in chloroform with fuming nitric acid or with a mixture of nitric and sulphuric acids, 1,2,3,4-tetramethyl-5,6-dinitrobenzene being formed (*Willstätter* and *Kubli, loc. cit.*; *L. I. Smith* and *S. A. Harris,* J. Amer. chem. Soc., 1935, **57,** 1289). Pentaethylbenzene behaves similarly to give 3,6-dinitrotetraethylbenzene. Hexamethyl- and hexaethylbenzenes also lose alkyl groups on nitration to give dinitro compounds (*Smith* and *C.O. Guss, ibid.*, 1940, **62,** 2635).

Other examples of the displacement of an alkyl group by a nitro group are the formation of small amounts of nitrotoluene in the nitration of cymene (*K. A. Kone et al.*, Ind. Eng. Chem., 1939, **31,** 257; 1957, **49,** 801; cf. *J. Alfthan,* Ber., 1920, **53,** 78), the elimination of alkyl groups in the nitration of the amyl- and butyl-xylenes (*R. de Capeller,* Helv., 1928, **11,** 426) and the replacement of an isopropyl group in the di- and tetra-isopropylbenzenes (*A. Newton,* J. Amer. chem. Soc., 1943, **65,** 2434). *Olah* and *S. J. Khun* (*ibid.*, 1964, **86,** 1067) discuss the dealkylating nitration of propylated and butylated alkylbenzenes using nitronium tetrafluoroborate; this nitrating agent avoiding possible oxidation.

The following nitrating agents have a limited use in the nitration of aromatic hydrocarbons; acetyl nitrate (*A. Pictet* and *A. Khotinsky,* Ber., 1907, **40,** 1163); inorganic nitrates in presence of acetic anhydride (*J. B. Menke,* Rec. Trav. chem., 1925, **44,** 141; 269; *J. R. Lacher et al.*, J. org. Chem., 1961, **26,** 2536); nitric anhydride in carbon tetrachloride (*A. N. Baryshnikova* and *A. I. Titov,* Doklady Akad. Nauk SSSR., 1957, **114,** 777); with benzoyl nitrate, durene, penta- and hexa-methylbenzenes are nitrated in the side chain (*Willstätter* and *Kubli, loc. cit.*). The usual method of introducing nitro groups directly into the side chain is to heat with dilute nitric acid at temperatures between 105 and 125°.

Small amounts of nitrophenols may be formed in the liquid-phase nitration of benzene and its homologues, the yield of nitrophenols being increased very considerably if the nitration is carried out in the presence of mercuric nitrate. Moderate yields of di- and tri-nitrophenols can be obtained from benzene whilst toluene gives nitrocresols and nitrohydroxybenzoic acids on nitration in presence of mercuric nitrate (*R. Wolffenstein* and *O. Boters,* Ber., 1913, **46,** 586; *F. H. Westheimer et al.*, J. Amer. chem. Soc., 1947, **69,** 773; *P. I. Petrovich,* Zhur.

priklad. Khim., 1959, **32**, 353). Oxidative nitration of toluene occurs with mixed acid in the presence of chromic acid, vanadium pentoxide or manganese dioxide, to give 2,4-dinitrobenzoic acid and intermediate products depending on conditions (*T. Urbánski et al.*, Biul. Wojskowej Akad. Tech., 1960, **9**, No. 97, 73; No. 98, 35).

Mechanism of nitration: The present knowledge of the nature of electrophilic aromatic nitration goes back to *H. Euler* (Ann., 1903, **330**, 280) who first suggested the nitronium ion $[NO_2]^{\oplus}$ as the active nitrating agent. Later work has provided convincing evidence that this is so; the nitronium ion being formed according to the following equation when nitric acid, reacting as a base, is dissolved in concentrated sulphuric acid:

$$HO \cdot NO_2 + 2\ H_2SO_4 \rightarrow [NO_2]^{\oplus} + [OH_3]^{\oplus} + 2\ [HSO_4]^{\ominus}$$

This conclusion is supported by several independent types of evidence among which may be noted cryoscopic data (*R. J. Gillespie et al.*, J. chem. Soc., 1950, 2504), ultra-violet and Raman spectra, the isolation of crystalline nitronium salts, kinetic, and conductivity measurements. A critical discussion of the significance of data from these sources in the light of further experimental evidence is given by *G. M. Bennett* and his coworkers (*ibid.*, 1946, 869, 875, 880; 1947, 474) and *C. K. Ingold* ("Structure and mechanism in Organic Chemistry", Cornell University Press, Ithaca, New York, 1953, Chapter V). Other strong Lewis acids such as boron trifluoride (*R. J. Thomas et al.*, Ind. Eng. Chem., 1940, **32**, 408), hydrochloric acid (*J. H. Simons et al.*, J. Amer. chem. Soc., 1941, **63**, 608), and nitronium tetrafluoroborate and analogous stable nitronium salts (*Olah et al.*, *J. chem. Soc.*, 1956, 4257; J. Amer. chem. Soc., 1961, **83**, 4564, 4571; 1962, **84**, 3687) also liberate the nitronium ion and catalyse the nitration, *e.g.*:

$$BF_3 + HNO_3 \rightarrow [NO_2]^{\oplus} + [HO \cdot BF_3]^{\ominus}$$

The nitronium ion is present even when the nitric acid is dissolved in aqueous or organic solvents such as nitromethane, acetic and dioxan, but such solutions give a much milder reaction. The conversion of the nitronium ion into an aromatic nitro group may be expressed thus:

$$ArH + [NO_2]^{\oplus} \longrightarrow Ar^{\oplus}\!<\!\begin{smallmatrix}H\\NO_2\end{smallmatrix} \longrightarrow Ar \cdot NO_2 + H^{\oplus}$$

and this view is strongly supported by the observations of *L. Melander* and of *Ingold* and coworkers (Nature, 1949, **163**, 599). It is however, not justified to conclude from this work that all nitrations depend on the participation of the nitronium ion. Indeed there are many, particularly nitrations in the vapour phase or with nitric anhydride in solution, which proceed by a radical mechanism.

For excellent discussions on aromatic nitrations see *Gillespie* and *D. J. Millen*

Quart. Reviews (London), 1948, **2**, 227; *Ingold, op. cit.*, p. 269; *de la Mare* and *Ridd*, "Aromatic Substitution, Nitration and Halogenation", Academic Press Inc., New York, 1959.

(iii) Sulphonation (*cf.* *p.* 71)

Nuclear sulphonation of the benzenoid hydrocarbons to form arenesulphonic acids is brought about by the action of sulphur trioxide derived from sulphuric acid:

$$2\ H_2SO_4 \rightarrow SO_3 + H^{\oplus}{}_3O + H^{\ominus}SO_4$$

and proceeds according to the equation:

$$ArH + SO_3 \rightleftharpoons Ar^{\oplus}\!\!<\!\!\begin{smallmatrix}H\\ SO^{\ominus}{}_3\end{smallmatrix} \rightleftharpoons ArSO^{\ominus}{}_3 + H^{\oplus}$$

In contrast to nitration the removal of the proton from the γ-complex is involved in the rate determining step and the reaction is reversible; the ease of desulphonation being attributed to the high stability of the complex (*R. Lantz*, Bull. Soc. chim. Fr., 1945, **12**, 1004). Aromatic hydrocarbons as a class are easily sulphonated. Sulphonation increases with methyl substitution and there is a pronounced difference in the ease of sulphonation shown by isomeric dialkylbenzenes, the *para* isomer being more difficult than the *ortho-meta* isomers. This difference has been used as a means of separating the isomeric xylenes.

Benzene reacts relatively slowly at room temperature with concentrated sulphuric acid (*M. Kilpatrick et al.*, J. phys. Chem., 1960, **64**, 1433; *J. Takenaka*, Kogyo Kagaku Zasshi, 1967, **70**, 476–487). On heating, an equilibrium is reached, the point of which is determined by the amount of sulphuric acid present. With equal volumes of benzene and sulphuric acid, about 80 % of the hydrocarbon is converted to sulphonic acid. If the conditions of sulphonation are so arranged that the water is removed continuously, either by using an apparatus of the type described by *H. Meyer* (Ann., 1923, **433**, 327) or by passing benzene vapour through heated sulphuric acid, it is possible to convert all the sulphuric acid into benzene sulphonic acid.

Toluene readily yields a mixture of *o*- and *p*-sulphonic acids with sulphuric acid at 25–35° and some *m*-sulphonic acid is also formed (*H. Cerfontain et al.*, Rec. Trav. chim., 1960, **79**, 935; 1964, **83**, 226; 1965, **84**, 551; *A. A. Spryskov*, Izvest. Vysshikh Ucheb. Zavedenii, Khim. i Khim. Tekhnol, 1961, **4**, 981). Di- and tri-methylbenzenes are sulphonated rapidly by stirring with concentrated sulphuric acid at room temperature. The rate of sulphonation of the trimethylbenzenes decreases in the order hemimellitene > mesitylene > pseudocumene (*Ya. I. Leitman* and *I. N. Diyarov*, Zhur. priklad. Khim., 1961, **34**, 1920; *Kilpatrick et al.*, J. phys. Chem., 1961, **65**, 530, 1189).

Dealkylation of tert-*butylbenzene* occurs when it is reacted with 72–91 % sulphuric acid at 5–35° (*Cerfontain et al.*, Rec. Trav. chim., 1963, **82**, 565). The behaviour of the *tetra-* and *penta-methylbenzenes* towards sulphuric acid is also abnormal in that pentamethylbenzene gives hexamethylbenzene and 1,2,3,4-tetramethylbenzene-5-sulphonic acid (prehnitenesulphonic acid). Durene similarly gives prehnitenesulphonic acid with some pseudocumenesulphonic acid, hexamethylbenzene and tar (*O. Jacobsen*, Ber., 1886, **19**, 1209; 1887, **20**, 896; 1888, **21**, 2814; *L. I. Smith*, Org. Reactions, 1942, **1**, 370; *Kilpatrick* and *M. W. Meyer*, J. phys. Chem., 1961, **65**, 1312; *F. Bohlmann* and *J. Riemann*, Ber., 1964, **97**, 1515). This is generally known as the Jacobson reaction. Durenesulphonic acid rearranges in polyphosphoric acid to give the sulphonic acids of prehnitene, pseudocumene, and hexamethylbenzene in low yields, the rearrangement being accompanied by desulphonation to durene and hexamethylbenzene (*E. N. Marvell* and *B. M. Graybill*, J. org. Chem., 1965, **30**, 4014).

Sulphonation may also be accomplished by sulphur trioxide as liquid, as vapour, with or without solvents (*e.g.* chloroform), or as an addition compound with one of several organic compounds such as pyridine, dioxane or dimethylformamide. Liquid sulphur dioxide is an excellent inert solvent for the sulphonation with sulphur trioxide. Sulphur trioxide is theoretically the most efficient sulphonating agent since only direct addition is involved in accordance with the equation:

$$RH + SO_3 \rightarrow RSO_3H$$

Diarylsulphone may also be formed. Until recently there have been practical difficulties in using sulphur trioxide and oleum made a suitable compromise. Fuming sulphuric acid brings about the formation of monosulphonic acids smoothly at room temperature (*C. M. Suter et al.*, J. Amer. chem. Soc., 1938, **60**, 538; *W. H. C. Rueggeberg et al.*, J. org. Chem., 1955, **20**, 455; *L. Leiserson et al.*, Ind. Eng. Chem., 1948, **40**, 508; *Spryskov* and *B. G. Gnedin*, Zhur. obshchei Khim., 1963, **33**, 1082; *Cerfontain et al.*, Rec. Trav. chim., 1964, **83**, 1103). Chlorosulphonic acid is an effective sulphonating agent; it is usual to employ an excess so that the product may be the sulphonyl chloride rather than the sulphonic acid. Sulphonation with chlorosulphonic acid may be accompanied by sulphone formation (*A. Rieche* and *W. Fischer*, Angew. Chem., 1957, **69**, 482; *F. Cuiban et al.*, Pharmazie, 1958, **13**, 407). Equilibrium becomes established in accordance with the equation:

$$R{\cdot}SO_3H + H{\cdot}SO_3Cl \rightleftharpoons R{\cdot}SO_2Cl + H_2SO_4$$

This equilibrium has been studied for toluene-*p*-sulphonic acid by *Spryskov* and *Y. L. Kuz'mina* (Zhur. obshchei Khim., 1951, **21**, 714) and by *I. Ognyanov* and *A. Zagorova* (Zhur. priklad. Khim., 1956, **29**, 1299). Fluorosulphonic acid has also been used in the same way for the preparation of sulphonyl fluorides *W. Steinkopf et al.*, J. pr. Chem., 1927, [ii], **117**, 1). Other sulphonating reagents which are used less frequently are sulphamic acid, alkali bisulphates, sodium trihydrogen sulphate, $NaH_3(SO_4)_2$, and combinations of sulphuric acid with phosphorus pentoxide, chlorosulphonic anhydride with aluminium trichloride, sulphuryl chloride with aluminium trichloride, and sulphuryl chloride with chlorosulphonic acid. Laboratory methods for the sulphonation of long-chain alkylbenzenes are described by *T. H. Liddicoet and S. A. Olund* (J. Amer. Oil Chemists' Soc., 1963, **40**, 5).

A second sulphonic group is introduced into arenesulphonic acids by heating with sulphuric acid at between 200 and 240° or with oleum. The sulphonation of toluene to disulphonic acid has been studied using 62 % fuming sulphuric acid at 100–210° (*Spryskov* and *T. I. Potapova*, Izvest. Vysshikh Ucheb. Zavedenii, Khim. i Khim. Tekhnol., 1962, **5**, 280). It has been established that the proportions of the isomeric disulphonic acids formed during the sulphonation may be determined by the duration of the reaction and the temperature. Thus benzenesulphonic acid gives *m*-disulphonic acid but if the sulphonation is prolonged a proportion of the *meta*- is converted to the *para*-sulphonic acid. A third sulphonic group is introduced by more drastic conditions (*Spryskov* and *S. P. Starkov*, Zhur. obshchei Khim., 1957, **27**, 2780; *A. P. Shestor*, *ibid.*, 1956, **26**, 1219). Benzenetrisulphonic acid is formed by heating the disulphonic acid with fuming sulphuric acid at 275° in presence of mercuric sulphate.

Sulphonation has been used for many years in the refining of petroleum distillates to produce mineral white oils, and in commercial processes based on benzenoid hydrocarbons. The monosulphonation of benzene is of commercial importance being the first stage of one process for the manufacture of phenol. The sulphonation of detergent alkylbenzene is also an important commercial process.

For further information on the sulphonation of benzene and the alkylbenzenes see *Suter*, "The Organic Chemistry of Sulphur", John Wiley and Sons Inc., New York, 1944; *Suter* and *A. W. Weston*, Org. Reactions, 1946, **3**, 141; *E. E. Gilbert*, "The Chemistry of Petroleum Hydrocarbons", Ed. *B. T. Brooks et al.*, Reinhold Publishing Corpn., 1955, Vol. III, p. 611.

(*iv*) *Friedel-Crafts reactions*

The alkylation of benzene and its homologues by the Friedel-Crafts reaction has been discussed in detail on p. 158. Under the influence of anhydrous aluminium chloride and other Friedel-Crafts type catalysts acyl and aldehydo groups can be introduced into the benzene nucleus by interaction between the hydrocarbon and the appropriate reactant, and this provides a convenient route to ketones and aldehydes. Other groups such as $-NH_2$, $-CONH_2$, $-CN$, $-CN{:}NOH$, $-SO_2R$ have been similarly introduced although generally with less satisfactory results. Alkylbenzenes with reactive groups in the side chain may be prepared by means of bifunctional reagents but the reaction may involve both active centres and additional reactions such as cyclialkylation (see p. 215) can also occur.

(i) *Acylation.* The substitution of an acyl group, RCO-, for hydrogen in the aromatic nucleus is known as *acylation.* The reaction with benzene and its homologues is readily brought about by acyl halides in the presence of Friedel-Crafts catalysts (*C. F. H. Allen,* Org. Synth., 1943, Coll. Vol. II, p. 3; *H. C. Brown* and *F. R. Jensen,* J. Amer. chem. Soc., 1958, **80**, 2291; *G. A. Olah et al., ibid.*, 1964, **86**, 2198):

$$\text{ArH} + \text{RCOCl} \xrightarrow{\text{cat.}} \text{ArCOR} + \text{HCl}$$

Acylation rates of benzene and mesitylene in aluminium chloride catalyzed reactions in nitromethane at 0^0 decrease in the order acetyl > propionyl > butyryl > isobutyryl > benzoyl > 2,4,6-trimethylbenzoyl chloride (*P. H. Gore* and *J. A. Hoskins,* Chem. Comm., 1966, 835). "Activated" acyl chlorides, *e.g.*, monochloroacetyl chloride, may react with hexa-alkylbenzenes, the acyl group replacing one of the alkyl groups (*V. A. Koptyug et al.,* Zhur. org. Khim., 1968, **4**, 662). A mixture of hydrogen chloride and carbon monoxide reacts as formyl chloride with benzene in the presence of aluminium chloride or bromide and gives benzaldehyde as the main product. This reaction is known as the Gattermann-Koch reaction and is a useful method for introducing the aldehydo group into an alkylbenzene (*L. Gattermann,* Ann., 1906, **347**, 347; *N. N. Crounse,* Org. Reactions, 1949, **5**, 290). An alternative method for introducing an aldehydo group into the benzene nucleus, also due to Gattermann, is the reaction of a mixture of hydrogen cyanide and hydrogen chloride with the aromatic hydrocarbon usually in the presence of aluminium chloride or zinc chloride (*W E. Truce, ibid.*, 1957, **9**, 37):

$$\text{ArH} + \text{HCN} + \text{HCl} \xrightarrow{\text{AlCl}_3} \text{ArCHO} + \text{NH}_4\text{Cl}$$

Ethoxalyl chloride, $EtO_2C{\cdot}COCl$, (see Vol. ID, p. 286), reacts with aromatic hydrocarbons to form aryl glyoxylic esters; this is a convenient route to the α-oxo-esters (*L. Boveault,* Bull. Soc. chim. Fr., 1896, [3] **15**, 1014; 1897, [3] **17**, 363, 940, 943). Oxalyl chloride reacts with benzene and monoalkylbenzenes, in the presence of aluminium chloride to give aryl-1,2-diketones (*W. Brawn,* G.P. 913891/1954) but with *symm.*-trialkylbenzenes good yields of 2,4,6-trialkylbenzoic acids are obtained (*G. A. Varvoglis* and *N. E. Alexandrou,* Chim. Chronika [Athens], 1961, **26**, No. 9, 137). Aromatic acid chlorides have been prepared by passing phosgene into the hydrocarbon and anhydrous aluminium chloride. The yields are generally poor. The acyl anhydrides may be used instead of the corresponding acyl halides. Saturated aliphatic acids also react with refluxing benzene and toluene in the presence of aluminium chloride to yield ketones, *e.g.* acetic acid gives acetophenone in good yield (*A. M. Glatz* and *Al. C. Razus,* Rev. Roum. Chim., 1966, **11**, 551). With aliphatic dibasic acid anhydrides, aroyl fatty acids, important intermediates in the synthesis of aromatic hydrocarbons, are formed (*E. Berliner,* Org. Reactions, 1949, **5**, 229; *M. S.*

Newman and *P. A. Scheurer*, J. Amer. chem. Soc., 1956, **78**, 5004; *D. D. Phillips* and *T. D. Hill, ibid.*, 1958, **80**, 3663):

$$ArH + (CH_2)_n\!\begin{matrix}<CO>\\<CO>\end{matrix}\!O \xrightarrow{\text{cat.}} ArCO(CH_2)_nCO_2H$$

The acylation reaction can be carried out in solvents such as nitromethane, nitrobenzene, carbon disulphide, etc. Aluminium chloride is commonly used as catalyst; it forms a 1:1 adduct with the acid chloride and is an essential part of the reaction. The active acylating species is probably the acylonium ion, $RCO^{\oplus}$, formed from the adduct (*K. Le Roi Nelson* and *H. C. Brown*, "The Chemistry of Petroleum Hydrocarbons", Ed. *B. T. Brooks et al.*, Reinhold Publishing Corpn., 1955, Vol. III, p. 465). In the reaction of acyl chlorides and anhydrides with hexa-alkylbenzenes one of the alkyl groups is displaced (*H. Hopff et al.*, Helv., 1960, **43**, 1473; 1965, **48**, 625). Acylation of aromatic hydrocarbons may also be effected by reaction with acetic alkane- or arene-sulphonic anhydrides, RSO_2OAc (R = alkyl, aryl), in the presence of Friedel-Crafts catalysts (*Olah* and *S. J. Kuhn*, J. org. Chem., 1962, **27**, 2667). The chlorides of sulphonic acids react in much the same way as the alkyl chlorides to give sulphones. The sulphonylation reaction is however less selective than acylation (*S. C. J. Olivier*, Rec. Trav. chim., 1913, **33**, 91; 1915, **35**, 166; *J. Boeseken* and *H. W. van Ockenburg*, *ibid.*, 1914, **33**, 317; *W. E. Truce* and *C. W. Vriesen*, J. Amer. chem. Soc., 1953, **75**, 5032; *G. Holt* and *B. Pagdin*, J. chem. Soc., 1960, 2508):

$$RSO_2Cl + ArH \xrightarrow{AlCl_3} RSO_2Ar + HCl$$

Methyl phenyl sulphoxide is formed from methanesulphinyl chloride and benzene in the presence of aluminium chloride (*I. B. Douglass* and *B.S. Farah*, J. org. Chem., 1958, **23**, 805). (Diphenyl sulphone)sulphonyl chlorides with benzene in the presence of aluminium chloride give products both of sulphonylation and of reduction to the sulphinic acid with concomitant formation of chlorobenzene (*E. C. Dart et al.*, J. chem. Soc., C, 1966, 1284).

(2) *Alkylation.* Substituted olefins, *e.g.*, halides, alcohols, esters, ketones, as well as dienes, can also be used as alkylating agents in the Friedel-Crafts reaction with benzene and its homologues. Reaction may occur with either or both functional groups and competing or additional reactions sometime occur such as cyclization, isomerization and polymerization (*R. Koncos* and *B. G. Friedman*, "Friedel-Crafts and related Reactions", Interscience Publishers, Ed. *Olah*, London, 1964, Vol. I, Part 1, p. 289; *L. R. C. Barclay*, *ibid.*, Vol. II, Part 2, p. 785). Thus in the reaction of the benzenoid hydrocarbons with dienes, alkenyl- and polyalkenyl-arenes, diarylalkanes and cyclized products are all formed. Mild conditions favour the formation of alkenylarenes usually with 1,4-addition; with increasing severity, dialkylation and cyclization predominate (*V. N. Ipatieff et al.*, J. Amer. chem. Soc., 1944, **66**, 816; *Yu. G. Mamadaliev*

et al., Azerb. Khim. Zhur., 1962, No. 5, 19; 1963, No. 4, 73; 1964, No. 2, 19). Thus the aluminium chloride catalyzed reaction of butadiene with benzene at 50° yields a mixture of 1,1-, 1,2-, and 1,3-diphenylbutanes; *sec*- and *n*-butylbenzenes are also formed (*T. Inukai*, J. org. Chem., 1966, **31**, 1124). Polyalkenylarenes are formed from benzenoid hydrocarbons by treatment with excess diene in the presence of boron trifluoride at 0–30° (*Ciba Ltd.*, B.P., 885,872/1961). Alkenyl halides react preferentially at the double bond in the presence of Brønsted acid catalysts, *e.g.*, sulphuric acid. With a Lewis type catalyst, *e.g.*, aluminium chloride, reaction occurs at both functional groups. Intermolecular alkylation also occurs to yield polycyclic aromatics. Thus 2-aryl-1-halogeno propanes have been prepared in good yields by alkylation of benzene with allyl halide in the presence of sulphuric acid at 50° (*Kh. Yu. Yuldashev* and *I. Tsukervanik*, Uzbeksk. Khim. Zhur., 1961, No. 6, 40; *Mamadaliev et al.*, Doklady Akad. Nauk Azerb. SSR., 1962, **18**, No. 2, 25; *A. D. Petrov et al.*, Neftekhimiya, 1962, **2**, 776) but in the presence of phosphoric acid–boron trifluoride, 1,2-diphenylpropane and higher alkylates are also formed (*Ya. M. Paushkin* and *I. Galal*, Doklady Akad. Nauk SSSR., 1962, **147**, 853). Tetraphenylethane is formed in high yield by condensation of benzene and trichloroethylene at 10° in the presence of aluminium chloride (*M. M. Guseinov et al.*, Doklady Akad. Nauk Azerb. SSR., 1966, **22**, No. 8, 38). Unsaturated alcohols react similarly to the alkenyl halides, proton-donating catalysts favouring addition to the double bond. Unsaturated carboxylic acids react with benzene and the alkylbenzenes in the presence of aluminium chloride, with addition to the double bond to form arylalkanoic acids. Addition to the β-carbon atom is the normal reaction and the alkylation is often accompanied by rearrangement of the double bond and replacement of an existing aryl group (*J. F. J. Dippy* and *J. T. Young*, J. chem. Soc., 1955, 3919). Arylalkenoic acids are formed with halogen substituted crotonic acids and benzene in the presence of aluminium chloride (*A. N. Nesmeyanov et al.*, Doklady Akad. Nauk SSSR., 1956, **111**, 114):

$$CH_2Cl{\cdot}CH{=}CH{\cdot}CO_2H + C_6H_6 \xrightarrow{AlCl_3} PhCH_2{\cdot}CH{=}CH{\cdot}CO_2H$$

Reaction with crotonyl chloride proceeds initially by acylation but addition to the double bond may also occur:

$$CH_3{\cdot}CH{=}CH{\cdot}COCl + C_6H_6 \xrightarrow{AlCl_3} CH_3{\cdot}CH{=}CH{\cdot}COPh + HCl$$

Alkylation with unsaturated esters proceeds without affecting the ester grouping. Alkylation of benzene with vinyl esters in the presence of aluminium chloride gives 9,10-dimethylanthracene as the major product (*J. M. Pepper et al.*, Canad. J. Chem., 1962, **40**, 122; 1963, **41**, 2103). Unsaturated ketones react with the benzenoid hydrocarbons in the presence of aluminium chloride by addition to the double bond. Thus benzene and toluene react with mesityl oxide in the presence of aluminium chloride to form methyl phenylisobutyl and methyl tolylisobutyl ketone (*S. D. Mekhtiev et al.*, Azerb. Khim. Zhur., 1967, No. 1, 13). Reaction of β-chlorovinyl ketones with benzene and its homologues in the presence of aluminium chloride involves the halogen atom. Thus β-chlorovinyl methyl ketone and mesitylene give 2,4,6-trimethylbenzylideneacetone (*Nesmeyanov et al.*, Izvest. Akad. Nauk SSSR., Otdel. khim. Nauk, 1956, 1197):

AcCH=CHCl + (Me, Me, Me) $\xrightarrow{AlCl_3}$ CH=CHAc (Me, Me, Me)

Ketene reacts to give methyl ketones in the presence of aluminium chloride (*R. E. Dunbar* and *R. T. Arndts*, Proc. N. Dakota Acad. Sci., 1960, **14**, 57). Diketene also condenses with the aromatic hydrocarbons to give 1,3-diketones. Thus with benzene in the presence of two molecules of aluminium chloride, benzoylacetone is obtained:

$CH_2{=}C{-}O$ / $CH_2{-}CO$ + (benzene) $\xrightarrow{AlCl_3}$ $CO{\cdot}CH_2Ac$ (phenyl)

The benzenoid hydrocarbons are readily cyanoethylated with vinyl cyanide in the presence of aluminium chloride (*A. D. Grebenyuk* and *I. P. Tsukervanik*, Zhur. obschchei Khim., 1955, **25**, 286).

Direct halogenoalkylation of benzene and the alkylbenzenes with a variety of bifunctional reagents is possible in the presence of Friedel-Crafts catalysts. Halogenoalkenes have already been mentioned and chloromethylation is described later (p. 217); other reagents include di- and poly-halogenoalkanes, halogenoalcohols, alkoxyhalogenoalkanes, halogenoalkyl sulphides and halogenoalkyl sulphates, *e.g.*

$$C_6H_6 + BrCH_2{\cdot}CH_2{\cdot}CHMeBr \xrightarrow[20^\circ]{AlCl_3} MeCHPh{\cdot}CH_2{\cdot}CH_2Br$$

Diarylalkanes and cycloalkylated products are also formed in these reactions (*Olah* and *W. S. Tolyyesi*, "Friedel-Crafts and Related Reactions", Ed. *Olah*, Interscience Publishers, London 1964, Vol. 2, Part II, p. 659; *F. A. Drahowzal*, *ibid.*, Part I, p. 417; *L. Schmerling et al.*, J. Amer. chem. Soc., 1956, **78**, 5406; 1957, **79**, 2636; *K. M. Shadmanov*, Uzbeksk. Khim. Zhur., 1961, No. 4, 70; *I. B. Tsukervanik* and *L. V. Burgrova*, Zhur. obschchei Khim., 1961, **31**, 2143; Zhur. org. Khim., 1965, **1**, 714; *D. L. Ransley*, J. org. Chem., 1966, **31**, 3595). The more severe alkylation of benzenoid hydrocarbons with bifunctional reagents leads to the attachment of a new ring to the nucleus and is known as *cyclialkylation*

(*H. A. Bruson* and *J. W. Kroeger*, J. Amer. chem. Soc., 1940, **62**, 36). The use of 1,4-dihalides in the cyclialkylation reaction is a convenient method of synthesising hydroaromatic hydrocarbons (*S. Yura* and *R. Oda*, J. Soc. Chem. Ind. Japan, 1943, **46**, 531; *K. Shishido* and *H. Nozaki*, *ibid.*, 1944, **47**, 516; *W. Reppe et al.*, Ann., 1955, **596**, 80). Halogen migration may occur in these reactions. Thus 1-methyltetralin and amylbenzene are found among the products of the reaction between benzene and 1,5-dibromopentane with aluminium chloride (*Nozaki et al.*, J. org. Chem., 1965, **30**, 1303). Cyclialkylation in the presence of aluminium chloride may also occur with glycols and their diacetates (*I. P. Labunskii* and *I. P. Tsukervanik*, Doklady Akad. Nauk SSSR., 1951, **80**, 369; *R. M. Lagidze* and *B. S. J. Potshverashvili*, Soobshch. Akad. Nauk Gruz. SSR., 1957, **19**, 429). An interesting example of cyclialkylation is the formation of 1,1,3,3,5-pentamethylindan by the reaction of *p*-cymene with *tert*-butanol in the presence of sulphuric acid (*M. J. Schlatter*, Amer. chem. Soc. Div. Pet. Chem. Symposium on Chemicals from Petroleum, 1956, Preprints No. 2, 77–82; see also *Ipatieff et al.*, J. Amer. chem. Soc., 1948, **70**, 2123):

Me Me CH Me + Me_3COH $\xrightarrow{H_2SO_4}$ Me Me Me Me Me

Benzene and its homologues react with alkynes in the presence of Friedel-Crafts catalysts to form *unsymm.* 1,1-diarylethanes, polycyclic aromatics, and polymeric material:

$$2\,C_6H_6 + CH{\equiv}CH \xrightarrow{AlCl_3} Ph_2CH{\cdot}CH_3 \xrightarrow[AlCl_3]{C_2H_2}$$ Me H Me H

Acetylene reacts with benzene in the presence of phosphoric acid–boron trifluoride to form 1,1-diphenylethane in 65 % yield (*O. W. Cook* and *V. J. Chambers*, *ibid.*, 1921, **43**, 334; *V. L. Vaiser*, Doklady Akad. Nauk, SSSR., 1953, **91**, 535). Methylacetylene and benzene in the presence of aluminium chloride give 2,2-diphenylpropane and the cyclized product XIV; some 2-phenylpropene is also formed (*I. P. Tsukervanik* and *K. Yu. Yuldashev*, Zhur. obshchei Khim., 1963, **33**, 3497). Functional derivatives of acetylene also react with aromatic hydrocarbons. 1,1,2-Triphenylpropane is formed in small yield when 3-chloropropyne reacts with benzene in the presence of aluminium chloride (*J. Cologne* and *Y. Infarnet*, Bull. Soc. chim. Fr., 1960, 1916). γ-Acetylenic glycols and their acetates react with the benzenoid hydrocarbons in the presence of aluminium chloride to form various polycyclic compounds, *e.g.* benzene and its homologues are alkylated with the diacetate of 3,6-dimethyl-4-octyne-3,6-diol in the presence of aluminium chloride to give compounds of the general formula XV (*Lagidze et al.*, Soobshch. Akad. Nauk Gruz. SSR., 1960, **25,** 19; 1962, **28,** 409; 1965, **37,** 311; Zhur. org. Khim., 1965, **1,** 1965). Reactions between acetylenes and aromatic hydrocarbons occur in the absence of Friedel-Crafts type catalysts. Thus, dicyanoacetylene reacts with benzene leading to the normal Diels-Alder adduct; however, the reaction is accelerated by aluminium chloride (*E. Ciganek*, Tetrahedron Letters, 1967, 3321).

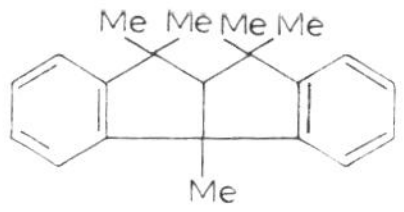

(XIV)

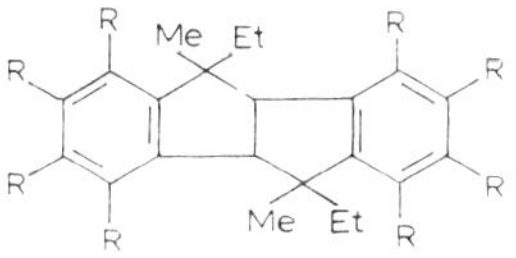

(XV)

Benzene and its homologues react normally with cyclo-olefins, cycloalkanols and cycloalkyl halides in the presence of Friedel-Crafts catalysts. Possible side reactions are migration of the double bond in the case of branched cyclo-olefins, and ring contraction (*Ipatieff et al.*, Org. Synth., 1943, Coll. Vol. II, p. 151; J. Amer. chem. Soc., 1950, **72,** 2772; *M. B. Turova-Pollyak* and *I. R. Davydova*, Zhur. obshchei Khim., 1956, **26,** 2710; *Schmerling*, U.S.P., 3,000,985/1958; *Mekhtiev* and *T. A. Pashaev*, Azerb. khim. Zhur., 1959, No. 2, 39; *N. G. Sidorova et al.*, Zhur. obshchei Khim., 1960, **30,** 1921; 1961, **31,** 2149; *J. Blackwell* and *W. J. Hickinbottom*, J. chem. Soc., 1963, 518; *S. V. Zavgorodnii* and *I. A. Nasyr*, Doklady Akad. Nauk SSSR., 1962, **145,** 1061).

Aldehydes react with benzene and its homologues in the presence of Friedel-Crafts catalysts, *e.g.*, sulphuric acid, phosphoric acid, boron trifluoride, hydrofluoric acid, zinc chloride etc., to give derivatives of diphenylmethane. Acetaldehyde (in the form of paraldehyde) gives 1,1-diarylethanes, $Ar_2CH{\cdot}CH_3$, while heptaldehyde gives 1,1-diarylheptanes in good yield. The mono-, di-, and tri-halogenoacetaldehydes react similarly (*A. Baeyer*, Ber., 1872, **5,** 1098; 1873, **6,** 222; 1874, **7,** 1190; *W. P. Duttenborg*, Ann., 1894, **279,** 324; *M. Delacre*, Bull. Soc. chim. Fr., 1895, [iii], **13,** 858; *G. B. Frankforter* and *W. Kritchevsky*, J. Amer. chem. Soc., 1914, **36,** 1511; *J. G. D. Schulz*, U.S.P., 3,002,034/1958).

(3) *Chloromethylation*, the reaction of an aromatic compound with formaldehyde and hydrogen chloride in the presence of a Friedel-Crafts catalyst, is of general application and an important reaction of the mononuclear aromatic hydrocarbons since the chloromethyl group may be converted to other groups such as $—CH_2OH$, —CHO, $—CH_2CN$, and $—CH_3$. Formaldehyde as an aqueous solution or in the form of one of its polymers may be used. Zinc chloride is usually employed to catalyze the reaction but phosphoric acid, phosphoryl chloride, aluminium chloride, stannic chloride, etc. have also been used. A chloromethyl ether or ester derivative such as bis chloromethyl ether $(ClCH_2)_2O$,

and chloromethyl acetate $ClCH_2O_2C{\cdot}CH_3$, is considered to be the intermediate reagent formed, and indeed, the chloromethyl group is introduced into the nucleus of aromatic hydrocarbons by the action of chloromethyl methyl ether in the presence of zinc chloride or stannic chloride:

$$C_6H_6 + ClCH_2OCH_3 \xrightarrow{\text{cat.}} PhCH_2Cl + CH_3OH$$

The earliest example of chloromethylation is the synthesis of benzyl chloride (*G. Grassi-Cristaldi* and *G. Maselli*, Gazz. chim. ital., 1898, **28**, II, 477):

$$C_6H_6 + CH_2O + HCl \xrightarrow{\text{cat.}} C_6H_5{\cdot}CH_2Cl + H_2O$$

A survey of the chloromethylation reaction is given by *R. C. Fuson* and *C. H. McKeever* (Org. Reactions, 1942, **1**, 63). More recent applications to the benzenoid hydrocarbons have been made by *I. N. Nazarov* and *A. V. Semenovikiĭ* (Izvest. Akad. Nauk, SSSR., Otdel. khim. Nauk, 1956, 1487); *R. Granger* and coworkers (Compt. rend., 1959, **249**, 2337); *A. A. Vansheĭdt* and coworkers (Zhur. priklad. Khim., 1961, **34**, 705); *K. Handrick* (Erdöl u. Kohle, 1966, **19**, 172); *G. W. Ayers* and *W. A. Krewer* (U.S.P., 3,284,518/1966; 3,294,850/1966); *G. I. Golivets et al.* (Zhur. priklad. Khim., 1968, **41**, 148). The chloromethylation of long-chain alkylbenzenes is described by *T. Kusano* and *T. Suzuki* (Kogyo Kagaku Zasshi, 1962, **65**, 223). *Amidomethylation* of benzene and alkylbenzenes to yield *N*-aralkylamides is analogous to chloromethylation; the hydrogen chloride being replaced by a nitrile (*C. L. Parris* and *R. M. Christenson*, J. org. Chem., 1960, **25**, 1888).

(4) *Amination.* Benzenoid hydrocarbons may be *aminated* in the presence of Friedel-Crafts catalysts with (a) (alkyl)hydroxylamines (*P. Kovacic et al.*, J. Amer. chem. Soc., 1962, **84**, 759; *Z. Yoshida et al.*, Kogyo Kagaku Zasshi, 1964, **67**, 76):

$$ArH + R_2NOR' \xrightarrow{AlCl_3} ArNR_2 + R'OH$$

(b) hydrazoic acid (*K. F. Schmidt*, Ber., 1924, **57**, 704; *G. M. Hoop* and *J. M. Tedder*, J. chem. Soc., 1961, 4685; *Kovacic et al.*, J. Amer. chem. Soc., 1964, **86**, 1588), or with (c) *N*-chloroamines (*F. Minisci et al.*, Tetrahedron Letters, 1966, 699; 2531; Chim. Ind. [Milan], 1966, **48**, 725; 1967, **49**, 252). Direct amination of alkylarenes with trichloroamine in the presence of aluminium halides yields *m*-alkylanilines as the major products, and in many cases presents an attractive alternative to the lengthy classical route to such compounds (*Kovacic et al.*, J. Amer. chem. Soc., 1964, **86**, 1650; 1965, **87**, 1262; 1966, **88**, 100, 1000; J. org. Chem., 1967, **32**, 585; Tetrahedron, 1967, **23**, 3965).

(5) *Amidation.* Amides of aromatic carboxylic acids are formed by reaction of carbamoyl chloride, HN_2COCl with aromatic hydrocarbons under the influence of anhydrous aluminium chloride (*L. Gatterman*, Ber., 1899, **32**, 1117; Ann.,

1888, **244**, 29; *F. Runge et al.*, Chem. Tech. [Berlin], 1956, **8**, 644). This reaction succeeds with substituted carbamoyl chlorides such as ethylphenylcarbamoyl chloride, PhEtNCOCl. Amides and anilides of aromatic acids result by treating the appropriate hydrocarbon with isocyanates in the presence of aluminium chloride. Similarly phenyl isothiocyanate gives thioanilides (*R. Leuckart*, Ber., 1885, **18**, 873; J. pr. Chem., 1890, [ii], **41**, 301; *M. T. Bogert* and *M. Meyer*, J. Amer. chem. Soc., 1922, **44**, 1568; *F. Effenberger* and *R. Gleiter*, Ber., 1964, **97**, 472; *R. W. Alder et al.*, J. chem. Soc., C, 1966, 52):

$$\text{ArH} + \text{RNCO} \xrightarrow{\text{AlCl}_3} \text{ArCONHR}$$

Thiocyanation of benzenoid hydrocarbons has been effected using lead thiocyanate in the presence of aluminium chloride (*E. Soderback*, Acta Chem. Scand., 1954, **8**, 1851).

(6) *Miscellaneous reactions.* Other miscellaneous Friedel-Crafts type reactions of benzenoid hydrocarbons include (a) *aminomethylation* with methylolamides (*R. O. Cinneide*, Nature, 1955, **175**, 47), (b) *hydroxyalkylation* with alkylene oxides (*V. R. Likhterov* and *V. S. Etlis*, Zhur. obshchei Khim., 1957, **27**, 2867; *T. Nakajima* and *S. Suga*, Kogyo Kagaku Zasshi, 1964, **67**, 1256), (c) *boronation* with boron tribromide (*W. Gerrard et al.*, Chem. and Ind., 1959, 1091), (d) *phosphination* with phosphorous trichloride (*R. A. Baldwin et al.*, J. org. Chem., 1961, **26**, 3547) and (e) *sulphination* with sulphur dioxide and hydrogen chloride (*E. Knoevenagel* and *J. Kenner*, Ber., 1908, **41**, 3315). The *linking of two aryl nuclei* by elimination of hydrogen is catalyzed by Friedel-Crafts catalysts and is known generally as the Scholl reaction (*Ipatieff* and *V. I. Komarewsky*, J. Amer. chem. Soc., 1934, **56**, 1926; *Kovacic* and *C. Wu*, J. org. Chem., 1961, **26**, 762):

Me, Me (m-xylene) $\xrightarrow{\text{FeCl}_3}$ Me, Me–biphenyl–Me, Me + other products

Mercuric fulminate reacts with benzene and aluminium chloride containing some hydrated salt, to give *benzaldoxime*; using freshly sublimed aluminium chloride, *benzonitrile* is formed (*R. Scholl*, Ber., 1899, **32**, 3492; *Scholl* and *F. Kacer*, *ibid.*, 1903, **36**, 322). Cyanogen bromide condenses with toluene in the presence of aluminium chloride to give a product which consists mainly of *p-toluonitrile.* This reaction appears to be more satisfactory with the phenol ethers (*P. Karrer* and *E. Zeller*, Helv., 1919, **2**, 482; 1920, **3**, 261). An alternative method described by *J. Houben* and *W. Fischer* (Ber., 1930, **63**, 2464; 1933, **66**, 339) consists in reacting trichloroacetonitrile, $CCl_3 \cdot CN$, with a hydrocarbon and aluminium chloride. A *ketimine* is formed which is split by treatment with alkali into the nitrile and chloroform. This reaction has been applied to toluene and to some of its homologues and also some phenols. Carbon dioxide reacts with methylbenzenes in the presence of aluminium bromide to give the corresponding *carboxylic acids* (*S. Fumasoni* and *M. Collepardi*, Ann. Chim. [Rome], 1964, **54** [11], 22).

For further information on the Friedel-Crafts and related reactions see: "Friedel-Crafts and Related Reactions," Vol. I–IV, Ed. *G. A. Olah*, Interscience, New York, 1963–1965; "Friedel-Crafts Reaction," *Olah* and *C. A. Cupas*, Kirk-Othmer Encycl. Chem. Technol., 2nd Edn., 1966, **10**, 135; "Alkylation of Aromatics by Bifunctional Compounds, Bibliographical Review," *C. Fabre* and *G. Bonavent*, Rev. Inst. franç. Petrole, 1967, **22**, 1829; 1968, **23**, 94.

(*v*) *Other substitution reactions*

Benzene and its homologues react with mercuric salts to give mono- and di-*mercuration products* (*L. I. Smith* and *F. L. Taylor*, J. Amer. chem. Soc., 1935, **57**, 2370; *F. H. Westheimer et al.*, *ibid.*, 1950, **72**, 4461; 1963, **85**, 2773; *H. C. Brown* and *C. W. McGary*, *ibid.*, 1955, **77**, 2300; *K. S. McMahon* and *K. A. Kobe*, Ind. Eng. Chem., 1957, **49**, 42):

$$\mathrm{ArH + HgX_2 \rightleftharpoons Ar{-}Hg{-}X + HX}$$

The highly reactive 2,4,6-trinitrobenzenediazonium ion couples with mesitylene, isodurene and pentamethylbenzene to form the corresponding 2,4,6-*trinitrobenzeneazoarene*; toluene and xylene give colour reactions but the products of coupling have not been isolated (*K. H. Meyer* and *H. Tochtermann*, Ber., 1921, **54**, 2283; *L. I. Smith* and *J. H. Poden*, J. Amer. chem. Soc., 1934, **56**, 2169).

Phenyl selenate is produced in good yield by reacting benzene with selenium trioxide in sulphur dioxide (*M. Schmidt* and *I. Wilhelm*, Ber., 1964, **97**, 872; *K. Dostal et al.*, Z. Chem., 1966, **6**, No. 4, 153).

The benzenoid hydrocarbons are readily *deuterated* by reaction with deuterium oxide, sulphate, amine etc., the exchange occurring predominantly in the same positions that are substituted in other electrophilic reactions (*W. M. Lauer et al.*, J. Amer. chem. Soc., 1958, **80**, 6433; 6437; *R. S. Bardasova et al.*, Tr. Gosudarst. Inst. priklad. Khim., 1960, No. 45, 111).

Benzene and its homologues may be *metalated* by reaction with alkali metals, alkali metal alkyls and aryls to give mono- and poly-metal aryls. Benzene reacts with *sec*-butylpotassium to give mainly phenylpotassium although some *sec*-butylbenzene is formed. With toluene the attack is exclusively on the methyl group but nuclear substitution increases with increase in size of the alkyl substituent and the orientation is predominantly *meta* and *para*. Thus reaction of *tert*-butylbenzene with ethylpotassium at 85° gives exclusively nuclear metalation. Transmetalation to more thermodynamically stable isomers occurs and the metalated products may be treated with carbon dioxide and Grignard type reactants to yield derivatives (*D. J. Bryce-Smith* and *E. E. Turner*, J. chem. Soc., 1953, 861; 1954, 1079; *A. A. Morton et al.*, J. org. Chem., 1955, **20**, 981; 1958, **23**, 1469; 1636; *R. A. Benkeser et al.*, J. Amer. chem. Soc., 1963, **85**, 3984; 1968, **90**, 4366; *C. D. Broaddus*, *ibid.*, 1966, **88**, 4174; J. org. Chem., 1970, **35**, 10).

Aryl and *alkyl free radicals* generated by decomposition of peroxides, aryldiazonium salts, *N*-nitrosoacetanilides, etc. are capable of reacting with benzene and the alkylbenzenes to produce both nuclear and side-chain substitution products. Such reactions are comparatively non-selective. The most important reaction of this type is the phenylation of the aromatic ring (*W. E. Bachmann* and *R. A. Hoffman*, Org. Reactions, 1944, **2**, 224; *C. S. Rondestvedt* and *H. S. Blanchard*, J. Amer. chem. Soc., 1955, **77**, 1769; *D. H. Hey et al.*, J. chem. Soc., 1962, 487; 1963, 5604; J. chem. Soc., C, 1967, 1153; Free Radical Chem., 1967, **2**, 47). Free radical methylation of benzene and its homologues has also been demonstrated (*E. L. Eliel et al.*, J. org. Chem., 1957, **22**, 859; *G. E. Corbett* and *G. H. Williams*, Proc. chem. Soc., 1961, 240). A novel method for arylating benzene consists of reacting a mixture of nitrobenzene and benzene at 600° for 9 seconds to give bi- and ter-phenyls (*E. K. Fields* and *S. Meyerson*, J. Amer. chem. Soc., 1967, **89**, 724).

A characteristic feature of the benzenoid hydrocarbons is the relative stability of the benzene nucleus to *high energy radiation*, although the nucleus is subject to attack by free radicals arising from the alkyl side chain (*A. J. Swallow*, "Radiation Chemistry of Organic Compounds", Pergamon Press, London, 1960).

6. Technical production and uses of the mononuclear hydrocarbons

There are two primary production sources of benzene and the alkylbenzenes, *viz.* coal and petroleum. The volume made from coal is dependent mainly on coke output and has remained relatively static over the years. Moreover, their availability in plentiful amounts from this source is limited to benzene. No such restrictions apply to their manufacture from petroleum, in fact, the potential production of the mononuclear hydrocarbons from crude oil far exceeds the foreseeable demand.

(a) From coal

Benzene and its homologues are produced from the gas and tar resulting as by-products from the carbonization of coal at high temperatures in gas work retorts and coke ovens. The presence of benzene in coal tar appears to have been recorded first by Leigh in 1842 and was later confirmed by *A. W. Hofmann* (Ann., 1845, **55**, 204); the isolation of benzene as well as some of its lower homologues in a pure state in any appreciable amount was first accomplished by *C. B. Mansfield* (J. chem. Soc., 1849, **1**, 244). The industrial exploitation of coal tar as a source of aromatic compounds, stimulated by the rise of the synthetic dye industry, led to a rapid development of the chemistry of aromatic compounds and their applications in industry, the arts and medicine.

The bulk of the aromatics produced from coal comes from by-product coke ovens. The gases and vapours leaving the coke oven contain the aromatic hydrocarbons. Coal tar, a black viscid liquid (8–12 gal./ton coal) which is condensed by a series of cooling operations, contains only a small part of the lower boiling aromatic hydrocarbons. The bulk (>95 %) of the benzene and alkylbenzenes produced from coke ovens is removed from the gas stream either by oil washing or by adsorption processes to give the light oil (2–4 gal./ton coal). Benzene is the major constituent (55–70 %) of the light oil and is recovered almost completely. Toluene and the xylenes are also largely recovered as pure products. They are easily separated from each other by distillation, impurities being removed by chemical treatment. The traditional method for purifying the light oil is to treat with sulphuric acid and distil. The thiophene content of the benzene so produced is high and this limits its application. A recent process for upgrading coke oven aromatics overcomes this problem and involves the catalytic hydrotreatment of the light oil. If desired the toluene and xylenes can be completely hydrodealkylated to benzene by a recycling operation (*A. K. Logwinuk et al.*, Ind. Eng. Chem., 1964, **56**, [4], 20). For other methods used in refining the light oil see *R. A. Fraser*, Chem. and Ind., 1961, 1678; and *G. Claxton*, The Gas World, 1962, 317. The mixed xylenes and other hydrocarbons in their boiling range may be marketed as such under the name "Light Solvent Naphtha" or separated into the various isomers by similar processes to those described in Section (b) below. Higher homologues of benzene are contained in "Heavy Solvent Naphtha," the distillate boiling above the xylenes.

In general, there is little possibility of improving the yield of light oil components when the overall economics of the coke oven process are considered. However, addition of heavy fuel oil to coal used in coking does give an increased yield of benzene (*W. Devecchi*, Schweiz. Ver. Gas-u. Wasserfach. Monats-Bull., 1952, **32**, 185). Coal hydrogenation is also a potential means of increasing the production of aromatics from this source (*E. E. Donath*, "Advances in Catalysis," 1956, **3**, 239).

For details concerning the reactions of coal carbonization see *P. H. Given et al.*, Bull. Brit. Coal Utilization Res. Assn., 1952, **16**, 245; *G. Huck*, Brennstoff-Chem., 1957, **38**, 51–4; and *W. Idris Jones*, J. Inst. Fuel, 1964, **37**, 3.

(b) From petroleum

The natural occurrence of benzene and its homologues in crude oil has already been discussed (p. 153); the total amount is enormous. At the present state of refining technology, however, the complexity of petroleum makes their isolation uneconomic, although toluene has been produced from certain crudes rich in natural toluene during national emergencies. The need for high octane fuels aroused the first real commercial interest in petroleum as a source of these aromatic hydrocarbons; benzene and the alkylbenzenes possessing the property of increasing the octane rating (see Vol. IA, p. 396) of gasolines (petrols). Major advances in the development of conversion processes were made during the second World War and many processes are now available for the production, recovery and purification of these aromatics from petroleum. Owing to developments in the automotive and aircraft industries the demand for quality gasolines levelled out by 1960. Capacity therefore became available for the production of pure aromatics and at a time when their demand as raw materials for plastics was expanding. Prior to 1950 the bulk of the pure aromatics was derived from coal. Since then the large expansion in aromatics production has been based almost entirely on petroleum sources and the petroleum industry is now the key supplier.

Benzene, toluene, the xylenes, ethylbenzene, cumene, pseudocumene, mesitylene and durene are all produced commercially in the petroleum industry by catalytic reforming operations. They can also be obtained from catalytic cracking operations but this route has not yet been developed appreciably. The basic reactions in catalytic reforming leading to the formation of aromatic hydrocarbons have been described elsewhere (p. 169). Several catalytic reforming processes are in use throughout the World but the majority are of the fixed bed type employing a supported platinum catalyst, catalyst regeneration being the main variation characterizing the process. *Platforming*, the pioneer process, is typical of catalytic reforming and a brief description of the process will serve as an illustration.

The usual feed stocks for platforming are straight run naphthas, preferably those rich in naphthenes (cycloalkanes), and will normally be prefractionated according to the final product desired, *e.g.* motor petrol, aviation petrol, or pure aromatics. The concentration of aromatics in petroleum naphthas is usually less than 15 % but the proportions of alkanes and cycloalkanes vary considerably depending on the crude source. In platforming a platinum catalyst (0·01–1 % platinum on alumina with combined halogen) is used. The charge and hydrogen-rich recycle gas are heated to 450–500° (500–700 psig) before entering the reactor section containing the catalyst. This section consists of several reactors with interposed reheaters to supply the heat required for the dehydrogenation reaction. As the reaction mixture passes through the reactors the slower hydrocracking and dehydrocyclization reactions are completed. The product coming from the last reactor is cooled and passed to the separator where gaseous and liquid components are separated, part of the gas being compressed and recycled. Light hydrocarbons (up to C_5 depending on the end use) are removed from the liquid component and the stabilized reformate, which is composed largely of aromatics and isoparaffins, is routed for petrol blending or further processing to make pure aromatic hydrocarbons. The pure aromatic hydrocarbons are obtained from the platformate by a solvent extraction process. For aromatics boiling in the gasoline (petrol) range the most widely used process is the Udex process, a solvent extraction process using aqueous diethylene glycol. This glycol has a high solubility for aromatic hydrocarbons and, because of the high boiling point of the glycol, the dissolved aromatics may be readily distilled from the solvent. Furthermore, the low solubility of the solvent in non-aromatics and its high solubility in water permits a clean separation of the solvent from the raffinate without distillation. The solvent-free aromatics leaving the stripping column are clay-treated to remove trace quantities of sulphur, olefinic and other impurities, and fractionally distilled to recover high purity benzene, toluene, and xylenes. High product purity is a characteristic of aromatic hydrocarbons produced by catalytic reforming because the bulk of impurities containing sulphur, nitrogen and oxygen are removed during the process. At the present time, over 6,000 million gallons of benzene are made annually by the Udex route. New and improved processes using sulpholane or *N*-methylpyrrolidone are gaining popularity. The re-forming process combines both the platforming and Udex extraction operations to produce either very high octane petrols or pure aromatic hydrocarbons (*H. W. Grote et al.*, Petroleum Refiner, 1955, **34**, 116; Chem. Eng. Progr., 1958, **54**, No 8, 43; *G. Voss*, Erdöl u. Kohle, 1962, **15**, 387; *F. S. Beardmore* and *W. C. G. Kosters*, J. Inst. Petroleum, 1963, **49**, 1; *R. Martin*, Petroleum Chem. Engr., 1966, **38** [2], 40–48).

Benzene, toluene and the xylenes are all isolated from catalytic reformed naphthas in large quantities by extraction processes referred to above. Ethylbenzene is also recovered from aromatized petrols for conversion to styrene. Other methylbenzenes produced commercially from platformate are cumene, pseudocumene, mesitylene and durene. Distillation, crystallization, isomerization, or extraction, or combinations of these processes, are used for their isolation. Pseudocumene, with a b.p. of $169 \cdot 4^0$, is separated commercially in high purity from other C_9 components by superfractionation. Durene cannot be easily recovered by distillation but it has an abnormally high melting point ($79 \cdot 2^0$) and is separated and purified by fractional crystallization after naphthalene has been removed by distillation. For further information on the manufacture of the polymethylbenzenes see *H. E. Cier* and *H. W. Earhart*, "Advances in Petroleum Chemistry and Refining", Ed. *J. J. McKetta*, Interscience Publishers, New York, 1964, Vol. VIII, p. 271.

Although the potential production of benzene, toluene, and the xylenes, by catalytic reforming far exceeds the present demand, the relative proportions in which they are produced by this means do not nearly coincide with market requirements, being appreciably short in benzene. It would of course be possible to restrict the boiling range of the reformer charge in order to produce mainly benzene. In practice, however, toluene and the xylenes can be produced at very little extra cost and refiners are reluctant to restrict the boiling range of the naphtha charge. With toluene prices well below those of benzene the problem of unbalanced production of this aromatic is now largely solved by hydrodealkylation (p. 161). Benzene produced by dealkylation of toluene and xylene now exceeds that from the coke ovens. For a description of this process see *G. F. Asselin*, *ibid.*, Vol. IX, p. 47. Benzene is also produced from "drip oils", a by-product from pyrolysis reactors at ethylene plants (*R. B. Stobaugh*, Hydrocarb. Process. Petrol. Refiner, 1955, **44**, 209; *J. Morrison*, Oil Gas Intern., 1968, **8**, 92–6).

Benzene is the most important aromatic hydrocarbon from a production viewpoint and has the most end uses. It is used mainly as the initial primary product for making styrene monomer, phenol, and Nylon, and in smaller amounts for the manufacture of detergents, aniline, nitrobenzene, maleic anhydride, resorcinol, biphenyl, the insecticides D.D.T. ("dichlorodiphenyltrichloroethane", see Vol. IC, p. 24) and *B.H.C.* ("benzene hexachloride", see Vol. IIB, p. 36), mono- and di-chlorobenzenes. Toluene is used largely in aviation and motor gasolines, for solvents, benzene production and for making diisocyanates required in the manufacture of urethane foams. Toluene is also used for making detergents, trinitrotoluene and numerous chemical intermediates. The largest outlet for mixed xylenes (*i.e.* the mixed C_8 hydrocarbons) is in petrol (gasoline) blending. They are also widely used as solvents. The demand for the separated isomers is increasing steadily. The main outlet for *o*-xylene is phthalic acid manufacture, *p*-xylene is required for producing terephthalic acid for synthetic fibres and films, *m*-xylene for making isophthalic acid, and ethylbenzene for styrene manufacture. The bulk of the pure xylene isomers come from petroleum. Ethylbenzene and *o*-xylene are separated commercially by distillation but the *m*- and *p*-xylenes require special fractional crystallization. The proportions in which the C_8 aromatic hydrocarbons are produced are out of balance with market requirements the *meta* isomer having the least demand. This problem has been overcome by isomerization processes (p. 162). Isomerization of the xylenes for maximum production of the more valuable *para* isomer is now an established process. The commercial applications for the higher polymethylbenzenes are mainly in the general plastics field, pseudocumene being required for the production of trimellitic anhydride and durene for making pyromellitic dianhydride.

The higher alkylbenzenes of commercial importance are made by alkylation of benzene with the appropriate olefin; the raw materials coming from petroleum or coal tar sources. The three large volume alkylbenzenes are ethylbenzene used for dehydrogenation to styrene, cumene used in the production of phenol, and dodecylbenzene used for making synthetic detergents. Ethylbenzene is made by alkylation of benzene with ethylene in the presence of aluminium chloride. Cumene is made commercially by alkylation of benzene with propylene. Dodecylbenzene is made by alkylation of benzene with propylene tetramer and mixed detergent alkylates are made from benzene and cracked wax olefins (see Vol. IA, p. 372), or monochloroparaffins, using aluminium chloride, hydrofluoric acid, or sulphuric acid as catalyst. For further information on these and other commercial alkylbenzenes produced by alkylation see *E. K. Jones*, Advances in Catalysis, 1958, **10**, 165; and *S. H. Patinkin* and *B. S. Friedman*, "Friedel-Crafts and Related Reactions", Ed. *G. A. Olah*, Interscience Publishers, London, 1964, Vol. 2, part I, p. 1.

Many other mononuclear aromatic hydrocarbons are potentially available from petroleum but await suitable technical advances and/or market developments for their commercial production.

For further information on the production and uses of aromatic hydrocarbons from coal and petroleum refer to reviews given by *A. H. Howland*, Proc. 7th World Petrol. Congress, 1967, **5**, 45; and *D. M. Samuel*, Chem. and Ind., 1968, 567.

7. Individual members of the mononuclear hydrocarbons

For physical properties see Table 2, p. 179.

Benzene, C_6H_6, is a clear colourless, highly refractive, flammable liquid, with a pleasant characteristic odour and burns with a smoky flame. The vapour when inhaled causes giddiness and ultimate insensibility. It was first isolated from pyrolysed sperm oil by M. Faraday in 1825 (*R. Kaiser*, Angew. Chem. intern. Edn., 1968, **7**, 345). Now that the bulk of commercial benzene is produced by catalytic reforming it is virtually free from sulphur. Much of the benzene derived from coal is also of comparable quality. The need for removing thiophene from commercial benzene therefore rarely exists today. The simplest and most satisfactory method of removing thiophene in the laboratory consists in shaking the impure benzene with Raney nickel (*J. Bougault et al.*, Bull. Soc. chim. Fr., 1940, [v], 7, 780). For a review of the production of pure benzene refer to *F. Trefny*, Erdöl u. Kohle, Erdgas, Petrochem., 1967, **20**, 629.

Benzene is sparingly soluble in water, 1·84 g/l at 0·8°, and 1·80 g/l at 24° (*D. M. Alexander*, J. phys. Chem., 1959, **63**, 1201). The solubility of water in benzene is of the order of 0·063 % v/w (*J. Wing* and *W. H. Johnston*, J. Amer. chem. Soc., 1957, **79**, 864). For the thermodynamic properties of benzene refer to *L. N. Canjar et al.*, Hydrocarb. Process. Petrol. Refiner, 1965, **44**, 297.

Toluene, *methylbenzene*, $C_6H_5{\cdot}CH_3$, a colourless, refractory, flammable liquid with a benzene-like odour, was first isolated by the distillation of resin from *Pinus maritima* (*M. Pelletier* and *P. Walther*, Ann. Chim., 1838, [ii], **67**, 278). It has also been obtained by the distillation of other resins such as tolu balsam and dragon's blood. The isomeric 5-methylene-1,3-cyclohexadiene (II), which is stable at —80°, has been obtained by pyrolysis of the ester I at 360°. It rapidly isomerizes at room temperature to toluene (*W. J. Bailey* and *R. A. Baylouny*, J. org. Chem., 1962, **27**, 3476):

CH_2OCO_2Et — 360° → CH_2

(I) (II)

Hydrocarbons C_8H_{10} comprise the three xylenes (dimethylbenzenes) and ethylbenzene. Their commercial separation has already been described (p. 224). Since they are invariably produced as a mixture their separation is important and the following techniques have been used: crystallization (*D. L. McKay et al.*, Chem. Eng. Progr., 1966, **62**, No. 11, 104; *A. Rabczuk* and *B. Kocko*, Przemysl Chem., 1968, **47**, 56); selective solid compound formation, *e.g. p*-xylene forms an equimolal *compound* with carbon tetrachloride which freezes at —32° (*E. J. Egan* and *R. V. Luthy*, Ind. Eng. Chem., 1955, **47**, 250); clathration (*R. N. Fleck* and *C. G. Wright*, U.S.P., 2,983,767/1961, 3,013,091/1961, and p. 203); sulphonation (*H. P. Hetzner* and *R. J. Miller*, U.S.P., 2,511,711/1950; *J. A. Spence* U.S.P., 2,943,211/1960; *W. Smolin*, U.S.P., 3,311,670/1967, and p. 209); extraction with $HF–BF_3$ (*D. A. McCaulay* and *A. P. Lien*, U.S.P., 2,528,892/1950); extractive distillation (*J. C. Chu et al.*, Ind., Eng. Chem., 1954, **46**, 754; *S. Suzuki*, U.S.P., 3,356,593/1967); azeotropic distillation (*R. L. Nelson*, Amer. chem. Soc. Div. Pet. Chem., Preprints, 1963, **8**, No. 1, 115); and conditioned membranes (*A. S. Michaels et al.*, Ind. Eng. Chem. Process Design Develop., 1962, **1**, 14). The use of the isomerization/disproportionation reaction to increase *p*-xylene production has already been mentioned (p. 162). Transalkylation between toluene and a polymethylbenzene for the production of *p*-xylene is of interest since the concentration of *p*-xylene produced at low conversions is much higher than normal thermodynamic values (*M. C. Hoff* and *McCaulay*, U.S.P., 2,954,414/1960). Oxidation is a particularly important reaction of the xylenes (p. 190). For a useful summary of the various processes for making aromatic dicarboxylic acids see Chem. Week. 1957, **80**, 32.

o-**Xylene**, 1,2-*dimethylbenzene*, is a clear colourless, toxic, flammable liquid with a characteristic odour. Nitration gives a mixture of 3- and 4-nitro-*o*-xylenes, and further nitration a mixture of 3,4-, 4,6-, and 4,5-dinitro-; and 3,4,5-, and 3,4,6-trinitro-*o*-xylenes (*A. W. Crossley* and *N. Renouf*, J. chem. Soc., 1909, **95**, 201; *K. A. Kobe* and *W. P. Prichett*, Ind. Eng. Chem., 1952, **44**, 1398).

m-**Xylene**, 1,3-*dimethylbenzene*, is a colourless liquid. It can be prepared in the pure state by the reaction of *m*-tolylmagnesium bromide with dimethyl sulphate, or by hydrolysis of the 4-sulphonic acid. Its formation by the decarboxylation of mesitylenic acid was important historically in that it provided proof of the orientation of the methyl groups and also of the toluic acid and isophthalic acid which are formed from it by further oxidation. On nitration it gives 4-nitro-*m*-xylene as the main product with a small amount of the 2-nitro-compound (*Kobe* and *H. Brennecke*, *ibid.*, 1954, **46**, 728). Further nitration gives a mixture of dinitro-compounds and finally 2,4,6-trinitro-*m*-xylene.

p-**Xylene**, 1,4-*dimethylbenzene*, exists as colourless plates. It is conveniently synthesized by reaction of dimethyl sulphate with *p*-tolylmagnesium halide or from *p*-xylidine (*G. R. Clemo et al.*, J. chem. Soc., 1929, 2375). Mono-substitution in the nucleus gives one product; disubstitution gives a mixture consisting largely of 2,3- and 2,6-disubstituted derivatives, *e.g.* nitration (*Kobe et al.*, Ind. Eng. Chem., 1950, **42**, 352; 356).

Ethylbenzene has been prepared by the Friedel-Crafts reaction from ethyl bromide or chloride and by the Fittig reaction from bromobenzene and ethyl bromide. Its preparation by the reaction of diethyl sulphate with phenylmagnesium bromide (*H. Gilman* and *R. E. Hoyle*, J. Amer. chem. Soc., 1922, **44**, 2623) or dimethyl sulphate with benzylmagnesium chloride is more satisfactory. Pure ethylbenzene has been made by hydrogenation of acetophenone at 200° with a copper–chromium catalyst (*E. Kuss et al.*, Naturwissenschaften, 1961, **48**, 73). For the nitration of ethylbenzene see *E. L. Cline* and *E. E. Reid*, J. Amer. chem. Soc., 1927, **49**, 3753, and *A. I. Titov*, Zhur. obshchei Khim., 1963, **33**, 1497.

Hydrocarbons C_9H_{12} comprise three trimethylbenzenes, three ethyltoluenes and two propylbenzenes. All occur in reformate; typical relative proportions (% wt.) being:

Isopropylbenzene	0·6
n-Propylbenzene	5·3
o-Ethyltoluene	9·3
m-Ethyltoluene	17·7
p-Ethyltoluene	8·8
Mesitylene	7·8
Pseudocumene	42·1
Hemimellitene	8·4

Only three isomers, *n*-propylbenzene, isopropylbenzene and pseudocumene, can be isolated in high purity and reasonable yield by distillation. Pseudocumene is the most prevalent aromatic in C_9 reformate and is separated commercially in high purity by this means.

Hemimellitene, 1,2,3-*trimethylbenzene,* has been prepared by reaction of methyl iodide with 2-bromo-*m*-xylene and sodium, and by distilling isodurylic acid. More satisfactory methods of preparation are hydrogenation of 2,3-dimethylbenzyl alcohol at 225° and 100–190 atmospheres using copper chromite catalyst (*L. I. Smith* and *I. J. Spillane,* J. Amer. chem. Soc., 1940, **62,** 2639) or reduction of 2,3-dimethylbenzyltrimethylammonium iodide with sodium amalgam (*S. W. Kantor* and *C. R. Hauser, ibid.*, 1951, **73,** 4122). It may be also made by alkylation of *o*-xylene with methanol at 475° using a synthetic zeolite as catalyst (*Y. Sidorenko et al.*, U.S.S.R. Pat., 186,418/1965; C.A., 1967, **66,** 115437c).

Pseudocumene, 1,2,4-*trimethylbenzene,* has been synthesized from *m*- and *p*-xylene by conversion into the corresponding dimethylbenzyl chloride and subsequent reduction (*H. Stephen et al.*, J. chem. Soc., 1920, **117,** 522; *L. Best,* Compt. rend., 1928, **186,** 373). Halogenation, nitration, and sulphonation take place in the 5-position preferentially: disubstitution products are chiefly 3,5-derivatives. For the side-chain halogenation of pseudocumene and durene see *A. Drechsler,* Z. Chem., 1965, **5,** 180. Oxidation by manganese dioxide, chromyl chloride or electrolytically yields 2,4-dimethylbenzaldehyde as the main product (*H. D. Law* and *F. M. Perkin,* J. chem. Soc., 1907, **91,** 263).

Mesitylene, 1,3,5-*trimethylbenzene,* occurs, together with pseudocumene, in coal tar solvent naphtha; its production from petroleum has already been mentioned. It is separated from the solvent naphtha fraction, b.p. 160–166°, by shaking with concentrated sulphuric acid and then steam distilling the sulphuric acid suspension of sulphonic acids. Fairly pure mesitylene passes over leaving a residue of pseudocumene sulphonic acid from which pseudocumene may be isolated by steam distillation from 50 % sulphuric acid (*Smith* and *O. W. Cass,* J. Amer. chem. Soc., 1932, **54,** 1606). Nitration proceeds very readily with the formation of di- and tri-nitro derivatives. The mononitro compound is prepared by careful nitration in acetic anhydride solution (*G. Powell* and *F. R. Johnson,* Org. Synth., 1943, Coll. Vol. II, p. 449), or by the elimination of the amino group from 4-amino-2-nitromesitylene. It is mainly substituted in the nucleus by chlorination in the dark (*G. Illuminati* and *F. Stegel*, Ric. Sci., rend., Sec. A, 1964, **7**, 458).

n-Propylbenzene is conveniently prepared by reaction of diethyl sulphate with benzylmagnesium chloride (*Gilman* and *W. E. Catlin*, Org. Synth., 1932, Coll. Vol. I, p. 458), or by reduction of propiophenone with amalgamated zinc (*W. N. White* and *C. D. Slater*, J. org. Chem., 1961, **26**, 3631). Other methods are the reduction of benzyl methyl ketone (*E. Clemmensen*, Ber., 1913, **46**, 1839; J. pr. Chem., 1910, [ii], **81**, 387) or the reaction of cyclopropane with benzene and sulphuric acid. For its behaviour on nitration see *O. L. Brady* and *R. N. Cunningham*, J. chem. Soc., 1934, 121.

Cumene, *isopropylbenzene*, $PhCHMe_2$, was first obtained by distilling cumic acid, $CO_2H \cdot C_6H_4 \cdot CHMe_2$, with lime (*Ch. Gerhardt* and *A. Cahours*, Ann., 1841, **38**, 80). It may be prepared by alkylation of benzene with isopropyl halide or propylene in the presence of aluminium chloride (*E. R. Kline et al.*, J. org. Chem., 1959, **24**, 1781; *G. A. Olah et al.*, J. Amer. chem. Soc., 1964, **86**, 1046). For the disproportionation of mono-, di-, and tri-isopropylbenzenes in the presence of aluminium chloride see *E. P. Babin*, Neftekhimiya, 1964, **4**, 21. Cumene and alkylcumenes react with sulphur at 185–200^0 in the presence of bases to give good yields of 4-aryl-1,2-dithiole-3-thiones (*E. K. Fields*, J. Amer. chem. Soc., 1955, **77**, 4255). The regulated oxidation of cumene by chromic acid or potassium permanganate in glacial acetic acid gives a small amount of dimethylphenylcarbinol, $Ph \cdot C(OH)Me_2$ (*E. Boedtker*, Bull. Soc. chim. Fr., 1901, [iii], **35**, 846). More vigorous oxidation leads successively to acetophenone (*H. Meyer* and *K. Bernhauer*, Monatsh., 1929, **53/54**, 724; 726) and benzoic acid. Bromination and chlorination are described by *W. Qvist* (Acta Acad. Aboensis, Math. Phys., 1936, **10**, No. 5) and nitration by *S. D. Mekhtiev* and *A. Sh. Novruzova* (Izvest. Akad. Nauk Azerb. SSR., Ser. Fiz.-Tekh. i Khim. Nauk, 1958, 85).

The three **ethyltoluenes** are prepared by reaction of the appropriate tolylmagnesium halide with diethyl sulphate (*Gilman* and *Hoyle, loc. cit.*), by reduction of the methyl tolyl ketones (*F. Eisenlohr* and *L. Schulz*, Ber., 1924, **57**, 1816; *Brady* and *J. N. E. Day*, J. chem. Soc., 1934, 117) or by alkylation of toluene (*Yu. G. Mamedaliev et al.*, Doklady Akad. Nauk Azerb. SSR., 1963, **19**, 13; *F. J. Soderquist* and *J. L. Amos*, U.S.P., 3,296,322/1967). Ethyltoluenes and ethylxylenes are of interest since they can be dehydrogenated to methylstyrenes (see *R. H. Boundy* and *R. F. Boyer*, "Styrene and its polymers, copolymers and derivatives", Reinhold Publishing Corp., New York, 1952).

Hydrocarbons $C_{10}H_{14}$. All the possible 22 isomers have been prepared in a state of purity (see Table 2). The amounts (% wt.) of the ten-carbon-atom aromatics in a typical reformate, which includes all the $C_{10}H_{14}$ alkylbenzenes, are given in Table 3 (*H. W. Earhart et al.*, Proc. 5th World Petrol. Congress, 1959, **4**, 1).

Prehnitene, 1,2,3,4-*tetramethylbenzene*, is prepared by the Jacobsen reaction (p. 210) from pentamethylbenzene and hydrolysis of the resulting prehnitenesulphonic acid. It is more convenient to use as starting point the mixture of

TABLE 3

PERCENTAGE OF C_{10} AROMATICS IN REFORMATE

tert-Butylbenzene	0·1
Isobutylbenzene	0·5
sec-Butylbenzene	0·1
m-Cymene	0·2
p-Cymene	0·7
o-Cymene	Trace
1,3-Diethylbenzene	1·6
1-Methyl-2-*n*-propylbenzene	1·8
1-Methyl-3-*n*-propylbenzene	3·0
1-Methyl-4-*n*-propylbenzene	2·8
n-Butylbenzene	3·6
1,2-Diethylbenzene	2·5
5-Ethyl-1,3-dimethylbenzene	4·0
1,4-Diethylbenzene	0·8
2-Methylindan	3·5
1-Methylindan	1·3
2-Ethyl-1,4-dimethylbenzene	4·7
4-Ethyl-1,3-dimethylbenzene	6·0
4-Ethyl-1,2-dimethylbenzene	9·6
2-Ethyl-1,3-dimethylbenzene	1·0
3-Ethyl-1,2-dimethylbenzene	4·1
1,2,4,5-Tetramethylbenzene (durene)	8·0
1,2,3,5-Tetramethylbenzene (isodurene)	12·7
5-Methylindan	2·9
4-Methylindan	13·4
1,2,3,4-Tetramethylbenzene (prehnitene)	5·3
Naphthalene	5·8

durene and isodurene obtained by methylation of xylene (*L. I. Smith*, Org. Synth., 1943, Coll. Vol. II, p. 248; *S. F. Birch et al.*, J. Amer. chem. Soc., 1949, **71**, 1362). It has also been made by the action of selenium oxide on α-pyronene (*W. Zacharewicz* and *A. Uzarewicz*, Roczniki Chem., 1960, **34**, 413).

Isodurene, 1,2,3,5-*tetramethylbenzene*, is obtained by the reduction of 2,4,6-trimethylbenzyl chloride (*W. J. Hickinbottom et al.*, J. Inst. Petroleum, 1949, **35**, 630), by the reaction of mesitylmagnesium bromide with dimethyl sulphate (*Smith* and *F. H. MacDougall*, J. Amer. chem. Soc., 1929, **51**, 3002; *Birch et al.*, *loc. cit.*), by precise fractionation of the tetramethylbenzenes from the methylation of xylene (*Birch, loc. cit.*), or by methylation of mesitylene (*K. S. Nisarbaeva* and *G. Kh. Khodzhaev*, Uzbeksk. khim. Zhur., 1963, **7**, 28). In the commercial production of durene from reformate the mother liquor from the crystallization process is high in isodurene.

Durene, 1,2,4,5-*tetramethylbenzene*, forms plate-like crystals with a camphoraceous odour subliming slowly at room temperature. It is isolated from the methylation product of xylene by taking advantage of its high melting point (*Smith, loc. cit.*). It has also been prepared by the reduction of 2,5-di(chloromethyl)-1,4-dimethyl-benzene (*J. von Braun* and *J. Nelles*, Ber., 1934, **67**, 1094), and by reaction of pseudocumene with formaldehyde followed by hydrogenation of the intermediate bis(trimethylphenyl)methane (*J. G. Hendrickson* and *F. T. Wadsworth*, Ind. Eng. Chem., 1958, **50**, 877). A process for making durene in high yield from pseudocumene, involving methylation, isomerization, disproportionation and transalkylation, has been developed in which all these reactions take place in one stage (*D. E. Brown et al.*, U.S.P., 2,945,899/1960).

Nitration of durene invariably gives the 3,6-dinitrodurene irrespective of the amount of nitric acid employed (*G. Illuminati* and *M. Illuminati*, J. Amer. chem. Soc., 1953, **75**, 2159). With dilute nitric acid, durylic acid and 2,5-dimethylterephthalic acid are produced. Pyromellitic acid is formed by oxidation with potassium permanganate while reaction with hydrogen peroxide or chromic acid gives duroquinone. 3-Chloro- and 3,6-dichloro-durene are formed by chlorination of durene in the cold. Heat and sunlight favour introduction of chlorine into the methyl groups.

Ethylxylenes. It is stated that 4-*ethyl*-o-*xylene* is formed together with other products when camphor is distilled with iodine or zinc chloride (*H. E. Armstrong* and *A. K. Miller*, Ber., 1883, **16**, 2259). All of the isomers have been prepared by reduction of the appropriate methyl xylyl ketone or by reaction of diethyl sulphate with a xylylmagnesium halide (*Birch et al., loc. cit.*). For the optimal conditions for alkylating *o*-, *m*-, and *p*-xylenes with ethylene in the presence of aluminium chloride see *S. I. Sadykh-Zade* and *A. K. Askerov*, Azerb. khim. Zhur., 1960, 39.

Diethylbenzenes. 1,3-*Diethylbenzene* is obtained by the action of ethylene on benzene and aluminium chloride; it is best separated from the 1,2- and 1,4-isomers through its sulphonic acid (*J. E. Copenhaver* and *E. E. Reid*, J. Amer. chem. Soc., 1927, **49**, 3157). *o*- and *p*-Diethylbenzenes are obtained pure by the reduction of the corresponding ethylphenyl methyl ketones, EtC_6H_4COMe (*Birch et al., loc. cit.*, p. 1367).

Methyl n-propylbenzenes. The three isomers are prepared by Clemmensen reduction of the ethyl tolyl ketones (*Birch et al., loc. cit.*,) or reduction of the propenyltoluenes (*F. Eisenlohr* and *L. Schulz*, Ber., 1924, **57**, 1817).

Cymenes, *methylisopropylbenzenes*. The most important of these is p-*cymene*; it occurs naturally in many essential oils such as Roman oil of cumin from which it was first isolated, oil of thyme, in some eucalyptus oils, Ceylon cinnamon oil, cypress oil, chenopodium oil, juniper berry oil, lemon oil, oil of limes, and other essential oils (*E. K. Nelson*, J. Amer. chem. Soc., 1920, **42**, 1207; *Y. Hirose et al.*, Nippon Zasshi, 1960, **81**, 1766; *B. A. Bernhard* and *J. R. Clarke*, Food Res., 1960, **25**, 389; J. Food Sci., 1961, **26**, 401; *C. A. Slater*, Chem. and Ind., 1961, 833). *p*-Cymene has been prepared from (+)-limonene by dehydrogenation over nickel at 280° (*W. Treibs* and *H. Schmidt*, Ber., 1927, **60**, 2340) or by heating with sulphur (*L. Ruzicka et al.*, Helv., 1922, **5**, 346); from dipentene by heating with concentrated sulphuric acid or phosphorus pentasulphide (*O. Wallach* and *W. Brass*, Ann., 1884, **225**, 311); from terpineol with dilute sulphuric acid (*Wallach, ibid.*, 1893, **275**, 106); in fair yield by heating camphor with phosphorus pentoxide (*F. Fittia, ibid.*, 1874, **172**, 307). The best conditions for the preparation of *p*-cymene from pinene are give by *E. Raymond* (Bull. Soc. chim. Fr., 1934, [v] **1**, 1470). *p*-Cymene is obtained in considerable quantities as a by-product in the manufacture of wood pulp (*A. S. Wheeler*, J. Amer. chem. Soc., 1920, **42**, 1843; *E. Profft* and *G. Buchmann*, Chem. Tech. [Berlin], 1955, **7**, 138). Vigorous oxidation of *p*-cymene yields terephthalic acid as the final product, but by controlled oxidation a number of intermediate products may be obtained, either from oxidation of the methyl group to give *p*-isopropylbenzaldehyde and *p*-isopropylbenzoic acid (cumic aldehyde and cumic acid) (*F. Fichter* and *J. Meyer*, Helv., 1925, **8**, 285), or methyl *p*-tolyl ketone and *p*-acetylbenzaldehyde (*H. N. Stephens*, J. Amer. chem. Soc., 1926, **48**, 1826, 2921; *H. Meyer* and *K. Bernhauer*, Monatsh., 1929, **53/54**, 729). 4-Carboxyphenyldimethylcarbinol, $CO_2H \cdot C_6H_4 \cdot C(OH)Me_2$, has been isolated using potassium permanganate as oxidizing agent (*R. Meyer*, Ann., 1883, **219**, 248), while *p*-methylacetophenone is obtained in 40 % yield by oxidation with dilute nitric acid (*J. Kulesza* and *W. Madalinski*, Przemysl Chem., 1955, **11**, 141). Nitration of *p*-cymene yields 4-isopropyl-1-methyl-2-nitrobenzene, an oil (*R. J. W. Le Fevre*, J. chem. Soc., 1933, 982). Further nitration gives the 2,6-*dinitro* compound, m.p. 54°, and other products including nitrotoluenes (*Wheeler et al.*, J. Amer. chem. Soc., 1921, **43**, 2613; 1927, **49**, 495; *M. Phillips, ibid.*, 1922, **44**, 1777).

m-*Cymene* is present in light resin oil (*W. Kelbe*, Ann., 1881, **210**, 10). It is obtained in a more or less impure state from some terpenes by the action of phosphorus pentasulphide or by heating with anhydrous zinc chloride (*Armstrong* and *Miller, loc. cit.*). The dehydrogenation of (+)-silvestrene over nickel–alumina gives *m*-cymene (*J. Herzenberg* and *S. Ruhemann, ibid.*, 1925, **58**, 2262) and it is formed by heating fenchene with phosphorus pentoxide at 120° (*Wallach*, Ann., 1893, **275**, 158; 1894, **284**, 324).

Pure specimens of *o*-, *m*- and *p*-cymenes have been prepared from the appropriate toluic ester by reaction with methylmagnesium bromide followed by dehydration of the resulting carbinol and subsequent hydrogenation of the *β*-tolylpropene which is thus formed (*Eisenlohr* and *Schulz, loc. cit.*, p. 1808; *Birch et al., loc. cit.*). The isomer distribution in the cymenes formed by Friedel-Crafts isopropylation of toluene with propylene or isopropyl bromide has been determined by *Olah* and coworkers (J. Amer. chem. Soc., 1964, **84**, 1046). Certain complexes of nickel thiocyanate with different benzylamines are selective for complexing with the isomeric cymenes and may be used for their separation (*P. de Radzitzky* and *J. Hanotier*, Ind. Eng. Chem. Process Design Develop., 1962, **1**, 10; Erdöl u. Kohle, 1962, **15**, 892).

n-**Butylbenzene** is obtained from benzyl chloride and *n*-propylmagnesium bromide, or better by Clemmensen reduction of *n*-butyrophenone (*Birch et al.*, J. Amer. chem. Soc., 1949, **71**, 1362).

sec-**Butylbenzene** is formed by the alkylation of benzene with *sec*-butyl alcohol (*H. Meyer* and *K. Bernhauer*, Monatsh., 1929, **53/54**, 721) and by the hydrogenation of 2-phenylbutene (*A. Klages*, Ber., 1902, **35**, 2642; *Birch et al.*, *loc. cit.*).

Isobutylbenzene is best prepared by the Clemmensen reduction of isobutyrophenone (*Birch et al., loc. cit.*).

tert-**Butylbenzene** is formed from benzene by the Friedel-Crafts reaction using *tert*-butyl chloride (*J. B. Shoesmith* and *A. Mackie*, J. chem. Soc., 1928, 2336) or by alkylation using isobutylene, or iso- or *tert*-butyl alcohol (*A. Verley*, Bull. Soc. chim. Fr., 1898, [iii], **19**, 72).

Monosubstitution in the four butylbenzenes occurs preferentially in the *para* position.

Hydrocarbons $C_{11}H_{16}$. **Pentamethylbenzene**, m.p. 54·3°, b.p. 230°, is prepared by passing methyl chloride into xylene and pure aluminium chloride at 95° (*L. I. Smith* and *F. J. Dobrovolny*, J. Amer. chem. Soc., 1926, **48**, 1415; *V. I. Mil'skii*, C.A., 1963, **59**, 11297a), or by the reduction of 1,3-di(chloromethyl)-2,4,6-trimethylbenzene (*Hickinbottom et al.*, J. Inst. Petroleum, 1949, **35**, 630; *M. Crawford* and *J. H. Magill*, J. chem. Soc., 1957, 3275). It has also been obtained by the rearrangement of 1,1,3,6-tetramethyl-4-methylene-cyclohexa-2,5-diene (*K. von Auwers* and *K. Ziegler*, Ann., 1921, **425**, 272). With concentrated sulphuric acid it gives prehnitene sulphonic acid and hexamethylbenzene (see p. 210; also *L. I. Smith* and *A. R. Lux*, J. Amer. chem. Soc., 1929, **51**, 2997).

Ethylmesitylene, b.p. 214° may be made by reaction of mesitylmagnesium bromide with ethyl sulphate or by hydrogenation of acetylmesitylene over copper chromite catalyst (*Crawford et al.*, J. chem. Soc., 1952, 4443; 1957, 3275). 5-*Ethylpseudocumene* has b.p. 207°; 3,5-*diethyltoluene*, b.p. 208°; n-*propyl*-p-*xylene*, b.p. 103°/30 mm, d_4^{20} 0·8665, n_D^{20} 1·4967, has been prepared by a Wurtz-Fittig synthesis while *isopropyl*-p-*xylene*, b.p. 102°/41 mm, d_4^{20} 0·8735, n_D^{20} 1·5007, results from the alkylation of *p*-xylene with either *n*- or iso-propyl alcohol in the presence of sulphuric acid (*T-H. Yang*, C.A., 1963, **59**, 11298f).

3-tert-**Butyltoluene**, b.p. 186°, is present in light oil from the distillation of pine resin. It is formed together with 4-tert-*butyltoluene* from toluene and *tert*-butyl chloride in the presence of aluminium chloride or ferric chloride (*J. B. Shoesmith* and *J. F. McGechen*, J. chem. Soc., 1930, 2232), or from toluene and isobutylene in the presence of synthetic aluminosilicate (*S. D. Mekhtiev et al.*, Azerb. khim. Zhur., 1965, 14). The 2,4,6-*trinitro*-derivative, m.p. 97°, is one of the artificial musks.

4-tert-**Butyltoluene** has been found in the essential oil of *Nepeta leucophylla* (*Y. P. Talwar et al.*, Soap, Perfumery Cosmetics, 1964, **37**, 45). It is the major product from the Friedel-Crafts alkylation of toluene with *tert*-butyl halide or isobutylene in nitromethane (*Olah et al.*, J. Amer. chem. Soc., 1964, **86**, 1060).

Pentylbenzenes. All eight isomers have been prepared. n-*Pentylbenzene* can be made by reaction of benzylmagnesium chloride with *n*-butyl-*p*-toluene sulphonate (*Gilman* and *J. Robinson*, Org. Synth., 1943, Coll. Vol. II, p. 47), hydrogenation of *n*-butyl phenyl ketone (*M. R. Musaev*, Azerb. khim. Zhur., 1960, 99), or from cinnamaldehyde and ethylmagnesium bromide followed by dehydration of the carbinol and hydrogenation (*Yu I. Kheruse* and *A. A. Petrov*, Zhur. obshchei Khim., 1961, **31**, 772). 3-*Methyl-1-phenylbutane*, b.p. 192–193°/725 mm, d_4^{20} 0·8529, n_D^{20} 1·4845, is prepared by reaction of benzylmagnesium chloride with isobutyl *p*-toluenesulphonate (*Gilman* and *N. J. Beaber*, J. Amer. chem. Soc., 1925, **47**, 522), and from 4-iodo-2-methylbutane, bromobenzene and sodium. 2-*Methyl-1-phenylbutane*, b.p. 189–190°/725 mm, d_4^{20} 0·8617, n_D^{20} 1·4898; 2-*phenylpentane*, b.p. 187–188°/725 mm, d_4^{20} 0·8641, n_D^{20} 1·4858; and 3-*methyl-2-phenylbutane*, b.p. 183–184°/725 mm, d_4^{20} 0·8632, n_D^{20} 1·4890, have all been prepared by Grignard synthesis from the appropriate carbonyl compound and reduction of the intermediate carbinol with hydrogen iodide. 2-*Methyl-2-phenylbutane*, b.p. 185–187°/725 mm, d_4^{20} 0·8685, n_D^{20} 1·4928, is obtained by alkylation of benzene with ethyl dimethyl carbinol in the presence of sulphuric acid. *Neopentylbenzene*, b.p. 180–181·5°/725 mm, d_4^{20} 0·8575, n_D^{20} 1·4862, has been made by a Wurtz-Grignard reaction from *tert*-butyl chloride and benzyl chloride and 3-*phenylpentane*, b.p. 185–185·5/725 mm, d_4^{20} 0·8632, n_D^{20} 1·4875, from ethylmagnesium bromide and ethyl benzoate followed by reduction with hydrogen iodide (*G. Kh. Khakimov* and *I. P. Tsukervanik*, Zhur. obshchei Khim., 1963, **33**, 493; *Kheruse* and *Petrov*, *loc. cit.*). For the aluminium chloride induced rearrangement of the eight pentylbenzenes see *R. M. Roberts* and *Y. W. Han*, J. Amer. chem. Soc., 1963, **85**, 1168.

Hydrocarbons $C_{12}H_{18}$. **Hexamethylbenzene,** m.p. 165°, b.p. 264°, is prepared by the action of methyl chloride on xylene or pentamethylbenzene in the presence of aluminium chloride at 95° (*Smith* and *Dobrovolny*, *loc. cit.*) or by the Jacobsen reaction from pentamethylbenzene (*Smith* and *Lux*, *loc. cit.*). It is formed by passing vaporised phenol, the cresols or xylenols with methyl alcohol over alumina at 400–440° (*E. Briner et al.*, Helv., 1924, **7**, 1049); the yield of hexamethylbenzene may reach 70 % (*N. M. Cullinane* and *S. J. Chard*, J. chem. Soc., 1945, 821). It is also obtained by heating the hydrochlorides of the xylidines with methyl alcohol at 300° (*D. H. Hey*, *ibid.*, 1931, 1586). More recently it has been made by contacting dimethylacetylene with a catalyst obtained by the reaction of vanadyl chloride with aluminium triisobutyl (*J. B. Armitage*, U.S.P., 3,082,269/1963). Hexamethylbenzene is formed by isomerization of hexamethylbicyclo[2·2·0]hexa-2,5-diene prepared from butyne (*W. Schaefer* and *H. Hellmann*, Angew. Chem. intern. Edn., 1967, **6**, 518). It is completely protonated in concentrated sulphuric acid. If kept for a few days in 96 % sulphuric acid it is gradually converted into pentamethylbenzylcarbonium ion (*H. M. Buck et al.*, Tetrahedron Letters, 1964, 2987; see also *R. Hulme* and *M. C. R. Syman*, Nature, 1965, **206**, 293; J. chem. Soc., A, 1966, 446). Oxidation of hexamethylbenzene occurs slowly to give mellitic acid, $C_6(CO_2H)_6$ (*C. Friedel* and *J. M. Crafts*, Compt. rend., 1880, **91**, 260) and also intermediate products such as prehnitene-1,2-dicarboxylic acid (*O. Jacobsen*, Ber., 1889, **22**, 1216).

The terminal product from the methylation of benzene and its methyl homologues is 1,1,2,3,4,5,6-heptamethylbenzonium ion. Loss of a proton from this gives 4-methylene-1,1,2,3,5,6-hexamethylcyclohexadiene-2,5 (*W.v. E. Doering et al.*, Tetrahedron, 1958, **4**, 178).

Ethyldurene, m.p. 54°, b.p. 120–124°/14 mm, is made by hydrogenation of 2,3,5,6-tetramethylacetophenone (*Crawford* and *Magill, loc. cit.*; *Ng. Ph. Buu-Hoi et al.*, Bull. Soc. chim. Fr., 1960, 1493).

1,2,4-**Triethylbenzene,** b.p. 93°/0·1 mm, n_D^{20} 1·5009, has been made from 1,2,4-trivinylcyclohexane by reaction with alkali metal (*F. Derichs et al.*, U.S.P. 3,166,604/1965). 1,3,5-*Triethylbenzene*, b.p. 211·2°, d_4^0 0·8772, is formed together with 1,2,4-triethylbenzene, by passing ethyl chloride into benzene in the presence of aluminium chloride. Further action gives 1,2,4,5-*tetraethyl-*, m.p. 10°, b.p. 246°, 1,2,3,5-*tetraethyl-*, m.p. –21°, b.p. 247°, *pentaethyl-* and *hexaethyl-benzenes* (*L. I. Smith* and *C. O. Guss*, J. Amer. chem. Soc., 1940, **62**, 2625). Thermodynamic data on the interconversion of the ethylbenzenes have been calculated by *A. E. Pinsker* (Khim. Prom., 1962, 486). The ethyl chloride may be replaced by ethylene with advantage for the formation of triethylbenzene (*C. H. Milligan* and *E. E. Reid*, J. Amer. chem. Soc., 1922, **44**, 206; *W. B. Dillingham* and *Reid, ibid.*, 1938, **60**, 2606).

Isopropylmesitylene, b.p. 99°/13·5 mm, d^{20} 0·8871, n_D^{20} 1·50969, can be made by alkylation of mesitylene with isopropyl alcohol in the presence of sulphuric acid (*B. Y. Ioffe* and *T-H. Yang*, Zhur. obshchei Khim., 1963, **33**, 2196). n-*Propylmesitylene* has b.p. 242°.

2-**Ethyl**-p-**cymene,** b.p. 210–212°, has been prepared by reduction of 2-acetyl-*p*-cymene (*T. Govindachari*, Tetrahedron, 1961, **12**, 105).

1,3-**Diisopropylbenzene,** b.p. 204°, is formed by reaction of propylene with benzene and aluminium chloride. Further action gives 1,2,4- and 1,3,5-triisopropylbenzenes (*T. M. Berry* and *Reid*, J. Amer. chem. Soc., 1927, **49**, 3148). This hydrocarbon is a potential raw material for the production of resorcinol (*B. Hammond* and *M. Thompson*, B.P., 857,113/1960).

1,4-**Diisopropylbenzene,** b.p. 210°, together with tri- and tetra-isopropylbenzenes and cumene result from the reaction of isopropyl alcohol or propylene with benzene in the presence of sulphuric acid (*H. L. Wunderly et al.*, J. Amer. chem. Soc., 1936, **58**, 1007; *V. N. Ipatieff et al., ibid.*, p. 919; *H. Meyer* and *K. Bernhauer*, Monatsh., 1929, **53/54**, 726). It has also been made by a modified Wittig reaction from triphenylphosphinomethylene and 1,4 diacetylbenzene (*A. A. D'onofrio*, J. appl. Polymer Sci., 1964, **8**, 521). It reacts with peroxides yielding polymers of $(p\text{-}Me_2CC_6H_4CMe_2\text{–})_n$ type arising through a series of radical recombinations; similar polymers are obtained with other polyalkylbenzenes (*V. V. Korshak et al.*, Vysokomolekulyarnye Soedineniya, 1961, **3**, 1332).

4-*n*-**Butyl**-1,2-**dimethylbenzene,** b.p. 82°/6 mm, d_4^{20} 0·8680, n_D^{20} 1·4975, is obtained in good yield by a Diels-Alder reaction with 2,3-dimethyl-1,3-butadiene and butylfumaric acid in acetic anhydride and heating the intermediate hydrophthalic anhydride with phosphorus pentoxide (*V. R. Skvarchenko et al.*, Zhur. obshchei Khim., 1962, **32**, 111). 4-sec-*Butyl*-1,2-*dimethylbenzene*, b.p. 105°/20 mm, d_4^{20} 0·8756, n_D^{20} 1·50136, results from the alkylation of *o*-xylene with *sec*-butyl alcohol in the presence of sulphuric acid (*Ioffe* and *Yang, loc. cit.*) but with isobutylene 4-tert-*butyl-1,2-dimethylbenzene,* b.p. 89°/14 mm, is obtained (*E. S. Pokzovskaya*, Trudy Inst. Nefti. Akad. Nauk SSSR., 1959, **13**, 5).

2-tert-**Butyl**-1,3-**dimethylbenzene,** b.p. 114–116°/16 mm, n_D^{25} 1·5150, is prepared from the bromodimethylbenzene by Grignard reaction with *tert*-butyl chloride (*A. W. Burgstahler et al.*, Tetrahedron Letters, 1965, 1625). 5-tert-*Butyl*-1,3-*dimethylbenzene,* b.p. 202°, is prepared by the Friedel-Crafts alkylation of *m*-xylene with isobutylene, diisobutylene or *tert*-butyl halide (*H. Bauer*, Ber., 1891, **24**, 2840; *H. M. Knight et al.*, J. org. Chem., 1963, **28**, 1218; *G. A. Olah et al.*, J. Amer. chem. Soc., 1964, **86**, 1060). The *trinitro* derivative, m.p. 97°, is one of the artificial musks.

sec-**Butyl-p-xylene,** b.p. 100°/14·5 mm, d_4^{20} 0·8724, n_D^{20} 1·49931, is obtained by alkylation of *p*-xylene with *sec*-butyl alcohol in the presence of sulphuric acid (*Ioffe* and *Yang, loc. cit.*).

3-tert-**Pentyltoluene**, b.p. 209°, may be made by the action of *tert*-pentyl-chloride on toluene using aluminium chloride (*J. C. Essner* and *E. Gossin*, Bull. Soc. chim. Fr., 1884, [ii], **42**, 213; *Knight et al., loc. cit.*).

n-**Hexylbenzene** may be prepared from *n*-pentyl bromide by a Grignard reaction with benzaldehyde followed by dehydration of the carbinol and hydrogenation. It has also been made by reaction of divinylacetylene with chlorobenzene followed by hydrogenation of the adduct; from bromobenzene and *n*-hexyl bromide with excess sodium; and from benzyl bromide and *n*-pentyl-*p*-toluenesulphonate (*T. W. Mears et al.*, J. Res. Nat. Bur. Stand. 1963, **67**, 475; *Yu. I. Kheruse* and *Petrov*, Zhur. obshchei Khim., 1963, **33**, 1111; *H. L. Holsopple*, C.A., 1964, **60**, 4029e).

2-**Phenylhexane,** b.p. 204–205°/725 mm, d_4^{20} 0·8610, n_D^{20} 1·4898, has been obtained from acetophenone and *n*-butylmagnesium bromide followed by reduction of the intermediate carbinol with hydrogen iodide. 3-*Phenylhexane,* b.p. 199–201·5°/725 mm, d_4^{20} 0·8615, n_D^{20} 1·4886, is prepared in a similar way from propiophenone (*G. Kh. Khakimov* and *I. P. Tsukervanik,* Zhur. obshchei Khim., 1963, **33**, 493).

Hydrocarbons $C_{13}H_{20}$. n-**Propyldurene**, m.p. 49–50°, b.p. 120–125°/30 mm, may be prepared by hydrogenation of propionyldurene (*M. Crawford* and *N. H. Magill,* J. chem. Soc., 1957, 3275).

n-**Butylmesitylene**, b.p. 137°/13 mm, d_4^{20} 0·8716, n_D^{20} 1·49955, and sec-*butylmesitylene,* b.p. 115°/17 mm, d_4^{20} 0·8865, n_D^{20} 1·50758, have both been made by alkylation of mesitylene with the appropriate butyl alcohol in the presence of sulphuric acid (*Ioffe* and *Yang, loc. cit.*).

p-sec-**Butylcumene,** b.p. 225·5°, d_4^{20} 0·8578, n_D^{20} 1·4890, has been obtained by alkylation of cumene with butylenes in the presence of boron trifluoride–phosphoric acid (*S. V. Zavgorodnii* and *G. P. Filinov,* Izvest. Vysshikh Ucheb. Zavendenii, Khim. i Khim. Tekhnol., 1961, 4, 792). 5-tert-**Pentyl**-m-**xylene** results from the alkylation of *m*-xylene with *tert*-pentyl chloride (*Knight et al., loc. cit.*).

2-p-**Tolylhexane** is prepared from *p*-methylacetophenone by Grignard reaction with *n*-butyl bromide followed by dehydration of the carbinol and hydrogenation (*M. S. Shvartsberg*, C.A., 1963, **58**, 2383h). 1-p-*Tolylhexane* has b.p. 248·1°, m.p. −26·9, d_4^{20} 0·8560, n_D^{20} 1·4878 (*A. L. Liberman*, Neftekhimiya, 1963, **3**, 825).

n-**Heptylbenzene** results from the reaction of bromobenzene, *n*-heptylbromide and excess sodium (*Holsopple, loc. cit.*). It has also been made by the Wolff-Kishner reduction of *n*-hexyl phenyl ketone (*F. Eisenlohr* and *L. Schultz*, Ber., 1924, **57**, 1808), and by hydrogenation of the adduct from vinylpropenylacetylene and chlorobenzene (*Kheruse* and *Petrov*, loc. cit.). 2-**Methyl**-5-**phenylhexane** has been made from 2-methyl-5-phenyl-3-hexyne-2,5-diol by dehydration and hydrogenation (*Shvartsberg, loc. cit.*).

Hydrocarbons $C_{14}H_{22}$. o-**Diethyl**-, m.p. 44·5–45·0°, m-**diethyl-**, m.p. 35·5–36·0°, and p-**diethyl-**, m.p. 101–102°, **-tetramethylbenzenes** have been prepared by reduction of the appropriate diacetyltetramethylbenzene to the carbinol followed by dehydration and hydrogenation of the resulting divinyltetramethylbenzene (*M. S. Newman et al.*, J. Amer. chem. Soc., 1964, **86**, 868). The **tetraethylbenzenes** result from the ethylation of benzene in the presence of aluminium chloride (*V. G. Plyusnin* and *T. I. Sukhorosova*, Tr. Inst. Khim. Acad. Nauk SSSR., Ural. Filial, 1960, 21). Both 1,2,3,5- and 1,2,4,5-tetra-ethylbenzenes undergo the Jacobsen rearrangement to 1,2,3,4-tetraethylbenzene (*Smith* and *Guss*, J. Amer. chem. Soc., 1940, **62**, 2631).

tert-**Butyl**-m-**cymene**, b.p. 227°/737 mm, d_4^{20} 0·8660, n_D^{20} 1·4950, and tert-**butyl**-p-**cymene,** b.p. 228°/730 mm, d_4^{15} 0·8876, n_D^{20} 1·4972 (*dinitro*-deriv., m.p. 132–133°, odour of musk), are both prepared from cymene by Friedel-Crafts reaction with *tert*-butyl alcohol and sulphuric acid (*H. Barbier*, Helv., 1932, **15**, 592).

o-**Di**-tert-**butylbenzene** and o-**neopentylcumene** have been made by step-wise reactions from 1,1,4,4-tetramethyl-2-tetralone (*Burgstahler* and *M. O. Abdel-Rahman*, J. Amer. chem. Soc., 1963, **85**,173). *o*-Di-*tert*-butylbenzene is a highly hindered molecule; nitration, acetylation, and other reactions are described by *Burgstahler* and coworkers (*ibid.*, 1964, **86**, 5281).

n-**Octylbenzene** is made by the alkylation of benzene with *n*-octyl bromide in the presence of aluminium bromide (*S. H. Sharman, ibid.*, 1962, **84**, 2945). 2-*Phenyloctane* results from the alkylation of benzene with octene-1 (*S. I. Faingol'd* and *Kh. Yu. Vuure*, Zhur. priklad. Khim., 1963, **36**, 2527). 2-*Benzyl*-2-*methylhexane*, b.p. 111°/6 mm, d_4^{20} 0·8717, n_D^{20} 1·4911; 2-*benzyl*-2,4-*dimethylpentane*, b.p. 86°/0·5 mm, d_4^{20} 0·8700, n_D^{20} 1·4930 (*A. D. Petrov*, Erdöl u. Kohle, 1958, **11**, 855).

Higher alkylbenzenes. 3-**Benzyl**-3,5-**dimethylhexane,** $C_{15}H_{24}$, b.p. 97°/0·5 mm, d_4^{20} 0·8888, n_D^{20} 1·4959 (*Petrov, loc. cit*).

3-**Benzyl**-2,3,5-**trimethylhexane,** $C_{16}H_{26}$, b.p. 107–108·5°/1·5 mm, d_4^{20} 0·8933, n_D^{20} 1·5034; 4-**benzyl**-2,4-**dimethylheptane,** b.p. 111·5°/3 mm, d_4^{20} 0·8811, n_D^{20} 1·4970 (*Petrov, loc. cit.*).

3,5-**Di**-tert-**butylcumene,** $C_{17}H_{28}$, b.p. 121–122°/12 mm, d_4^{20} 0·8720, n_D^{20} 1·4860, may be made by hydrogenation of 3,5-di-*tert*-butylisopropenylbenzene obtained by the Grignard reaction (p. 174) (*H. D. Scharf* and *F. Doering*, Ber., 1967, **100**, 1761). **Di-sec-butylmesitylene**, b.p. 159°/18 mm, d_4^{20} 0·8958, n_D^{20} 1·51277, has been made by alkylation of mesitylene with *sec*-butyl alcohol in the presence of sulphuric acid (*Ioffe* and *Yang*, *loc. cit.*). 4-**Benzyl**-2,4,6-**trimethylheptane**, b.p. 133–135°/5 mm, d_4^{20} 0·8968, n_D^{20} 1·5005; 4-**benzyl**-2,4-**dimethyloctane**, b.p. 150·5–151·5°/6 mm, d_4^{20} 0·8833, n_D^{20} 1·4989; 2-**methyl**-2-**phenyldecane**, m.p. − 93°, b.p. 146·5°/2 mm, d_4^{20} 0·8665, n_D^{20} 1·4881 (*Petrov*, *loc. cit.*).

1,3,5-**Tri-n-butylbenzene**, $C_{18}H_{30}$, b.p. 143°/3 mm, d_4^{20} 0·8609, n_D^{20} 1·4870 has been made in 92 % yield by the catalytic trimerization of 1-hexyne (*V. O. Reikhsfel'd et al.*, Zhur. obshchei Khim., 1962, **32**, 653). Both 1,2,4-*tri*-tert-*butylbenzene*, m.p. 49–50°, and 1,3,5-*tri*-tert-*butylbenzene*, m.p. 72–72·5°, are obtained by catalytic cyclotrimerization of 3,3-dimethyl-but-1-yne (*idem.*, Zhur. org. Khim., 1966, **2**, 961). 1,3,5-Tri-*tert*-butylbenzene also results from the alkylation of *m*-di-*tert*-butylbenzene with *tert*-butyl chloride in the presence of aluminium chloride (*S. Watarai*, Bull. chem. Soc. Japan, 1963, **36**, 747). 1,2,4-Tri-*tert*-butylbenzene (III) has been converted by uv radiation to the isomer IV containing the bicyclo[2·2·0]hexadiene structure of Dewar benzene (p. 186). The alicyclic isomer reverts to III on heating at 200° (*S. P. Pappas* and *E. E. van Tamelen*, J. Amer. chem. Soc. 1962, **84**, 3789; see also *K. E. Wilzbach* and *L. Kaplan*, ibid., 1965, **87**, 4004):

Bu^t Bu^t Bu^t → Bu^t Bu^t Bu^t

(III) (IV)

The **phenyldodecanes**, are widely used in detergents. A mixture of secondary phenyldodecanes in which the 2-isomer predominates, is produced when benzene is alkylated with 1-dodecene in the presence of a Friedel-Crafts catalyst. In the absence of Lewis base the yield of internal isomers increases (*J. J. Shook*, U.S.P., 3,365,509/1968). The position of the phenyl group on the alkyl chain is an important factor in determining the surface active properties of their sulphonates and most of the isomers have been prepared. The physical properties of some of these isomers are given in Table 4. 1-*Phenyl-dodecane* has been made by the aluminium chloride-catalyzed acylation of benzene with dodecanoyl chloride and Wolff-Kishner reduction of the ketone. The isomeric phenyl *n*-dodecanes have been prepared from the appropriate alkyl aryl ketone by the normal Grignard–dehydration–reduction method (*A. C. Olson*, Ind. Eng. Chem., 1960, **52**, 833; *Petrov*, *loc. cit.*).

Physical properties of some isomeric phenyl-*n*-tetradecanes, $C_{20}H_{34}$, benzyl-*n*-tridecanes and *p*-di-*n*-alkylbenzenes are also given in Table 4. The *n*-tetradecanes can be made by the normal Grignard–dehydration–reduction route,

TABLE 4

PHYSICAL PROPERTIES OF SOME ISOMERIC $C_{18}H_{30}$, $C_{20}H_{34}$ AND $C_{26}H_{46}$ ALKYLBENZENES

	M.p.°	B.p.°/mm	n_D^{20}
1-Phenyldodecane		166·1–166·7/5	1·4824
2-Phenyldodecane		154·0–154·4/5	1·4817
3-Phenyldodecane		172·2–176·1/15	1·4818
4-Phenyldodecane		164–166/10	1·4810
5-Phenyldodecane		155–161/10	1·4812
6-Phenyldodecane		144·4–148·3/5	1·4813
2,4-Dimethyl-2-phenyldecane		136–137/5	1·4850
5,9-Dimethyl-2-phenyldecane		134–135/2	1·4815
2,2,4,6,6-Pentamethyl-4-phenylheptane		133–135/14	1·4930
1-Phenyltetradecane	8·7	203/12	1·4815
2-Phenyltetradecane	−1·5	197–198/12	1·4810
3-Phenyltetradecane	−3·0	193–194/12	1·4815
4-Phenyltetradecane	−4·0	191–192/12	1·4818
5-Phenyltetradecane	−5·1	188–189/12	1·2820
6-Phenyltetradecane	−6·2	186–187/12	1·4821
7-Phenyltetradecane	−7·2	184–185/12	1·4823
2-Benzyltridecane	−15	196/11	1·4816
3-Benzyltridecane	—	199/14	1·4831
4-Benzyltridecane	—	192/12	1·4824
5-Benzyltridecane	—	195/14	1·4821
6-Benzyltridecane	—	187/11	1·4820
7-Benzyltridecane	—	189/11	1·4818
1-Methyl-4-tridecylbenzene	14·5	196/8	1·4835
1-Dodecyl-4-ethylbenzene	0·0 to 0·3	194/8	1·4842
1-Propyl-4-undecylbenzene	−9·2 to −9·0	193/8	1·4839
1-Butyl-4-decylbenzene	−12·2 to −12·0	192/8	1·4839
1-Amyl-4-nonylbenzene	−14·8 to −14·5	192/8	1·4841
1-Hexyl-4-octylbenzene	−13·0 to −12·6	192/8	1·4840
p-Di-*n*-heptylbenzene	− 7·2 to −7·9	192/8	1·4840
1-Ethyl-4-octadecylbenzene	33	—	1·4702
1-Butyl-4-hexadecylbenzene	21	—	1·4705
1-Hexyl-4-tetradecylbenzene	21	—	1·4703
1-Dodecyl-4-octylbenzene	21	—	1·4700
p-Di-*n*-decylbenzene	29–30	—	1·4709
p-Di-(1-propylheptyl)benzene	—	150/0·05	1·4686
p-Di-(4-propylheptyl)benzene	—	191/0·1	1·4728
(±)-*p*-Di-(1-methylnonyl)benzene	39–40	184/0·5	1·4699
1,2,3,5-Tetra-amylbenzene	—	175/0·4	1·4772

the *n*-tridecanes *via* acylation of benzene with the appropriate acyl chloride, and the *p*-di-*n*-alkylbenzenes by acylation of the alkylbenzene with the desired acyl chloride followed by reduction of the intermediate ketone (*F. Asinger et al.*, J. pr. Chem., 1963, 153). 3,4,5,6-Tetramethyl-1,2-dineopentyl-benzene is an example of hindered rotation; rotation about the ar-CH_2 bonds can be followed by n.m.r. (*D. T. Dix et al.*, Tetrahedron Letters, 1966, 517).

6-(2,4,5-**Trimethylphenyl**)**pentadecane,** $C_{24}H_{42}$, m.p. <—60°, d^{20} 0·8665, n_D^{20} 1·4915 (*Petrov et al.*, Neftekhimiya, 1963, **3,** 465). **Hexa-n-propylbenzene,** m.p. 102·5–103°, has been made in good yield by propylation of benzene with *n*-propyl chloride, and by catalytic cyclotrimerization of di-*n*-propylacetylene (*H. Hopff* and *A. Gati*, Helv., 1965, **48,** 509).

The isomeric alkylbenzenes, $C_{26}H_{46}$, listed in Table 4, have been made by acylation of the appropriate alkylbenzene with the appropriate acyl chloride and reduction or by the Grignard–dehydration–reduction route (*P. Studt*, Ann., 1965, **685,** 49; 1966, **693,** 90).

p-**Di**-n-**dodecylbenzene,** $C_{30}H_{54}$, m.p. 42°, and p-**di**-n-**tetradecylbenzene,** $C_{34}H_{62}$, m.p. 53° (*J. Jovavonic et al.*, *ibid.*, 1966, **696,** 55).

Chapter 3

Halogen Derivatives of Benzene and its Homologues

W. J. FEAST AND W. K. R. MUSGRAVE

Introduction

A very great number of halogenated derivatives of benzene and its homologues have been made and characterized. The material included in this chapter is intended to give a wide survey rather than a comprehensive one; however, in order to assist the more demanding reader, reference to detailed compilations and reviews has been made where possible. At the time of writing the subject matter is an area of some activity and the interested reader will find reference to continuing specialist reports (see text) rewarding.

Nomenclature. Substitutive nomenclature is invariably used in naming nuclear halogenated derivatives of benzene and its homologues; thus C_6H_5Cl is chlorobenzene, C_6H_5IO iodosylbenzene, etc. Radicofunctional names are used for aralkyl halides and the general class of nuclear halogenated arenes may be referred to as aryl halides. In certain cases additive names, *e.g.* stilbene dibromide, $C_6H_5 \cdot CHBr \cdot CHBr \cdot C_6H_5$, are used. The conventional symbols Ar and Ph are used for aryl and phenyl, respectively.

1. Nuclear halogen derivatives

(a) Methods of preparation

A number of approaches to the synthesis of benzenoid compounds having halogen atoms bound to the nucleus have proved successful. The most widely used approach has a benzenoid compound as the starting material and involves the replacement of hydrogen or a functional group by halogen (methods *i* to *vii*, pp. 243 *et seq.* below). A second route involves the synthesis of the halogenated aromatic nucleus from a more saturated structure, usually by dehydrohalogenation or dehalogenation of polyhalogenocyclohex-anes, -enes, or -dienes (methods *viii* and *ix*). The other methods described, although successfully applied in the cases cited, have not found widespread use.

Preparative details for well established methods of synthesis of halogeno-aromatic compounds are to be found in a number of specialist texts and reviews. In particular *Houben-Weyl*, "Methoden der Organischen Chemie", (George Thieme Verlag, Stuttgart) gives details of the preparation of fluoro compounds (see *E. Forche*, Vol. 5/3, 1962, pp. 1–502), chloro compounds (*W. Hahn* and *R. Stroth*, Vol. 5/3, 1962, pp. 511–1078), bromo compounds (*A. Roedig*, Vol. 5/4, 1960, pp. 13–516), and iodo compounds (*Roedig*, Vol. 5/4, 1960, pp. 517–678). Preparations of fluoro compounds are given by *A. E. Pavlath* and *A. J. Leffler* ("Aromatic Fluorine Compounds", Reinhold, New York, 1962, Chapter 1), by *M. Hudlicky* ("Chemistry of Organic Fluorine Compounds", Pergamon, Oxford, 1961, p. 136) and by *W. A. Sheppard* and *C. M. Sharts* ("Organic Fluorine Chemistry", Benjamin, New York, 1969). The series "Organic Reactions" provides reviews of synthetic and mechanistic aspects of well established methods, and "Organic Syntheses" gives detailed instructions for the preparation of individual compounds. Detailed literature compilations such as those by *E. H. Huntress* ("Organic Chlorine Compounds", J. Wiley and Sons, Inc., New York, 1948) and *H. P. Braendlin* and *E. T. McBee* ("Halogenation in Friedel-Crafts and Related Reactions", Vol. 3, *G. A. Olah*, Ed., Interscience Publishers, New York, 1964, Chap. 43) are useful aids to literature searches.

The mechanistic aspects of halogenation reactions have attracted considerable attention and have been comprehensively reviewed by *P. B. D. de la Mare* and *J. H. Ridd* ("Aromatic Substitution", Butterworths, London, 1959), *R. O. C. Norman* and *R. Taylor* ("Electrophilic Substitution in Benzenoid Compounds", Elsevier, Amsterdam, 1965), and *L. M. Stock* ("Aromatic Substitution Reactions", Prentice-Hall, Englewood Cliffs, N.J., 1968). More detailed treatments of particular reactions are to be found in "Advances in Physical Organic Chemistry", *V. Gold*, Ed., Academic Press, New York, *e.g. Stock* and *H. C. Brown*, "A Quantitative Treatment of Directive Effects in Aromatic Substitution", Vol. I, 1963, p. 35, and *H. Zollinger*, "Hydrogen Isotope Effects in Aromatic Substitution Reactions", Vol. II, 1964, p. 163; and in "Progress in Physical Organic Chemistry", *A. Streitwieser* and *R. W. Taft*, Ed., Interscience Publishers, New York, *e.g. E. Baciocchi* and *G. Illuminati*, "Electrophilic Aromatic Substitution and Related Reactions in Polyalkylbenzene Systems", Vol. 5, 1967, p.1, and *E. Berliner*, "Electrophilic Aromatic Substitution", Vol. 2, 1964, pp. 253–321. Publications on reaction mechanisms are reviewed annually in "Organic Reaction Mechanisms", *B. Capon* and *C. W. Rees*, Eds., Interscience, New York, 1966 *et seq.* Many of these reviews cited for their mechanistic content also give practical details of syntheses.

(i) Replacement of aromatic hydrogen by halogen

Molecular chlorine, bromine and iodine can effect nuclear halogenation in benzenoid compounds which are activated towards electrophilic attack, the order of reactivity being chlorine > bromine > iodine. Chlorination and bromination are usually carried out in a polar solvent and give products with halogen substitution at sites *ortho* and *para* to the activating substituent, *para* substitution usually predominating. Successful iodination requires the addition of an oxidizing agent such as nitric acid, iodic acid, sodium persulphate or hydrogen peroxide. Iodination of aromatic amines goes to completion probably because the hydrogen iodide is removed, as the salt, as it is formed. Iodination generally gives a product with the iodine in the *para* position with respect to the activating substituent (see *de la Mare* and *Ridd, loc. cit.*, pp. 105–111). The reaction of molecular fluorine with aromatic compounds does not produce fluorinated aromatic products. The reaction is apparently a free radical process and the materials isolated are tetrafluoromethane and other fragmentation products, uncharacterized polymeric material, and poor yields of polyfluorocyclohexanes (*J. M. Tedder*, "The Fluorination of Organic Compounds using Elementary Fluorine" in "Advances in Fluorine Chemistry", Volume 2, Eds. *M. Stacey, J. C. Tatlow* and *A. G. Sharpe*, Butterworths, London, 1961, pp. 129–131).

The addition of a catalyst such as iron, ferric chloride, aluminium chloride, iodine or pyridine makes molecular chlorination and bromination of less reactive benzenoid compounds conveniently possible. The use of iron and ferric chloride in chlorination is described by *P. G. Harvey et al.* (J. appl. Chem., 1954, **4**, 325), for example, toluene gives 2,3,4,5,6-pentachlorotoluene. Careful choice of catalyst and operating conditions gives a considerable degree of control over the reaction, thus *T. Kametani et al.* (J. pharm. Soc. Japan, 1968, **88**, 950) report the chlorination of benzene to *p*-dichlorobenzene in 70 % yield using a catalyst of iron, sulphur and lead oxide. The function of the catalyst is to increase the electrophilicity of the halogen; the detail of the mechanisms probably varies in different systems but at some stage in the reaction sequence the catalyst polarizes the inter-halogen bond. If this polarization is complete prior to reaction, then the halogenating reagent is a positively charged species.

Many preparative methods involve the formation of so-called positive halogen; positive halogenating species may also be involved, *e.g.* in iodination in the presence of oxidizing agents where the oxidant serves the function of removing the hydrogen iodide formed during the reaction and of liberating a positively charged iodinating species, possibly $I^{\oplus}$. Positive species active in iodination such as $I_3^{\oplus}$ and $I_2^{\oplus}$ can be obtained in 20 % and 60 % oleum, respectively, and mono- or poly-iodination of deactivated substrates can be effected by choice of the appropriate reagent (see *J. Arotsky, R. Butler* and *A. C. Darby*, Chem. Comm., 1966, 650, and references therein). Chlorination and bromination reactions involving positive halogen are accomplished by using acidified solutions of hypohalous acids, the active entity being $X^{\oplus}$ or $H_2OX^{\oplus}$. *N*-Chloro- and *N*-bromo-amides and imides are also used to effect nuclear halogenation. These reactions are very sensitive to reaction conditions and the reagent structure. Usually under aprotic conditions, side-chain(free radical mechanism) halogenation occurs; in polar conditions, and often with added catalyst, nuclear halogenation is observed probably through the agency of molecular bromine and the catalyst. *A. Henne* (J. Amer. chem. Soc., 1951, 73, 1103) has shown that while *N*-bromosuccinimide reacts with toluene to give exclusively benzyl bromide, perfluoro-*N*-bromosuccinimide under the same conditions gives exclusively *ortho*- and *para*-bromotoluene; *J. D. Park* (*ibid.*, 1952, 74, 2189) has shown that this difference in behaviour between hydrocarbon and fluorocarbon analogues holds good in the case of *N*-bromoacetamide and *N*-bromotrifluoroacetamide also.

Acyl hypohalites form useful halogenating agents for aromatic compounds provided they do not undergo the Hunsdiecker Reaction very rapidly as is the case with trifluoroacetyl hypobromite and hypoiodite (*C. V. Wilson*, Organic Reactions, Vol. 9, Ch. 5, 1957). Thus silver trifluoroacetate and iodine do not form appreciable amounts of trifluoromethyl iodide below 100^0 and when heated with benzene an 85 % yield of iodobenzene is obtained (*R. N. Haszeldine* and *Sharpe*, J. chem. Soc., 1952, 993). Excellent yields of bromo- and iodo-benzenes containing methyl, halogen, methoxyl, amino, and carboxyl groups can be obtained, the halogen entering the *para* or *meta* positions according to the directing effect of the substituent group which must not have too great a deactivating effect or no substitution occurs at all. It appears that little *ortho* substitution ever occurs. This reagent resembles the silver perchlorate–halogen reagent but has the advantages that the trifluoroacetic acid is volatile, its silver salt is more soluble in organic solvents than is silver perchlorate, and the explosion hazard is avoided.

An alternative method of generating acetyl hypohalites is from the reaction of mercuric acetate with a solution of the halogen in acetic acid. Detailed kinetic investigations with acetyl hypoiodites and hypobromites generated in this way indicate that the rate-determining step involves both hypohalite and hydrocarbon and that the reaction is polar in character

(*E. M. Chen*, *R. M. Keefer* and *J. L. Andrews*, J. Amer. chem. Soc., 1967, 89, 428 and references therein):

$$Hg(OCO{\cdot}CH_3)_2 + X_2 \rightarrow Hg(OCO{\cdot}CH_3)X + CH_3{\cdot}CO_2X$$

$$CH_3{\cdot}CO_2X + ArH \rightarrow ArX + CH_3{\cdot}CO_2H$$

Many investigators have sought examples of electrophilic fluorination but it was not until recently that *D. H. R. Barton* and co-workers (Chem. Comm., 1968, 806) published the first example of such a reaction. Organic fluoroxy-compounds, *e.g.* trifluorofluoroxymethane (CF_3OF), react with activated aromatic compounds in a halogenated solvent at low temperature to give the products of electrophilic aromatic fluorination. Thus, for example, salicylic acid reacts with trifluorofluoroxymethane at 0^0 in $CHCl_3$ giving a mixture of 5- and 3-fluorosalicylic acids (70 %).

A new method of preparing aromatic bromides and iodides appears promising (*A. Mckillop et al.*, Tetrahedron Letters, 1969, 1623, 2423 and 2427); an aromatic compound reacts first with thallium(III) trifluoroacetate and subsequent addition of bromine in carbon tetrachloride or of aqueous potassium iodide gives the aryl bromide or iodide in good yield.

The interaction of xenon difluoride with benzene gives low yields of fluoro- and difluoro-benzenes (*M. J. Shaw*, *H.H. Hyman* and *R. Filler*, J. Amer. chem. Soc., 1969, 91, 1563).

When aryl radicals are generated in polyhalogenomethanes, halogen abstraction predominates over hydrogen abstraction, often exclusively, and aryl halides are formed (*J. I. G. Cadogan*, *D. H. Hey* and *P. G. Hibbert*, J. chem. Soc., 1965, 39 and references therein).

Free radical halogenation of unsubstituted aromatic compounds, *i.e.* direct replacement of aromatic hydrogen by halogen atoms, has been postulated to account for the change in isomer distribution above the so-called "transition-point" in vapour-phase aromatic halogenations (*G. H. Williams*, "Homolytic Aromatic Substitution", Pergamon Press, Oxford, 1960, p. 120 and references therein). Vapour-phase chlorination and bromination of aromatic compounds, particularly halogenobenzenes and benzotrifluoride, have been investigated by *E. C. Kooyman* and co-workers and substitution patterns markedly different from those obtained in well characterized electrophilic substitution reactions were observed (*Kooyman*, Adv. Free Radical Chem., 1965, 1, 137).

Irradiation of solutions of aromatic nitro compounds in concentrated hydrochloric acid with u.v. light ($>$290 nm) results in reduction of nitro to amine and replacement of aromatic hydrogens by chlorine:

NO_2, OH → NH_2, Cl, Cl, Cl, OH (71%) + OH, Cl, Cl, Cl, Cl, OH

This remarkable result has been attributed to nucleophilic attack by $Cl^{\ominus}$ on the photochemically excited protonated nitrobenzene species (*R. L. Letsinger* and *G. G. Wubbels*, J. Amer. chem. Soc., 1966, **88**, 5041).

(ii) Replacement of carboxylic acid function by halogen

Carboxyl groups can be replaced by bromine or iodine using the Hunsdiecker Reaction (*Wilson, loc. cit.*). The reaction is not so general in the aromatic as in the aliphatic series and the variable yields are apparently dependent on temperature:

$$C_6H_5 \cdot CO_2Ag + Br_2 \rightarrow C_6H_5Br + CO_2 + AgBr$$

Derivatives of benzoic acid possessing electron-attracting groups give satisfactory yields of halogenobenzenes. Reactions are usually carried out in carbon tetrachloride solution.

Aroyl chlorides are thermally decarbonylated giving low yields of aryl chlorides (*A. Maille*, Compt. rend., 1925, **180**, 1111). The use of chlorotris(triphenylphosphine)rhodium as catalyst makes the method generally applicable to the synthesis of aryl halides; for use in the synthesis of aryl fluorides see *G. A. Olah* and *P. Kreienbühl*, J. org. Chem., 1967, **32**, 1614; aryl chlorides and bromides, *J. Blum*, Tetrahedron Letters, 1966, 1605 and J. Amer. chem. Soc., 1967, **89**, 2338; and iodides, *Blum, H. Rosenman* and *E. D. Bergman*, J. org. Chem., 1968, **23**, 1928.

(iii) Replacement of phenolic OH and thiophenolic SH by halogen

The preparation of aryl halides from phenols and phosphorus pentahalides is of little value, the main products being the triaryl phosphates, except in the case of the nitrophenols (*Engelhardt* and *Latschinov*, Ber., 1870, **3**, 98). If, however, a triaryloxyphosphorus dichloride is first prepared, then used to prepare the appropriate tetra-aryloxyphosphorus monochloride and this is subjected to thermal decomposition, a good yield of the appropriate aryl chloride is formed as follows:

$$3\,ArOH + PCl_5 \xrightarrow{\text{heat}} (ArO)_3PCl_2 + 3\,HCl$$

$$(ArO)_3PCl_2 + Ar'OH \xrightarrow{\text{heat}} (ArO)_3(Ar'O)PCl + HCl$$

$$(ArO)_3(Ar'O)PCl \xrightarrow{200^{\circ}-360^{\circ}} Ar'Cl + (ArO)_3PO$$

For the preparation of aryl bromides or iodides the mixed tetra-aryloxy-phosphorus monochloride is converted into the mono-bromide or -iodide by halogen exchange with ethyl bromide or methyl iodide. For good yields in all cases Ar′ must contain a more powerful electron-attracting group than does Ar. The reaction fails with dihydric phenols (*D. G. Coe, H. N. Rydon* and *B. L. Tonge,* J. chem. Soc., 1957, 232).

Aryl fluorides may be prepared in good yield from phenols and thiophenols by a two step process in which they are first converted to aryl fluoroformates or thioformates by the action of carbonyl chloride fluoride. Pyrolysis of these products eliminates carbon dioxide or carbonyl sulphide to give the aryl fluoride (*K. O. Christie* and *A. E. Pavlath,* J. org. Chem., 1965, 30, 3170; 1966, 31, 559):

$$\mathrm{ArXH + COClF \rightarrow FCOXAr\ (X = O\ or\ S)}$$

$$\mathrm{FCOXAr \xrightarrow{\Delta} ArF + COX}$$

The pyrolysis step is more efficient in the presence of a platinum catalyst and the reaction is applicable to some substituted and dihydric phenols.

(*iv*) *Replacement of an amino group by halogen*

The diazotization of arylamines followed by replacement of the diazonium group by halogen provides a good route to many aryl halides. Chlorine and bromine can be introduced by the Sandmeyer reaction (*H. H. Hodgson,* Chem. Reviews, 1947, 40, 251):

$$\mathrm{ArN^{\oplus}{\equiv}N\ X^{\ominus} \xrightarrow[HX]{Cu_2X_2} ArX + N_2}$$

or by the Gatterman reaction (*L. A. Bigelow,* Org. Synth., Coll. Vol. I, 1941, p. 133):

$$\mathrm{CH_3C_6H_4N^{\oplus}{\equiv}N\ Br^{\ominus} \xrightarrow[HBr]{Cu\ powder} CH_3C_6H_4Br}$$

Iodine is introduced by decomposing the diazonium salt in the presence of aqueous potassium iodide (*H. J. Lucas* and *E. R. Kennedy,* Org. Synth., Coll. Vol. II, 1943, p. 351):

$$\mathrm{C_6H_5N_2^{\oplus}\ Cl^{\ominus} + KI \rightarrow C_6H_5I + N_2 + KCl}$$

The most general method used to introduce fluorine is by the decomposition of diazonium fluoroborates, a process known as the Schiemann Reaction

(*A. Roe*, "Organic Reactions", Vol. 5, Chap. 5, 1949; *H. Suschitzky*, "Advances in Fluorine Chemistry", Vol. 4, Chap. 1, Butterworths, London, 1965):

$$ArN^{\oplus}{\equiv}N\ BF_4^{\ominus} \rightarrow ArF + N_2 + BF_3$$

By repetition of the cycle: nitration, reduction of nitro compound, and decomposition of the diazonium borofluoride, on benzene it is possible to prepare 1,2,4,5-tetrafluorobenzene (*G. C. Finger et al.*, J. Amer. chem. Soc., 1951, 73, 145) but further treatment of this compound with nitrating mixture gives 2,5-difluoroquinone.

Variations of the Schiemann reaction are surveyed in detail by *Suschitzky* (*loc. cit.*) and recent improvements in the yields and the range of compounds to which the reaction is applicable are reported by *C. Sellers* and *Suschitzky* (J. chem. Soc. C, 1968, 2317).

The various mechanisms invoked to account for the replacement of a diazonium group by an halogen atom have been reviewed (*H. Zollinger*, "Azo and Diazo Chemistry", Chap. 7, Interscience Publishers, Inc., New York, 1961, and *Suschitzky*, *loc. cit.*).

Alternative one step processes for the replacement of arylamine groups by halogen have been developed; *Cadogan*, *D. A. Roy* and *D. M. Smith* (J. chem. Soc. C, 1966, 1249) have shown that a selection of aromatic amines react with pentyl nitrite in the presence of bromoform to give the corresponding aryl bromide in reasonably good yield (50–70 %). The analogous reaction with chloroform was less successful.

A one step conversion of arylamines into the corresponding chlorides and bromides is achieved by reaction with the cupric halide/nitric oxide complex:

$$CuX_2NO + ArNH_2 \rightarrow CuX + ArX + N_2 + H_2O$$

The reaction proceeds in good yield at room temperature and is applicable to a wide range of substituted arylamines, yields being best in the benzenoid series (*W. Brackama* and *P. J. Smit*, Rec. Trav. chim., 1966, 85, 857).

(v) *Replacement of nitro, chlorosulphonyl and substituted amino groups by halogen*

When nitro compounds or sulphonyl chlorides are heated in sealed tubes with phosphorus pentachloride or thionyl chloride the nitro or chlorosulphonyl group is replaced by chlorine (*H. Meyer*, Monatsh., 1915, 36, 721). This reaction has been applied to the preparation of di-, tri- and tetrachloro compounds (*J. Pollak* and *A. Wienerberger*, *ibid.*, 1914, 35, 1472;

J. Schmidt and *H. Wagner*, Ann., 1912, **387**, 164; *S. C. J. Olivier*, Rec. Trav. chim., 1918, **37**, 313; *R. Anschütz* and *E. Molineus*, Ann., 1918, **415**, 62; *W. Davies* and *E. H. C. Hickox*, J. chem. Soc., 1922, **121**, 2648).

Expected straightforward electrophilic halogenation can sometimes result in displacment of functional groups; *e.g.*, *C. R. Harrison* and *J. F. W. McOmie* (J. chem. Soc. C, 1966, 997) report that 4-acetylamino-3-nitroanisole reacts with bromine in acetic acid to give 2,3,4,5-tetrabromoanisole:

$$\text{4-AcNH-3-}NO_2\text{-}C_6H_3OCH_3 \xrightarrow{Br_2/CH_3\cdot CO_2H} \text{2,3,4,5-}Br_4C_6H(OCH_3)$$

(*vi*) *Replacement of one halogen by another*

Treatment of polyhalogeno aromatic compounds with alkali metal fluorides in polar aprotic solvents, in fused salt systems, or alone, leads to the replacement of other halogens by fluorine and is a useful method of preparing polyfluoro, chloropolyfluoro, and bromopolyfluoro aromatic compounds.

In solvents such as dimethylformamide, *N*-methylpyrrolidone, and sulpholane, mixtures of chlorofluoro derivatives with little, if any, perfluorocompound are obtained from the reaction of potassium fluoride with hexachlorobenzene (*J. T. Maynard*, J. org. Chem., 1963, **28**, 112; Fr. Pat., 1,360,917/1964 to *Nat. Smelting Co. Ltd.*):

$$C_6Cl_6 \xrightarrow[230\text{–}240^0/18\ h]{\text{sulpholane: KF}} C_6ClF_5 + C_6Cl_2F_4 + C_6Cl_3F_3$$

G. W. Holbrook, L. A. Loree and *O. R. Pierce* (J. org. Chem., 1966, **31**, 1259) report that in the exchange fluorination of hexabromo- and hexachlorobenzene in polar aprotic solvents, the first three bromines or chlorines are exchanged relatively easily but thereafter products containing hydrogen are obtained.

When perchloro or perbromo aromatic compounds and potassium fluoride are heated together in an autoclave at autogenous pressure, perfluoro compounds are obtained (*N. N. Vorozhtsov, G. G. Yakobson et al.*, Izvest. Akad. Nauk S.S.S.R., Ser. Khim., 1963, 1524; J. gen. Chem. U.S.S.R., 1965, **33**, 1161, and 1966, **36**, 152, 2124, and 2128; Belg. Pat., 659,239/1965 to *Imp. Smelting Corp. Ltd.*):

$$C_6Cl_6 \xrightarrow[KF]{450\text{–}500^0} C_6F_6 + C_6ClF_5 + C_6Cl_2F_4 + C_6Cl_3F_3$$

$$C_6Cl_5\cdot CF_3 \xrightarrow[KF]{560\text{–}590^0} C_6F_5\cdot CF_3\ (53\,\%\ \text{yield})$$

The reaction can also be run in a flow system using a potassium fluoride chloride melt at 780° and nitrogen as carrier gas (*H. C. Fieldings et al.*, J. chem. Soc. C, 1966, 2142):

$$1{,}3{,}5\text{-}C_6Cl_3F_3 \xrightarrow[780^0]{KF/KCl} C_6F_6\ (37\,\%) \text{ and } C_6ClF_5\ (40\,\%)$$

An alternative method of exchanging one halogen for another has been known for some time, thus chlorine replaces bromine in bromobenzene in the absence of halogen carriers and the reaction is catalysed by light (*A. Eibner*, Ber., 1903, 36, 1229; see also *R. K. Sharma* and *N. Kharasch*, Angew. Chem. internat. Edn., 1968, 7, 36; and *B. Milligan et al.*, J. Amer. chem. Soc., 1967, **89**, 4081).

(*vii*) *Perchlorination and bromination*

Poly-chlorination can be carried out using sulphuryl chloride in the presence of a catalyst consisting of aluminium trichloride and sulphur monochloride (*O. Silberrad*, J. chem. Soc., 1922, **121**, 1015). Stepwise chlorination occurs and the extent of chlorination depends on the amount of sulphuryl chloride used and the temperature.

Chlorobenzene gives successively *o*- and *p*-dichloro-, 1,2,4-trichloro-, 1,2,4,5- and 1,2,3,5-tetrachloro-, and hexachloro-benzenes. When fully halogenated side-chains are already attached to the nucleus it is not possible to attain complete substitution into the nucleus either by using the catalyst as described by Silberrad or with the more conventional catalysts, but a more active modification of the Silberrad catalyst will cause complete chlorination of benzotrichloride and 1,4-bistrichloromethylbenzene but not of 1,3-bistrichloromethyl- or 1,3,5-tristrichloromethyl benzenes because in the last two cases there is greater deactivation of the ring system. This method has been intensively developed and applied by *M. Ballester* and co-workers (J. Amer. chem. Soc., 1966, **88**, 957; U.S. Report, ARL67-0167, Aug. 1967; Pure appl. Chem., 1967, **15**, 123 and references therein).

Excellent yields of perbromo aromatic compounds are reported from the extremely rapid, room temperature bromination with dibromoisocyanuric acid in concentrated sulphuric acid, oleum or fluorosulphonic acid (*W. Gottardi*, Monatsh., 1969, **100**, 42).

(*viii*) *Dehydrohalogenation routes to halogenated benzenes*

Photochemical chlorination of benzene gives a mixture of 1,2,3,4,5,6-hexachlorocyclohexanes which can be dehydrochlorinated by alkali to trichlorobenzenes.

Fluorination of benzene with high valency metallic fluorides or complex metal fluorides (*M. Stacey* and *J. C. Tatlow* in "Advances in Fluorine Chemistry", Eds. *Stacey, Tatlow* and *Sharpe,* Butterworths, London, 1960, Vol. 1, p. 166; and *P. L. Coe, R. G. Plevey* and *Tatlow,* J. chem. Soc. C, 1969, 1060) gives polyfluoro-cyclohexanes and -cyclohexenes which on dehydrofluorination with alkali yield fluorinated benzenes. Thus, hexa- and penta-fluorobenzenes can be prepared by fluorinating benzene over cobalt trifluoride to a mixture of the nona-, and octa-fluorocyclohexanes. Separation of these followed by dehydrofluorination with concentrated aqueous caustic potash gives mixtures of polyfluorocyclohexadienes and hexa- and penta-fluorobenzenes, respectively (*J. A. Godsell, Stacey* and *Tatlow,* Nature, 1956, 178, 199; *E. Nield, R. Stephens* and *Tatlow,* J. chem. Soc., 1959, 159):

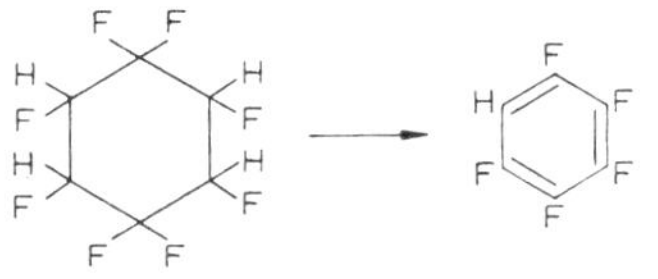

(*ix*) *Dehalogenation routes to halogenobenzenes*

The defluorination of perfluoroalicyclic compounds over iron, nickel, or iron oxide gives the corresponding perfluoro aromatic compounds. Perfluorocyclohexane requires a temperature of about 600° but the presence of a perfluoroalkyl side-chain allows the reaction to proceed at a lower temperature (*Tatlow et al.*, Nature, 1959, 183, 586 and 588; Fr. Pat., 1,357,911/1963):

$$CF_3 \cdot C_6F_{11} \xrightarrow[\text{Fe gauze}]{500°C} CF_3 \cdot C_6F_5$$

Polyfluoro-cyclohexenes and -cyclohexadienes obtained by the dehydrofluorination of deca-, nona- and octa-fluorocyclohexanes as described in method (*viii*) can also be defluorinated in this way to give hexa-, penta- and the isomeric tetra fluorobenzenes (*P. L. Coe, C. R. Patrick* and *Tatlow,* Tetrahedron, 1960, 9, 240).

A modification of this dehalogenation process is to treat hexachlorobenzene with either chlorine trifluoride or fluorine gas and so obtain a mixture of chlorofluorocyclohexanes. On dehalogenation of the mixture over stirred iron powder at 330° a mixture of hexafluoro- and chlorofluoro-benzenes is obtained, the chlorine atoms having been removed preferentially (*G. M. Brooke et al.*, J. chem. Soc., 1964, 729):

$$C_6Cl_6 + 3F_2 \xrightarrow{CCl_2F \cdot CClF_2} C_6Cl_xF_{12-x} \xrightarrow[300°]{Fe} \begin{cases} C_6F_6,\ C_6ClF_5 \\ C_6Cl_2F_4,\ C_6Cl_3F_3 \end{cases}$$

$$(x = 5, 6 \text{ and } 7)$$

(x) Pyrolysis of lower molecular weight materials

A method for the production of hexafluorobenzene only, involves the pyrolysis of tribromofluoromethane (*Y. Desirant*, Bull. classe sci., Acad. roy. Belg., 1955, **41**, 759; *L. A. Wall et al.*, J. Res. Nat. Bur. Stand., Sect. A., 1961, **65A**, 239 and references therein):

$$6\,CBr_3F \xrightarrow[\text{Pt tube}]{630\text{–}640^0} C_6F_6 \text{ and } 9\,Br_2$$

This starting material is relatively inaccessible; others such as $CHFCl_2$, $CHClF{\cdot}CH_2F$, and $CHClF{\cdot}CHClF$ give very low conversions to C_6F_6.

(xi) Trimerization of alkynes

Halogenoalkynes are unstable and investigators generally report a considerable explosion hazard, nevertheless several successful trimerization reactions of halogenoalkynes to halogenobenzene derivatives have been reported; chloroacetylene is reported to trimerize in the gas phase when diluted with nitrogen and under the influence of light or heat, the product being 1,3,5-trichlorobenzene (*E. H. Ingold*, J. chem. Soc., 1924, **125**, 1535). Dichloroacetylene yields hexachlorobenzene when a solution in diethyl ether is heated and illuminated (*E. Ott* and *G. Dittus*, Ber., 1943, **76B**, 80). Fluoroacetylene yields some 1,2,4-trifluorobenzene when stored at low pressure in the gas phase and in the dark (*W. J. Middleton* and *W. H. Sharkey*, J. Amer. chem. Soc., 1959, **81**, 803). *H. G. Viehe* and co-workers (see Angew. Chem. internat. Edn., 1965, **4**, 746 and references therein) have shown that *tert*-butylfluoroacetylene undergoes spontaneous exothermic oligomerization in the liquid phase giving mainly trimers and tetramers. The primary trimeric products are:

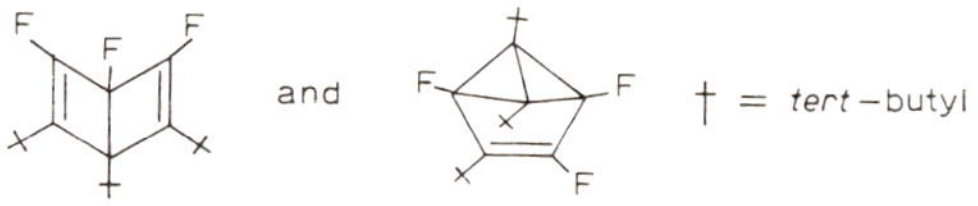

and   † = *tert*-butyl

which are readily isomerized thermally to 1,2,3- and 1,2,4-tri-*tert*-butyltrifluorobenzenes.

(xii) Synthesis of aryl halides via *dihalogenocarbene addition to 1,3-dienes*

Addition of dihalogenocarbenes to cyclopentadiene followed by ring expansion and elimination of hydrogen halide yields chloro- and fluoro-benzene:

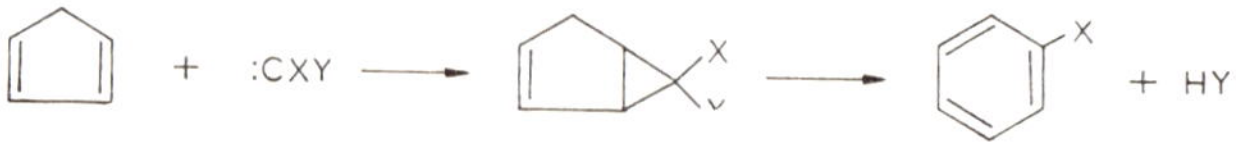

+ :CXY ⟶ ⟶ + HY

A. P. ter Borg and *A. F. Bickel* (Rec. Trav. chim., 1961, **80**, 1217) report that a suspension of cyclopentadienylsodium and cyclopentadiene in pentane reacts with chloroform to yield chlorobenzene (23 %). A better yield (45·5 %) was obtained when a mixture of chloroform and cyclopentadiene (1:1 molar ratio) was pyrolysed at 450° in a gas flow reactor (*J. W. Engelsma, ibid.*, 1965, **84**, 187). Similarly pyrolysis of materials which give rise to chlorofluoro- or difluoro-carbene, such as $CHClF_2$, CBr_2F_2, CCl_2F_2, $CHCl_2F$, C_2F_4, in the presence of cyclopentadiene yields fluorobenzene (*O. M. Nefedov* and *A. A. Ivashenko*, Bull. Acad. Sci. U.S.S.R., 1968, 445). These authors also report (Izvest. Akad. Nauk S.S.S.R. ser. Khim., 1968, 1417) that the reaction of difluorocarbene with 1,3-butadiene yields a mixture of *ortho-*, *meta-* and *para*-difluorobenzenes:

$$CHClF_2 + CH_2{=}CH{-}CH{=}CH_2 \xrightarrow[650^\circ]{740\text{ mm.}} C_6H_4F_2 \text{ (48 \% yield)}.$$

(*xiii*) *Synthesis of halogenated benzene derivatives* via *Diels-Alder adducts of halogenated dienes*

2,3,4,5-Tetrachlorocyclopentadienone acetals react with acetylene derivatives and give polychlorinated benzene derivatives either directly or on thermolysis of the Diels-Alder adducts (*R. W. Hoffmann* and *H. Hauser*, Tetrahedron, 1965, **21**, 891; and *D. M. Lemal, E. P. Grosselink* and *S. D. McGregor*, J. Amer. chem. Soc., 1966, **88**, 582):

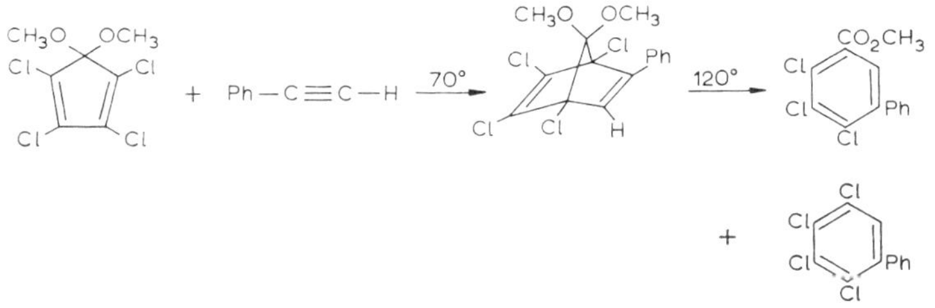

Similarly pyrolysis of the Diels-Alder adducts of perfluorocyclopentadiene (*R. E. Banks et al.*, J. chem. Soc. C, 1967, 1608) and perfluorocyclohexa-1,3-diene (*L. P. Anderson, W. J. Feast* and *W. K. R. Musgrave, ibid.*, 1969, 211) gives tetrafluorobenzene derivatives.

(*b*) *Properties and reactions*

The monohalogeno derivatives of benzene are colourless, neutral liquids each having a distinct and characteristic odour. They do undergo some free-radical reactions such as the formation of organometallic compounds under very similar reaction conditions to the alkyl chlorides, bromides and iodides; their nuclear hydrogen atoms can be replaced on treatment with electrophiles; but their outstanding chemical characteristic is the stability of the halogen to nucleophiles in comparison with that in an alkyl halide. Thus they are indifferent to prolonged boiling with aqueous or alcoholic reagents such as alkali, ammonia, or potassium cyanide under conditions which would certainly bring about replacement of the halogen in an alkyl chloride, bromide or iodide. If, however, the halogen atom is activated by the presence of some other group in the nucleus or very vigorous conditions are used, sometimes in the presence of a cuprous salt as catalyst, then nucleophilic substitutions, or elimination–addition processes whose overall effect is the same as nucleophilic substitution, can be made to occur.

(*i*) *Reaction with metals*

Aryl bromides and iodides form **Grignard reagents** easily by standard methods; aryl chlorides are less reactive but if small amounts of a more reactive halide, *e.g.* ethylene dibromide, are added at intervals to activate the magnesium surface, good yields can be obtained (*M. S. Kharasch* and *O. Reinmuth*, "Grignard Reactions of Nonmetallic Substances", Constable and Co. Ltd., London, 1954). Alternatively, tetrahydrofuran can be used as solvent instead of diethyl ether, when reaction occurs more readily. All of these arylmagnesium halides undergo normal reactions. Aryl fluorides do not form Grignard reagents. Di-Grignard reagents from aromatic dihalides are difficult to prepare particularly in ethereal solution; tetrahydrofuran as solvent results in better yields. Grignard reagents formed from *o*-dihalogenobenzenes easily form aryne intermediates (*I. T. Millar* and *H. Heaney*, Quart. Reviews, 1957, **11**, 109):

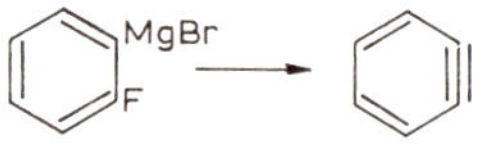

and those from polyhalides are best prepared at low temperatures (-10^0 to -60^0)

$$C_6F_5Br + Mg \xrightarrow[-30^oC]{THF} C_6F_5MgBr \rightarrow C_6F_5\cdot(C_6F_4)_n\cdot C_6F_5$$

or they polymerise (*E. Nield, R. Stephens* and *J. C. Tatlow*, J. chem. Soc., 1959, 166; *E. J. P. Fear, J. Thrower* and *M. A. White*, 19th Int. Cong. Pure and appl. Chem. London, 1963). The latter are often better prepared by an exchange process which makes use of the fact that any hydrogen atoms left in the molecule are acidic because of the electron attraction of the halogen atoms:

$$C_6HF_5 + PhMgBr \longrightarrow C_6F_5MgBr + C_6H_6$$

1,2,4,5-Tetrafluorobenzene forms a di-Grignard reagent in this way (*R. J. Harper, E. J. Soloski* and *C. Tamborski*, J. org. Chem., 1964, **29**, 2385).

Aryl-lithium compounds can be made by the direct action of lithium on the aryl chloride, bromide or iodide, preferably in an atmosphere of nitrogen because of the ready reaction of the product with oxygen. Often the reaction is slow but is faster in ethers than in hydrocarbon solvents such as pentane. However, many aryl-lithiums attack ether links (that in tetrahydrofuran more readily than that in diethyl ether) and it is often more convenient to prepare and store the organometallic compound in a hydrocarbon even though this reduces its reactivity. Aryl fluorides do not give aryl-lithiums because the more highly electronegative fluorine makes the *ortho* hydrogen sufficiently acidic to allow metallation to occur. This is followed by loss of lithium fluoride and then addition of the *o*-fluoroaryl-lithium to the benzyne intermediate:

$$C_6H_5F \xrightarrow{Li} o\text{-}FC_6H_4Li \longrightarrow C_6H_4 \longrightarrow o\text{-}FC_6H_4C_6H_5$$

Organo-lithium compounds are often more readily formed by halogen-metal exchange between the aryl halide and, preferably, butyl-lithium. *p*-Bromophenyl-lithium is formed in 90 % yield in this way (*Millar* and *Heaney, loc. cit.*):

$$p\text{-}C_6H_4Br_2 + BuLi \rightarrow p\text{-}BrC_6H_4Li + BuBr$$

As is the case for Grignard reagents, polyfluoro-, and presumably other polyhalogeno-aryls easily form lithium derivatives by hydrogen-metal exchange, particularly when tetrahydrofuran is used as the solvent, because of the relatively great acidity of the residual hydrogen atoms in the molecule. Because the metallating ability of an alkyl-lithium is greater than that of a Grignard reagent, it is possible to metallate both 5 and 6 positions of 1,2,3,4-tetrafluorobenzene with *n*-butyl-lithium (*Harper et al., loc. cit.*):

$$1,2,3,5\text{-}F_4C_6H_2 \xrightarrow[-70^\circ\ \text{THF}]{n\text{-BuLi}} 1,2,3,5\text{-}F_4C_6H(Li) + 1,2,3,5\text{-}F_4C_6Li_2 \xrightarrow{CO_2} 1,2,3,5\text{-}F_4C_6H(CO_2H) + 1,2,3,5\text{-}F_4C_6(CO_2H)_2$$

Organo-lithium compounds usually undergo the same reactions as Grignard reagents but do so more readily. Details of the use of aryl halides in their preparation can be found in *G. Wittig*, "Syntheses with Organo-lithium Compounds, Newer Methods of Preparative Organic Chemistry", Interscience Publishers, New York, 1948; *R. G. Jones*, "Halogen-Metal Interconversion Reaction with Organo-lithium Compounds", Organic Reactions, Vol. 6, Chap. 7, John Wiley, New York, 1951; *G. E. Coates, M.L. H. Green* and *K. Wade*, "Organo-Metallic Compounds", 3rd Edn., Methuen, London, 1967.

Sodium and **potassium aryls** are much more reactive than the lithium or magnesium derivatives. They are not much used in organic synthesis. For their preparation see Vol. III B, Chap. 10.

The coupling of aryl and alkyl halides or of two aryl halides by the Wurtz-Fittig or Fittig reactions, respectively, is not of great preparative significance. Biaryls and alkaryls are more conveniently prepared by making the Grignard reagent first and then treating it, alone or in the presence of another organic halide, with cobaltous chloride (*Kharasch et al.*, J. Amer. chem. Soc., 1943, **65**, 491–507; *Kharasch* and *Reinmuth, loc. cit.*), or by using the Ullman technique. The latter reaction has been reviewed by *P. E. Fanta* (Chem. Reviews, 1946, **38**, 139; 1964, **64**, 613) who discusses the relative reactivities of the halogens (I > Br > Cl; F does not react), the influence of other groups present, solvent effects, temperature, the nature of the copper catalyst, and the mechanism. Mixed coupling of aryl and alkyl iodides through the agency of reagents formally designated as lithium di-*n*-alkyl coppers gives excellent yields (*E. J. Corey* and *G. H. Posner*, J. Amer. chem. Soc., 1968, **90**, 5616):

$$C_6H_5I + (n\text{-}C_4H_9)_2CuLi \xrightarrow[\text{(ii)} + n\text{-}C_4H_9I,\ 0^\circ,\ 3{\cdot}5\ \text{h}]{\text{(i) } 0^\circ,\ 5\ \text{h}} n\text{-}C_4H_9{\cdot}C_6H_5\ (75\,\%)$$

(ii) Dehalogenation

Dehalogenation of the monohalogeno-hydrocarbons is effected by hydrogenation over reduced nickel in the vapour phase at 270° (*J. B. Sabatier* and *A. Mailhe*, Compt. rend., 1904, **138**, 246); in the liquid phase by hydrogenation in presence of alkali with palladium–calcium carbonate (*M. Busch* and *H. Stöve*, Ber., 1916, **49**, 1069); with colloidal palladium (*K. W. Rosenmund* and *F. Zetsche, ibid.*, 1918, **51**, 578) or with reduced nickel (*C. Kelber, ibid.*, 1917, **50**, 305). Using palladium–calcium carbonate, biaryl formation may occur; the conditions which determine this course of the reaction have been investigated by *F. R. Mayo* and *M. D. Harwitz* (J. Amer. chem. Soc., 1949, **71**, 776) who give references to earlier work. Dehalogenation has also been effected with greater or less success by sodium or sodium amalgam and alcohol (*A. Stepanow*, J. Soc. phys.-chem. Russe, 1905, **37**, 15; *R. Löwerherz*, Z. physik. Chem., 1902, **40**, 415). Chloro- and bromo-aromatic compounds are photochemically reduced when isopropanol solutions are irradiated (*J. T. Pinkey* and *R. G. D. Rigby*, Tetrahedron Letters, 1969, 1267 and 1271). The process is not always straightforward replacement of halogen by hydrogen and in some cases alkyl aryl ethers are formed. In polyhalogeno compounds, in addition to catalytic dehalogenation (*R. E. Florin, W. J. Pummer* and *L. A. Wall*, J. Res. Nat. Bur. Stand., 1959, **62**, 119, and *G. G. Yakobson, N. N. Vorozhtsov et al.*, J. gen. Chem. U.S.S.R., 1966, **36**, 150),

$$C_6F_5Cl \xrightarrow{H_2/Pd} C_6F_5H$$

some of the halogen atoms can be removed by nucleophilic attack (see later) using hydride ion from lithium aluminium hydride (*G. M. Brooke, J. Burdon* and *Tatlow*, J. chem. Soc., 1962, 3253):

$$C_6F_5H \xrightarrow[\text{reflux, 53 hrs}]{LiAlH_4,\ \text{ether}} \underset{(90\%)}{1,2,4,5\text{-}C_6F_4H_2} + \underset{(3\%)}{1,2,3,5\text{-}C_6F_4H_2} + \underset{(7\%)}{1,2,3,4\text{-}C_6F_4H_2}$$

(*iii*) *Electrophilic substitution*

Electrophilic substitution in the halogenobenzenes, which results in replacement of nuclear hydrogen atoms, is brought about by the general methods already given for benzene (see pp. 204 *et seq.*) and its homologues, *para* derivatives being formed predominantly. The theory of the process is discussed in Vol. I A, pp. 284 *et seq.*, and in this volume pp. 71 *et seq.*). Highly halogenated aromatic compounds are unreactive in electrophilic substitution, but normal reaction products can be obtained by using forcing conditions. Thus, for example, Friedel-Crafts alkylation of pentafluorobenzene gives the expected products but the reaction is conducted under autogenous pressure at 150° (*W. F. Beckert* and *J. U. Lowe*, J. org. Chem., 1967, **32**, 1212). Similarly conventional nitration procedures fail for pentafluorobenzene but good yields (82 %) of pentafluoronitrobenzene are obtained using a nitrating mixture of boron trifluoride and fuming nitric acid in tetramethylene sulphone (sulpholane) (*P. L. Coe, A. A. Jukes* and *Tatlow*, J. chem. Soc., C, 1966, 2323).

(iv) Nucleophilic substitution

Nucleophilic substitution results in replacement of the halogen atoms of the halogenobenzenes and can take place by a bimolecular process when the halogen atoms are activated by other groups attached to the benzene ring. Alternatively it can occur by an elimination–addition (benzyne) mechanism for non-activated halides although some of these can be substituted by both mechanisms under certain reaction conditions. The bimolecular substitution mechanism is typified by the reaction of the 2,4-dinitrophenyl halides with nucleophiles when the order of displacement generally, though not invariably, observed is F $\gg$ Cl $\sim$ Br $\sim$ I. This order indicates the formation of a rate-determining transition state in which the breaking of the C-halogen bond has not made significant progress; for detailed discussion of mechanisms see *J. F. Bunnett*, Quart. Reviews, 1958, **12**, 1; *S. D. Ross*, Prog. phys. org. Chem., 1963, **1**, 31, and *F. Pietra*, Quart. Reviews, 1969, **23**, 504:

F, NO_2, NO_2 + $Nu^{\ominus}$ ⇌ [F, Nu, N, O, O, N, O, O]$^{\ominus}$ → Nu, NO_2, NO_2

In the polyhalogenobenzenes, the other halogen atoms activate the nucleus to this type of attack because of their electron attractive (*I*σ) effect and this explains the great ease of nucleophilic substitution in polychloro- (*A. L. Rocklin*, J. org. Chem., 1956, **21**, 1478; *A. F. Holleman*, Rec. Trav. chim., 1920, **39**, 736; *T. de Crauw, ibid.*, 1931, **50**, 753; *L. S. Kobrina Yakobson* and *Vorozhtsov*, Zhur. obshch. Khim. 1965, **35**, 137 and 142) and polyfluoro-benzenes (*Tatlow*, Endeavour, 1963, **22**, 89; *Wall et al.*, J. Res. Nat. Bur. Stand., 1963, **67A**, 481; *Burdon* and *W. B. Hollyhead*, J. chem. Soc., 1965, 6326). The reactions between hexabromo- or hexaiodo-benzenes and Grignard reagents to give polyaryl- and polyalkyl-benzenes (*T. A. Geissman* and *R. C. Mallet*, J. Amer. chem. Soc., 1939, **61**, 1788; *J. F. Durand*, Compt. rend., 1930, **191**, 1960) may be explained partly on the same basis. An explanation of the orientation of nucleophilic substitution in polychloro- and polyfluoro-benzenes has been suggested by *Burdon* (Tetrahedron, 1965, **21**, 3373).

The "benzyne mechanism", which can operate only in the presence of strong bases since only these can remove protons from the benzene nucleus, is common but not exclusive among those aryl halides with no activating group. Thus, treatment of *o*-chlorotoluene with potassamide in liquid ammonia gives a mixture of *o*- and *m*-toluidines (*J. D. Roberts et al.*, J. Amer. chem. Soc., 1956, **78**, 611):

CH_3, Cl $\xrightarrow{-HCl}$ CH_3 $\xrightarrow{NH_3}$ CH_3, NH_2 + CH_3, NH_2

The mechanism is confirmed by the fact that compounds with no hydrogen atoms *ortho* to the halogen do not react with potassiumamide in liquid ammonia. The intermediate formation of benzyne from fluorobenzene *via* the lithium derivative has already been mentioned. Other modifications of this preparation involve halogen-metal interchange by *o*-halogenofluorobenzenes and butyl-lithium with subsequent elimination of lithium fluoride (*H. Gilman et al.*, *ibid.*, 1940, **62**, 2327; 1955, **77**, 3919; 1956, **78**, 2217; 1957, **79**, 2625; *Wittig* and *L. Pöhmer*, Ber., 1956, **89**, 1334). Tetrafluorobenzyne is an intermediate in the reactions between pentafluorophenyllithium and bromine and furan (*Coe, Stephens* and *Tatlow*, J. chem. Soc., 1962, 3227; *D. D. Callender, Coe* and *Tatlow*, Tetrahedron, 1965, **22**, 419). The halogenobenzenes and halogenotoluenes are hydrolysed readily by aqueous sodium hydroxide at temperatures between 250° and 350° by either the benzyne mechanism or by nucleophilic substitution depending on the exact reaction conditions. Thus *p*-iodotoluene reacts with *N*-NaOH at 250° almost completely by direct substitution whereas chlorobenzene at 340° is hydrolysed by both mechanisms (*A. Bottini* and *Roberts*, J. Amer. chem. Soc., 1957, **79**, 1458).

By one mechanism or the other, aromatic halogen has been replaced using other halide, cyanide, thio- and seleno-cyanate, azide, hydroxide, alkoxide, phenoxide, carboxylate, sulphide, mercaptide, sulphite, sulphinate, and arsenite ions, ammonia, amines, carbanions and organometallic compounds. Sometimes cuprous salts, cuprous oxide, or copper are used as catalysts as in a commercial process for preparing phenol from chlorobenzene (*W. J. Hale* and *E. G. Britton*, Ind. Eng. Chem., 1928, **20**, 114). In these cases it is probable that the cuprous ion forms a complex with the halide and nucleophilic substitution then occurs. The field of nucleophilic aromatic substitution up to 1951 has been surveyed by *Bunnett* and *R. E. Zahler* (Chem. Reviews, 1951, **49**, 273); the theory of both mechanisms has been discussed (*Bunnett*, Quart. Reviews, 1958, **12**, 1; *E. S. Gould*, "Mechanism and Structure in Organic Chemistry", Holt, Rinehart and Winston Inc., New York, 1959;

and *C. K. Ingold,* "Structure and Mechanism in Organic Chemistry", G. Bell and Sons Ltd., London 1953). The elimination–addition (benzyne) mechanism is reviewed by *Bunnett* (J. chem. Ed., 1961, **38**, 278); *Heaney,* (Chem. Reviews, 1962, **62**, 81); and *R. W. Hoffman* in "Dehydrobenzene and Cycloalkynes", Weinheim, Verlag Chemie, 1967.

A complex, provisionally assigned the structure

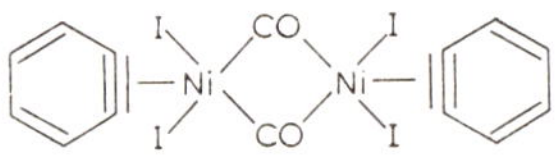

is obtained from the reaction of 1,2-diiodobenzene with tetracarbonyl nickel (*E. W. Gowling, S. F. A. Keitle* and *G. M. Sharples,* Chem. Comm., 1968, 21).

(*v*) *Photochemical reactions*

Photochemical investigations of halogenobenzenes have been theoretically and synthetically useful. The spectrum of phenyl radicals, obtained by flash photolysis of halogenobenzenes, has been recorded (*G. Porter* and *B. Ward,* Proc. chem. Soc., 1964, 288). Photochemical halogenation (*R. L. Letsinger* and *G. G. Wubbels,* J. Amer. chem. Soc., 1966, **88**, 5041) and halogen exchange (*R. K. Sharma* and *N. Kharasch,* Angew. Chem. internat. Edn., 1968, **7**, 36) have been mentioned previously; the review by the latter authors indicates the range and value of photochemical reactions of aryl iodides. Photosubstitution reactions are described by *E. Havinga, R. O. de Jongh* and *M. E. Kronenberg* (Helv., 1967, **50**, 2550). The isomerization of hexafluorobenzene to the Dewar isomer (75 %) on photolysis at 2537Å in the vapour phase (*G. Camaggi, F. Gozzo* and *G. Cevidalli,* Chem. Comm., 1966, 313 and J. chem. Soc., C, 1969, 489; and *I. Haller,* J. chem. Phys., 1967, **47**, 1117), contrasts with other photoisomerizations of aromatics where benzvalene and fulvene isomers predominate:

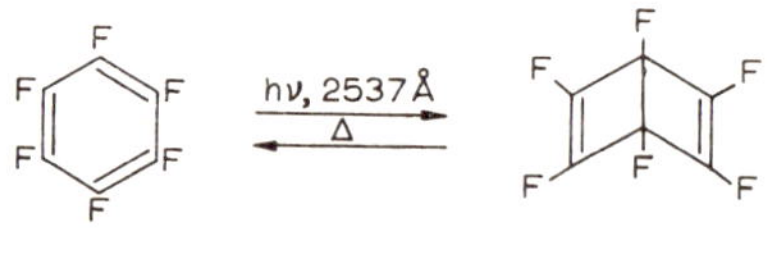

(*c*) *Halogenobenzenes*

(*i*) *Fluorobenzenes*

Some physical constants are given in Table 1. References for the preparation of these and many other fluorinated aromatic compounds can be found in

"Aromatic Fluorine Compounds" by *A. E. Pavlath* and *A. L. Leffler*, Reinhold Pub. Corp., New York, 1962; "Aromatic Fluorine Compounds" by *A. K. Barbour, M. W. Buxton* and *G. Fuller* in the Kirk-Othmer Encyclopaedia of Chemical Technology, Interscience Publishers, New York, 1966, and "Chemistry of Organic Fluorine Compounds" by *M. Hudlicky*, Pergamon Press, London, 1961.

TABLE 1

FLUOROBENZENES

Compound	M.p.°	B.p.°	$d_4^{t°}$	$n_D^{t°}$
Fluorobenzene	−41·9	85	d_4^{15} 1·0309	n_D^{20} 1·4667
o-Difluorobenzene	−34	91	d_4^{25} 1·1496	n_D^{20} 1·4452
m-Difluorobenzene	−59·3	83	d_4^{20} 1·1572	n_D^{20} 1·4410
p-Difluorobenzene	−13	89	d_4^{20} 1·1716	n_D^{20} 1·4421
1,2,3-Trifluorobenzene	−13·5	95		n_D^{20} 1·4230
1,2,4-Trifluorobenzene	−35	88		
1,3,5-Trifluorobenzene	−5·5	75·5	d_4^{20} 1·277	n_D^{20} 1·4140
1,2,3,4-Tetrafluorobenzene	−42	95	d_4^{25} 1·422	n_D^{20} 1·4069
1,2,3,5-Tetrafluorobenzene	−48	83	d_4^{25} 1·410	n_D^{25} 1·4011
1,2,4,5-Tetrafluorobenzene	4	90	d_4^{25} 1·424	n_D^{20} 1·4045
Pentafluorobenzene	−48	85	d_4^{20} 1·531	n_D^{25} 1·3881
Hexafluorobenzene	5	80	d_4^{20} 1·6182	n_D^{25} 1·3761

Vapour pressures and critical constants of C_6F_6, C_6F_5Br and C_6F_5Cl and a comparison with those of C_6H_6, C_6H_5F, C_6H_5Cl and C_6H_5Br are given by *F. D. Evans* and *P. F. Tiley* (J. chem. Soc., B, 1966, 134). References to work on the ^{19}F N.M.R. spectra of fluorobenzenes and their derivatives are given by *J. W. Emsley, J. Feeney* and *L. H. Sutcliffe* ("High Resolution Nuclear Magnetic Resonance Spectroscopy", Vol. II, Pergamon Press, Oxford, 1965) and *J. Homer* and *L. F. Thomas* (J. chem. Soc., B, 1966, 141). The influence of the fluorine atom on further substitution in the aromatic nucleus has been discussed by *D. T. Clark, J. N. Murrell* and *J. M. Tedder* (J. chem. Soc., 1963, 1250), *Murrell* ("Quantitative Evaluation of Substituent Effects by Electronic Spectroscopy", Royal Inst. of Chem. Lecture Series 1963, No. 2), *R. O. C. Norman*, ("Quantitative Aspects of Aromatic Substitution", *ibid.*) and *J. Burdon* (Tetrahedron, 1965, **21**, 3373).

Fluorobenzene, C_6H_5F, is best prepared by the thermal decomposition of benzenediazonium fluoroborate, PhN_2BF_4 (*D. T. Flood*, Org. Synth., Coll. Vol. II, 1943, p. 295; *A. Roe*, Org. Reactions, Vol. V, Chapter 4) or by the decomposition of the diazonium fluoride in anhydrous hydrogen fluoride (*P. Osswald* and *O. Scherer*, G. P. 600,706/1934; C. A., 1934, **28**, 7260). Electrophilic substitution in *ortho* and (mainly) *para* positions occurs much more readily than for other halogenobenzenes, in the case of chlorination being more rapid than for benzene itself. This is because the combination of $I\pi$ and mesomeric effects effectively counterbalance the deactivating $I\sigma$ effect. For a comparison of the rates of halogenation under electrophilic and free radical conditions see *G. Olah, Pavlath* and *G. Varsanyi*, J. chem. Soc., 1957, 1823. A nuclear fluorine atom has little effect on the Friedel and Crafts reaction and rarely influences the functioning of other groups present in the nucleus as in the formation of Grignard reagents, condensation, hydrolysis, esterification, etherification, etc.

(*ii*) *Chlorobenzenes*

Some physical constants of the chlorobenzenes are given in Table 2. References for the preparations and reactions of these and many other chlorinated aromatic compounds can be found in "Organic Chlorine Compounds" by *E. H. Huntress*, Chapman and Hall, London, 1948.

TABLE 2

CHLOROBENZENES

Compound	M.p.⁰	B.p.⁰	$d_4^{t^0}$	$n_D^{t^0}$
Chlorobenzene	—45·2	132·0	d_4^{15} 1·1117	n_D^{51} 1·5275
o-Dichlorobenzene	—17·2	179·2	d_4^{25} 1·2973	$n_D^{20\cdot4}$ 1·5485
m-Dichlorobenzene	—26·3	172	d_4^{25} 1·2799	$n_D^{20\cdot9}$ 1·5457
p-Dichlorobenzene	53·0	174·5	d^{55} 1·2495	$n_D^{19\cdot9}$ 1·5267
1,2,3-Trichlorobenzene	52·4	218		
1,2,4-Trichlorobenzene	16·6	213	d_{25}^{25} 1·4634	n_D^{25} 1·5524
1,3,5-Trichlorobenzene	63	208		
1,2,3,4-Tetrachlorobenzene	47·5	254		
1,2,3,5-Tetrachlorobenzene	51	246		
1,2,4,5-Tetrachlorobenzene	138	244		
Pentachlorobenzene	87	276		
Hexachlorobenzene	229	326		

Chlorobenzene, C_6H_5Cl, is conveniently prepared by the chlorination of benzene using a halogen carrier. For the industrial manufacture of chlorobenzene, iron powder is used; in the laboratory preparation, aluminium or iron or their anhydrous chlorides are suitable carriers. Further chlorination yields *p*-dichlorobenzene with some *o*-dichlorobenzene, and by prolonged chlorination, 1,2,4-trichloro- and 1,2,4,5-tetrachloro-benzene are the main products. Chlorination of chlorobenzene in the vapour phase at 500–600⁰ gives a large proportion of *m*-dichlorobenzene (*J. P. Wibaut et al.*, Rec. Trav. chim., 1937, **56**, 65, 815). Nitration occurs normally to give *o*- and *p*-nitro- and 2,4-dinitro-chlorobenzenes. For a comparison of the relative rates of nitration of the monohalogenobenzenes see *M. L. Bird* and *C. K. Ingold* (J. chem. Soc., 1938, 918). Chlorobenzene behaves normally in the Friedel-Crafts reaction to yield *p*-substituted alkyl- and arylbenzenes; it is chloromethylated to give *p*-chlorobenzyl chloride and condenses with aldehydes in presence of concentrated sulphuric acid.

(*iii*) *Bromobenzenes*

Physical constants are given in Table 3.

TABLE 3

BROMOBENZENES

Compound	M.p.°	B.p.°	$d_4^{t°}$	$n_D^{t°}$
Bromobenzene	−30·6	156·2	d_4^{20} 1·4946	n_D^{20} 1·5572
o-Dibromobenzene	+ 6·7	225	$d_4^{20\cdot4}$ 1·956	n_D^{20} 1·6081
m-Dibromobenzene	− 7	220	d_4^{20} 1·952	$n_D^{18\cdot7}$ 1·6088
p-Dibromobenzene	+87·3	220		
1,2,3-Tribromobenzene	88			
1,2,4-Tribromobenzene	44			
1,3,5-Tribromobenzene	119	271		
1,2,3,4-Tetrabromobenzene	47·5			
1,2,3,5-Tetrabromobenzene	98			
1,2,4,5-Tetrabromobenzene	173			
Pentabromobenzene	159			
Hexabromobenzene	326			

Bromobenzene is conveniently prepared by the direct bromination of benzene using a halogen carrier. On further bromination 1,4-dibromobenzene is formed together with some of the *ortho* isomer and both yield 1,2,4-tribromo- and then 1,2,4,5-tetrabromo-benzene by the further action of bromine (*M. Copisarow*, *ibid.*, 1921, **119**, 447). On sulphonation, by heating with concentrated sulphuric acid, there may be a migration of the halogen on prolonged heating. Thus bromobenzene gives 3,5-dibromobenzenesulphonic acid together with other products. *p*-Dibromobenzene similarly yields some tetrabromo- and hexa-bromo-benzene under drastic conditions (*J. Herzig*, Monatsh., 1881, **2** 192; *R. R. Baxter* and *F. D. Chattaway*, J. chem. Soc., 1915, **107**, 1815) A convenient procedure for perbromination was described on p. 250.

(*iv*) *Iodobenzenes*

Iodobenzene is conveniently prepared from aniline by diazotization and treatment with potassium iodide: yield >80 %. Sulphonation of iodobenzene with fuming sulphuric acid yields *p*-iodobenzenesulphonic acid. At higher temperature sulphonation gives *p*-diiodobenzene and benzenesulphonic acid (*G. S. Neumann*, Ann., 1887, **241**, 39; *J. Troeger* and *F. Hurdelbrinck*, J. pr. Chem., 1902, [ii], **65**, 82; *M. Boyle*, J. chem. Soc., 1909, **95**, 1683; *E. H. Huntress* and *F. H. Carten*, J. Amer. chem. Soc., 1940, **62**, 511). The formation of polyiodo compounds in presence of sulphuric acid appears to be a characteristic of iodo aromatic compounds. The formation of hexaiodobenzene, C_6I_6, is an extreme example; it is obtained in about 80% yield by the action of iodine and 60% oleum on benzene at 170–180°. It is similarly formed from benzoic and terephthalic acids (*J. F. Durand* and *M. Mancet*, Bull. Soc. chim. Fr., 1935, [v], **2**, 665; *E. Rupp*, Ber., 1896, **29**, 1630; see also *J. Arotsky*, *R. Butler* and *A. C. Darby*, Chem. Comm., 1966, 650 and refs. therein).

Iodobenzene, m.p. —31°, b.p. 188·5°, d_4^{25} 1·823, $n_D^{17.8}$ 1·6213. o-*Diiodobenzene*, m.p. 26·7°, b.p. 286°; m-*diiodobenzene*, m.p. 35°, b.p. 285°; p-*diiodobenzene*, m.p. 129°, b.p. 285°. 1,2,3-*Triiodobenzene*, m.p. 116°; 1,2,4-*triiodobenzene*, m.p. 91·5°; 1,3,5-*triiodobenzene*, m.p. 184°. 1,2,3,4-*Tetraiodobenzene* m.p. 136°; 1,2,3,5-*tetraiodobenzene*, m.p. 148°; 1,2,4,5-*tetraiodobenzene*, m.p. 254°. *Pentaiodobenzene*, m.p. 172°. *Hexaiodobenzene*, m.p. 340–350°.

(d) Iodosyl, iodyl and iodonium compounds*

The preparation, reactions and properties of these compounds have been reviewed by *R. B. Standin* (Chem. Reviews, 1943, **32**, 249) and more recently by *D. F. Banks* (*ibid.*, 1966, **66**, 243).

A characteristic reaction of iodobenzene and its homologues is the addition of chlorine to give (**dichloroiodo**)**arenes** (older nomenclature iodoarene dichloride; aryliododichlorides) of the general formula $RICl_2$ (R = aryl) The earlier literature favoured an ionic structure, $RI^{\oplus}Cl\ Cl^{\ominus}$, for these compounds but more recent determinations of conductivity, molecular weight, and dipole moment indicate that any dissociation is strictly molecular (*Banks, loc. cit.*):

$$RICl_2 \rightleftharpoons RI + Cl_2$$

X-ray structure determinations confirms the covalent nature of the compounds. Substituted iodobenzenes also give (dichloroiodo)arenes in the same way, although an accumulation of electronegative groups in the nucleus tends to prevent their formation (*C. Willgerodt* and *K. Wilcke*, Ber., 1910, **43**, 2746). The reaction is brought about by passing chlorine into a solution of the iodo compound in an inert solvent, *e.g.* chloroform, glacial acetic acid or light petroleum (*Willgerodt et al.*, *ibid.*, 1893, **26**, 1533; 1896, **29**, 1568; 1910, **43**, 2641; J. pr. Chem., 1905, [ii], **71**, 540).

The (dichloroiodo)arenes are relatively unstable, decomposing on heating or on prolonged exposure to light thus:

$$C_6H_5ICl_2 \rightarrow p\text{-}ClC_6H_4I + HCl$$

* Under the I.U.P.A.C. Rules these terms have replaced the older terms "iodoso" and "iodoxy", respectively.

They are converted to **iodosyl compounds**, RIO, by the action of aqueous alkali (*H. J. Lucas, E. R. Kennedy* and *M. W. Formo*, Org. Synth., Coll. Vol. III, p. 483, 1955) or aqueous pyridine (*G. Ortoleva*, Gazz. chim. ital , 1900, **30**, II, 1).

Iodosyl compounds have also been prepared by the direct oxidation of iodo compounds preferably by peracetic acid in acetic acid (*J. Böeseken* and *C. Schneider*, Proc. koninkl. Nederland. Akad. Wetenschap., 1930, **33**, 827; 1932, **35**, 1140). They behave as the anhydrides of the hypothetical base $RI(OH)_2$ and their salts, of general formula RIX_2, can be prepared by dissolving the neutral iodosyl compounds in the appropriate acids ($X = NO_3$, Cl, F, HCO_2, $CH_3 \cdot CO_2$, $PhCO_2$; $X_2 = CrO_4$). This method has some limitations and a better preparative route is *via* a double displacement on the dichloroiodo compound in acetonitrile solution (*N. W. Alcock* and *T. C. Waddington*, J. chem. Soc., 1963, 4103);

$$C_6H_5ICl_2 + 2\ AgX \rightarrow C_6H_5IX_2 + 2\ AgCl$$

(Difluoroiodo)arenes have been prepared from the iodosyl compound by the action of hydrogen fluoride in acetic acid (*B. S. Garvey, L. F. Halley* and *C. F. H. Allen*, J. Amer. chem. Soc., 1937, **59**, 1827) and from (dichloroiodo)-arenes by reaction with mercuric oxide and aqueous hydrogen fluoride in methylene chloride (*W. Carpenter*, J. org. Chem., 1966, **31**, 2688):

$$ArICl_2 + HgO + 2HF \rightarrow ArIF_2 + H_2O + HgCl_2$$

(Dichloro- and difluoro-iodo)arenes have been used respectively as chlorinating (*D. D. Tanner* and *P. B. van Bostelen*, J. org. Chem., 1967, **32**, 1517) and fluorinating (*Carpenter, loc. cit.*) reagents.

The iodosyl compounds are oxidising agents and can be estimated quantitatively by the liberation of iodine from acidified iodide solutions:

$$C_6H_5IO + 2HI = C_6H_5I + I_2 + H_2O$$

They readily oxidise thiols to disulphides (*L. Hellerman et al.*, J. Amer. chem. Soc., 1941, **63**, 2551).

Iodosyl compounds are oxidized by hypochlorite or peracids such as perbenzoic or peracetic to **iodyl compounds,** RIO_2. Iodyl compounds are also formed by the dismutation of iodosyl compounds on boiling with water, by rapid steam distillation (*H. J. Lucas* and *E. R. Kennedy*, Org. Synth., Coll. Vol. III, p. 485, 1955), or on prolonged storage:

$$2\ PhIO \rightarrow PhI + PhIO_2$$

They are also formed conveniently by the oxidation of iodo compounds by permonosulphuric acid, Caro's reagent (*E. Bamberger* and *A. Hill*, Ber., 1900, 33, 533; *I. Masson et al.*, J. chem. Soc., 1935, 1669).

They are amphoteric substances and iodylbenzene, $PhIO_2$, forms crystalline salts with strong inorganic acids; $PhIO_2,H_2SO_4$ has been analysed and there is good evidence for a perchlorate and nitrate (*Masson et al., loc. cit.*). A fluoro compound (difluoroiodosyl)benzene, $PhIOF_2$ is formed from iodylbenzene and hot concentrated hydrofluoric acid, presumably by elimination of water from the salt which is the initial product (*R. F. Weinland* and *W. Stille*, Ber., 1901, 34, 2631). Iodylbenzene also behaves as a weak acid giving easily hydrolysable alkali metal salts of the base "iodylbenzene hydroxide", $PhIO_2OH$, known as iodylates or iodoxylates (*Masson et al., loc. cit.*). By prolonged action of alkali at room temperature, or on boiling, the iodyl group is replaced by hydrogen so that iodylbenzene yields benzene and sodium iodate; *o*-iodylbenzoic acid and *p*-iodyl-nitrobenzene behave similarly (*C. Hartmann* and *V. Meyer*, Ber., 1893, **26**, 1727; *D. Vorländer* and *K. Büchner*, *ibid.*, 1925, **58**, 1291). The iodyl compounds, like iodosyl compounds and their salts, are oxidising agents which are quantitatively reduced by adding iodide ion and acidifying:

$$C_6H_5IO_2 + 4\ HI \rightarrow C_6H_5I + 2\ H_2O + 2\ I_2$$

In the absence of nitrous acid, iodylbenzene can be nitrated to give an excellent yield of the *meta*-nitro derivative (*Masson et al., loc. cit.*). Iodyl compounds should be handled with extreme care because they explode on impact and deflagrate on touching with a hot wire.

If a mixture of iodyl- and iodosyl-benzene in aqueous suspension is heated with silver oxide, the aqueous solution becomes strongly alkaline due to the formation of **diphenyliodonium hydroxide**:

$$PhIO + PhIO_2 + OH^{\ominus} \rightarrow [Ph_2I]^{\oplus}OH^{\ominus} + IO_3^{\ominus}$$

The addition of a soluble iodide to the solution precipitates diphenyliodonium iodide (*C. Hartmann* and *V. Meyer*, Ber., 1894, **27**, 502; *A. J. Lucas* and *E. R. Kennedy*, Org. Synth., Coll. Vol. III, 1955, p. 355).

Iodonium salts of the type

$$[(C_6H_4X)_2I]^{\oplus}Y^{\ominus},\ (X = H, Me, Br, Cl, \text{or } I;\ Y = HSO_4)$$

are formed by the reaction of C_6H_5X with a solution of iodine pentoxide, I_2O_5, in concentrated sulphuric acid. Nitrobenzene and benzenesulphonic acid give iodosyl compounds (*Masson* and *W. E. Hanby*, J. chem. Soc., 1938, 1699; *Masson* and *E. Race*, *ibid.*, 1937, 1718).

The action of concentrated sulphuric acid on iodosyl compounds gives iodine-substituted iodonium sulphates (*Hartmann* and *Meyer*, Ber., 1894, **27**, 427; *Willgerodt et al.*, *ibid.*, 1900, **33**, 842; 1905, **38**, 1474):

2PhIO ⟶ (I-C₆H₄)(C₆H₅)I⊕ HSO₄⊖

2,2′-Biphenyleneiodonium iodide has been prepared from 2,2′-di-iodosobiphenyl or its tetrachloride by keeping them in water for several months and treating the aqueous solution with sulphur dioxide (*L. Mascarelli*, Gazz. chim. ital., 1913, **43**, I, 26). The 4,4′-diethoxycarbonyl derivative also has been made by treating the diazonium salt obtained from 2,2′-diamino-4,4′-diethoxylcarbonylbiphenyl with sodium iodide (*N. E. Searle* and *R. Adams*, J. Amer. chem. Soc., 1933, **55**, 1649).

The iodonium hydroxides are strong bases which form stable salts when neutralised by acids. The iodonium salts decompose on heating:

$$(CH_3C_6H_4)_2I^{\oplus}I^{\ominus} \xrightarrow{155^\circ} 2\ CH_3C_6H_4I$$

(*J. J. Lucas et al.*, J. Amer. chem. Soc., 1936, **58**, 157);

$$\begin{matrix}p\text{-anisyl}\\ \text{Ph}\end{matrix}\!\!>\!I^{\oplus}Br^{\ominus} \xrightarrow{180^0} p\text{-}CH_3O\cdot C_6H_4I + C_6H_5Br$$

(*R. B. Sandin et al.*, *ibid.*, 1937, **59**, 2014).

(**Dichloroiodo**)**benzene**, (***phenyliododichloride***), $PhICl_2$, yellow needles from chloroform, decomposes between 110^0 and 135^0 to give some *p*-chloroiodobenzene; this decomposition occurs also on keeping, particularly on exposure to light.

It is reduced by warm alcohol and liberates an equivalent of iodine from acidified potassium iodide solution to give iodobenzene. (Dichloroiodo)benzene reacts with silver acetylide to give the iodonium chloride $(PhIC{\equiv}CH)^{\oplus}Cl^{\ominus}$ (*Willgerodt*, Ber., 1895, **28**, 2110); with sodiomalonic ester to give iodobenzene and tetra-ethyl ethanetetracarboxylate (*H. H. Hodgson*, Proc. Camb. Phil. Soc., 1908, **14**, 547).

(*Dinitratoiodo*)*benzene*, $PhI(NO_3)_2$, yellow plates decomp. 105–106^0; (*diacetoxyiodo*)*benzene*, $PhI(OAc)_2$, prisms, m.p. 156–157^0; (*chromatoiodo*)*benzene*, $PhICrO_4$, yellow precipitate, explodes at 65–67^0 (*Willgerodt*, Ber., 1892, **25**, 3498).

Iodosylbenzene, PhIO, an amorphous yellow solid, exploding near 210^0, is formed from (dichloroiodo)benzene by the action of aqueous alkali or aqueous pyridine. It is reduced by aqueous sulphur dioxide or dilute hydroiodic acid to iodobenzene. Oxidation by hypochlorite or peracids gives **iodylbenzene**, $PhIO_2$, long needles, sparingly soluble in most solvents; explodes at about 235^0.

Diphenyliodonium iodide, $(Ph_2I)^{\oplus}I^{\ominus}$, yellow needles, m.p. 175–176^0, decomposing to give iodobenzene (*V. Meyer*, Ber., 1894, **27**, 1592).

For further examples see *Banks loc. cit.*

(e) *Mixed halogenobenzenes*

Dihalogenobenzenes with dissimilar substituents are best obtained *via* the

TABLE 4

MIXED DIHALOGENOBENZENES

Compound	*ortho*		*meta*		*para*	
	M.p.^{0}	B.p.^{0}	M.p.^{0}	B.p.^{0}	M.p.^{0}	B.p.^{0}
Chloro-fluoro	−42	138[1]	—	—	−28	130[1]
Bromo-fluoro	—	57/22 mm[2]	—	151[3]	− 8	152[3,5]
Fluoro-iodo	−41·5	188[4]	—	—	−18) −27·2)	183[4]
Bromo-chloro	−12	204[6]	−21	196[6]	67	196[6,7]
Chloro-iodo	—	110/16 mm[8]	—	230[8]	57	227[9]
Bromo-iodo	2	257[6]	−9·3	252[6]	92	251[6]

1 *C. K. Ingold* and *C. C. N. Vass*, J. chem. Soc., 1928, 423, 2265.
2 *E. Bergmann et al.*, Z. physik. Chem., 1930, **B,10,** 120.
3 *G. Schiemann* and *R. Pillarsky*, Ber., 1931, **64,** 1343.
4 *I. J. Rinkes*, Chem. Weekblad, 1918, **16,** 206.
5 *C. M. Suter* and *A. W. Weston*, J. Amer. chem. Soc., 1941, **63,** 602.
6 *J. Narbutt*, Ber., 1919, **52,** 1031.
7 *H. W. Schwechten*, *ibid.*, 1932, **65,** 1605.
8 *A. F. Holleman*, Rec. Trav. chim., 1915, **34,** 223.
9 *L. Birckenbach* and *G. Goubeau*, Ber., 1932, **65,** 399.

diazonium compounds from the appropriate halogen-substituted anilines (*H. S. Fry* and *I. W. Grote,* J. Amer. chem. Soc., 1926, **48,** 711). Some of the *para* compounds can be prepared conveniently by direct substitution. Table 4 gives physical constants of some of these compounds. Many more highly halogenated compounds are known containing halogen atoms of more than one kind, a great number being prepared during the intensive investigation of halogen exchange reactions (cf. p. 249 *et seq.*).

(*f*) *Halogen derivatives of alkylbenzenes*

Table 5 gives the constants of the **monohalogenotoluenes**. All may be conveniently prepared *via* the diazonium compounds from the corresponding toluidines (cf. *C. S. Marvel* and *S. M. McElvain,* Org. Synth., Coll. Vol. I, 1932, p. 163; *L. A. Bigelow, ibid.,* p. 130). Pure *m*-bromotoluene is best prepared by the elimination of the amino group from 3-bromo-*p*-toluidine (*Bigelow et al., loc. cit.,* p. 128). *o*- and *p*-Chloro- and -bromo-toluenes are also formed by the direct halogenation of toluene in presence of a catalyst followed by precise fractionation. Further chlorination of the chlorotoluenes is described by *J. B. Cohen et al.* (J. chem. Soc., 1901, **79,** 1111; 1902, **81,** 1325; 1904, **85,** 1274). *m*-Chlorotoluene has been made from 1-methylcyclohex-1-en-3-one (*A. Klages* and *E. Knoevenagel,* Ber., 1894, **27,** 3019; cf. *Klages, ibid.,* 1896, **29,** 310).

The halogenotoluenes can undergo further nuclear substitution by halogenation, nitration and sulphonation and the alkyl group may be halogenated.

TABLE 5

MONOHALOGENOTOLUENES, $CH_3C_6H_4X$

Compound	M.p.°	B.p.°	$d^{t°}$	$n_D^{t°}$
o-Fluorotoluene	—	114	$d^{17\cdot3}$ 1·001	$n_D^{17\cdot3}$ 1·4716
m-Fluorotoluene	—	116	$d^{13\cdot4}$ 0·9972	—
p-Fluorotoluene	—	117	d_4^{16} 1·0007	—
o-Chlorotoluene	−36·5	158·4	d_4^{20} 1·073	$n_D^{20\cdot2}$ 1·5247
m-Chlorotoluene	−48·0	161·6	d_4^{20} 1·0722	$n_D^{18\cdot7}$ 1·5225
p-Chlorotoluene	8·0	162·3	d_4^{20} 1·0697	$n_D^{19\cdot0}$ 1·5199
o-Bromotoluene	−28	181·8	d_{15}^{15} 1·4309	—
m-Bromotoluene	−40	184	d_4^{20} 1·4099	—
p-Bromotoluene	28	184·6	d_4^{20} 1·3998	—
o-Iodotoluene	—	204	d^{20} 1·698	—
m-Iodotoluene	28	213(204)	d^{20} 1·698	—
p-Iodotoluene	35	211	—	—

They may be oxidized to halogen-substituted benzaldehydes and benzoic acids.

More highly halogenated toluenes are listed in Table 6 and halogenated xylenes in Table 7.

TABLE 6

DI- AND POLY-HALOGENOTOLUENES

Compound	Fluoro-	Chloro-		Bromo-		Iodo-	
	B.p.°	M.p.°	B.p.°	M.p.°	B.p.°	M.p.°	B.p.°
2,3-Dihalogenotoluene	117–119[12]	—	204[1]	28[4]		31	
2,4-Dihalogenotoluene	114–115[12]	—	195[1,6]	—	244[4,5]		296
2,5-Dihalogenotoluene	117[12]	5	199[1,7]	—	236[4]	31	
2,6-Dihalogenotoluene	112[12]	—	197[1,2]	—	243[4,5]	42	
3,5-Dihalogenotoluene	—	26	201[1]	39[5]			
2,3,4-Trihalogenotoluene	—	41	232[3,6]	44[4]		92[11]	
2,3,5-Trihalogenotoluene	—	46	231[3]	54[4]		73[11]	
2,3,6-Trihalogenotoluene	—	46[3]	—	60[4]		80[10]	
2,4,5-Trihalogenotoluene	—	82[3,6]	—	112[4]		118[11]	
2,4,6-Trihalogenotoluene	—	34[3]		66[4]		105[11]	
3,4,5-Trihalogenotoluene	—	43[3]	—	89[4]		122	
2,3,4,5-Tetrahalogenotoluene	—	98[4,6]		111		285[11]	
2,3,4,6-Tetrahalogenotoluene	—	92[6,8]		108		125[10]	
2,3,4,5,6-Pentahalogenotoluene	115–118[13]	218[2,6]		288[9]		340[11]	

1 *J. B. Cohen* and *H. D. Dakin*, J. chem. Soc., 1901, **79**, 1114.
2 *W. Davies*, *ibid.*, 1921, **119**, 873.
3 *Cohen* and *Dakin*, *ibid.*, 1902, **81**, 1328.
4 *Cohen* and *P. K. Dutt*, *ibid.*, 1914, **105**, 501.
5 *S. C. J. Olivier*, Rec. Trav. chim., 1926, **45**, 303.
6 *O. Silberrad*, J. chem. Soc., 1925, **127**, 2680.
7 *T. de Crauw*, Rec. Trav. chim., 1931, **50**, 773.
8 *Cohen* and *Dakin*, J. chem. Soc., 1906, **89**, 1454.
9 *M. D. Bodroux*, Ann. Chim., 1929, [x], **11**, 546.
10 *H. L. Wheeler*, Amer. chem. J., 1910, **44**, 135.
11 *idem*, *ibid.*, p. 501.
12 *K. Inukai* and *Y. Maki*, J. chem. Soc. Japan, 1956, **59**, 1160; C. A., 1958, **52**, 13265.
13 *A. K. Barbour et al.*, J. chem. Soc., 1961, 808.

For **bromochlorotoluenes** see *J. B. Cohen* and *C. J. Smithells*, J. chem. Soc., 1914, **105**, 1908; **chloro-** or **bromo-iodotoluenes** see *F. B. McAlister* and *J. Kenner*, *ibid.*, 1928, 1915; *Cohen* and *J. Miller*, *ibid.*, 1904, **85**, 1627; **bromofluoro-, chlorofluoro-** and **fluoroiodo-toluenes** see "Aromatic Fluorine Compounds" by *A. E. Pavlath* and *A. Leffler*, Reinhold, New York, 1962.

p-**Chlorocumene**, b.p. 125°/20 mm (*H. Meyer*, Monatsh., 1929, **53/54**, 741); p-*bromocumene*, b.p. 216° (*L. Bert*. Bull. Soc. chim. Fr., 1925, [iv], **37**, 1265); p-*iodocumene*, b.p. 234° (*E. Schreiner*, J. pr. Chem., 1910, [ii], **81**, 562); p-*fluorocumene*, b.p. 151–153° (*A. A. Petrov* and *A. V. Tumanova*, Zhur. obshchei. Khim., 1956, **26**, 2744; C. A., **51**, 7325). m-*Bromocumene*, b.p. 95°/20 mm (*R. D. Haworth*, J. chem. Soc., 1939, 1302).

5-**Chlorohemimellitene**, m.p. 70°, b.p. 213°; 5-*bromohemimellitene*, m.p. 73°, b.p. 233°; 5-*iodohemimellitene*, m.p. 37°, b.p. 256°. 4,5,6-*Trichlorohemimellitene*, m.p. 217° (*A. W. Crossley* and *J. S. Hills*, *ibid*., 1906, **89**, 882); 4,5,6-*tribromohemimellitene*, m.p. 245° (*S. F. Birch* and *W. S. G. Norris*, *ibid*., 1926, 2551).

5-**Chloropseudocumene**, m.p. 70°, b.p. 213° (*A. Huender*, Rec. Trav. chim., 1915, **34**, 8); 5-*bromopseudocumene*, m.p. 73°, b.p. 233°; 5-*iodopseudocumene*, m.p. 37°, b.p. 256° (*K. Elbs* and *A. Jaroslawzew*, J. pr. Chem., 1913, [ii], **88**, 93); 5-*fluoropseudocumene*, m.p. 27°, b.p. 174–175° (*O. Wallock* and *F. Heusher*, Ann., 1888, **243**, 219). 3,5,6-*Trichloropseudocumene*, m.p. 197° (*G. Schultz*, Ber., 1909, **42**, 3604); 3,5,6-*tribromo-pseudocumene*, m.p. 231° (*Birch* and *Norris*, *loc. cit.*). For other monohalogenopseudocumenes see *L. I. Smith*, J. Amer. chem. Soc., 1936, **58**, 8.

2-**Chloromesitylene**, b.p. 205° (*W. Davies* and *E. S. Wood*, J. chem. Soc., 1928, 1126); 2-*bromomesitylene*, m.p. 0°, b.p. 225° (Org. Synth., Coll. Vol. II, 1943, p.95); 2-*iodomesitylene*, m.p. 31° (*R. L. Datta*, J. Amer. chem. Soc., 1921, **43**, 315); 2-*fluoromesitylene*, m.p. —36·7°, b.p. 167–169° (*E. T. McBee* and *R. E. Leech*, Ind. Eng. Chem., 1947, **39**, 393). 2,4-*Dichloromesitylene*, m.p. 59° (*idem*, *ibid*., 1944, **36**, 1011); 2,4-*dibromomesitylene*, m.p. 64°, 2,4-*diiodomesitylene*, m.p. 82°, 2,4-*difluoromesitylene*, m.p. —18·5°, b.p. 168–169°. 2,4,6-*Trichloro*-, m.p. 203°, -*tribromo*-, m.p. 226°, -*triiodo*- m.p. 280°, 2,4,6-*trifluoro-mesitylene*, m.p. 68°, b.p. 169° (*G. C. Finger et al.*, J. Amer. chem. Soc., 1951, **73**, 149).

Mono- and 2,5-di-chloro derivatives of *p*-cymene are described by *E. Gysin* (Helv., 1926, **9**, 66), mono- and di-bromo derivatives by *M. T. Bogert* and *J. R. Tuttle* (J. Amer. chem. Soc., 1916, **38**, 1353).

Monohalogen derivatives of 1,3-diethylbenzene, see *J. E. Copenhaver* and *E. E. Reid* (*ibid*., 1927, **49**, 3157) and *H. R. Snyder et al.* (*ibid*., 1941, **63**, 3280). 5-*Bromo*-, m.p. 26°, and 5,6-*dibromo*-, m.p. 205°, -1,2,3,4-*tetramethylbenzene* (*L. I. Smith*, *ibid*., 1929, **51**, 3004).

2. Alkylbenzenes with halogen in the side-chain

(a) Methods of preparation

(1) Halogenation of alkylbenzenes, in the absence of catalysts which promote nuclear substitution (see pp. 243 *et seq.*), occurs in the side-chain preferentially. The reaction is favoured by exposure to white or ultra-violet light and by boiling the liquid which is being halogenated. Although *o*-xylene will only give *o*-dichloromethylbenzotrichloride, presumably because of

TABLE 7

HALOGENATED XYLENES

Compound	Fluoro-		Chloro-		Bromo-		Iodo-	
	M.p.°	B.p.°	M.p.°	B.p.°	M.p.°	B.p.°	M.p.°	B.p.°
3-Halogeno-*o*-xylene				193[1]		213[2]		111/11 mm.[3]
4-Halogeno-*o*-xylene				191[1]		215[3]		111/11 mm.
3,4,-Dihalogeno-*o*-xylene			9	234[4]	7	277		
3,5-Dihalogeno-*o*-xylene			4	226[5]				
3,6-Dihalogeno-*o*-xylene			68	227[4]				
4,5-Dihalogeno-*o*-xylene			76	240[4]	88	278[7,9]		
3,4,5-Trihalogeno-*o*-xylene			96[6]		105[8]			
3,4,6-Trihalogeno-*o*-xylene			48[6]		86[8]			
3,4,5,6-Tetrahalogeno-*o*-xylene			227[6]			257–261[5,10]		
2-Halogeno-*m*-xylene				187[11]				
4-Halogeno-*m*-xylene		143–144[26]		188[12,13]		203[15]		111/14 mm.
5-Halogeno-*m*-xylene		145[27]		191[13]				107/11 mm.
2,4-Dihalogeno-*m*-xylene				221[11]	—8	269[16]		
4,5-Dihalogeno-*m*-xylene					11	257		
4,6-Dihalogeno-*m*-xylene			68	222[14]		68–72[16,17]	72	
2,4,5-Trihalogeno-*m*-xylene			95[11]		87[19]			
2,4,5,6-Tetrahalogeno-*m*-xylene			219[11]		248–251[10,18]		128	
2-Halogeno-*p*-xylene	—6[28]			183		205[20]		217[20]
2,5-Dihalogeno-*p*-xylene			71[21]		75[22]			
2,6-Dihalogeno-*p*-xylene					36[24]			
2,3,5-Trihalogeno-*p*-xylene			96[24]		89[25]			
2,3,5,6-Tetrahalogeno-*p*-xylene		143–144[29]	223[24]		253[18]			

1 *A. Krüger*, Ber., 1885, **18**, 1755.
2 *K. v. Auwers*, Ann., 1919, **419**, 116.
3 *Auwers, ibid.*, 1921, **422**, 164.
4 *L. E. Hinkel et al.*, J. chem. Soc., 1928, 1876; 1934, 1946.
5 *Hinkel, ibid.*, 1928, 2532.
6 *Idem, ibid.*, 1920, **117**, 1302.
7 *S. Coffey*, Rec. Trav. chim., 1923, **42**, 433.
8 *F. M. Jaeger* and *J. J. Blanksma, ibid.*, 1906, **25**, 354.
9 *W. H. Mills* and *I. G. Nixon*, J. chem. Soc., 1930, 2524.
10 *A. W. Crossley* and *N. Renouf, ibid.*, 1921, **119**, 274.
11 *I. G. Farbenind.*, G.P. 491,220/1927.
12 *V. Grignard et al.*, Ann. Chim., 1915, [ix], **4**, 45.
13 *A. Klages* and *E. Knoevenagel*, Ber., 1924, **27**, 3019.
14 *A. Claus*, J. pr. Chem., 1890, [ii], **41**, 556.
15 *G. T. Morgan* and *E. A. Coulson*, J. chem. Soc., 1929, 2208.
16 *O. Jacobsen*, Ber., 1888, **21**, 2824.
17 *R. L. Datta*, J. Amer. chem. Soc., 1916, **38**, 2550.
18 *S. F. Birch* and *W. S. Norris*, J. chem. Soc., 1926, 2549.
19 *Jaeger* and *Blanksma*, Rec. Trav. chim., 1906, **25**, 360.
20 *Morgan* and *Coulson*, J. chem. Soc., 1929, 2211.
21 *A. S. Wheeler* and *M. Moyse*, J. Amer. chem. Soc., 1924, **46**, 2574.
22 *Wheeler* and *E. W. Constable, ibid.*, 1923, **45**, 2000.
23 *E. Bures*, Chem. Ztbl., 1937, II, 3743.
24 *Bures, ibid.*, 1929, I, 507.
25 *Jaeger* and *Blanksma*, Rec. Trav. chim., 1906, **25**, 362.
26 *G. Balz* and *G. Schiemann*, Ber., 1927, **60B**, 1186.
27 *N. Lofgren, B. Lundquist* and *S. Lindstrom*, Acta. chim. Scand., 1955, **9**, 1079.
28 *R. L. Fern* and *C. A. van der Werf*, J. Amer. chem. Soc., 1950, **72**, 4809.
29 *A. K. Barbour et al.*, J. chem. Soc., 1961, 808.

steric hindrance, toluene, *m*- and *p*-xylenes, mesitylene and ethylbenzene give products with fully chlorinated side-chains (*P. G. Harvey et al.*, J. appl. Chem., 1954, **4**, 319). Some nuclear chlorination occurs at the same time and some viscous condensation products are also produced. The chlorination of the side-chains of polyalkylbenzenes containing methyl, ethyl and isopropyl groups has also been described by *E. T. McBee* and his co-workers (Ind. Eng. Chem., 1947, **39**, 384–399) who in some cases added water to the reaction mixture and who increased the temperature of reaction progressively as chlorination proceeded. More conveniently the alkylbenzene is boiled with sulphuryl chloride in presence of benzoyl peroxide or some other diacyl peroxide (*M. S. Kharasch* and *H. C. Brown*, J. Amer. chem. Soc., 1939, **61**, 2142; cf. *Kharasch, Brown* and *T. H. Chao, ibid.*, 1940, **62**, 3435):

$$C_6H_5\cdot CH_3 + SO_2Cl_2 \xrightarrow[15\ min]{(PhCOO)_2} C_6H_5\cdot CH_2Cl\ (80\%\ \text{yield}) + SO_2 + HCl$$

This process proceeds rapidly in the dark.

Another efficient reagent for free-radical chlorination is *tert*-butyl hypochlorite which reacts with toluene at 40° in the presence of azobisisobutyronitrile to give benzyl chloride in 84 % yield (*C. Walling* and *B. B. Jacknow, ibid.*, 1960, **82**, 6108). The degree of halogenation can be controlled by regulating the amount of halogenating agent used, but some mono- and tri-chloro compound will, for example, contaminate a di-chloro derivative. Similarly *tert*-butyl hypoiodite can be used to effect free-radical substitution of benzylic hydrogens by iodine. The reagent is prepared by reaction of *tert*-butyl hypochlorite with mercuric iodide at 0° in a halogenated solvent. Addition of the hydrocarbon to the solution followed by irradiation gives useful yields of iodinated product (*D. D. Tanner* and *G. C. Gidley, ibid.*, 1968, **90**, 808):

$$HgI_2 + 2\,(CH_3)_3COCl \rightarrow 2\,(CH_3)_3COI + HgCl_2$$

$$(CH_3)_3COI + C_6H_5{\cdot}CH_3 \rightarrow C_6H_5{\cdot}CH_2I\ (34\,\%) + (CH_3)_3COH$$

Bromination of the side-chains of toluene and its derivatives has been carried out using elemental bromine in carbon disulphide or carbon tetrachloride solution with either white or ultra-violet light (*J. R. Sampey, F. S. Fawcett* and *B. A. Morehead, ibid.*, 1940, **62**, 1839) and by heating alkyl benzenes with carbon tetrabromide in sealed tubes at 150–180° for about eight hours. In this way toluene was converted, stepwise, into benzyl bromide, benzal bromide and benzotribromide; ethylbenzene was converted into α-phenyl-α-bromoethane and α-phenyl-α,β-dibromoethane (*W. H. Hunter* and *D. E. Edgar, ibid.*, 1932, **54**, 2025). *N*-Bromosuccinimide converts toluene into benzyl bromide in 64 % yield when the reaction is carried out in carbon tetrachloride and in the presence of benzoyl peroxide (*H. Schmid* and *P. Karrer*, Helv., 1946, **29**, 573). The use of this reagent is reviewed by *C. Djerassi* (Chem. Reviews, 1948, **43**, 271) and by *L. Horner* and *E. H. Wenkelmann* ("Newer Methods of Preparative Organic Chemistry", Vol. III, Academic Press, New York, 1964).

(2) Replacement of the OH group in the aryl alcohols by halogen by the action of hydrogen halides, thionyl chloride or phosphorus halides:

$$PhCH_2OH \rightarrow PhCH_2Cl$$

(3) Reaction of aromatic hydrocarbons with formaldehyde and hydrogen chloride in presence of zinc chloride gives nuclear chloromethyl derivatives. The formaldehyde may be replaced by chloromethyl ether and the catalyst by stannic chloride. The applicability of this reaction to benzenoid compounds

has been studied by *H. Stephen et al.* (J. chem. Soc., 1920, **117**, 510). Arylbromomethanes have been prepared similarly. A survey of chloromethylation is given by *R. C. Fuson* and *C. H. McKiever* (Org. Reactions, Vol. I, 1942, p. 63). Brief mention of bromo- and iodo-methylation is also made by these authors. Iodomethylation with iodomethyl methyl ether is described by *S. V. Rogozhin et al.* (Bull. Acad. Sci. U.S.S.R., 1966, 1448).

(4) Arylolefins will add hydrogen halides or halogens to give mono- and di-halogeno compounds (*J. J. Drysdale* and *W. D. Phillips*, J. Amer. Chem. Soc., 1957, **79**, 319; *L. I. Smith* and *M. M. Falkof*, Org. Synth., Coll. Vol. III, p. 350, 1955):

$$PhCH{=}CF_2 + Br_2 \rightarrow PhCHBr{\cdot}CBrF_2$$

$$PhCH{=}CHPh + Br_2 \rightarrow PhCHBr{\cdot}CHBrPh$$

At high dilution in halogenated solvent below -78^0, aryl olefins add fluorine, the product being predominantly *cis* (*R. F. Merritt*, J. Amer. chem. Soc., 1967, **89**, 609):

$$PhCH{=}CH{\cdot}CH_3 + F_2 \xrightarrow[(5\text{ mm pressure }F_2)]{-78^0,\text{ halosolvent}} PhCHF{\cdot}CHF{\cdot}CH_3$$

$$CH_2{=}C_6H_3F{=}CH_2 + I_2 \longrightarrow ICH_2{-}C_6H_3F{-}CH_2I$$

(*M. Szwarc*, J. Polym. Sci., 1951, **6**, 319);

$$PhCH{=}CH_2 + HBr \rightarrow PhCHBr{\cdot}CH_3$$

(*F. Ashworth* and *G. N. Burkhardt*, J. chem. Soc., 1928, 1798).

(5) Replacement of the other halogens by fluorine or iodine is a convenient method for the preparation of fluoro compounds and many iodo compounds (see Vol. I A, p. 482).

(6) Many dihalogeno compounds have been prepared by reaction of aromatic aldehydes and alkyl aryl ketones with phosphorus pentachloride, thionyl chloride or certain acyl chlorides:

$$PhCHO \rightarrow PhCHCl_2$$

For references to the preparative methods see Vol IA, p. 482. Aromatic aldehydes, ketones, and acids can be converted into fluorides of the types

$ArCHF_2$, $ArCF_2Ar$, $ArCF_2Alk$, $ArCF_3$ on heating with sulphur tetrafluoride under pressure. Other fluorides such as HF, BF_3, AsF_3 act as catalysts (*W. C. Smith*, J. Amer. chem. Soc., 1960, **82**, 543):

$$C_6H_5\cdot CO_2H + SF_4 \xrightarrow[\text{temp.}]{\text{room}} C_6H_5\cdot COF \rightarrow C_6H_5\cdot CF_3 + HF + SOF_2$$

If phenylsulphur trifluoride is used instead of sulphur tetrafluoride the reactions can be carried out in glass apparatus (*W. A. Sheppard, ibid.*, 1962, **84**, 3058):

$$C_6H_5\cdot CHO + PhSF_3 \xrightarrow{100^0} \underset{(70\text{–}80\,\%)}{C_6H_5\cdot CHF_2} + C_6H_5SOF$$

(7) Decarbonylation (analogous to p. 246) is also a possible synthetic route (see *J. Tsuji et al.*, Tetrahedron Letters, 1965, 4565):

$$PhCH_2\cdot COCl \xrightarrow{1\,\%\ Pd/C,\ 220^0,\ 3h} PhCH_2Cl\ (42\,\%)$$

(8) Dihalogenoalkanes undergo Friedel-Crafts reaction with arenes but the usual products are without halogen since the primary product is generally more reactive than the starting materials (*G. A. Olah* and *S. J. Kuhn*, J. org. Chem., 1964, **29**, 2317; and *Olah* (Ed.), "Friedel-Crafts and Related Reactions", Pt.I, Vol. 2, 1964). However, using mild conditions with aluminium chloride or hydrogen fluoride as catalyst it is sometimes possible to stop the reaction after the first step (*D. L. Ransley*, J. org. Chem., 1966, **31**, 3595):

$$ClCH_2\cdot CHCl\cdot CH_2\cdot CH_3 + C_6H_6 \xrightarrow[3\ h]{AlCl_3\ (6.6\ \text{moles}\ \%)} \underset{(80\cdot 2\,\%)}{CH_3\cdot \overset{\overset{\displaystyle Ph}{|}}{C}H\cdot CH_2\cdot CH_2Cl}$$

(b) Properties and reactions

These compounds are exemplified by benzyl chloride, $PhCH_2Cl$, benzylidene chloride, $PhCHCl_2$,benzotrichloride $PhCCl_3$, 1-chloro-1-phenylethane, $PhCHCl\cdot CH_3$, and styrene dibromide $PhCHBr\cdot CH_2Br$. They are sharply distinguished from the nuclear halogeno compounds in that the halogen has the general reactions of that in alkyl halides. Thus the halogen in benzyl chloride is readily replaced by a large variety of groups, OH, OAlk, NH_2, SH, CN, NO_2. The $CHCl_2$ group in benzylidene chloride can be hydrolysed to CHO and the CCl_3 group in benzotrichloride to CO_2H. Some reactions are discussed below.

(i) Benzyl halides

Benzyl fluoride, $PhCH_2F$, m.p. —35°, b.p. 139·9°, $d_4^{25.3}$ 1·0228, $n_D^{25.3}$ 1·4892, in contrast to the other benzyl halides is not lachrymatory; it is a mobile liquid with a benzene-like odour. It is prepared by the action of mercuric fluoride on benzyl bromide in chloroform. A number of substituted benzyl fluorides are prepared similarly (*J. Bernstein et al.*, J. Amer. chem. Soc., 1948, **70**, 2310). Alternatively it can be prepared in 59 % yield by fluoromethylation of benzene using paraformaldehyde and anhydrous hydrogen fluoride at about 80° in the presence of zinc chloride (*Olah* and *A. Pavlath,* Acta chim. Acad. Sci. Hung., 1953, **3**, 425; C. A., 1955, **49**, 2384) or by heating benzyl bromide with potassium fluoride in glycol solution (*K. Fukin* and *N. Kitano,* J. chem. Soc., Japan. ind. Chem. Sect., 1955, **58**, 352). It polymerises readily in presence of catalysts. It is not changed by boiling with zinc dust and alcohol; it is only partially hydrolysed to benzyl alcohol after boiling for 6 hours with 10 % aqueous alkali carbonate; benzyl ethyl ether is formed in poor yield after boiling it for 1 hour with alcoholic sodium ethoxide. Nitration with acetic anhydride and nitric acid yields chiefly 4-nitrobenzyl fluoride with some 2- and 3-nitro compounds (*C. K. Ingold* and *E. H. Ingold,* J. chem. Soc., 1928, 2249).

Benzyl chloride, $PhCH_2Cl$, is a lachrymatory liquid, m.p. —41°, b.p. 179·3°, d_4^{15} 1·1043, $n_D^{17.4}$ 1·5391. It is prepared conveniently by passing the necessary amount of chlorine into boiling toluene exposed preferably to strong sunlight or ultra-violet light (see *J. Silberrad et al., ibid.*, 1925, **127**, 1724, 2449).

It is obtained by the action of paraformaldehyde or aqueous formaldehyde, hydrochloric acid and zinc chloride on benzene (*Stephen et al., ibid.*, 1920, **117**, 518; *G. Blanc,* Bull. Soc. chim. Fr., 1923, [iv], **33**, 314; *L. Bert,* Compt. rend., 1928, **186**, 373). Conditions for the manufacture of benzyl chloride by this method are given by *A. Ginsberg et al.* (Ind. Eng. Chem., 1946, **38**, 478). Monochlorodimethyl ether can be used instead of formaldehyde and hydrogen chloride (*M. Sommelet,* Compt. rend., 1913, **157**, 1445). Another method of formation is the reaction of sulphuryl chloride with toluene at about 105–110° (*A. Wohl,* G.P. 139,552/1901).

Reactions. Benzyl chloride is slowly hydrolysed by boiling water, more quickly by alkalis, to benzyl alcohol. Measurements of the rate of hydrolysis by aqueous solvents are recorded by *G. Harker* (J. chem. Soc., 1924, **125**, 500) and *S. C. J. Olivier et al.* (Rec. Trav. chim., 1922, **41**, 304, 639; 1929, **48**, 234). Benzyl esters are formed by heating benzyl chloride with the sodium salts of the appropriate acids; benzyl ethers by reaction with sodium alkoxides. Reaction with the sodium salts of phenols in acetone or alcohol proceeds normally to give aryl benzyl ethers; in non-dissociating solvents such as toluene the benzyl group enters the nucleus (*L. Claisen et al.*, Ann., 1925, **442**, 210; *W. F. Short* and *M. L. Stewart,* J. chem. Soc., 1929, 554).

Benzyl chloride is converted into mono-, di- and tri-benzylamines by reaction with ammonia; it reacts similarly with primary and secondary amines and with tertiary amines forms benzylammonium salts. Benzyl cyanide is formed

TABLE 8

BENZYL HALIDES WITH HALOGEN SUBSTITUENTS IN THE NUCLEUS

Compound	Fluoro		Chloro		Bromo		Iodo	
	M.p.°	B.p.°/mm	M.p.°	B.p.°/mm	M.p.°	B.p.°/mm	M.p.°	B.p.°/mm
o-Halogenobenzyl fluoride				58·5/9[1]		84–85/18[1]		
m-Halogenobenzyl fluoride				65/11[1]		83/13[1]		68/2·5[1]
p-Halogenobenzyl fluoride			3–4[1]	72–73/16[1]	33[1]	80–81·5/10[1]	52[1]	75/4[1]
o-Halogenobenzyl chloride		68/16[2]		94/15[2]		111/15[2]	29[6]	
m-Halogenobenzyl chloride		68/15[2]		110/25[3]	23	112/15	27[6]	
p-Halogenobenzyl chloride		76/20[2]	30[4]		36	137/37[2]	53[2,6]	
o-Halogenobenzyl bromide				103/13[5]	31[5]		55[6]	
m-Halogenobenzyl bromide				120/24[5]	40[5]		50[6]	
p-Halogenobenzyl bromide			51[5]		63[5]		80[6]	
o-Halogenobenzyl iodide		91/8[7]			47[5]			
m-Halogenobenzyl iodide		98/12[7]			42[5]			
p-Halogenobenzyl iodide	34–35[7]	102/9			73[5]			

TABLE 9

METHYLBENZYL HALIDES

Compound	Fluoride	Chloride	Bromide	Iodide
o-Methylbenzyl		b.p. 198°	m.p. 21°; b.p. 102°/11 mm	m.p. 33–34°
m-Methylbenzyl	b.p. 48°/8·5 mm	196°	97–98°/8 mm	
p-Methylbenzyl	m.p. 20°, b.p. 47°/8 mm	80°/2 mm	100°/9 mm	46–47°

References Table 8

1 *J. Bernstein, J. S. Roth* and *W. T. Miller*, J. Amer. chem. Soc., 1948, **70,** 2310.
2 *G. M. Bennett* and *B. Jones*, J. chem. Soc., 1935, 1818.
3 *S. C. J. Olivier*, Rec. Trav. chim., 1922, **41,** 420; *J. Kenner* and *E. Witham*, J. chem. Soc., 1921, **119,** 1460.
4 *G. Blanc*, Bull. Soc. chim. Fr., 1923, [iv], **33,** 317.
5 *J. B. Shoesmith* and *R. H. Slater*, J. chem. Soc., 1926, 219.
6 *Olivier*, Rec. Trav. chim., 1923, **42,** 522.
7 *P. S. Varma, K. S. Venkataraman* and *P. M. Nilkantiah*, J. Indian chem. Soc., 1944, **21,** 112; C. A., 1945, **39,** 1395.

by heating it with an aqueous or alcoholic solution of potassium cyanide (*M. Gomberg* and *C. C. Buchler*, J. Amer. chem. Soc., 1920, **42,** 2059). It forms benzyl derivatives with β-keto-esters and 1,4-diketones (*G. T. Morgan* and *C. J. A. Taylor*, J. chem. Soc., 1925, **127,** 801).

Benzyl chloride forms a Grignard compound easily by reaction with magnesium in ether. The most suitable conditions are given by *H. Gilman* and *R. J. Vanderwal* (Bull. Soc. chim. Fr., 1929, [iv], **45,** 642). For the reaction of benzyl chloride with Grignard compounds see *M. S. Kharasch* and *O. Reinmuth*, "Grignard Reactions of Nonmetallic Substances", Constable, London, 1954.

Benzyl bromide, m.p. —3·9°, b.p. 210°, is formed by the bromination of boiling toluene; by the action of bisbromomethyl ether on benzene and zinc chloride at room temperature (*Stephen et al.*, J. chem. Soc., 1920, **117,** 520); or from toluene and *N*-bromosuccinimide (*H. Schmid* and *P. Karrer*, Helv., 1946, **29,** 573). It is slowly decomposed by water at room temperature, more rapidly on heating. For measurements of rate of replacement of the halogen in halogenobenzyl bromides and ω-bromoxylenes see *J. B. Shoesmith et al.* (J. chem. Soc., 1924, **125,** 2281; 1926, 221).

Benzyl iodide, m.p. 24°, is strongly lachrymatory. It is best prepared from benzyl chloride and sodium iodide in acetone or methyl alcohol (*F. C. Whitmore* and *E. N. Thurman*, J. Amer. chem. Soc., 1929, **51,** 1497; *G. H. Coleman* and *C. R. Hanser*, *ibid.*, 1928, **50,** 1196).

The constants of some nuclear halogenated benzyl chlorides and bromides are given in Table 8 and, in Table 9, those of methylbenzyl halides.

For observations on the preparation of pure specimens of some of these compounds and measurements of reactivity see *Bennett* and *Jones* (*loc. cit.*).

2,4-, 2,6- and 3,5-*Dibromobenzyl chlorides*, m.p. 33°, 66° and 50°, respectively (*Olivier*, Rec. Trav. chim., 1926, **45,** 302). 2,4,6-*Tribromobenzyl chloride*, m.p. 153° (*M. Henraut*, Bull. Soc. chim. Belg., 1924, **33,** 132).

(*ii*) *Benzylidene halides; benzotrihalides*

Benzylidene fluoride, $PhCHF_2$, b.p. 139·9°, n_D^{20} 1·4578, is formed by heating benzylidene chloride with antimony pentafluoride (*T. van Hove*, Bull. Acad. roy. Sci. Belg., 1913, 1078) or by reduction of benzochlorodifluoride with sodium amalgam in alcohol or by hydrogenation (*F. Swarts*, *ibid.*, 1920, 408). It is hydrolysed by sulphuric acid at 200° to benzaldehyde.

Benzylidene chloride, $PhCHCl_2$, a lachrymatory liquid, m.p. —16·4°, b.p. 205·2°, d^{16} 1·295, $n_D^{19.4}$ 1·5515, is prepared by the chlorination of toluene at its boiling point. It is also formed by heating toluene with two molecular proportions of phosphorus pentachloride at 190–195° (*A. Colson* and *H. Gautier*, Ann. Chim., 1887, [vi], **11**, 19); by the reaction of benzaldehyde with phosphorus pentachloride (*Cahours*, Ann., 1849, **70**, 39) or thionyl chloride (*F. Loth* and *A. Michaelis*, Ber., 1894, **27**, 2548; cf. *P. Hoering* and *F. Baum*, *ibid.*, 1908, **41**, 1918) or with phosgene at 120–130° (*R. Kempf*, J. pr. Chem., 1870, [ii], **1**, 412) or oxalyl chloride at 130–140° (*H. Staudinger*, Ber., 1909, **42**, 3976).

Reactions. It is hydrolysed to benzaldehyde by heating with water at 140–160° or more conveniently with aqueous calcium hydroxide or other alkaline reagents or with acids. Reaction with sodium ethoxide in alcohol gives benzaldehyde diethyl acetal, $PhCH(OEt)_2$; sodium sulphide gives β-tristhiobenzaldehyde; for hydrogenation using nickel or palladium-calcium carbonate see *M. Busch* and *H. Stöve*, *ibid.*, 1916, **49**, 1068), *W. Borsche* and *G. Heimbürger* (*ibid.*, 1915, **48**, 457). Nitration by nitric acid or by nitric acid in acetic anhydride gives a mixture of *m*- (~35 %), *p*- (~40 %) and *o*- (~25 %) nitro derivatives (*B. Flürscheim* and *E. L. Holmes*, J. chem. Soc., 1928, 1622; *A. F. Holleman et al.*, Rec. Trav. chim., 1914, **33**, 25).

Benzylidene bromide, $PhCHBr_2$, b.p. 156°/23 mm, is a fuming liquid easily hydrolysed by water to hydrobromic acid and benzaldehyde. It is prepared by the action of phosphorus tribromide on benzaldehyde (*T. Curtius* and *E. Quendenfeldt*, J. pr. Chem., 1898, [ii], **58**, 389).

Benzotrichloride, m.p. —4·8°, b.p. 220·8°/760 mm, *d* 1·38, is best prepared by the chlorination of boiling toluene containing 2 % of phosphorus trichloride (*F. Swarts*, Bull. Soc. chim. Belg., 1922, **31**, 376). It is hydrolysed to benzoic acid under a variety of conditions, by hot water, concentrated sulphuric acid, or dilute aqueous alkali. With one molar proportion of water and a little ferric chloride it is hydrolysed to benzoyl chloride. Nitration gives ~65 % of *m*-nitro derivative (*Holleman et al.*, Rec. Trav. chim., 1914, **33**, 32; *Flürscheim* and *Holmes*, J. chem. Soc., 1928, 1613).

By reaction with antimony pentafluoride *ω,ω-dichloro-ω-fluorotoluene*, $PhCCl_2F$, b.p. 178–180°, is formed. At higher temperatures *ω-difluoro-ω-chlorotoluene*, $PhCF_2Cl$, b.p. 142·6° and *ω,ω,ω-trifluorotoluene* (**benzotrifluoride**), b.p. 103°, $n_D^{13.3}$ 1·4149, are formed (*W. T. Miller* and *A. H. Fainberg*, J. Amer. chem. Soc., 1957, **79**, 4164). Yields of benzotrifluoride can be improved and the formation of tars avoided by using anhydrous hydrogen fluoride at ~100° and 15 atm. pressure instead of antimony trifluoride (*J. H. Brown, C. W. Suckling* and *W. B. Whalley*, J. chem. Soc., 1949, 595). Benzotrifluoride is stable to water and aqueous alkali and is not altered by heating with iron or copper at 350°. Hydrolysis to benzoic acid is effected by heating with an excess of concentrated hydrobromic acid at 150° in presence sulphur dioxide. The *ortho*- and *para*-hydroxyl and amino substituted benzotrifluorides are readily hydrolysed to benzoic acid derivatives by aqueous sodium hydroxide whereas the *meta*-isomers are resistent to refluxing alkali, successful hydrolysis requiring prolonged heating with concentrated sulphuric acid. In contrast, when the reaction is carried out under ultraviolet irradiation the *ortho-* and *meta-*isomers are readily hydrolysed in acid media, whereas the *para-*isomer reacts only slowly to give non-acidic products (*E. Heilbronner et al.*, Tetrahedron Letters, 1968, 3845). Nitration by concentrated nitric acid gives *m*-nitrobenzotrifluoride (*Swarts, loc. cit.*). The trifluoromethyl group of benzotrifluoride is readily converted to the more reactive trichloro- and tribromo-methyl groups by the action of the appropriate aluminium halide (*L. M. Yagupol'skii* and *N. V. Kondratenko* (J. gen. Chem. U.S.S.R., 1967, **37**, 1686, and references therein).

For nuclear halogenated benzylidene chlorides and benzotrichlorides see, *inter alia, T. de Crauw*, Rec. Trav. chim., 1931, **50**, 773; *V. Villiger*, Ber., 1928, **61**, 2598; *L. Anschütz*, Ann., 1927, **454**, 99. For the fluorides see *Pavlath* and *Leffler*, "Aromatic Fluorine Compounds", Reinhold, New York, 1962. Nucleophilic substitution in these compounds has been discussed by *S. M. Shein et al.* (J. gen. Chem. U.S.S.R., 1968, **38**, 480, 485; and 1965, **35**, 1944).

(*iii*) *Higher halogenophenylalkanes*

1-**Chloro**-1-**phenylethane**, *α-phenylethyl chloride*, $PhCHCl{\cdot}CH_3$, b.p. 81°/17 mm. is prepared by the chlorination of boiling ethylbenzene in sunlight (*E. E. Turner et al.*, J. chem. Soc., 1927, 1159, 1163), or more conveniently from the corresponding alcohol by the action of hydrogen chloride or thionyl chloride (*A. M. Ward, ibid.*, 1927, 452). It is hydrolysed by water at 50° to α-phenylethyl alcohol.

1-*Bromo*-1-*phenylethane*, $PhCHBr{\cdot}CH_3$, b.p. 96°/16 mm, is prepared by the bromination of boiling ethylbenzene and by the action of hydrogen bromide on the alcohol or on styrene (*Ashworth* and *Burkhardt, loc. cit.*).

2-**Chloro**-1-**phenylethane**, *β-phenylethyl chloride*, $PhCH_2{\cdot}CH_2Cl$, b.p. 96°/23 mm, is obtained from the corresponding alcohol by the action of hydrogen chloride or thionyl chloride (*Ward, loc. cit.*).

2-*Bromo*-1-*phenylethane*, b.p. 109°/22 mm, is prepared by the action of hydrogen bromide or phosphorus pentabromide on the alcohol (*R. P. Linstead* and *L. T. D. Williams, ibid.*, 1926, 2745; *J. B. Shoesmith* and *R. J. Connor, ibid.*, 1927, 1771).

1,1-**Dichloro**-1-**phenylethane**, $PhCCl_2{\cdot}CH_3$, readily eliminates hydrogen chloride to give α-chlorovinylbenzene, $PhCCl{:}CH_2$. Further action of alcoholic potash yields phenylacetylene (*J. U. Nef*, Ann., 1899, **308**, 266).

1,2-**Dichloro**-1-**phenylethane**, *styrene dichloride*, $PhCHCl{\cdot}CH_2Cl$, b.p. 115°/15 mm, from styrene and chlorine in chloroform (*H. Biltz*, Ann., 1897, **296**, 275). For 2,2-dichloro-1-phenylethane and tri-, tetra- and penta-chloro-1-phenylethane see *Biltz* (*loc. cit.*).

1,2-**Dibromo**-1-**phenylethane**, *styrene dibromide*, m.p. 73°, from bromine and styrene in chloroform (*W. L. Evans* and *L. H. Morgan*, J. Amer. chem. Soc., 1913, **35**, 56). For its conversion to α-bromostyrene and phenylacetylene see *M. Bourguel* (Ann. Chim., 1925, [x], **3**, 228); *Ashworth* and *Burkhardt* (*loc. cit.*).

1,2-**Dibromo**-1-**phenylpropane**, $PhCHBr{\cdot}CHBr{\cdot}CH_3$, m.p. 66° (*R. C. Huston* and *D. D. Sager*, J. Amer. chem. Soc., 1926, **48**, 1957; *E. Spath* and *G. Koller*, Ber.,

1925, **58**, 1269), when heated with methyl alcohol at 100° gives 2-bromo-1-methoxy-1-phenylpropane.

2,3-*Dibromo-1-phenylpropane*, $PhCH_2 \cdot CHBr \cdot CH_2Br$, b.p. 115°/5 mm (*Huston* and *Sager, loc. cit.*; *M. Porcher*, Bull. Soc. chim. Fr., 1922, [iv], 31, 339).

1,2-*Dibromo-2-phenylpropane*, $PhCBrMe \cdot CH_2Br$, b.p. 114°/7 mm (*P. Ramart* and *P. Amagat*, Ann. Chim., 1927, [x], **8**, 304).

For the (chlorobutyl)benzenes see *J. von Braun et al.*, Ber., 1917, **50**, 53; 1910, **43**, 2846; 1912, **45**, 397, 2177, 2517; *P. A. Levene* and *L. A. Mikeska*, J. biol. Chem., 1926, **70**, 362; *J. B. Conant* and *W. R. Kirner*, J. Amer. chem. Soc., 1924, **46**, 242; *S. S. Rosander* and *C. S. Marvel, ibid.*, 1928, **50**, 1495; *I. E. Muskat* and *K. A. Huggins, ibid.*, 1929, **51**, 2500.

Aryl derivatives of group VIII metal compounds, prepared from group VIII metal salts and mercury and tin aryls, react with olefins in the presence of cupric halides to form 2-arylalkyl halides, the yields being better for the chloro compounds than for the bromo compounds (*R. F. Heck*, J. Amer. chem. Soc., 1968, **90**, 5538):

$$PhHgCl + LiCl + 2\,CuCl_2 \xrightarrow[\text{20 h, HOAc/H}_2\text{O/Li}_2\text{PdCl}_4\text{ Cat.}]{C_2H_4,\ \text{2 atmospheres,}} PhCH_2 \cdot CH_2Cl\ (70\%)$$

(*iv*) *Compounds with more than one side-chain*

In general these compounds may be prepared by extension of the methods discussed above. In addition, trimerizations of alkynes have also been reported (*J. F. Harnis et al.*, J. org. Chem., 1960, **25**, 633 and *H. C. Brown et al., ibid.*, 1960, **25**, 634):

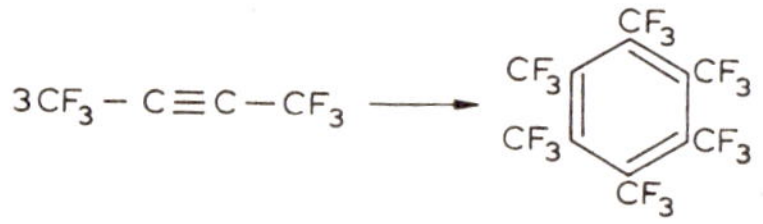

$$3CF_3-C\equiv C-CF_3 \longrightarrow C_6(CF_3)_6$$

The hexakis(trifluoromethyl)benzene undergoes ready photochemical valence-bond isomerization to the Dewar, prismane and benzvalene isomers, which exhibit surprising stability (*D. M. Lemal, J. V. Staros* and *V. Austel*, Abstract 45, First North East Regional Meeting of the American Chemical Society, 1968; *idem.*, J. Amer. chem. Soc., 1969, **91**, 3373; and *M. G. Barlow, R. N. Haszeldine* and *R. Hubbard*, Chem. Comm., 1969, 202).

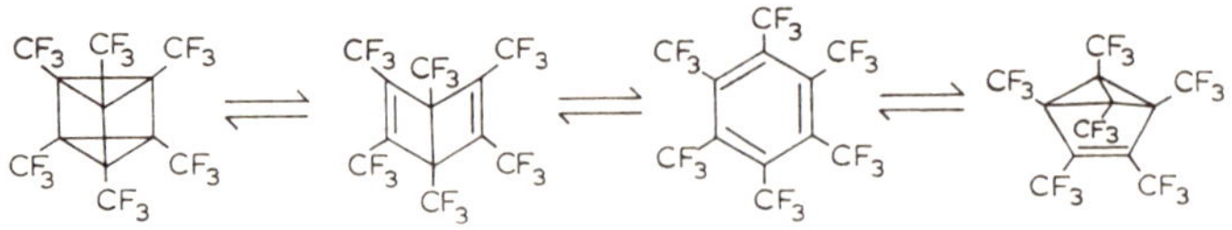

Hexafluorobut-2-yne can also react with arenes to give 1,2-bis-(trifluoromethyl)arenes (*C. C. Krespan, B. C. McKusick* and *T. L. Cairns*, J. Amer. chem. Soc., 1961, **83**, 3428; and *R. S. H. Liu, ibid.*, 1968, **90**, 215):

$$C_6H_6 + CF_3{-}C{\equiv}C{-}CF_3 \longrightarrow C_6H_4(CF_3)_2 + C_2H_2$$

The reaction proceeds *via* a 1,4-cycloaddition product, 2,3-bis(trifluoromethyl) bicyclo[2.2.2]octa-2,5,7-triene, which can be isolated in low yield, together with the product of further addition–elimination reactions 1,2,4,5-tetrakis-(trifluoromethyl)benzene.

3. Benzene homologues halogenated in both nucleus and side-chain

Photochemical chlorination of the side-chains of benzene homologues in which all of the nuclear hydrogen atoms have already been replaced by chlorine, proceeds readily but it is not possible to carry it to completion. Methyl groups are converted to dichloromethyl and ethyl to isomeric forms of trichloro-ethyl (*P. G. Harvey et al.*, J. appl. Chem., 1954, **4**, 319; *M. Ballester, C. Molinet* and *J. Castaner*, J. Amer. chem. Soc., 1960, **82**, 4254). These workers could not repeat an earlier reported preparation of perchloromesitylene (*E. T. McBee* and *R. E. Leech*, Ind. Eng. Chem., 1947, **39**, 393) by photochlorination. Similarly, up to 1954 it was not possible, using ordinary nuclear chlorination catalysts, to complete the chlorination of the nucleus of benzene homologues with fully chlorinated side-chains. One hydrogen, in the *ortho* position, could not be removed without chlorinolysis of the side-chain (*Harvey et al.*, J. appl. Chem., 1954, **4**, 325). It was in this year that *Ballester* and his co-workers (see p. 250) introduced the improved Silberrad catalyst ($AlCl_3 + S_2Cl_2$) with sulphuryl chloride and prepared perchlorotoluene. Thereafter several perchloroalkaryl compounds with saturated and unsaturated side-chains were made by Ballester's group but *m*-xylene and mesitylene have resisted complete chlorination.

Perchloro-alkyl groups forming all or part of the side-chains of the perchloroalkaryl compounds are easily hydrolysed by concentrated sulphuric acid at 100^0 to carboxyl groups:

$$C_6Cl_5{\cdot}CCl{=}CCl{\cdot}CCl_3 \rightarrow C_6Cl_5{\cdot}CCl{=}CCl{\cdot}CO_2H$$

Prolonged treatment with chlorine, under either nuclear chlorination conditions or photolysis, usually causes chlorinolysis with the production of hexachlorobenzene:

$$C_6Cl_5 \cdot CCl_3 \xrightarrow{Cl_2 + h\nu (\text{or catalyst})} C_6Cl_6 + CCl_4$$

Treatment with iodide ion in acetic acid or stannous chloride in dioxane causes dechlorination, sometimes with coupling, to give products with unsaturated side-chains:

$$2\ C_6Cl_5 \cdot CCl_3 \xrightarrow[\text{or } Sn^{2\oplus}]{I^{\ominus}} C_6Cl_5 \cdot CCl{=}CCl \cdot C_6Cl_5$$

Those with trichloromethyl side-chains will condense with trichloroethylene to give perchloropropenylbenzenes (see later).

Perfluoro-homologues of benzene have been prepared by the defluorination of the corresponding perfluoro-homologues of cyclohexane by passing them, with nitrogen, over iron gauze heated to temperatures varying between 300° and 600° (*B. Gething et al.*, Nature, 1959, 183, 590); by heating compounds such as pentachlorobenzotrifluoride with potassium fluoride (*A. K. Barbour, M. W. Buxton*, and *G. Fuller*, "Aromatic Fluorine Compounds", Kirk-Othmer Encyclopaedia of Chemical Technology, Interscience Publishers, New York; 1966; and *J. P. Kolenko et al.*, J. gen. Chem. U.S.S.R., 1967, 37, 1606) and also by the polyfluoroalkylation of fluorinated aromatic systems; this reaction, resulting from the nucleophilic attack of a perfluoroalkyl anion generated from a perfluoroalkene and an alkali metal fluoride on the fluorinated aromatic compound can be regarded as the fluorocarbon chemistry analogue of the Friedel-Crafts reaction of hydrocarbon chemistry (*R. D. Chambers et al.*, J. chem. Soc., C, 1968, 2221):

$$\text{(F)(CF}_3\text{)C}_6\text{F}_{10} \xrightarrow[\text{Fe gauze}]{500^\circ} C_6F_5 \cdot CF_3$$

$$C_6Cl_5 \cdot CF_3 \xrightarrow{KF} C_6F_5 \cdot CF_3$$

$$C_6F_6 + R_fCF{=}CF_2 \xrightarrow{F^{\ominus}} C_6F_5 \cdot CF(CF_3)R_f$$

(R_f = perfluorinated alkyl)

Trifluoromethyl groups in these compounds, which are attached to the nucleus, can be hydrolysed by concentrated sulphuric acid to carboxyl groups. The fluorine atoms in the nucleus are highly susceptible to attack by nucleophilic reagents such as lithium aluminium hydride, hydrazine, ammonia, sodium methoxide, methyl-lithium, substitution occurring in the *para* position with respect to a perfluoroalkyl group when this is possible and in the *ortho* position as the next best alternative.

(a) Polyhalogenated benzene homologues

All of the data below on chlorinated compounds have been taken from the papers of *Ballester* and his co-workers which have already been referred to. Except where stated, compounds described result from direct chlorination, by Ballester's method, of the appropriate starting material with fully halogenated side-chains and non-halogenated nucleus.

2,3,4,5-**Tetrachloro-1-trichloromethylbenzene,** m.p. 122–123°. Hydrolysis with concentrated sulphuric acid gives 2,3,4,5-*tetrachlorobenzoic acid*, m.p. 190–191° (*Harvey et al.*, J. appl. Chem., 1954, **4**, 325).

2,3,5,6-**Tetrachloro-1-trichloromethylbenzene,** m.p. 68–69°. Hydrolysis with concentrated sulphuric acid gives 2,3,5,6-*tetrachlorobenzoic* acid, m.p. 180–181°.

Perchlorotoluene, m.p. 71·5–72·5°. Hydrolysis gives pentachlorobenzoic acid in 98 % yield. Photo- or nuclear chlorination gives hexachlorobenzene. Photolysis with u.v. light and in carbon tetrachloride solution gives **perchloroethylbenzene** presumably *via* a perchlorobenzyl radical. Condensation of perchlorotoluene and trichloroethylene using a mixture of anhydrous aluminium chloride, anhydrous calcium chloride in methylene chloride saturated with dry hydrogen chloride gives cis-**perchloropropenylbenzene**, m.p. 146–148°, the trans *isomer*, m.p. 163–165°, and a substance $C_{11}HCl_{13}$, m.p. 183–185°.

$$C_6Cl_5{\cdot}CCl_3 + CHCl{=}CCl_2 \longrightarrow \underset{(55\%)}{(Cl)(C_6Cl_5)C{=}C(Cl)(CCl_3)} + \underset{(16\%)}{(Cl)(C_6Cl_5)C{=}C(CCl_3)(Cl)}$$

Treatment with iodide ion in acetic acid or stannous chloride in dioxane gives cis-**perchlorostilbene**, m.p. 230·5–231·5° (48 %), the trans *isomer* (37 %), m.p. 400°, and some (9 %) α-*H*-heptachlorotoluene* (*Ballester* and *J. Rosa*, Tetrahedron, 1960, **9**, 156); in this respect, perchlorotoluene differs from benzotrichloride and partly chlorinated benzotrichlorides which, under the same reaction conditions, give α,α,α′,α′-tetrachlorobibenzyl and its partly chlorinated derivatives, respectively.

Perchloro-p-xylene, m.p. 153–154·5°. Hydrolysis gives *tetrachloroterephthalic acid*, m.p. 335° (decomp.). It is thermochromic, becoming colourless at —20°. With iodide ion in acetic acid it gives a quantitative yield of **perchloro-p-xylylene,** m.p. 168–168·5°, which reverts to the starting material on photochlorination with chlorine and white light in acetic acid containing hydrogen chloride:

* For indicated hydrogen mode of nomenclature see I.U.P.A.C. Rule A 21.6.

$$Cl_3C\text{-}C_6Cl_4\text{-}CCl_3 \underset{Cl_2,\,h\nu}{\overset{I^\ominus}{\rightleftharpoons}} Cl_2C{=}C_6Cl_4{=}CCl_2$$

Condensation with trichloroethylene under the same conditions as for perchlorotoluene gives four isomeric perchloro-1,4-bispropenylbenzenes in about 30 % yield.

2,5-**Dichloro**-1,4-**bistrichloromethylbenzene,** m.p. 202–204°. Hydrolysis gives 2,5-*dichloroterephthalic acid,* m.p. 305°.

1,3,5-**Trichloro**-2,4,6-**trisdichloromethylbenzene,** m.p. 178·5–180·5°, is obtained by the photochlorination of 1,3,5-trichloromesitylene together with 1,2,3,5-**tetrachloro**-4,6-**bis-dichloromethylbenzene,** m.p. 99–101°.

Perchlorostyrene, m.p. 92·5–96·5°, is obtained in high yield by treating perchloroethylbenzene in acetic acid with iodide ion:

$$C_6Cl_5\cdot C_2Cl_5 \xrightarrow{I^\ominus} C_6Cl_5\cdot CCl{=}CCl_2$$

It is also obtained by nuclear chlorination of pentachloroethylbenzene or α,β,β-trichlorostyrene:

$$C_6H_5CCl{=}CCl_2 \xrightarrow[AlCl_3,\,S_2Cl_2]{SO_2Cl_2} C_6Cl_5\cdot C_2Cl_5 + C_6Cl_5\cdot CCl{=}CCl_2$$

cis- and trans-**Perchloropropenylbenzene** prepared as described above give, on hydrolysis, the same *perchlorocinnamic acid,* m.p. 206–208°. On treatment with ferrous chloride or stannous chloride in dioxane they each give the same mixture of two isomers of *perchloro*-1,6-*diphenylhexatriene* with m.ps. 304–304·5° and 141–146°. Presumably the reaction occurs *via* the perchloro-3-phenyl-2-propenyl radical.

$$C_6Cl_5\cdot CCl{=}CCl\cdot CCl_3 \rightarrow C_6Cl_5\cdot CCl{=}CCl\cdot CCl_2\cdot \rightarrow (C_6Cl_5\cdot CCl{=}CCl\cdot CCl_2)_2$$
$$\rightarrow C_6Cl_5\cdot CCl{=}CCl\cdot CCl{=}CCl\cdot CCl{=}CCl\cdot C_6Cl_5$$

Irradiation of either of the two isomers with ultra-violet light gives a third *isomer,* m.p. 335–336°, which is also the main product when either of the original isomers is heated to 500°.

Perchlorobis-1,4-propenylbenzenes. Four isomers are obtained from perchloro-*p*-xylene as described above. On hydrolysis they give isomers of perchloro-*p*-phenylphenylenediacrylic acid.

Perfluorotoluene, b.p. 102–103°, n_D^{19} 1·3680, is prepared by defluorination of perfluoromethylcyclohexane at 500° or by heating pentachlorobenzotrifluoride with potassium fluoride at 560–590°. With concentrated sulphuric acid at 160°

for 12 hours it gives *pentafluorobenzoic acid*, m.p. 104–105°. With lithium aluminium hydride, methyllithium, potassium benzenethiolate, hydrazine, ammonia, sodium hydrogen sulphide and sodium ethoxide it gives $CF_3 \cdot C_6F_4X$ where X is H, Me, SPh, $NHNH_2$, NH_2, SH or OEt and is *para* with respect to the CF_3 group (*D. J. Alsop, J. Burdon* and *J. C. Tatlow*, J. chem. Soc., 1962, 1801; *Burdon, W. B. Hollyhead* and *Tatlow, ibid.*, 1965, 5152).

Perfluoroethylbenzene, b.p. 114–115°, n_D^{19} 1·3620, is prepared by defluorination of perfluoroethylcyclohexane at 490°; with nucleophiles such as those above substitution occurs in position 4 (*B. R. Letchford, C. R. Patrick* and *Tatlow, ibid.*, 1964, 1776).

Perfluoro-o-xylene, b.p. 128°, n_D^{19} 1·3670, is prepared by defluorination of perfluoro-1,2-dimethylcyclohexane at 460°; or alternatively by pyrolysis of the Diels-Alder adduct of hexafluorobutyne-2 with perfluorocyclohexa-1,3-diene (*L. P. Anderson, W. J. Feast* and *W. K. R. Musgrave*, J. chem. Soc., C, 1969, 211). Hydrolysis with fuming sulphuric acid (20 % of SO_3) at 150° for 12 hours gives *tetrafluorophthalic* acid, m.p. 153–154° (*Gething, Patrick* and *Tatlow, ibid.*, 1961, 1574). With nucleophiles, substitution in the ring occurs in the position *para* to one of the trifluoromethyl groups (*E. V. Aroskar et al., ibid.*, 1964, 2975).

Perfluoro-m-xylene, b.p. 119–120°, n_D^{17} 1·3621, is prepared by defluorination of perfluoro-1,3-dimethylcyclohexane at 460°. Hydrolysis with fuming sulphuric acid (20 % SO_3) gives *tetrafluoroisophthalic* acid, m.p. 203° (decomp.) (*P. Robson et al., ibid.*, 1964, 5748); with nucleophiles, substitution in the ring occurs *ortho* to one trifluoromethyl group and *para* to the other (*Aroskar et al., ibid.*, 1965, 2658).

Perfluoro-p-xylene, b.p. 122, n_D^{17} 1·3621, is prepared by defluorination of perfluoro-1,4-dimethylcyclohexane at 460°. Hydrolysis with oleum gives *tetrafluoroterephthalic acid*, m.p. 283–284° (*Gething, Patrick* and *Tatlow, loc. cit.*). Nucleophilic substitution occurs *ortho* to one of the trifluoromethyl groups (*Aroskar, et al.*, J. chem. Soc., 1964, 2975).

Octafluorostyrene, b.p. 121°, n_D^{22} 1·3550, is prepared in about 15 % yield by the defluorination of decafluoroethylbenzene at 600° (*Letchford et al.*, Chem. and Ind., 1962, 1472) or, in much better yield by preparing a pentafluorophenylcopper complex from pentafluorophenylmagnesium bromide and cuprous chloride and causing this to react with trifluorovinyl iodide (*C. Tamborski, R. de Pasquale* and *E. J. Soloski*, Abstracts of papers at 5th Int. Symp. on Fluorine Chemistry, Moscow, 1969):

$$C_6F_5MgBr + CuCl \rightarrow C_6F_5Cu \text{ complex} \xrightarrow{CF_2=CFI} C_6F_5 \cdot CF{=}CF_2$$

Chapter 4

Nuclear Hydroxy Derivatives of Benzene and its Homologues

A. R. FORRESTER AND J. L. WARDELL

1. Monohydric phenols

The term "phenol", specifically signifying the monohydroxy derivative of benzene, is applied generally to all derivatives of benzene and its homologues having nuclear hydroxyl groups. According to the number of hydroxyl groups present the compounds are termed mono-, di-, tri-hydric phenols, etc.

Simple phenols exist exclusively in the enolic form I, rearrangement to a keto form II being accompanied by a decrease in resonance energy of about 35 kcal·mole^{-1}. Spectroscopic measurements indicate that dihydric and

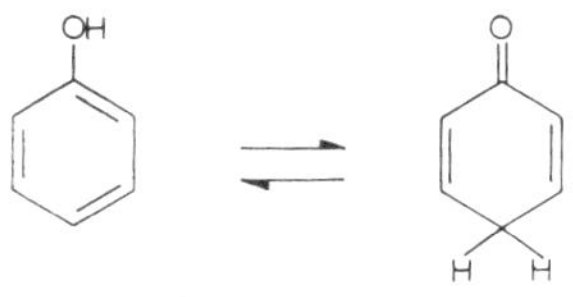

(I) (II) 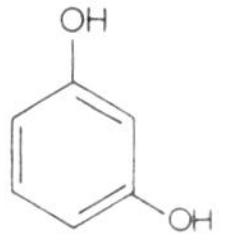 (III) 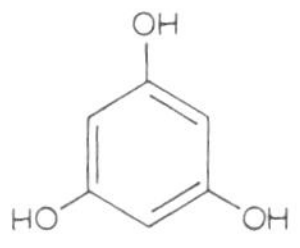 (IV)

trihydric phenols such as resorcinol (III) and phloroglucinol (IV) also exist (in the solid state and in solution) in the enolic form although the latter will react in the oxo form in alkaline solution giving, for example, a trioxime. Certain phenols with strongly electron-withdrawing *ortho-* or *para-*substituents, however, do exist as an equilibrium mixture of tautomeric forms, *e.g.*, *p*-nitrosophenol (V) (see p. 346) (*R. H. Thomson*, Quart. Reviews, 1956, 10, 27; *V. V. Ershov* and *G. A. Nikiforov*, Russ. chem. Reviews, 1966, 35, 817).

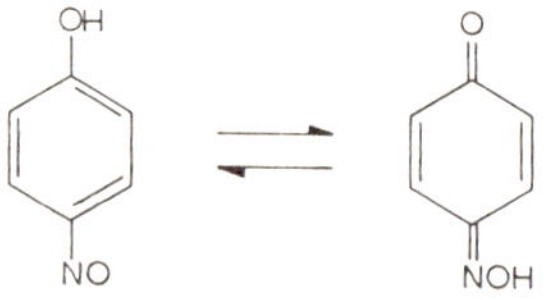

(V)

Occurrence. Although complex phenols are widespread in Nature, simple phenols are relatively uncommon. Phenol itself is found in mammalian urine, in pine needles and cones, and in the essential oil of tobacco leaves, while *p*-cresol, 4-allylphenol, thymol and carvacrol have been isolated from the essential oils of plants and alkylnitrophenols from the odoriferous oil from rocks and clays (*I. J. Bear* and *Z. H. Kranz*, Austral. J. Chem., 1969, **22**, 2343). Of the dihydric phenols, quinol is the most abundant; the trihydric phenol phloroglucinol has been found free only recently. Crude petroleum also contains some simple phenols (*J. B. Harborne* and *N. W. Simmonds* in "Biochemistry of Phenolic Compounds", Academic Press, London, 1964, Ch. 3). Thermal decomposition of lignin gives a variety of phenols (*F. E. Brauns* and *D. A. Brauns*, "The Chemistry of Lignins", Academic Press, London, 1960, supp. Vol., Ch. 22). The commercial usefulness of this method is at present under investigation. Simple phenols are present in the tar formed in the dry distillation of peat and coal, coal tar being an important industrial source of phenol and its homologues (cf. p. 309).

(a) Methods of preparation

Only methods in which a hydroxyl (or acyloxyl) group is introduced into the benzene ring are described at this stage. Preparations in which substituted phenols undergo reactions not involving the hydroxyl group are discussed later. The industrial preparation of phenol has been reviewed by *G. Szonyi* in "Homogeneous Catalysis", American Chemical Society, Washington, 1968, p. 53.

(*i*) *From sulphonic acids*

Fusion of an arenesulphonic acid with alkali at about 300° gives the corresponding phenol, generally in good yield. The reaction is a simple S_N2-type aromatic substitution (*S. Oae et al.*, Bull. chem. Soc. Japan, 1966, **39**,

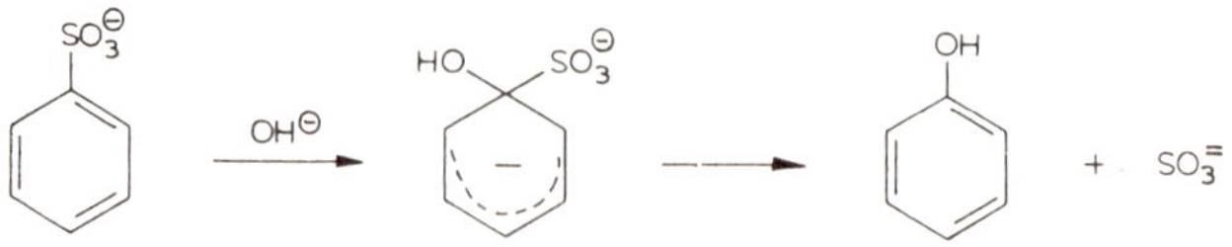

$$C_6H_5SO_3^{\ominus} \xrightarrow{OH^{\ominus}} [C_6H_5(OH)(SO_3^{\ominus})]^{-} \longrightarrow C_6H_5OH + SO_3^{=}$$

1212) and a typical procedure is described by *W. W. Hartman* (Org. Synth., Coll. Vol. 1, 1964, p. 175). The industrial importance (*e.g., O. Tyrer,* U.S.P., 2,451,996/1948) of this method is now decreasing (but see *S. W. Englund, R. S. Aries* and *D. F. Othmer,* Ind. Eng. Chem., 1953, **45**, 189). Fusion of arylsulphones with alkali also yields phenols (*Oae* and *N. Furukawa,* Bull· chem. Soc. Japan, 1966, **39**, 2260).

(ii) From aryl halides

This is an extremely important industrial method for the production of phenol. In the continuous process developed by Dow, chlorobenzene is hydrolysed by 10–20% aqueous caustic soda at 350–400° under pressure in a steel reactor (*L. A. Angello* and *W. H. Williams,* Ind. Eng. Chem., 1960, **52**, 894). In the related Raschig process, phenol is produced in two steps

$$C_6H_6 + HCl + \tfrac{1}{2}O_2 \rightarrow C_6H_5Cl + H_2O \qquad [1]$$

$$C_6H_5Cl + H_2O \rightarrow C_6H_5OH + HCl \qquad [2]$$

by passing a mixture of benzene, air, and hydrogen chloride through an aluminium–copper–iron hydroxide catalyst at about 200°. The hydrogen chloride produced at step [2] is recycled (*W. Mathes,* Angew. Chem., 1939, **52**, 591; *S. Raschig G.m.b.H.,* G.P., 588,649/1933; *R. M. Crawford,* Chem. Eng. News, 1947, **25**, 235).

Aryl halides with *ortho-* or *para*-electron-withdrawing substituents (NO_2, etc.) may be hydrolysed with aqueous alkali or converted into aryl ethers (which may be dealkylated to phenols) with alkoxide, under relatively mild conditions (*J. F. Bunnett* and *R. E. Zahler,* Chem. Reviews, 1951, **49**, 273). Other groups (NH_2) similarly activated may also be easily replaced (*Bunnett,* Quart. Reviews, 1958, **12**, 1).

(iii) From hydroperoxides

Decomposition of hydroperoxides of alkylbenzenes with acid gives phenol in good yield (*H. Hock* and *S. Lang,* Ber., 1944, **77**, 257; *M. S. Kharasch,*

$$\mathrm{Me_2C(Ph)\!-\!O\!-\!OH} \xrightarrow{H^{\oplus}} \mathrm{Me_2C(Ph)\!-\!O\!-\!\overset{\oplus}{O}H_2} \xrightarrow{-H_2O} \mathrm{Me_2\overset{\oplus}{C}\!-\!O\!-\!Ph}$$

$$\xrightarrow[-H^{\oplus}]{H_2O} \mathrm{Me_2C(OH)\!-\!O\!-\!Ph} \longrightarrow \mathrm{Me_2C{=}O} + \mathrm{HOPh}$$

A. Fono and *W. Nudenberg*, J. org. Chem., 1950, **15**, 748; *E. S. E. Hawkins*, "Organic Peroxides", Spon, London, 1961, p. 91). The acid decomposition of cumyl hydroperoxide at 45–65° is considered to be the most economical commercial process for the preparation of phenol (*Hawkins, loc. cit.*, p. 345; *B. V. Aller, R. H. Hall* and *R. N. Lacey*, B.P., 629,429/1949).

(iv) From aromatic carboxylic acids

Oxidative decarboxylation of the cupric salts of aromatic acids or of the acids in the presence of other cupric salts at 200–350°, with or without oxygen, yields either phenols or esters depending upon the solvent (*W. G. Toland*, J. Amer. chem. Soc., 1961, **83**, 2507; *W. W. Kaeding*, J. org. Chem., 1961, **26**, 3144). The manufacture of phenol from toluene *via* benzoic acid utilises this reaction (*Toland*, U.S.P., 2,762,838/1951; *Kaeding, R. O. Lindblom* and *R. G. Temple*, Ind. Eng. Chem., 1961, **53**, 805). Basic cupric salts of benzoic and toluic acids at 200–220° yield the corresponding *o*-hydroxy acids (*Kaeding* and *A. T. Shulgin*, J. org. Chem., 1962, **27**, 3551). At higher temperatures these decarboxylate rapidly to phenols (*B. R. Brown, D. L. Hammick* and *A. J. B. Scholefield*, J. chem. Soc., 1950, 778).

(v) From diazonium salts

Hydrolysis of diazonium salts to phenols in hot acid solution (*H. Zollinger*, "Azo and Diazo Chemistry", Interscience, New York, 1961, 138; *J. P. Lambooy*, J. Amer. chem. Soc., 1950, **72**, 5327) is considered to proceed by an S_N1-type mechanism (*D. F. Detar* and *A. R. Ballantine*, *ibid.*, 1958, **78**, 3916), the initially-formed phenyl cation combining rapidly with water (but see *E. S. Lewis* and *R. E. Holliday*, *ibid.*, 1966, **88**, 5043):

$$ArN_2^{\oplus}X^{\ominus} \rightarrow Ar^{\oplus} + N_2 + X^{\ominus} \xrightarrow{H_2O} ArOH$$

Hydrolysis with alkali is unsatisfactory because the diazonium salt couples with the phenoxide-anion formed.

(vi) From Grignard reagents

Autoxidation of an aryl Grignard reagent or aryl-lithium gives a mixture of products which includes the corresponding phenol usually in 10–50% yield (*G. Sosnovsky* and *J. H. Brown*, Chem. Reviews, 1966, **66**, 527; *T. G. Brilkina* and *V. A. Shushunov*, Russ. chem. Reviews, 1966, **35**, 613):

$$ArLi + O_2 \rightarrow ArOOLi$$

$$ArOOLi + ArLi \rightarrow 2\ ArOLi \xrightarrow{H_2O} 2\ ArOH$$

Grignard reagents may be converted into phenols more efficiently by treatment either with ethyl borate followed by hydrogen peroxide (*M. F. Hawthorne,* J. org. Chem., 1957, **22,** 1001) or with *tert*-butyl hydroperoxide (*S. O. Lawesson* and *N. C. Yang,* J. Amer. chem. Soc., 1959, **81,** 4230):

$$ArMgX \xrightarrow{B(OEt)_3} ArB(OH)_2 \rightarrow ArOH + H_3BO_3$$

(vii) Autoxidation of benzene

This subject has been reviewed recently (*E. T. Enisov* and *D. I. Metelitsa,* Russ. chem. Reviews, 1968, **37,** 656). The yield of phenol formed by oxidation of benzene with air in the gas phase is enhanced by using atomic oxygen (43% yield) (*F. Porter,* U.S.P., 2,392,875/1946) and by the addition of homogeneous catalysts such as iodine (yield 56%) (*R. A. Harman,* U.S.P., 2,382,148/1946) or more easily oxidised organic compounds such as cyclohexane (62% yield) (*M. Donald* and *M. Darlington,* Ind. Chem., 1958, **34,** 5).

Oxidation by air in aqueous solution with or without irradiation (of various types) is strongly catalysed by copper and iron salts. For example, phenol is formed in ~60% yield when mixtures of benzene and aqueous cupric sulphate are heated to 304^0 (*J. B. Braunworth,* U.S.P., 2,976,329/1961). The industrial usefulness of these oxidations is still being explored. Benzene may be economically converted into phenol, however, by first reducing it to cyclohexane, oxidising the cyclohexane to cyclohexanol and dehydrogenating the cyclohexanol. In a commercial process a mixture of cyclohexane and benzene is used in proportions such that the hydrogen produced in the oxidation stage is utilised in the reduction stage (*A. S. Banciu,* Chem. Proc. Eng., 1967, 31).

(viii) Oxygenation of aromatic compounds

(1) *Using peroxides and peroxycarbonates.* Electrophilic oxygenation of alkylbenzenes and alkyl aryl ethers has been achieved using hydrogen peroxide (*D. H. Derbyshire* and *W. A. Waters,* Nature, 1950, **165,** 401; *J. D. McClure* and *P. H. Williams,* J. org. Chem., 1962, **27,** 24; *K. Omura* and *T. Matsuura,* Tetrahedron, 1970, **26,** 255), aroyl peroxides (*D. Z. Denney, T. M. Valega* and *D. B. Denney,* J. Amer. chem. Soc., 1964, **86,** 46; *J. T. Edward, H. S. Chang* and *S. A. Samad,* Canad. J. Chem., 1962, **40,** 804); *P. Kovacic, C. G. Reid* and *M. J. Brittain,* J. org. Chem., 1970, **35,** 2152) and peroxycarbonates (*Kovacic* and *M. E. Kurz, ibid.,* 1966, **31,** 2011, 2459; *G. A. Razuvaev, N. A. Kartashova* and *N. S. Boguslavskaya,* J. gen. Chem. U.S.S.R., 1964, **34,** 2108) in the presence of Brönsted or Lewis (usually aluminium chloride) acids. Di-isopropyl peroxydicarbonate ($Pr^iOCOO \cdot OCOOPr^i$)

decomposes homolytically, however, if cupric chloride is the catalyst (*Kovacic et al.*, J. Amer. chem. Soc., 1968, **90**, 1818). Thermal decomposition of *m*-nitrobenzenesulphonyl peroxide in aromatic solvents without acid (thought to be an ionic process) gives good yields of esters (56–70%) from which the phenols are easily generated (*R. L. Dannley* and *G. E. Corbett*, J. org. Chem., 1966, **31**, 153).

(2) *Using peroxy acids.* Cationic hydroxylation of alkylbenzenes, (*R. D. Chambers, P. Goggin* and *W. K. R. Musgrave*, J. chem. Soc., 1959, 1804), aryl ethers (*McClure* and *Williams*, J. org. Chem., 1962, **27**, 627) and other aromatic compounds (*A. J. Davidson* and *R. O. C. Norman*, J. chem. Soc., 1964, 5404) with trifluoroperoxyacetic acid gives, in some cases, the corresponding phenol in moderate yield. Better results have been obtained using trifluoroperoxyacetic acid : boron trifluoride etherate, mesitylene, for example, gives mesitol in 88% yield (*H. Hart* and *C. A. Buehler*, J. org. Chem., 1964, **29**, 2397).

(3) *Using Fenton's reagent.* Although Fenton's (ferrous ion and hydrogen peroxide) and Udenfriend's (ferrous ion, oxygen or hydrogen peroxide, ascorbic acid and ethylenediaminetetra-acetic acid) reagents are useful model systems for the study of biological hydroxylating processes their synthetic use is limited (*Norman* and *J. R. L. Smith* in "Oxidases and Related Redox Systems", Wiley, New York, 1965, Vol. 1, p. 131; *L. S. Boguslavskaya*, Russ. chem. Reviews, 1965, **34**, 503 and refs. cited therein). Usually, yields of monohydric phenols are low because the hydroxylating agent(HO·) reacts with other substituent (alkyl) or with the initial product (*G. H. Williams*, "Homolytic Aromatic Substitution", Pergamon, London, 1960, p. 110; *Smith* and *Norman*, J. chem. Soc., 1963, 2897). Acetanilide, however, has been *ortho*-hydroxylated in 46% yield by this method (*C. T. Clark, R. D. Downing* and *J. B. Martin*, J. org. Chem., 1962, **27**, 4698). Hydroxylation by irradiation of the substrate dissolved or suspended in water (*G. Stein* and *J. Weiss*, J. chem. Soc., 1949, 3245 and *Williams*, *loc. cit.*, p. 110) or by treatment with pernitrous acid (*R. B. Heslop* and *P. L. Robinson*, J. chem. Soc., 1954, 1271) also give low yields of phenols.

(*ix*) *By thallation of aromatic hydrocarbons*

In this method of considerable promise, the hydrocarbon, *e.g.* toluene, is first treated with thallium(III) trifluoroacetate and the resultant arylthallium di-trifluoroacetate treated successively with lead tetra-acetate in trifluoroacetic acid and triphenylphosphine to form the aryl trifluoroacetate, which is eventually hydrolysed with a base to the phenol: in this way toluene can be converted into *p*-cresol in 62% yield (*E. C. Taylor et al.*, J. Amer. chem. Soc., 1970, **92**, 3522).

$$ArH + Tl(O\cdot CO\cdot CF_3)_3 \xrightarrow[20^0]{C_3F\cdot CO_2H} Ar\text{–}Tl(O\cdot CO\cdot CF_3)_2 \xrightarrow[\text{(ii)}Ph_3P]{\text{(i)}Pb(O\cdot CO\cdot CF_3)_4}$$

$$[ArO\cdot CO\cdot CF_3] \xrightarrow{NaOH} ArOH$$

(*x*) *Other methods*

Phenols are formed in many other ways but generally these are less useful either because the starting materials are not readily available or the yields are poor, *e.g.*, by heating 2,5- or 3,5-dienoic acids with acidic or basic catalysts (*G. P. Chiusoli* and *G. Agnès*, Proc. chem. Soc., 1963, 310; Ital. P. 695,436/1965); from methylated hexoses by treatment with sodium in liquid ammonia (*G. V. Davydova* and *N. N. Shorygina*, Doklady Chem., 1964, 154, 12); by ozonolysis of cinnamic esters (*E. Spath, M. Pailer* and *G. Gergelly*, Ber., 1940, 73, 795); by the photolysis of a wide range of compounds, particularly unsaturated cyclic ketones (*J. Streith, B. Danner* and *C. Signalt*, Chem. Comm., 1967, 979; *A. Small*, J. Amer. chem. Soc., 1964, 86, 2091; *H. F. Zimmerman* and *D. I. Schuster*, *ibid.*, 1962, 84, 4527); and by condensation of 1,3-diketones with ketonic esters (*V. Prelog, O. Metzler* and *O. Jeger*, Helv., 1947, 30, 675).

(*b*) *Properties*

(*i*) *Spectra*

Introduction of a hydroxyl group into the benzene ring shifts the u.v. absorption bands to longer wavelength and intensifies them; phenol absorbs at 210·5 nm (ε = 6,200) and 270 nm (ε = 1,450). The bathochromic shift which occurs on addition of base has been utilised in the determination of phenols (*D. Harvey* and *E. G. Penketh*, Analyst, 1957, 82, 498). The effect of substituents and solvent on the positions and intensities of these bands is discussed by *J. C. Dearden* and *W. F. Forbes* (Canad. J. Chem., 1959, 37, 1294).

The i.r. spectra of monohydric phenols, measured in dilute solution in a solvent with which there is little hydrogen bonding (CCl_4), normally show strong absorption in the range 3650–3590 cm^{-1} due to O–H stretching vibrations. Electron-withdrawing substituents in the *meta-* or *para*-positions decrease and electron-donating substituents increase ν_{max} relative to that of phenol; ν_{max} may be related to both the Hammett σ-values of the substituents and the pK_a values of the phenols. Polar *ortho*-substituents (NO_2, CHO) or solvents with which there is strong hydrogen bonding (acetone) lower ν_{max} to 3570–3450 cm^{-1}; the magnitude of the shift can be used to measure the strength of the hydrogen-bond formed. Phenols with a *tert*-

butyl *ortho*-substituent show two bands in the 3600 cm^{-1} region attributable to different conformational arrangements of the OH and Bu^t groups (*L. J. Bellamy*, "The Infra-Red Spectra of Complex Molecules", Methuen, London, 1957, Ch. 6, and "Advances in Infrared Group Frequencies", Methuen, London, 1968; *G. C. Pimentel* and *A. L. McClellan*, "The Hydrogen Bond", Reinhold, New York, 1960).

The chemical shift of the hydroxyl proton in substituted phenols lies in the range $\tau = -1{\cdot}0$ to $1{\cdot}5$ (measured in dimethyl sulphoxide). The shifts bear a linear relationship to the Hammett σ-values of the substituents even when the substituents are in the *ortho*-position (*J. G. Traynham* and *G. A. Knesel*, J. org. Chem., 1966, 31, 3350).

In the mass spectra of phenols the parent ion usually gives rise to the base peak although the M–1 peak may also be intense. Peaks corresponding to loss of CO and CHO are also prominent (*J. H. Beynon, R. A. Saunders* and *A. E. Williams*, "The Mass Spectra of organic Molecules", Elsevier, Amsterdam, 1968, p. 153).

(ii) Acidity

Phenol and its lower homologues are weak acids soluble in aqueous alkali but not in bicarbonate solution and are thus easily distinguished from simple aromatic carboxylic acids; the p*K*a values of phenol and benzoic acid are 9·98 and 4·18, respectively. Electron-withdrawing substituents, particularly if they are situated in the *ortho*- and/or *para*-positions enhance, and electron-donating groups suppress, the acid character of phenols. For example, *p*-nitrophenol has p*K*a = 7·15 and *p*-methoxyphenol p*K*a = 10·21. 2,4,6-Trinitrophenol, commonly known as picric acid (p*K*a = 0·71), is readily soluble in bicarbonate solution. Intramolecular hydrogen bonding also tends to reduce the acidity of phenols. The solubility of phenol homologues in alkali decreases as the size and number of the alkyl substituents increases. Phenols with bulky *para*-substituents are generally soluble only in hot concentrated alkali while polyalkylated phenols (pentamethylphenol) or phenols with bulky *ortho*-substituents are insoluble even on heating. Such phenols, termed "crypto" phenols, may be dissolved in "Claisen's alkali" [35% solution of potassium hydroxide in aqueous methanol (1:4)]. If the *ortho*-alkyl substituents are extremely large then the reactivity of the hydroxyl group may be completely suppressed, 2,4,6-tri-*tert*-pentylphenol, for example, will not react with sodium in liquid ammonia (*G. H. Stillson et al.*, J. Amer. chem. Soc., 1945, 67, 303; 1946, 68, 722).

The salts of phenols are termed phenolates or alternatively phenoxides; *e.g.* PhONa is sodium phenolate or sodium phenoxide in IUPAC nomenclature.

(c) Reactions of the hydroxyl group

(i) O-Alkylation and acylation

Phenols may be *O*-alkylated giving ethers (p. 318) and *O*-acylated giving esters (p. 330). Usually phenols are esterified by treatment with an acid chloride or anhydride since they do not react directly with carboxylic acids. Esters (*p*-nitrobenzoates and *p*-toluenesulphonates), ethers (aryloxyacetic acids) and aryl urethanes (PhOCONHAr, formed by treatment with aryl isocyanates) are also used to characterise phenols.

(ii) Substitution

The conversion of phenols into aryl chlorides by treatment with phosphorus pentachloride is effective only for nitrophenols (*H. A. Engelhardt* and *P. Latschinow,* Ber., 1870, **3**, 97). Other phenols give, among other products, triarylphosphates. *H. N. Rydon* and his coworkers (J. chem. Soc., 1957, 323) however, have developed a method in which the phenol is first heated with a triaryloxyphosphorus dihalide [usually *p*-(*tert*-butyl)phenyl] giving a tetra-aryloxyphosphorus monohalide which is then pyrolysed. Bromides and iodides may also be prepared by this route.

$$\mathrm{Ar'OH} + \mathrm{(ArO)_3PCl_2} \xrightarrow[-\mathrm{HCl}]{} \mathrm{(Ar'O)(ArO)_3PCl} \rightarrow \mathrm{Ar'Cl} + \mathrm{(ArO)_3PO}$$

In a related method an aryloxytriphenylphosphorus halide is formed from the phenol and triphenylphosphine dihalide, thermal decomposition of which gives the aryl halide in excellent yield (*L. Horner et al.*, Ber., 1962, **95**, 523; *G. A. Wiley et al.*, J. Amer. chem. Soc., 1964, **86**, 964; *J. P. Schaefer* and *J. Higgins,* J. org. Chem., 1967, **32**, 1607).

$$\mathrm{ArOH} + \mathrm{Ph_3PX_2} \xrightarrow[-\mathrm{HX}]{} \mathrm{(ArO)Ph_3PX} \rightarrow \mathrm{ArX} + \mathrm{Ph_3PO}$$

Bulky *ortho*-substituents alter the course of this reaction (*D. G. Lee,* Chem. Comm., 1968, 1554).

Although phenols may be converted into thiols in moderate yield by heating with phosphorus pentasulphide the reaction is of little synthetic value and is rarely used (*N. Hoshi* and *M. Saito,* Jap. P., 4717/1952). Treatment of phenols with thiols at 180° in the presence of hydrochloric acid gives the corresponding sulphides (*Oae* and *R. Kiritani,* Bull. chem. Soc. Japan, 1965, **38**, 1381). Conversion of phenols into thiols *via* their dialkylthiocarbamates has been described by *M. S. Newman* and *H. A. Karnes* (J. org. Chem., 1966, **31**, 3980) and into the corresponding sulphonic acids by *J. E. Cooper* and *J. M. Paul* (*ibid.*, 1970, **35**, 2046).

(iii) Oxidation

The extensive use of phenols as antioxidants for oils and fats (*G. Scott*, "Atmospheric Oxidation and Antioxidants", Elsevier, Amsterdam, 1965) and the realisation that many naturally occurring compounds such as lignin and certain alkaloids are the result of oxidative coupling of relatively simple phenols *in vivo*, has led to numerous and intensive studies in this field. The subject has been reviewed several times and only a broad outline of the scope and mechanisms of these oxidations will be attempted here. Further and more extensive cover is given by the following: *D. H. R. Barton* and *T. Cohen* ("Festschrift A. Stoll", Birkhäuser, Basle, 1957, p. 117); *R. G. R. Bacon* and *H. A. O. Hill* (Quart. Reviews, 1965, 19, 95); *A. I. Scott* (*ibid.*, 1965, 19, 1); *H. Musso* (in "Oxidative Coupling of Phenols", Arnold, London, 1967, p. 1); *Thomson* (in "Biochemistry of Phenolic Compounds", Academic Press, London, 1964, p. 1).

The products formed on oxidising monohydric phenols depend upon the position and nature of the ring substituents, the oxidizing agent and the conditions. Although a vast amount of information has been accumulated it is still not always easy to predict the course that a particular oxidation will take and reference should be made to the reaction list compiled by *Musso* (see above) for exceptional cases. Generally, however, for phenols with a free *ortho*- or *para*-position, one-electron oxidants give initially, resonance-stabilised aroxyl radicals (VI) (detected by e.s.r., see *Waters et al.*, J. chem. Soc., 1964, 213, 408) which may (a) dimerise (C–C) to give dihydroxybiphenyls or diphenoquinones, (b) couple (O–C) to give diaryl ethers or polymeric ethers or (c) couple with a second mole of oxidant giving a dienone which may decompose to a benzoquinone:

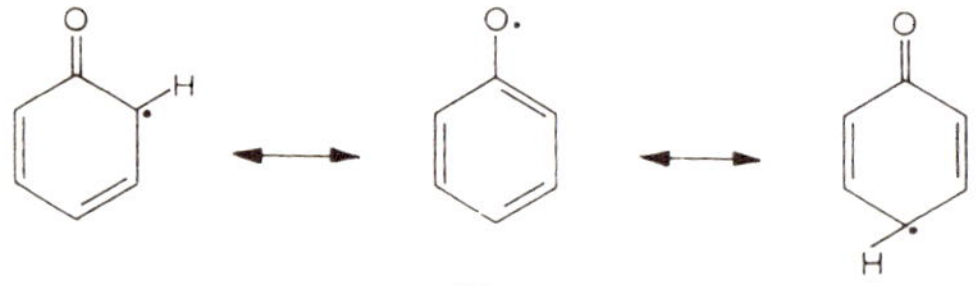

(VI)

Oxidations using alkaline ferricyanide or ferric chloride usually lead to C–C dimers, 2,6-dimethylphenol, for example, gives the diphenoquinone VII (*Bacon* and *A. R. Izatt*, J. chem. Soc., C, 1966, 791) and *p*-cresol the (*ortho-ortho*) dimer VIII and Pummerer's ketone (X) (*C. L. Chen, W. J. Connors* and *W. M. Shinker*, J. org. Chem., 1969, 34, 2966):

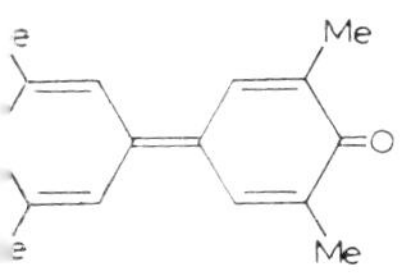

(VII) 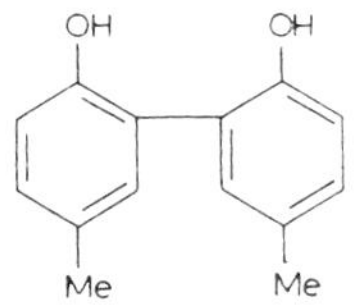 (VIII) 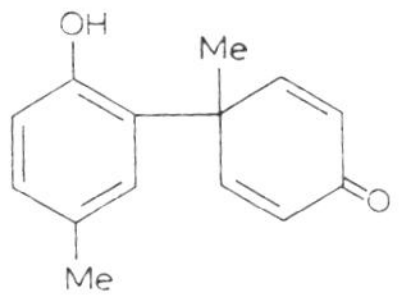 (IX) 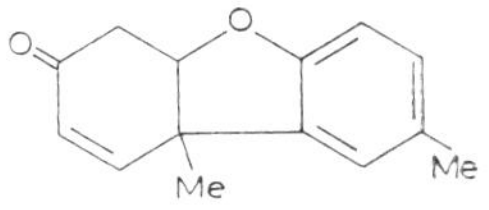 (X)

The latter product is structurally related to the naturally occurring usnic acid and is formed *via* the unsymmetrical dimer IX (*Barton, A. M. Deflorin* and *O. E. Edwards*, J. chem. Soc., 1956, 530). Many other oxidants including enzymes (*W. W. Westerfield* and *C. Lowe*, J. biol. Chem., 1942, **145**, 463) give similar results (*Musso, loc. cit.*).

Polyether formation is favoured using an excess of a metal oxide (Ag_2O, PbO_2, MnO_2) (*E. McNelis*, J. org. Chem., 1966, **31**, 1255) or by autoxidation, in the presence of a cuprous chloride–pyridine catalyst (*G. F. Endres et al.*, *ibid.*, 1966, **31**, 549).

Fremy's salt (*H.-J. Teuber* and *W. Rau*, Ber., 1952, **86**, 1036 and subsequent papers in this series), organic nitroxides (*A. R. Forrester* and *R. H. Thomson*, J. chem. Soc., C, 1966, 1844), perchloryl fluoride (*A. S. Kende* and *P. MacGregor*, J. Amer. chem. Soc., 1961, **83**, 4197), chromyl chloride (*J. A. Strickson* and *M. Leigh*, Tetrahedron, 1968, **24**, 5145) and hydroperoxides (*E. C. Horswill* and *K. U. Ingold*, Canad. J. Chem., 1966, **44**, 263) are useful reagents for the conversion of phenols to quinones. A common mechanism operates and is illustrated here for Fremy's salt:

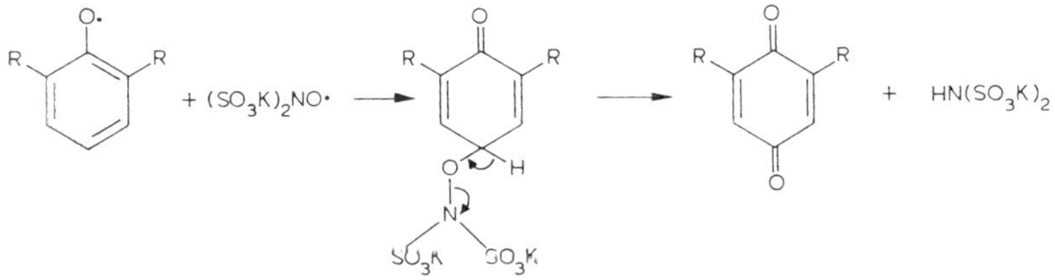

$$+ (SO_3K)_2NO\cdot \longrightarrow \longrightarrow + HN(SO_3K)_2$$

Other reagents such as trifluoroperoxyacetic acid (*Chambers, Goggin* and *Musgrave*, J. chem. Soc., 1959, 1804), peracetic acid (*D. Bryce-Smith* and *A. Gilbert*, J. chem. Soc., 1964, 873), nitric acid (*M. F. Ansell, B. W. Nash* and *D. A. Wilson*, *ibid.*, 1963, 3028), lead tetra-acetate (*G. W. K. Cavell et al.*, *ibid.*, 1954, 2785) and potassium periodate (*B. Sklarz*, Quart. Reviews, 1967, **21**, 3 and refs. cited therein) also give quinones but these reactions are almost certainly ionic.

Homolytic oxidation of 2,4,6-tri-*tert*-alkyl-, and triaryl-phenols and other tri-substituted phenols with bulky *ortho*-substituents gives relatively stable radicals whose properties have been intensively studied, mainly by *E. Müller* and his collaborators (*Müller, K. Ley* and *W. Kiedaisch,* Ber., 1954, 87, 1605 and subsequent papers in this series; *K. Dimroth, A. Berndt* and *H. Perst,* Org. Synth., 1970, 49, 116). Many of these radicals dimerise in the solid state giving quinol ethers, *e.g.*, XI, and combine with other reactive radicals (NO_2, $\cdot$OOH, O_2) giving cyclohexadienone derivatives

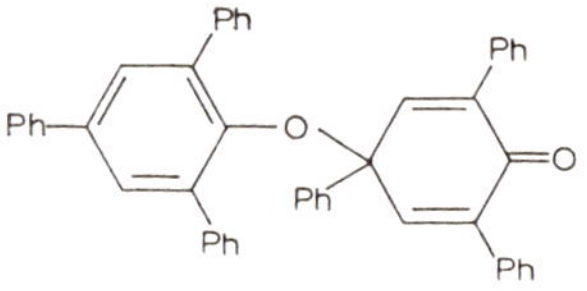

(XI) 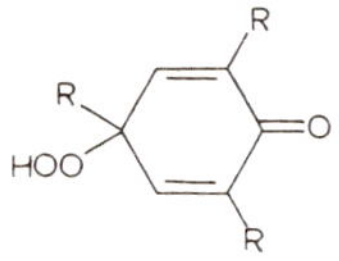 (XII)

such as XII which may react further. If a *para*-methyl substituent is present either dimerisation to a dihydroxydiphenylethane (XIII) and hence to a stilbene quinone (XIV) (*C. D. Cook, W. G. Nash* and *H. R. Flanagan,* J. Amer. chem. Soc., 1955, 77, 1783) or oxidation of the alkyl to a formyl group may occur (*H.-D. Becker,* J. org. Chem., 1965, 30, 982) depending upon the oxidant. 2,6-Di-*tert*-butyl-4-vinylphenols also give C–C dimers of type XV. Here coupling occurs *via* the β-carbon atom, and so provides a model system for lignan biosynthesis (see *H. Erdtman,* Svensk Kem. Tids., 1935, 47, 223; *N. J. Cartwright* and *R. D. Howarth,* J. chem. Soc., 1944, 535; *Müller,*

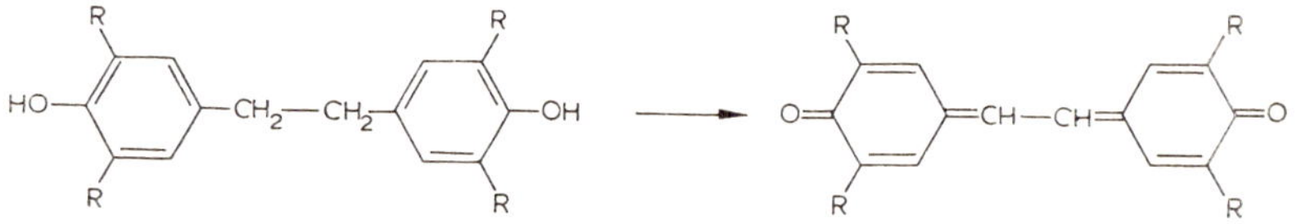

(XIII) (XIV)

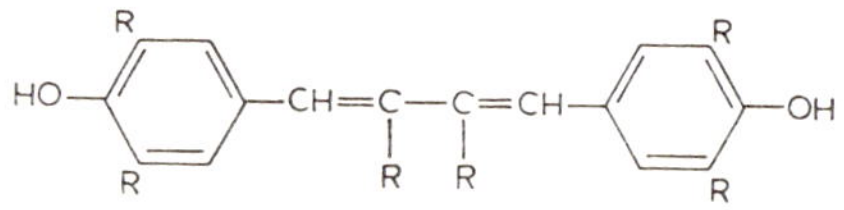

(XV) (XVI)

R = *e.g. tert*-Butyl

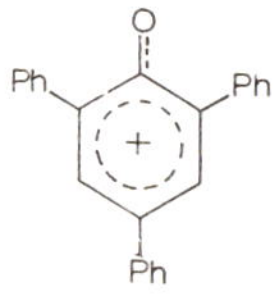

H. D. Spanagel and *A. Rieker,* Ann., 1965, 681, 141). 2,4,6-Triphenylphenoxyl may be oxidised further to a stable phenoxonium cation XVI (*K. Dimroth, W. Umbach* and *H. Thomas,* Ber., 1967, 100, 132). Photochemical oxidation of 2,6-di-alkylphenols using benzophenone as a sensitiser leads to either substituted triphenylcarbinols, fuchsones or, if a trace of acid is present, tetraphenylmethanes (*Becker*, J. org. Chem., 1967, **32**, 2115, 2124, 2131, 2136, 2140). Full accounts of the chemistry of stable aroxyl radicals is given by *Forrester, J. M. Hay* and *Thomson* ("Organic Chemistry of Stable Free Radicals", Academic Press, London, 1968, Ch. 7) and *E. R. Altwicker* (Chem. Reviews, 1967, **67**, 475).

Vigorous oxidation of phenols generally leads to rupture of the benzene ring; oxidation of *tert*-alkylated phenols with permanganate, for example, gives fatty acids. Phenol may also be degraded by ozonisation in ethyl acetate to formic acid, carbon dioxide, glyoxal and oxalic acid (*E. Bernatek* and *C. Frengen*, Acta Chem. Scand., 1961, **15**, 471) and by chlorination followed by alkaline hydrolysis to trichloroethylene and dichloromaleic acid (p. 151).

(*iv*) *Reduction*

(1) *To cyclohexanol derivatives.* Hydrogenation in the vapour or liquid phase over a suitable catalyst (often platinum or nickel) yields cyclohexanones, cyclohexanols or cyclohexanes (*H. A. Smith* and *B. L. Stump*, J. Amer. chem. Soc., 1961, **83**, 2739; *N. I. Shuikin* and *L. A. Erivanskaya*, Russ. chem. Reviews, 1960, **29**, 309; see also Vol. IIB, pp. 1 and 44). Reduction of phenol with sodium in liquid ammonia also gives cyclohexanone (*A. J. Birch* and *H. Smith*, Quart. Reviews, 1958, **12**, 17).

(2) *To benzene.* In a recently described procedure phenol is converted into an arylheterocyclic ether by treatment with, for example, a chlorotetrazole, which is then hydrogenated over 5% palladium on charcoal (*W. J. Musliner* and *J. W. Gates*, J. Amer. chem. Soc., 1966, **88**, 4271). This method is claimed to be milder and more general than previous ones which involved cleavage of diaryl ethers, or phosphate esters with sodium in liquid ammonia or distillation of the phenol with zinc dust (*W. H. Pirkle* and *J. L. Zabriskie*, J. org. Chem., 1964, **29**, 3124; *Y. K. Sawa, N. Tsuji* and *S. Maeda*, Tetrahedron, 1961, **15**, 144, 154; *G. W. Kenner* and *N. R. Williams*, J. chem. Soc., 1955, 522). *P. N. Rylander* ("Catalytic Hydrogenation over Platinum Metals", Academic Press, New York, 1967, p. 331) has reviewed the hydrogenation of phenols and phenyl ethers.

(*d*) *Reactions of the aromatic ring*

Electrophilic substitution occurs more readily than with benzene because of the $+M$ effect of the hydroxyl group and may be effected by reagents which have no action on benzene and its homologues. For example, phenol is nitrated by treatment with dilute nitric acid (p. 340) and sulphonation (p. 374) and halogenation (p. 334) are equally easy. The substituent enters mainly at the *ortho*- and *para*-positions relative to the hydroxyl group.

(1) *Reaction with diazo compounds.* Phenols react with diazo compounds in the absence of free mineral acid to give hydroxyazo compounds (p. 366).

(2) *Nitrosation.* With nitrous acid phenols in aqueous or alcoholic solution give nitrosophenols (p. 346).

(3) *Mercuration.* Treatment of phenol with mercuric acetate in boiling water followed by addition of sodium chloride yields *o*-chloromercuriphenol (*F. C. Whitmore* and *E. R. Hanson*, Org. Synth., Coll. Vol. 1, 1964, p. 161).

(4) *Alkylation.* Homologues of phenol are prepared by the alkylation of phenol wih alkenes, alcohols or alkyl halides in the presence of a catalyst (*N. I. Shuikin* and *E. A. Viktorova*, Russ. chem. Reviews, 1960, **29**, 560; *S. N. Patinkin* and *B. S. Friedman*, "Friedel Crafts and Related Reactions", Interscience, New York, 1964, Vol. II, pt. 1, p. 75; *A. V. Topchiev, S. V. Zavgorodnii* and *Y. M. Paushkin*, "Boron Trifluoride and its compounds as catalysts in organic chemistry", Pergamon, London, 1959). The most widely used catalysts with alkenes are concentrated sulphuric acid, boron trifluoride, metal halides ($AlCl_3$, $SnCl_4$, $ZnCl_2$) and cation-exchange resins (*B. Loev* and *J. T. Massengale*, J. org. Chem., 1957, **22**, 988) of which the first-named is the most common (*G. H. Stillson, D. W. Sawyer* and *C. K. Hunt*, J. Amer. chem. Soc., 1945, **67**, 303). The alkyl group, which may isomerise (primary → secondary → tertiary) during the course of the reaction, enters mainly the *ortho*- and *para*- (predominantly) positions although at low temperatures ethers may also be formed. *o*-Alkylphenols are best prepared using aluminium phenoxide as catalyst under carefully controlled conditions (*J. Kolka et al.*, J. org. Chem., 1957, **22**, 642). It should be remembered, however, that alkylation of phenols is a reversible process. For example, *m*- and *p*-cresol may be separated by converting them into their di-*tert*-butyl derivatives which are then fractionally distilled and dealkylated by heating in the presence of a catalyst.

Alkylation with alcohols in the vapour phase is effected in the presence of alumina or natural clays. In the liquid phase, phosphoric acid or aluminium chloride are usually the catalysts of choice although zinc chloride or sulphuric acid may also be used. Although alcohols are convenient alkylating agents in the laboratory, alkenes, because of their cheapness and availability in quantity, are normally used in industrial processes.

Ring alkylation of phenols with alkyl or benzyl halides is usually carried out in the presence of aluminium chloride or a halogen acid. Heating is often necessary. Formation of alkylphenols by treatment of phenolate anions with alkyl halides is frequently complicated by ether and cyclohexadienone formation. These dienones may be formed even when there is a free *para*-position, *e.g.*, as in XVII (*D. Y. Curtin* and *A. R. Stein*, Org. Synth., 1966, **46**, 115), and rearrange to phenols on treatment with acid. The factors

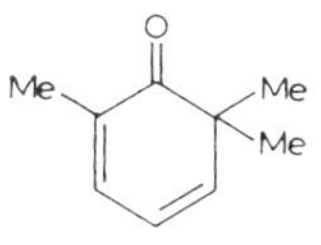

(XVII) 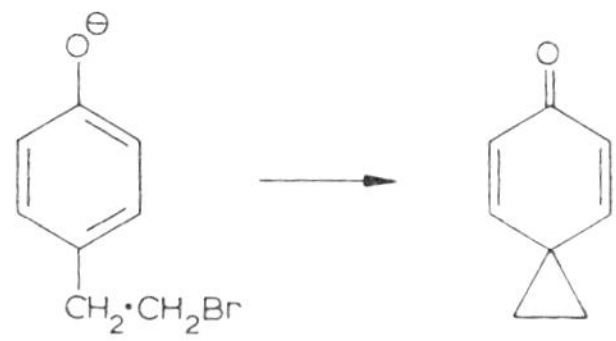 (XVIII) (XIX)

which govern *C*- and *O*-alkylations are not yet fully understood in spite of intensive investigation. Generally, however, *C*-alkylation is favoured by structural features such as bulky *ortho*-substituents and by working in "heterogeneous systems", *i.e.*, where there is much undissolved phenolate salt present or where there are ion-aggregates in solution (*N. Kornblum et al.*, J. Amer. chem. Soc., 1961, **83**, 3668 and 1959, **81**, 2705; *Curtin* and *D. H. Dybvig*, *ibid.*, 1962, **84**, 225). In homogeneous systems, hydrogen bonding solvents encourage *C*-alkylation (*Kornblum et al.*, *ibid.*, 1963, **85**, 1141, 1148). An interesting example of an intramolecular alkylation (XVIII → XIX) in which a highly reactive spirodienone is formed has been described by *R. Baird* and *S. Winstein*, *ibid.*, 1963, **85**, 567).

(5) *Formylation.* This may be achieved either by heating the phenol (or ether) with a mixture of hydrogen cyanide, hydrogen chloride and aluminium chloride and then hydrolysing the resulting aldimine (Gattermann aldehyde synthesis; *W. E. Truce*, Org. Reactions, 1957, **9**, 37) or by heating under reflux an alkaline solution of the phenol and chloroform (Reimer-Tiemann synthesis; *H. Wynberg*, Chem. Reviews, 1960, **60**, 169). For further details of these and other methods of preparing phenolic aldehydes see *V. I. Minkin* and *G. N. Dorofeenko*, Russ. chem. Reviews, 1960, **29**, 599 and also C.C.C. Vol. IIIC.

In addition to phenolic aldehydes, cyclohexadienones such as XX may also be formed in the Reimer-Tiemann reaction with certain *para*-substituted phenols. Similar products are formed by heating *para*-alkylphenols with

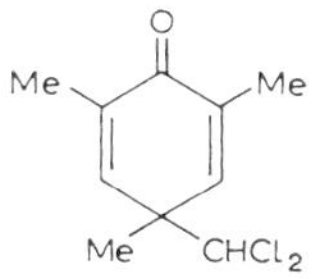

(XX) 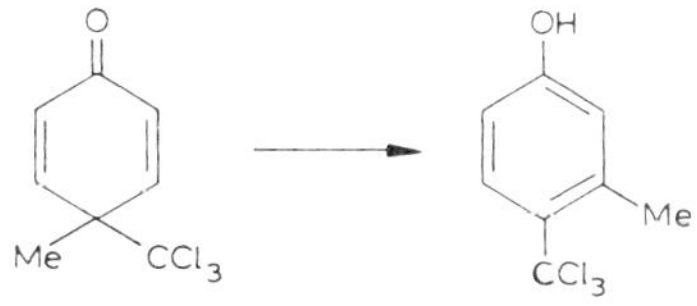 (XXI) (XXII)

carbon tetrachloride in the presence of aluminium chloride (Zincke-Suhl reaction; *M. S. Newman* and *A. G. Pinkus,* J. org. Chem., 1954, **19**, 978). *p*-Cresol gives the dienone XXI which undergoes a dienone-phenol rearrangement to XXII in acid solution (*V. V. Ershov, A. A. Volodkin* and *G. N. Bogdanov,* Russ. chem. Reviews, 1963, **32**, 75). Other rearrangements of this and related cyclohexadienones are described by *Newman, D. Pawellek* and *S. Ramachandran,* J. Amer. chem. Soc., 1962, **84**, 995 and other papers in this series.

(6) *Hydroxymethylations.* Phenol reacts with formaldehyde in aqueous alkaline solution at low temperature to give a mixture of *o*- and *p*-(hydroxymethyl)phenols. Unless the reaction is strictly controlled di- and tri-(hydroxymethyl)phenols, diphenylmethane derivatives and resinous products are also formed.

The condensation of phenols with formaldehyde in the presence of base or acid to form polymeric resinous materials, the so-called Bakelite resins (after *L. H. Baekeland,* their discoverer, Ind. Eng. Chem., 1909, **1**, 149) is of great technical importance. The resins (phenoplasts) were initially used for the production of insulating parts but their use has since been extended to the manufacture of many moulded and laminated materials and varnishes.

Polymerisation is usually carried out in two stages; the first (A) is catalysed by acid or base and gives a polymer of low molecular weight which is soluble and fusible, and the second (B) gives a lightly cross-linked product which is insoluble but fusible. Sometimes a third stage (C) is used to produce a highly cross-linked insoluble, infusible polymer. If base is used to catalyse the first stage the mono-, di-, and tri-(hydroxymethyl)phenols formed initially (*J. H. Freeman* and *C. W. Lewis,* J. Amer. chem. Soc., 1954, **76**, 2080) condense to some extent, with or without elimination of formaldehyde, giving a mixture of mainly di-, tri- and poly-nuclear diphenylmethane derivatives termed *resoles* (*L. M. Yeddanapalli* and *D. J. Francis,* Makromol. Chem., 1962, **55**, 74):

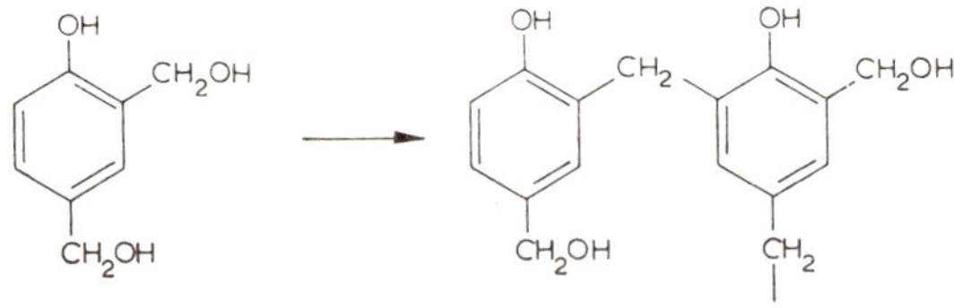

At stage B the resoles are neutralised or made slightly acidic and heated to about 150° when cross-linking occurs with the formation of *resites* containing mainly methylene and ether linkages (*H. Hanus* and *E. Fuchs,* J. pr. Chem., 1939, **153**, 327):

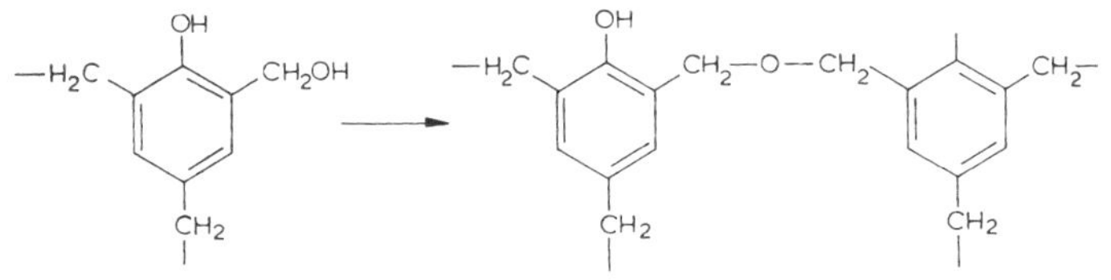

Quinone methides are probably formed as intermediates in this stage.

If acid is used to catalyse stage A the (hydroxymethyl)phenols condense to give diphenylmethane derivatives coupled mainly *via* the *para*-position (*J. S. Rodia* and *Freeman*, J. org. Chem., 1959, **24**, 21).

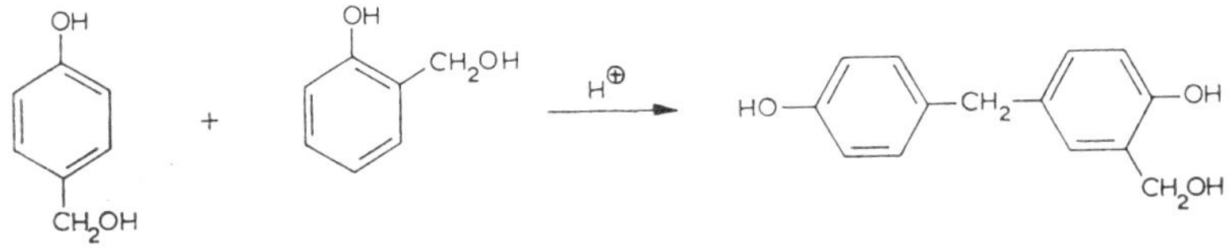

The oligomers of average molecular weight ~600, thus produced, are termed *novolacs*. Uncontrolled cross-linking is prevented at this stage by using phenol: formaldehyde ratios of slightly greater than 1. At stage B a catalyst, usually hexamethylenediamine, and heat are required to effect cross-linking because of the small number of free hydroxymethyl groups. At high phenol nucleus: hexamethylenediamine ratios no nitrogen is incorporated into the polymer and the cross-links are almost exclusively methylene groups. When the ratio is 0.5:1, however, nitrogen is incorporated to the extent of 10% and many of the linkages are secondary amino groups (CH_2—NH—CH_2) (*G. Zigeuner* and *W. Schaden*, Monatsh., 1950, **81**, 1017). At higher temperatures azomethine links may also be formed.

Clearly, by varying such parameters as the structure of the phenol, ratio of reactants, temperatures, pH, and catalysts, a vast number of resins with differing physical properties may be produced. Full accounts of the preparation, structure and technology of these resins are given elsewhere (*J. F. Walker*, "Formaldehyde", Reinhold, New York, 1944; *N. J. L. Megson*, "Phenolic Resin Chemistry", Academic Press, New York, 1956; *R. W. Lenz*, "Organic Chemistry of Synthetic High Polymers", Wiley, New York, 1967, Ch. 5).

Thiomethylation of phenols in the *ortho*-position may be achieved using a mixture of dicyclohexylcarbodi-imide, dimethyl sulphoxide, and an acid (*M. G. Burdon* and *J. G. Moffat*, J. Amer. chem. Soc., 1966, **88**, 5855; *T. Durst*, in "Advances in Organic Chemistry", Vol. 6, Interscience, New York, 1969, p. 285).

(7) *Carbonation.* In the Kolbe-Schmitt reaction (*A. S. Lindsey* and *H. Jeskey,* Chem. Reviews, 1957, 57, 583) an alkali metal phenolate is heated under pressure with carbon dioxide or carbon monoxide and carbonate (*Y. Vashura* and *T. Nogi,* Chem. and Ind., London, 1968, 77; Bull. chem. Soc. Japan, 1969, 42, 2070). *ortho*-Hydroxy acids, probably formed as

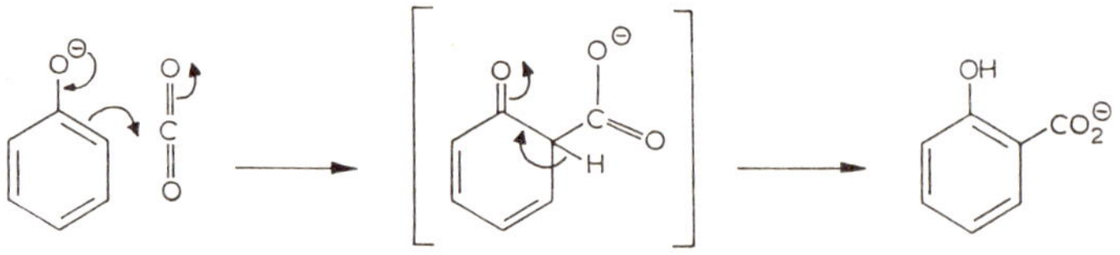

shown, predominate below 200⁰ but at higher temperatures and particularly if the potassium salt is used the *para*-isomer is favoured (*C. A. Buehler* and *W. A. Cate,* Org. Synth., Coll. Vol. II, 1963, p. 341). In an alternative but frequently less useful method the phenol is heated with carbon tetrachloride in alkaline solution in the presence of copper or copper salts (*G. Hasse,* Ber., 1877, 10, 2185).

(8) *Reactions with other carbonyl compounds.* With acetone in the presence of dry hydrogen chloride phenol gives the 4,4′-isopropylidenebisphenol, commonly known as "bisphenol A". A large number of such products have been prepared and used as antioxidants and monomers in condensation polymerisations (*e.g., J. Kahovec* and *J. Popisil,* Coll. Czech. Chem. Comm., 1969, 34, 2843; 1968, 33, 1709).

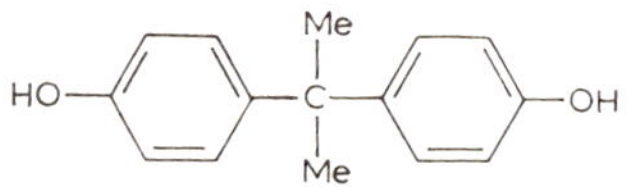

4,4′-Isopropylidenebisphenol

Condensation of phenols with *β*-oxo-esters yields either coumarins (Pechmann-Duisberg reaction) or chromones (Simonis reaction) depending on the dehydrating agent (usually sulphuric acid or phosphorus pentoxide (*S. Sethna* and *R. Phadke,* Org. Reactions, 1953, 7, 2; *E. H. Woodruff,* Org. Synth., Coll. Vol. III, 1963, p. 581). 4-Hydroxycoumarins are formed by heating phenols with malonic acid in the presence of zinc chloride or phosphorus oxychloride (*V. R. Shah, J. L. Bose* and *R. C. Shah,* J. org. Chem., 1960, 25, 677). Acid-catalysed condensation of glyoxal with phenol gives dihydrobenzfuran derivatives (XXIII) (*E. C. M. Coxworth,* Canad. J. Chem., 1967,

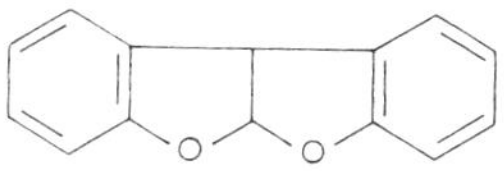

(XXIII) 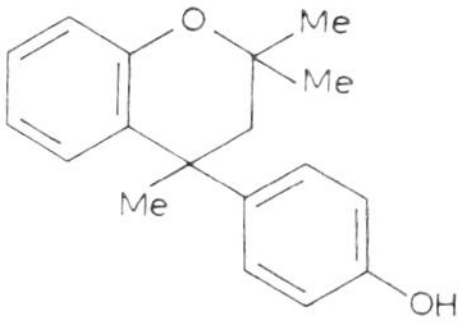 (XXIV)

45, 1777) and of mesityl oxide with phenol gives Dianin's compound (XXIV) (*W. Baker et al.*, J. chem. Soc., 1956, 2010, 2018). The latter product forms inclusion compounds with a large number of organic solvents. The above reactions are merely a selection of the condensations which phenols undergo with simple carbonyl compounds. Further examples are summarised by *J. C. McGowan, J. M. Anderson* and *N. C. Walker* (Rec. Trav. chim., 1964, 83, 597; see also C.C.C. 1st Edn., Vol. IVB).

Reaction with phthalic anhydride in the presence of a catalyst such as sulphuric acid or zinc chloride gives either phthaleins or hydroxyanthraquinones. Aurin dyes are formed by condensation of phenols with, for example, carbon tetrachloride or formic acid in the presence of zinc chloride.

(9) *Hydroxylation and acyloxylation.* Free-radical hydroxylation by irradiation in an aqueous solvent or using Fenton's reagent (*J. H. Merz* and *W. A. Waters*, J. chem. Soc., 1949, 2427) gives exclusive *ortho-para* hydroxylation generally in low yields (but see *K. Omura* and *T. Matsuura*, Tetrahedron, 1968, 24, 3475). Better results are obtained by regulated oxidation using potassium persulphate in alkaline solution (*K. Elbs*, J. pr. Chem., 1893, 48, 179; *S. M. Sethna*, Chem. Reviews, 1951, 49, 91). The second hydroxyl group enters into the *para*-position or if this is occupied into the *ortho*-position. Kinetic studies have shown that this is a bimolecular reaction between phenoxide and persulphate anions (*E. J. Behrman* and *P. P. Walker*, J. Amer. chem. Soc., 1962, 84, 3454). Exclusive *ortho*-hydroxylation may be achieved by converting the phenol into an aryloxynitrobenzophenone XXV which is hydroxylated (hydrogen peroxide/sulphuric acid) to a phenolic ether XXVI and then cleaved to a catechol by heating with piperidine (*J. D. Loudon*, "Progress in Organic Chemistry", Butterworths, London, 1961, Vol. 5):

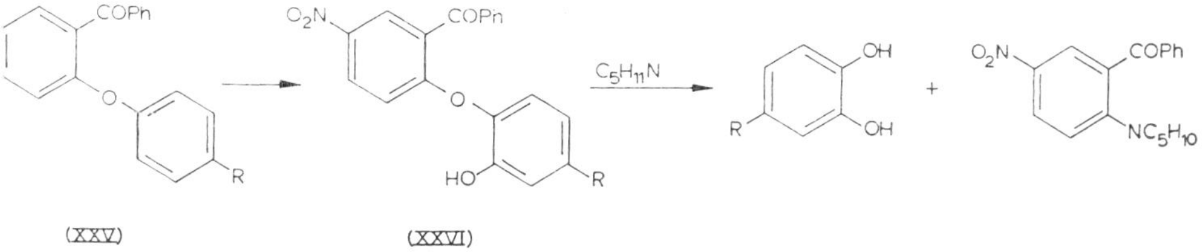

(XXV) (XXVI)

Acyloxylation using benzoyl or acetyl peroxide leads, after saponification, to dihydric phenols. Benzoyloxylation, but not acetoxylation, occurs *ortho*- to the hydroxyl group (*S. L. Cosgrove* and *Waters*, J. chem. Soc., 1951, 388; *F. Wessly* and *E. Schinzelle*, Monatsh., 1953, 84, 969; *D. H. R. Barton et al.*, Chem. Comm., 1969, 550). Lead tetra-acetate in acetic acid converts phenols into quinol acetates, *e.g.*, *o*-cresol gives XXVII and toluquinone. The former rearranges to a resorcinol derivative XXVIII on treatment with sulphuric

(XXVII) (XXVIII)

acid–acetic anhydride. Other phenols behave similarly giving quinol acetates and/or diacetates, *e.g.*,

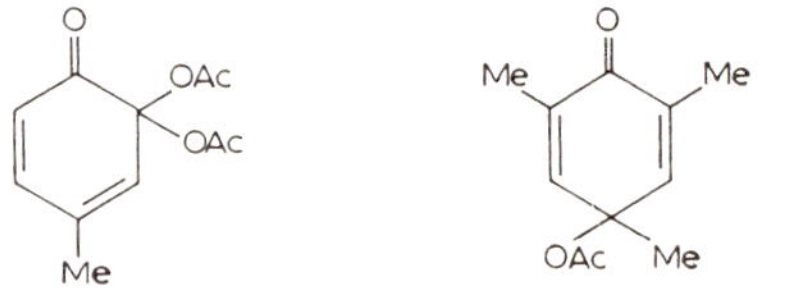

which undergo dienone-phenol rearrangements before and/or after mild hydrolysis. Much of the work concerning these interesting and synthetically useful dienones is summarised by their discoverer *Wessly* (J. roy. Inst. Chem., 1959, 424).

(10) *Ring expansion.* Lithium phenolate reacts with chlorocarbene, generated from methylene chloride and methyl-lithium giving tropone and 2-methyl-3,5-cycloheptadienone (XXIX) (*G. L. Closs* and *L. E. Closs*, J. Amer. chem. Soc., 1961, 83, 599):

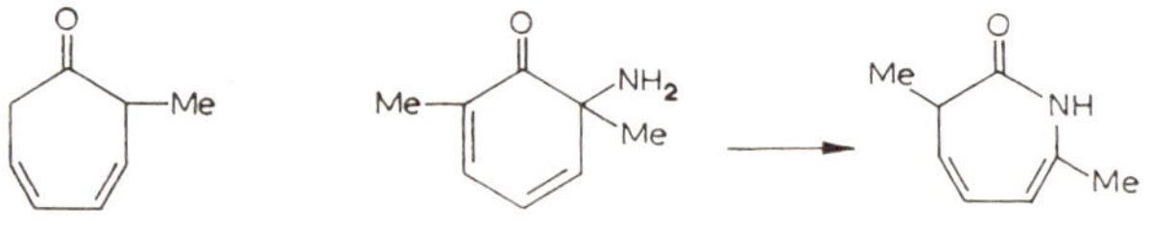

(XXIX) (XXX) (XXXI)

The sodium salt of 2,6-dimethylphenol undergoes ring expansion to the 1,3-dihydro-2*H*-azepin-2-one (XXXI) on treatment with chloramine. The reaction is considered to proceed *via* the dienone XXX (*L. A. Paquette*, *ibid.*, 1963, 85, 3288; Org. Synth., 1964, 44, 41).

(11) *Colour reactions.* Solutions of phenols turn green-blue on treatment with ferric chloride and red with nitrous acid containing a little mercuric nitrate (Millon's test) (*Z. Tamura* and *Y. Kido,* Chem. Pharm. Bull. Japan, 1968, **16**, 1808). 2,6-Dichloroquinone-4-chloroimine condenses with phenols, unsubstituted in the *para*-position, to give coloured indophenols (Gibb's test). Indophenols are also formed by the Liebermann reaction in which nitrite is added to a solution of the phenol in concentrated sulphuric acid. The indophenols dissolve in concentrated sulphuric acid or in alkali, forming blue solutions. In addition to the above reagents, diazonium salts, potassium permanganate, antimony pentachloride and phosphomolybdic acid are some of the other reagents which have been used as chromatographic sprays to detect phenols. Many of these reagents may be used also to determine phenols quantitatively. Fuller accounts of the colour reactions are given by *H. D. Gibbs* (Chem. Reviews, 1927, **3**, 291–319), *F. Feigl* ("Spot Tests in Organic Analysis", Elsevier, Amsterdam, 1966, p. 179–188) and *M. K. Seikel* ("Biochemistry of Phenolic Products", Academic Press, London, 1964, p. 33).

In addition to their use in the manufacture of resins, synthetic polyamide fibres, dyes, drugs, herbicides, insecticides and as antioxidants, phenol and its derivatives are important bactericides. A survey of the bactericidal action of phenols is given by *C. M. Suter* (Chem. Reviews, 1941, **28**, 269) and of phenolic disinfectants by *D. Jerchel* and *H. Oberheiden* (Angew. Chem., 1955, **67**, 145).

(e) Phenol and its homologues

Phenol, also known as *carbolic acid,* d^{25} 1·05446, n_D^{45} 1·54027, crystallises in colourless prisms, but develops a red tint on exposure to light and air presumably due to autoxidation to catechol and quinones. The bond angles and distances have been determined by *H. Gillier-Pandraud* (Bull. Soc. chim. Fr., 1967, 1988). Phenol has a characteristic odour, is corrosive to the skin and is toxic.

Crude phenol was first isolated from coal tar by Runge in 1834 and named carbolic acid. *A. Laurent* (Ann., 1842, **43**, 203) obtained phenol in a somewhat purer state from coal tar and called it phenic acid or phenyl hydrate. It was first prepared from benzene by heating aniline hydrochloride with silver nitrite (*Hunt,* Jahresb., 1849, 391). Its preparation by the alkaline fusion of benzene-sulphonic acid (*A. Wurtz,* Compt. rend., 1867, **64**, 749; *A. Kekulé, ibid.*, p. 752) is still used for the industrial preparation of phenol. Lister first made use of its germicidal properties in 1867.

Of the one and a half million tons of phenol required annually (1967 figure) only 10% is obtained from coal tar. The remainder is manufactured mainly by cumene oxidation (50%), the Raschig and Dow processes, and by sulphonation (*A. S. Banciu,* Chem. Proc. Eng., 1967, 31).

Phenol is soluble in about 15 parts of water at 20° (*C. R. Bailey*, J. chem. Soc., 1925, **127**, 1952), forms a *monohydrate*, m.p. 17·2°, and is volatile in steam. It is readily soluble (in alcohol and ether) and forms complexes with many organic solvents (*e.g.* acetone). It dissolves in dilute solutions of sodium or potassium hydroxide with the formation of alkali metal phenoxides which may be precipitated by addition of a large excess of alkali. *Sodium phenoxide* has m.p. 60° and *potassium phenoxide* has m.p. 104°. Compounds of potassium phenoxide with water or phenol of crystallisation have been described. *Calcium salt*, $Ca(OPh)_2 \cdot 2H_2O$; *barium salt*, $Ba(OPh)_2$. When phenol is heated to 180° in 10 *M*-hydrochloric acid in $H_2^{18}O$ considerable oxygen exchange occurs (*S. Oae et al.*, Bull. chem. Soc. Japan, 1964, **37**, 770, and 1966, **39**, 1961).

Addition of bromine water to dilute solutions of phenol gives 2,4,6-*tribromophenol* (m.p. 93°). With 1-naphthyl isocyanate it gives a *urethane* (m.p. 136–137°) and with *p*-toluenesulphonyl chloride an *ester* (m.p. 95°). Alkylation of phenol using methanol in the presence of alumina gives, rather surprisingly, hexamethylbenzene (*N. M. Cullinane, S. J. Chard* and *C. W. C. Dawkins*, Org. Synth., Coll. Vol. IV, 1963, p. 520; *A. P. Krysin* and *V. A. Koptyug*, Bull. Acad. Sci., U.S.S.R., 1969, 1479).

Homologous phenols. The general reactions of the homologues of phenol are similar to those of phenol and have already been described. On account of the ease of oxidation of the aromatic nucleus the oxidation of alkyl substituents requires special conditions. The methyl group of cresols, xylenols and other substituted phenols may be oxidised by fusion with caustic potash, preferably with the addition of lead dioxide or copper oxide (*D. Todd* and *A. E. Martell*, Org. Synth., 1960, **40**, 48). Alternatively, the hydroxyl group may be protected by converting it into an ester (sulphonate or acetate) group and the alkyl group oxidised to a formyl (with manganese dioxide and sulphuric acid) or carboxyl group (with permanganate).

Halogenation and nitration follow the usual course, although if conditions are suitable halogenocyclohexadienones and perhalides may be formed. It is also possible to halogenate the side chain giving hydroxybenzyl halides if the appropriate halogenating agent is used. The toxicity of the homologues of phenol decreases and the germicidal value increases as the alkyl groups become more complex. Thus, the cresols have a phenol coefficient of about 2 and amyl-*m*-cresol one of about 250 (*C. E. Coulthard, J. Marshall* and *F. L. Pyman*, J. chem. Soc., 1930, 280). Homologues of phenol are sometimes characterised by their "hydroxyl value" (or number) which corresponds to the number of mg of potassium hydroxide equivalent to the amount of acid used on esterification of 1 g of alkylphenol. The hydroxyl value of phenol is 596, those of alkylphenols being correspondingly lower.

Cresols, $CH_3 \cdot C_6H_4OH$. All three isomers are present (1–2% total) in coal tar, *o*- and *m*-cresol in almost equal proportions and *p*-cresol in about one quarter to a fifth of the total cresol content. They arise in the middle or carbolic fraction, after initial fractionation of the tar, and are separated from the neutral components present by extraction with alkali. Further careful fractionation gives phenol and o-*cresol* (b.p. 191^0) but m- and p-*cresols* (b.p. 202^0 and 201^0, respectively) are not separable in this way. These isomers may be separated by sulphonation followed by fractional hydrolysis of their sulphonic acids by superheated steam (*H. Bruckner*, Z. angew. Chem., 1928, **41**, 1043). Other methods of separation depend upon differences in their capacity to form addition compounds with, for example, urea or oxalic acid (*G. Darzens*, Compt. rend., 1931, **192**, 1657; *L. Rappen* and *H. Wille*, U.S.P., 3,193,585/1965; *K. H. Engel*, U.S.P. 2,095,801/1937; *D. F. Gould*, U.S.P., 2,099,109/1937; *H. G. Rosencrantz* and *H. Winter*, G.P. (East), 48,625/1966; *J. Fleischer* and *E. Mier*, G.P., 1,215,726/1966), fractional crystallisation of their phosphate esters (*M. Kotake*, Jap.P., 133,250/1939) or alkylation to give products which are separable and may be dealkylated subsequently (*W. Weinrich*, Ind. Eng. Chem., 1943, **35**, 264; *D. R. Stevens*, *ibid*., p. 655; *C. H. Young* and *P. A. Munday*, U.S.P., 2,969,401/1961).

Cresols are also formed during petroleum cracking and are isolated with phenol and xylenols by extraction with alkali. This mixture termed "cresylic acid" or "creosote" is an important industrial source of cresols (*H. L. Coonradt* and *B. W. Rope*, Ind. Eng. Chem., P.R.O. section, 1963, **2**, 317; *D. C. Jones*, *J. A. Kohlbeck* and *M. B. Neunorth*, *ibid*., 1963, **2**, 217).

Creosote is used as a wood preservative and disinfectant; a dilute solution in soapy water is called "lysol". Individually the cresols are of technical importance in the manufacture of fine chemicals, plastics, dyestuffs, insecticides and herbicides. For details of the i.r. spectra of cresols and xylenols, and other alkylphenols, see *D. H. Whiffen* and *H. W. Thompson*, J. chem. Soc., 1945, 268, *D. D. Shrewsbury*, Spectrochim. Acta, 1960, **16**, 1294, and *T. A. Kletz* and *W. C. Price*, J. chem. Soc., 1947, 644.

o-*Cresol*, d_4^{20} 1·0465, n_D^{40} 1·5372, p*K*a 10·20. The main product of halogenation and nitration is usually the 4-substituted 2-methylphenol but nitration in the presence of nitrous acid gives a mixture (1:1) of 4- and 6-nitrocresols (*S. Veibel*, Ber., 1930, **63**, 2074). 2-Methyl-4,6-dinitrophenol is prepared in a continuous process using 70% nitric acid (*W. Mathes* and *L. Seelman*, G.P., 805,645/1951). Halogenation in the presence of sulphuric acid gives good yields of the 6-halogeno isomers (*R. C. Huston* and *A. H. Neeley*, J. Amer. chem. Soc., 1935, **57**, 2176). On heating *o*-cresol at 134^0 with aluminium chloride, isomerisation to a mixture of *m*- and *p*-cresols occurs (*H. P. Meissner* and *F. E. French*, *ibid*., 1952, **74**, 1000). Fusion with caustic soda, preferably with the addition of lead dioxide (*C. Graebe* and *H. Kraft*, Ber., 1906, **39**, 794), or treatment with palladous chloride (*G. W. Cooke* and *J. M. Gulland*, J. chem. Soc., 1939, 872) gives salicylic acid. The action of carbon dioxide on the sodium salt gives mainly 2-hydroxy-3-methylbenzoic acid at 200^0 (*O. Baine et al*., J. org. Chem., 1954, **19**, 510); at higher temperatures 4-hydroxy-5-methylbenzoic acid is formed. Oxidation with Fremy's salt gives toluquinone in 82% yield (*H. J. Teuber* and *W. Rau*, Ber., 1952, **86**, 1036).

m-*Cresol*, d_4^{20} 1·0336, n_D^{22} 1·5352, p*K*a 10·01, is formed by dealkylation of thymol or by Clemmensen reduction of *m*-hydroxybenzaldehyde. Nitration with nitrogen dioxide gives either the 4,6-dinitro or the 2,4,6-trinitro derivative depending upon the temperature. 4-Chloro-3-methylphenol is prepared by treatment of *m*-cresol with sulphuryl chloride in the absence of solvent (*P. P. T. Sah* and *H. H. Anderson*, J. Amer. chem. Soc., 1941, **63**, 3164). Its chlorination and bromination are described by *Huston* and *Neeley* (*loc. cit.*) and by *G. P. Gibson* (J. chem. Soc., 1926, 1424).

p-*Cresol*, d_4^{20} 1·0347, n_D^{40} 1·5319, p*K*a 10·17, is prepared from *p*-toluenesulphonic acid by alkali fusion (*W. W. Hartman*, Org. Synth., Coll. Vol. I, 1964, p. 175) or from *p*-toluidine (*W. J. Hickinbottom*, "Reactions of Organic Compounds", Longmans, London, 1957, p. 484). Nitration with dilute nitric acid gives 4-*methyl*-2-*nitrophenol* (m.p. 33°). Bromination in chloroform gives either 2-bromo-, or 2,6-dibromo-4-methylphenol depending upon the amount of bromine used. 2,3,6-Tribromo-4-methylphenol can be prepared only when iron is present as a halogen carrier and the preparation of the tetrabromo-compound requires the presence of aluminium chloride. Oxidation of *p*-cresol with a number of reagents (*e.g.*, ferric chloride, potassium persulphate, hydrogen peroxide/ferrous sulphate) gives a mixture of Pummerer's ketone, and dimeric and trimeric phenols (see p. 298). In the presence of aluminium chloride–hydrogen chloride *p*-cresol rearranges to *m*-cresol (80%) (*L. A. Fury* and *D. E. Pearson*, J. org. Chem., 1965, **30**, 2301).

Xylenols $(CH_3)_2 \cdot C_6H_3OH$. The physical constants of the various isomers are given in Table 1. All of the isomers are present in the higher boiling fractions of cracked petroleum (200–260°) and all, except 2,3-dimethylphenol (*o*-3-xylenol), are present in coal tar, 2,5-dimethylphenol (*p*-xylenol) being the most abundant in petroleum and 3,5-dimethylphenol (*m*-xylenol) being the most abundant in coal tar. Because of the difficulty in separating the isomers (*Stevens*, Ind. Eng. Chem., *loc. cit.*) specific xylenols are now often synthesised. 2,4-, 2,5- and 3,4-Dimethylphenols may be converted into 3,5-dimethylphenol in high yield by heating with aluminium chloride–hydrogen chloride (*Fury* and *Pearson*, *loc. cit.*).

Ethylphenols. See Table 1 for physical constants and refs. to preparations. All are present in the xylenol fraction of coal tar and may be separated by alkylation followed by fractional distillation (*Stevens*, *loc. cit.*). Their conversion into phenol by thermal dealkylation over acidic catalysts has been investigated (*P. Schneider*, *M. Kraus* and *V. Bažant*, Coll. Czech. chem. Comm., 1960, **26**, 1636 and *P. H. Given*, J. appl. Chem., 1957, **7**, 182). I.R. and U.V. data for the isomers are given by *P. Demerseman et al.* (Bull. Soc. chim. Fr., 1962, 1700).

C_9-Phenols $C_9H_{11}OH$. These comprise the trimethyl-, ethylmethyl- and the propyl-phenols.

Trimethylphenols. See Table 1 for physical properties and refs. to preparations. Generally, they are prepared from the corresponding diazonium salts by treatment with acid. Oxidation of mesitol (2,4,6-trimethylphenol) has been most

(*text continued on p. 316*)

TABLE 1

PHYSICAL CONSTANTS OF PHENOL AND ITS HOMOLOGUES

Phenol	M.p. °C	B.p. °C
Phenol	41	181·7
C_7H_7OH		
o-Cresol	30·5	191
m-Cresol	11·8	202
p-Cresol	34·8	201
C_8H_9OH		
2,3-Dimethylphenol[1] (*o*-3-xylenol)	75	218
2,4-Dimethylphenol[2] (*m*-4-xylenol)	26	211·5
2,5-Dimethylphenol[3] (*p*-xylenol)	75	211·5
2,6-Dimethylphenol[4] (*m*-2-xylenol)	45	203
3,4-Dimethylphenol[5] (*o*-4-xylenol)	65	227
3,5-Dimethylphenol[6] (*m*-5-xylenol)	64	221·5
o-Ethylphenol[7]	—	206
m-Ethylphenol[8]	−4	214
p-Ethylphenol[7]	48	218
$C_9H_{11}OH$		
2,3,4-Trimethylphenol[9]	69 or 81	—
2,3,5-Trimethylphenol[10,21]	94	—
2,3,6-Trimethylphenol[11]	62	—
2,4,6-Trimethylphenol[12,14,21] (mesitol)	68	—
2,4,5-Trimethylphenol[13] (*ψ*-cumenol)	70	—
3,4,5-Trimethylphenol[14,24] (hemimellitol)	81	—
o-Propylphenol[15]	—	225
m-Propylphenol[16]	—	228
p-Propylphenol[15]	21	230
o-Isopropylphenol[17]	16	212
m-Isopropylphenol[18]	26	228
p-Isopropylphenol[19]	61	229
$C_{10}H_{13}OH$		
2-Isopropyl-3-methylphenol[20]	70–71	229
2-Isopropyl-4-methylphenol[20]	36–37	223
2-Isopropyl-5-methylphenol[20] (thymol)	51	232
2-Isopropyl-6-methylphenol[20]	−14·5	226
3-Isopropyl-2-methylphenol[20]	82–83	—
3-Isopropyl-4-methylphenol[20]	39–40	—
3-Isopropyl-5-methylphenol[20]	50	241
5-Isopropyl-2-methylphenol[20] (carvacrol)	3	236
4-Isopropyl-3-methylphenol[20]	112–113	246
4-Isopropyl-2-methylphenol[20]	8·6 and 28·5	231

(*continued*)

Table 1 (*continued*)

Phenol	M.p. °C	B.p. °C
2,3,4,5-Tetramethylphenol[21,24] (prehnitol)	87	—
2,3,4,6-Tetramethylphenol[22,31] (isodurenol)	79–81	—
2,3,5,6-Tetramethylphenol[21,23,24] (durenol)	119	248
o-Butylphenol[25]	—	236
m-Butylphenol[26]	—	136/20 mm.
p-Butylphenol[27]	22	249
o-(*tert*-Butyl)phenol[28]	−7	224
m-(*tert*-Butyl)phenol[29]	41	237
p-(*tert*-Butyl)phenol[30]	97	239
$C_{11}H_{15}OH$		
Pentamethylphenol[31]	126	267
4-Butyl-2-methylphenol[32]	124	—
6-Butyl-2-methylphenol[32]	14	118/15 mm.
6-Butyl-3-methylphenol[32]	18	133/15 mm.
2-Butyl-4-methylphenol[32]	19	125/15 mm.
4-(*tert*-Butyl)-2-methylphenol[33]	28	236
5-(*tert*-Butyl)-2-methylphenol[29]	32	93/2 mm.
6-(*tert*-Butyl)-2-methylphenol[33]	27	226
4-(*tert*-Butyl)-3-methylphenol[34]	72	153/20 mm.
5-(*tert*-Butyl)-3-methylphenol[35]	47	127/11 mm.
6-(*tert*-Butyl)-3-methylphenol[36]	23	122/15 mm.
2-(*tert*-Butyl)-4-methylphenol[37]	47	112/10 mm.
p-*tert*-Pentylphenol[30]	95	267
Higher phenols		
2,6-Di-*tert*-butyl-4-methylphenol[38]	70	133/11 mm.
2,6-Di-*tert*-butyl-4-ethylphenol[38]	44	134/10 mm.
2,4,6-Tri-*tert*-butylphenol[38]	131	136/10 mm.
4-Methyl-2,6-di-*tert*-pentylphenol[38]	—	140/5 mm.
2,4,6-Tri-*tert*-pentylphenol[38]	—	126/2 mm.
p-Hexylphenol[39]	29	155/15 mm.
p-Heptylphenol[32,40]	16	164/15 mm.
p-Octylphenol[41]	—	113/3 mm.
p-(*tert*-Octyl)phenol[42]	83	287
p-Nonylphenol[40,42]	42·5	301
4-(*p*-Hydroxyphenyl)-2,2,4,4-tetramethylbutane	84	283
3-Pentadecylphenol[44] (hydrogingol)	51	231/8 mm.
p-Dodecylphenol[43]	66	322

References Table 1

1 *W. F. Short, H. Stromberg* and *A. E. Wiles,* J. chem. Soc., 1936, 319; *L. I. Smith* and *J. W. Opie,* J. org. Chem., 1941, **6,** 427.
2 *W. Cocker* and *C. Lipman,* J. chem. Soc., 1947, 533; *W. Harmsen,* Ber., 1880, **13,** 1558.
3 *E. Noelting, O. N. Witt* and *S. Forel, ibid.,* 1885, **18,** 2664; *V. Rericha* and *M. Protiva,* Chem. Listy, 1951, **45,** 157.
4 *R. B. Carlin* and *R. P. Landerl,* J. Amer. chem. Soc., 1950, **72,** 2762.
5 *G. Baddeley,* J. chem. Soc., 1944, 330.
6 *E. C. Horning,* J. Amer. chem. Soc., 1945, **67,** 1421; *L. A. Fury* and *D. E. Pearson,* J. org. Chem., 1965, **30,** 2301; *K. von Auwers* and *E. Borsche,* Ber., 1915, **48,** 1698.
7 *P. Schneider, M. Kraus* and *V. Bažant,* Coll. Czech. chem. Comm., 1961, **26,** 1636.
8 *J. Kenner* and *F. S. Statham,* J. chem. Soc., 1935, 299.
9 *K. von Auwers* and *F. Wieners,* Ber., 1925, **58,** 2815; *N. P. Buu-Hoi, P. Jacquignon* and *O. Roussel,* Bull. Soc. chim. Fr., 1965, 322.
10 *von Auwers* and *K. Saurwein,* Ber., 1922, **55,** 2372.
11 *G. T. Morgan* and *A. E. J. Pettet,* J. chem. Soc., 1934, 418.
12 *H. Hart* and *C. A. Buehler,* J. org. Chem., 1964, **29,** 2397; *C. W. Porter* and *F. H. Thurber,* J. Amer. chem. Soc., 1921, **43,** 1194.
13 *von Auwers* and *J. Marwedel,* Ber., 1895, **28,** 2902.
14 *W. E. Doering* and *F. M. Beringer,* J. Amer. chem. Soc., 1949, **71,** 2221.
15 *L. H. Farinholt, W. C. Hardin* and *D. Twiss, ibid.,* 1933, **55,** 3383.
16 *W. H. Hartung* and *F. S. Crossley, ibid.,* 1934, **56,** 158.
17 *F. J. Sowa, H. D. Hinton* and *J. A. Nieuwland, ibid.,* 1932, **54,** 3694.
18 *H. Gilman et al.,* J. org. Chem., 1954, **19,** 1067.
19 *J. R. Stevens* and *R. H. Beutel,* J. Amer. chem. Soc., 1941, **63,** 308; *L. Bert,* Bull. Soc. chim. Fr., 1925, 37, 1397.
20 *M. S. Carpentier* and *W. M. Easter,* J. org. Chem., 1955, **20,** 401.
21 *H. Dakshinamruty* and *M. Santappa, ibid.,* 1962, **27,** 1839.
22 *Hart* and *Buehler, loc. cit.*
23 *C. E. Ingham* and *G. C. Hampson,* J. chem. Soc., 1939, 981.
24 *J. S. Fitzgerald,* J. applied Chem., 1955, **5,** 289.
25 *P. Demerseman et al.,* Bull. Soc. chim. Fr., 1963, 2559.
26 *Schneider, Kraus* and *Bažant,* Coll. Czech. chem. Comm., 1962, **27,** 9.
27 *J. B. Niederl et al.,* J. Amer. chem. Soc., 1937, **59,** 1113.
28 *Hart, ibid.,* 1949, **71,** 1966.
29 *Carpentier, Easter* and *T. F. Wood,* J. org. Chem., 1951, **16,** 586.
30 *R. C. Huston* and *T. Y. Hsieh,* J. Amer. chem. Soc., 1936, **58,** 439.
31 *D. H. Hey,* J. chem. Soc., 1931, 1581.
32 *C. E. Coulthard, J. Marshall* and *F. L. Pyman, ibid.,* 1930, 280.
33 *Hart* and *E. A. Haglund,* J. org. Chem., 1950, **15,** 396.
34 *D. R. Stevens* and *R. S. Bowman,* U.S.P., 2,560,666/1948.
35 *B. M. Dubinin* and *N. E. Kozhevnikova,* C.A., 1951, **45,** 9500.
36 *R. Stroh, R. Seydel* and *W. Hahn,* Angew. Chem., 1957, **69,** 699.
37 *Kraus* and *Bažant,* Coll. Czech. chem. Comm., 1963, **28,** 1877.
38 *G. H. Stillson, D. W. Sawyer* and *C. K. Hunt,* J. Amer. chem. Soc., 1945, **67,** 303.
39 *D. W. Goheen* and *W. R. Vaughan,* J. org. Chem., 1958, **23,** 891.
40 *G. Sandulesco* and *A. Girard,* Bull. Soc. chim. Fr., 1930, **47,** 1300.
41 *E. Profft* and *G. Rietz,* J. prakt. Chem., 1960, **11,** 94.
42 *B. Loev* and *J. T. Massengale,* J. org. Chem., 1957, **22,** 988.
43 *A. F. McKay et al.,* Canad. J. Chem., 1960, **38,** 2042.
44 *H. J. Backer* and *N. H. Haack,* Rec. Trav. chim., 1941, **60,** 661.

closely studied. The *para* methyl group is usually attacked; the final product, which may be a benzyl alcohol, aldehyde, acid, benzoquinone, dihydroxybibenzyl or stilbenequinone, depends upon the oxidising agent (*C. Graebe* and *H. Kraft*, Ber., 1906, **39**, 794; *A. F. Bickel* and *E. C. Kooyman*, J. chem. Soc., 1953, 3211; *W. A. Waters et al., ibid.*, 1954, 243 and 1951, 1726; *H. Booth* and *B. C. Saunders, ibid.*, 1956, 940).

Propylphenols. *p*-Isopropylphenol is present in eucalyptus oils (*A. R. Penfold*, Proc. roy. Soc. N.S. Wales, 1928, **61**, 179). Oxidation of *p*-propylphenol with alkaline ferricyanide gives an analogue of Pummerer's ketone (*C. G. Haynes*, *A. H. Turner* and *Waters*, J. chem. Soc., 1956, 2823).

C_{10}-Phenols $C_{10}H_{13}OH$. The two best known are thymol, 2-isopropyl-5-methyl phenol, and carvacrol, 5-isopropyl-2-methylphenol, both of which occur naturally in some essential oils.

Thymol, which has a powerful odour of thyme, crystallises in large colourless plates. It occurs, together with cymene, in oil of thyme from *Thymus vulgaris* and in the oils from *Ptychotis ajowan* and *Monarda punctata*. It may be manufactured by condensing *m*-cresol with isopropyl chloride, isopropyl alcohol or propylene in the presence of the appropriate catalyst (*M. S. Carpenter*, B.P., 551,624/1943; U.S.P., 2,286,953/1942; *S. Skraup*, U.S.P., 2,103,736/1937; *G. Leston*, U.S.P., 3,267,155, 3,267,152 and *T. Hokama*, U.S.P., 3,267,153/1966). Alternatively, nitration of *p*-cymene with subsequent reduction to aminocymene and diazotisation, alkali fusion of *p*-cymene-3-sulphonic acid (*R. M. Cole*, U.S.P., 1,378,939/1921 and *R. H. McKee*, U.S.P., 1,449,121/1923), dehydrobromination of dibromomenthone or oxidation of piperitone with ferric chloride may also be used. It is an antiseptic.

It may be hydrogenated to di-menthol in high yield (*A. L. Barney* and *H. B. Hass*, Ind. Eng. Chem., 1944, **36**, 85). Oxidation with ferric chloride yields the (*para-para*) coupled product (*H. Cousin* and *H. Hérissey*, Bull. Soc. chim. Fr., 1908, [iv], **3**, 586) while Fremy's salt gives thymoquinone in 98% yield (*H.-J. Teuber* and *W. Rau*, Ber., 1953, **86**, 1036). Iodine and caustic soda convert thymol into 3,3′-di-iodo-5,5′-di-isopropyl-2,2′-dimethyl-diphenoquinone which has been used in surgery under the name of Aristol (*J. Messinger* and *G. Vortmann, ibid.*, 1889, **22**, 2312). Reduction to *p*-cymene may be effected in high yield by hydrogenolysis of the corresponding phenyltetrazotyl ether (*W. J. Musliner* and *J. W. Gates*, J. Amer. chem. Soc., 1966, **88**, 4271).

Carvacrol, occurs in the combined state in the oil of some species of *Satureja* and *Origanum* such as *S. hortensis* and *O. hiritum*. It is obtained (a) by heating carvone with phosphoric acid or sulphuric acid in methanol (*G. Büchi* and *R. E. Erickson, ibid.*, 1954, **76**, 3493); (b) by heating camphor with catalytic quantities of iodine; (c) from *p*-cymene-2-sulphonic acid by alkali fusion (*G. Udaka* and *M. Imai*, J. pharm. Soc. Japan, 1943, **63**, 242; *McKee, loc. cit.*); and (d) by heating 2-bromo-*p*-cymene in ethylene glycol with alkali (*W. Strubell* and *H. Baumgaertel*, Arch. Pharm., 1956, **289**, 719; *idem*, G.P. (East), 14,460/1958). Treatment with *tert*-butyl hypochlorite yields a dichloro-derivative (*D. Ginsburg*, J. Amer. chem. Soc., 1951, **73**, 2723) and oxidation with Fremy's salt gives thymoquinone (73%) (*Teuber* and *Rau, loc. cit.*). Carvacrol and many of its halogenated derivatives are powerful antiseptics.

p-(tert-**Butyl**)**phenol** is manufactured by the action of isobutene or of di-isobutene on phenol in the presence of aluminium chloride (*R. P. Perkins* and *H. S. Nutting*, U.S.P., 2,091,565/1937) or sulphuric acid (*Monsanto Chem. Co.*, B.P., 452,335/1937); a small amount of *o*-(*tert*-butyl) phenol is also formed (*J. I. de Jong*, Rec. Trav. chim., 1964, **83**, 469). *B. Loev* and *J. T. Massengale* (J. org. Chem., 1957, **22**, 988) describe a simple laboratory preparation in which a cation-exchange resin is used as catalyst.

Thousands of tons of this chemical are produced annually, mainly for conversion into formaldehyde resins. It is also used as a stabiliser, an anticracking agent for rubber, and an antiseptic. Oxidation with cupric *o*-toluate gives the *o,o*-coupled dihydro dimer (*W. W. Kaeding, ibid.*, 1963, **28**, 1063).

C_{11}-**Phenols** $C_{11}H_{15}OH$. **Pentamethylphenol** gives no colour with ferric chloride and is not soluble in aqueous alkali.

6-tert-**Butyl**-3-**methylphenol** is almost insoluble in water and dilute alkali but forms a crystalline hydrate. It is an effective antioxidant and is an intermediate in the production of musk ambrette (6-*tert*-butyl-3-methyl-2,4-dinitro-anisole) which is used as a fixative in the perfume industry (*A. E. Chichibabin*, Bull. Soc. chim. Fr., 1935, 497).

2-tert-**Butyl**-4-**methylphenol** is used to prepare antioxidants for rubber.

p-tert-**Amylphenol** condenses with formaldehyde giving resins which are widely used in varnish and paint manufacture. Products with ethylene oxide are used as surface-active agents.

Higher phenols. *G. H. Stillson et al.* (J. Amer. chem. Soc., 1945, **67**, 303) have described a number of alkylated phenols which are only sparingly soluble in aqueous alkali, the acidic nature of the hydroxyl group being suppressed by the presence of bulky alkyl groups in the 2- and 6-positions.

2,6-**Di**-tert-**butyl**-4-**methylphenol.** Oxidation of this phenol to a 3,5,3′,5′-tetra-*tert*-butylstilbenequinone by one-electron oxidants has been extensively studied (*L. M. Strigun, L. S. Vartanyan* and *N. M. Emanuel*, Russ. chem. Reviews, 1968, **37**, 421). Mild oxidation with chromic acid or with bromine yields the corresponding 4-formylphenol (*G. M. Coppinger* and *T. W. Campbell*, J. Amer. chem. Soc., 1953, **75**, 734). It is used extensively as an antioxidant.

2,4,6-**Tri**-tert-**butylphenol.** Oxidation with a variety of oxidants gives the relatively stable blue 2,4,6-tri-(*tert*-butyl)phenoxyl radical (*E. Müller* and *K. Ley*, Ber., 1954, **87**, 922). Reactions of this radical are reviewed in the refs. given on p. 300.

p-(tert-**Octyl**)**phenol** is sparingly soluble in aqueous alkali but is soluble in alcoholic potash. It is prepared by the reaction of phenol with di-isobutylene (2,2,4-trimethylpent-1- and -2-enes) in the presence of sulphuric acid (*J. B. Niederl*, Ind. Eng. Chem., 1938, **30**, 1269) or hydrogen fluoride (*J. H. Simons* and *S. Archer*, J. Amer. chem. Soc., 1940, **62**, 451). If the conditions of the reactions become vigorous *p*-(*tert*-butyl)phenol is formed. An account of its reactions is given by *Niederl* (*loc. cit.*). For other *p*-(*tert*-octyl)phenols see *R. C. Huston et al.* (J. Amer. chem. Soc., 1948, **70**, 1474).

Nonyl- and **dodecyl-phenols** are manufactured on a large (several thousand tons per year) scale for use in the detergent industry. The phenols are condensed with ethylene oxide giving long chain polyethers which are used as nonionic detergents and as wetting and emulsifying agents.

2. Functional derivatives of phenols

This section deals with the ethers of phenols (aryl ethers) and the phenyl esters (aryl esters) of organic and inorganic acids. The subject matter is confined almost entirely to the derivatives of phenol which are known sufficiently well to illustrate the methods of formation and reactions of these compounds.

(a) Phenyl ethers

(i) Methods of preparation

(1) Alkyl and alkenyl aryl ethers are obtained by treatment of an alkyl or alkenyl halide with an alkali metal areneolate (aryl oxide), preferably in warm alcoholic solution. Alternatively, the phenol and alkyl or alkenyl iodide may be dissolved in acetone and the solution heated under reflux with solid potassium carbonate for some hours (*W. N. White et al.*, J. Amer. chem. Soc., 1958, 80, 3271; *G. N. Vyas* and *N. M. Shah*, Org. Synth., Coll. Vol. IV, 1963, p. 836). Aryloxy-acids, alcohols, ketones, etc. may be similarly prepared using the appropriate halogen-substituted acid, alcohol, ketone, etc. Factors which affect *O- versus C*-alkylation in such reactions have been investigated by *N. Kornblum* and his coworkers (J. Amer. chem. Soc., 1955, 77, 6269; 1959, 81, 2705). Some of these factors have already been discussed (p. 297).

(2) Conversely, aryl halides do not react readily with alkoxides unless the halogen atom is activated by the presence of one or more electron-withdrawing groups preferably in the *ortho-* and/or *para*-positions (*e.g., L. McMaster* and *A. C. Magill, ibid.*, 1928, 50, 3038). The preparation of *tert*-butyl phenyl ether from bromobenzene and potassium *tert*-butoxide in dimethyl sulphoxide solution is not a general method since an aryne intermediate is thought to be involved (*D. J. Cram, B. Rickborn* and *G. R. Knox, ibid.*, 1960, 82, 6412). The ease of replacement of halogen in *o*- and *p*-halogenonitrobenzenes is F > Cl > Br > I (*B. A. Bolto, J. Miller* and *V. A. Williams,* J. chem. Soc., 1955, 2926). Suitably activated nitro groups may also be replaced and 3,5-dinitroanisole is formed in 77% yield from 1,3,5-trinitrobenzene and sodium methoxide (*F. Reverdin*, Org. Synth., Coll. Vol. I, 1964, p. 219). Transetherification of nitroaryl ethers such as trinitroanisole

by treatment with, for example, ethoxide proceeds *via* the red "Meisenheimer complex" I. (For a review see *E. Buncel, A. R. Norris* and *K. E. Russell*, Quart. Reviews, 1968, **22**, 123):

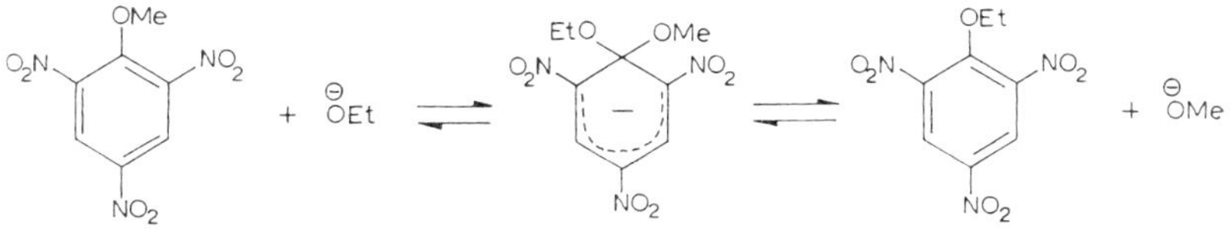

Trinitroanisole + $\overset{\ominus}{O}Et$ ⇌ (I) ⇌ + $\overset{\ominus}{O}Me$

(3) *O*-Alkylation of phenols may be satisfactorily achieved using dialkyl sulphates or the alkyl esters of sulphonic acids (*F. Ullmann*, Ann., 1903, **327**, 104; *D. A. Shirley* and *W. H. Reedy*, J. Amer. chem. Soc., 1951, **73**, 458).

(4) *O*-Methylation may be effected using an ethereal solution of diazomethane. Weakly acidic phenols require the presence of a catalyst such as fluoboric acid (*M. Neeman et al.*, Tetrahedron, 1959, **6**, 36).

(5) When phenols are heated with primary alcohols in the presence of dicyclohexylcarbodi-imide (D.C.C.) with or without solvent in a sealed tube the corresponding ether is formed in good yield (*E. Vowinkel*, Angew. Chem., 1963, **2**, 218). The reaction proceeds *via* an initial alcohol–D.C.C. adduct II, which is attacked by aryl oxide as indicated (*F. L. Bach*, J. org. Chem., 1965, **30**, 1300; *Vowinkel*, Ber., 1966, **99**, 1479):

(II) ⟶ $ArOCH_2R + (C_6H_{11}NH)_2CO$

Secondary and tertiary alcohols give poor yields of ethers.

(6) Moderately good yields of alkyl aryl ethers are obtained by treatment of aryl Grignard reagents or, preferably, aryl-lithiums with a dialkyl peroxide (*G. A. Baramki, H. S. Chang* and *J. T. Edward*, Canad. J. Chem., 1962, **40**, 441). Better results are achieved using an aryl Grignard reagent and a perbenzoate ester (*S. O. Lawesson* and *N. C. Yang*, J. Amer. chem. Soc., 1959, **81**, 4230; Org. Synth., 1967, **41**, 91). This method is particularly useful for preparing aryl *tert*-butyl ethers.

(7) Diaryl ethers are prepared by heating an aryl bromide with an alkali metal aryl oxide in the presence of copper powder (*Ullmann* and *P. Sponagel*, Ber., 1905, **38**, 2211; *H. E. Ungnade*, Chem. Reviews, 1946, **38**, 405). Alternatively the phenol and aryl halide may be heated with cuprous oxide in

collidine (*R. G. R. Bacon* and *H. A. O. Hill*, J. chem. Soc., 1964, 1108). In some cases the salt of an aryl sulphonic acid has been used instead of an aryl bromide in the Ullmann reaction (*E. H. Nollau* and *L. C. Daniels*, J. Amer. chem. Soc., 1914, **36**, 1885).

(8) Reaction of phenols with diaryliodonium salts at 100° in the presence of alkali (*F. M. Beringer et al., ibid.*, 1953, **75**, 2708) and thermal decomposition of diaryl carbonates (*H. Witt, H. Holtschmidt* and *E. Müller*, Angew. Chem., 1970, **9**, 67) also yields diaryl ethers.

(9) Special methods of limited application are:

(a) Decomposition of diazonium salts in the presence of alcohols frequently gives significant quantities of ethers but the ratio of ether to hydrocarbon produced depends upon a number of factors of which the structure of the alcohol and the acidity of the solution are particularly important (*D. F. De Tar* and *T. Kosuge, ibid.*, 1958, **80**, 6072).

(b) Phenols may be alkylated with alcohols or alkenes over alumina or silica. Mixtures of *O*- and *C*-alkylated products are formed, the former predominating below 200° (*N. M. Cullinane* and *W. C. Davies*, B.P., 600,835 and 600,837/1948).

(c) A mixture of phenol and alcohol when passed over thoria at 420° yields an ether (*P. Sabatier* and *A. Mailhe*, Compt. rend., 1910, **151**, 359).

(d) Generation of an aryne in an alcoholic solvent usually leads to the formation of an alkyl aryl ether (*J. F. Bunnett et al.*, J. Amer. chem. Soc., 1966, **88**, 5250).

(10) *Polymeric ethers*. (a) *Polyaryl ethers* of type III are prepared by autoxidative coupling of hindered phenols using a copper salt and amine as catalyst or by oxidation of bromophenols with ferricyanide (*A. S. Hay*, J. Pol. Sci., 1962, **58**, 581; *W. A. Batte, N. S. Chu* and *C. C. Price*, in "Macromolecular Syntheses", Wiley, New York, 1963, p. 76). Poly(2,6-dimethyl-1,4-phenylene)

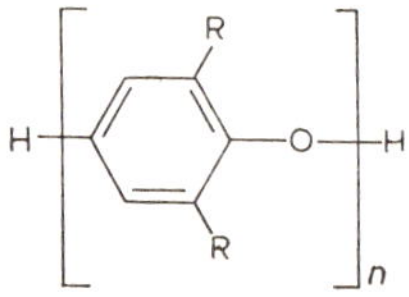

(III)

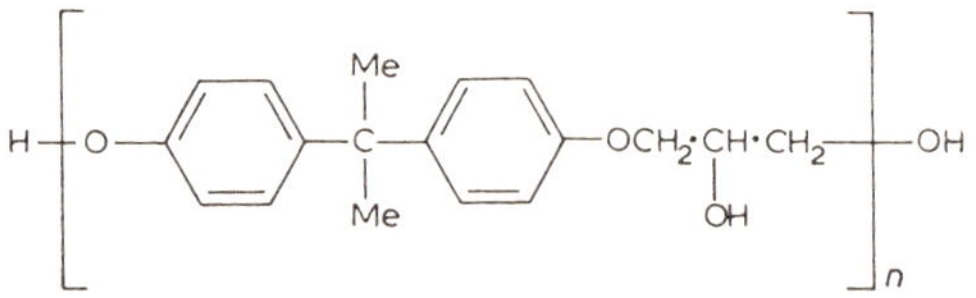

(IV)

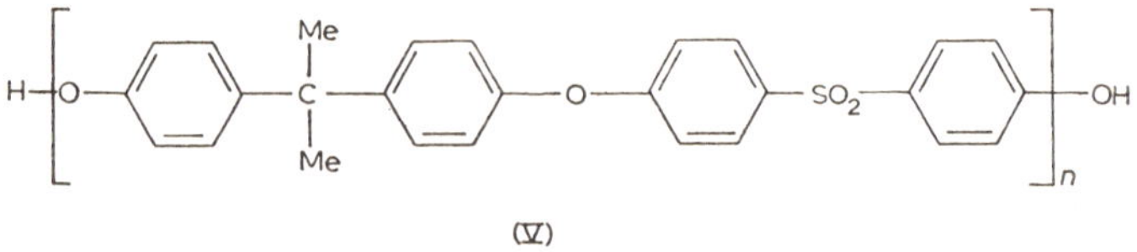

(V)

ether (**III**; R = Me) is marketed under the name PPO and is used in the manufacture of insulating materials.

(b) *Polyhydroxy ethers* (phenoxy resins), *e.g.*, **IV** are formed from dihydric phenols, epichlorohydrin and sodium hydroxide in dimethyl sulphoxide solution (*J. Wynstra*, U.S.P., 3,277,051/1966; *Union Carbide*, U.S.P., 3,305,528/1967). The related polysulphones, *e.g.* V, are similarly prepared from a dihydric phenol, 4,4'-dichlorodiphenyl sulphone and alkali (*Union Carbide Corp.*, Dutch P., 6,408,130/1965). Both polymers are useful for a variety of purposes including the manufacture of food containers and electronic equipment.

(*ii*) *Properties and reactions of ethers*

(1) *Spectra*. Diaryl and alkyl aryl ethers show relatively strong i.r. absorption in the 1270–1200 region attributable to C–O stretching vibrations. Methoxybenzenes have an easily detectable band at 2850 cm^{-1} (C–H) (*K. Nakanishi*, "Infrared Absorption Spectroscopy", Holden Day, San Francisco, 1962). The chemical shifts of the O–CH protons normally lie in the range τ 6·1–6·4 and there is a reasonably good correlation between these shifts and the Hammett σ-parameters of *meta*- or *para*-substituents (*C. Heathcock*, Canad. J. Chem., 1962, 40, 1865). In the mass spectra of diaryl ethers the base peak is usually that of the parent-ion, fission of one of the C–O bands being a major decomposition route of this ion. The parent-ions of alkyl phenyl ethers, predictably, give less intense peaks. The base peak is usually of mass 94 (except for methoxybenzenes) and is attributed to the ion VI formed by a "McLafferty rearrangement" of the parent-ion as indicated (*J. H. Beynon, R. A. Saunders* and *A. E. Williams*, "The Mass Spectra of Organic Molecules", Elsevier, London, 1968, p. 160).

$$\left[\text{PhO–CH}_2\text{–CH(H)R}\right]^{\oplus\cdot} \longrightarrow \left[\text{C}_6\text{H}_6\text{O}\right]^{\oplus\cdot} \text{(VI)} + \text{RCH=CH}_2$$

(VI)

(2) *With bases*. Alkyl aryl ethers are relatively stable to amines and to aqueous alkali. With alcoholic alkali there is some decomposition into aryl oxide and alcohol. Thus, anisole gives phenol (7%) after heating at 180–200° for 7 h with 20% alcoholic potash (*G. K. Hughes* and *E. O. P. Thomson*, Proc. roy. Soc. N.S.W., 1949, 83, 269). This resistance is diminished if electron-withdrawing substituents are present in the *ortho*- and/or *para*-positions. *o*- and *p*-Nitroanisoles are slowly converted into nitrophenols by

refluxing with aqueous alkali, 2,4-dinitro- and 2,4,6-trinitro-anisoles being hydrolysed even more easily. The alkoxyl group in the alkyl poly-nitroaryl ethers may be replaced by hydrazino, amino, and hydroxylamino groups. (For a review see *R. L. Burwell*, Chem. Reviews, 1954, **54**, 615). Very strong bases such as sodamide (*Bunnett* and *T. K. Brotherton*, Chem. Ind. London, 1957, 80), sodium hydride and lithium diphenylphosphide (*F. B. Mann* and *M. J. Pragnell*, *ibid.*, 1964, 1386) also cleave phenyl ethers. Thiolates in hot dimethylformamide demethylate aryl methyl ethers (*G. I. Feutrill* and *R. N. Mirrington*, Tetrahedron Letters, 1970, 1327).

(3) *With acids.* Ethers, by virtue of the unshared pair of electrons on their oxygen atoms, are Lewis bases and are protonated in strongly acidic media. n.m.r. measurements of alkyl aryl and diaryl ethers in strong acids such as hydrogen fluoride have shown, however, that protonation may occur on C and/or on O giving the ions VII and VIII, respectively (*D. M. Brower*,

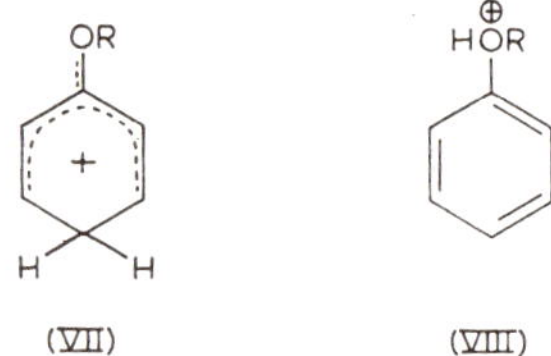

E. L. Mackor and *C. McLean*, Rec. Trav. chim., 1966, **85**, 108; *T. Birchall et al.*, Canad. J. Chem., 1964, **42**, 1433). Cleavage of alkyl aryl ethers by heating with hydriodic acid or preferably with constant boiling hydrobromic acid (*D. M. W. Anderson et al.*, Analyst, 1963, **88**, 353) is thought to proceed by nucleophilic attack by halide ion on an intermediate oxonium species such as VIII. (Detailed discussion of the mechanism is given by *E. Staude* and *F. Patat*, in "The Chemistry of the Ether Linkage", Interscience, New York, 1966, Ch. 2.) The above reaction with hydrogen iodide is the basis of the Zeisel method for the determination of alkyl aryl ethers. The alkyl iodide formed is distilled into alcoholic silver nitrate and the excess of silver nitrate is back-titrated. Diaryl ethers because of their lower basicities are not cleaved by hydrogen halides (*N. Orlow* and *W. Ipatiew*, Ber., 1927, **60**, 1963). The hydrohalides of bases such as pyridine or aniline may also be used to de-alkylate phenolic ethers (*V. Prey*, *ibid.*, 1941, **74**, 1219).

Lewis acids such as aluminium halides (*C. Hartmann* and *L. Gattermann*, *ibid.*, 1892, **25**, 3531) and boron halides (*W. Gerrard* and *M. F. Lappert*, Chem. Reviews, 1958, **58**, 1081; *J. F. W. McOmie* and *M. L. Watts*, Chem. Ind. London, 1963, 1658) yield complexes with ethers which, when heated, decompose to aryloxy metal halides from which the phenols may be obtained by treatment with water:

$$ArOR + MX_3 \rightarrow RX + ArOMX_2 \xrightarrow{H_2O} ArOH + MX_2OH$$

(4) *With alkali metals.* Both alkyl aryl and diaryl ethers are readily cleaved by heating with dispersed sodium or by treatment with sodium–potassium alloy at room temperature (*E. Müller* and *W. Bunge*, Ber., 1936, **69**, 2164; *P. Schorigin, ibid.*, 1924, **57**, 1627; *K. W. Müller*, U.S.P. 2,862,035/1958; *Prey*, Ber., 1943, **76**, 156). Fission of ethers may often be achieved more efficiently by treatment with an alkali metal in the presence of biphenyl (*D. H. Eargle*, J. org. Chem., 1963, **28**, 1703; *J. J. Eisch* and *A. M. Jacobs, ibid.*, 1963, **28**, 2145), the effective reagent with lithium being the dilithiobiphenyl. Alkali metal alkyls also cleave some ethers (*R. L. Letsinger* and *A. Bobko*, J. Amer. chem. Soc., 1953, **75**, 2649).

(5) *Electrophilic aromatic substitution.* Nuclear substitution of aryl ethers occurs in the *ortho*- and *para*-positions, the ring being more reactive than that of benzene.

(a) *Halogenation.* No catalyst is necessary for bromination or chlorination and two or three halogens may be introduced into the ring. Extensive measurements of the rates of halogenation have been made (*e.g.*, see *H. C. Brown* and *L. M. Stock, ibid.*, 1960, **82**, 1942; *B. Jones* and *E. N. Richardson*, J. chem. Soc., 1956, 3939; *D. R. Harvey* and *R. O. C. Norman, ibid.*, 1961, 3604). Disproportionate amounts of *ortho*-halogeno isomers are usually formed in these reactions. If a halogen carrier such as aluminium trichloride or tribromide is used, dealkylation occurs with the formation of a tetra- or penta-halogenophenol (*A. Bonneaud*, Bull. Soc. chim. Fr., 1910, **7**, 776). Chlorination of alkyl aryl ethers with sulphuryl chloride leads to α-chlorination of the alkyl group (*F. G. Bordell* and *B. M. Pitt*, J. Amer. chem. Soc., 1955, **77**, 572). Chlorination of alkyl aryl ethers containing deactivating ring substituents such as chlorine in the presence of phosphorus pentachloride at about 200° gives mono-, di-, or tri-α-chlorinated products (*H. J. Barber, R. F. Fuller* and *M. B. Green*, J. applied Chem. London, 1953, **3**, 409; *H. Gross* and *W. Bürger*, Org. Synth., 1970, **49**, 16). A mixture of odine and diborane readily cleaves aromatic ethers (*L. H. Long* and *G. F. Freeguard*, Nature, 1966, **207**, 403).

(b) *Nitration.* Nitration occurs smoothly giving either mono-, di- or tri-nitroaryl ethers; anisole is about 70 times more reactive than benzene to nitration (*N. C. Deno* and *R. Stein*, J. Amer. chem. Soc., 1956, **78**, 578). Relative reactivity and isomer distribution studies have been reported by *C. A. Bunton, G. J. Minkoff* and *R. I. Reed* (J. chem. Soc., 1947, 1416) and *M. J. S. Dewar* and *D. S. Urch* (*ibid.*, 1958, 3079). Under suitable conditions *K. H. Meyer* and *H. Gottlieb-Billroth* (Ber., 1919, **52**, 1476) obtained highly coloured "diaryloxoammonium salts" (IX) by treatment of alkyl aryl ethers with mixed acids. These salts are readily reduced to diaryl nitroxides by a variety of reagents.

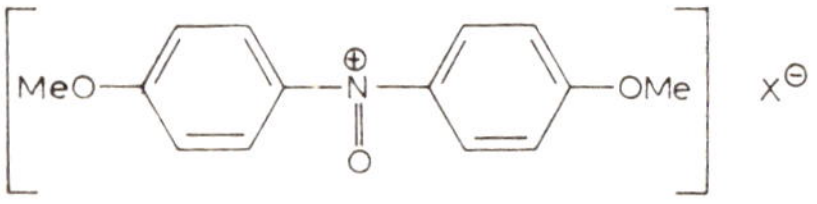

(IX)

(c) *Diazo coupling.* Alkyl aryl ethers couple with reactive diazo compounds, preferably in acetic acid solution to give azo compounds (*Meyer et al.*, Ber., 1914, 47, 1741). In some cases dealkylation accompanies the coupling reaction (*J. F. Bunnett* and *G. B. Hoey,* J. Amer. chem. Soc., 1958, 80, 3142).

(d) *Acylation and alkylation.* Alkyl aryl ethers undergo normal Friedel-Craft acylations and alkylations in the presence of polyphosphoric acid (*P. D. Gardiner, ibid.*, 1954, 76, 4550). Aluminium chloride usually dealkylates ethers and hence its use in such reactions is not recommended unless the substituted phenol is required (*M. Kulka, ibid.*, 1954, 76, 5469). Carboxylation of anisole by anodic oxidation of a methanolic solution of anisole containing sulphuric acid (1%) has been reported (*V. D. Parker,* Chem. Ind. London, 1968, 1363).

(6) *Rearrangements.* (a) *Claisen rearrangement.* This subject has been reviewed several times (*D. S. Tarbell,* Org. Reactions, 1946, 2, 1; *S. J. Rhoads,* in "Molecular Rearrangements", Interscience, New York, 1963, Vol. 1, p. 660; *D. L. Dalrymple, T. L. Kruger* and *White,* in "The Chemistry of the Ether Linkage", Interscience, New York, 1967, p. 635).

The reaction was discovered by *L. Claisen* (Ber., 1912, 45, 3157) who showed that allyl *o*-anisyl ether rearranged on heating to 2-allyl-6-methoxyphenol. Unsaturated ethers with double bonds in other positions did not isomerise under these conditions. The following points briefly summarise

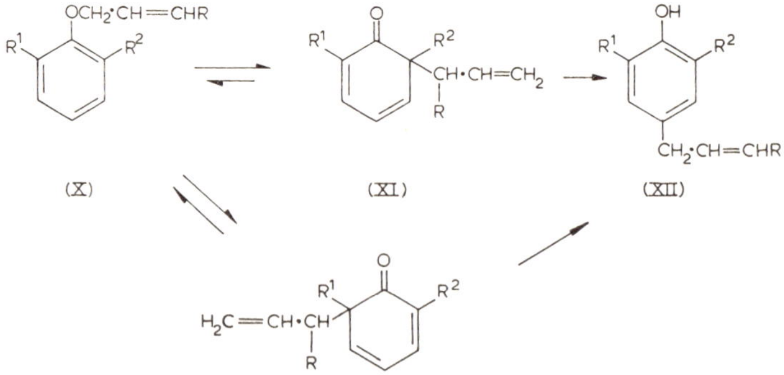

(X) (XI) (XII)

the numerous and intensive studies of the mechanism of the reaction which is outlined below.

(i) In the "*ortho*-rearrangement" (*i.e.* X → XI) the alkyl group is reversed while in the "*para*-rearrangement" it is not. Studies in which the α- or γ-positions of the allyl group were labelled with alkyl groups or ^{14}C confirmed this result.

(ii) The rearrangements are intramolecular, proceeding *via* cyclic transition states. No scrambled products are formed when a mixture of two allyl aryl ethers is heated.

(iii) The intermediate formation of *ortho*-dienones XI (which rapidly isomerise to phenols when R^1 or R^2 = H) in the "*para*-rearrangement" has been established by trapping the dienone XI ($R^1 = R^2$ = Me) as its Diels-Alder adduct with maleic anhydride. The dienone XI ($R^1 = R^2$ = Me) has also been synthesised and shown to rearrange to the phenol XII ($R^1 = R^2$ = Me) and the ether X ($R^1 = R^2$ = Me).

(iv) The initial step is reversible. Allyl-2,6-di(methallyl)phenyl ether and 2-allyl-6-methallylphenyl methallyl ether each give the same mixture of phenols on heating.

Abnormal products are frequently encountered. These are usually the result of the displacement of a ring substituent, isomerisation of the normal product of rearrangement or the migration of the allyl group to an *ortho*- or *para*-vinyl side-chain.

(b) *Acid-catalysed rearrangements.* Alkyl aryl ethers rearrange to *ortho*- and/or *para*-alkyl phenols in the presence of Lewis acids such as aluminium chloride or boron trifluoride. Unlike the Claisen rearrangement both intra- and inter-molecular migration of the alkyl group is thought to occur, the extent of each being dependent on a number of factors such as the catalyst and the solvent (*Dewar* and *N. A. Puttman,* J. chem. Soc., 1960, 959; 1959, 4086, 4090; *R. A. Spanninger* and *J. L. von Rosenberg,* Chem. Comm., 1970, 795) Acid-catalysed rearrangement of allyl ethers gives products similar to those obtained from the thermal Claisen rearrangement (*Gerrard, Lappert* and *H. B. Silver,* Proc. chem. Soc., 1957, 19); *o*-methyldiaryl ethers rearrange to *o*-benzylphenols at ~400° (*A. Factor et al.,* J. org. Chem., 1970, **35**, 57).

(c) *Wittig rearrangement.* Aryl benzhydryl ethers and other ethers with relatively acidic protons yield isomeric carbinols on treatment with strong bases such as aryl-lithiums, sodamide or alkoxides (*G. Wittig, H. Döser* and *I. Lorenz,* Ann., 1949, **562**, 192). A summary of work on the mechanism of this reaction is given by *Dalrymple, Kruger* and *White* (*loc. cit.*).

$$Ph_2CH—O—Ar \xrightarrow{PhLi} Ph_2C^{\ominus}—O—Ar \rightarrow Ph_2\underset{\substack{|\\ Ar}}{C}—O^{\ominus} \rightarrow Ph_2\underset{\substack{|\\ Ar}}{C}—OH$$

(7) *Reactions induced by light.* Photolysis of aryl ethers leads to rearranged products (cf., photo-Fries rearrangement on p. 330). Thus, benzyl phenyl ether gives mainly *p*-benzylphenol, and diphenyl ether gives *p*-hydroxybiphenyl (*M. S. Kharasch, G. Stampa* and *W. Nudenberg,* Science, 1959, **116,** 309; *D. P. Kelly, J. T. Pinhey* and *R. D. G. Rigby,* Austral. J. Chem., 1969, **22,** 977; Tetrahedron Letters, 1966, 5953). Phenoxyacetic acids behave similarly (*Kelly* and *Pinhey, ibid.,* 1964, 3427). When anisole is photolysed in the presence of good nucleophiles such as ethoxide, displacement of the methoxyl group occurs to a small extent (*J. A. Baltrop, N. J. Bruce* and *A. Thomson,* J. chem. Soc., C, 1967, 1142). The methoxyl groups in photoactivated *m*-nitroanisoles are much more labile and may be displaced by hydroxide and cyanide ions and by amines (*E. Havinga et al.,* Rec. Trav. chim., 1966, **85,** 56, 275; *R. L. Letsinger* and *J. H. McCain,* J. Amer. chem. Soc., 1966, **88,** 2884; 1969, **91,** 6425). In some cases the nitro group in *p*-nitroanisole may be similarly replaced.

(8) *Oxidation and reduction.* Oxidation of alkyl aryl ethers has been reviewed recently (*O. C. Musgrave,* Chem. Reviews, 1969, **69,** 499). Oxidation may lead to the formation of biaryls, hydroxylated or acyloxylated products, or quinones depending upon the oxidising agent and the conditions. Homolytic oxidants frequently oxidise the side-chain.

Reduction of alkyl aryl ethers with sodium in liquid ammonia yields *cyclo*-hexenones (*A. J. Birch,* Quart. Reviews, 1950, **4,** 69; *H. Smith,* "Chemistry in Non Aqueous Ionising Solvents", Interscience, New York, 1963, Vol. 1, pt. 2, p. 245). Lithium aluminium hydride in the presence of catalytic quantities of nickel chloride hydrogenolyses alkyl aryl ethers to phenols and alkenes (*V. L. Tweedie* and *M. Cuscurida,* J. Amer. chem. Soc., 1957, **79,** 5463). *E. J. Corey et al.* (*ibid.,* 1969, **91,** 4782) have recently described a novel method of reducing a phenolic ring. Electrophilic substitution of the first-formed aryloxy acid gives a cationic intermediate which is trapped by an internal nucleophile (the carboxyl group) giving a cyclohexadiene which may then be reduced to a cyclohexane derivative:

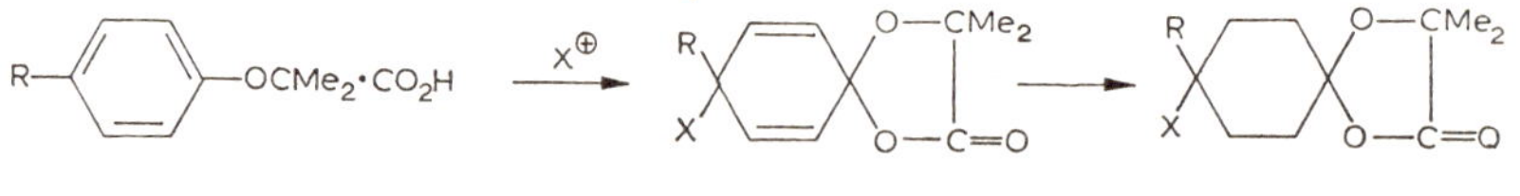

(*iii*) *Alkyl aryl ethers*

Anisole, methyl phenyl ether, m.p. —37°, b.p. 153–158°, d_4^0 1·012, n_D^{22} 1·5156, is conveniently prepared by the reaction of phenol with dimethyl sulphate in aqueous alkaline solution (*G. S. Hiers* and *F. D. Hager,* Org. Synth., Coll. Vol. 1, 3rd Edn., 1964, p. 58) or by treating phenol with dimethyl ether and methanol over alumina at about 200⁰ (*N. M. Cullinane* and *W. C. Davies*, B.P., 600,835 and 600,837/1948). Vigorous bromination in the presence of aluminium bromide gives pentabromophenol (*A. Bonneaud*, Bull. Soc. chim. Fr., 1910, **7**, 776). *p*-Iodoanisole is formed by the action of iodine in the presence of mercuric oxide (*F. F. Blicke* and *F. D. Smith*, J. Amer. chem. Soc., 1928, **50**, 1229). Nitration and sulphonation occur readily. Photolysis of aromatic iodo compounds in anisole gives reasonably good yields of the corresponding biaryls (*W. Wolf* and *N. Kharasch*, J. org. Chem., 1961, **26**, 283). It is used as an antioxidant and as a stabiliser for vinyl polymers.

Phenetole, ethyl phenyl ether, m.p. —30^0, b.p. 172^0, d_4^0 0·9852, n_D^{19} 1·5084. *Phenyl* n-*propyl ether*, b.p. 189^0. *Isopropyl phenyl ether*, b.p. 176·8^0. *tert-Butyl phenyl ether*, b.p. 61·8^0 (*W. T. Olson et al.*, J. Amer. chem. Soc., 1947, **69**, 2451). n-*Butyl phenyl ether*, b.p. 210^0 (*A. I. Vogel*, J. chem. Soc., 1948, 616). *Isobutyl phenyl ether*, b.p. 196^0 (*I. B. Douglass* and *F. B. Dains*, J. Amer. chem. Soc., 1934, **56**, 719). *Isopentyl phenyl ether*, b.p. 225^0 (*E. Waser et al.*, Helv., 1929, **12**, 418).

Methyl o-*tolyl ether*, b.p. 172^0, *methyl* m-*tolyl ether*, b.p. 176·5^0, *methyl* p-*tolyl ether*, b.p. 176·7^0 (*Olson et al., loc. cit.*); *ethyl* o-*tolyl ether*, b.p. 190^0 (*G. Baddeley* and *M. A. Vickars*, J. chem. Soc., 1958, 4665).

(*iv*) *Unsaturated alkyl aryl ethers*

Phenyl vinyl ether, b.p. 158^0, is best prepared by distilling phenoxyethyl bromide over powdered potassium hydroxide (*S. G. Powell* and *R. Adams*, J. Amer. chem. Soc., 1920, **42**, 646), or from phenol and vinyl chloride in the presence of bases (*I.G. Farbenind.*, G.P., 513,679/1927), or from phenol and ethylene dichloride and sodium ethoxide (*idem*, G.P., 525,188/1929). On pyrolysis it yields phenol and the diphenyl acetal of acetaldehyde (*W. M. Lauer* and *M. A. Spielman*, J. Amer. chem. Soc., 1933, **55**, 1572) or phenol and acetylene (*H. V. R. Iengar* and *P. D. Ritchie*, J. chem. Soc., 1957, 2556) depending upon the conditions.

2-*Bromovinyl phenyl ether*, PhOCH:CHBr, b.p. 98^0/23 mm, is formed from acetylene tetrabromide and potassium phenoxide. Alcoholic potash converts it into *phenoxyacetylene*, an unstable oil, b.p. 75^0/35 mm, which forms metal acetylides such as $(PhOC{:}C)_2Cu$ (*M. Slimmer*, Ber., 1903, **36**, 289). A review of ethynyl ethers was published in 1960 (*J. F. Arens*, Advances in Organic Chemistry, Vol. 2, Interscience, New York, 1960, p. 117).

Allyl phenyl ether, b.p. 191–192^0, 73^0/11 mm, is a liquid with a geranium-like odour. It is formed by heating phenol and allyl bromide in acetone containing suspended potassium carbonate. When boiled it gradually isomerises to 2-allylphenol, the temperature gradually rising until at 220^0 the rearrangement is complete (*Claisen, loc. cit.*; Ann., 1919, **418**, 69). This reaction is discussed more fully on p. 324.

3-*Phenoxyprop*-1-*yne, phenyl* 2-*propynyl ether*, $PhOCH_2{\cdot}C{:}CH$, b.p. 98^0/23 mm, is obtained by the reaction of propargyl bromide on phenol in alcoholic alkali (*M. M. Kreevoy, H. B. Charman* and *D. R. Vinard*, J. Amer. chem. Soc.,

1961, **83**, 1978). Heating in diethylaniline gives 3-chromene (*I. Iwai* and *J. Ide*, Chem. Pharm. Bull. Japan, 1963, **11**, 1042).

(*v*) *Alkyl aryl ethers with substituents in the alkyl group*

2-**Bromoethyl phenyl ether,** m.p. 39°, b.p. 240–250° (decomp.), 125–130°/18 mm, is prepared by the reaction of 1,2-dibromoethane with sodium phenoxide. 1,2-Diphenoxyethane is also formed. 3-*Bromopropyl phenyl ether*, m.p. 7–8°, b.p. 136–142°/20 mm, is similarly prepared from dibromopropane (*C. S. Marvel* and *A. L. Tanenbaum*, Org. Synth., Coll. Vol. 1, 1964, p. 435). 2-*Chloroethyl phenyl ether*, m.p. 28°, b.p. 220°, 101°/12 mm (*G. R. Clemo* and *W. H. Perkin*, J. chem. Soc., 1922, **121**, 642). 3-*Chloropropyl phenyl ether*, m.p. 12°, b.p. 240°, 127–129°/13 mm (*R. E. Lyle, E. J. De Witt* and *I. C. Pattison*, J. org. Chem., 1956, **21**, 61). 2-*Iodoethyl phenyl ether*, m.p. 31° and 3-*iodopropyl phenyl ether*, m.p. 12°, b.p. 155°/16 mm, may be prepared from the corresponding bromides by heating with sodium iodide in alcohol (*J. von Braun*, Ber., 1913, **46**, 1782). The halogen atoms in the above ethers are easily replaced by CN, NH_2, NO_2, OR and OAr by standard methods.

2-**Hydroxyethyl phenyl ether,** or **glycol monophenyl ether,** b.p. 237°, 165°/80 mm, is prepared by the reaction of sodium phenoxide with ethylene chlorohydrin, or by the action of ethylene oxide on phenol (*R. A. Smith*, J. Amer. chem. Soc., 1940, **62**, 994; *I.G. Farbenind.*, U.S.P., 1,976,677/1933). 3-*Hydroxypropyl phenyl ether*, $HOCH_2{\cdot}CH_2{\cdot}CH_2OPh$, b.p. 158–160°/25 mm (*Powell*, J. Amer. chem. Soc., 1923, **45**, 2708). 2,3-*Dihydroxypropyl phenyl ether*, *glycerol α-phenyl ether*, $PhOCH_2{\cdot}CH(OH){\cdot}CH_2(OH)$, m.p. 70°, b.p. 176°/16 mm, is conveniently prepared by reaction of glycerol α-chlorohydrin with sodium phenoxide in alcohol (*T. S. Wheeler* and *F. G. Willson*, Org. Synth., Coll. Vol. 1, 1964, p. 296). 2-*Phenoxypropane*-1,3-*diol*, $CH_2OH{\cdot}CHOPh{\cdot}CH_2OH$, m.p. 68° (*S. W. Chaikin*, J. Amer. chem. Soc., 1948, **70**, 3522). 1,3-*Diphenoxypropan-2-ol*, *glycerol* 1,3-*diphenyl ether*, m.p. 82°, is prepared by heating sodium phenoxide with 1,3-dichloropropan-2-ol at 100–120° (*A. Fairbourne, G. P. Gibson* and *D. W. Stephens*, J. chem. Soc., 1932, 1965) or with epichlorhydrin (*T. W. Evans* and *K. E. Marple*, U.S.P., 2,351,025/1941).

Diphenoxymethane, m.p. 15°, b.p. 295° (*W. Baker*, J. chem. Soc., 1931, 1765). 1,2-*Diphenoxyethane*, *ethylene glycol diphenyl ether*, m.p. 97–98° (*A. C. Cope*, J. Amer. chem. Soc., 1935, **57**, 572). 1,3-*Diphenoxypropane*, m.p. 61° (*J. A. King* and *F. H. McMillan*, *ibid.*, 1945, **67**, 336). 1,4-*Diphenoxybutane*, m.p. 98° (*L. W. Deady, R. D. Topsom* and *J. Vaughan*, J. chem. Soc., 1965, 5718). 1,5-*Diphenoxypentane*, m.p. 46° (*von Braun* and *E. Kamp*, Ber., 1937, **70**, 973). 1,6-*Diphenoxyhexane*, m.p. 83°; 1,7-*diphenoxyheptane*, m.p. 55° (*A. Müller* and *W. Vanc*, *ibid.*, 1944–46, **77–79**, 669).

Phenoxy-substituted amines, aldehydes, ketones and acids are generally prepared from the corresponding halogen-substituted derivatives by reaction with sodium phenoxide.

1-**Amino-2-phenoxyethane, 2-phenoxyethylamine,** b.p. 228°; 3-*phenoxypropylamine*, b.p. 241°, 126°/15 mm (*von Braun*, *ibid.*, 1937, **70**, 979). 4-*Phenoxybutyl-*

amine, b.p. 254^0, 146–148^0/17 mm (*D. G. Doherty, R. Shapira* and *W. T. Burnett*, J. Amer. chem. Soc., 1957, **79**, 5667). cis- and trans-2-*Phenoxycyclopropylamine*, b.p. 67^0/1 mm and m.p. 122–124^0, respectively (*J. Finkelstein, E. Chiang* and *J. Lee*, J. med. Chem., 1965, **8**, 432).

Phenoxyacetaldehyde, b.p. 215^0, 83^0/4 mm, is best prepared by oxidation of glycerol α-phenyl ether (*R. J. Speer* and *H. R. Mahler*, J. Amer. chem. Soc., 1949, **71**, 1133) (*semicarbazone*, m.p. 145^0; 2,4-*dinitrophenylhydrazone*, m.p. 138^0, *diethyl acetal*, b.p. 254^0; *oxime*, m.p. 95^0). **Phenoxyacetone,** b.p. 230^0, 128^0/23 mm, is prepared from phenoxyacetyl chloride by reaction with methylzinc iodide (*Brettle*, J. chem. Soc., 1956, 1891) (*semicarbazone*, m.p. 172^0). On treatment with concentrated sulphuric acid it yields 3-methylbenzofuran (*R. Stoermer*, Ber., 1895, **28**, 1253; 1902, **35**, 3553).

Phenoxyacetic acid, m.p. 96^0, is formed in poor yield from the reaction of chloroacetic acid with phenol in aqueous alkaline solution and is best prepared by condensation of phenol and ethyl chloroacetate followed by hydrolysis of the resulting phenoxyacetate (*Brettle, loc. cit.*). Homologues of phenol usually react smoothly with chloroacetic acid to give highly crystalline derivatives which have been recommended for the characterisation of phenols (*A. I. Vogel*, "Elementary Practical Organic Chemistry", Longmans, New York, 1966, Pt. II, p. 326; *C. F. Koelsch*, J. Amer. chem. Soc., 1931, **53**, 304). Aryloxyisobutyric acids are prepared from chloretone, [$Me_2C(OH)\cdot CCl_3$], alkali and phenol in acetone solution (*E. J. Corey, S. Barcza* and *G. Klotmann, ibid.*, 1969, **91**, 4782).

2,4-Dichloro- and 2,4,5-trichloro-phenoxyacetic acids are still the most extensively used selective weed killers (*K. A. Hassall*, "World Crop Production", Iliffe, London, 1969, Vol. 2, p. 220). 2,4-Dichloro-6-hydroxyphenoxyacetic acid, however, shows no plant growth regulating properties (*G. W. K. Cavill* and *D. L. Ford*, J. chem. Soc., 1954, 565).

(*vi*) *Diaryl ethers*

Diphenyl ether, m.p. 28^0, b.p. 252^0, forms long needles with a floral odour. It is obtained commercially as a by-product in the production of phenol by the Dow and Raschig processes (p. 291) (*W. J. Hale* and *E. C. Britton*, U.S.P., 1,882,884/1933). It may also be prepared by an Ullmann reaction using bromobenzene and potassium phenoxide (*F. Ullmann* and *P. Sponagel*, Ber., 1905, **38**, 2211; Ann., 1906, **350**, 83). The pyrolysis of the aluminium salt of phenol is also claimed to give a good yield of diphenyl ether (*A. N. Cook*, J. Amer. chem. Soc., 1906, **28**, 608). It is unaffected by alkali, acid and chromium trioxide in acetic acid at room temperature. Because of its stability diphenyl ether is used as a component of high-temperature heat-transfer fluids and is marketed under the names Dowtherm A and B, which are eutectics with biphenyl and naphthalene, respectively. It may be di-lithiated in the 2 and 2′ positions with butyllithium in tetrahydrofuran (*H. Gilman* and *W. J. Trepka*, J. org. Chem., 1961, **26**, 5202) and on photolysis yields dibenzofuran (*H. Stegemeyer*, Naturwiss., 1966, **53**, 582).

Decafluorodiphenyl ether, m.p. 72^0, is prepared by treatment of hexafluorobenzene with sodium pentafluorophenolate in dimethylacetamide solution (*R. J. De Pasquale* and *C. Tamborski,* J. org. Chem., 1967, **32,** 3163).

(b) Esters of phenols

(i) Aryl esters of carboxylic acids

Esters are prepared from the corresponding phenol and acid chloride or anhydride (*F. D. Chattaway,* J. chem. Soc., 1931, 2495). The use of pyridine as solvent is recommended (*T. G. Bonner* and *P. McNamara,* J. chem. Soc., B, 1968, 795). They are not formed by the normal process of esterification with the free carboxylic acid and phenol unless either trifluoroacetic anhydride (*E. J. Bourne et al.,* J. chem. Soc., 1949, 2976; *R. C. Parish* and *L. M. Stock,* J. org. Chem., 1965, **30,** 927), trifluoromethanesulphonic anhydride (*T. Gramstad* and *R. N. Haszeldine,* J. chem. Soc., 1957, 4069), or polyphosphoric acid (*A. R. Bader* and *A. D. Kontowicz,* J. Amer. chem. Soc., 1953, **75,** 5416) is used as a catalyst. Dicyclohexylcarbodiimide has also been used as a condensing agent (*S. Neelakantan, R. Padmasini* and *T. R. Seshadri,* Tetrahedron, 1965, **21,** 3531).

Other methods of less synthetic importance include the Baeyer-Villiger oxidation of aromatic ketones (*C. H. Hassall,* Org. Reactions, 1966, **9,** 73), acyloxylation of aromatic hydrocarbons and ethers with nitric acid–acetic anhydride (*A. Fischer et al.,* J. chem. Soc., 1964, 3687), peroxides or peroxy acids (see p. 293) or acyloxylation by electrolysis in acetic acid solution containing sodium acetate (*L. Eberson* and *K. Nyberg,* Acta chem. Scand., 1964, **18,** 1568).

Their reactions are generally similar to those already described for alkyl esters (*Rodd,* C.C.C., 2nd Edn., Vol. 1C, p. 144). An important difference, however, is that aryl esters react with certain metal halides (aluminium trichloride, titanium tetrachloride) to give mixtures of *o*- and *p*-hydroxyaryl ketones, a reaction generally known as the Fries rearrangement (*K. Fries* and *G. Finck,* Ber., 1908, **41,** 4271). The ratio of *ortho-* to *para*-acylphenol formed depends largely upon the structure of the ester, the temperature and the solvent. A full account of the scope of the reaction is given by *A. H. Blatt* (Org. Reactions, 1948, **1,** 342) and a typical experimental procedure by *E. Miller* and *W. H. Hartung* (Org. Synth., Coll. Vol. II, 1966, p. 543). The mechanism is uncertain although it does seem to be at least partly intramolecular (see *M. J. S. Dewar,* in "Molecular Rearrangements", Interscience, New York, 1963, Vol. 1, p. 295 and *A. Gerecs,* in "Friedel Crafts and Related Reactions", Interscience, London, 1964, Vol. III, p. 299; *Dewar* and *L. Hart,* Tetrahedron, 1970, **26,** 973). Similar rearrangements occur on irradiation of aryl esters (photo-Fries reaction). Both cyclic and free radical-cage mechanisms have been proposed to account for the expected and unexpected products obtained (*e.g.*, see *J. S. Bradshaw, A. L. Loveridge* and *L. White,* J. org. Chem., 1968, **33**, 4127), the greater weight of evidence favouring the latter (*A. C. Day,* Ann. Reports B, 1967, **64**, 185; 1968, **65**, 209).

Phenyl formate, b.p. 90°/30 mm, is prepared in 85% yield by treatment of phenol with formic acid–acetic anhydride mixture. The method is a general one for the preparation of formate esters (*W. Stevens* and *A. van Es,* Rec. Trav. chim., 1964, **83**, 1294; 1965, **84**, 1247). *Phenyl orthoformate,* $CH(OPh)_3$, m.p. 76°, is formed from chloroform and potassium phenoxide (*H. Baines* and *J. E. Driver,* J. chem. Soc., 1924, **125**, 907). *Phenyl acetate,* b.p. 195°, gives mainly *p*-hydroxyacetophenone on warming with aluminium chloride in nitrobenzene (*K. W. Rosenmund* and *W. Schurr,* Ann., 1928, **460**, 56). When no solvent is used equal quantities of *o*- and *p*-hydroxyacetophenones are formed (*D. B. Bruce, A. J. S. Sorrie* and *R. H. Thomson,* J. chem. Soc., 1953, 2403). Selective reduction to phenylacetaldehyde may be accomplished using lithium tri-*tert*-butoxyaluminohydride at 0° (*P. M. Weissman* and *H. C. Brown,* J. org. Chem., 1966, **31**, 283). *Phenyl orthoacetate,* $MeCH(OPh)_3$, m.p. 98° (*S. M. McElvain* and *B. Fajardo-Pinzon,* J. Amer. chem. Soc., 1945, **67**, 650); *phenyl propanoate,* b.p. 80°/8 mm; *phenyl butanoate,* b.p. 222°; *phenyl pentanoate,* b.p. 120°/16 mm.

Diphenyl oxalate, m.p. 136°, b.p. 191°; *diphenyl malonate,* m.p. 50° (*C. A. Bischoff* and *A. von Hedenström,* Ber., 1902, **35**, 3437, 3452); *diphenyl succinate,* m.p. 120° (*G. H. Daub* and *W. S. Johnson,* Org. Synth., Coll. Vol. IV, 1962, p. 390); *diphenyl fumarate,* m.p. 161°; *diphenyl maleate,* m.p. 73° (*Bischoff* and *von Hedenström, loc. cit.*, p. 4084).

(ii) *Aryl esters of carbonic acid*

Phenyl carbonic acid, (*phenyl hydrogen carbonate*), $PhO{\cdot}CO_2H$, is not known. Its sodium salt has been obtained by the action of carbon dioxide on sodium phenoxide under pressure. It is decomposed by water and when heated at 120° under pressure it rearranges to sodium salicylate (*R. Schmitt,* J. pr. Chem., 1885, **31**, 397). The esters of phenylcarbonic acid, *e.g.*, $PhO{\cdot}CO_2Et$, are formed by the action of chloroformic esters on phenols (*Gilman* and *S. E. Kirby,* J. Amer. chem. Soc., 1932, **54**, 344). They decompose on heating with the formation of either alkyl phenyl ethers or at 500° to alkenes and phenols (*P. D. Ritchie,* J. chem. Soc., 1935, 1054).

Phenyl carbamate, "phenylurethane", $PhO{\cdot}CO{\cdot}NH_2$, m.p. 143°, is prepared from phenol and carbamoyl chloride (*R. Avenarius,* Z. Angew. Chem., 1923, **36**, 165) and by acid hydrolysis of phenyl cyanate (*D. Martin,* Angew. Chem., 1964, **76**, 311).

Phenyl N-*phenylcarbamate,* $PhO{\cdot}CO{\cdot}NHPh$, m.p. 124°, is formed by treatment of phenol with phenyl isocyanate. Phenols are frequently characterised in this way (*N. G. Gaylord* and *J. J. O'Brien,* Rec. Trav. chim., 1955, **74**, 218). *Phenyl* N,N-*diphenylcarbamate,* m.p. 105°, is formed from diphenylcarbamoyl chloride and phenol (*J. Herzig*, Ber., 1907, **40**, 1833) and *phenyl allophanate*, PhO·CO·NHCO·NH_2, m.p. 178°, from an excess of carbamoyl chloride on phenol (*Avenarius, loc. cit.*).

Diphenyl carbonate, $(PhO)_2CO$, m.p. 80°, b.p. 168°/15 mm, is prepared by passing phosgene (carbonyl chloride) into an aqueous solution of sodium phenoxide (*Bischoff* and *von Hedenström, loc. cit.*, p. 3431) or, less conveniently, by heating phenol and phosgene at 150°, by boiling phenol and carbon tetrachloride with zinc oxide and zinc chloride (*M. Gomberg* and *H. R. Snow*, J. Amer. chem. Soc., 1925, **47**, 198), by reaction of phenoxymagnesium bromide with phosgene (*S. T. Bowden* and *J. John*, J. chem. Soc., 1939, 314), and from an aryloxy nitrile and phenoxide as shown (*D. Martin* and *S. Rackow*, Ber., 1965, **98**, 3662):

$$ArO^{\ominus} + ArO{-}CN \longrightarrow (ArO)_2C{=}NH \longrightarrow (ArO)_2C{=}O$$

It is more easily hydrolysed than phenyl benzoate (*G. D. Cooper* and *B. Williams*, J. org. Chem., 1962, **27**, 3717). With sodium hydroxide at 200° it yields sodium salicylate and with ammonia, urea (*H. Eckenroth* and *K. Kock*, Ber., 1894, **27**, 3410). Diphenyl carbonate is used to prepare linear polycarbonates which are high melting polymers suitable for films, fibres and coatings. The carbonate and a dihydric phenol, *e.g.*, bisphenol A [2,2-bis(*p*-hydroxyphenyl)-propane] are allowed to react at 150° and 300° until a product of the correct molecular weight is obtained. In an alternative process the phenol in alkaline solution is treated with phosgene (*H. Schnell*, Ind. Eng. Chem., 1959, **51**, 157).

O,O-**Diphenyl thiocarbonate**, PhO·CS·OPh, m.p. 106°, is prepared by reaction of thiocarbonyl chloride (thiophosgene) with two moles of phenol in benzene–pyridine solution (*A. Schonberg et al.*, Ber., 1930, **63**, 178; Ann., 1930, **483**, 107). Such esters have been little studied and their chemistry is summarised by *E. R. Reid* ("Organic Chemistry of Bivalent Sulphur", Vol. IV, Chemical Publishing Co., New York, 1962, p. 138). On heating they undergo an interesting rearrangement with O to S migration yielding *O,S*-diaryl thiocarbonates (*H. R. Al-Kazimi*, *D. S. Tarbell* and *D. Plant*, J. Amer. chem. Soc., 1955, **77**, 2479):

$$PhO{\cdot}CS{\cdot}OPh \rightarrow PhS{\cdot}CO{\cdot}OPh$$

(*iii*) *Aryl esters of inorganic acids*

Phenylsulphurous acid, phenyl hydrogen sulphite, is not known in the free state. Its sodium salt, PhO·SO·ONa, is formed by the action of sulphur dioxide on sodium phenoxide; by reaction with methyl iodide it forms methyl phenyl sulphite, PhO·SO·OMe.

Diphenyl sulphite, $(PhO)_2SO$, b.p. 185°/15 mm, is prepared in 87% yield by heating thionyl chloride with chlorobenzene at 130° (*W. E. Bissinger* and *F. E.*

Kung, ibid., 1948, **70**, 2664). It is rapidly hydrolysed to phenol in alkaline solution and gives a sulphoxide on treatment with a Grignard reagent. For other reactions see *H. F. van Woerden*, Chem. Reviews, 1963, **63**, 557.

Phenylsulphuric acid, phenyl hydrogen sulphate, $PhOSO_2OH$, is known only as it salts. The *potassium salt*, $PhOSO_2 \cdot OK$, is not easily soluble in cold water, but is freely soluble in hot. It is present in the urine of herbivorous animals and also in that of man after ingestion of phenol. The *sodium salt*, $PhOSO_3Na,3H_2O$, is more soluble than the potassium salt. The salts may be prepared (a) by treatment of phenol with sulphur trioxide-amine complexes with or without a solvent below 100^0 (*P. Baumgarten*, Ber., 1924, **59**, 1976; *W. B. Hardy et al.*, J. Amer. chem. Soc., 1951, **73**, 3094; 1952, **74**, 5212); (b) by reaction of chlorosulphonic acid with phenol in the presence of pyridine or dimethylaniline (*R. D. Haworth* and *A. Lapworth*, J. chem. Soc., 1924, **125**, 1299; *J. Feigenbaum* and *C. A. Neuberg*, J. Amer. chem. Soc., 1941, **63**, 3529); (c) from sulphuryl chloride and an alkali phenoxide in benzene (*M. Battegay*, Compt. rend., 1932, **194**, 1505).

The salts of the arylsulphuric acids are in general stable in aqueous alkaline solution undergoing slow hydrolysis on heating to 100^0. In the presence of aqueous acid, hydrolysis to phenol and sulphuric acid is rapid.

Triphenyl phosphite, $(PhO)_3P$, m.p. $+22^0$, b.p. $210^0/11$ mm, is prepared by reaction of phenol with phosphorus trichloride (*H. B. Gottlieb*, J. Amer. chem. Soc., 1932, **54**, 748). It reacts with alcohols in the presence of either hydrogen halide or halogen to give alkyl halides:

$$(PhO)_3P + ROH + HX \rightarrow RX + (PhO)_2P^{\oplus}(O^{\ominus})H + PhOH$$

$$(PhO)_3P + ROH + X_2 \rightarrow RX + (PhO)_2P^{\oplus}(O^{\ominus})X + PhOH$$

(*H. N. Rydon et al.*, J. chem. Soc., 1953, 2224; 1954, 2281). Its use in the conversion of phenols into aryl halides has already been described (see p. 297). With ozone, triphenyl phosphate is formed (*Q. E Thompson*, J. Amer. chem. Soc., 1961, **83**, 845). *Phenoxyphosphorus dichloride*, $PhOPCl_2$, b.p. $90^0/11$ mm, and *diphenoxyphosphorus chloride*, $(PhO)_2PCl$, b.p. $172^0/11$ mm, are prepared by heating triphenyl phosphite with phosphorus trichloride at 150^0 (see *Rydon*, Chem. Soc. Special Publication, 1957, No. 8, p. 61). *Triphenylphosphite dibromide*, $(PhO)_3PBr_2$, is used to convert acetylenic alcohols into bromides (*D. K. Black et al.*, Tetrahedron Letters, 1963, 483).

Triphenyl phosphate, $(PhO)_3PO$, m.p. 49^0, b.p. $254^0/14$ mm, is best prepared by the action of phosphoryl chloride on phenol in aqueous alkaline solution. With phenylmagnesium bromide it gives phenol and triphenylphosphine oxide in low yield (*H. Gilman* and *C. C. Vernon*, J. Amer. chem. Soc., 1926, **48**, 1063). *Diphenyl phosphate*, $(PhO)_2PO(OH)$, m.p. 71^0, *dihydrate*, m.p. 51^0, and *monophenyl phosphate*, $PhOPO(OH)_2$, m.p. 99^0, are hydrolysed by base with increasing difficulty. Their *chlorides*, $PhOPOCl_2$, b.p. 240^0; and $(PhO)_2POCl$, b.p. $215^0/21$ mm, are formed by heating phenol with phosphoryl chloride at 110^0

(*H. F. Freeman* and *C. W. Colver, ibid.*, 1938, **60**, 750; *R. H. A. Plimmer* and *W. T. N. Burch*, J. chem. Soc., 1929, 292). These chlorides are used for the preparation of monoalkyl and symmetrical dialkyl phosphates (*D. M. Brown*, in "Advances in Organic Chemistry", 1963, Vol. 3, p. 87).

Tetraphenyl orthosilicate, m.p. 48°, is prepared by heating silicon tetrachloride with phenol (*B. Smith*, Acta chem. Scand., 1955, **9**, 1337) or anisole (*R. Schwarz* and *W. Kuchen*, Ber., 1956, **89**, 169). *Monophenoxysilicon trichloride*, b.p. 70°/9 mm, *diphenoxysilicon dichloride*, $(PhO)_2SiCl_2$, b.p. 156–157°/13 mm, and *triphenoxysilicon monochloride*, b.p. 212°/9 mm, are also formed in the former reaction.

Triphenyl borate, $B(OPh)_3$, b.p. 158°/0·05 mm, m.p. 92–93°, and other triaryl borates are obtained by heating phenols with boric acid or boric oxide with continuous removal of the water produced (*L. H. Thomas*, J. chem. Soc., 1946, 820, 823). Alternatively, the phenol or phenolic ester may be treated with boron trichloride at —60 to —80° (*T. Colclough, W. Gerrard* and *M. F. Lappert, ibid.*, 1955, 907, 4987). The *pyridine complexes of phenoxyboron dichloride*, m.p. 98–102°, and *diphenoxyboron chloride*, m.p. 116–118° are obtained by the interaction of boron trichloride and triphenyl borate at —70° in methylene chloride followed by addition of pyridine.

3. Halogenophenols

(a) Methods of preparation

(1) Direct chlorination, bromination, and iodination occur particularly easily and may be effected using either acetic acid, carbon tetrachloride, chloroform or nitromethane as solvent (*S. V. Zubarev et al.*, Russ. J. org. Chem., 1968, **4**, 1769). The halogen enters the *ortho*- and/or *para*-positions and mono-, di-, or tri-halogenophenols may be prepared as required. Iodination is reversible and an oxidising agent must also be present to remove the hydrogen iodide formed. Direct fluorination cannot be controlled. Bromination and chlorination occur more vigorously in aqueous solution and 2,4,6-tribromo- and 2,4,6-trichloro-phenols are precipitated quantitatively on addition of bromine and chlorine water, respectively, to phenol. Prolonged treatment with bromine or chlorine in the presence of a halogen carrier brings about the introduction of four or more halogens into the nucleus. Aromatic halogenation is discussed by *P. D. B. de la Mare* on pp. 71 *et seq.* and more detailed accounts are given by *de la Mare* and *J. H. Ridd* ("Aromatic Substitution", Butterworths, London, 1969) and *R. O. C. Norman* and *R. Taylor* ("Electrophilic Substitution in Benzenoid Compounds", Elsevier, New York, 1965). In addition to the halogens a number of other reagents may be used and some of the more useful of these are mentioned below.

Although it has been reported that *tert*-butyl hypochlorite chlorinates phenols mainly in the *ortho*-positions (*D. Ginsberg,* J. Amer. chem. Soc., 1951, **73**, 2723), *D. R. Harvey* and *Norman* (J. chem. Soc., 1961, 3604) obtained *ortho*:*para* ratios of nearly 1:1 using this reagent. Their results were similar to those recorded for chlorinations in which the chlorinium ion, Cl^+, was involved but were in contrast with those obtained for direct chlorinations in carbon tetrachloride (2·8:1) and in molten phenol (1:1·7). Sulphuryl chloride with or without a catalyst such as aluminium chloride (*P. P. T. Sah* and *H. H. Anderson,* J. Amer. chem. Soc., 1941, **63**, 3164) and *N*-chloroamides, first used by *K. J. P. Orton et al.* (J. chem. Soc., 1911, **99**, 1185), are also useful. Selective *para*-chlorination has been achieved using cupric chloride in dimethylformamide (*E. M. Kosower et al.,* J. org. Chem., 1963, **28**, 630). Chlorination of phenol in cold alkaline solution yields 2,2,4-trichloro-1,3-dihydroxycyclopent-4-ene-1-carboxylic acid (I) which

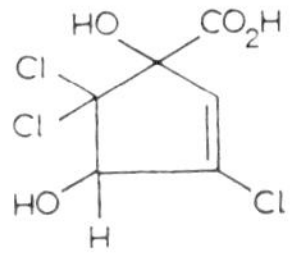

(I)

is a starting material for the synthesis of caldariomycin (*A. W. Burgstahler, T. B. Lewis* and *M. O. Abdel Rahman, ibid.,* 1966, **31**, 3516) and exhaustive chlorination in acetic acid yields chlorinated cyclohexanone derivatives (*L. Vollbracht et al.,* Tetrahedron, 1968, **24**, 6265; *E. Morita* and *M. W. Dietrich,* Canad. J. Chem., 1969, **47**, 1943).

Although iodine monochloride is an iodinating agent, iodine monobromide (*W. Militzer,* J. Amer. chem. Soc., 1938, **60**, 256) and bromine chloride (*C. O. Obenland,* J. chem. Ed., 1964, **41**, 566) readily brominate phenols and have advantages over bromine in certain cases. Dioxan dibromide is another useful mild brominating agent (*L. A. Yanovaskaya, A. P. Terent'ev* and *L. I. Belen'kii,* C.A., 1953, **47**, 8032). *N*-Bromosuccinimide preferentially brominates phenol in the *para*-position (*A. Wohl* and *K. Jaschinowski,* Ber., 1919, **52**, 51). Thionyl bromide (but not thionyl chloride) readily halogenates phenols at room temperature (*S. D. Saraf,* Canad. J. Chem., 1969, **47**, 2803). A procedure for the selective *ortho*-bromination of phenol is described by *D. E. Pearson, R. D. Wysong* and *C. V. Breder* (J. org. Chem., 1967, **32**, 2358). Certain *ortho*-bromophenols which are difficult to obtain by direct means, may be prepared by hydrogen bromide-catalysed isomerisation of the corresponding *para*-isomer (*E. J. O'Bara, R. J. Balsley* and *I. Strare, ibid.,* 1970, **35**, 16).

Mercuric oxide, iodic acid, potassium persulphate, and hydrogen peroxide in the presence of a strong acid (*L. Jurd,* Austral. J. sci. Research, 1949, **2A,** 595; 1950, **3A,** 587) are among the most commonly used oxidising agents in iodinations. Iodination with iodine and silver perchlorate is thought to involve the iodinium ion (*L. Birckenbach* and *J. Goubeau,* Ber., 1932, **65,** 395). Iodine monochloride has been used to iodinate both phenols (*B. Jones* and *E. N. Richardson,* J. chem. Soc., 1953, 713) and phenolic acids (*G. H. Woollett* and *W. W. Johnson,* Org. Synth., Coll. Vol. II, 1966, p. 343). The morpholine–iodine complex (*P. Charbier, J. Seyden-Penne* and *A. M. Fouace,* Compt. rend., 1957, **245,** 174) and di-iododimethylhydantoin (*O. O. Orazi, R. A. Corral* and *H. A. Bertorello,* J. org. Chem., 1965, **30,** 1101) are other useful mild iodinating agents.

In order to produce the required halogenophenol it is frequently necessary to introduce blocking groups such as —SO_3H or —CO_2H which are removed at a later stage. The preparation of *o*-bromophenol (*R. C. Huston* and *M. M. Ballard,* Org. Synth., Coll. Vol. II, 1966, p. 97) and 2,6-dichlorophenol (*D. S. Tarbell, J. W. Wilson* and *P. E. Fanta,* Org. Synth., Coll. Vol. III, 1963, p. 267) are illustrative. *o*-Iodophenol is best prepared by treatment of *o*-chloromercuriphenol with iodine (*F. C. Whitmore* and *E. R. Hanson,* Org. Synth., Coll. Vol. I, 1964, p. 326).

(2) From halogen-substituted amines by hydrolysis of their diazonium salts. For examples see the preparation of mono-, di- and tri-fluorophenols (*G. C. Finger et al.,* J. Amer. chem. Soc., 1959, **81,** 94) and 3-bromo-4-hydroxytoluene (*H. E. Ungnade* and *E. F. Orwoll,* Org. Synth., Coll. Vol. III, 1963, p. 130).

(3) From the diazonium salts of aminophenols by treatment with cuprous chloride or bromide (Sandmeyer reaction), potassium iodide, or borofluoric acid followed by thermal decomposition of the resultant diazonium borofluoride (Balz-Schieman reaction). For examples see the preparation of *p*-iodophenol (*F. B. Dains* and *F. Eberly,* Org. Synth., Coll. Vol. II, 1966, 355). Yields of products obtained from the Sandmeyer and Balz-Schieman reaction of aminophenols are frequently low (*P. H. Cheek, R. H. Wiley* and *A. Roe,* J. Amer. chem. Soc., 1949, **71,** 1863) and it is usually advantageous to use the corresponding amino ethers and to dealkylate the products (*C. M. Suter, E. J. Lawson* and *P. G. Smith, ibid.,* 1939, **61,** 161).

(4) Replacement of one halogen of a di-, tri-, or poly-halogenobenzene by the action of alcoholic alkali has been used to prepare certain tri-, tetra-, and penta-halogenophenols (*A. F. Holleman,* Rec. Trav. chim., 1918, **37,** 103, 201; 1920, **39,** 435; *J. de Crauw, ibid.,* 1929, **48,** 1063), *p*-fluorophenol (*M. M. Boudakian et al.,* J. org. Chem., 1961, **26,** 4641) and pentafluorophenol

TABLE 2

HALOGENOPHENOLS

Position(s) of substituents	Fluorophenols		Chlorophenols		Bromophenols		Iodophenols	
	m.p.°	b.p.°	m.p.°	b.p.°	m.p.°	b.p.°	m.p.°	b.p.°
ortho-	16	152	9	171	5·5	195	40	187
meta-	14	178	33	214	33	237	40	—
para-	48	186	43	220	64	238	92	—
2,3-di-	32	54/25 mm.	57	—	68	—	—	—
2,4-di-	22·5	150	45	—	40	—	72	—
2,5-di-	42	—	58	—	73	—	99	—
2,6-di-	46	60/17 mm.	65	—	56	—	68	—
3,4-di-	—	85/20 mm.	68	—	79	—	83	—
3,5-di-	55	—	68	—	81	—	104	—
2,3,4-tri-	—	69/43 mm.	80	—	95	—	—	—
2,3,5-tri-	29	—	62	—	95	—	114	—
2,3,6-tri-	—	—	58	—	—	—	—	—
2,4,5-tri-	42	—	68	—	87	—	—	—
2,4,6-tri-	50	—	70	—	96	—	156	—
3,4,5-tri-	—	—	101	—	129	—	—	—
2,3,4,5-tetra-	—	143	116	—	123	—	225	—
2,3,4,6-tetra-	—	—	70	—	114	—	170	—
2,3,5,6-tetra-	30	146·5	115	—	161	—	—	—
penta-	29	143	189	—	229	—	—	—

(*J. M. Birchall* and *R. N. Haszeldine*, J. chem. Soc., 1959, **13**; *E. J. Forbes et al.*, *ibid.*, 2019).

(5) From lithium halogenoaryls by conversion into the corresponding borinic ester with trimethyl borate followed by oxidation with hydrogen peroxide. Tetrafluoro- and pentafluoro-phenols have been prepared in this way (*G. M. Brooke et al.*, Tetrahedron Letters, 1965, 2991; J. chem. Soc., C, 1967, 869).

The melting and boiling points of mono- and more highly halogenated phenols containing halogen of one type only are given in Table 2.

(b) Properties and reactions

The halogenophenols undergo the general reactions of phenols described on p. 297 *et seq.*

(i) Acidity

The acid character of phenol (p*K*a 9·98) is enhanced by the presence of halogen in the nucleus, *e.g.*, *p*-fluoro-, *p*-chloro-, *p*-bromo-, and *p*-iodo-phenols have p*K*a = 9·95, 9·38, 9·36, 9·31, respectively. Trichloro- and tribromophenols are sufficiently acidic to decompose aqueous alkali metal carbonates and pentachlorophenol can be titrated in dilute aqueous alcohol with standard alkali and thymol blue (*T. S. Carswell* and *H. K. Nason*, Ind. Eng. Chem., 1938, **30**, 622). Pentafluorophenol (p*K*a 5·5) is about as acidic as benzoic acid but is less so than pentachlorophenol (p*K*a = 5·26) thus indicating the extent to which the $+M$ effect of the fluorine atoms offsets their powerful $-I$ effect (*Birchall* and *Haszeldine*, J. chem. Soc., 1959, 3653). A list of the acid strengths of halogenophenols is given by *A. Albert* and *G. L. Serjeant* ("Ionisation Constants of Acids and Bases", Methuen, London, 1962, p. 130) and *P. J. Pearce* and *R. J. J. Simkins* (Canad. J. Chem., 1968, **46**, 241).

(ii) Replacement of halogen

Fusion of the monohalogenophenols with alkali brings about replacement of the halogen by hydroxyl. Resorcinol is the main product from each of the monochlorophenols presumably arising *via* a benzyne intermediate (cf., *A. T. Bottini* and *J. D. Roberts*, J. Amer. chem. Soc., 1957, **79**, 1458). By using aqueous alkali at about 170–195° a normal replacement without migration of the hydroxyl group occurs (*C. F. Boehringer* and *Söhne*, G.P., 269,544/1914; 284,533, 268,266/1915; *A.G. f. Anilinfabrikation*, G.P., 349,794/1914).

Hydrogenation using a nickel catalyst in the presence of alkali removes halogen as hydrohalide from the chloro- and bromo-phenols (*C. Kelber*, Ber., 1922, **54**, 2255; cf., *M. Busch* and *W. Schmidt*, *ibid.*, 1930, **62**, 2612). In the tribromophenols, one halogen may be removed by zinc dust and acetic acid (*M. Kohn* and *J. Pfeiffer*, Monatsh., 1927, **48**, 218). When bromo-, and bromochloro-phenols are treated with aluminium chloride in benzene bromines *ortho* and *para* to the hydroxyl group are replaced by hydrogen but chlorines are unaffected (*Kohn* and *S. Reichmann*, J. org. Chem., 1947, **12**, 213). *p*-Bromophenol with aluminium chloride and hydrogen halide co-catalyst rearranges to *m*-bromophenol (54%) (*L. A. Fury* and *D. E. Pearson*, *ibid.*, 1965, **30**, 2301). Photo-dehalogenation of phenols and phenolic ethers has been described by *J. T. Pinhey* and *R. D. G. Rigby* (Tetrahedron Letters, 1969, 1267, 1271). Irradiation of *p*-halogenophenols in the presence of cyanide ion gives *p*-cyanophenols which, in turn, may be converted into *p*-hydroxybenzaldehydes by irradiation in alkaline solution (*K. Omura* and *T. Matsuura*, Chem. Comm., 1969, 1374, 1516). In the mass spectrometer halogenophenols readily lose halogen and carbon monoxide (*T. L. Folk* and *L. G. Wideman*, Chem. Comm., 1969, 491).

(iii) Nitration

Nitration of halogenophenols (or ethers) frequently leads to anomalous products (*D. V. Nightingale*, Chem. Reviews, 1947, **40**, 117). The halogen may be replaced by a nitro group or be oxidatively eliminated giving a quinone. In some cases the displaced halogen may re-enter the ring to give an apparently rearranged product. 2,4,6-Tribromophenol, for example, reacts with nitric acid to give a mixture of 6-bromo-2,4-dinitro-, 2,6-dibromo-4-nitro-phenols and 2,6-dibromobenzoquinone.

(iv) Formation of cyclohexadienones

Bromination of 2,4,6-trisubstituted phenols yields products usually considered to be cyclohexadienones (but see *E. Schulek, K. Burger* and *E. Körös*, Acta Chem. Acad. Sci. Hung., 1959, **21**, 67). For example, 2,4,6-tribromophenol gives the relatively stable dienone II previously named, "tribromophenol bromide" [ν_{max} 6·0 μ (conjugated CO), λ_{max} 280 mμ (cyclohexa-2,5-dienone)] (*J. A. Price*, J. Amer. chem. Soc., 1955, **77**, 5436) which readily loses a halogen atom on treatment with proton donors such as alcohol, phenol or aniline (*R. Benedikt*, Ber., 1879, **12**, 1005). Sterically hindered phenols such as 2,4,6-tri-*tert*-butylphenol yield more stable dienones which give ethers with alcohols (*E. Müller, K. Ley* and *W. Kiedaisch*, *ibid.*, 1954, **87**, 1605) and stable phenoxyls with metals or iodide (*C. D. Cook* and *R. C. Woodworth*, J. Amer. chem. Soc., 1953, **75**, 6242). Bromocyclohexadienones readily undergo rearrangement on treatment with acid, on heating, or, in

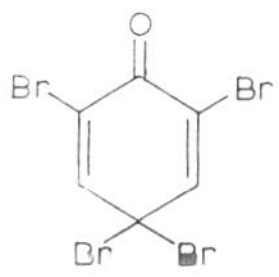

(II) 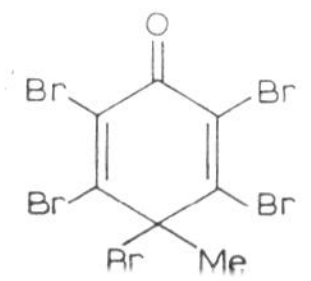 (III) 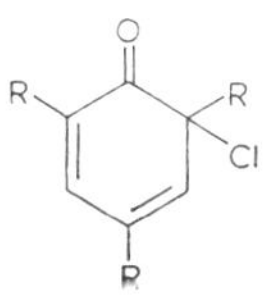 (IV)

some cases, even on standing. Thus, II gives 2,3,4,6-tetrabromophenol and III gives 2,3,5,6-tetrabromo-4-bromomethylphenol (*Th. Zincke*, J. pr. Chem., 1897, **56**, 157). Elimination of a *p*-*tert*-butyl group or of a *p*-CH_2OH, —CHO, and —CO_2H during bromination also occurs presumably *via* a cyclohexadienone intermediate (*L. E. Forman* and *W. C. Sears*, J. Amer. chem. Soc., 1954, **76**, 4977; *I. W. Ruderman*, Ind. Eng. Chem., Analyt., 1946, **18**, 735; *Burgstahler, P. L. Chien* and *Abdel Rahman*, J. Amer. chem. Soc., 1964, **86**, 5281).

Chlorination of 2,4,6-trisubstituted phenols also yields cyclohexadienones but it is thought that these are 2,4-cyclohexadienones (IV) (λ_{max} 350–380 nm). Pentachlorophenol, for example, gives a hexachloro-2,4-cyclohexadienone which rearranges to a hexachloro-2,5-cyclohexadienone on standing (*R. Fort,* Ann. Chim. Fr., 1959, **13**, 203). For further details of this interesting group of compounds see *V. V. Ershov, A. A. Volod'kin* and *G. N. Bogdanov,* Russ. chem. Reviews, 1963, **32**, 75.

(v) Oxidation

Oxidation of 2,4,6-trihalogenophenols usually leads to the loss of the 4-halogeno group and the formation of polyaryl ethers (*H. S. Blanchard, H. C. Finkbeiner* and *G. A. Russell,* J. Polymer Sci., 1962, **58**, 469; *W. H. Hunter et al.,* J. Amer. chem. Soc., 1932, **54**, 2456; 1916, **38**, 1761). Similar polymers are obtained by heating the potassium salt of 2,4,6-tribromophenol in benzene (*H. A. Torrey* and *Hunter, ibid.,* 1911, **33**, 194). The thermal stability of these polymers has been examined by *J. M. Cox, B. A. Wright* and *W. W. Wright* (J. appl. Polymer Sci., 1965, **9**, 513). Oxidation of pentabromo- and pentachloro-phenols with lead dioxide or fuming nitric acid at -25^0 gives dienone dimers of type V (*Müller, A. Rieker* and *W. Beckert,* Z. Naturforsch., 1962, **17b**, 567) which may be oxidised further to quinones (*W. Chang,* J. org. Chem., 1962, **27**, 2921).

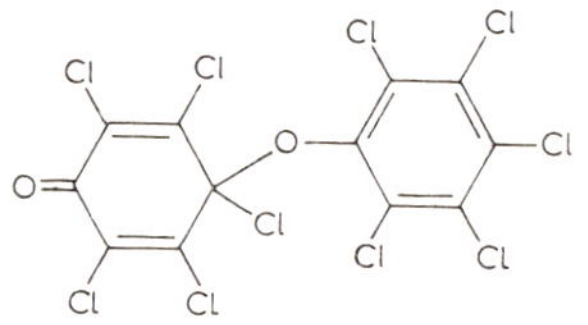

(V)

4. Phenols carrying substituents attached through nitrogen

(a) Nitrophenols

(i) Methods of preparation

(1) Phenols are easily nitrated by dilute nitric acid to mononitrophenols. In this way *S. Veibel* (Ber., 1930, **63**, 1582, 2074) obtained a mixture of approximately equal quantities of *o*- and *p*-nitrophenols, which were conveniently separated by distillation of the *ortho* isomer in steam. More concentrated nitric acid or a mixture of nitric and sulphuric acids gives di- and tri-nitro compounds. Higher yields of *o*-nitrophenols are obtained

when an organic solvent, often acetic acid, is used (*F. Arnall*, J. chem. Soc., 1924, **125**, 811). Other nitrating agents which may be used include dinitrogen tetroxide (also $BF_3 \cdot N_2O_4$) and acetyl nitrate. Nitration of aniline with dinitrogen tetroxide surprisingly gives 2,4-dinitrophenol in 25% yield. High yields of *o*-nitrophenols may be obtained by treatment of aryl chloroformates wih silver nitrate in acetonitrile (*M. J. Zabik* and *R. D. Schuetz*, J. org. Chem., 1967, **32**, 300). For a review of nitration see *A. V. Topchiev* ("Nitration of Hydrocarbons and Other Organic Compounds", Pergamon, London, 1959). Frequently-encountered side reactions are, oxidation with the formation of benzoquinones or diphenoquinones and replacement of a ring substituent by a nitro group (*G. A. Zlobina* and *V. V. Ershov*, Bull. Acad. Sci., 1964, 1570; *T. J. Barnes* and *W. J. Hickinbottom*, J. chem. Soc., 1961, 953; *D. V. Nightingale*, Chem. Reviews, 1947, **40**, 117). Nitration of sterically hindered (polyalkyl or polyhalogeno) phenols often gives nitro-2,5-cyclohexadienones; 2,4,6-tri-*tert*-butylphenol, for example, on treatment with nitric acid at 0^0 readily gives a nitrodienone which loses nitrogen dioxide on heating (*K. Ley* and *E. Müller*, Ber., 1956, **89**, 1402).

(2) Nitration of phenol with dilute nitric acid is strongly catalysed by the presence of nitrite. Some of the product at least arises from nitrosation (by $NO^{\oplus}$) followed by oxidation by nitric acid with regeneration of the nitrous acid consumed. The *ortho*:*para* ratio is extremely sensitive to the amount of nitrous acid present and to the solvent (*C. K. Ingold et al.*, J. chem. Soc., 1950, 2628, 2657). Nitrophenols have also been prepared from nitrosophenols by oxidation with, for example, hydrogen peroxide.

(3) Di- and tri-nitrophenols have been prepared by nitration of aromatic hydrocarbons with nitric acid in the presence of mercuric nitrate. Presumably, mercuration precedes nitration in such cases (*Topchiev, loc. cit.*). Direct hydroxylation of *m*-dinitro- and 1,3,5-trinitro-benzene with ferricyanide has also been reported.

(4) Nitroanilines are easily converted into nitrophenols by boiling their diazonium salts with acid (*P A. S. Smith*, "Open Chain Nitrogen Compounds", Benjamin, New York, 1966, Pt. 2, p. 282).

(5) Hydrolyses of nitroanilines, nitrophenylhydrazines, nitrophenyl ethers and halogenonitrobenzenes in which the nitro groups arc *ortho* or *para* to the leaving group give good yields of nitrophenols (*J. F. Bunnett*, Quart. Reviews, 1958, **12**, 1).

(*ii*) *Properties and reactions*

Acidity. Nitrophenols are much stronger acids than alkylphenols and liberate carbon dioxide from bicarbonate solution. The salts so-formed are more intensely coloured than the parent phenols. Phenol, *o*-nitrophenol,

2,4-dinitrophenol and 2,4,6-trinitrophenol have p*K*a values of 9·98, 7·23, 4·01 and 0·71, respectively. The deep yellow colour produced on salt formation is considered to be due to mesomeric anions, *e.g.*, I in which the quinonoid form is dominant in the resonance hybrid (see *L. A. Cohen* and *W. M. Jones*, J. Amer. chem. Soc., 1963, **85**, 3400 for comparative U.V. data). *A. Hantzsh* and *H. Gorke* (Ber., 1906, **39**, 1073) prepared two series of ethers from a number of nitrophenols (but not *p*-nitrophenol); the normal

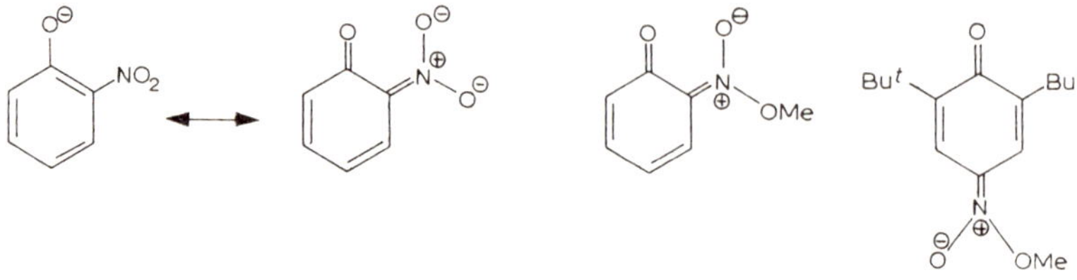

NO_2 Bu^t Bu^t OMe OMe

(I) (II) (III)

ethers, such as nitroanisole which are almost colourless, and highly coloured unstable quinonoid ethers, *e.g.*, II, which were readily hydrolysed. More recently *Cohen* and *Jones* (*loc. cit.*) have isolated the orange quinonoid ether III and shown that it is stable at temperatures below that of its m.p. The reason why *m*-nitrophenols also yield highly coloured salts is less obvious (*N. V. Sidgwick*, "The Organic Chemistry of Nitrogen", Clarendon, Oxford, 3rd Edn., 1966, pp. 400–406).

The physical characteristics of *o*-nitrophenol differ markedly from those of its *meta* and *para* isomers. The *ortho* compound is bright yellow, volatile in steam and boils at about 214°. Its isomers are almost colourless, non-volatile in steam and boil at about 280–300°. These differences are attributed to the formation of a chelate ring by hydrogen bonding in the *ortho* compound for which there is ample spectroscopic evidence (*C. N. R. Rao*, in "Chemistry of the Nitro- and Nitroso-Groups", Interscience, London, 1969, Pt. 1, pp. 112–116).

(*iii*) *Mononitrophenols*

o-**Nitrophenol,** yellow crystals with a distinctive odour, m.p. 45°, b.p. 214°. It can be prepared, in addition to the ways already described by (a) nitration of phenol-4-sulphonic acid and subsequent hydrolysis of the resultant 2-nitrophenol-4-sulphonic acid, (b) hydrolyses of *o*-chloro-, -bromo-, -nitro-, or -aminonitrobenzene with base, (c) hydroxylation of nitrobenzene with alkali. Halogenation, nitration and sulphonation proceed normally, the substituent entering the 4-position although a little of the 2,6-disubstituted compound may also be formed. Treatment with mercuric nitrate gives 6-acetoxymercuri-2-nitrophenol. (*H. H. Hodgson*, J. Amer. chem. Soc., 1927, **49**, 2840).

m-*Nitrophenol*, m.p. 96°, is prepared by the diazo reaction from *m*-nitroaniline (*R. H. F. Manske*, Org. Synth., Coll. Vol. I, 1964, p. 404). Nitration gives a mixture of 2,3-, 2,5- and 3,4-dinitrophenols and further nitration yields 2,3,4,6-tetranitrophenol. Photolysis in the presence of hydrochloric acid gives 3-amino-2,4,6-trichlorophenol in 37% yield (*R. L. Letsinger* and *G. G. Wubbels*, J. Amer. chem. Soc., 1966, **88**, 5041).

p-*Nitrophenol*, m.p. 114°, is dimorphous (*Sidgwick*, J. chem. Soc., 1915, **107**, 672). It is prepared by (a) direct nitration of phenol, (b) nitration of the *p*-toluenesulphonic ester of phenol, followed by hydrolysis, (c) oxidation of *p*-nitrosophenol with nitric acid, (d) hydrolyses of *p*-chloro-, -bromo-, -nitro-, or -aminonitrobenzenes, and (e) hydroxylation of nitrobenzene.

Bromination gives 2,6-*dibromo-4-nitrophenol*, m.p. 141° (*W. W. Hartman* and *J. B. Dickey*, Org. Synth., Coll. Vol. II, 1966, p. 173), reduction of which with tin and hydrochloric acid followed by treatment with sodium hypochlorite gives 2,6-dibromoquinone-4-chloroimide (*Hartman, Dickey* and *J. G. Stampfli, ibid.*, p. 175). Chloromethylation by methylal in the presence of hydrochloric acid gives a 69% yield of 2-chloromethyl-4-nitrophenol (*C. A. Buehler, F. K. Kirchner* and *G. F. Deebel*, Org. Synth., Coll. Vol. III, 1963, p. 468). It gives esters with benzoxycarbonyl amino acids which readily react with esters of amino acids to give products with a new peptide bond (*M. Bodanszky* and *V. du Vigneaud*, J. Amer. chem. Soc., 1959, **81**, 5688).

(*iv*) *Alkyl nitroaryl ethers*

Alkyl nitroaryl ethers are formed by alkylation of nitrophenols in the usual way with alkyl iodides or sulphates, or by treatment of nitroaryl halides with alkoxide or by direct nitration of alkyl aryl ethers (*J. Allan* and *R. Robinson*, J. chem. Soc., 1926, 376).

o-, m-, and p-**Nitroanisoles** have m.p. 9°, b.p. 272°; m.p. 38°, b.p. 258° and m.p. 54°, b.p. 259°, respectively, while o-, m-, and p-*nitrophenetoles* have m.p. 2°, b.p. 275°; m.p. 45°, and m.p. 60°, b.p. 283°, respectively. *p*-Nitroanisole on irradiation in the presence of cyanide ion gives 2-cyano-4-nitroanisole in high yield (*R. L. Letsinger* and *R. R. Hautala*, Tetrahedron Letters, 1969, 4205) and in acetonitrile gives principally *p*-nitroanisole and 4-methoxy-2-nitrophenol (*L. B. Jones, J. C. Kudrna* and *J. P. Foster*, Tetrahedron Letters, 1969, 3263).

For the preparation of aryl nitroaryl ethers see *M. Allen* and *R. Y. Moir*, Canad. J. Chem., 1959, **37**, 1799, and *G. A. Neville* and *Moir, ibid.*, 1969, **47**, 2787.

(*v*) *Dinitrophenols*

2,4-**Dinitrophenol,** m.p. 114°, is prepared by (a) alkaline hydrolysis of 2,4-dinitrochlorobenzene (*Hickinbottom*, "Reactions of Organic Compounds", Longmans, London, 1957, p. 95), (b) hydroxylation of *m*-dinitrobenzene with ferricyanide or with 30% oleum (*K. Lauer*, J. pr. Chem., 1935, **142**, 310), (c) oxidative nitration of benzene in the presence of mercuric nitrate (*W. E. Bachmann et al.*, J. org. Chem., 1948, **13**, 390) and (d) dinitration of phenol. Reduction with ammonium sulphide gives 2-amino-4-nitrophenol (*Hartman* and *H. L. Silloway*, Org. Synth., Coll. Vol. III, 1963, p. 82).

3,5-*Dinitrophenol*, m.p. 122^0, is formed by acid hydrolysis of 3,5-dinitroanisole (*Sidgwick* and *T. W. J. Taylor*, J. chem. Soc., 1922, **121**, 1853), which is easily obtained from 1,3,5-trinitrobenzene by the action of sodium methoxide in methanol (*F. Reverdin*, Org. Synth., Coll. Vol. I, 1964, p. 219).

2,3-*Dinitrophenol*, m.p. 145^0; 2,5-*dinitrophenol*, m.p. 105^0 and 3,4-*dinitrophenol*, m.p. 137^0 (*K. H. Pausacker* and *J. G. Scroogie*, J. chem. Soc., 1955, 1897; *Sidgwick* and *W. M. Aldous*, *ibid.*, 1921, **119**, 1001); 2,6-*dinitrophenol*, m.p. 64^0 (*H. Zinner et al.*, Ber., 1959, **92**, 407).

(*vi*) *Trinitrophenols*

Picric acid, 2,4,6-**trinitrophenol.** *Historical.* Picric acid was first obtained by *Woulfe* (1771) from indigo and later by *Welter* (1799) by treatment of silk with nitric acid and was known for a long time as *Welter's* bitter yellow. *Dumas* analysed the acid and gave it the name picric acid from the Greek *pikros*, bitter. It was recognised as a derivative of phenol by *Laurent* (1848).

Methods of preparation. It is prepared by (a) nitration of phenol, *o*- and *p*-nitrophenols and of 2,4- and 2,6-dinitrophenols as well as suitable nitrophenolsulphonic acids (see p. 378) and *p*-hydroxybenzoic acid (*Y. Takayama* and *Y. Tsubuku*, Bull. chem. Soc. Japan, 1942, **17**, 109), (b) oxidative nitration of benzene with nitric acid and mercuric nitrate (*E. E. Aristoff et al.*, Ind. Eng. Chem., 1948, **40**, 1281), (c) hydrolysis of chloro-2,4,6-trinitro- or bromo-2,4,6-trinitro-benzene with warm alkali, and (d) oxidation of 1,3,5-trinitrobenzene with ferricyanide in sodium carbonate solution (*P. Hepp*, Ann., 1882, **215**, 344).

Properties and reactions. Picric acid crystallises from hot water as yellow leaflets or prisms. It explodes violently when heated or subjected to shock and has been used as a military high explosive (lyddite). It is soluble in about 160 parts of cold water and is freely soluble in hot water. Its aqueous solutions have been used to dye silk and wool yellow with a greenish tint. With alkali or carbonates it yields salts which are more sensitive to shock than the free acid. The *potassium salt* is soluble in 260 parts and the *sodium salt* is soluble in 10 parts of water at 15^0.

(1) Picric acid forms 1:1 molecular complexes (charge transfer complexes) with hydrocarbons. These have a characteristic m.p. and are useful for purposes of identification, separation, and purification of polycyclic hydrocarbons. They should not be confused with amine picrates, however, which are salts formed by proton transfer (*R. Foster*, "Organic Charge Transfer Complexes", Academic Press, London, 1969). Both amine and hydrocarbon picrates can be used for the determination of molecular weights by measuring the intensity of the absorption at 380 nm (*K. G. Cunningham*, *W. Dawson* and *F. S. Spring*, J. chem. Soc., 1951, 2305).

(2) The hydroxyl group of picric acid may be replaced by chlorine by reaction

with phosphorus pentachloride, arenesulphonyl chlorides (*Hickinbottom*, "Reactions of Organic Compounds", Longmans, London, 1957, p. 96) or thionyl chloride in dimethylformamide (*H. Eilingsfeld, M. Seefelder* and *H. Weidinger*, Angew. Chem., 1960, **72,** 836) to give picryl chloride, 1-chloro-2,4,6-trinitrobenzene.

(3) Reduction with alcoholic ammonium sulphide or with zinc dust and ammonia yields picramic acid, 6-amino-2,4-dinitrophenol, which is an important intermediate for the preparation of certain azo-dyes (*Hodgson* and *E. R. Ward*, J. chem. Soc., 1945, 663).

(4) It is degraded to chloropicrin, CCl_3NO_2, by treatment with chlorine in calcium hydroxide solution (*E. D. G. Frahm*, Rec. Trav. chim., 1931, **50,** 1125).

2,4,6-**Trinitroanisole,** m.p. 65^0, and 2,4,6-*trinitrophenetole*, m.p. 78^0, may be prepared by nitration of anisole and phenetole, respectively, or by treatment of picryl chloride with the corresponding alkoxide. Treatment of either 2,4,6-trinitroanisole with ethoxide or 2,4,6-trinitrophenetole with methoxide gives the same red Meisenheimer complex (*K. L. Servis*, J. Amer. chem. Soc., 1965, **87,** 5495). Both form charge transfer complexes with hydrocarbons (*Foster, loc. cit.*).

2,3,6-**Trinitrophenol**, m.p. 117^0, and 2,4,5-**trinitrophenol**, m.p. 96^0, are among the products of further nitration of *m*-nitrophenol (*J. J. Blanksma*, Rec. Trav. chim., 1902, **21,** 261; *P. S. Varma* and *D. A. Kulkarni*, J. Amer. chem. Soc., 1925, **47,** 143).

2,3,4-, 2,3,5-, 3,4,5- and 2,4,5-**Trinitroanisoles** have m.p.s of 155^0, 104^0, 120^0 and 107^0, respectively.

(vii) *Tetra- and penta-nitrophenols*

2,3,4,6-**Tetranitrophenol,** m.p. 140^0, is formed by the nitration of *m*-nitrophenol (*Blanksma, loc. cit.*; *G. F. van Duin* and *B. C. van Lennep*, Rec. Trav. chim., 1920, **39,** 162). With boiling water it yields 2,4,6-trinitroresorcinol, with alcoholic ammonia 3-amino-2,4,6-trinitrophenol and with hydrogen halides, 3-halogeno-2,4,6-trinitrophenols (*E. Yu. Orlova et al.*, C.A., 1965, **63,** 16239).

Pentanitrophenol, yellow crystals, m.p. 190^0 (decomp.), is formed by nitration of 3,5-dinitrophenol. With boiling water it yields trinitrophloroglucinol (*Blanksma, loc. cit.*).

(viii) *Nitro derivatives of homologues of phenol*

These may be prepared by the general methods used for the nitrophenols. In the nitration of homologues of phenol, the nitro group is directed, in general, to positions *ortho* and *para* to the hydroxyl. Thus *o*-cresol gives a mixture of 2-methyl-4- and -6-nitrophenols on nitration and 2-methyl-4,6-dinitrophenol on further nitration.

4-**Methyl**-2-**nitrophenol** (o-*nitro*-p-*cresol*), m.p. 33^0, is formed by the nitration of *p*-cresol in benzene (*Hickinbottom, loc. cit.*, p. 149) or by the alkaline hydrolysis of 4-amino-3-nitrotoluene at 130^0 for 12 hours (*L. Gindraux*, Helv., 1929, **12,** 921). 4-*Methyl*-3-*nitrophenol*, m.p. 76^0, is formed when di-*p*-tolyl carbonate is nitrated and the product hydrolysed (*M. Copisarow*, J. chem. Soc., 1929, 251) or by the diazo reaction from 4-methyl-3-nitroaniline (*F. Kuffner, G. Lenneis* and *H. Bauer*, Monatsh., 1960, **91**, 1152).

2-**Methyl**-4-**nitro**- and 2-**methyl**-6-**nitro**-**phenol**, m.p. 96° and 70°, respectively, are prepared by nitration of *o*-cresol (*G. P. Gibson*, J. chem. Soc., 1925, **127**, 42). 2-*Methyl*-3-*nitrophenol*, m.p. 94°, is prepared from 2-methyl-3-nitroaniline by the diazo reaction (*Kuffner et al., loc. cit.*). 2-*Methyl*-5-*nitrophenol*, m.p. 118° is obtained by the diazo reaction from the corresponding toluidine (*F. Ullmann* and *R. Fitzenkam*, Ber., 1905, **38**, 3787).

2-*Methyl*-4,6-*dinitrophenol*, m.p. 86°, is formed by nitrating *o*-cresol or by heating 2,3,5-trinitrotoluene with sodium acetate solution (*O. L. Brady, S. W. Hewetson* and *L. Klein*, J. chem. Soc., 1924, **125**, 2400).

3-**Methyl**-2-, -4-, and -6-**nitrophenols** have m.p. 41°, 129°, and 56°, respectively. They are all formed when *m*-cresol is nitrated with nitric acid in acetic acid at low temperature, the 2-nitro compound being produced in minimal quantity however (*Hickinbottom, loc. cit.*, p. 149; *Kuffner, Lenneis* and *Bauer, loc. cit.*; *Gibson*, J. chem. Soc., 1923, **123**, 1269; *R. D. Howarth* and *A. Lapworth, ibid.*, p. 2982). 3-*Methyl*-5-*nitrophenol*, m.p. 60°, is prepared by the diazo reaction on the corresponding amine (*Haworth* and *Lapworth, loc. cit.*).

3-*Methyl*-4,6-*dinitrophenol*, m.p. 74°, and 3-*methyl*-2,4-*dinitrophenol*, m.p. 133°, are both formed by nitration of 3-methyl-4-nitrophenol and may be separated by extraction with cold benzene in which only the former is readily soluble (*I. Pastorek*, C.A., 1965, **63**, 5547h).

3-*Methyl*-2,4,6-*trinitrophenol*, m.p. 109°, is formed from *o*-nitrosotoluene by warming with a nitrous acid–nitric acid mixture at 95° (*F. H. Westheimer, E. Segel* and *R. Schramm*, J. Amer. chem. Soc., 1947, **69**, 773).

The discovery that many alkyldinitrophenols were useful insecticides and herbicides (*F. Tattersfield* and *C. T. Gimingham*, J. Soc. chem. Ind., 1927, **46**, 368T; *G. H. Coleman* and *G. A. Griess*, U.S.P., 2,365,056/1944) has led to the investigation of a large number of such compounds. 2-Alkyl-4,6-dinitro- and 4-alkyl-2,6-dinitro-phenols are the most active and are conveniently prepared by nitration of the appropriate 2- or 4-alkylphenol in an organic solvent by heating under reflux with nitric acid (35%) (*M. Pianka* and *J. D. Edwards*, J. chem. Soc., C, 1967, 2281; *G. G. S. Sutton, M. E. D. Hillman* and *J. G. Moffatt*, Canad. J. Chem., 1964, **42**, 480).

4-tert-**Butyl**-3-**methoxy**-2,6-**dinitrotoluene**, m.p. 85°, is used in perfumery as a synthetic musk.

(b) Nitrosophenols

Nitrosophenols exist in solution as tautomeric mixtures of quinone-monoximes (IV) and nitrosophenols (V) in which the former usually predominates (*vide infra*). They give the expected reactions of both forms and may be prepared from phenols or quinones.

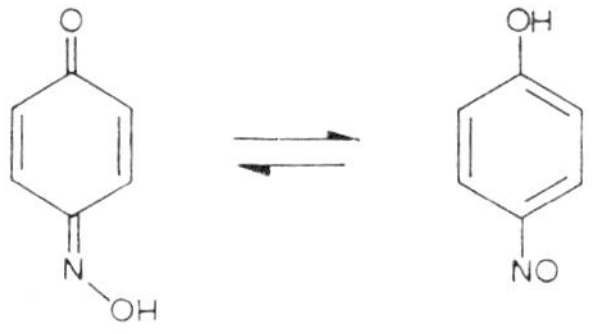

(IV) (V)

(i) Methods of preparation

(1) Nitrosation of phenols with nitrous acid in alkaline solution usually gives good yields of nitrosophenols although in some cases the nitrophenol is the main product (*Hodgson* and *J. S. Wignall*, J. chem. Soc., 1927, 2216). Generally, the substituent enters the *para*-position but if this is blocked either *o*-nitrosophenols are formed (*Hodgson* and *A. Kershaw*, *ibid.*, 1929, 1553; *S. Veibel*, Ber., 1930, **63**, 1577) or the *para*-substituent is displaced (*R. A. Henry*, J. org. Chem., 1958, **23**, 648; *K. M. Ibne-Rasa*, J. Amer. chem. Soc., 1962, **84**, 4962). Nitrosylsulphuric acid or nitrosyl chlorides have been used instead of nitrous acid (*R. Nietzki* and *A. L. Guitermann*, Ber., 1888, **21**, 428; *N. Dost* and *C. M. Dijs*, Dutch P., 87,355/1958; C.A., 1959, **53**, 17057h).

(2) *o*-Nitrosophenols are best prepared directly from the parent aromatic hydrocarbons by oxidation with hydrogen peroxide, in the presence of hydroxylamine and a copper salt (Baudisch reaction) (*O. Baudisch*, J. Amer. chem. Soc., 1941, **63**, 622; *G. Cronheim*, J. org. Chem., 1947, **12**, 1, 7, 20). Best yields of *o*-nitrosophenol are obtained at pH 2·5–3·5; above pH 4 catechol is also formed (*Tanimoto*, Bull. chem. Soc. Japan, 1970, **93**, 139). The mechanism of this reaction is not fully understood (*K. Murayama*, *I. Tanimoto* and *R. Goto*, *ibid.*, 1967, **32**, 2516; 1970, **43**, 1182). Careful reduction of *o*-nitrophenols has also been used.

(3) *p*-Nitrosophenols may be obtained from the corresponding quinones by reaction with hydroxylamine (*H. Goldschmidt*, Ber., 1884, **17**, 213; *R. K. Norris* and *S. Sternhell*, Austral. J. Chem., 1969, **22**, 935).

(4) Alkaline hydrolysis of *p*-nitrosoanilines or *p*-nitrosoalkylarylamines yields *p*-nitrosophenols. This method has also been used for the preparation of amines (*C. W. L. Bevan*, *J. Hirst* and *A. J. Foley*, J. chem. Soc., 1960, 4543).

(5) *Norris* and *Sternhell* (*loc. cit.*; Austral. J. Chem., 1966, **19**, 841) have described the preparation of a number of *p*-nitrosophenols by the following reaction sequence; reduction of alkyl nitroaryl ethers to alkyl aminoaryl ethers, oxidation with potassium persulphate [*E. Havinga et al.* (Rec. Trav.

chim., 1958, 77, 746) used Caro's acid] and, finally, hydrolysis of the ensuing alkyl nitrosoaryl ether with acid.

(6) Photolysis of 2,4-dinitroaryloxy acids yields 4-nitro-2-nitrosophenols which are not easily accessible by other routes (*P. H. McFarlane* and *D. W. Russell*, J. chem. Soc., D, 1969, 475).

(ii) Properties and reactions

(1) *Spectra and tautomerism.* Detailed u.v., i.r., n.m.r. and X-ray studies (*C. Romers, C. B. Shoemaker* and *E. Fischmann*, Rec. Trav. chim., 1957, 76, 490) of *p*-nitrosophenols clearly show that the vast majority exist mainly in the quinonoid form. In particular, the presence of one H—O stretching vibration at about 3563 cm^{-1} in the i.r. spectrum of *p*-nitrosophenol (*A. W. Baker*, J. phys. Chem., 1958, **62**, 744; *D. Hadži*, J. chem. Soc., 1956, 2725), the similarity of its u.v. spectrum to that of the methyl ether of the quinonoid form (*A. Schors, A. Kraaijeveld* and *Havinga*, Rec. Trav. chim., 1955, 74, 1243) and the proton splitting pattern in its n.m.r. spectrum, provide convincing evidence. *Norris* and *Sternhell* (*loc. cit.*), who have recently summarised the bulk of the spectroscopic evidence, have estimated by n.m.r. measurements that *p*-nitrosophenol in dioxan exists to the extent of 83% as the quinone oxime. This result agrees with the earlier calculations of *H. H. Jaffe* (J. Amer. chem. Soc., 1955, 77, 4448) that the quinonoid form is 4·6 kcal/mole more stable than the aromatic form. The introduction of alkyl substituents generally reduces the percentage of nitrosophenol even further but groups such as methoxycarbonyl (in the 2- or 6-positions) which can form intramolecular hydrogen bonds with the phenolic hydroxyl group disturb the equilibrium in favour of the nitrosophenol. n.m.r. studies have further shown that in solution isomerisation (*syn* ⇌ *anti*) of the oxime group occurs and is accelerated by the addition of water or hydrochloric acid and retarded by the addition of trifluoroacetic acid.

o-Nitrosophenols, by comparison, have been shown by u.v. (*A. Burawoy et al.*, J. chem. Soc., 1955, 3727) and i.r. (*P. M. Boll*, Acta Chem. Scand., 1958, **12**, 1777) measurements to exist mainly in the nitrosophenol form. This has been attributed to intramolecular hydrogen bonding. 5-Methoxy-2-nitrosophenol, however, has been obtained in two crystalline forms which it is claimed correspond to the two tautomeric forms (*Burawoy et al., loc. cit.*). A similar claim for 3-chloro-4-nitrosophenol was disproved (*Schors* and *Havinga*, Rec. Trav. chim., 1951, 70, 59).

Tautomerism of nitrosophenols has been reviewed by *V. V. Ershov* and *G. A. Nikiforov*, Russ. chem. Reviews, 1966, 35, 817.

(2) *Salt formation.* Nitrosophenols are acidic and form salts which are generally highly coloured. The negative charge in the anions of *p*-nitroso-

phenols is considered to be localised mainly on the oxygen of the oxime groups (*Norris* and *Sternhell, loc. cit.*). *o*-Nitrosophenols readily form deeply coloured, water-soluble co-ordination complexes with metal ions such as $Cu^{\oplus\oplus}$ (*Cronheim, loc. cit.*; *H. Shimura*, J. chem. Soc. Japan, 1955, **76**, 867). The alkylmercuri derivatives of *p*-nitrosophenol also exhibit tautomerism (*A. N. Nesmeyanov* and *D. N. Kraftsov*, Doklady Akad. Nauk S.S.S.R., 1960, **135**, 331).

(3) *Oxidation and reduction*. Oxidation to the corresponding nitrophenol is effected by hydrogen peroxide or alkaline ferricyanide (*E. Nölting* and *D. Kohn*, Ber., 1884, **17**, 370; *Hodgson* and *F. H. Moore*, J. chem. Soc., 1925, **127**, 2260). Nitric acid may also be used but with this reagent nitration may occur (*Nölting* and *Kohn, loc. cit.*). A trimeric product has also been obtained by oxidation with ferricyanide (*M. S. Kharasch* and *B. S. Joshi*, J. org. Chem., 1962, **27**, 651).

Nitrosophenols are reduced to the corresponding aminophenols by a wide variety of reagents. These include tin and hydrochloric acid, hydrogenation over a metal catalyst (*C. F. Winans* and *H. Adkins*, J. Amer. chem. Soc., 1933, **55**, 2051) and sodium hydrosulphite (*W. R. Vaughan* and *G. K. Finch*, J. org. Chem., 1956, **21**, 1201). Reaction of *p*-acetamidobenzenesulphinic acid and *p*-nitrosophenol affords the aminosulphone VI (*S. Pickholz*, J. chem. Soc., 1946, 1058).

(4) *Alkylation*. Methylation of *p*-nitrosophenol with ethereal diazomethane (*Hodgson, ibid.*, 1932, 1395) or dimethyl sulphate in alkaline solution, or methylation of its silver salt with methyl iodide (*ibid.*, 1931, 1494) gives the same product as that obtained by treatment of benzoquinone with *O*-methylhydroxylamine (*J. L. Bridge*, Ann., 1893, **277**, 79), *i.e.*, the methyl ether of *p*-benzoquinone oxime. The dinitrone VII is also formed in the

NH_2, OH, SO_2, NHAc

(VI)

$(p\text{-}MeOC_6H_4\overset{\oplus}{N}(O^{\ominus})\!=\!CH\text{—})_2$

(VII)

reaction with diazomethane (cf., reactions of nitrosobenzene). Alkylation of *p*-nitrosophenol by alcohols, in the presence of sulphuric acid, however, gives alkyl *p*-nitrosoaryl ethers in high yield (*Schors, Kraaijeveld* and *Havinga, loc. cit.*; *J. T. Hays, E. H. de Butts* and *H. L. Young*, J. org. Chem., 1967, **32**, 153). The conversions of *p*-nitrosophenol into 4-amino-3,5-dichlorophenol by treatment with hydrogen chloride in ether and into 4-amino-3,5-dichloroanisole with hydrogen chloride in methanol, reported much earlier by *G. Bargellini* and his coworkers (Gazz., 1929, **59**, 16; Atti Accad. Lincei, 1928, [6], **8**, 404), are obviously related to the above alkylation.

(5) *Acylation.* Acylation of *p*-nitrosophenol with acetic anhydride gives a mixture of the *p*-nitrosophenolic ester and benzoquinone oxime ester, the latter rearranging to the former on crystallisation. Substituted nitrosophenols appear to behave similarly but a re-investigation using n.m.r. spectroscopy is required (*P. Ramart-Lucas et al.*, Bull. Soc. chim. Fr., 1949, 905). The claim that the benzenesulphonic ester undergoes a Beckmann rearrangement has been disproved and the product shown to be 4,4′-dihydroxyazoxybenzene (*R. A. Raphael* and *E. Vogel,* J. chem. Soc., 1952, 1958; *N. J. Leonard* and *J. W. Curry,* J. org. Chem., 1952, **17**, 1071).

(6) *Nucleophilic addition reactions. p*-Nitrosophenol reacts with hydroxylamine hydrochloride (*Nietzki* and *F. Kehrman,* Ber., 1887, **20**, 613), semicarbazide hydrochloride (*J. Thiele* and *W. Barlow,* Ann., 1898, **302**, 331), *p*-nitrophenylhydrazine (*W. Borsche, ibid.*, 1905, **343**, 199; 1907, **357**, 182) and oxalyldihydrazine (*idem, ibid.*, 1929, **475**, 129) to give products derived from the quinone oxime form. Phenylhydrazine in neutral solvent, however, reduces it to *p*-aminophenol (*O. Fischer* and *L. Wacker,* Ber., 1888, **21**, 2609).

(7) *Other reactions. p*-Nitrosophenol is converted into *p*-nitrosoaniline by heating with ammonium acetate and ammonium chloride in the presence of ferric chloride (*J. Willenz,* J. chem. Soc., 1955, 2049). With aniline in acetic acid it yields *p*-hydroxyazobenzene (*C. Kimich,* Ber., 1875, **8**, 1027), with aniline hydrochlorideazophenine (2,5-dianilino-*p*-benzoquinoneanil) (*O. W. Witt* and *E. G. P. Thomas,* J. chem. Soc., 1883, **43**, 112) and with phenol in the presence of 70% sulphuric acid an indophenol. Condensation of alkyl ethers of *p*-nitrosophenol with primary aromatic amines gives *p*-nitrosodiphenylamines in high yield (*Hays, Young* and *H. H. Espy,* J. org. Chem., 1967, **32**, 158).

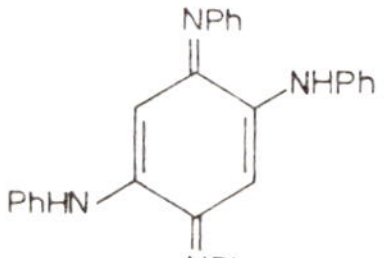

NPh NHPh PhHN NPh

Azophenine

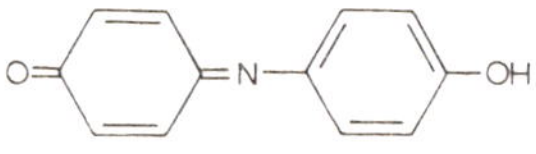

O= =N— —OH

Indophenol

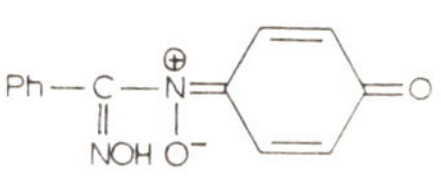

Ph—C—N(⊕)= =O NOH O⁻

(VIII)

Condensation of *p*-nitrosophenols with nitrile oxides yields tautomeric quinone-imine *N*-oxides VIII (*F. Minisci, R. Galli* and *A. Quilco,* Tetrahedron Letters, 1963, 785). Diazonium nitrates may be produced from *p*-nitrosophenol or alkyl *p*-nitrosoaryl ethers by reaction with nitrous acid and nitrogen dioxide, respectively (*E. Bamberger*, Ber., 1918, **51**, 634; *L. Horner* and *F. Hübenett*, *ibid.*, 1952, **85**, 804). Diazonium salts are also thought to be formed as intermediates when *p*-nitrosophenol and halogenated nitrosophenols react with hydroxylamine salts and cuprous halide–hydrohalide mixtures yielding *p*-halogenophenols (*Hodgson* and *W. H. H. Norris*, J. chem. Soc., 1949, S181).

The ready condensation of nitrosophenols with phenols [cf., *Lieberman*'s nitroso reaction (Ber., 1874, **7**, 247) is now considered to be the origin of the "diazo resins" which are formed when diazonium salts are decomposed in hot acid solution containing nitrous acid (*H. Gies* and *E. Pfeil*, Ann., 1952, **578**, 11).

(*iii*) *Individual nitrosophenols and their derivatives*

p-**Nitrosophenol** forms colourless needles from boiling water, yellowish white needles from acetone–benzene. It dissolves fairly easily in water and most organic solvents (except hydrocarbons). In polar solvents such as water, alcohol, or ether, the solution is green, while in other solvents it is much yellower. Values reported for its m.p. vary over a wide range, 126–138°. A paper which describes the purification of commercial samples of *p*-nitrosophenol gives 135° (*Hays, de Butts* and *Young*, *loc. cit.*). Solutions in dilute sulphuric acid are unstable (*T. Suzawa* and *H. Hiyama*, C.A., 1956, **50**, 227d).

It is best prepared from phenol and nitrous acid but is also formed by the other methods described previously.

The following *p*-nitrosophenols (most of which exist mainly in the quinone-oxime form) are prepared by standard methods: *2-methyl-4-nitrosophenol*, m.p. 135° (*Bridge* and *W. C. Morgan*, Amer. J. Chem., 1898, **20**, 766); *2-ethyl-4-nitrosophenol*, m.p. 101°; *2-isopropyl-4-nitrosophenol*, m.p. 110°; *2-*tert*-butyl-4-nitrosophenol*, m.p. 105° (*Norris* and *Sternhell*, *loc. cit.*); *3-methyl-4-nitrosophenol*, m.p. 155° (*Goldschmidt* and *H. Schmid*, Ber., 1884, **17**, 2060); *2,6-dimethyl-4-nitrosophenol*, m.p. 172°; *2,6-di-*tert*-butyl-4-nitrosophenol*, m.p. 211°; *3,5-dimethyl-4-nitrosophenol*, m.p. 183°; *3,5-diethyl-4-nitrosophenol*, m.p. 130° (*Vaughan* and *Finch*, *loc. cit.*); *2-methoxy-6-methyl-4-nitrosophenol*, m.p. 198°, (*J. F. W. McOmie* and *I. M. White*, J. chem. Soc., 1955, 2619); *2-chloro-*, *2-bromo-*, *2-iodo-* and *2-fluoro-4-nitrosophenols*, m.p. 148°, 151°, 160° and 144° (*Hodgson* and *D. E. Nicholson*, *ibid.*, 1940, 810); *2-acetyl-6-methyl-4-nitrosophenol*, m.p. 114°; *2-cyano-4-nitrosophenol*, m.p. 150°; *2-methylthio-4-nitrosophenol*, m.p. 158° (*Norris* and *Sternhell*, *loc. cit.*); *nitrosothymol*, m.p. 164° (*E. Kremers*, *N. Wakeman* and *R. M. Hixon*, Org. Synth., Coll. Vol. I, 1964, p. 511); *2,6-dimethoxy-4-nitrosophenol*, m.p. 219° (*H. I. Bolker* and *F. L. Kung*, Canad. J. Chem., 1969, 47, 2109); *nitrosocarvacrol*, m.p. 153° (*A. Klager*, Ber., 1899, 32, 1516).

o-**Nitrosophenol**, forms pale greenish yellow needles which are highly volatile and have a piercing odour. Its solutions in organic solvents are green. The cupric salt, $Cu(C_6H_4O_2N)_2$, is red, the sodium salt, $Na(C_6H_4O_2N)$, forms reddish green shimmering leaflets and the ferric salt, $Fe(C_6H_4O_2N)_3$, is greenish black.

It is best prepared by the Baudisch reaction (*Maruyama, Tanimoto* and *Goto, loc. cit.*). For other preparations see *O. Baudisch et al.* (Ber., 1912, **45**, 1171; 1915, **48**, 1662; 1918, **51**, 1058). *Cronheim* (*loc. cit.*) gives details of the colours, solubilities and spectral properties of the metal complexes of many substituted *o*-nitrosophenols but, unfortunately, inadequately describes the preparation of the parent nitrosophenols. 5-*Methoxy-2-nitrosophenol*, m.p. 130–140^0 and 168^0 (see p. 348) (*F. Henrich* and *H. Eisenach*, J. pr. Chem., 1904, 70, 332).

m-**Nitrosophenol**, m.p. 104–105^0 (decomp.), is prepared by oxidation of the corresponding hydroxylamine with ferric chloride (*L. Alfonso* and *D. N. Kravtsov*, Doklady Chem., 1968, **169**, 216).

Alkyl nitrosoaryl ethers. p-*Nitrosoanisole*, m.p. 25^0; p-*nitrosophenetole*, m.p. 35^0, *butyl* p-*nitrosophenyl ether*, m.p. —6^0 (*Hays, de Butts* and *Young, loc. cit.*); 3-*chloro-4-nitrosoanisole*, m.p. 59^0 and 3,5-*dichloro-4-nitrosoanisole*, m.p. 125^0 (*Hodgson et al.*, J. chem. Soc., 1927, 2216; 1929, 1553); m-*nitrosoanisole*, m.p. 48^0 (*Baudisch* and *R. Fürst*, Ber., 1915, **48**, 1665); o-*nitrosoanisole*, m.p. 103^0 (*A. Baeyer* and *E. Knorr*, *ibid.*, 1902, **35**, 3034) is obtained from *o*-anisidine by oxidation with Caro's acid. 5-*Dimethylamino-2-nitrosoanisole*, m.p. 131^0 (*L. Fieser* and *H. T. Thompson*, J. Amer. chem. Soc., 1939, **61**, 376).

(c) *Aminophenols*

(i) *Methods of preparation*

(1) Aminophenols are obtained by reduction of nitrophenols (p. 340), nitrosophenols (p. 349) and hydroxyazo compounds. A wide variety of reducing agents, *e.g.*, sodium dithionite and zinc dust, have been used; details of many of these are given by *R. Schröter* (in Houben Weyl's "Methoden der Organischen Chemie", Vol. XI/I, George Thieme Verlag, Stuttgart, 1957, Chapter 4). By using sodium or ammonium sulphide it is generally possible to reduce di- and tri-nitrophenols to aminonitrophenols, *e.g.*, 2,4-dinitrophenol may be reduced to 2-amino-4-nitrophenol in this way (*W. W. Hartman* and *H. L. Silloway*, Org. Synth., Coll. Vol. III, 1963, p. 82) [*cf.*, reduction of 2,4-dinitro-1-naphthol to 2-nitro-4-amino-1-naphthol by stannous chloride (*H. H. Hodgson* and *E. W. Smith*, J. chem. Soc., 1935, 671)].

(2) *p*-Aminophenols are formed by rearrangement of arylhydroxylamines under the influence of dilute sulphuric acid (Bamberger rearrangement) (*E. Bamberger*, Ann., 1921, **424**, 297). The rearrangement is intermolecular (*H. E. Heller, E. D. Hughes* and *C. K. Ingold*, Nature, 1951, **168**, 909) and *p*-phenetidine is formed when the reaction is carried out in ethanol (*Y. Yukawa*, J. chem. Soc. Japan, Pure Chem. Sect., 1950, **71**, 603). In a variation

of this method a nitro compound is reduced electrochemically in dilute sulphuric acid. *o*-Acylhydroxylamines rearrange on heating, or in some cases spontaneously, mainly to *o*-acyloxyamines from which the corresponding *o*-aminophenol may be obtained by hydrolysis (*G. T. Tisue, M. Grassmann* and *W. Lwowski*, Tetrahedron, 1968, **24**, 999, and refs. cited therein). Phenylhydroxylamine-*O*-sulphonic acid rearranges to *o*-aminophenol by a cyclic ionic process (*E. Boyland* and *R. Nery*, J. chem. Soc., 1962, 5217).

(3) Replacement of the halogen atom in halogenophenols by the action of ammonia in the presence of a cupric salt also yields aminophenols. Unless the halogen atom is activated by *ortho* or *para* nitro groups the replacement requires forcing conditions and this method is seldom used in the laboratory. *m*-Aminophenol may be formed by replacement of one of the hydroxyl groups of resorcinol by the amino group by heating with aqueous ammonia and ammonium chloride under pressure at 200°.

(4) Direct amination of phenols has received scant attention. Treatment of phenol with chloramine has been reported to give a little *p*-aminophenol (*F. Raschig*, Z. angew. Chem., 1907, **20**, 2065) while sodium phenoxide gives *o*-aminophenol (*W. Theilacker* and *E. Wegner*, Angew. Chem., 1960, **72**, 127). 2,6-Dialkylphenoxides give ring-expanded products with chloramine (see p. 308). *N*-Benzoyloxypiperidine reacts with phenol to give *N*-(2-hydroxyphenyl)piperidine (*P. Kovacic, R. P. Bennett* and *J. L. Foote*, J. org. Chem., 1961, **26**, 3013).

(5) Ortho-hydroxylation of aromatic amines is effected by treatment with potassium persulphate followed by acid hydrolysis of the resulting *o*-aminophenyl sulphate (*Boyland* and *P. Sims*, J. chem. Soc., 1954, 980). Udenfriend's reagent hydroxylates aniline in the *para*-position (*S. Udenfriend et al.*, J. biol. Chem., 1954, **208**, 731, 741). *N*-Alkylarylamines are oxidised by benzoyl peroxide to *o*-benzamidophenols (*J. T. Edward*, J. chem. Soc., 1954, 1464).

(6) Substituted *m*-aminophenols are formed when *o*-quinol acetates are treated with secondary amines (*F. Langer, E. Zbiral* and *F. Wessely*, Monatsh., 1959, **90**, 623).

o-**Aminophenol**, m.p. 174°, is not very soluble in cold water and is difficultly soluble in benzene. It forms complexes of varying stability with a wide variety of metal ions (*D. D. Perrin*, J. chem. Soc., 1961, 2244). Crystallographic data is given by *W. C. McCrone* (Anal. Chem., 1949, **21**, 531). It is usually prepared by reduction of *o*-nitrophenol. A wide variety of reducing agents has been used, including sodium dithionite, zinc dust and boiling water, sodium borohydride and palladised charcoal (*T. Neilson, H. C. S. Wood* and *A. G. Wylie*, J. chem. Soc., 1962, 371), and hydrazine and palladised charcoal (*P. M. G. Bavin*, Canad. J. Chem., 1958, **36**, 238). *G. Bing, F. Kaufler* and *R. A. Kreiger* (Australian

P., 205,525/1957) describe its preparation from *o*-chlorophenol by treatment with ammonia under pressure in the presence of a copper catalyst. *o*-Dialkylaminophenols may be obtained from 2-acylfurans *via* the enamine of type IX.

m-*Aminophenol*, m.p. 122^{0}, is fairly soluble in hot water and very soluble in ether and alcohol. Crystallographic data is given by *J. Krc* and *A. Hinch* (Anal. Chem., 1956, **28**, 137). It is prepared by (a) reduction of *m*-nitrophenol (see refs.

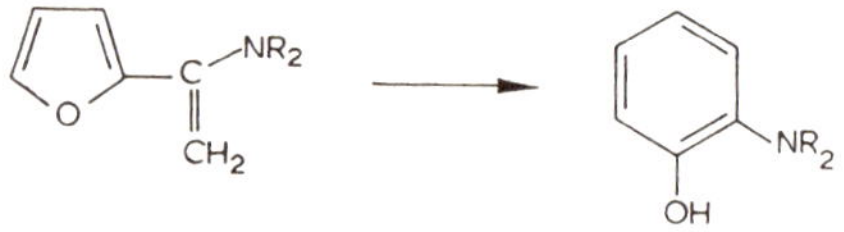

(IX)

to reduction of *o*-nitrophenol); (b) alkali fusion of metanilic (*R. Meyer* and *W. Sundmacher*, Ber., 1899, **32**, 2112) or of *p*-amino-*o*-hydroxybenzoic acids (*Sumner Chem. Co. Inc.*, B.P., 722,074/1954); (c) heating resorcinol with ammonia and ammonium bisulphite at 100^{0} (*Bad. Anilin u. Soda Fabrik*, G.P., 117,471/1899) or with aqueous ammonia and ammonium chloride at 200^{0} (*Leonhard & Co.*, G.P., 49,060/1889). With benzenediazonium chloride it gives a mixture of mono-, bis-, and tris-azo compounds (*T. S. Gore* and *P. K. Inamdar*, Indian J. Chem., 1969, **7**, 437).

p-*Aminophenol*, m.p. 184^{0} (decomp.), is soluble in 90 parts of water at 0^{0} but is insoluble in chloroform and benzene. The crystal structure is given by *C. J. Brown* (Acta Cryst., 1951, **4**, 100). It is prepared by (a) reduction of *p*-nitrophenol (*L. Spiegler*, U.S.P., 2,947,781/1960), *p*-nitrosophenol (*J. H. Boyer* and *S. E. Ellzey*, J. Amer. chem. Soc., 1960, **82**, 2525) or 4,4′-dihydroxyazobenzene; (b) heating chloro- or bromo-phenol with aqueous ammonia and a little copper sulphate (*H. E. Podall* and *W. E. Foster*, J. org. Chem., 1958, **23**, 280; *A. G. Anilinfabrikation*, B.P., 4044/1908); (c) reduction of nitrobenzene in acid solution either by catalytic hydrogenation, electrochemical or chemical reduction (aluminium or zinc and acid) (*R. G. Benner*, B.P., 1,112,678/1965; *F. R. Bean*, U.S.P., 2,446,519/1948, and many other refs. in the patent literature); (d) dealkylation of alkyl *p*-aminophenyl ethers with strong acid (*I. G. Farbenind.*, *A.G.*, B.P., 293,792/1928).

(*ii*) *Properties and reactions of aminophenols*

(1) *Acidity*, The acidity of phenols is depressed by the presence of an amino group. The p*K*a values of phenol and *p*-aminophenol are 9·98 and 10·3, respectively. The aminophenols behave as weak bases (*p*-aminophenol p*K*b = 5·5) giving salts with mineral acids.

(2) *Oxidation*. *p*-Aminophenol is rapidly oxidised by air. It is a strong reducing agent and is well-known under the name Rodinal as a photographic developer; it may be used also for dyeing furs (*J. F. Corbett*, J. chem. Soc., B,

1969, 207 and later papers in this series). Oxidation by silver oxide yields *p*-benzoquinoneimine which is unstable to light and acids, and polymerises in solution (*R. Willstätter et al.*, Ber., 1904, **37**, 1494; 1909, **42** 1902) but has been trapped as its Diels-Alder adduct with cyclopentadiene (*C. J. Sunde, J. G. Erickson* and *E. K. Raunio*, J. org. Chem., 1948, **13**, 742). *N*- and *C*-Alkyl homologues also decompose rapidly but *N*-aryl-, *N*-acyl-, and *N*-chloroquinoneimines are sufficiently stable to be isolated (*R. Adams* and *J. H. Looker*, J. Amer. chem. Soc., 1951, **73**, 1145; *Adams* and *W. Reifschneider*, Bull. Soc. chim. Fr., 1958, 23). The last mentioned imine is prepared by oxidation of *p*-aminophenol with bleaching powder (*Willstätter* and *E. Meyer*, Ber., 1904, **37**, 1494). In aqueous solution *p*-amino- and *p*-alkylaminophenols are oxidised by dichromate to *p*-benzoquinone (*Willstätter et al.*, *ibid.*, 1905, **38**, 2244; 1909, **42**, 2166). Controlled oxidation gives *p*-aminophenoxyl the E.S.R. spectrum of which has been recorded (*H. Stegmann* and *K. Scheffler*, Z. Naturforsch., 1964, **19b**, 537).

Photo-oxidation (*H. Ogawa* and *S. Natori*, Chem. Pharm. Bull. Japan, 1968, **16**, 1709) or oxidation of *o*-aminophenol by one-electron oxidants such as ferric chloride (*W. Schäfer*, Progr. in org. Chem., 1964, **6**, 135), enzymes (*L. R. Morgan, D. M. Weimorts* and *C. C. Aubert*, Biochim. Biophys. Acta, 1965, **100**, 393) or simply by autoxidation on silica thin layer plates gives the corresponding 2-aminophenoxazin-3-one (X) (*N. N. Gerber*, Canad. J. Chem., 1968, **46**, 790). In certain cases the corresponding 2-hydroxyphenoxazin-3-one (XI) may also be formed (*Schäfer, loc. cit.*). Oxidation

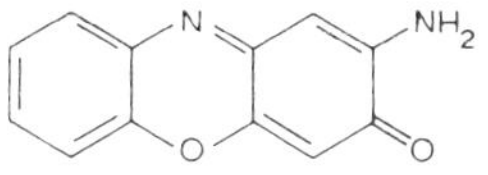

(X) 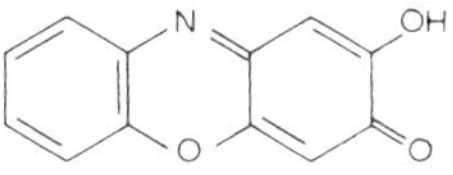 (XI)

of *o*-aminophenol with nitroso compounds, ferricyanide, or by heating with ethanolic potassium hydroxide yields the benzoxazinophenoxazine (triphenoxdioxazine) (XII) (*M. Khalifa*, J. chem. Soc., 1960, 2779):

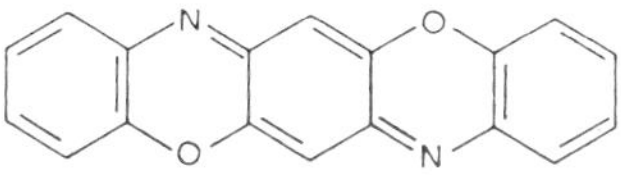

(XII)

(3) *Acylation* of *o*-aminophenols generally leads to *N*-acylated products but if an excess of reagent is used *O,N*-diacylated compounds are formed.

O- to *N*-Acyl migration occurs extremely easily and *o*-benzamidophenol may be obtained by reduction of *o*-nitrophenyl benzoate (*A. Einhorn* and *B. Pfyl*, Ann., 1900, **311**, 34). Hydrolyses of the "mixed" diacyl compounds (XIII; R = Me, R′ = Ph and R = Ph, R′ = Me) in aqueous alkali yields

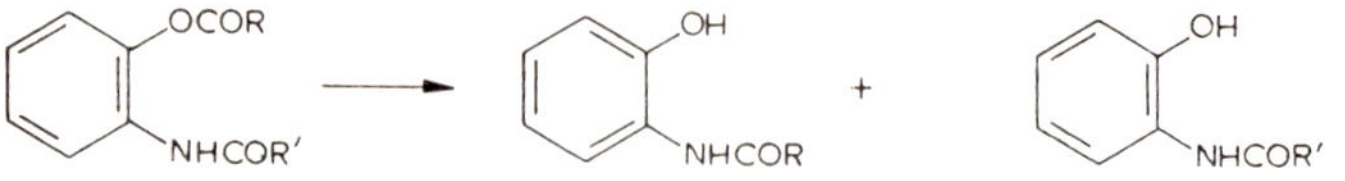

OCOR, NHCOR′ (XIII) → OH, NHCOR + OH, NHCOR′

a mixture of both *o*-acetamido- and *o*-benzamido-phenols. These "mixed" diacyl derivatives on treatment with either alcohol, pyridine, or water and heat, give an equilibrium mixture of the two diacylated isomers (*A. L. Lerosen* and *E. D. Smith*, J. Amer. chem. Soc., 1948, **70**, 2705; 1949, **71**, 2815; cf., *F. Bell*, J. chem. Soc., 1931, 2962). Although *o*-aminophenyl esters are generally too unstable to be isolated (but see ref. 6 in *L. H. Amundsen* and *C. Ambrosio*, J. org. Chem., 1966, **31**, 731), *o*-aminophenyl ethyl carbonate (XIV) has been isolated and shown to rearrange slowly in aqueous acidic

OCO_2Et, NH_2 (XIV) → OH, $NHCO_2Et$

solution to ethyl *o*-hydroxyphenylcarbamate (*J. H. Ransom*, Amer. chem. J., 1900, **23**, 1). Other examples of this and related migrations are given by *N. N. Crounse* and *L. C. Raiford* (J. org. Chem., 1945, **10**, 419); *Raiford* and *Lerosen* (J. Amer. chem. Soc., 1945, **67**, 2163); *Amundsen* and *Ambrosio* (*loc. cit.*).

Acylation of *p*-aminophenols with, for example, ketene (*M. Bergmann* and *F. Stern*, G.P., 453,577/1925) yields the corresponding *N*-acylated products which usually darken on storage. Very pure *p*- (and *m*-)hydroxyacetanilide have been prepared in high yield by reductive acetylation of *p*- (and *m*-)nitrophenol using palladium as catalyst (*M. Freifelder*, J. org. Chem., 1962, **27**, 1092). *p*-Aminophenyl esters are normally prepared by reduction of the corresponding *p*-nitrophenyl esters but a surprising *O*- to *N*-acetyl migration has been reported on catalytic reduction of *p*-nitrophenyl acetate on platinum (*R. Feldstein, M. H. Aldridge* and *B. H. Alexander*, *ibid.*, 1961, **26**, 1656). Selective *O*-benzoylation of aminophenols using sodium benzoyl thiosulphate ($PhCOS_2O_3Na$) under carefully controlled conditions has been claimed by *A. Ito* (C.A., 1962, **58**, 6732).

o-**Hydroxyacetanilide,** m.p. 210°; o-*hydroxybenzanilide,* m.p. 167°; o-*formamidophenol,* m.p. 129°, is prepared by heating *o*-aminophenol with formic acid (*E. Bamberger,* Ber., 1903, **36,** 2042); o-*aminophenyl benzoate,* m.p. 79°; o-*acetamidophenyl acetate,* m.p. 122°; o-*benzamidophenyl benzoate,* m.p. 185°.

m-*Hydroxyacetanilide,* m.p. 149°; m-*hydroxybenzanilide,* m.p. 174°; m-*aminophenyl benzoate,* m.p. 172°; m-*acetamidophenyl acetate,* m.p. 100°; m-*benzamidophenyl benzoate,* m.p. 153°.

p-*Hydroxyacetanilide,* m.p. 169°; p-*hydroxybenzanilide,* m.p. 217°; p-*aminophenyl benzoate,* m.p. 154°; p-*aminophenyl benzenesulphonate,* m.p. 101°; p-*acetamidophenyl acetate,* m.p. 151°; p-*benzamidophenyl benzoate,* m.p. 235°.

(4) *Alkylation.* All possible mono-, di-, and tri-methylated aminophenols are known. Direct methylation of aminophenols with dimethyl sulphate under neutral conditions leads to the formation of *N,N*-dimethylaminophenols (*M. L. Crossley et al.,* J. Amer. chem. Soc., 1952, **74,** 573), which may also be obtained from the corresponding diazonium salts (*V. O. Süs,* Ann., 1947, **557,** 237; *J. Dejonge* and *R. Dijkstra,* Rec. Trav. chim., 1949, **68,** 426) or by dealkylation of *N,N*-dimethylanisidines with hydrogen iodide (*F. G. Bordwell* and *P. J. Boutan,* J. Amer. chem. Soc., 1956, **78,** 87). *N*-Monoalkylation may be achieved by heating the aminophenol with the appropriate alkyl bromide (1 mole) (*Crossley, loc. cit.*) or alcohol in the presence of Raney nickel (*J. Horyna* and *O. Cerny,* Coll. Czech. chem. Comm., 1956, **21,** 906). Reduction of the condensation products from phthalimide, formaldehyde and aminophenols also yield *N*-methylaminophenols; *N*-methylanisidines are similarly prepared (*M. Sekiya* and *K. Ito,* Chem. Pharm. Bull. Japan, 1966, **14,** 1007). Trimethylammonium formate converts *p*-dimethylaminophenol (and its ethers) into the corresponding *N*-methyl-*N*-formyl derivatives (*M. Sekiya, M. Tomie* and *N. J. Leonard,* J. org. Chem., 1968, **33,** 318).

The anisidines and other alkyl aminoaryl ethers are frequently obtained by reduction of the corresponding nitro compounds if they are available (*J. N. Ashley et al.,* J. chem. Soc., 1959, 897). *p*-Anisidine and related ethers are prepared by the Bamberger rearrangement of phenylhydroxylamine in the appropriate alcohol containing sulphuric acid (*Bamberger,* Ber., 1900, **33,** 3600). For *m* anisidine it is more convenient to methylate *m*-acetylaminophenol or *m*-aminophenol under alkaline conditions (*P. K. Kadaba* and *S. P. Massie,* J. org. Chem., 1957, **22,** 333; *S. Solomon, C. H. Wang* and *S. G. Cohen,* J. Amer. chem. Soc., 1957, **79,** 4104). Although methylation of *o*-anisidine has been reported to give only *N,N*-dimethyl-*o*-anisidine (*J. W. Cook, J. D. Loudon* and *P. McCloskey,* J. chem. Soc., 1952, 3904; *Bordwell* and *Boutan, loc. cit.*), *H-H. Stroh* and *G. Westphal* (Ber., 1964, **97,** 83) claim that the monomethylamino compound can also be obtained in this way. Methylation of *p*-anisidine with diazomethane in the presence of boron trifluoride gives a mixture of the mono- and di-methylanisidines (*E. Müller, H. Hüber-Emden* and *W. Rundel*, Ann., 1959, **623**, 34). When *o*- or *m*- or *p*-halogenoanisoles are treated with sodamide or lithium dialkylamides the corresponding *m*-anisidine or *m-N,N*-dimethylanisidine is the major product, formed by way of a benzyne intermediate (*H. Gilman* and *R. H. Kyle*, J. Amer. chem. Soc., 1952, **74**, 3027; *J. D. Roberts et al., ibid.*, 1956, **78**, 611). *N*-Methylanisidines are usually obtained by methylation of the sodium salts of acetanisidides with dimethyl sulphate followed by hydrolysis with sulphuric acid (*Cook, Loudon* and *McCloskey, loc. cit.*; *M. Julia* and *J. Lenzi*, Bull. Soc. chim. Fr., 1962, 1051) or by reduction of *N*-formylanisidines with lithium aluminium hydride (*M. Bory* and *M. C. Mentzer, ibid.*, 1953, 814; *F. Bennington, R. D. Morin* and *L. C. Clark*, J. org. Chem., 1958, **23**, 19).

The reaction of *o*-aminophenol with chloroacetic acid is pH-dependent; *O*-alkylation may be avoided by working at pH <8. Mono- and di-*N*-alkylated products are formed, the latter readily lactonising to the 2-phenomorpholone (XV). Conversely, the salt of *o*-aminophenoxyacetic acid lactamises to the 3-phenomorpholone (XVI) on acidification (*H. H. Freedman* and *A. E. Frost, ibid.*, 1958, **23**, 1292):

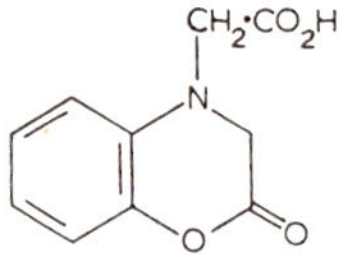

(XV) 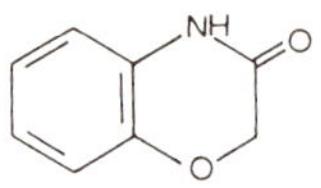 (XVI)

Other aminophenoxyacetic acids have been prepared by *W. A. Jacobs* and *M. Heidelberger* (J. Amer. chem. Soc., 1917, **39**, 2188).

o-**Anisidine,** m.p. 3°, b.p. 218°; o-*phenetidine*, b.p. 232°; *allyl 2-aminophenyl ether*, b.p. 130°/10 mm; *2-aminodiphenyl ether*, m.p. 45°, b.p. 173°/14 mm. m-*Anisidine*, b.p. 243°; m-*phenetidine*, b.p. 240°; *3-aminodiphenyl ether*, m.p. 37°. p-*Anisidine*, m.p. 57°, b.p. 243°, (p-*methoxyacetanilide*, m.p. 127°); p-*phenetidine*, b.p. 254° (p-*ethoxyacetanilide*, m.p. 135°, is used as an antipyretic under the name Phenacetin). *Diacetyl*-p-*phenetidine*, m.p. 54°, is formed by the prolonged action of boiling acetic anhydride on *p*-phenetidine. p-*Ethoxyphenylsuccinimide*, Pyrantin, m.p. 155°, is also used as an antipyretic. N-(p-*Ethoxyphenyl*)*urea*, m.p. 174°, is freely soluble in hot water and has an intensely sweet taste. It is known commercially as the sweetening agent Dulcin.

o-**Methylaminophenol,** m.p. 86°, has been used, mixed with quinol, as a photographic developer. *2-Hydroxydiphenylamine*, m.p. 70° is obtained by the action of acetyl or benzoyl peroxide on diphenylamine (*P. B. Denney* and *D. Z. Denney, ibid.*, 1960, **82**, 1389). m-*Methylaminophenol*, b.p. 170°/12 mm; m-*ethylamino-*

phenol, m.p. 62°, b.p. 176°/12 mm; p-*methylaminophenol*, m.p. 85°; its sulphate is the photographic developer Metol; 4-*hydroxydiphenylamine*, m.p. 70°, b.p. 330°, is formed by heating quinol with aniline and zinc chloride at 180–185° (*A. E. Bradfield, L. H. N. Cooper* and *K. J. P. Orton*, J. chem. Soc., 1927, 2854), or by reaction of *p*-aminophenol with bromobenzene and cuprous iodide (*N. O. Witt*, G.P., 187,870/1906). It is widely used as an antioxidant. o-*Dimethylaminophenol*, m.p. 44°, b.p. 200° (*J. von Braun*, Ber., 1916, **49**, 1101). m-*Dimethylaminophenol*, m.p. 85°, b.p. 268° and m-*diethylaminophenol*, m.p. 78°, b.p. 278° have been obtained by heating resorcinol with the appropriate dialkylamine and its sulphite in aqueous solution at 125° (*Bad. Anilin u. Sodafabrik*, G.P., 121,683/1901). m-*Dimethylaminophenol* is easily soluble in acid and alkali and shows strong reducing properties. The mono- and di-alkylaminophenols react with phthalic and succinic anhydrides, in the presence of sulphuric acid at 170°, yielding technically valuable rhodamine dyes (*S. Wawzonek*, in "Heterocyclic Compounds", Ed. *R. C. Elderfield*, Wiley, New York, 1950, Vol. 2, p. 419; *I. S. Ioffe et al.*, Zhur. obschei Khim., 1962, **32**, 1477, 1480, 1485; C.A., 1963, **58**, 1563, and subsequent papers in this series). The position *para* to the dialkylamino group is reactive and condenses with aldehydes and phosgene (*R. Möhlau* and *P. Koch*, Ber., 1894, **27**, 2895; *F. von Meyenburg, ibid.*, 1896, **29**, 501). p-*Dimethylaminophenol*, m.p. 78°.

N-Methyl-o-anisidine, b.p. 230°; N-*methyl*-m-*anisidine*, m.p. 37°, b.p. 125°/13 mm; N-*methyl*-p-*anisidine*, b.p. 121°/15 mm; N,N-*dimethyl*-o-*anisidine*, b.p. 211°; N,N-*dimethyl*-m-*anisidine*, b.p. 118°/6 mm. (*H. P. Crocker* and *B. Jones*, J. chem. Soc., 1959, 1808); N,N-*dimethyl*-p-*anisidine*, m.p. 49°, b.p. 126°/20 mm.

(5) *Condensation reactions. o*-Aminophenols and their derivatives are useful starting materials for the syntheses of phenoxazones, phenoxazines, benzoxazoles and thiobenzoxazoles. Thus, phenoxazine is formed when *o*-aminophenol and its hydrochloride are heated to 200–224° or when *o*-aminophenol is heated with catechol at 270° (*F. Kehrman* and *A. A. Niel*, Ber., 1914, **47**, 3102), the catechol acting only as a proton donor (*J. de Antoni*,

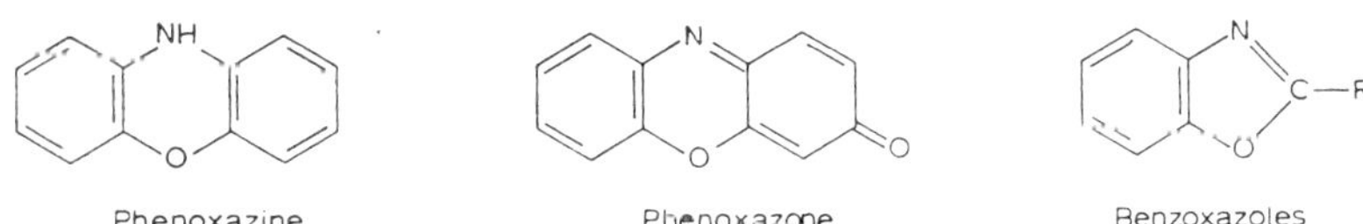

Bull. Soc. chim. Fr., 1963, 2871). Better yields of phenoxazine may be obtained by the autocondensation of *o*-aminophenol in the presence of iodine (*P. Müller, N. P. Buu-Höi* and *R. Rips*, J. org. Chem., 1959, **24**, 37). Picryl chloride, 2,4,6-trinitroanisole and 4,5-dichloro-1,3-dinitrobenzene may be condensed with *o*-aminophenol yielding 2,4-dinitrophenoxazine (*G. S. Turpin*, J. chem. Soc., 1891, **59**, 722; *F. Ullmann* and *S. M. Sane*, Ber., 1911, **44**, 3730; *H. Musso* and *P. Wager, ibid.*, 1961, **94**, 2551; *Musso, ibid.*, 1963,

96, 1927; *H. Beecken* and *Musso, ibid.*, 1961, 94, 601). Oxidative coupling of *o*-aminophenols to phenoxazones has already been mentioned. Oxidative condensation of *o*-aminophenols with *o*-dihydric phenols (*A. Butenandt et al.*, Ann., 1957, 602, 72) or condensation of *o*-aminophenols with hydroxybenzoquinones is the most widely used method for the preparation of phenoxazones (*Kehrmann et al., ibid.*, 1893, 26, 2375; 1896, 29, 2076; *Schäfer* and *H. Schlude*, Tetrahedron Letters, 1968, 2161). A summary of more recent work is given by *Schäfer*, in Progr. in org. Chem., 1964, 6, 135.

The *N*-acyl derivatives and *N,O*-diacyl derivatives of *o*-aminophenols are readily converted on heating into benzoxazoles. The parent compound (R = H) is prepared by distillation of *o*-formamidophenol (*Bamberger*, Ber., 1903, 36, 2052) and 2-phenylbenzoxazole by heating *o*-benzamidophenol. Heating *o*-aminophenol with *s*-triazine also yields benzoxazole (*C. Grundmann* and *A. Kreutzberger*, J. Amer. chem. Soc., 1955, 77, 6559). Many other acid derivatives (amides, nitriles, anilides) and aldehydes (*V. V. Somayajulu* and *N. V. Subba Rao*, Proc. Ind. Acad. Sci. Sect. A, 1964, 59, 396) react with *o*-aminophenols to give benzoxazoles; phosgene gives 2-hydroxybenzoxazole (R = OH) and halogenophosphazenes, *e.g.*, $(NPCl_2)_3$, give trimeric cyclised products (*H. R. Allcock* and *R. L. Kugel*, J. Amer. chem. Soc., 1969, 91, 5452). With diethyl oxalate, however, no ring closure occurs, instead 2,2'-dihydroxyoxanilide is formed (*E. Campaigne* and *J. E. van Verth*, J. org. Chem., 1958, 23, 1344). *J. W. Cornforth* (in "Heterocyclic Compounds", Ed. *R. C. Elderfield*, Wiley, New York, 1957, Vol. 5, p. 418) has reviewed the preparation of benzoxazoles from *o*-aminophenols.

Aminophenols condense in the usual way with aldehydes and ketones to give imines (Schiff bases) (*F. G. Pope*, J. chem. Soc., 1908, 93, 532). Those from *o*-aminophenol are the most interesting since they may be cyclised to benzoxazoles by heating or by oxidation (*F. J. Kreysa et al.*, J. Amer. chem. Soc., 1951, 73, 1155; *F. F. Stephens* and *J. D. Bower*, J. chem. Soc., 1949, 2971; 1950, 1722).

(6) *Electrophilic substitution.* The hydroxyl- and the amino-groups are both powerful electron-donating groups and when their relative positions in the ring are such that their effects are in opposition it is not always easy to predict which isomeric trisubstituted benzene will be formed. In strongly acidic media the amino or alkylamino group is protonated and the orientating power of the hydroxyl group prevails. Acylation of the amino and hydroxyl groups, which is frequently necessary to protect these groups from oxidation by, for example, "mixed acid" (nitric/sulphuric acid) reduces their electron-donating power. Hence, preferential acylation may be used in certain cases to ensure the formation of the required isomer. For examples see the preparations of the aminonitrophenols.

On sulphonation with concentrated sulphuric acid the sulphonic acid groups enters the *ortho*- and/or *para*-positions relative to the hydroxyl group, *e.g.*, *m*-aminophenol yields 2-amino-4-hydroxybenzenesulphonic acid (*A. L. Miller et al.*, J. Amer. chem. Soc., 1949, 71, 3559). Sulphur trioxide–amine complexes may either sulphate the hydroxyl group or sulphamate the amino group depending upon the amount of amine present (*Boyland et al.*, J. chem. Soc., 1958, 532; 1953, 3623).

Aminonitrophenols. **2-Amino-3-nitrophenol,** m.p. 217^0 (*H. King, ibid.*, 1927, 1058; *H. Zinner et al.*, Ber., 1959, **92,** 407); *2-amino-4-nitrophenol,* m.p. 143^0 (*W. W. Hartman* and *H. L. Silloway,* Org. Synth., Coll. Vol. III, 1966, p. 82), [*hydrate,* m.p. 80–90^0]; its *n*-propyl ether is reported to be 4,000 times as sweet as cane sugar (*J. J. Blanksma* and *P. W. van der Weyden,* Rec. Trav. chim., 1940, **59,** 629); *2-amino-5-nitrophenol,* m.p. 204^0 (*L. F. Hewitt* and *King,* J. chem. Soc., 1926, 817), (*ethyl ether,* m.p. 91^0); *2-amino-6-nitrophenol,* m.p. 112^0, [*ethyl ether,* m.p. 37^0 (*P. E. Verkade* and *P. H. Witjens,* Rec. Trav. chim., 1943, **62** 204; 1946, **65,** 361; *Zinner et al., loc. cit.*)]. *3-Amino-2-nitrophenol,* m.p. 140^0, formed by reduction of 2,3-dinitrophenol with stannous chloride (*G. W. Amery* and *J. F. Corbett,* J. chem. Soc., C, 1967, 1053); *3-amino-4-nitrophenol,* m.p. 185–186^0; *3-amino-5-nitrophenol,* m.p. 165^0, and *3-amino-6-nitrophenol,* m.p. 162^0 (*R. Meldola et al.*, J. chem. Soc., 1906, **89,** 923); *4-amino-2-nitrophenol,* m.p. 128^0 (*A. Girard,* Bull. Soc. chim. Fr., 1924, [iv], **35,** 772; *Blanksma* and *E. M. Petri,* Rec. Trav. chim., 1947, **66,** 365). *4-Amino-3-nitrophenol,* m.p. 154^0, from the nitration of *p*-aminophenyl acetate (*Girard, loc. cit.*), [*methyl ether,* m.p. 123^0 (*E. E. Fanta* and *D. S. Tarbell,* Org. Synth., Coll. Vol. III, 1963, p. 661)]. Most of the *N*- and *O*-monomethyl-, *N,O*- and *N,N*-dimethyl- and *N,N,O*-trimethyl derivatives of the above phenols have been described (*Zinner et al., loc. cit.*; *Amery* and *Corbett, loc. cit.*; *Hewitt* and *King, loc. cit.*; *J. D. Bower* and *F. F. Stephens,* J. chem. Soc., 1951, 325). U.V. and visible spectroscopic data are given by *Corbett* (Spectrochim. Acta, 1967, **23A,** 2315).

Aminodinitrophenols. **2-Amino-4,6-dinitrophenol** (**picramic acid**), dark red needles, m.p. 175^0, is formed by reduction of picric acid under a wide variety of conditions, *e.g.*, alcoholic ammonium sulphide or sodium sulphide, iron and aqueous ferric chloride (*R. F. Lyons* and *L. T. Smith,* Ber., 1927, **60,** 173), hydrazine in ethanol (*S. Kubota, K. Nara* and *S. Onishi,* J. pharm. Soc. Japan, 1956, **76,** 801) or by metal-catalysed hydrogen transfer (*E. A. Braude, R. P. Linstead* and *K. R. H. Woolridge,* J. chem. Soc., 1954, 3586). Further reduction yields 2,6-diamino-4-nitrophenol and 2,4,6-triaminophenol (*F. L. English,* Ind. Eng. Chem., 1920, **12,** 994). Picramic acid is used widely for the preparation of metachrome mordant dyes (*K. Venkataraman,* "The Chemistry of Synthetic Dyes", Vol. 1, Academic, New York, 1952). It condenses with picryl chloride or 2,4,6-trinitroanisole to give the corresponding phenoxazine (*A. Misslin* and *A. Bau,* Helv., 1917, **2,** 311). *2-Amino-3,5-dinitrophenol,* m.p. 218^0 (*L. Horner, U. Schwenk* and *E. Junghanns,* Ann., 1953, **579,** 212). *4-Amino-2,5-dinitrophenol,* dark violet needles, m.p. 166^0 (*Girard, loc. cit.*). *4-Amino-2,3-dinitrophenol* is unstable; 4-*amino*-2,6-*dinitrophenol* (*isopicramic acid*), m.p. 170°; 4-*amino*-3,5-*dinitrophenol*, m.p. 231° (*R. Meldola et al.*, J. chem. Soc., 1907, **91**, 1474; 1914, **105**, 410, 2073); 3-*amino*-2,4-*dinitrophenol*, m.p. 220°; 3-*amino*-2,6-*dinitrophenol*, m.p. 225°; 3-*amino*-4,6-*dinitrophenol*, m.p. 231° (*F. Bell*, *ibid.*, 1931, 2343).

Diazonium salt formation. Hydroxybenzenediazonium salts may be prepared from the corresponding aminophenols by diazotisation in aqueous sulphuric acid solution but difficulties can be encountered when the aminophenol is easily oxidised or is of low solubility. Numerous alternative procedures have been reported to overcome such difficulties and these include the use of nitrogen oxides in an organic solvent (*A. Hantzsch* and *W. B. Davidson*, Ber., 1896, **29**, 1522) and nitrosyl sulphuric acid in concentrated sulphuric acid (*H. H. Hodgson* and *S. Birtwell*, J. chem. Soc., 1943, 321). *J. M. Tedder* and *G. Theaker* (*ibid.*, 1958, 2573) have prepared diazonium salts directly from phenols by treatment with an excess of nitrous acid or its derivatives. *o*- and *p*-Hydroxybenzenediazonium salts have also been obtained by nucleophilic substitution of suitably substituted halogeno-, methoxyl- and nitro-benzenediazonium salts. This reaction has been utilised in the preparation of diazophenols, by *Meldola* and his co-workers (*ibid.*, 1902, **81**, 988 and other papers in this series; *Orton et al.*, *ibid.*, 1907, **91**, 1554; *G. T. Morgan* and *J. W. Porter*, *ibid.*, 1915, **107**, 645; *H. Bart*, Ann., 1922, **429**, 97) and is of considerable industrial importance in the manufacture of hydroxyazo-dyes which form lakes with heavy metal cations (*K. H. Saunders*, "The Aromatic Diazo Compounds and Their Technical Application", Arnold, London, 1949).

Crystalline diazonium salts have been isolated by diazotising the hydrochloride or sulphate of the aminophenol under anhydrous conditions. These salts couple with amines and phenols in the normal way but the replacement of the diazo-group by hydroxyl may not occur smoothly. Extensive use is made of hydroxyarenediazonium salts in the dye industry; the monographs by *H. Zollinger* ("Azo and Diazo Chemistry", Interscience, New York, 1961) and *Saunders* (*loc. cit.*) should be consulted for details.

Di- and triaminophenols. 2,4-**Diaminophenol,** m.p. 78–80°, is easily the most important. It is rapidly darkened by exposure to air and is easily soluble in acids and alkalis. Its salts are used under the name of Amidol as developers in photography and the free base is used in the preparation of hair dyes. 2,4-Diaminophenol is prepared by reduction of 2,4-dinitrophenol catalytically, or with metals and acid (*B. Jaeckel*, Ber., 1950, **83**, 578) or by metal-catalysed hydrogen transfer (*Braude, Linstead* and *Woolridge, loc. cit.*). It is also formed by electrolytic reduction of *m*-dinitrobenzene or *m*-nitroaniline in sulphuric acid solution (*B. B. Dey et al.*, Indian P., 39,426/1950; *H. V. K. Udupa, G. S. Subramanian* and *K. S. Udupa*, Indian P., 79,212/1964; *L. Gattermann*, Ber., 1893, **26**, 1844). The hydrochloride is oxidised by aqueous ferric chloride to 2-aminobenzoquinoneimine (*Kehrmann* and *H. Prager*, *ibid.*, 1906, **39**, 3437). 2,6-*Diaminophenol* is unstable and readily reduces silver nitrate solution. It is prepared by a reduction of 2,6-dinitrophenol with zinc and sulphuric acid (*E. Fromm* and *R. Ebert*, J. pr. Chem., 1924, **107**, 75). 2,5- and 4,5-*Diaminophenols* are prepared from the corresponding nitro compounds (*R. Lantz* and *E. Michel*, Bull. Soc. chim. Fr., 1964, 534, 538; *Kehrmann*, Ber., 1897, **30**, 2096; 1898, **31**, 2403). 2,3-*Diaminophenol*, m.p. 170°; 3,4-*diaminophenol*, m.p. 172° (*S. K. Freeman* and *P. E. Spoerri*, J. org. Chem., 1951, **16**, 438). 3,5-*Diaminophenol*, m.p. 170°, is prepared from phloroglucinol and aqueous ammonia (*J. Pollak*, Monatsh., 1893, **14**, 425) or by heating 2,4,6-triaminobenzene with aqueous acid (*J. Kreuger* and *R. L. Hayes*, U.S.P., 2,447,020/1948).

2,4,6-**Triaminophenol** is extremely unstable but gives salts, with three equivalents of hydrochloric or hydroiodic acid, which are more stable. It is formed by reduction of picric acid or its salts (*G. Zemplĕn* and *J. Schwartz*, Acta Chim. Acad. Sci. Hung., 1953, **3**, 487; *Bamberger*, Ber., 1883, **16**, 2400). With ferric chloride its hydrochloride gives an intensely blue solution from which brownish blue needles of diamino-*p*-benzoquinoneimine hydrochloride separate.

(*d*) *Diazophenols* (*quinone diazides or diazoquinones*)

Treatment of *o*- and *p*-hydroxybenzene diazonium chlorides with base or dehydrohalogenation with silver oxide yields *o*- and *p*-diazophenols, respectively. These compounds are usually given quinonoid structures but they exist as mesomeric hybrids of XVII ↔ XVIII and XIX ↔ XX. Evidence for their structure will be discussed in the subsequent chapter in Vol. IIIB, dealing with diazo compounds. The considerable amount of work done on this interesting class of compounds has been reviewed (*L. A. Kazitsyna, B. S. Kikot* and *A. V. Upadysheva*, Russ. chem. Reviews, 1966, **35**, 388) and only a brief summary of their chemistry will be given here.

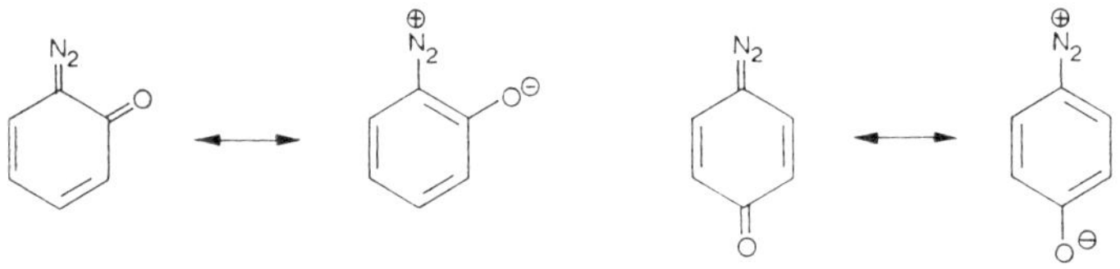

(XVII) ↔ (XVIII) (XIX) ↔ (XX)

(*i*) *Methods of formation and preparation*

(1) The conversion of a hydroxybenzenediazonium salt into a diazophenol is usually effected either by neutralisation with base or, more simply, by dilution with water (*O. Süs*, Ann., 1953, **579**, 133; *K. B. Whetsel, G. F. Hawkins* and *F. E. Johnson*, J. Amer. chem. Soc., 1956, **78**, 3360). If the diazophenol is very soluble in water then it may be obtained by shaking an alcoholic solution of the diazonium salt with silver oxide, adding ether, and then cooling (*L. C. Anderson et al.*, *ibid.*, 1945, **67**, 955; *J. D. C. Anderson et al.*, J. chem. Soc., 1949, 2082).

(2) Aromatic sulphonohydrazides react with *o*- and *p*-quinones to give the corresponding hydrazones, which, on treatment with base (or acid), yield diazophenols (*W. Ried* and *R. Dietrich*, Ann., 1961, **649**, 57; Ber., 1962, **94**, 387; *L. Horner* and *W. Durckheimer*, *ibid.*, 1962, **95**, 1206). In a related procedure a phenoxide is treated with *p*-toluenesulphonyl azide (*J. M. Tedder* and *B. Webster*, J. chem. Soc., 1960, 4417).

(ii) Properties and reactions

(1) Diazophenols are bases which are converted into diazonium salts on treatment with acid (*V. V. Ershov* and *G. A. Nikiforov*, Doklady Chem., 1964, **158**, 1101). Thus, their chemistry in acid solution is that of the diazonium salts.

(2) They form complexes with a large number of metal chlorides ($ZnCl_2$, $HgCl_2$) in weakly acidic media. These have the general formula (*Kazitsyna et al.*, *ibid.*, 1964, **159**, 1109):

$$[(HOC_6H_4N_2)_m OC_6H_4N_2]^{m\oplus} [MCl_{n+m}]^{m\ominus}$$

(3) Diazophenols readily react at nucleophilic sites in a number of compounds. The best known examples of such reactions are the azo-coupling reactions with phenols and amines. Generally the *para* isomer couples more readily than the *ortho* isomer and electron-withdrawing substituents accelerate the rate of coupling (*L. A. Kazityna, N. D. Klyneva* and *K. V. Romanova*, Doklady Chem., 1968, **181**, 955). Reaction also occurs with aliphatic diazo compounds. *o*-Diazophenols yield derivatives of benzhydro-oxadiazines (XXI) and *p*-diazophenols give the *N*-substituted diazophenol (XXII) (*R. Huisgen* and *R. Fleischmann*, Ann., 1959, **623**, 47):

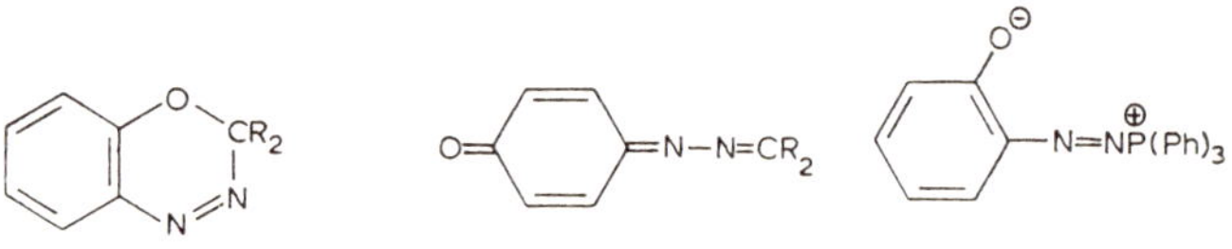

(XXI) (XXII) (XXIII)

With triphenylphosphine they yield quinonetriphenylphosphazines (XXIII) (*Ried* and *H. Appel*, Z. Naturforsch., 1960, **15B**, 684; Ann., 1961, 646, 83) which may be used as dyes and with alkyl Grignard reagents, alkaneazophenols are formed (*G. A. Nikiforov, A. A. Efremenko* and *V. V. Ershov*, Bull. Acad. Sci., U.S.S.R., 1967, 2574).

(4) *o*-Diazophenols react with ketenes to give the oxadiazepines XXIV or

XXV depending upon whether one or two moles of the ketene are used (*Ried et al., ibid.*, 1963, **666**, 113, 135, 144; 1965, **681**, 45, 52):

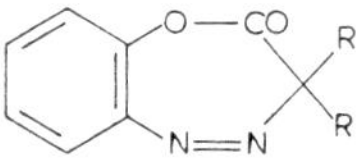

(XXIV) 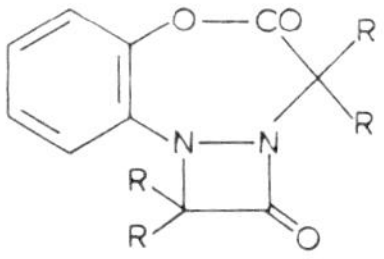 (XXV)

(5) *Photochemistry.* Diazophenols are extremely sensitive to light. *o*-Diazophenols undergo an interesting and synthetically useful ring contraction on irradiation to give cyclopentadienecarboxylic acids (*Süs, ibid.*, 1944, **556**, 65, 85; *Süs, H. Steppan* and *Dietrich, ibid.*, 1958, **617**, 20):

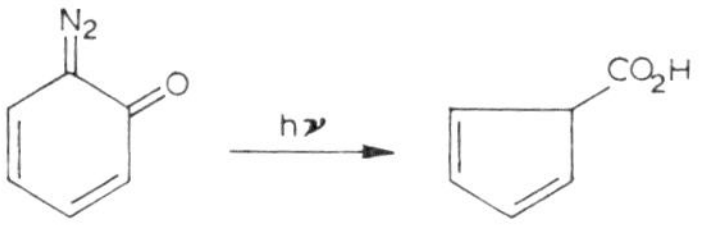

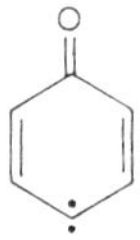

(XXVI)

Heating causes a similar change (*P. Yates* and *E. W. Robb,* J. Amer. chem. Soc., 1957, **79**, 5760). The products obtained by thermolysis or photolysis of *p*-diazophenols depend upon the solvent. In tetrahydrofuran and other ethers 1:1 alternating copolymers are formed (*J. K. Stille* and *P. Cassidy,* J. pol. Sci. [B], 1963, 563; *Stille, Cassidy* and *L. Plummer,* J. Amer. chem. Soc., 1963, **85**, 1318); in aromatic solvents *p*-arylphenols (*T. Kunitake* and *C. C. Price, ibid.*, 1963, **85**, 761) or polymers (*M. J. S. Dewar* and *A. N. James,* J. chem. Soc., 1958, 917; *Süs, K. Möller* and *H. Heis,* Ann., 1956, **598**, 123) are obtained; in chlorinated hydrocarbons the corresponding diazonium salts are regenerated and in ketonic solvents epoxyspirodienones are produced (*B. V. Svirdov, Nikoferov* and *Ershov,* Bull. Acad. Sci., U.S.S.R., 1969, 1732). In the formation of *p*-arylphenols, carbenes are thought to be the reactive intermediates (*Dewar* and *K. Narayanaswami,* J. Amer. chem. Soc., 1964, **86**, 2422). e.s.r. spectra of the triplet XXVI, obtained by irradiation of *p*-diazophenol, has been recorded by *E. Wasserman* and *R. W. Murray* (*ibid.*, 1964, **86**, 4203). A brief summary of the use of diazophenols in the manufacture of light-sensitive papers is given by *Zollinger* (*loc. cit.*, p. 172).

o-**Diazophenol,** o-**benzoquinone diazide,** bright yellow crystals, m.p. 66° (decomp.), from *o*-hydroxybenzenediazonium chloride and silver oxide in amyl alcohol (*J. D. C. Anderson, R. G. W. Lefevre* and *I. R. Wilson,* J. chem. Soc., 1949, 2082) or from *o*-benzoquinone and tosyl hydrazide in methylene chloride

(*Horner* and *Dürckheimer, loc. cit.*). It is a readily soluble in water and organic solvents and couples with alkaline β-naphthol.

p-**Diazophenol,** p-*benzoquinone diazide,* crystallises with four water molecules of crystallisation, m.p. 39⁰. It is easily soluble in water and alcohol, sparingly soluble in benzene or ether. It couples with alkaline β-naphthol and evolves nitrogen slowly on boiling with water. Its preparation from the diazonium salt is described by *L. C. Anderson* and *M. J. Roedel* (J. Amer. chem. Soc., 1945, **67**, 955).

2,6-**Dimethyl-4-diazophenol,** m.p. 121⁰ (*Kunitake* and *Price, loc. cit.*) forms a hydrated complex with boron trifluoride. 2,5-*Dimethyl-4-diazophenol,* m.p. 120⁰ (*Ried* and *Dietrich,* Ber., 1961, **94**, 387). 2,6-, 3,5- and 2,5-*Dichloro-4-diazophenols,* m.p. 141⁰, 108⁰ and 133⁰, respectively (*Ried* and *Dietrich,* Ann., 1961, **649**, 57). 2,6-*Dibromo-4-diazophenol,* golden yellow, m.p. 150⁰ (decomp.), is prepared by the action of bromine water on diazotised *p*-aminophenol (*C. Böhmer,* J. pr. Chem., 1881, **24**, 449). 2-*Nitro-4-diazophenol* forms yellow leaflets which explode at about 17⁰ (*Morgan* and *Porter, loc. cit.*). 2,3,5-*Trinitro-4-diazophenol* forms an explosive, yellow, microcrystalline powder.

5-*Methyl-*, 4,5-*dimethyl-* and 4,6-*dimethyl-2-diazophenols,* m.p. 71⁰, 63⁰ and 84⁰, respectively, are prepared from the corresponding quinones (*Horner* and *Dürckheimer, loc. cit.*).

3,5-*Dibromo-* and 2,6-*dibromo-2-diazophenols,* m.p. about 130⁰ and 103⁰, respectively, are prepared from the corresponding diazonium salts (*Orton, loc. cit.*). *Tetrachloro-2-diazophenol,* m.p. 123⁰, dimerises with loss of nitrogen on treatment with halogens at room temperature (*Ried,* Angew. Chem., 1966, **5**, 257; *Huisgen* and *Fleischmann, loc. cit.*).

4-*Nitro-2-diazophenol,* yellow precipitate which explodes at about 120⁰, and 4,6-*dinitro-2-diazophenol,* yellow plates, m.p. 158⁰, are prepared from the corresponding diazonium salts (*Morgan* and *Porter, loc. cit.*). The latter has also been prepared by oxidation of picramic acid with dichromate (*T. Urbanski et al.,* C.A., 1964, **60**, 2798).

(e) *Azoxyphenols (hydroxyazoxybenzenes)*

(i) *Methods of preparation*

Usually, azoxyphenols are prepared by methods which are applicable to azoxybenzenes (*K. H. Schundehütte,* in Houben Weyl's "Methoden der Organischen Chemie", Vol. X/III, Georg Thieme Verlag, Stuttgart, 1965, p. 749).

(1) Oxidation of the corresponding azophenol with peracetic acid (*A. Angeli,* Gazz., 1916, 46, [ii], 106) or perbenzoic acids is probably the most useful method. Oxidation of a monosubstituted or an unsymmetrically disubstituted azobenzene does, of course, produce two structural isomers (geometrical isomers are also possible but usually the *trans*-form is much

more stable) which have to be separated. These isomers are usually distinguished by the prefixes α- and β-, or by numbering the ring positions as indicated.

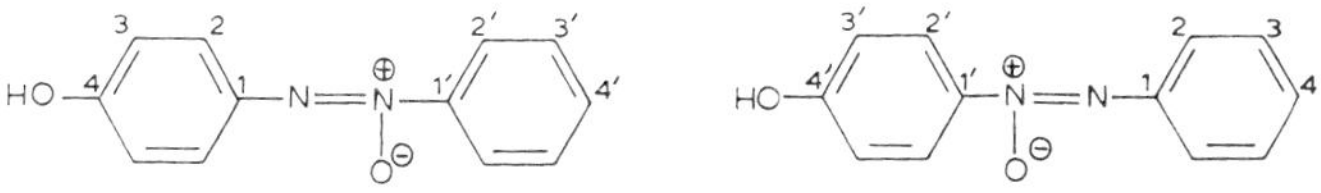

4-Hydroxyazoxy benzene
α-isomer

4'-Hydroxyazoxybenzene
β-isomer

The relative ease of oxidation of the nitrogen atoms in azobenzenes depends upon the position and nature of the ring substituents (*A. Risaliti*, *ibid.*, 1963, **93**, 585).

(2) In another method of fairly general application nitrosophenols are treated with arylhydroxylamines:

$$HOC_6H_4NO + HONHPh \rightarrow HOC_6H_4\overset{\oplus}{N}(\text{—}O^{\ominus})\text{=}N\text{—}Ph$$

(3) Azoxyphenols have also been obtained from the corresponding amino-azoxybenzenes by diazotisation and hydrolysis in the usual way (*Risaliti*, *loc. cit.*).

o-**Hydroxyazoxybenzenes.** Both isomers, golden yellow needles, m.p. 76° and 108°, are formed together with azoxybenzene and other products by the action of alkali on nitrosobenzene (*E. Bamberger*, Ber., 1900, **33**, 1939).

p-*Hydroxyazoxybenzenes*; α-form, m.p. 156°; β-form, m.p. 117°, are prepared from the corresponding azophenol or by condensation of phenylhydroxylamine with nitrosophenol (*Angeli*, *loc. cit.*). The forms differ in their behaviour towards bromine, permanganate and nitric acid (*B. Valori*, Atti Accad. Lincei, 1915 [v], **23**, II, 291; *Angeli*, Gazz., 1921, **51**, [i], 35).

2,2'-**Dihydroxyazoxybenzene**, m.p. 153°, is prepared by reduction of *o*-fluoronitrobenzene with glucose in alkaline solution (*A. Pavlath* and *I. Kuhn*, Acta Chim. Acad. Sci. Hung., 1955, **7**, 65).

4,4'-*Dihydroxyazoxybenzene*, reddish yellow needles, m.p. 234°, is obtained either from the corresponding azo-compound (*Angeli*, Gazz., 1921, **51**, [i], 35) or from *p*-nitrosophenol by treatment with benzenesulphonyl chloride in pyridine (*R. A. Raphael* and *E. Vogel*, J. chem. Soc., 1952, 1958).

Ethers of azoxyphenols. 2-**Ethoxyazoxybenzenes,** α-form m.p. 72°, β-form m.p. 75°, are obtained by oxidation of the corresponding azo compound. 4-*Methoxyazoxybenzenes*, α-form m.p. 67°, β-form m.p. 43°. Both are prepared by oxidation of the appropriate azobenzene (*C. S. Hahn* and *H. H. Jaffe*, J. Amer. chem. Soc., 1962, **84**, 949). An alternative preparation of the α-form is given by *T. E.*

Stevens (J. org. Chem., 1964, **29**, 311). 4-*Ethoxyazoxybenzenes*, α-form m.p. 74°, β-form m.p. 135°. Unequivocal assignment of the structures of the products obtained by oxidation of 4-ethoxyazobenzene was achieved by synthesis of both isomers from the corresponding indazole oxides by oxidation, followed by decarboxylation of the resultant azoxybenzene-2-carboxylic acids (*L. C. Behr*, J. Amer. chem. Soc., 1954, **76**, 3672).

2,2′-**Dimethoxyazoxybenzene**, m.p. 81°, prepared by reduction of *o*-nitro- or *o*-nitroso-anisole in alkaline solution (*P. H. Gore* and *O. H. Wheeler*, *ibid.*, 1956, **78**, 2160) is reported to exist in two forms. The higher melting form (m.p. 117°) is less stable giving the other isomer on heating. Dipole moment measurements indicate that the two forms are geometrical isomers (*E. Müller*, Ann., 1932, **495**, 132). 3,3′-*Dimethoxyazoxybenzene*, m.p. 51° (*I. Yoshioka* and *H. Otomasu*, J. pharm. Soc. Japan, 1956, **76**, 1051). 4,4′-*Dimethoxyazoxybenzene* is formed by (a) the action of alcoholic solutions of sodium or magnesium alkoxides or sodium hydroxide on *p*-nitroanisole (*L. Gattermann* and *A. Ritschke*, Ber., 1890, **23**, 1738); (b) reduction of nitroanisole with glucose in alkaline solution (*H. W. Galbraith, E. F. Degering* and *E. F. Hitch*, J. Amer. chem. Soc., 1951, **73**, 1323) or with sodium lead alloy (*K. Tabei* and *M. Yamaguchi*, Bull. chem. Soc. Japan, 1967, **40**, 1538); (c) oxidation of *p*-anisidine with sodium perborate (*S. M. Mehta* and *M. V. Vakilwala*, J. Amer. chem. Soc., 1952, **74**, 563) or peracid (*D. Lefort, C. Four* and *A. Pourchez*, Bull. Soc. chim. Fr., 1961, 2378); or (d) by condensation of the appropriate nitroso compound and *p*-anisylhydroxylamine (*A. Rising*, Ber., 1904, **37**, 43). I.R. (*W. Maier* and *G. Englert*, Z. Electrochem., 1958, **62**, 1020), N.M.R. (*D. Webb* and *H. Jaffe*, Tetrahedron Letters, 1964, 1875) and U.V. (*P. H. Gore* and *O. H. Wheeler*, J. Amer. chem. Soc., 1956, **78**, 2160; *M. P. Grammaticakis*, Bull. Soc. chim. Fr., 1951, 951) data on this dialkoxyazoxybenzene have been recorded. 4,4′-Dimethoxyazoxybenzene and its homologues (*C. Weygand* and *R. Gabler*, J. pr. Chem., 1940, [ii], **155**, 332) yield liquid crystals which have been used extensively as solvents for N.M.R. and E.S.R. studies (*G. R. Luckhurst*, Quart. Reviews, 1968, **22**, 179; *Luckhurst et al.*, Mol. Phys., 1966, **11**, 49). 4,4′-*Dimethoxyazoxybenzene* melts at 118° to an anisotropic liquid which becomes clear at 135°; in this range a nematic mesophase is formed. 4,4-*Diethoxyazoxybenzene* melts at 138° and gives a clear liquid at 168°.

(*f*) *Azophenols or hydroxyazobenzenes*

(*i*) *Methods of preparation*

(1) Diazonium salts couple with phenols to give hydroxyazobenzenes

$$PhN_2^{\oplus}X^{\ominus} + C_6H_5OH \rightarrow PhN{=}N{\cdot}C_6H_4{\cdot}OH + HX$$

The above reaction which is of fundamental importance in the manufacture of azo dyes has been well studied (*H. Zollinger*, "Azo and Diazo Chemistry",

Interscience, London, 1961; *K. H. Schündehütte,* in Houben Weyl's "Methoden der Organischen Chemie", Vol. X/III, Georg Thieme Verlag, Stuttgart, 1965, p. 219).

The diazonium cation is a weak electrophile and couples with phenols only in weakly alkaline solution. The reaction is bimolecular and there is no kinetic isotope effect except in special cases (*R. Ernst, O. A. Stamm* and *Zollinger,* Helv., 1958, **41**, 2274). The rate of coupling decreases, as the pH of the solution is lowered and the concentration of phenoxide anion decreases, and also when the pH is increased and the equilibrium moves to the right. Electron-withdrawing substituents increase the reactivity of the diazonium

$$PhN_2^{\oplus} + {}^{\ominus}OH \rightleftharpoons PhN_2OH$$

salt and electron-donating groups increase that of the phenol; normally diazonium salts do not couple with ethers. Substitution occurs predominantly *para* to the hydroxyl group and in some cases a *para* substituent in the phenol is eliminated (*H. Wittmann,* Monatsh., 1962, **93**, 1; *J. H. Freeman* and *C. E. Scott,* J. Amer. chem. Soc., 1955, **77**, 3384). When an excess of diazonium salt is used di- and tri-substitution may occur and bis- and tris-azo compounds result (*E. Grandmougin* and *H. Freimann,* Ber., 1907, **40**, 2662). Resorcinol reacts particularly easily in this way (*H. F. Hodson, Stamm* and *Zollinger,* Helv., 1958, **41**, 1816). With unreactive phenols coupling can occur on oxygen with formation of diazo ethers, *e.g.*, ArN_2O-$C_6H_4NO_2$, which are unstable, rearranging on warming to *o*-hydroxyazophenols. *O*-Azo compounds have also been prepared from picric acid and other phenols such as mesitol and pentamethylphenol which have no free coupling positions (*O. Dimroth et al.,* Ber., 1908, **41**, 4012; *K. von Auwers, ibid.,* 1908, **41**, 4304).

(2) *Wallach rearrangement.* On warming azoxybenzenes with concentrated sulphuric acid hydroxyazobenzenes are formed (*O. Wallach, ibid.,* 1881, **14**, 2617). *p*-Hydroxyazobenzenes are the usual products (but see *Hahn* and *Jaffe,* J. Amer. chem. Soc., 1962, **84**, 946) but the *ortho*-isomers are obtained if both *para* positions are blocked (*J. Singh et al.,* Canad. J. Chem., 1963, **41**, 499). With unsymmetrically substituted azoxybenzenes in which both *para* positions are available (but not for azoxybenzene itself) the oxygen usually shows a preference for the ring adjacent to the $-\underset{\displaystyle O^{\ominus}}{\overset{|}{N^{\oplus}}}=$ group (*Singh et al., loc. cit.*). The mechanism of this rearrangement is still a controversial subject but the view that the symmetrical dication $PhN^{\oplus}{\equiv}N^{\oplus}Ph$ is the key intermediate is a popular one (*E. Buncel et al.,* Chem. Comm., 1969, 765). A summary of the mechanistic work is given by *D. Duffey* and

E. C. Hendley (J. org. Chem., 1968, **33**, 1918). In the related photo-Wallach rearrangement *o*-hydroxyazobenzenes are formed, the oxygen entering exclusively the more distant ring. An intramolecular mechanism involving a 5-membered intermediate seems reasonable in this case (*G. M. Badger* and *R. G. Buttery*, J. chem. Soc., 1954, 2243; *R. Tanikaga*, Bull. chem. Soc. Japan, 1968, **41**, 2151); for a review of the photo-Wallach rearrangement see *G. G. Spence, E. C. Taylor* and *O. Buchardt*, Chem. Reviews, 1970, **70**, 231.

(3) Symmetrical dihydroxyazobenzenes may be obtained from the diazonium salt of the corresponding aminophenol by treatment with a cuprous salt.

(4) Condensation of nitrosophenols with primary aromatic amines is useful for preparing unsymmetrical azo compounds.

(5) Reduction of aromatic nitrophenols with a wide variety of reagents, many of which are given by *Schundehütte* (*loc. cit.*), is used for symmetrical azo compounds.

(6) Benzoquinones condense with many nitrophenylhydrazines to give hydrazones which tautomerise to azophenols. Phenylhydrazine, however, reduces *p*-benzoquinones:

$$O_2N{-}C_6H_4{-}NHN{=}C_6H_4{=}O \rightleftharpoons O_2N{-}C_6H_4{-}N{=}N{-}C_6H_4{-}OH$$

(7) A number of hydroxynitroazobenzenes have been obtained by condensation of nitroareneazomalonic dialdehydes with aliphatic ketones in the presence of base (*D. Leuchs*, Ber., 1965, **98**, 1335):

$$O_2N{-}C_6H_4{-}N{=}N{-}CH(CHO)_2 + RCH_2{-}CO{-}CH_2R \longrightarrow O_2N{-}C_6H_4{-}N{=}N{-}C_6H_2R_2{-}OH$$

(ii) Tautomerism of azophenols

Although some hydroxyazobenzenes give reactions of both phenols and ketones, U.V. (*A. Burawoy* and *J. T. Chamberlain*, J. chem. Soc., 1952, 2310; 1952, 3734; *R. Kuhn* and *F. Bär*, Ann., 1935, **516**, 143), I.R. (*D. Hadzi*, J. chem. Soc., 1956, 2143) and N.M.R. (*K. Venkataraman et al.*, Tetrahedron Letters, 1966, 3897) measurements show that almost all exist in the phenolic form. For the reactions of certain azophenols with semicarbazide and 2,4-dinitrophenylhydrazine see *W. Borsche et al.* (Ann., 1929, **472**, 201) and *R. Willstätter et al.* (*ibid.*, 1929, **477**, 161). 4-Hydroxy-3-methyl-2′,4′-dinitroazobenzene seems to be exceptional since it has been shown (*Venkataraman et al., loc. cit.*) by N.M.R. to exist in both forms thus supporting an earlier chemical observation that 4-hydroxy-2',4'-dinitroazobenzene, unlike the other azophenols examined, underwent Diels-Alder addition with cyclopentadiene (*W. Lauer* and *S. Miller*, J. Amer. chem. Soc., 1935, **57**, 520). For further discussion of this subject see *V. V. Ershov* and *G. A. Nikiforov* (Russ. chem. Reviews, 1966, **35**, 817).

o-**Hydroxyazobenzene,** orange red needles, m.p. 83^0, is a minor product of the rearrangement of azoxybenzene on warming with sulphuric acid or of the coupling of phenol with benzenediazonium salts (*Bamberger*, Ber., 1900, **33**, 3188) and is separated from the main product by steam distillation. In the photo-Wallach rearrangement of azoxybenzene, however, it is the principal product. Its formation by dehydration of cyclohexa-1,2-dione monophenylhydrazone is described by *T. S. Gore* and *P. K. Inamdar* (Indian J. Chem., 1968, **6**, 19). Photochemical disproportionation of nitrosobenzene also gives *o*-hydroxyazobenzene (*M. L. Scheinbaum*, J. org. Chem., 1964, **29**, 2200).

Numerous I.R. (*Hadzi, loc. cit.*; *A. G. Catchpole, W. B. Foster* and *R. S. Holden*, Spectrochim. Acta, 1962, **18**, 1353) studies have shown that in *o*-hydroxyazobenzene and its homologues there is strong intramolecular hydrogen bonding between the OH and the more distant nitrogen atom. With metal ions ($Cu^{\oplus\oplus}$, $Ni^{\oplus\oplus}$, $Co^{\oplus\oplus}$, etc.) it forms chelates. The *copper complex*, m.p. 222^0, is the most stable and may be utilised in the purification of *o*-hydroxyazobenzene (*K. Ueno*, J. Amer. chem. Soc., 1957, **79**, 3066; *G. Haefelinger* and *E. Bayer*, Naturwiss., 1964, **51**, 136; *H. D. K. Drew et al.*, J. chem. Soc., 1938, 292; 1939, 823). o-*Methoxyazobenzene*, m.p. 41^0, is formed from nitrobenzene and *o*-anisidine (*Bamberger*, Ber., 1900, **33**, 3188). With phenylmagnesium bromide it yields 3-methoxybenzidine dihydrochloride (48%) and 2-methoxy-2',6'-diphenylazobenzene (34%) (*A. Risaliti* and *S. Bozzini*, Ann. Chim. Rome, 1964, **54**, 685).

2,2'-**Dihydroxyazobenzene,** golden yellow plates, m.p. 172^0, is obtained in 53% yield when the diazonium salt of *o*-aminophenol is added to a solution of a cuprous salt. The product is isolated as its copper chelate from which the azophenol is formed on treatment with acid (*D. C. Freeman* and *C. E. White*, J. org. Chem., 1956, **21**, 379). It has also been obtained by melting *o*-nitrophenol with alkali and a little water (*Willstätter* and *M. Benz*, Ber., 1906, **39**, 3492) and by hydroxylating the copper chelate of *o*-hydroxyazobenzene with hydrogen peroxide ("oxidative coppering") (*Z. Yoshida, K. Kazama* and *R. Oda*, C.A., 1962, **57**, 13922; cf., *H. Pfitzner* and *H. Baumann*, Angew. Chem., 1958, **70**, 232). The strong intramolecular hydrogen bonds which exist in this and homologous compounds have been studied by N.M.R. (*L. W. Reeves*, Canad. J. Chem., 1960, **38**, 748). 2,2'-Dihydroxyazo compounds form many stable copper and chromium complexes which are important dyes (*Zollinger*, "Azo and Diazo Chemistry", Interscience, New York, 1961, p. 351–360 and refs. cited therein; *E. J. Gonzales* and *H. B. Jonassen*, J. inorg. and nuclear Chemistry, 1962, **24**, 1595 and other papers by these authors; *G. Schetty*, Helv., 1970, **53**, 1437). 2,2'-*Dimethoxyazo-*

benzene, m.p. 153^0, is obtained from *o*-anisidine by autoxidation in strongly alkaline solution (*L. Horner* and *J. Dehnert*, Ber., 1963, **96**, 786), *o*-anisylmagnesium bromide and the diazonium salt of *o*-anisidine (*Y. Nomura et al.*, Bull. chem. Soc. Japan, 1964, **37**, 967) and from *o*-nitroanisole and sodium methoxide in methanol.

3-**Hydroxyazobenzene,** m.p. 114–117^0, amber coloured crystals, is usually obtained by dealkylating 3-*methoxyazobenzene*, m.p. 23^0, which results from the deamination of 4-amino-3-methoxyazobenzene (*P. Jacobson* and *F. Hönigsberger*, Ber., 1903, **36**, 4093) or by condensation of nitrobenzene with *m*-anisidine in the presence of base (*M. Martynoff*, Compt. rend., 1946, **223**, 747).

3,3′-**Dihydroxyazobenzene,** brown crystals, m.p. 205^0. Reduction of *m*-nitrophenol with zinc in alkaline solution is the most convenient preparative method (*P. Ruggli* and *M. Hinovker*, Helv., 1934, **17**, 396). Fusion of *m*-nitrophenol with alkali and a little water (*Willstätter* and *Benz*, *loc. cit.*) and reduction of *m*-nitrophenol with silica powder have also been used (*R. Meier* and *F. Böhler*, *ibid.*, 1956, **89**, 2301). It can be nitrated to 3,3′-*dihydroxy*-2,4,6,2′,4′,6′-*hexanitroazobenzene*, a yellow red powder, melting at 238^0 explosively, hydrolysis of which with aqueous potassium carbonate yields phloroglucinol and nitro- and hydroxylamino-phloroglucinol, according to the conditions (*K. Elbs* and *F. Schliephake*, J. pr. Chem., 1922, **104**, 282). 3,3′-*Dimethoxyazobenzene*, m.p. 76^0, is prepared by reduction of the corresponding azoxy compound (*Th. Rotarski*, Ber., 1908, **41**, 865) or of 4-bromo-3-nitroanisole with lithium aluminium hydride, in which case bromine is eliminated (*J. Dutta* and *R. N. Biswas*, J. chem. Soc., 1963, 2387; see also *Nomura et al.*, *loc. cit.*).

4-**Hydroxyazobenzene,** orange red needles, m.p. 152^0, is obtained by (a) coupling diazotised aniline with phenol; (b) rearrangement of azoxybenzene under the influence of sulphuric acid (*C-S. Hahn* and *H. H. Jaffe*, J. Amer. chem. Soc., 1962, **84**, 949); and (c) by heating diazoaminobenzene with phenol (*K. Heumann* and *L. Oeconomides*, Ber., 1887, **20**, 372). Mass spectral (*J. H. Bowie, C. E. Lewis* and *R. G. Cooks*, J. chem. Soc., B, 1967, 621), I.R. (*R. Kübler, W. Lüttke* and *S. Weckherlin*, Z. Electrochem., 1960, **64**, 650; *Hadzi*, *loc. cit.*) and U.V. (*Hahn* and *Jaffe*, *loc. cit.*; *Kuhn* and *Bär*, *loc. cit.*) data on this compound are well-documented.

Acylation and alkylation of the hydroxyl group are effected in the usual ways (*R. Pfister* and *F. Häfliger*, Helv., 1957, **40**, 395; *S-J. Yeh* and *Jaffe*, J. Amer. chem. Soc., 1959, **81**, 3274) and the potassium salt has been recommended as a reagent for the identification of alkyl halides (*E. O. Woolfolk, E. Donaldson* and *M. Payne*, J. org. Chem., 1962, **27**, 2653). The methyl carbamates of 4-hydroxyazobenzene and its homologues are used as insecticides (*W. W. Kaeding*, U.S.P., 3,009,909/1960). Nitration and bromination of 4-hydroxyazobenzene give the expected substitution products (*J. T. Hewitt et al.*, J. chem. Soc., 1900, **77**, 99 and 810) but under different conditions a diazonium salt or its decomposition products are formed (*M. P. Schmidt*, J. pr. Chem., 1912 [ii], **85**, 235; *G. Charrier* and *G. Ferrari*, Gazz., 1914, **44**, [i], 170).

The products formed on coupling phenol with benzenediazonium salts depend upon the relative proportions of the reactants and the conditions. In addition to the hydroxyazobenzenes, 2,4-*bisbenzeneazophenol*, m.p. 123°, and 2,4,6-*tris*-benzeneazophenol, m.p. 215°, have been isolated (*E. Grandmougin* and *H. Freimann*, Ber., 1907, **40**, 2662; *G. Heller* and *O. Nötzel*, J. pr. Chem., [ii] 1907, **76**, 58).

4-*Methoxy*-, m.p. 156°, 4-*benzyloxy*-, m.p. 116°, 4-*acetoxy*-, m.p. 89°, -*azobenzenes*. 4′-*Hydroxy*-2-*nitroazobenzene*, m.p. 162°, 4′-*hydroxy*-2,4-*dinitroazobenzene*, m.p. 204°, and 4′-*hydroxy*-2,4,6-*trinitroazobenzene*, m.p. 194° (*W. Borsche*, Ann., 1907, **357**, 171).

4,4′-**Dihydroxyazobenzene,** greenish brown crystals, m.p. 215°, is prepared by (a) alkaline fusion of *p*-nitrophenol; (b) treatment of diazotised *p*-aminophenol with a copper(I) salt (*E. R. Atkinson et al.*, J. Amer. chem. Soc., 1945, **67**, 1513); and (c) reduction of *p*-diazophenol (*Willstätter* and *Benz*, *loc. cit.*; Ber., 1907, **40**, 1578). The product obtained from (c) forms reddish brown leaflets and is thought to be a different crystalline form (β) from that obtained by methods (a) and (b) (*A. H. Cook* and *D. G. Jones*, J. chem. Soc., 1939, 1309). 4-*Hydroxy*-4′-*methoxyazobenzene*, m.p. 144°; 4,4′-*dimethoxyazobenzene*, m.p. 166° (*Yeh* and *Jaffe*, J. Amer. chem. Soc., 1959, **81**, 3274).

Dihydroxyazobenzenes in which both hydroxyl groups are in the same ring are usually prepared by coupling the appropriate diazonium salt with a dihydric phenol. Thus, resorcinol can couple in the 2-, 4-, or 6-positions and the product obtained depends mainly upon the conditions (pH) and the proportions of the reactants (*T. S. Gore* and *Venkataraman*, Proc. Ind. Acad. Sci., 1951, **34A**, 368; *Hodson, Stamm* and *Zollinger*, *loc. cit.*). Ethers of resorcinol couple with reactive diazonium salts giving hydroxyazobenzenes. Dealkylation is considered to occur after coupling in such cases (*J. Haginina* and *J. Murakoshi*, C.A., 1951, **46**, 7068).

2,4-**Dihydroxyazobenzene,** dark red needles, m.p. 170°; 2,4-*bisbenzeneazoresorcinol*, red needles, m.p. 121°; 4,6-*bisbenzeneazoresorcinol*, brown red needles, m.p. 217°; 2,4,6-*trisbenzeneazoresorcinol*, brown needles, m.p. 254°.

(g) Hydrazophenols or hydroxyhydrazobenzenes

These are usually obtained by reduction of the corresponding azo compounds with either zinc and acetic acid or ammonium sulphide. The ease of reduction is directly proportional to the electron-donating power of the substituents and when the hydroxyl group(s) are in the *ortho* and/or *para* positions further reduction to the aminophenol occurs. This may be prevented by acylation of the hydroxyl group(s) before reduction. With 3- and 3,3′-dihydroxyazobenzene there are no such problems (*M. Khalifa*, J. chem. Soc., 1960, 1854; *Khalifa* and *W. H. Linnell*, J. org. Chem., 1959, **24**, 853). Condensation of acylated or alkylated hydrazophenols with alkylmalonyl chlorides yields 3,5-dioxopyrazolidine derivatives which are of value in the treatment of rheumatoid arthritis (*R. Pfister* and *F. Häfliger*, Helv., 1957, **40**, 395; *Linnell* and *Khalifa*, J. chem. Soc., 1959, 1315).

4-**Ethoxyhydrazobenzene,** m.p. 86°; 4-*acetoxyhydrazobenzene*, m.p. 117°; 3-*ethoxyhydrazobenzene*, m.p. 75° (*P. Jacobson* and *F. Hönigsberger*, Ber., 1903,

36, 4112); *2-benzyloxyhydrazobenzene*, m.p. 66° (*Pfister* and *Häfliger*, *loc. cit.*); *4,4'-diacetoxyhydrazobenzene*, m.p. 140°; *4,4'-diethoxyhydrazobenzene*, m.p. 119°; *3,3'-diethoxyhydrazobenzene*, m.p. 85°; *3,3'-dibenzyloxyhydrazobenzene*, m.p. 108°; *2,2'-dimethoxyhydrazobenzene*, m.p. 102°.

5. Phenolsulphonic acids

(a) Methods of preparation

Phenol is readily sulphonated to give phenolmonosulphonic acids and under more vigorous conditions, phenoldi- and phenoltri-sulphonic acids (*H. Cerfontain*, "Mechanistic Aspects in Aromatic Sulphonation and Desulphonation", Interscience, New York, 1968; *E. E. Gilbert*, "Sulphonation and Related Reactions", Interscience, New York, 1965).

(1) Using an excess of 85–100% sulphuric acid at temperatures below 120°, a mixture of *o*- and *p*-phenolsulphonic acids is formed; the *ortho*:*para* ratio decreases with increasing temperature and acid concentration. These isomers can be separated by crystallisation of their barium salts as the *ortho* isomer is less soluble. The *para* isomer is obtained from the filtrate as the magnesium salt (*J. Obermiller*, Ber., 1907, **40**, 3623). Alternatively separation can be achieved chromatographically. (*E. Grebenovsky*, Z. anal. Chem., 1965, **213**, 412).

The yield of the *meta* isomer is small under kinetically controlled con-

TABLE 3

SULPHONATION OF PHENOL BY SULPHURIC ACID

%H_2SO_4	Temp. °C	*ortho* %	*meta* %	*para* %	Ref.
98	20	49		51	1
98	70	25		75	1
98	120	11		89	1
83	160 (5 h)		1·7	98	2
83	160 (50 h)		21	79	2
83	180 (15 h)		17	83	2
98	180 (15 h)		28	72	2
83	209 (5 h)		26	74	2
83	209 (20 h)		38	62	2

1 *H. Cerfontain*, Mechanistic aspects in aromatic sulphonation and desulphonation, Interscience, New York, 1968.

2 *B. T. Karavaev* and *A. A. Spryskov*, J. gen. Chem. U.S.S.R., 1963, **33**, 1840.

ditions, *e.g.*, only 0·1% is produced at 20° with 98% acid (*Y. Muramoto*, C.A., 1956, 50, 9946), but increases on increasing the temperature, reaction time, and acid concentration (see Table 3). With sufficiently long reaction times, the isomers equilibrate; for example, in 55% sulphuric acid at 130° after 200 hours, the equilibrium mixture contains 53% *meta*, the remainder being the *para* isomer apart from a little *ortho* isomer and disulphonic acid (*R. N. Khelevin* and *A. A. Spryskov*, C.A., 1969, 70, 114756). The isomeration occurs *via* a desulphonation–resulphonation mechanism. 4,4′-Dihydroxydiphenyl sulphone and related compounds are also produced, especially at higher temperatures.

Fuming sulphuric acid is used to prepare phenol-di- and -tri-sulphonic acids; for conditions, see references to the individual acids.

(2) Phenol is also sulphonated by chlorosulphonic acid in dichloroethane or carbon disulphide (*Spryskov* and *B. G. Gnedin*, J. Russ. org. Chem., 1965, 1, 1983) and by liquid sulphur trioxide to give mono-, 2,4-di- and 2,4,6-trisulphonic acids at 50°, 95° and 120°, respectively (*B. K. Davison* and *L. F. Byrne*, B.P., 820,659/1960).

(3) Other methods of preparation include the hydrolysis of diazonium salts of anilinesulphonic acids, and alkaline fusion or hydrolysis of benzenedisulphonic acids or halogenobenzenesulphonic acids.

(b) *Properties and reactions*

Phenolsulphonic acids are soluble in alcohol and water and are very hygroscopic.

(1) Phenolsulphonic acids are desulphonated by heating in aqueous mineral acid. The ease of hydrolysis of the monosulphonic acids is *ortho* > *para* > *meta* (*Cerfontain, loc. cit.*).

(2) Alkaline fusion of *o*- and *m*-phenolsulphonic acids proceeds normally to give dihydric phenols, but the *para* compound behaves abnormally giving phenol and complex phenolic material rather than hydroquinone (*L. R. Buzbee*, J. org. Chem., 1966, 31, 3289).

(3) Phosphorus pentachloride reacts with sodium *m*-phenolsulphonate to form the corresponding sulphonyl chloride but with the *ortho* and *para* isomers phosphorodichloridates, such as I, are initially formed; these give dichlorobenzenes on heating. *o*- and *p*-Phenolsulphonyl chlorides can be obtained if the hydroxyl group is initially protected by acylation. Phenol-di- and -tri-sulphonic acids on reaction with phosphorus pentachloride produce chlorobenzene-di- and -tri-sulphonyl chlorides (*C. M. Suter*, "The Organic Chemistry of Sulphur", Wiley, New York, 1944):

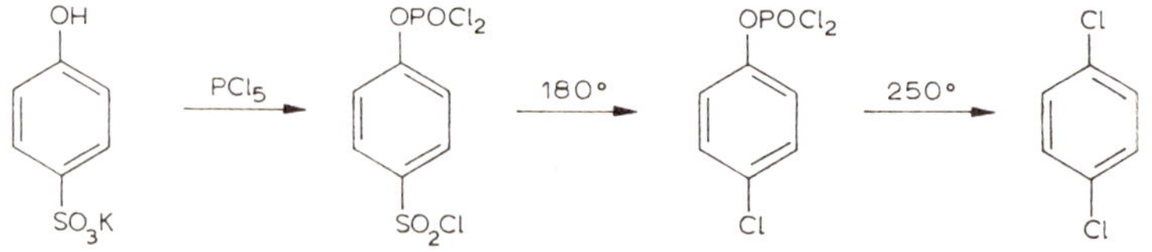

(I)

(4) The hydroxyl group is alkylated by alkyl halides or sulphates in alkaline solution (*M. H. Carr* and *H. P. Brown*, J. Amer. chem. Soc., 1947, 69, 1170) and acetylated by the usual methods.

(5) Both direct substitution into the ring and replacement of the sulphonyl group by electrophilic species occur readily.

(c) Individual compounds

(*i*) *Phenolsulphonic acids*

o-**Phenolsulphonic acid** gives an intense violet colour with ferric chloride. Phenol, potassium phenoldisulphonate and other products are formed on heating its potassium salt at 300° (*J. Obermiller*, Ber., 1910, **43**, 1413). Bromine water produces 2,4,6-tribromophenol and nitrous acid forms the 4-diazo compound. o-*Phenolsulphonyl chloride*, m.p. 72°, *anilide*, m.p. 127° (*R. Anschütz* and *C. Zymandl*, Ann., 1918, **415**, 66).

m-**Phenolsulphonic acid** is prepared from benzene-*m*-disulphonic acid and aqueous alkali at 250° (*F. Willson* and *K. H. Meyer*, Ber., 1914, **47**, 3160), from diazotised metanilic acid (*Obermiller*, Ann., 1911, **381**, 114) and from resorcinol and bisulphite (*V. N. Ufimtsov*, Zhur. Priklad. Khim., 1947, **20**, 1199). It gives a violet colour with ferric chloride.

p-**Phenolsulphonic acid** is obtained by sulphonating phenol with the pyridine-sulphur trioxide adduct at 170° and with sulphuric acid (*P. Baumgarten*, Ber., 1926, **59**, 1982), or by hydrolysis of diazotised sulphanilic acid and *p*-chlorobenzenesulphonic acid in aqueous alkali (*Willson* and *Meyer*, *loc. cit.*). It produces a weak violet colour with ferric chloride and is oxidised by manganese dioxide and sulphuric acid to *p*-benzoquinone. Halogenation produces 2,6-di-bromo- and 2,6-di-iodo-phenol-4-sulphonic acids, 2,4,6-tribromo- (*L. G. Cannell*, J. Amer. chem. Soc., 1957, **79**, 2927) and -trichloro-phenol (*R. L. Datta* and *H. K. Miller*, *ibid.*, 1919, **41**, 2028). 1,3,5-Tribromocyclohexadien-4-onesulphonic acid (II), is an intermediate in bromination. For nitration, see p. 378.

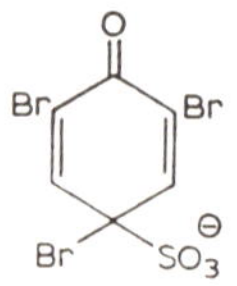

(II)

p-*Phenolsulphonyl fluoride*, m.p. 77° (*W. Steinkopf*, J. pr. Chem., 1927, **117**, 1), *anilide*, m.p. 141° (*R. Anschütz* and *E. Molineus*, Ann., 1918, **415**, 51).

Phenol-2,4-disulphonic acid prepared by sulphonation of phenol (*R. King*, J. chem. Soc., 1921, **119**, 2105; *W. Davies* and *E. S. Wood*, *ibid.*, 1928, 1128), gives a weak blue-red colour with ferric chloride. Halogenation readily produces 6-bromophenol-2,4-disulphonic acid and 2,4,6-trihalogenophenols. See p. 378 for nitration. *Phenol-2,4-disulphonyl chloride*, m.p. 89°, *dianilide*, m.p. 205°.

Phenol-2,4,6-trisulphonic acid is obtained by heating phenol with fuming sulphuric acid at 100° (*J. Pollak*, Monatsh., 1918, **39**, 193; *M. Margueyrol* and *P. Carré*, Bull. Soc. chim. Fr., 1920, **27**, 200). It gives an intense blue-red colour with ferric chloride. Chlorination leads to 2,4,6-trichlorophenol and nitric acid gives picric acid. *Phenol-2,4,6-trisulphonyl chloride*, m.p. 193°, *trianilide*, m.p. 247° (*Davies* and *Wood*, *loc. cit.*).

Scheme 1.
Nitration of phenolsulphonic acids.

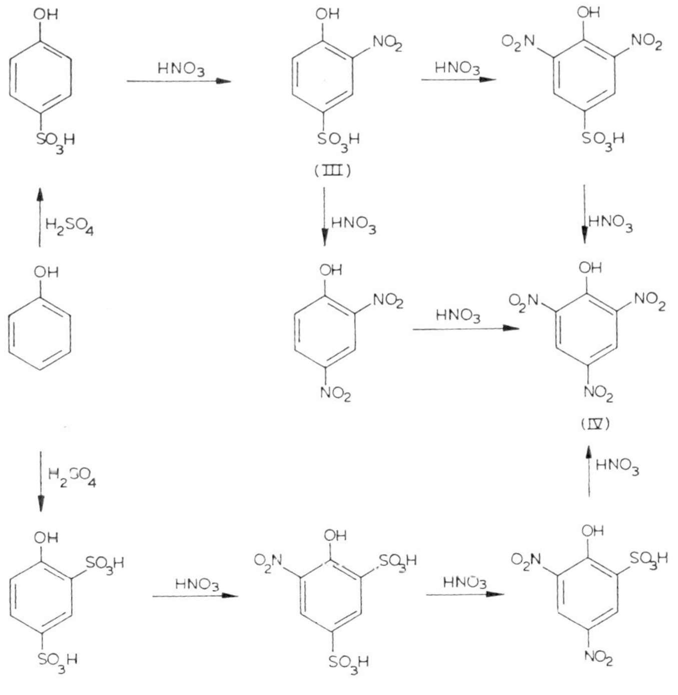

(*ii*) *Ethers*

These are prepared by alkylation of phenolsulphonic acids, by sulphonation of appropriate alkyl phenyl ethers (*M. E. Hultqvist et al.*, J. Amer. chem. Soc., 1951, **73**, 2558; *R. D. Haworth* and *A. Lapworth*, J. chem. Soc., 1924, **125**, 1299) or by oxidation of alkoxyarenethiols, diaryl disulphides or arenesulphinic acids (*L. Gatterman*, Ber., 1899, **32**, 1136). o-*Anisolesulphonyl chloride*, m.p. 56^{0}, *amide*, m.p. 169^{0}; o-*phenetolesulphonyl chloride*, m.p. 65^{0}, *amide*, m.p. 163^{0} (*Gatterman*, *loc. cit.*). m-*Anisolesulphonyl chloride*, b.p. 156^{0}/20 mm; *amide*, m.p. 128^{0} (*K. Fries* and *E. Engelbertz*, Ann., 1915, **407**, 210). p-*Anisolesulphonyl chloride*, m.p. 41^{0}, *amide*, m.p. 110^{0}; p-*phenetolesulphonyl chloride*, m.p. 37^{0}, *amide*, m.p. 151^{0} (*Carr* and *Brown*, *loc. cit.*).

(*iii*) *Nitrophenolsulphonic acids*

Nitrophenolsulphonic acids may be prepared by sulphonating nitrophenols or by nitrating phenolsulphonic acids. Scheme 1 gives the sequence of reactions, leading, eventually to the formation of picric acid (IV) on nitrating certain phenolsulphonic acids [*King*, *loc. cit.*; *F. Olsen* and *J. C. Goldstein*, Ind. Eng. Chem., 1924, **16**, 66; *K. Lesniak* and *T. Urbanski*, in "Nitro Compounds" (Proc. internat. Symp. Nitro Compounds, Warsaw, 1963) Ed. *T. Urbanski*, Pergamon, N.Y. 1964, p. 61].

4-**Nitrophenol**-2-**sulphonic acid** is formed by sulphonating *p*-nitrophenol (*R. Gnehm* and *O. Knecht*, J. pr. Chem., 1906, [ii], **73**, 519), by nitrating *o*-phenolsulphonic acid (*J. Post* and *C. Stuckenberg*, Ann., 1880, **205**, 45) and by the alkaline hydrolysis of 6-chloro-3-nitrobenzenesulphonic acid (*King*, *loc. cit.*).

2-*Nitrophenol*-4-*sulphonic acid* (III), is obtained by sulphonating *o*-nitrophenol (*H. E. Armstrong*, Z. Chem., 1871, 321; *A. Kekulé*, *ibid.*, 1867, 641) and by nitrating *p*-phenolsulphonic acid. The conditions for the latter reaction and for the formation of 4,6-*dinitrophenol*-2- and 2,6-*dinitrophenol*-4-*sulphonic acids* and 6-*nitrophenol*-2,4-*disulphonic acid* are given by *King* (*loc. cit.*).

Halogenation of nitrophenolsulphonic acids leads to replacement of the sulphonyl group as well as to direct substitution *e.g.*, III, gives V and VI, with bromine (*E. Sakellarios*, Ber., 1922, **55**, 2846; *J. Post* and *F. Brackebusch*, Ann. 1880, **205**, 88):

(V) (VI)

(*iv*) *Aminophenolsulphonic acids*

4-**Aminophenol**-2-**sulphonic acid** is prepared by sulphonation of *p*-aminophenol (*R. Bauer*, Ber., 1909, **42**, 2106), reduction of 4-nitrophenol-2-sulphonic acid (*F. R. Bean*, U.S.P., 2,446,519/1948), reductive hydroxylation of *m*-nitrobenzenesulphonic acid (*Gatterman*, Ber., 1893, **27**, 1927; *Bean*, U.S.P., 2,525,515/1950) or ammonolysis of 4-chlorophenol-2-sulphonic acid (*E. Bamberger*, Ber., 1901, **33**, 3600).

5-**Aminophenol**-2-**sulphonic acid** from sulphonation of *m*-aminophenol (*A. L. Miller et al.*, J. Amer. chem. Soc., 1949, **71**, 3559; *J. Büchi, R. Lieberherr* and *M. Flury*, Helv., 1951, **34**, 2076).

6-**Aminophenol**-2-**sulphonic acid** is prepared by reduction of 6-nitrophenol-2-sulphonic acid (*King*, J. chem. Soc., 1921, **119**, 1415).

4-**Aminophenol**-3-**sulphonic acid** by reducing 4-nitrophenol-3-sulphonic acid and by the reduction of nitrobenzene with sodium sulphite in aqueous bicarbonate solution (*A. Seyevitz* and *Vignat*, Compt. rend., 1922, **174**, 297). Energetic oxidation gives *p*-benzoquinone. It reduces ammoniacal silver nitrate and is used as a photographic developer.

6-**Aminophenol**-3-**sulphonic acid** from *o*-aminophenol *via* benzoxazolone (*J. F. Gaunt, F. M. Rowe* and *J. B. Speakmann*, J. Soc. Dyers and Colourists, 1947, **63**, 48) or 2-methylbenzoxazole (*E. R. Riegel, H. W. Post* and *E. E. Reid*, J. Amer. chem. Soc., 1929, **51**, 505).

2-**Aminophenol**-4-**sulphonic acid** by reducing the corresponding 2-nitro compound (*S. Hashimoto* and *J. Sunamoto*, C.A., 1964, **60**, 437; *H. Pelster, C. Koenig* and *R. Pulter*, F.P., 1,481,040/1967) or by sulphonating *o*-aminophenol (*J. Post*, Ann., 1880, **205**, 52).

(*v*) *Homologues of phenolsulphonic acids*

Cresolsulphonic acids. Sulphonation of *o*-cresol by sulphuric acid can lead to three isomers depending on the time and the temperature (*Gilbert, loc. cit.*):

TABLE 4

SULPHONATION OF *O*-CRESOL

$C_6H_4OH(1)Me(2)$-

Hours	Temperature	6-SO_3H	5-SO_3H	4-SO_3H
24	—10°	0	0	100
—	20°	47	0	53
12	20–60°	0	25	75
1	150	0	87	13

3-*Methylphenol*-4- and 3-*Methylphenol*-6-*sulphonic acids* are produced by sulphonation of *m*-cresol. Higher temperatures favour the formation of the former (*Haworth* and *Lapworth, loc. cit.*). 4-*Methylphenol*-2-*sulphonic acid*, m.p. 54°, and 3,4-*dimethylphenol*-6-*sulphonic acid* from *p*-cresol and 3,4-dimethylphenol, respectively (*R. Anschutz* and *L. Hodenius*, Ann., 1918, **415**, 74; *G. P. Mueller* and *W. S. Pelton*, J. Amer. chem. Soc., 1949, **71**, 1504).

Uses. Phenolsulphonic acids and their salts have industrial uses as monomers and co-monomers for sulphonated phenol-formaldehyde resins, both water-soluble (synthetic tannin agents) and water-insoluble (ion-exchange resins) are produced. Other industrial uses are as intermediates in dyestuff manufacture, as plasticizers for concrete and as antiseptics and antiperspirants.

6. Dihydric phenols

Many dihydric phenols, especially the *ortho-* and *para*-dihydric phenols, are found in plants or are obtained by the breakdown of naturally occurring substances. Catechol and derivatives occur in wood tar, lignin, and other wood products, melanins and in sources containing tannic acids. Few resorcinol derivatives occur naturally; some 3,5-dihydroxystilbenes are known and other resorcinol derivatives are found in lichens and in fungal metabolites. Hydroquinone occurs in the form of the glucoside, arbutin, which is widely distributed in nature.

(a) Methods of preparation

(1) Replacement of the appropriate substituents by hydroxyl in (a) mono-halogenophenols and phenolsulphonic acids, (b) arenedisulphonic acids or halogen-substituted sulphonic acids, (c) dihalogen-substituted aromatic hydrocarbons, is used. These substitutions are normally brought about by alkali fusion. Rearrangement to the thermodynamically most stable *meta*-dihydric phenols can arise from fusion of *ortho-* and *para*-disubstituted derivatives at high temperatures.

(2) *ortho-* and *para*-Dihydric phenols are obtained by the reaction of *ortho-* and *para*-hydroxy-aldehydes and -ketones with hydrogen peroxide in aqueous alkaline solution.

(3) Acid-catalysed decomposition of bis-(hydroperoxyalkyl)benzenes gives dihydric phenols.

(4) *ortho-* and *para*-Dihydric phenols result from reduction of *ortho-* and *para*-benzoquinones.

(5) Oxidation of anilines and monohydric phenols is used for *ortho-* and especially *para*-dihydric phenols.

(6) Other methods include, (a) decarboxylation of dihydroxybenzoic acids, (b) hydrolyses of diazonium salts of aminophenols, (c) dealkylation of mono- and di-ethers of dihydric phenols.

(b) Properties and reactions

The simpler dihydric phenols have no characteristic odour. They are more soluble in water than the monohydric phenols and at 20^0 are very soluble in alcohol and ether but poorly so in benzene and carbon tetrachloride.

TABLE 5

SOLUBILITY IN g/l AT 20°

	Catechol	Resorcinol	Hydroquinone
Water	312	584	58
Ethanol	582	610	v.sol.
Benzene	8	22	0·2
Carbon tetrachloride	1		

They are weak dibasic acids: the p*K*a values (first dissociation) for catechol, resorcinol and hydroquinone are, at 20°, 9·25, 9·20 and 9·91, respectively. The dihydric phenols are much more sensitive to oxidation than the monohydric phenols and are useful reducing agents. The *ortho*- and *para*-isomers in particular are readily oxidised by atmospheric oxygen in alkaline solution. Halogenation, sulphonation, nitration, alkylation, acylation, Kolbe and other substitutions are particularly easily effected. The ease of oxidation of catechol and hydroquinone prevents some direct substitutions, such as mercuration and iodination; indirect methods are available, however.

ortho-Dihydric phenols are distinguished from the other isomers by the formation of chelate complexes on reaction with polybasic acids and certain metal hydroxides. They also form cyclic ethers. The *ortho*- and *para*-isomers yield the corresponding *ortho*- and *para*-benzoquinones by suitable methods of oxidation. *meta*-Dihydric phenols containing free 2- and 4-positions condense with anhydrides of 1,2-dicarboxylic acids to give fluoresceins.

(*i*) *Spectra*

The following spectral studies have been made; u.v. (*J. C. Dearden* and *W. F. Forbes*, Canad. J. Chem., 1959, **37**, 1294); i.r. (*A. Hidalgo* and *C. Otero*, Spectrochim. Acta, 1960, **16**, 528; *D. K. Mukherjee* and *S. B. Banerjee*, Indian J. Phys., 1967, **41**, 108); mass spectra (*I. Aczel* and *H. E. Lumpkin*, Anal. Chem., 1960, **32**, 1819); proton magnetic resonance (*J. C. Schug* and *J. C. Deck*, J. chem. Phys., 1962, **37**, 2618).

(*ii*) *Dihydric phenols and tautomerism*

A dihydric phenol should show a greater tendency than phenol to tautomerise since the energy difference between the oxo and enol forms is smaller (*G. Wheland*, *ibid*., 1933, **1**, 731; *R. H. Thomson*, Quart. Reviews, 1956, **10**, 27; *V. V. Ershov* and *G. A. Nikiforov*, Russian Chem. Reviews, 1966, **35**, 817). However, there is no clear evidence for the existence of equilibria between enol and oxo forms of dihydric phenols in solution. The oxo form of hydroquinone, cyclohex-2-ene-1,4-dione (I), has been prepared. The crystalline solid is stable over prolonged periods at 10°. In non-polar aprotic solvents rearrangement to the enol form is slow, but is rapid in aqueous or alcoholic solutions. The reverse reaction does not occur; the stability of the cyclohexenedione is due to kinetic and not thermodynamic factors (*E. W. Garbisch,* J. Amer. chem. Soc., 1965, **87**, 4971). Some substituted hydroquinone derivatives, such as II from addition of halogens to *p*-benzoquinone (*T. H. Clark*, Amer. Chem. J., 1892, **14**, 553; *R. K. Norris* and *S. Sternhell*, Australian Chem. J., 1966, **19**, 617; Chem. Comm., 1965, 608) have also been isolated. II is converted totally into the enol form in acid solution: again this is an irreversible reaction.

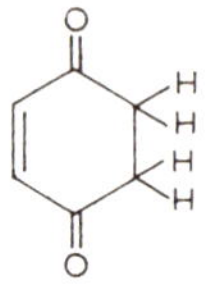

X = Br, Cl

(I) 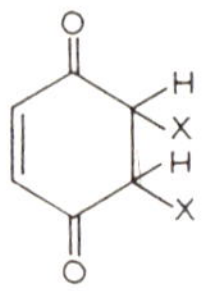 (II)

There is abundant chemical evidence to show that dihydric phenols can, react *via* the oxo form especially in basic solution, from which the presence of the mesomeric anions can be inferred, *e.g.*, for resorcinol:

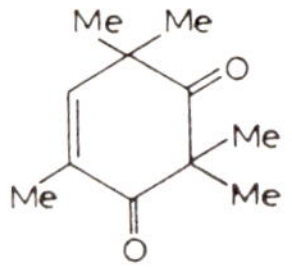

(III)

Thus, resorcinol, like α,β-unsaturated carbonyl compounds, is readily reduced by sodium amalgam to cyclohexane-1,3-dione (*G. Merling*, Ann., 1894, **278**, 20) and it also reacts with methyl iodide in base to give *C*-alkylated products, such as III (*A. R. Stein*, Canad. J. Chem., 1965, **43**, 1493, 1508).

(c) ortho-*Dihydric phenols; Catechol and its derivatives*

Catechol *pyrocatechol, o-dihydroxybenzene*, 1,2-$C_6H_4(OH)_2$, m.p. 105°, b.p. 245°, is present in many plant and tree extracts and is obtained from the barks, sapwoods, heartwoods, leaves, kino exudates, etc. ("Wood Extractives", Ed. *W. E. Hillis*, Academic Press, New York, 1962; see also an article on polyhydric phenols and their derivatives in tannins and colouring matter by *M. A. Buchanan*, in "Wood Chemistry", Ed. *L. E. Wise* and *E. C. Jahn*, Reinhold, New York, 1952, p. 618). Catechol and other phenols including alkylcatechols are obtained from lignins in a variety of ways, including the use of catalytic hydrogenation and alkali fusion (*I. A. Pearl*, "The Chemistry of Lignin", Arnold, London, 1967; "Lignin Structure and Reactions", American Chemical Society, 1966; *B. L. Browning*, "Methods of Wood Chemistry", Interscience, New York, 1967, Vol. II, part IV; *F. E.* and *D. A. Braun*, "The Chemistry of Lignin", Academic Press, New York, 1960). Catechol was first obtained in 1839 by distilling catechin.

(i) *Methods of preparation*

(1) Catechol is obtained by the alkali fusion of *o*-phenolsulphonic acid, *o*-chlorophenol and *o*-bromophenol. Thus, heating *o*-chlorophenol with anhydrous sodium hydroxide in a nitrogen atmosphere gives catechol and considerable amounts of the *m*-isomer, resorcinol (*K. W. Muller* and *D. Delfs*, G.P., 1,040,563/1958). Other industrial processes for the production of catechol from *o*-chlorophenol use barium and sodium hydroxides, lead and cuprous chlorides or calcium chloride and hydroxide instead of sodium hydroxide (U.S. Dept. Com. Office Tech. Serv. PB. Report 58747; 63959).

(2) The oxidation of salicylaldehyde by hydrogen peroxide in alkaline solution is a very useful laboratory method for the preparation of catechol (*H. D. Dakin*, Amer. chem. J., 1909, **42**, 477; Org. Synth., Coll. Vol. I, 1964, p. 149).

(3) Heating guaiacol, the monomethyl ether of catechol, with concentrated hydrobromic acid gives catechol (*H. T. Clarke* and *E. R. Taylor*, *ibid.*, p. 150). Several other acids such as hydrochloric, hydriodic, aluminium chloride (references cited by *Clarke* and *Taylor*) and orthophosphoric acid (*V. A. Shishkin* and *D. V. Tishchenko*, U.S.S.R.P., 165,465/1964) are also used.

(4) The direct hydroxylation of phenol by aqueous hydrogen peroxide on irradiation at 2537 Å produces catechol ($< 26\%$) as well as hydroquinone (14%) and some trihydric phenols (*K. Omura* and *T. Matsuura*, Tetrahedron, 1968, **24**, 3475). Alternatively, oxidation of phenol at 50–150° by performic or peracetic acid in the presence of phosphoric acid can be used; again mixtures of catechol and hydroquinone result (*A. Marlard*, F.P., 1,479,354/1967).

(5) Oxidation of 1,4-diethyl-1,2,3,4-tetrahydronaphthalene by oxygen in the presence of potassium carbonate gives the dihydroperoxide. This is decomposed by acids to produce catechol (*D. P. Young*, U.S.P., 2,797,249/1957):

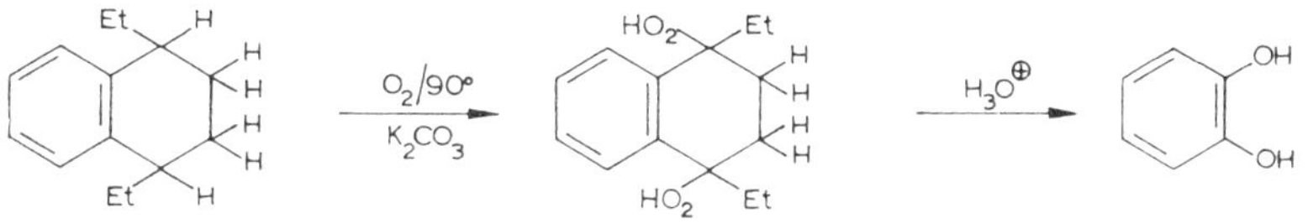

(6) Other methods include (a) dry distillation of protocatechuic acid (*H. Kunz-Krause* and *P. Manicke*, Ber., 1920, **53**, 190), (b) copper sulphate catalysed hydrolysis of diazotised *o*-aminophenol (*Soc. chim. des Usines du Rhône*, G.P., 167,211/1904) and (c) dehydrogenation of 1,2-cyclohexanediol on palladium/potassium carbonate coated carbon (*Y. M. Pauskin, E. M. Buslova* and *S. A. Nizova*, C.A., 1969, **71**, 123799).

(ii) *Properties and reactions*

(1) *Colour reactions*. An aqueous solution of catechol with ferric chloride gives a green colour changing to dark red or violet on addition of sodium carbonate. Catechol reduces ammoniacal silver solutions at room temperature and alkaline cupric salts (Fehling's solution) on warming. On exposure to air, its alkaline

solution becomes green then blue and finally black (see below). Another distinctive reaction of catechol is the formation of a white insoluble lead "salt", $C_6H_4O_2Pb$.

(2) *Oxidation of catechol and its derivatives.* The first step in the oxidation of catechol is the formation of the semiquinone radical IV. In strongly alkaline solutions, the resonance-stabilised symmetrical anion V, is formed, whereas in strongly acidic media, the diprotonated radical VI, is produced. Electron spin resonance spectra of these radicals in solutions of different pH have been recorded (*J. C. P. Smith* and *A. Carrington*, Mol. Phys., 1967, **12**, 439). The alkali metal salts of the semiquinone radical-anion from 4,6-di-*tert*-butylcatechol are stable enough to be isolated (*K. Ley* and *E. Muller*, Angew. Chem., 1958, **70**, 469). Reactions of stable *o*-semiquinone radical ions have been reviewed (*A. R. Forrester, J. M. Hay* and *Thomson*, "Organic Chemistry of Stable Free Radicals", Academic Press, London, 1968, p. 334). Further oxidation leads immediately to the *o*-benzoquinone VII. Various oxidants are available for the simple oxidation to *o*-benzoquinone; these include ferricyanide, Fremy's salt, *o*-chloranil (*H. Musso*, Angew. Chem. internat. Edn., 1963, **2**, 723) and ceric sulphate (*R. Brockhaus*, Ann., 1968, **712**, 214). For the mechanism of the autoxidation of 3,5-di-*tert*-butylcatechol, see *C. A. Tyson* and *A. E. Martell*, J. phys. Chem., 1970, **74**, 2601):

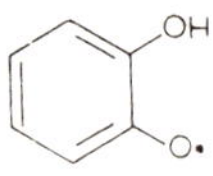

(IV) 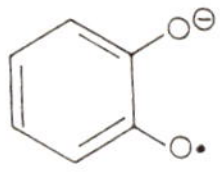 (V) 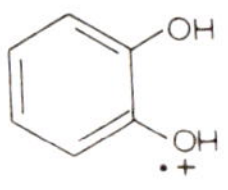 (VI) 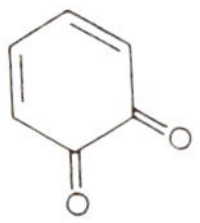 (VII)

Dimeric *o*-quinones VIII, are formed if silver oxide in acetone is the oxidant (*J. Harley-Mason* and *A. H. Laird*, J. chem. Soc., 1958, 1718). Many other carbon-carbon coupled products, such as 3,4,3',4'-tetrahydroxybiphenyl (IX), can be produced, *e.g.*, by irradiating at 2537 Å an aqueous solution of catechol (*H. I. Joschak* and *S. I. Miller*, J. Amer. chem. Soc., 1966, **88**, 3273), and by

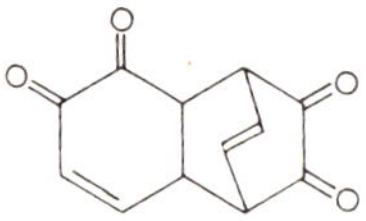

(VIII) 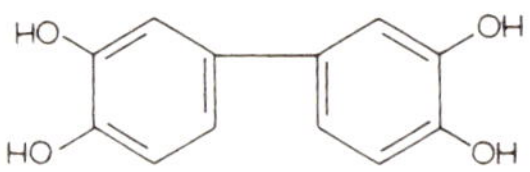 (IX)

the use of silver nitrate, ferric chloride or the enzyme polyphenol oxidase as the oxidant (*W. G. C. Forsyth et al.*, Biochim. Biophys. Acta, 1957, **25**, 155; 1960, **37**, 322). The oxygen-carbon coupled product X is produced by aerial oxidation in alkaline solution (*F. R. Hewgill, T. J. Stone* and *W. A. Waters*, J. chem. Soc., 1964, 408; *Forsyth et al., loc. cit.*; *Musso* in "Oxidative Coupling of Phenols", Ed. *W. I. Taylor* and *A. R. Battersby*, Arnold, London, 1967; *A. I. Scott*, Quart. Reviews, 1965, **19**, 11):

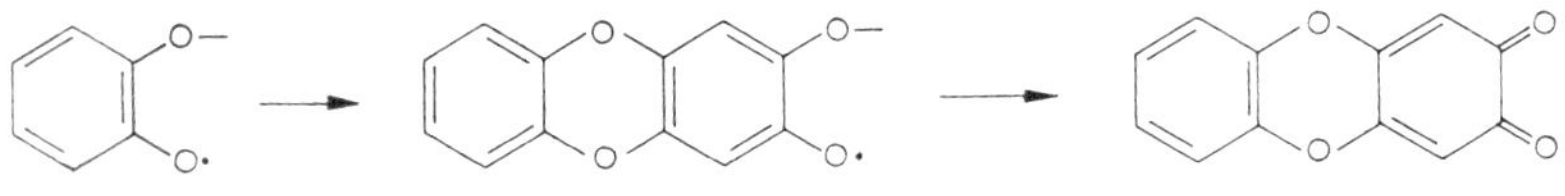

(X)

Further oxidation and coupling can lead to polymeric material; thus coniferyl alcohol gives a compound resembling natural spruce lignin on oxidation either by enzymes or manganese dioxide (*K. Freudenberg*, Pure appl. Chem., 1962, **5**, 9).

More drastic oxidation causes ring opening; both enzymes (*e.g. M. Nozaki et al.*, "Oxidation of Organic Compounds", Vol. III, American Chemical Society, Special Pub., 1968, p. 242) and chemical oxidants, *e.g.*, ozone and peracids (*L. B. Wingard*, Dissertation Abs., 1966, **26**, 5942; *A. Wacek* and *R. Fiedler*, Monatsh., 1949, **80**, 170) give *cis,cis*-muconic acid (XI):

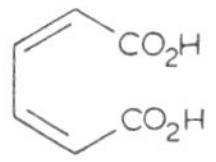

CO_2H
CO_2H

(XI)

(3) *Cyclic derivatives*. Catechol with its *ortho* hydroxyl groups forms a variety of cyclic compounds. These include the cyclic esters of organic acids and of di- and poly-basic inorganic acids, such as those of phosphorus, boron and sulphur. For phosphorus acids, for example, the following types of compounds have been reported; XIIa, XIII, XIV (n = 1) and XV from phosphorus trichloride, XVI from phosphoryl chloride, XIIb from methylphosphorus dichloride and the remainder from phosphorus pentachloride (see Scheme 2).

For boron, several analogous compounds (XII → XV) are known (see *M. J. S. Dewar, V. P. Kubba* and *R. Pettit*, J. chem. Soc., 1958, 3076; *W. Gerrard, M. F. Lappert* and *B. A. Mountfield, ibid.*, 1959, 1529; *H. Steinberg*, "Organoboron Chemistry", Vol. I, Interscience, New York, 1964, Chaps. V and VI).

The cyclic esters, XIX, XX, XXI and the cyclic ethers, XXII, XXIII and XXIV, (Scheme 3), are described later (see pp. 386).

Catechol forms an extensive series of heterocyclic condensation products; *e.g.*, *o*-phenylenediamine gives dihydrophenazine, *o*-aminophenol, phenoxazine, and *o*-aminothiophenol, phenothiazine (*D. E. Pearson*, in "Heterocyclic Compounds" Vol. 6, Ed. *R. C. Elderfield*, Wiley, New York, 1957):

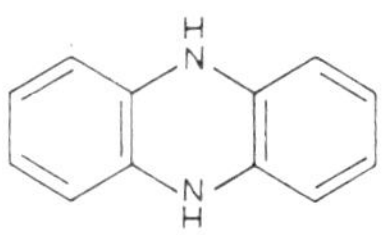

Dihydrophenazine

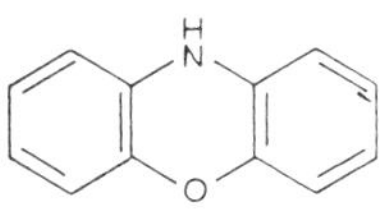

Phenoxazine

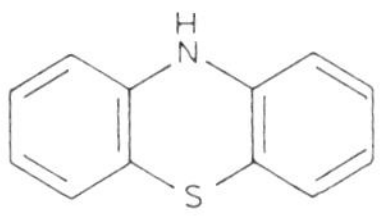

Phenothiazine

Scheme 2

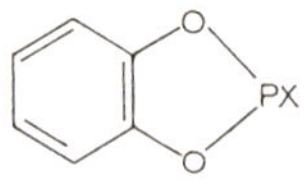

(a) X = Cl, m.p. 30° (ref. 1)
Br, b.p. 105°/13 mm (ref. 2)
(b) X = Me, m.p. 2° (ref. 8)

(XII)

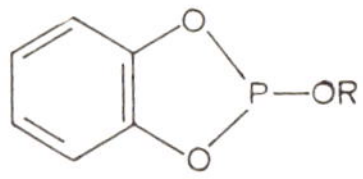

R = Me, b.p. 77°/15 mm
Et, b.p. 83°/11 mm
o-HOC_6H_4, m.p. 117° (ref. 3)

(XIII)

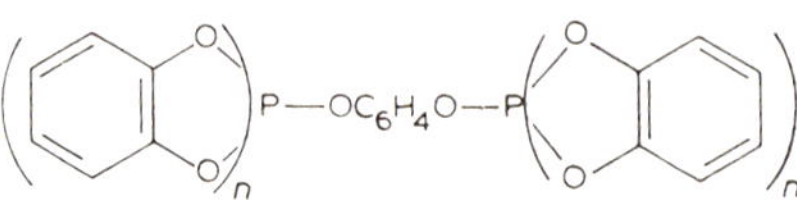

n = 1, b.p. 255°/16 mm (ref. 4)
n = 2, (ref. 5)

(XIV)

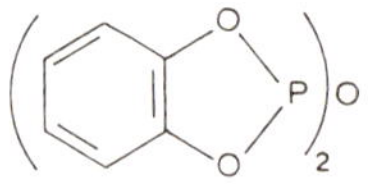

m.p. 72° (ref. 1)

(XV)

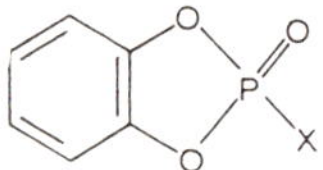

X = OH, b.p. 225°/3 mm (ref. 6)
Cl, m.p. 58° (ref. 5)
Br, m.p. 35-39° (ref. 2)

(XVI)

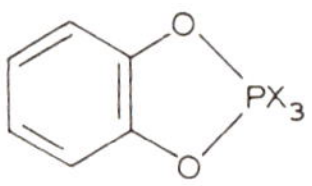

X = Cl, m.p. 62° (ref. 5,7)
Br, m.p. 67-70° (ref. 2)

(XVII)

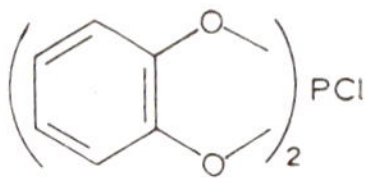

m.p. 166° (ref. 5)

(XVIII)

References (for Scheme 2)

1 *P. C. Crofts, J. H. H. Markes* and *H. N. Rydon*, J. chem. Soc., 1958, 4250.
2 *H. Gross* and *U. Karsch*, J. pr. Chem., 1965, [iv], **29**, 315.
3 *L. Anschütz* and *H. Walbrecht, ibid.*, 1932, [ii], **133**, 65.
4 *Anschütz et al.*, Ber., 1943, **76**, 218.
5 *Idem*, Ann., 1927, **454**, 71.
6 *E. Cherbuliez, M. Schwarz* and *J. P. Leber*, Helv., 1951, **34**, 841.
7 *Gross* and *J. Gloede*, Ber., 1963, **96**, 1387.
8 *M. Wieber* and *W. R. Hoos*, Monatsh., 1970, **101**, 776.

Scheme 3

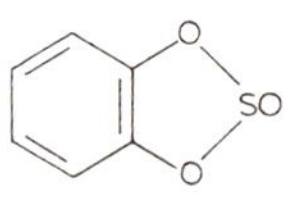

(XIX)

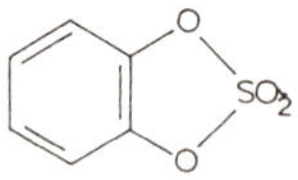

(XX)

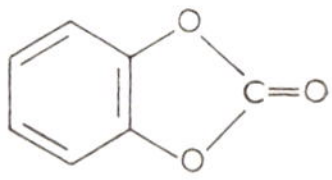

(XXI)

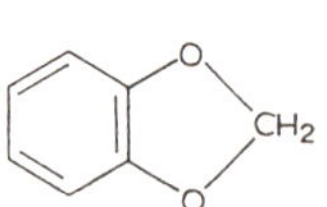

(XXII)

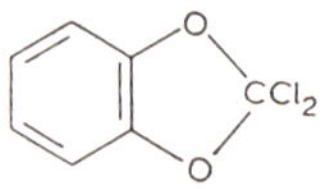

(XXIII)

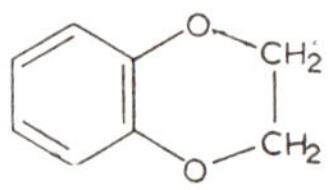

(XXIV)

Catechol forms *chelate complexes* with elements having a high affinity for oxygen; these elements include Ti, Zr, Hf, V, Nb, Cr, Mo, As, Sn and Sb (*F. P. Dwyer* and *D. P. Mellor*, "Chelating Agents and Metal Chelates", Academic Press, New York, 1964, p. 106). Arsenic gives the complexes, $H[As(C_6H_4O_2)_2]\cdot 4H_2O$ and $H[As(C_6H_4O_2)_3]\cdot 5H_2O$. Use has been made of the quinine and cinchonidine salts to resolve the latter complex into optical enantiomers (*F. G. Mann* and *J. Watson*, J. chem. Soc., 1947, 505; *J. H. Craddock* and *M. M. Jones*, J. Amer. chem. Soc., 1961, **83**, 2839; for substituted catechols, see *T. H. Larkins* and *Jones*, Inorg. Chem., 1963, **2**, 142).

(4) *Substitutions*. *C*-Acylation, with acyl halides, acid anhydrides and acids, and *C*-alkylation, with alkanes, alcohols and alkenes, in the presence of the usual acid catalysts, proceed readily to 4-mono- and di-alkyl derivatives ("Friedel-Crafts and Related Reactions", Ed. *G. A. Olah*, Vol. II and III, Interscience, New York, 1964). In alkylations with alkenes, if aluminium or an aluminium alkoxide is used instead of the usual catalysts, then 3-mono- and 3,6-dialkylcatechols are formed. A catechyl aluminium alkoxide complex is probably formed and an intramolecular reaction ensues to give the *ortho*-substituted product in contrast with the normal intermolecular Friedel-Craft reaction (*R. Stroh*, *R. Seydel* and *W. Hahn*, in "Newer Methods of Preparative Organic Chemistry", Ed. *W. Foerst*, Vol II, Academic Press, New York, 1963, p. 337; *G. G. Ecke* and *A. J. Kolka*, U.S.P., 2,831,817/1958).

The acid-catalysed Fries-rearrangement of esters gives 4-acylcatechols ("Friedel-Crafts and Related Reactions", Vol. III), whereas the photo-Fries rearrangement produces comparable yields of the 3- and 4-acylcatechols (*J. C. Anderson* and *C. B. Reese*, Proc. chem. Soc., 1960, 217; J. chem. Soc., 1963, 1781). Indirect methods of preparation have been developed for 3-acylcatechols (*e.g.*, see *W. I. Awad*, *M. F. El-Neweihy* and *S. F. Selim*, J. org. Chem., 1958, **23**, 1783; *W. R. Boehme* and *W. G. Scharpf*, *ibid.*, 1961, **26**, 1692).

The aldehyde group has been introduced using the Reimer-Tiemann reaction and by the reaction of an orthoformate and catechol in the presence of a Lewis acid (*H. Gross*, *A. Rieche* and *G. Matthey*, Ber., 1963, **96**, 308). Simply heating

H(COEt)$_3$ + OH / OH (catechol) $\xrightarrow{AlCl_3}$ HO / HO (C$_6$H$_3$)–CH(OEt)$_2$ $\xrightarrow{H_3O^{\oplus}}$ HO / HO (C$_6$H$_3$)–CHO

catechol in aqueous ammonium carbonate at 140^0 gives protocatechuic acid (see also *O. Baine et al.*, J. org. Chem., 1954, **19**, 510).

Sulphonation of catechol produces mono- and di-sulphonic acids (*J. N. Ray* and *M. L. Dey*, J. Chem. Soc., 1920, **117**, 1405; *C. Gentsch*, Ber., 1910, **43**, 2018; *J. Pollak* and *E. Gebauer-Fülnegg*, Monatsh., 1926, **47**, 109).

(*iii*) *Halogeno derivatives of catechol*

3-**Fluoro-**, m.p. 71^0, and 4-**fluoro-catechol,** m.p. 90–91^0, are formed by the thermal decomposition of diazonium borofluorides of dimethoxyanilines and subsequent demethylation (*J. Corse* and *L. L. Ingraham*, J. org. Chem., 1951, **16**, 1345). *Tetrafluorocatechol*, m.p. 67·5–69⁰ (*J. Burdon, V. A. Damodaran* and *J. C. Tatlow*, J. chem. Soc., 1964, 763).

Controlled chlorination of catechol with sulphuryl chloride gives the desired degree of substitution. Chlorine readily produces tetrachlorocatechol, *via* 4-chloro-, 4,5-dichloro- and 3,4,5-trichloro-catechols (*H. Schwarz*, Z. chem., 1967, **7**, 18). Further chlorination produces tetrachloro-*o*-benzoquinone and a hexachlorocyclohexenedione. Direct bromination can give tetrabromocatechol. The m.p. of chloro- and bromo-derivatives are given in Table 6.

4-*Iodocatechol* has been obtained from 4-aminocatechol diacetate (*E. Fourneau* and *J. Druey*, Compt. rend., 1934, **199**, 870).

TABLE 6

HALOGENOCATECHOLS

	Chloro		Bromo	
	M.p. ⁰C.	Ref.	M.p. ⁰C.	Ref.
3-Mono-	46–48	1	40·5–41·5	4
4-Mono-	91	1	87	5
3,4-Di-	—		73–74	6
3,5-Di-	83–84	2	58–60	7
4,5-Di-	116–117	1	119–121	8
3,4,5-Tri-	106–109	1, 7	139–141	8
3,4,6-Tri-	—		106	10
Tetra	193–194	3	190	9, 11

References

1 *R. Willstatter* and *H. E. Müller*, Ber., 1911, **44**, 2182.
2 *H. D. Dakin*, Amer. chem. J., 1909, **42**, 477.
3 *T. Zincke* and *F. Küster*, Ber., 1888, **21**, 2719.
4 *H. S. Mason*, J. Amer. chem. Soc., 1947, **69**, 2241.
5 *K. W. Rosenmund* and *W. Kuhnhenn*, Ber., 1923, **56**, 1262.
6 *J. Frejka* and *B. Sefranek*, Coll. Czech. chem. Comm., 1936, **8**, 130.
7 *H. Cousin*, Ann. Chim., 1898, [vii], **13**, 480.
8 *M. Kohn*, J. Amer. chem. Soc., 1951, **73**, 480.
9 *Zincke*, Ber., 1887, **20**, 1776.
10 *Kohn* and *L. Steiner*, J. org. Chem., 1947, **12**, 30.
11 *C. L. Jackson* and *W. Koch*, Amer. chem. J., 1901, **26**, 10.

(*iv*) *Nitro derivatives of catechol*

Catechol and nitric acid in ether solution form a mixture of 3-**nitrocatechol,** m.p. 86⁰, and 4-**nitrocatechol,** m.p. 174⁰, whereas sodium nitrite and sulphuric acid produces only the 4-isomer and, when in excess, 3,4-*dinitrocatechol*, m.p.

147–148° (*D. H. Rosenblatt, E. Epstein* and *M. Levitch*, J. Amer. chem. Soc., 1953, **75**, 3277). Demethylation of 3,6-dinitroguaiacol gives 3,6-*dinitrocatechol*, m.p. 164·5–165·5° (*P. M. Heertjes et al.*, J. chem. Soc., 1955, 1313) and of 4,5-dinitroveratrole, 4,5-*dinitrocatechol*, m.p. 166·5–167·5° (*J. Ehrlich* and *M. T. Bogert*, J. org. Chem., 1957, **12**, 522). 3,5-*Dinitrocatechol*, m.p. 164° (*F. C. Laxton, E. B. R. Prideaux* and *W. H. Radford*, J. chem. Soc., 1925, **127**, 2499).

(v) *Amino derivatives of catechol*

The usual method of preparation is by reduction of the corresponding nitro compound and the following have been prepared in this way (unless otherwise stated).

3-**Aminocatechol,** m.p. 142–145°, *hydrochloride*, m.p. 198–203° (*J. H. Boyer* and *L. R. Morgan*, J. org. Chem., 1961, **26**, 1654; *L. Horner* and *K. Sturm*, Ann., 1957, **608**, 128).

4-*Aminocatechol*, m.p. 124–125°, is obtained by reduction of the 4-nitro and 4-nitroso compounds and of the coupled product of guaiacol and diazotised sulphanilic acid (*R. F. Collins* and *M. Davis*, J. chem. Soc., C, 1966, 366; *N. L. Drake, H. G. Harris* and *C. B. Jaeger*, J. Amer. chem. Soc., 1948, **70**, 168; *R. E. Harman* and *J. Cason*, J. org. Chem. 1952, **17**, 1058) or by demethylation of 4-aminoguaiacol (*W. A. Jacobs, M. Heidelberger* and *I. P. Rolf*, J. Amer. chem. Soc., 1919, **41**, 458). Autoxidation of 4-aminocatechol proceeds readily to give the dimer XXV, while 3-aminocatechol gives polymers based on units of XXVI (*Horner* and *Sturm*, *loc. cit.*).

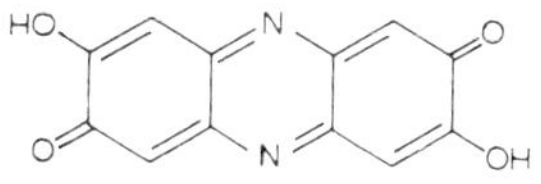

(XXV) 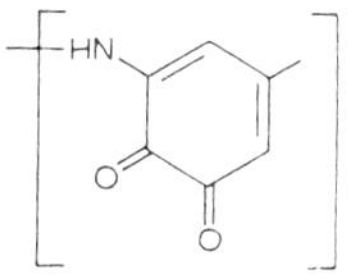 (XXVI)

(vi) *Catechol ethers*

O-Alkylation readily proceeds with alkyl halides, alkyl sulphates and *p*-toluenesulphonyl esters especially in alkaline solution. A number of catechol ethers, such as eugenol, safrole, apiole, piperonal and papaverine, occur in Nature.

Guaiacol, *catechol monomethyl ether*, 1,2-$C_6H_4(OH)(OMe)$, m.p. 28°, b.p. 205°, $d_4^{21.4}$ 1·1287, n_D^{20} 1·5417, is present in wood tars from which it can be obtained by distillation. Lignin and its derivatives and guaiacum resin are other sources of guaiacol and its alkyl derivatives. (See p. 382 for references.)

Methods of preparation of guaiacol. Guaiacol is obtained by the monomethylation of catechol, dimethyl sulphate in an alkaline solution being the preferred reagent, especially if nitrobenzene is also present to form a two-phase solvent system (*H. Bredereck, I. Henning* and *W. Rau*, Ber., 1953, **86**, 1085; *Bredereck* and *Hennig*, G.P., 874,445/1953). Hydrolysis of the diazonium salt of *o*-anisidine is also extensively used (*M. A. Phillips*, Chem. Age, 1956, **74**, 631; *G. H. Herbst*,

G.P., 1,148/236/1963; *W. J. Hickinbottom,* "Reactions of Organic Compounds", Longmans, London, 1957).

Properties and reactions of guaiacol. Oxidation. An alcoholic solution of guaiacol gives with ferric chloride a green colour, which turns to reddish-violet on addition of carbonate. Guaiacol and other monoalkyl ethers of catechol are readily oxidised to *o*-benzoquinone and alcohol by periodate and bismuthate (*E. Adler* and *R. Magnusson,* Acta chem. Scand., 1959, **13,** 505; *Adler, I. Falkehag* and *B. Smith, ibid.,* 1962, **16,** 529) while lead tetra-acetate (*F. Wessely* and *J. Koltan,* Monatsh., 1953, **84,** 291) and Fremy's salt (*H. J. Teuber* and *Rau,* Ber., 1953, **86,** 1036) produce methoxy-*p*-benzoquinone. When a *para*-substituent is present, the latter oxidants give a 3-methoxy-*o*-quinone. Further oxidation can occur with periodate.

Coupled oxidation products, *e.g.,* 4,4′-dihydroxy-3,3′-dimethoxybiphenyl (XXVII), 4-hydroxy-2′,3-dimethoxydiphenyl ether (XXVIII) and a polymeric material, $[C_6H_3O(OMe)]_n$ are obtained if either silver oxide, ferricyanide or persulphate is the oxidant (*B. O. Lindgren,* Acta chem. Scand., 1960, **14,** 2089; see also *F. R. Hewgill* and *B. S. Middleton,* J. chem. Soc., C, 1967, 2316; *Hewgill* and *S. L. Lee, ibid.,* 1969, 2080):

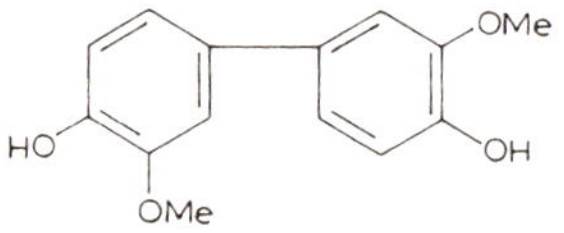

OMe HO OMe OH
(XXVII)

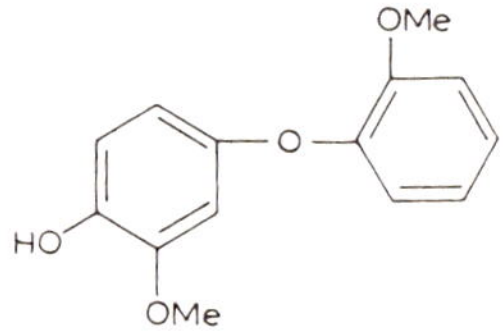

OMe O HO OMe
(XXVIII)

Substitution of guaiacol. Guaiacol may be mono- and di-sulphonated (*A. Rising,* Ber., 1906, **39,** 3685; *L. Paul, ibid.,* p. 4093); it undergoes the Reimer-Tiemann reaction to give vanillin and 2-hydroxy-3-methoxybenzaldehyde (*F. Tiemann* and *P. Koppe, ibid.,* 1881, **14,** 2015); it can be mono-acylated (*J. Iwas* and *K. Tomino,* J. pharm. Soc. Japan, 1956, **76,** 808) and alkylated (*Ng. Ph. Buu-Hoi et al.,* J. org. Chem., 1952, **17,** 243, 1122; *R. H. Rosenwald,* J. Amer. chem. Soc., 1952, **74,** 4602). Its esters undergo the Fries rearrangement (*O. Dann* and *G. Mylius,* Ann., 1954, **587,** 1) and its allyl ether rearranges to *o*-eugenol (*C. F. H. Allen* and *J. W. Gates,* Org. Synth., Coll. Vol. III, 1963, p. 418). 2-Methoxy-4-nitrosophenol is formed by the action of sodium nitrite in acetic acid, or alkyl nitrite in sodium ethoxide solution (*A. Pfob,* Monatsh., 1898, **18,** 409).

Substituted guaiacols. [OH = 1; OMe = 2, in accord with Chemical Abstracts usage].

Halogenoguaiacols. 5-*Fluoroguaiacol,* b.p. 94°/16 mm (*Corse* and *Ingraham, loc. cit.*). Direct chlorination by chlorine in aqueous solution gives 5-chloro-, 4,5-dichloro- and trichloro-guaiacols (*K. Sato* and *H. Mikawa,* Bull. chem. Soc. Japan, 1960, **33,** 1736) and in acetic anhydride 4,5,6-trichloro- and tetrachloro-guaiacols (*R. Fort, J. Sleziona* and *L. Denivelle,* Bull. Soc. chim. Fr., 1955, 810) (see Table 7).

Nitroguaiacols. 3-*Nitroguaiacol,* m.p. 68·5–69·5°, is obtained, by hydrolysis of nitroguaiacol acetate (*A. E. Oxford,* J. chem. Soc., 1926, 2004). 4-*Nitro-,* m.p. 101–102°, and 5-*nitro-guaiacol,* m.p. 104–105°, are prepared by base and acid hydrolyses, respectively, of 4-nitroveratrole (*D. Cardwell* and *R. Robinson, ibid.,* 1951, **107,** 255); 6-*nitroguaiacol,* m.p. 61° (*A. Klemenc,* Monatsh., 1912, **33,** 701). Further nitration of the 3-isomer produces 3,6-*dinitro-,* m.p. 69–70°, and 3,4-*dinitro-guaiacol,* m.p. 110°, (*Oxford, loc. cit.*); the 4-isomer gives 4,6-*dinitro-,* m.p. 123°, and the 5-isomer, 4,5-*dinitro-,* m.p. 172°, and 5,6-*dinitro-guaiacol,* m.p. 205–208° decomp. 3,5-*Dinitroguaiacol,* m.p. 80°, is obtained by hydrolysis of the corresponding carbonate (*F. Pollecoff* and *Robinson,* J. chem. Soc., 1918, **113,** 645). 3,4,6-*Trinito-,* m.p. 108–109°, 4,5,6-*trinito-,* m.p.

TABLE 7

HALOGENOGUAIACOLS

	Chloro		Bromo		Iodo
	M.p. °C.	Ref.	M.p. °C.	Ref.	M.p. °C.
3-Mono-	32–33	1	—	—	—
4-Mono-	16–17	2	119/5 mm	6	43°
5-Mono-	36–37	2	65	7	87–88°
6-Mono-	54	2	62–63	5, 8	—
3,5-Di	63–64	2	—		—
4,5-Di-	71·5–73	3	95	7	—
4,6-Di-	64–65	2	64–65	8	92
4,5,6-Tri-	111	4	117	9	—
3,5,6-Tri-	103	11	—		—
Tetra-	121	4	162–163	10	—

References

1 *A. Angeletti* and *M. Pirona,* C.A., 1937, **31,** 678.
2 *J. P. Brown* and *E. B. McCall,* J. chem. Soc., 1955, 3681.
3 *M. Matell,* Acta chem. Scand., 1955, **9,** 1017.
4 *R. Fort, J. Sleziona* and *L. Denivelle.* Bull. Soc. chim. Fr., 1955, 810.
5 *P. W. Robertson,* J. chem. Soc., 1908, **93,** 788.
6 *B. Riegel* and *H. Wittcoff,* J. Amer. chem. Soc., 1946, **68,** 1913.
7 *E. M. Hindmarsh, I. Knight* and *R. Robinson,* J. chem. Soc., 1917, **111,** 940.
8 *N. L. Drake, H. C. Harris* and *C. B. Jaeger,* J. Amer. chem. Soc., 1948, **70,** 168.
9 *A. Zangirolami,* Gazz., 1932, **62,** 570.
10 *C. L. Jackson* and *H. A. Torrey,* Amer. chem. J., 1898, **20,** 395.
11 *H. Koopman,* Tetrahedron Letters, 1965, 513.

144° (*Oxford, loc. cit.*) and 3,5,6-*trinito-guaiacol,* m.p. 129° (*Pollecoff* and *Robinson, loc. cit.*).

For nitration of other catechol ethers, see *Oxford* (*loc. cit.*).

5-**Aminoguaiacol,** 4-*aminocatechol* 1-*methyl ether,* m.p. 131–133°; 4-*amino-guaiacol,* 4-*aminocatechol* 2-*methyl ether,* m.p. 117° (*R. G. Fargher,* J. chem. Soc. 1920, **117,** 865). Oxidation of 4-aminoguaiacol with dichromate gives methoxy-*p*-benzoquinone (*Harmon* and *Cason, loc. cit.*).

3-*Aminoguaiacol,* 3-*aminocatechol* 2-*methyl ether,* m.p. 95–96°; 6-*amino-guaiacol,* 3-*aminocatechol* 1-*m thyl ether,* m.p. 83·5° (*Oxford, loc. cit.*).

Veratrole, *catechol dimethyl ether,* m.p. 22·5°, b.p. 207°, is obtained by methylation of catechol or guaiacol in aqueous alkaline solution with dimethyl sulphate (*W. H. Perkin* and *C. Weizmann,* J. chem. Soc., 1906, **89,** 1649). The m.ps. of some veratrole derivatives are given in Table 8. Oxidation of veratrole by chloranil in sulphuric acid gives 2,3,6,7,10,11-hexamethoxytriphenylene (*I. M. Matheson, O. C. Musgrave* and *C. J. Webster,* Chem. Comm., 1965, 278). For a general review on oxidation of alkyl aryl ethers, see *Musgrave* (Chem. Reviews, 1969, **69,** 499).

TABLE 8

VERATROLE DERIVATIVES

	Chloro		Bromo		Iodo		Nitro	
	M.p. or b.p. °C.	Ref.	M.p. or b.p. °C.	Ref.	M.p. °C.	Ref.	M.p. °C.	Ref.
3-Mono-	98/3 mm	1	23	5	45	9	65	11
4-Mono-	242/760	2	157/30 mm	6	34	10	97	12
4,5-Di-	86	2	92	7	132	4	131	13
3,4,5-Tri-	68	3	84	8	—		145	13
Tetra-	88	4	118	4	—		—	

References

1 *E. D. Hornbaker* and *A. Burger,* J. Amer. chem. Soc., 1955, **77,** 5314.
2 *A. Peratoner* and *G. Ortoleva,* Gazz., 1898, **28,** [i], 229.
3 *H. Cousins,* Ann. chim., 1903, [vii], **29,** 58.
4 *F. Bruggeman,* J. pr. Chem., 1896, [ii], **53,** 251.
5 *H. S. Mason,* J. Amer. chem. Soc., 1947, **69,** 2241.
6 *Ng. Ph. Buu-Hoi,* Ann., 1944, **556,** 1.
7 *H. W. Underwood, O. L. Baril* and *G. C. Toone,* J. Amer. chem. Soc., 1930, **52,** 4087.
8 *J. Frejka* and *B. Sefranek,* Coll. Czech. chem. Comm., 1936, **8,** 130.
9 *F. Mauthner,* J. pr. Chem., 1937, [ii], **149,** 328.
10 *D. E. Janssen* and *C. V. Wilson,* Org. Synth., Coll. Vol. IV, 1963, 547.
11 *W. Baker* and *H. A. Smith,* J. chem. Soc., 1931, 2542.
12 *N. L. Drake, H. C. Harris* and *C. B. Jaeger,* J. Amer. chem. Soc., 1948, **70,** 168.
13 *J. Ehrlich* and *M. T. Bogert,* J. org. Chem., 1947, **12,** 522.

4-Aminoveratrole, 4-aminocatechol dimethyl ether, m.p. 90^0 (*Fargher, loc. cit.*), can also be made by reaction of veratronitrile with hypochlorite (*J. B. Buck* and *W. S. Ide*, Org. Synth., Coll. Vol. II, 1966, p. 45).

3-Aminoveratrole, 3-aminocatechol dimethyl ether, b.p. 137^0/10 mm (*H. Vermeulen*, Rec. Trav. chim., 1929, **48**, 971).

Catechol monoethyl ether, m.p. 28^0, b.p. 211^0; *catechol ethyl methyl ether,* b.p. 213^0 (*J. Allan* and *Robinson*, J. chem. Soc., 1926, 376); *catechol diethyl ether,* m.p. 44^0. *Catechol monoallyl ether,* b.p. 82^0/5 mm, and *diallyl ether,* b.p. 107^0/5 mm, are obtained from catechol and allyl bromide (*Perkin* and *V. M. Trikojus, ibid.*, 1927, 1663); both ethers undergo the Claisen rearrangement on heating. *Catechol monophenyl ether, 2-hydroxydiphenyl ether,* HOC_6H_4OPh, m.p. 104–105^0, is formed by demethylating its methyl ether (*H. E. Ungnade* and *K. T. Zilch*, J. org. Chem., 1950, **15**, 1108). *Catechol methyl phenyl ether,* m.p. 77–78^0, is prepared by heating guaiacol, bromobenzene and copper in aqueous alkali (*Ungnade* and *E. F. Orwoll*, Org. Synth., Coll Vol. III, 1963, p. 566) or preferably in collidine solution (*R. G. R. Bacon* and *O. J. Stewart*, J. chem. Soc., 1965, 4953). *Catechol diphenyl ether,* $C_6H_4(OPh)_2$, m.p. 93^0, is produced by heating *o*-dibromobenzene with sodium phenoxide and copper powder at 240^0 (*F. Ullmann* and *P. Sponagel*, Ann., 1906, **350**, 83).

Catechol methylene ether (XXII, p. 386), b.p. 172^0, is formed by reaction of methylene chloride and catechol either in sodium ethoxide in alcohol (*B. N. Ghosh*, J. chem. Soc., 1915, **107**, 1588) or hydroxide in dimethyl sulphoxide (*W. Bonthrone* and *J. W. Cornforth, ibid.*, C, 1969, 1202). Its halogenation (*I. G. H. Jones* and *Robinson, ibid.*, 1917, **111**, 903; *A. M. B. Orr, Robinson* and *M. M. Williams, ibid.*, p. 946) and nitration (*Perkin, Robinson* and *F. Thomas, ibid.*, 1909, **95**, 1977; *E. Marieli* and *E. Boi*, Gazz., 1904, **34**, [ii], 174; 1909, **39**, [ii], 179) have been described. On heating with phosphorous pentachloride, **catechol dichloromethylene ether** (XXIII, p. 386) is formed (*G. Barger*, J. chem. Soc., 1908, **93**, 563). The latter is also produced from catechol carbonate and phosphorus pentachloride (*H. Gross, A. Rieche* and *E. Hoft*, Ber., 1961, **94**, 544). It is used in a synthesis of arenecarboxylic acids from aromatic hydrocarbons (*Gross, J. Rusche* and *M. Mirsch, ibid.*, 1963, **96**, 1382).

Catechol ethylene ether, 1,4-benzodioxane, (XXIV), b.p. 214^0, is formed by heating catechol with ethylene dibromide, alkali and copper powder (*Ghosh, loc. cit.*). For reactions and general chemistry of 1,4-benzodioxane, see *R. C. Elderfield* (in "Heterocyclic Chemistry", Vol. VI, Ed. *Elderfield*, Wiley, New York, 1957, p. 69).

(*vii*) *Catechol esters*

Catechol sulphite [*thionyl catechol* (XIX, p. 386)], b.p. 138^0/105 mm, results from the reaction between catechol and thionyl chloride in pyridine (*A. Green*, J. chem. Soc., 1927, 500). It reacts with acetic acid containing a little pyridine to give catechol monoacetate and with acetic anhydride to give catechol diacetate. **Catechol sulphate** (XX, p. 386) m.p. 34–35^0, is produced from catechol, sulphuryl chloride and pyridine in petroleum ether (*E. T. Kaiser, I. R. Katz* and *T. F.*

Wulfers, J. Amer. chem. Soc., 1965, **87**, 3781). Alkaline hydrolysis gives *o*-hydroxyphenyl sulphate.

Catechol carbonate (XXI, p. 386), m.p. 118°, b.p. 227°, is produced from phosgene and catechol in cold aqueous alkaline solution (*R. S. Hanslick, W. F. Bruce* and *A. Mascitti*, Org. Synth., Coll. Vol. IV, 1963, p. 788). Other methods of preparation include the reaction of catechol with chloroformate ester and the hydrolysis of catechol dichloromethylene ether (*Barger, loc. cit.*).

Catechol monoacetate, m.p. 57–58°; **diacetate**, m.p. 64° (*Green, loc. cit.*); *monobenzoate*, m.p. 130–131° (*H. Bredereck* and *H. Heckh*, Ber., 1958, **91**, 1314); *dibenzoate*, m.p. 84°; *oxalate*, $C_6H_4O_2(CO)_2$, m.p. 185° (*Ghosh, loc. cit.*).

(*viii*) *Homologues of catechol*

2,3-**Dihydroxytoluene,** *isohomocatechol*, 3-*methylcatechol*, m.p. 68°. The acetate is obtained together with that of methylhydroquinone by the thermal rearrangement of 2-acetyl-2-methylcyclohexa-2,4-dienone, prepared from *o*-cresol by oxidation with lead tetra-acetate (*E. Zbiral, F. Wessely* and *E. Lahrmann*, Monatsh., 1960, **91**, 331). It has also been prepared by the demethylation of its methyl ethers (*R. Majima* and *Y. Okazaki*, Ber., 1916, **49**, 1482). 3-*Methylcatechol* 1-*methyl ether*, m.p. 42°; 2-*methyl ether hemihydrate*, m.p. 39°.

3,4-*Dihydroxytoluene, homocatechol*, 4-*methylcatechol*, m.p. 65°, b.p. 130°/11 mm, occurs, together with its ethers, in wood tars and has been obtained from lignins. It is prepared by reduction of protocatechuic aldehyde (*W. Schepss, ibid.*, 1913, **46**, 2564; *K. W. Rosenmund* and *G. Jordan, ibid.*, 1925, **58**, 160), by the oxidation of *p*-cresol with peracids, or by electrolytic means (*C. G. Henderson* and *R. Boyd*, J. chem. Soc., 1910, **97**, 1659; *F. Fichter* and *F. Ackermann*, Helv., 1919, **2**, 583).

TABLE 9

n-ALKYLCATECHOLS

	3-Isomers			4-Isomers		
	M.p. °C	B.p. °C/mm	Ref.	M.p. °C	B.p. °C/mm	Ref.
Methyl-	68			65	130/11	
Ethyl-		130/15	1		130/9	4
n-Propyl-	72	112/1·5	2		155/12	4
n-Butyl-	33		3	42	143/5	4
n-Pentyl-	34		3		155/7	4

References

1 *W. Mosiman* and *J. Tambor*, Ber., 1916, **49**, 1261.
2 *J. English* and *G. W. Barber*, J. Amer. chem. Soc., 1949, **71**, 3310.
3 *R. D. Haworth* and *D. Woodcock*, J. chem. Soc., 1946, 999.
4 *E. Miller et al.*, J. Amer. chem. Soc., 1938, **60**, 7; *S. Mamura, H. Okubo* and *H. Kaneta*, C.A., 1957, **51**, 14617.

4-*Methylcatechol* 1-*methyl ether, creosol,* b.p. 222^0 (*R. Schwarz* and *H. Hering*, Org. Synth., Coll. Vol. IV, 1963, p. 203); *dimethyl ether, homoveratrole,* m.p. 24^0. These ethers are readily available by Clemmensen reduction of vanillin, and its methyl ether.

Dimethylcatechols, are prepared by *ortho*-hydroxylation of dimethylphenols (*J. D. Loudon* and *J. A. Scott,* J. chem. Soc., 1953, 265), the Dakin oxidation of hydroxy-benzaldehydes and -acetophenones (*W. Baker et al., ibid.,* 1953, 1615) and thermal rearrangement of 2-acetylcyclohexa-2,4-dienones (*Zbiral, Wesseley* and *Lahrmann, loc. cit.*).

3,4-*Dimethylcatechol,* m.p. 84–85^0; 3,5-*dimethylcatechol,* m.p. 70–71^0; 3,6-*dimethylcatechol,* m.p. 101^0; 4,5-*dimethylcatechol,* m.p. 85–86^0.

3,4,5-**Trimethylcatechol** has been obtained by reducing the *o*-benzoquinone (*L. Horner* and *E. Geyer,* Ber., 1965, **98,** 2016); tetramethylcatechol has been synthesised from prehnitol (*Horner* and *W. Spietschka,* Ann., 1952, **579,** 159).

4-*n*-**Alkylcatechols** are best obtained by Clemmensen reduction of the acyl derivatives (*E. Miller et al.,* J. Amer. chem. Soc., 1938, **60,** 7; *S. Mamura, H. Okubo* and *H. Kaneta,* C.A., 1957, **51,** 14617). Table 9 lists the m.ps. and b.ps. of some 4-*n*-alkyl- and 3-*n*-alkyl-catechols.

Other catechols can be obtained from natural sources, *e.g.* 3-*pentadecylcatechol, tetrahydrourushiol,* m.p. 58^0, and 3-*heptadecylcatechol,* m.p. 63^0, are formed by hydrogenating the unsaturated phenols derived from Japanese lacquer (*G. Bertrand, H. J. Backer* and *N. H. Haack,* Bull. Soc. chim. Fr., 1939, 1670).

Uses of catechol and its derivatives. Among the interesting uses for catechol and its derivatives may be mentioned the preparation of alizarin by condensation with phthalic anhydride; in medicine the compound of bismuth and tetrabromocatechol is used as an astringent, antiseptic and deodorant; in photography catechol is used as a fine grain developer. Catechol and its derivatives have antioxidant properties.

(*d*) meta-*Dihydric phenols: resorcinol and its derivatives*

Resorcinol, m-*dihydroxybenzene,* m.p. 110^0, b.p. 281^0, was first obtained in 1864 by the alkaline fusion of *Galbanum* and *Asafetida* resins. Dry distillation or alkaline fusion of Brazil wood also gives resorcinol.

(*i*) *Methods of preparation*

1. Resorcinol is formed by alkaline fusion of benzene-*m*-disulphonic acid, *m*-halogenophenols and *m*-phenolsulphonic acid (*e.g., W. R. Cake,* U.S.P., 2,856,437/1958). Rearrangement of *o*- and *p*-disubstituted benzenes during alkaline fusion can also lead to resorcinol (*e.g., K. W. Muller* and *D. Delfs,* G.P., 1.040,563/1958); see also *S. Oae, N. Furukawa* and *T. Asari,* Bull. Soc. Chem. Japan, 1969, **42,** 177).

2. Acid-catalysed decomposition of *m*-di-isopropylbenzene dihydroperoxide, obtained by oxidation of *m*-di-isopropylbenzene with oxygen, is a well-patented preparation of resorcinol (*e.g., D. I. H. Jacobs,* U.S.P., 2,736,753/1956).

(ii) *Properties and reactions*

(1) *Colour reactions.* An aqueous solution of resorcinol gives a dark-violet colour on addition of ferric chloride solution. It reduces both ammoniacal silver nitrate and Fehlings solution on warming. On exposure to air, a pink tint develops.

(2) *Oxidation.* Resorcinol cannot give *ortho-* or *para*-benzoquinonoid products on two-electron oxidation. The behaviour of a resorcinol derivative in the first oxidation step is simply that of a monohydric phenol whose ionisation potential is lowered by the second hydroxyl group. The E.S.R. spectrum of the relatively unstable *m*-benzene semiquinone radical has been observed (*T. J. Stone* and *W. A. Waters,* J. chem. Soc., 1964, 4302). Subsequent oxidation and reaction gives di- and poly-meric compounds, *e.g.* oxidation of orcinol, 5-methylresorcinol, by alkaline ferricyanide, ferric chloride or lead dioxide gives mainly polymeric material of type XXIX, formed by both C–C and C–O coupling, and a small amount of 2,2′,4,4′-tetrahydroxy-6,6′-dimethylbiphenyl (XXX). If oxygen or hydrogen peroxide is the oxidant, then high yields of bibenzo-, mono-, and -di-quinones, XXXI, and XXXII, are formed:

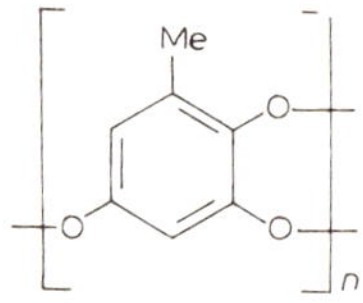

(XXIX)

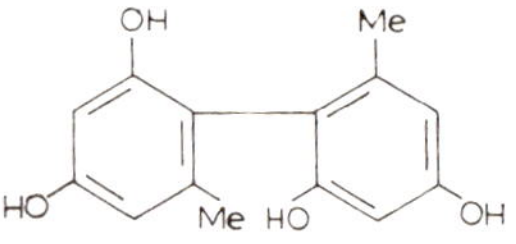

(XXX) 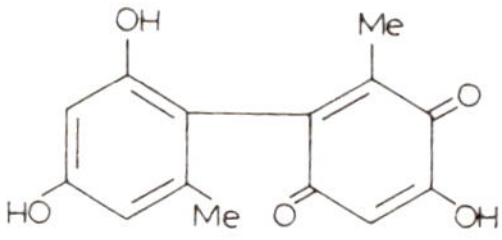 (XXXI) 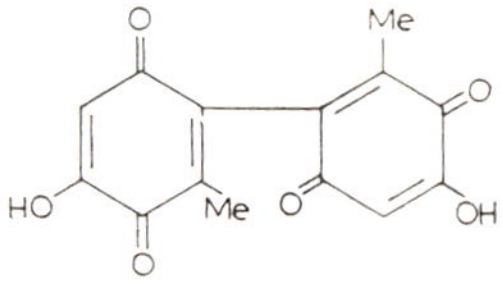 (XXXII)

Autoxidation of resorcinol and its derivatives in the presence of ammonia produces orceine and litmus dyes. These consist of a mixture of 7-hydroxy- and 7-amino-phenoxazones (XXXIII; R = H: resorcinol; R = Me: orcinol) (*H. Musso et al.*, Angew. chem., 1961, **73**, 665; Ber., 1963, **96**, 1579):

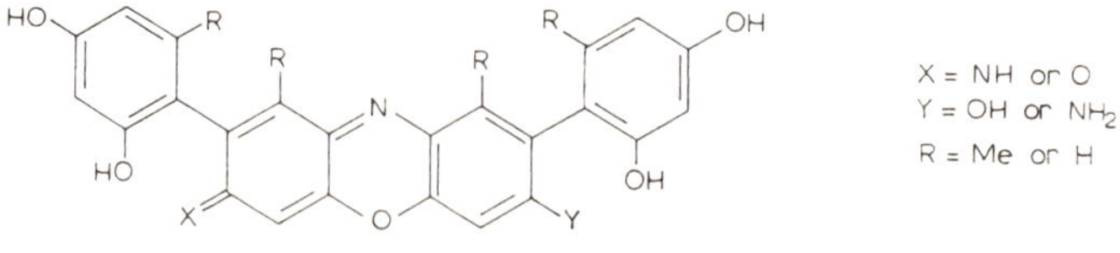

X = NH or O
Y = OH or NH_2
R = Me or H

(XXXIII)

Aerial oxidation of sterically hindered, 4,6-di-*tert*-butylresorcinol, at -20^0 gives the hydroperoxide XXXIV, which produces, on warming, the hydroxyquinone XXXV, and other products (*Musso* and *D. Maassen*, Ann., 1965, **689**, 93):

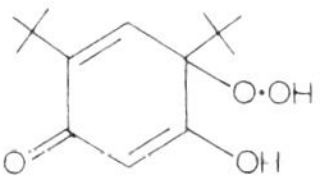

(XXXIV) 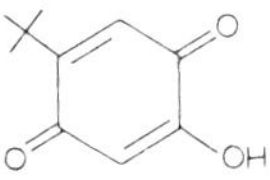 (XXXV)

For other references, see *Musso* (Angew. Chem. internat. Edn., 1963, **2**, 729; in "Oxidative Coupling of Phenols", Ed. *W. I. Taylor* and *A. R. Battersby*, Arnold, London, 1967) and *A. I. Scott* (Quart. Reviews, 1965, **19**, 1).

(3) *Substitution.* Nucleophilic substitution by ammonia gives *m*-aminophenol (*Bad. Anilin u. Sodafabrik*, G.P., 126,136/1918), by secondary amines 3-dialkylaminophenols and *m*-bis-dialkylaminobenzenes (*F. Effenberger et al.*, Ber., 1970, **103**, 1456), and by bisulphite *m*-phenolsulphonic acid (by way of 1,3-dihydroxycyclohexane-1,3,5-trisulphonic acid) (*J. Allan* and *J. Podstata*, Coll. Czech. chem. Comm., 1966, **31**, 3563, 3571, see also p. 376).

Electrophilic substitution is particularly easy and can lead to polysubstitution unless controlled conditions are used. *C*-Alkylation and *C*-acylation, including the Houben-Hoesch synthesis, occur readily. Preparations of 2- and 4-acylresorcinols are described in Organic Syntheses (*A. Russell* and *J. F. Frye*, Org. Synth., Coll. Vol. III, 1964, p. 281; *S. R. Cooper*, p. 761); more forcing conditions result in the formation of 4,6-di- and 2,4,6-tri-acylresorcinols (*H. J. Teuber, D. Cornelius* and *U. Wölcke*, Ann., 1966, **696**, 116; *D. B. Limaye* and *R. G. Chitale*, C.A., 1951, **45**, 7049). The Fries rearrangement of diesters also leads to mono- and di-acylresorcinols (*R. D. Desai* and *C. K. Marant*, Proc. Indian Acad. Sci., 1949, **29A**, 269). The aldehyde group has been introduced into the 4-position by the Reimer-Tiemann, Gatterman, and Vilsmeier reactions and also by reaction with alkyl dichloromethyl ethers and orthoformates. β-Resorcylic acid, resorcinol-4-carboxylic acid, is formed by reaction with carbon dioxide and potassium bicarbonate (*M. Nierenstein* and *D. A. Clibbens*, Org. Synth., Coll. Vol. II, 1966, p. 557); higher temperatures produce the 4,6-dicarboxylic acid (*R. F. Bann* and *R. C. Conn*, U.S.P., 2,703,812/1955). For reviews on topics in this paragraph, see "Friedel-Crafts and Related Reactions", (Vols. II and III, Ed. *G. A. Olah*, Interscience, New York, 1964).

A sulphonation study has shown that 2-mono-, 4-mono-, 2,4-di-, 4,6-di- and 2,4,6-trisulphonic acids can be produced (*Podstata* and *Z. J. Allan*, Coll. Czech. chem. Comm., 1967, **32**, 3004).

(4) *Condensation reactions.* Phthalic anhydride condenses with resorcinol to give fluorescein; other 1,2-dicarboxylic acids react similarly. Resorcinol and its derivatives condense with a variety of compounds to form flavones, coumarins, pyrones, etc. ("Heterocyclic Chemistry", Vol. II, Ed. *R. C. Elderfield*, Wiley, New York, 1951).

(iii) Halogenoresorcinols

2,4,6-Tribromo- and trichloro-resorcinols are easily produced. Further halogenation leads to 1,3,3,5,5-*pentachlorocyclohex*-1-*ene*-2,4-*dione*, m.p. 94–96°, and 2,2,4,4,5,6,6-*heptachlorocyclohexane*-1,3-*dione*, m.p. 56·5–58·5° (*T. Zincke* and *S. Rabinowitsch*, Ber., 1890, **23**, 3766; 1891, **24**, 912; see also *C. v. d. Stelt, B. G. Suurmond* and *W. T. Nauta*, Rec. Trav. chim., 1954, **73**, 1022) and the penta-bromo-compound (*F. Trischler* and *K. Szivos*, C.A., 1966, **65**, 12854).

Several procedures control halogenation to the mono- or di-substitution stage (see Table 10). These include the use of (1) sulphuryl chloride (*M. L. Moore, A. A. Day* and *C. M. Suter*, J. Amer. chem. Soc., 1934, **56**, 2456); (2) bromine and iodine monochloride complexes with dioxane (*L. A. Yanokskaya, A. P. Terentov* and *L. J. Belenkii*, Zhur. obshchei Khim., 1952, **22**, 1594; 1954, **24**, 1265); (3) bromine in pyridine (*K. W. Rosenmund* and *W. Kuhnhenn*, Ber., 1923, **56**, 1262); (4) dichlorourea, alone, or with potassium bromide or iodide for chlorination, bromination or iodination, respectively (*M. V. Likhosherstov*, J. gen. Chem. USSR., 1933, **3**, 164, 172, 177). Alternative methods include brominating resorcylic acids and subsequently removing the carboxyl group (*G. P. Rice*, J. Amer. chem. Soc., 1926, **48**, 3125) or debrominating tribromoresorcinol with sodium bisulphite or stannous chloride (*T. L. Davis* and *V. F. Harrington*, *ibid.*, 1934, **56**, 129).

TABLE 10

HALOGENORESORCINOLS

	Chloro		Bromo		Iodo	
	M.p. °C.	Ref.	M.p. °C.	Ref.	M.p. °C.	Ref.
2-Mono-	97	1	102	6	100	10
4-Mono-	105	2	100–102	7	67	11
5-Mono-	117	3	87	3	92	3
2,4-Di-	85	4	93	8	—	
4,6-Di-	112	2	110–112	9	145	11
2,4,6-Tri-	83	5	110	10	154	12

References

1 *H. W. Wanzlick* and *S. Mohrmann*, Ber., 1963, **96**, 2257.
2 *M. L. Moore, A. A. Day* and *C. M. Suter*, J. Amer. chem. Soc., 1934, **56**, 2456.
3 *H. H. Hodgson* and *J. S. Wignall*, J. chem. Soc., 1926, 2826.
4 *P. A. Pectynin* and *A. S. Kuchina*, J. gen. Chem., USSR, 1947, **17**, 278.
5 *R. Bendedikt*, Monatsh., 1883, **4**, 224.
6 *G. P. Rice*, J. Amer. chem. Soc., 1926, **48**, 3125.
7 *R. B. Sandin* and *R. A. McKee*, Org. Synth., Coll. Vol. II, 1966, p. 100.
8 *T. L. Davis* and *V. F. Harrington*, J. Amer. chem. Soc., 1934, **56**, 129.
9 *K. W. Rosenmund* and *W. Kuhnhenn*, Ber., 1923, **56**, 1262.
10 *M. V. Likhosherstov*, J. gen. Chem., USSR, 1933, **3**, 164, 172, 177.
11 *B. H. Nicolet* and *J. R. Sampey*, J. Amer. chem. Soc., 1927, **49**, 1796.
12 *E. Roberts*, J. chem. Soc., 1923, **123**, 2707.

(*iv*) *Nitrosoresorcinols*

4-**Mono-** and 2,4-**di-nitrosoresorcinols** are formed from resorcinol and amyl nitrite in alcoholic alkali (*F. Henrich*, Ber., 1902, **35**, 4191; *F. L. Gilbert, F. C. Laxton* and *E. B. R. Prideaux*, J. chem. Soc., 1927, 2295); the latter also using aqueous nitrous acid (*W. R. Orndorff* and *M. L. Nichols*, J. Amer. chem. Soc., 1923, **45**, 1536). Both compounds are oxidised to nitroresorcinols by hydrogen peroxide or nitric acid (*Gilbert, Laxton* and *Prideaux, loc. cit.*; *A. Gay-Lussac* and *J. Meniger*, Poudres, 1958, **40**, 7; C.A., 1961, **55**, 3489) and reduced to amino-resorcinols by stannous chloride and hydrochloric acid.

Reaction of 2,4-dinitrosoresorcinol with hydroxylamine gives cyclohexane-1,2,3,4-tetraone tetra-oxime (*F. Kehrmann* and *J. Messinger*, Ber., 1890, **23**, 2815; *R. Nietzki* and *F. Blumenthal*, *ibid.*, 1897, **30**, 181). Tautomerism of nitroso-phenols has been discussed (p. 348).

(*v*) *Nitroresorcinols*

2-**Nitroresorcinol,** m.p. 84^0, b.p. 234^0, is prepared by the nitration of resorcinol-4,6-disulphonic acid and subsequent desulphonation. This is achieved simply by adding a cooled paste of resorcinol in oleum to nitric acid and oleum and then steam distilling (*M. S. Carpenter, V. M. Easter* and *T. F. Wood*, J. org. Chem., 1956, **16**, 586). 4-*Nitroresorcinol*, m.p. 115^0, is obtained by oxidation of 4-nitroso-resorcinol and by alkaline hydrolysis of 1,3-diamino-4-nitrobenzene (*Allan* and *J. A. Langer*, Chem. Listy, 1951, **45**, 281). 5-*Nitroresorcinol*, m.p. 157–159^0 (*J. M. Grosheintz* and *H. O. L. Fischer*, J. Amer. chem. Soc., 1948, **70**, 1479).

2,4-**Dinitroresorcinol,** m.p. 147–149^0. 4,6-*Dinitroresorcinol*, m.p. 214^0, is produced by nitration of resorcinol diacetate (*D. N. Mukerji*, J. chem. Soc., 1922, **121**, 545; *W. Borsche* and *E. Feske*, Ber., 1928, **61**, 690).

2,4,6-**Trinitroresorcinol,** *styphnic acid*, m.p. 175^0, is prepared by nitration of resorcinolsulphonic acids (*V. Merz* and *G. Zetter*, *ibid.*, 1879, **12**, 2035; *Borsche* and *Feske, loc. cit.*) and by oxidation of 2,4,6-trinitrocresol with dichromate (*H. G. Adolph, J. C. Dacons* and *M. J. Kamlet*, Tetrahedron, 1963, **19**, 801). It is a strong acid with p*K*a in acetone of 1·74 and 4·86 (*P. J. Pearce* and *R. J. J. Simkins*, Canad. J. Chem., 1968, **46**, 241) and like picric acid, forms addition compounds.

Tetranitroresorcinol, m.p. 152^0 (*J. J. Blanksma*, Rec. Trav. chem., 1908, **27**, 35) is hydrolysed by water to trinitrophloroglucinol.

The i.r. spectra of 2-nitro- (*J. C. Dearden* and *W. F. Forbes*, Canad. J. Chem., 1960, **38**, 1852), 2,4-dinitro- and 2,4,6-trinitro-resorcinol (*F. Pristera et al.*, Anal. Chem., 1960, **32**, 495), show there is strong hydrogen bonding between the nitro and hydroxyl groups (see also p. 342).

(*vi*) *Aminoresorcinols*

2-**Aminoresorcinol,** *hydrochloride*, m.p. 240–241^0 (*H. D. Dell et al.*, Ann., 1967, **709**, 70); *monomethyl ether*, m.p. 123^0 (*R. K. Smalley* and *H. Suschitzky*, J. chem.

Soc., 1963, 5571), *dimethyl ether*, m.p. 75–77^0 (*B. T. Stephanov*, Zhur. obshchei Khim., 1962, **32**, 3741).

4-*Aminoresorcinol, triacetate*, m.p. 114^0, is prepared by ammonolysis of hydroxy-hydroquinone at room temperature (*R. Lantz*, Compt. rend., 1960, **251**, 2045, 2984) and by reduction of 4-nitro-, 4-nitroso- and 4-azo-resorcinols (*F. Heinrich* and *B. Wagner*, Ber., 1902, **35**, 4195; *R. Meyer* and *H. Kreis*, *ibid*., 1883, **16**, 1329; *E. Gebauer-Fülnegg* and *H. A. Beatty*, J. Amer. chem. Soc., 1927, **49**, 1361). 2-*Amino*-5-*methoxyphenol*, m.p. 128–130^0, *hydrochloride*, m.p. 214^0 (*J. Hill* and *G. R. Ramage*, J. chem. Soc., 1964, 3709). 4-*Amino*-3-*methoxyphenol* (*F. R. Bean*, U.S.P., 2,525,515/1950). 2,4-*Dimethoxyaniline*, m.p. 39^0, 2,4-*dimethoxyacetanilide*, m.p. 116^0 (*J. D. Loudon* and *J. Ogg*, J. chem. Soc., 1955, 739).

5-*Aminoresorcinol*, m.p. 146–152^0, is prepared from phloroglucinol and ammonia (*J. Pollak*, Monatsh., 1893, **14**, 419); *triacetate*, m.p. 121^0; 3,5-*dihydroxy-acetanilide*, m.p. 213^0 (*F. Sorm* and *L. Novotny*, Chem. Listy, 1955, **49**, 901). 3,5-*Dimethoxyaniline*, m.p. 46^0, b.p. 120^0/0·2 mm; (-*acetanilide*, m.p. 157^0; *hydrochloride*, m.p. 209^0), is prepared from 3,5-dimethoxybenzamide and hypo-chlorite (*N. B. Dean* and *W. B. Whalley*, J. chem. Soc., 1954, 4638), from bromo-2,4-dimethoxybenzene and sodamide (*R. A. Benseker et al*., J. Amer. chem. Soc., 1958, **80**, 5289) or by reduction of the corresponding nitro compound (*S. H. Oakeshott* and *S. G. P. Plant*, J. chem. Soc., 1927, 484).

(*vii*) *Ethers and esters*

Resorcinol mono- and **di-methyl ethers** b.p. 244^6 and 217^0, respectively (*W. H. Perkin et al*., J. chem. Soc., 1926, 941). Particularly good yields of mono-ethers are obtained by using dialkyl sulphate in a two-phase solvent system composed of nitrobenzene and aqueous alkali (*H. Bredereck, I. Hennig* and *W. Rau*, Ber., 1953, **86**, 1085). Irradiation, using a mercury arc, of *m*-chlorophenol in alcoholic solution produces *m*-alkoxyphenols (*J. T. Pinhey* and *R. D. G. Rigby*, Tetra-hedron Letters, 1969, 1271). Use of alkyl halides under basic conditions also produces ethers but forcing conditions lead to *C*-alkylated products (*J. W.* and *R. H. Cornforth* and *R. Robinson*, J. chem. Soc., 1942, 682; *A. R. Stein*, Canad. J. Chem., 1965, **43**, 1493; 1508).

Resorcinol monoethyl ether, b.p. 246^0 (*Bredereck, Hennig* and *Rau, loc. cit.*); *diethyl ether*, m.p. 12^0, b.p. 235^0; *resorcinol monophenyl ether*, b.p. 143^0/2·0 mm, is obtained from *m*-phenoxyanisole, *m*-phenoxyaniline or from *m*-bromobenzene, copper and phenol (*K. J. Sax et al*., J. org. Chem., 1960, **25**, 1590, see also *E. Klarmann et al*., J. Amer. chem. Soc., 1931, **53**, 3397); *diphenyl ether*, m.p. 61^0 (*F. Ullmann* and *P. Sponagel*, Ann., 1906, **350**, 83).

Resorcinol monoethers behave essentially as simple monohydric phenols on oxidation, *e.g.*, *m*-methoxyphenol and Fremy's salt or periodate give methoxy-*p*-benzoquinone (*H. J. Teuber* and *Rau*, Ber., 1953, **86**, 1036; *E. Adler* and *R. Magnusson*, Acta chem. Scand., 1959, **13**, 505).

The nitration (*M. S. Carpenter, Easter* and *Wood*, J. org. Chem., 1951, **16**, 586) and controlled monobromination of resorcinol ethers have been described (*D. C. Schlegel, C. D. Tipton* and *K. L. Reinhart, ibid*., 1970, **35**, 849).

Resorcinol monoacetate, curesol, b.p. 283°, is best obtained (in 70% yield) from resorcinol, acetic anhydride and alkali (*S. S. Israelstam* and *J. D. M. Simpson*, J. South Africa chem. Inst., 1956, **9**, 92); *monobenzoate*, m.p. 132°, is prepared (in 90% yield) from resorcinol and benzoyl chloride in alkali (*Bredereck* and *H. Heckh*, Ber., 1958, **91**, 1314); *resorcinol diacetate*, b.p. 273°; *dibenzoate*, m.p. 117°.

(viii) Homologues of resorcinol

The most important of these is orcinol, 5-methylresorcinol. **Orcinol** occurs in fungal metabolites and lichens of the species *Rocella* and *Lecanora*, partly as orsellinic acid (2,4-dihydroxy-6-methylbenzoic acid) and partly as erythrin (erythritol orsellinate). It is obtained from orsellinic acid by heating or by alkaline hydrolysis. It is prepared by alkali fusion of suitable 3,5-disubstituted toluenes (*R. H. C. Neville* and *A. Winther*, Ber., 1882, **15**, 2976; *M. S. Kisteneva* and *M. S. Rozndestvenskii*, Zhur. Priklad. Khim., 1949, **20** 1108). A convenient synthesis starts from ethyl crotonate and ethyl acetoacetate (*R. M. Ankar* and *A. H. Cook*, J. chem. Soc., 1945, 311).

Orcinol forms a *monohydrate*, m.p. 56°; it is very soluble in alcohol and ether. Its aqueous solution gives a blue colour with ferric chloride. Orcinol undergoes all of the substitution reactions of resorcinol. For example, chlorine gives *trichloro-orcinol*, m.p. 127°; the calculated quantity of sulphuryl chloride produces 4-*chloro-orcinol*, m.p. 104°; 2,4-*dichloro-orcinol*, m.p. 121°, is obtained by chlorinating ethyl orsellinate followed by decarboxylation. Nitration gives 2-*nitro*-, m.p. 120°; 4-*nitro*-, m.p. 125–127°; 2,4-*dinitro*-, m.p. 164°; and 2,4,6-*trinitro*-3,5-*dihydroxytoluene*, m.p. 171° (decomp.) (*H. Musso* and *H. Beecken*, Ber., 1959, **92** 1416). *Orcinol monomethyl ether*, m.p. 62°, b.p. 259°; *dimethyl ether*, b.p. 244°.

The m.p. of other methyl homologues of resorcinol are given in Table 11. Methyl groups have been introduced into the resorcinol nucleus by a combination of formylation (using zinc chloride and hydrochloric acid) and Clemmenson reduction (*e.g.*, see *W. B. Whalley*, J. chem. Soc., 1949, 3038, 3278; *D. J. Cram*, J. Amer. chem. Soc., 1950, **72**, 595, and also *A. Bauer-Benedikt* and *H. Punzengruber*, Monatsh., 1950, **81**, 772). Another important preparation is the acid-catalysed rearrangement of a 2-acetoxy-2-methylcyclohexa-3,5-dienone or a 4-acetoxy-4-methylcyclohexa-2,5-dienone, formed by oxidation of an *o*- and *p*-cresol, respectively, with lead tetra-acetate (*W. Metlesics, F. Wessely* and *H. Budzikiewicz*, Tetrahedron, 1959, **6**, 345; *S. Goodwin* and *B. Witkop*, J. Amer. chem. Soc., 1957, **79**, 179).

4-n-*Alkylresorcinols*, best prepared by Clemmenson reduction of the acyl compounds (*T. B. Johnson* and *W. W. Hodge*, J. Amer. chem. Soc., 1913, **35**, 1014; *A. R. L. Dohme, E. H. Cox* and *E. Miller*, *ibid.*, 1926, **48**, 1688) are bactericides; 4-n-*hexylresorcinol*, m.p. 68–71°, is used as an antiseptic. 2-*n*-Alkylresorcinols are prepared similarly and also by reaction of 2-lithioresorcinol dimethyl ether with alkyl halides (*Dell et al.*, Ann., 1967, **709**, 63, 70).

5-*Alkylresorcinols* are prepared from 5-alkyl-3-hydroxy-4-methoxycarbonyl-

TABLE 11

HOMOLOGUES OF RESORCINOL

(a) *Methyl homologues*

	M.p. °C.	Ref.		M.p. °C.	Ref.
2-Mono-	120	1	4,6-Di-(*m*-xylorcinol)	127	6
4-Mono- (cresorcinol)	104	2	2,4,5-Tri-	148	4
5-Mono- (orcinol)	107		2,4,6-Tri- (mesorcinol)	151	7
2,4-Di-	110	3	4,5,6-Tri-	165	8
2,5-Di- (β-orcinol)	163	4	Tetra-	157	4
4,5-Di- (*o*-xylorcinol)	136	5			

(b) *Other homologues*

	2-		4-		5-	
	M.p. °C.	Ref.	M.p. °C.	Ref.	M.p. °C.	Ref.
Ethyl-	97	9	98	12	96	14
n-Propyl-	101	10	78	12	83 (Divarinol)	15
n-Butyl-	88	10	47	13	82	16
n-Pentyl-	55	11	72	13	49 (Olivetol)	16

References

1 *P. R. Dinganka*r and *T. S. Gore*, Indian J. Chem., 1964, **2**, 294.
2 *J. C. Bell, W. Bridge* and *A. Robertson*, J. chem. Soc., 1937, 1542.
3 *W. Baker et al., ibid.*, 1949, 2834.
4 *W. B. Whalley, ibid.*, 1949, 3278.
5 *D. J. Cram*, J. Amer. chem. Soc., 1948, **70**, 4240.
6 *Cram* and *F. W. Cranz, ibid.*, 1950, **72**, 595.
7 *J. W* and *R. H. Cornforth* and *R. Robinson*, J. chem. Soc., 1942, 682.
8 *A. Robertson* and *Whalley, ibid.*, 1949, 3038.
9 *F. Takacs*, Monatsh , 1964, **95**, 961.
10 *H. D. Dell et al.*, Ann., 1967, **709**, 63, 70.
11 *R. Adams, C. K. Cain* and *B. R. Baker*, J. Amer. chem. Soc., 1940, **62**, 2201.
12 *T. B. Johnson* and *W. W. Hodge, ibid.*, 1913, **35**, 1014.
13 *A. R. L. Dohme, ibid.*, 1926, **48**, 1688.
14 *J. P. Brown et al.*, J. chem. Soc., 1940, 859.
15 *F. Mauthner*, J. pr. Chem., 1924, [ii], **108**, 275.
16 *R. M. Ankar* and *A. H. Cook*, J. chem. Soc., 1945, 311.

cyclohex-2-enones and bromine in acetic acid. The cyclohexenones are conveniently prepared from 1-alkenyl acetates and ethylmalonic esters in the presence of sodium methoxide (*A. Brossi, A. Focella* and *S. Teitel,* G.P., 2,002,815/1970).

Uses. Resorcinol and its derivatives find limited use in the dyestuffs industry, *e.g.*, eosins and azo dyes; as synthetic tannin agents and resins; as antiseptics (especially 4-*n*-hexylresorcinol) and for other medical purposes.

(*e*) para-*Dihydric phenols*: *hydroquinone and its derivatives*

Hydroquinone, p-*dihydroxybenzene, quinol,* is so named since it is readily formed by reduction of *p*-benzoquinone. Hydroquinone occurs widely in nature in the form of its β-glucopyranoside, *arbutin,* m.p. 200°, $[\alpha]_D$ —64·3°, from which it is obtained by acid hydrolysis. Arbutin is present in leaves of many plants and trees such as bearberry (*Arbutus uva-ursi*), cranberry (*Vaccinium vitisidaea*), whortleberry (*V. myrtillus*). It also occurs in other species such as *Chinaphila, Pirola, Gaultheria, Rhodcdendrum aureum. Methylarbutin,* m.p. 160° $[\alpha]_D$ —63°, is found in *P. incarnata*. Hydroquinone was first obtained in 1820 by Pelletier and Caventov from the distillation of quinic acid.

(*i*) *Methods of preparation*

(1) Alkaline hydrolyses of *p*-halogenophenols (*e.g., K. W. Muller* and *D. Delfs,* G.P., 1,040,563/1958) and *p*-dichlorobenzene (*S. J. Lloyd* and *A. M. Kennedy,* U.S.P., 1,849,844/1933) lead to hydroquinone and some resorcinol.

(2) Electrolytic oxidation of benzene in the presence of sulphuric acid, using a lead dioxide anode and lead cathode produces *p*-benzoquinone, which is reduced to hydroquinone (*H. von Bramer* and *A. C. Ruggles,* U.S.P., 2,043,912/1936).

(3) Aniline is oxidised by manganese dioxide in sulphuric acid to *p*-benzoquinone, which is reduced to hydroquinone by iron in water (*H. v. Bramer* and *J. W. Zabriskie,* U.S.P., 1,998,177/1935; *Fuji Photo Film Co.,* F.P., 1,371,138/1964).

(4) Oxidation of *p*-hydroxyacetophenone or *p*-hydroxybenzaldehyde by hydrogen peroxide in alkaline solution is used (*H. D. Dakin,* Amer. chem. J., 1909, **42**, 477).

(5) Phenol is oxidised to hydroquinone by persulphate (*W. Baker* and *N. C. Brown,* J. chem. Soc., 1948, 2303), performic or peracetic acids in the presence of phosphorous acid at 50–150° (*A. Marlard,* F.P., 1,479,354/1967) and by irradiating an aqueous solution of phenol containing hydrogen peroxide (*K. Omura* and *T. Matsuura,* Tetrahedron, 1968, **24**, 3475). The last two methods also produce catechol.

(6) The acid-catalysed decomposition of *p*-di-isopropylbenzene dihydroperoxide, obtained by oxidation of *p*-di-isopropylbenzene, is an industrial process (*D. I. H. Jacobs,* B.P., 775,813/1956; *W. Webster,* B.P., 727/498/1955).

(7) Acetylene, carbon monoxide and a hydrogen donor, such as water or

hydrogen, in the presence of a transition metal compound, *e.g.*, $Ru_3(CO)_{12}$ at $\sim 280^0$, give hydroquinone (*B. W. Howk* and *J. C. Sauer*, U.S.P., 3,055,949/1962; *Lonza Ltd.*, B.P., 1,119,520/1968).

(8) Hydroquinone is produced by the hydrolysis of the diazonium salt of *p*-aminophenol (*J. de Jonge* and *R. Dijkstra*, Rec. Trav. chim., 1949, **68**, 426).

(9) *p*-Benzoquinone is reduced by several reducing reagents, including sulphur dioxide, lithium aluminium hydride, hydrazine, stannous chloride and iron.

(ii) *Properties and reactions*

Hydroquinone exists in three crystalline forms, α, β and γ (*W. A. Casperi*, J. chem. Soc., 1926, 2944; 1927, 1093; *D. F. Evans* and *R. E. Richards*, *ibid.*, 1952, 3932). The common form is the α-modification, m.p. $172 \cdot 3^0$; the β-modification is the one which forms clathrate compounds with small molecules (*H. M. Powell*, *ibid.*, 1950, 468). Hydroquinone and *p*-benzoquinone form an equimolar charge-transfer complex, *quinhydrone*, m.p. 171^0, which is used as the basis of the quinhydrone electrode for measuring hydrogen-ion activity. Hydroquinone reduces Fehling's solution and ammoniacal silver nitrate in the cold.

(1) *Oxidation.* Hydroquinones, like catechols, are extremely easily oxidised to the corresponding benzoquinones by iodine, silver oxide, ferricyanide, chloranil, etc. (*H. Musso*, Angew. Chem., internat. Edn., 1963, **2**, 723; in "Oxidative Coupling of Phenols", Ed. *W. I. Taylor* and *A. R. Battersby*, Arnold, London, 1967; *A. I. Scott*, Quart. Reviews, 1963, **19**, 1). The initial products, the *p*-semiquinone radical-anions have been well-studied, more so than the *o*-semiquinones (*I. C. P. Smith* and *A. Carrington*, Mol. Phys., 1967, **12**, 439; *Carrington*, Quart. Reviews, 1963, **17**, 67). The following equilibrium is rapidly attained and the equilibrium constant has been measured (*C. A. Bishop* and *L. K. J. Tong*, J. Amer. chem. Soc., 1965, **87**, 501):

$$\text{}^{\ominus}O\text{-}C_6H_4\text{-}O^{\ominus} + O{=}C_6H_4{=}O \rightleftharpoons 2\ \cdot O\text{-}C_6H_4\text{-}O^{\ominus}$$

Autoxidation of hydroquinone in alkaline solution gives *p*-benzoquinone initially; hydroxy-*p*-benzoquinone has been isolated as a further oxidation product (*W. Flaig* and *J. C. Salfeld*, Naturwiss., 1960, **47**, 516; *M. Eigen* and *P. Matthies*, Ber., 1961, **94**, 3309). Nucleophiles may add during the course of the oxidation, *e.g.*, oxidation in the presence of bisulphite produces mono- and 2,5-di-sulphonic acids (*T. H. James* and *A. Weissberger*, J. Amer. chem. Soc., 1939, **61**, 442). The coupled oxidation product 2,2′,5,5′-tetrahydroxybiphenyl, is formed by irradiating an aqueous solution of hydroquinone (*H. I. Joschek* and *S. I. Miller, ibid.*, 1966, **88,** 3273). Like catechol, humic acid-like products can result from a succession of coupling reactions.

(2) *Substitution.* Electrophilic substitution, although easy, does not proceed as readily as with resorcinol. Both *C*-alkylation (mono- and 2,5-di-substitution) and *C*-acylation have been variously studied ("Friedel-Crafts and Related Reactions", Vol. II and III, Ed. *G. A. Olah,* Interscience, N.Y., 1964). The aldehyde group is introduced by the Reimer-Tiemann reaction (*H. Wynberg,* Chem. Reviews, 1960, **60,** 169) and the Kolbe-Schmidt reaction gives both mono- and di-carboxylic acids (*O. Baine et al.*, J. org. Chem., 1954, **19,** 510); the mono-carboxylic acid is formed simply by heating with aqueous potassium bicarbonate at 130^0 (*C. Senhofer* and *P. Sarlay,* Monatsh., 1881, **2**, 448). The Fries rearrangement of diesters produces monoacylhydroquinones (*G. C. Amin* and *N. M. Shah,* Org. Synth., Coll. Vol. III, 1963, p. 280). Hydroquinone can be monosulphonated (*A. Quilico,* Gazz., 1927, **57,** 793) and disulphonated (*H. Kauffmann,* Ber., 1907, **40,** 838). Diazo compounds do not couple with hydroquinone but oxidise it to *p*-benzoquinone (*K. J. P. Orton* and *R. W. Everatt,* J. chem. Soc., 1908, **93,** 1021).

(3) *Addition reactions.* 1,4-Diels Alder addition, a rare occurrence in benzenoid systems, has been achieved with maleic anhydride to give XXXVI, (*R. C. Cookson* and *N. S. Wariyar, ibid.*, 1957, 327):

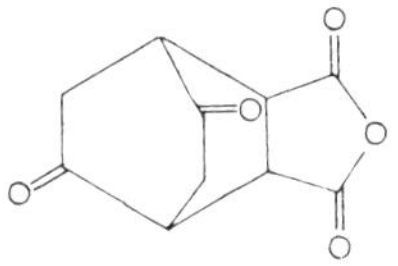

(XXXVI)

(*iii*) *Ethers and esters*

Hydroquinone monomethyl ether, p-**methoxyphenol,** m.p. 53–55^0, b.p. 241^0, is obtained by the controlled action of dimethyl sulphate on hydroquinone (*H. Bredereck, I. Hennig* and *W. Rau,* Ber., 1953, **86,** 1085). *Monoethyl ether,* p-*ethoxyphenol,* m.p. 64–66^0, is similarly prepared. Alternatively, hydrolyses of dialkyl *p*-hydroxyphenyl phosphates can be used (*F. Ramirez, E. H. Chen* and *S. Dershowitz,* J. Amer. chem. Soc., 1959, **81,** 4338). *Hydroquinone dimethyl ether,* p-*dimethoxybenzene,* m.p. 57^0, b.p. 205^0 (*G. M. Dyson, H. J. George* and *R. F. Hunter,* J. chem. Soc., 1927, 436). Treatment of *p*-dibromobenzene with alkoxide ion in the presence of cuprous iodide in refluxing collidine gives *para*-dialkoxybenzenes (*R. G. R. Bacon* and *S. C. Rennison, ibid.*, C, 1969, 312). *Hydroquinone ethyl methyl ether,* m.p. 37^0; *diethyl ether,* m.p. 72^0.

Hydroquinone monophenyl ether, p-*hydroxydiphenyl ether,* m.p. 84^0, is formed by demethylation of p-*methoxydiphenyl ether,* b.p. 186^0/32 mm, obtained from *p*-methoxyphenol and bromobenzene in the presence of alkali and copper powder at 240^0 (*T. R. Lea* and *R. Robinson, ibid.*, 1926, 411); *hydroquinone diphenyl ether,* m.p. 77°, is obtained from phenol and dibromobenzene at 180° also in the presence of copper (*F. Ullmann* and *P. Sponagel,* Ann., 1906, **350,** 83).

Oxidation of hydroquinone monoethers by periodate or bismuthate produces *p*-benzoquinone and the alcohol or phenol. The monophenyl ether also gives XXXVII. Fremy's salt yields *p*-alkoxy- and *p*-aroyl-*o*-quinones (*Adler* and *Magnusson, loc. cit.* and references cited therein; *Adler, I. Falkehag* and *B. Smith,* Acta chem. Scand., 1962, **16,** 529). For other oxidations see *C. G. Haynes, A. H. Turner* and *W. A. Waters* (J. chem. Soc., 1956, 2823) and *D. F. Bowman, F. R. Hewgill* and *B. R. Kennedy* (*ibid.*, C, 1966, 2274).

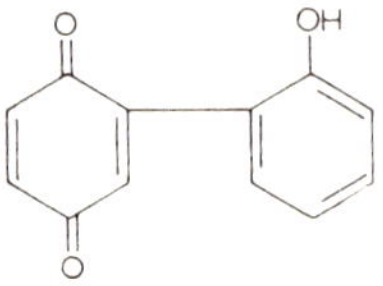

(XXXVII)

Hydroquinone monoacetate m.p. 65–66° (*C. A. Bunton* and *J. Hellyer,* J. Amer. chem. Soc., 1967, **89,** 6252); *monobenzoate,* m.p. 164° (*Bredereck* and *H. Heckh,* Ber., 1958, **91,** 1314); *diacetate,* m.p. 121° (*W. W. Prichard,* Org. Synth., Coll. Vol. III, 1963, p. 452); *dibenzoate,* m.p. 199°.

TABLE 12

HALOGENOHYDROQUINONES

	Chloro		Bromo		Iodo	
	M.p. °C.	Ref.	M.p. °C.	Ref.	M.p. °C.	Ref.
Mono-	105	1, 2	113–115	3	115–116	6
2,3-Di-	144–145	1, 3	—		—	
2,5-Di-	172	1, 3	186–187	5	—	
2,6-Di-	157–158	1, 3	166–167	2	144	7
Tri-	136	1	135	2	—	
Tetra-	230–232	1, 4	244–246	5	258	8

References

1 *J. B. Conant* and *L. F. Fieser,* J. Amer. chem. Soc., 1923, **45,** 2194.
2 *A. P. Terentev, A. N. Grinev* and *A. B. Terentev,* Zhur. obshchei Khim., 1954, **24,** 1433.
3 *T. Akita,* C.A., 1962, **57,** 9711.
4 *R. H. Thomson,* J. chem. Soc., 1953, 1196.
5 *J. Boehm* and *W. Zamlynski,* C.A., 1967, **67,** 53817.
6 *D. E. Kvalnes,* J. Amer. chem. Soc., 1934, **56,** 667.
7 *P. Block* and *G. Powell, ibid.*, 1942, **64,** 1070.
8 *C. L. Jackson* and *E. K. Bolton, ibid.*, 1914, **36,** 1483.

(iv) *Halogenohydroquinones*

(1) Chloro- and bromo-hydroquinones are readily formed by the addition of hydrogen halide to *para*-benzoquinones in solvents such as chloroform and ether (*J. B. Conant* and *L. F. Fieser*, J. Amer. chem. Soc., 1923, **45**, 2194 and references cited therein; *A. P. Terentev, A. N. Grinev* and *A. B. Terentev*, Zhur. obshchei Khim., 1954, **24**, 1433). Use of formaldehyde in acetic acid in the above reaction results in addition of Cl_2 rather than HCl (*R. H. Thomson*, J. chem. Soc., 1953, 1196).

(2) For direct halogenation of hydroquinones either chlorine, bromine or sulphuryl chloride is used (*Conant* and *Fieser*, *loc. cit.*; *G. F. Rodgers*, U.S.P., 2,748,173/1956; *J. E. Kovack*, U.S.P., 3,143,576/1964; *J. Boehm* and *W. Zamlynski*, C.A., 1967, **67**, 53817). A milder brominating agent is the bromine–dioxane complex (*L. A. Yanovskaya, A. P. Terentev* and *L. I. Belenkii*, C.A., 1953, **47**, 8032).

(3) Reduction of halogeno-*p*-benzoquinones by hydrazine hydrate (*T. Akita*, *ibid.*, 1962, **57**, 9711), sulphur dioxide (*H. E. Ungnade* and *K. T. Zilch*, J. org. Chem., 1950, **16**, 64) or stannous chloride (*Conant* and *Fieser*, *loc. cit.*) is also effective.

Trifluoro-, m.p. 112–113°, and **tetrafluoro-hydroquinone**, m.p. 168–169°, have been described (*E. Nield* and *J. C. Tatlow*, Tetrahedron, 1960, **8**, 38).

The m.ps. of chloro-, bromo- and iodo-hydroquinones are given in Table 12.

(v) *Nitrohydroquinones*

Direct nitration is not usually employed owing to the ease of oxidation of hydroquinone. Instead, esters and ethers are nitrated and subsequently hydrolysed; the monobenzenesulphonate ester has proved particularly useful (*E. M. Kampouris*, J. chem. Soc., C, 1967, 1235). All possible dinitrohydroquinones and their methyl ethers were obtained by the elegant use of the relative directing effects of the hydroxyl and benzenesulphonate groups (Table 13). Oxidation of *o*-nitrophenol by ammonium persulphate gives nitrohydroquinone (*W. Ruske*, Ann., 1957, **610**, 156).

TABLE 13

NITROHYDROQUINONES

	Nitrohydroquinone M.p. °C.	Monomethyl ether M.p. °C.	Dimethyl ether M.p. °C.
2-Mono-	132	1-ether 85 4-ether 79	71
2,3-Di-	98–100	162	183
2,5-Di-	201–203	165	201
2,6-Di-	135–136	1-ether 152	111
2,3,5-Tri-	—	1-ether 111 4-ether 78	98
Tetra-	—	116	179

(vi) Aminohydroquinones

These are prepared by reduction of the nitro-compounds. 2-**Aminohydroquinone hydrochloride** (*R. Lantz* and *E. Michel*, Bull. Soc. chim. Fr., 1964, 534); 4-*methyl ether* (*W. J. Close, B. D. Tiffany* and *M. A. Spielman*, J. Amer. chem. Soc., 1949, **71**, 1265); *dimethyl ether*, m.p. 80° (*A. Blackhall* and *Thomson*, J. chem. Soc., 1954, 3916).

(vii) Homologues of hydroquinone

Reduction of the appropriate alkyl-*p*-benzoquinone is a general method. Such quinones are prepared by oxidation of *C*-methylanilines (*K. Sato, Y. Fujima* and *A. Yamada*, Bull. chem. Soc. Japan, 1968, **41**, 442) and methyl-*p*-aminophenols (*N. Zenker* and *E. Jorgensen*, J. org. Chem., 1959, **24**, 1351; *J. L. G. Nilsson, H. Sievertsson* and *H. Selander*, Acta Pharm. Suecica, 1968, **5**, 215; C.A., 1968, **69**, 106132).

Direct oxidation by persulphate of a methylphenol with a free *para* position gives the hydroquinone (*W. Baker* and *N. C. Brown*, J. chem. Soc., 1948, 2303). Oxidation of *p*-alkylphenols with peracids in acidic solution also leads to alkylhydroquinones. The intermediates, the 4-alkyl-4-hydroxycyclohexa-2,5-dienones, can be isolated in neutral solution and rearrange with both acid and base catalysts to the alkylhydroquinones (*S. Goodwin* and *B. Witkop*, J. Amer. chem. Soc., 1959, **79**, 179 and references cited therein). Acid-catalysed rearrangements of *p*-alkylphenylhydroxylamines also produce hydroquinones (Bamberger reaction: *H. J. Shine*, "Aromatic Rearrangements", Elsevier, Amsterdam, 1967, p. 812; *K. Sato* and *Y. Fujima*, C.A., 1967, **67**, 11307).

Other preparations involve a combination of the Mannich reaction and hydrogenation to introduce methyl groups into the nucleus (*S. Abe* and *Sato*, J. org. Chem., 1963, **28**, 1928; Japan P., 23,558/1964; *W. J. Burke et al.*, J. org. Chem., 1961, **26**, 4669). Methyl- and dimethyl-acetylenes react with carbon monoxide in the presence of a carbonylation catalyst to give dimethyl- and tetramethyl-hydroquinones (*W. Reppe* and *H. Vetter*, Ann., 1953, **582**, 133; Netherlands P., 6,602,880/1966).

Syntheses from aliphatic compounds have been devised *e.g.*, 2,5-dimethyl-

TABLE 14

HYDROQUINONE HOMOLOGUES

Hydroquinone	M.p. °C.	Hydroquinone	M.p. °C.
Methyl-	125		
2,3-Dimethyl-	218	Ethyl-	115
2,5-Dimethyl-	211	*n*-Propyl-	88
2,6-Dimethyl-	148	*n*-Butyl-	87
Trimethyl-	167	*n*-Pentyl-	86
Tetramethyl-	226		

hydroquinone from biacetyl and sodium bisulphite (*W. Hafner*, G.P., 1,220,437/1966, see also *M. Mühlstadt* and *O. Scholz*, Ber., 1964, **97**, 1).

n-Alkylhydroquinones are formed by reduction of the acyl analogues (*A. Vanderberghe* and *J. F. Willems*, Bull. Soc. chim. Belges, 1965, **74**, 397; *E. C. Armstrong et al.*, J. Amer. chem. Soc., 1960, **82**, 1928), by reaction of trialkylboranes with *p*-benzoquinones (*M. F. Hawthorne* and *M. Reintjes, ibid.*, 1964, **86**, 951; 1965, **87**, 4584) and by reduction of *p*-alkylnitrobenzenes with aluminium in sulphuric acid (*F. R. Bean*, U.S.P., 2,533,203/1950).

Uses. Hydroquinone and its derivatives are used principally as photographic developers; other uses include antioxidants for fats and oils and inhibitors of polymerisation.

7. Polyhydroxybenzenes

All possible tri-, tetra-, penta- and hexa-hydroxybenzenes are known. Derivatives of these compounds occur widely in plants as glucosides, chromones, coumarins, flavanoids, lignins and tannins. The most important and best-studied compounds are pyrogallol and phloroglucinol. Polyhydroxybenzenes and their ethers are all extremely reactive towards electrophiles and alkyl, acyl, formyl, carboxyl ("Friedel-Craft and Related Reactions", Vol. II and III, Ed. *G. A. Olah*, Interscience, N.Y., 1964), halogeno and nitro groups are readily introduced. Nucleophilic substitutions such as ammonolysis and hydroxylation also occur easily. Oxidation, especially of those polyhydroxybenzenes with hydroxyl groups *ortho* or *para* to each other, is a particularly easy process; most polyhydroxybenzenes are oxidised by air, and solids and solutions become coloured on exposure to it. Phloroglucinol and other compounds with *meta*-hydroxyl groups, show many of the reactions of oxo compounds but there is no structural and spectroscopic evidence for the existence of oxo–enol equilibria. Some oxo forms, for example, that of 1,2,3,4-tetrahydroxybenzene, can be isolated.

(a) 1,2,3-Trihydroxybenzenes

Pyrogallol, 1,2,3-trihydroxybenzene, is incorporated in tannins, flavones and alkaloids and is obtained from wood tar distillates (*J. G. Gatsis*, U.S.P., 2,934,567/1960). *K. Scheele* first obtained pyrogallol in 1876 by the dry distillation of gallic acid, 3,4,5-trihydroxybenzoic acid, extracted from nutgalls.

(i) Methods of preparation

(1) Decarboxylation of gallic acid is still extensively used. This may be achieved, apart from by dry distillation, by heating gallic acid with water with aqueous alkaline earth oxides (*Nitrilefabrik*, G.P., 335,153/1916) or tertiary base (*P. N. Zemenko*, U.S.S.R. Pat., 105,427/1957).

(2) Pyrogallols are produced by *ortho*-hydroxylation of catechols (*J. D. Loudon et al.*, J. chem. Soc., 1953, 269; 1954, 1134).

(3) Rearrangement of 2,2-diacetoxycyclohexa-3,5-dienone, either by heating or with an acid catalyst, such as boron trifluoride or sulphuric acid in acetic anhydride, gives pyrogallol triacetate. Substituted pyrogallols are similarly prepared from dienones with free 6-positions (*F. Wessely et al.*, Monatsh., 1964, **95**, 533; 1960, **91**, 117; *S. Goodwin* and *B. Witkop*, J. Amer. chem. Soc., 1957, **79**, 179).

(*ii*) *Reactions and properties*

Pyrogallol, m.p. 132⁰, p*K*a 9·28, 11·34 at 20⁰, is soluble in water, alcohol and ether but only sparingly soluble in benzene and carbon tetrachloride. The I.R., U.V. and N.M.R. spectra have been obtained (*L. P. Kuhn* and *R. E. Bowman*, Spectrochim. Acta, 1961, **17**, 650; *L. Jurd*, J. Amer. chem. Soc., 1956, **78**, 3445; *J. C. Schug* and *J. C. Deck*, J. chem. Phys., 1962, **37**, 2618). Owing to its *ortho*-hydroxyl groups, pyrogallol forms chelates and is useful in analytical chemistry (*e.g.*, *A. Pakhomova*, *V. L. Rudyakova* and *I. A. Sheka*, Russ. J. inorg. Chem., 1966, **11**, 621; *E. C. Hunt* and *R. A. Wells*, Analyst, 1954, **79**, 345). It also forms cyclic ethers, *e.g. pyrogallol methylene ether*, m.p. 65⁰, and esters *e.g. pyrogallol carbonate*, m.p. 135⁰.

Oxidation. Pyrogallol is a powerful reducing agent and precipitates mercury, silver and gold from solutions of their salts. With ferric chloride, a blue colour is formed which turns brown on standing. Its aqueous alkaline solutions rapidly absorb oxygen and are used to remove oxygen from gases. Hydroxybiphenyl has been isolated from such solutions (*H. Erdtman*, Proc. roy. Soc., 1933, **143A**, 196).

The e.s.r. spectrum of its semiquinone radical-anion has been obtained (*H. Rein* and *O. Ristau*, Z. phys. Chem. Leipzig, 1968, **239**, 115; *M. Adams*, *M. S. Blois* and *R. H. Sands*, J. chem. Phys., 1958, **28**, 774).

Oxidation of pyrogallol by isoamyl nitrite and acetic acid in ethanol gives I, (*H-J. Teuber*, *P. Heinrich* and *M. Dietrich*, Ann., 1966, **696**, 64; *A. G. Perkin* and *A. B. Steven*, J. chem. Soc., 1906, **89**, 802), whereas purpurogallin (II) is produced by a variety of other oxidants, of which one of the most convenient and satisfactory is sodium iodate (*T. W. Evans* and *W. M. Dehn*, J. Amer. chem. Soc., 1930, **52**, 3647; for the mechanism, see *L. Horner*, *K. H. Weber* and *W. Dürckheimer*, Ber., 1961, **94**, 2881; *J. C. Salfeld* and *E. Baume*, *ibid.*, 1964, **97**, 307). 4-Halogeno- and 4-alkyl-, but not 5-alkyl- and 4,6-dialkyl-pyrogallols, are similarly oxidised to purpurogallin derivatives (*e.g. A. Critchlow et al.*, Tetrahedron, 1967, **23**, 2829). Other tropolones, *e.g.* III, are obtained by co-oxidation with an *o*-benzoquinone or a catechol (*Horner et al.*, Ber., 1964, **97**, 312):

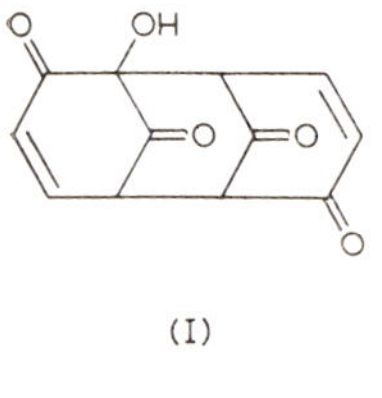

(I) 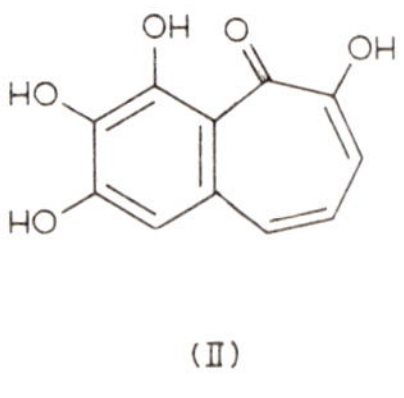 (II) 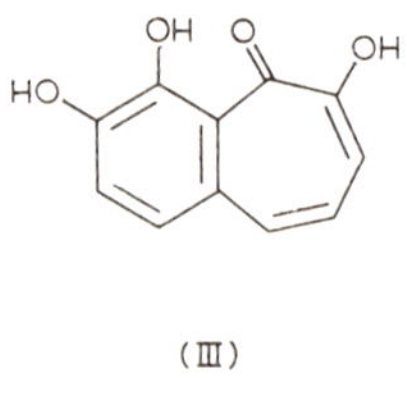 (III) 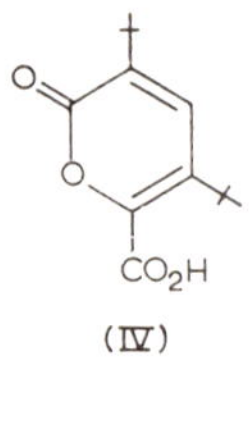 (IV)

Pyrogallols undergo ring cleavage on oxidation, *e.g.*, 4,6-di-*tert*-butylpyrogallol is oxidised to the lactonic acid, IV, by ferricyanide (*T. W. Campbell,* J. Amer. chem. Soc., 1951, **73**, 4190).

(*iii*) *Ethers and esters*

The Dakin reaction is used to obtain **pyrogallol 1-methyl ether,** m.p. 41°, b.p. 136°/22 mm (*A. R. Surrey,* Org. Synth., Coll. Vol. III, 1964, p. 759) and 1,2-*dimethyl ether,* b.p. 116°/12 mm (*E. Profft* and *G. Rietz,* J. pr. Chem., 1960, [iv], **11**, 94); 2-*methyl ether,* m.p. 87°, b.p. 155°/15 mm (*E. Spath* and *H. Schmid,* Ber., 1941, **74**, 193). Methylation of pyrogallol in alkaline solution by methyl iodide gives the 1,3-*dimethyl ether,* m.p. 55°, b.p. 262°, while dimethyl sulphate produces the *trimethyl ether,* m.p. 47° (*E. Chapman, A. G. Perkin* and *R. Robinson,* J. chem. Soc., 1927, 3015).

The 1-methyl ether is oxidised by silver oxide to 3-methoxy-*o*-benzoquinone (*R. Willstatter* and *F. Muller,* Ber., 1911, **44**, 2171). Chemical and enzymic oxidants produce coerulignone, 3,5,3′,5′-tetramethoxy-4,4′-diphenoquinone, from the 1,3-dimethyl ether (*B. C. Saunders* and *B. P. Stark,* Tetrahedron, 1967, **23**, 1867).

Pyrogallol monoacetate, m.p. 85°; 1,2-*diacetate,* m.p. 115°; *triacetate,* m.p. 163°. *Pyrogallol 1-benzoate,* m.p. 138°, *tribenzoate,* m.p. 90° (*H. Bredereck* and *H. Heckh,* Ber., 1958, **91**, 1314). Nitrous acid and 1,2,3-triethoxybenzene give 2,6-dimethoxy-1,4-benzoquinone (*H. T. Bolker* and *F. L. Kung,* J. chem. Soc., 1969, 2298 and refs. cited therein).

(*iv*) *Halogenopyrogallols*

Direct halogenation readily gives trichloro- and tribromo-pyrogallol and can proceed further *e.g.*, to give 1,2,2,6-tetrabromocyclohexene-3,4,5-trione (*F. J. Moore* and *R. M. Thomas,* J. Amer. chem. Soc., 1917, **39**, 974). Controlled use of *N*-chloro- and *N*-bromo-succinimide and *tert*-butyl hypochlorite gives the 4-, 4,6-di- and 4,5,6-tri-halogeno derivatives of pyrogallol and its 1,3-dimethyl and trimethyl ethers (*D. Friedman* and *D. Ginsberg,* J. org. Chem., 1958, **23**, 16). Use of sulphuryl chloride and halogenation of pyrogallol carbonate are other methods of controlling substitution (*Horner* and *S. Gowecke,* Ber., 1961, **94**, 1267).

4-**Chloropyrogallol,** m.p. 162° (*Horner* and *Gowecke, loc. cit.*), 1,3-*di-* and *tri-methyl ethers,* b.p. 175°/18 mm and 252°, respectively; 5-*chloropyrogallol,* m.p. 166°, 1-*methyl ether,* m.p. 67°, is obtained from hydrogen chloride and 3-methoxy *o*-benzoquinone, *trimethyl ether,* m.p. 72° (*Horner* and *Gowecke, loc. cit.*). 4,6-*Dichloropyrogallol,* m.p. 129°, *trimethyl ether,* b.p. 80°/0·1 mm; 4,5,6-*trichloropyrogallol,* m.p. 177°, 1,3-*di-* and *tri-methyl ethers,* m.p. 119° and 52°.

4-**Bromopyrogallol,** m.p. 119°, 1,3-*di-* and *tri-methyl ethers,* b.p. 88°/0·4 mm, and 95°/0·5 mm; 5-*bromopyrogallol* 1,3-*dimethyl ether,* m.p. 106°, is obtained by reductive dehalogenation of the tribromo compound (*M. Kohn* and *L. Steiner,* J. org. Chem., 1947, **12**, 30), *trimethyl ether,* m.p. 78°.

4,6-**Dibromopyrogallol,** m.p. 155°, 1,3-*di-* and *tri-methyl ethers,* m.p. 127° and b.p. 118°/0·4 mm, respectively; 4,5,6-*tribromopyrogallol,* m.p. 168°, 1-, 1,3-*di-* and *tri-methyl ethers,* m.p. 111°, 134° and 73° respectively.

(v) Nitropyrogallols

5-**Nitropyrogallol,** m.p. 203^0, is obtained by the nitration of pyrogallol tribenzyl ether (*T. S. Gardner, E. Wenis* and *J. Lee*, J. org. Chem., 1950, **15,** 841). 4-*Nitropyrogallol,* m.p. 162^0, and 4,6-*dinitropyrogallol,* m.p. 208^0, are similarly produced from pyrogallol carbonate (*A. Einhorn, J. Cobliner* and *H. Pfeiffer,* Ber., 1904, **37,** 100). 4-*Nitropyrogallol* 1-, 3-, 1,2-*di*-, 1,3-*di*- and *tri-methyl ethers,* m.p. 127^0, 122^0, 102^0, 67^0 and 44^0, respectively; 4,6-*dinitropyrogallol* 1,2-*di*-, 1,3-*di*- and *tri-methyl ethers,* m.ps. 76^0, 162^0 and 85^0, respectively (*E. Spath* and *E. Dobrovolny,* Ber., 1938, **71,** 1831; *W. Baker* and *H. A. Smith,* J. chem. Soc., 1931, 2542; *K. Brand* and *H. Collischorn,* J. pr. Chem., 1921, [ii] **103,** 345). *Trinitro*-1,2,3-*trimethoxybenzene,* m.p. 128^0.

(vi) Homologues of pyrogallol

Methylpyrogallols are prepared by rearrangement of diacetoxycyclohexadienones (p. 308), *ortho*-hydroxylation of catechols (p. 409), reaction of acetic anhydride with an *o*-benzoquinone and subsequent hydrolysis (*Horner et al.,* Ann., 1955, **597,** 1; Ber., 1964, **97,** 312) and reduction of the appropriate benzaldehydes. The best method for *n*-alkyl derivatives is the Clemmenson reduction of the acyl compounds (*M. C. Hart* and *E. H. Woodruff,* J. Amer. chem. Soc., 1936, **58,** 1957). Some alkyl derivatives occur naturally ,*e.g.*, 5-n-*propylpyrogallol,* m.p. 78^0, which is found in *Hamaline diracerata.* 4-, 5-, 4,5-*Di*-, 4,6-*di*- and 4,5,6-*tri-methylpyrogallol,* m.ps. 142^0, 126^0, 154^0, 122^0 and 165^0, respectively. 4-*Ethyl*-, m.p. 108^0, 4-n-*propyl*-, m.p. 110^0, 4-n-*butyl*-, m.p. 88^0, 4-n-*pentyl-pyrogallol,* m.p. 90^0.

(b) 1,2,4-*Trihydroxybenzenes*

Hydroxyhydroquinone, 1,2,4-**trihydroxybenzene,** m.p. 140^0, *triacetate,* m.p. 96^0.

(i) Methods of preparation

(1) Treatment of *p*-benzoquinones with acetic anhydride in the presence of an acid gives substituted 1,2,4-triacetoxybenzenes (*H. S. Wilgus* and *J. W. Gates,* Canad. J. Chem., 1967, **45,** 1975; *J. M. Blatchly, J. F. W. McOmie* and *J. B. Searle,* J. chem. Soc., C, 1969, 1350, 1353; for 1,2,4-triacetoxybenzene, see *E. B. Vliet,* Org. Synth., Coll. Vol. I, 1964, p. 317). For the hydrolyses of the triacetates, see *M. Healey* and *Robinson* (J. chem. Soc., 1934, 1625).

(2) Dakin oxidation of 2,4- or 3,4-dihydroxy-benzaldehydes or -acetophenones yields hydroxyhydroquinones (*H. Dakin,* Amer. chem. J., 1909, **42,** 495; *W. Baker,* J. chem. Soc., 1934, 1684).

(3) Thermal rearrangements of 2,2-diacetoxy-6-alkylcyclohexa-3,5-dienones produce substituted hydroxyhydroquinones (*H. Budlikiewicz, W. Metlesks* and *F. Wessely,* Monatsh., 1960, **91,** 117).

(4) 1,2,4-Trialkoxybenzenes are formed from the tetra-*O*-alkyl ethers of 1,2-*O*-isopropylidene-*myo*-inositol (cf. Vol. IIB, pp. 59–60) by reaction with

potassium *tert*-butoxide in dimethyl sulphoxide (*P. A. Gent* and *R. Gigg*, J. chem. Soc., C, 1970, 2253).

(*ii*) *Properties and reactions*

Hydroxyhydroquinone, p*K*a 9·08, 11·82 at 20° is soluble in water and alcohol but sparingly so in benzene and chloroform. It gives with ferric chloride a brown colour, changed to blue and then red by the cautious addition of sodium bicarbonate solution. Its alkaline solutions absorb oxygen and rapidly darken. The e.s.r. spectra of the semiquinone radical anion of hydroxyhydroquinone and its mono-, di- and tri-methyl homologues have been recorded (*K. A. K. Lott, E. L. Short* and *D. N. Waters, ibid.*, B, 1969, 1232). Hydroxy-*p*-benzoquinone (*M. S. Mason*, J. biol. Chem., 1948, **181**, 803) and the quinone V, are initial oxidation products (*H. Musso et al.*, Ann., 1964, **676**, 10), further carbon-carbon coupling eventually producing humic acid-type material (*J. E. LuValle*, J. Amer. chem. Soc., 1952, **74**, 2907). Several other oxidation products, including cyclopentane derivatives, have been obtained (*J. F. Corbett*, J. chem. Soc., C, 1967, 611; see also *ibid.*, 1970, 2101).

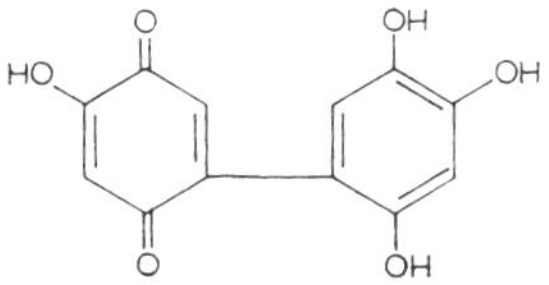

(V)

Methyl ethers are available from the appropriate anilines (by hydrolyses of the diazonium salts) or by the Dakin oxidation. 1,2,4-*Trihydroxybenzene* 1-, 2- and 4-*methyl ethers*, m.p. 71°, 90°, and 49°, respectively; 1,2-, 1,4- and 2,4-*dimethyl ethers*, m.p. 80°, b.p. 93°/7 mm and m.p. 28° respectively; *trimethyl ether*, m.p. 19°, b.p. 251°. 1,2,4-*Trihydroxybenzene*, 1-*mono*-, 4-*mono*- and tri-*benzoates*, m.p. 170°, 140° and 120°, respectively.

Direct halogenation produces 1,2,5,5-tetrahalogenocyclohexene-3,4,6-trione, which is reduced by stannous chloride to the trihalogenohydroxyhydroquinone (*T. Zincke* and *E. Weishaut*, Ann., 1924, **437**, 86). For other halogeno and nitro derivatives see *Blatchly, McOmie* and *Searle*, (*loc. cit.*), *A. Oliverio* and *G. Castelfranchi* (Gazz., 1952, **82**, 109; 1950, **80**, 267, 276) and *H. W. Dorn, W. H. Warren* and *J. L. Bullock* (J. Amer. chem. Soc., 1939, **61**, 144). Replacement of hydroxyl by an amino group is easily accomplished at room temperature (*R. L. Lantz* and *E. Michel*, Bull. Soc. chim. Fr., 1961, 2402).

(*c*) 1,3,5-*Trihydroxybenzenes*

Phloroglucinol, 1,3,5-*trihydroxybenzene*, was first obtained in 1855 by *Hlasiwetz* by hydrolysis of the glycoside, phoretin, found in the bark of fruit trees. Subsequently it has been isolated from many other naturally occurring substances, *e.g.*, flavones, catechin, catechu, kino and gamboge.

(i) Methods of preparation

An important industrial process is the reductive decarboxylation of 2,4,6-trinitrobenzoic acid (prepared from 2,4,6-trinitrotoluene by oxidation with dichromate) by iron, or tin, and hydrochloric acid to 1,3,5-triaminobenzene and subsequent hydrolysis (*M. L. Kastens* and *J. F. Kaplan*, Ind. Eng. Chem., 1950, **42,** 402). The laboratory preparation is described by *H. T. Clarke* and *W. W. Hartman* (Org. Synth., Coll. Vol. I, 1964, p. 455). 1,3,5-Trinitrobenzene (*E. Varo* and *J. N. Vickers*, B.P., 1,106,088/1964), picryl chloride (*P. M. Heertjes*, Rec. trav. Chim., 1959, **78,** 452) and 3,5-diaminonitrobenzene (*I.G. Farbwerke Hoechst A.-G.*, B.P., 1,012,782/1965) are also used as starting materials.

Other methods include the acid-catalysed decomposition of the trihydroperoxide of 1,3,5-tri-isopropylbenzene (*F. Seidel, M. Schulze* and *H. Baltz*, J. pr. Chem., 1956, [iv], **3,** 278; *A. F. Shepard*, B.P., 751,598/1958) and condensations of aliphatic compounds, *e.g.*, of malonyl chloride with acetone (*T. Komninos*, Bull. Soc. chim. Fr., 1918, **23,** 449) and of diethyl malonate with sodium (*D. Ullrich* and *J. Seiffert*, G.P. [East], 24,998/1963).

(ii) Properties and reactions

Phloroglucinol, anhydrous, m.p. 219^0, *dihydrate*, m.p. 113–116^0, p*K*a 7·97 and 9·23 at 20^0, is a colourless, odourless, sweet-tasting compound not very soluble in cold water but fairly soluble in alcohol and ether.

(1) *Oxidation.* Its aqueous solution gives a violet colour with ferric chloride, reduces Fehling's solution and precipitates gold, silver and platinum from solutions of their salts. Alkaline solutions of phloroglucinol absorb oxygen from air but much less readily than those of pyrogallol. Its reactivity towards periodate is also smaller (*D. E. Pennington* and *D. M. Ritter*, J. Amer. chem. Soc., 1947, **69,** 187). Oxidation of 3-methyl-, and of 3,5-dimethyl-phloroacetophenone by ferricyanide and ferric chloride, respectively, give the compounds VI and VII (*D. H. R. Barton, A. M. Deflorin* and *O. E. Edwards*, J. chem. Soc., 1956, 530; *H. Davies, H. Erdtman* and *M. Nilsson*, Tetrahedron Letters, 1966, 2491):

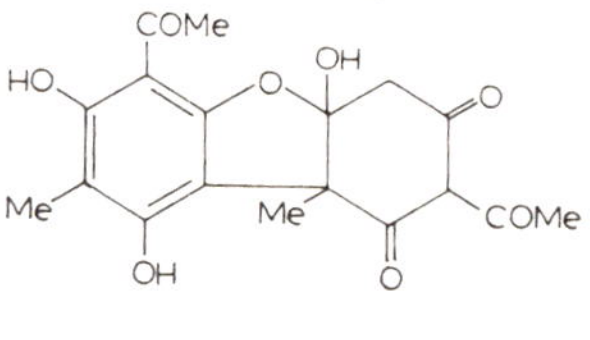

(VI)

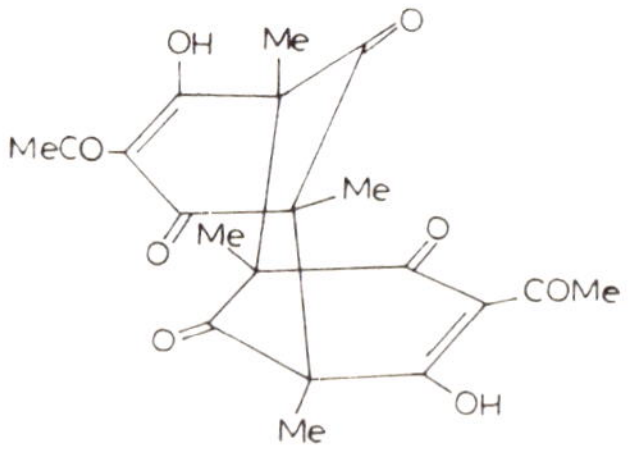

(VII)

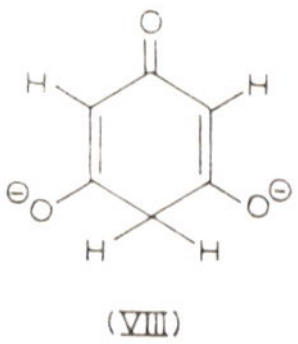

(VIII)

(2) *Tautomerism.* The u.v. (*T. W. Campbell* and *G. M. Coppinger,* J. Amer. chem. Soc., 1951, **73,** 2708), i.r. and n.m.r. (*R. J. Highet* and *T. J. Batterham,* J. org. Chem., 1964, **29,** 475) and X-ray spectra (*K. Maartman-Moe,* Acta Cryst., 1965, **19,** 155) are consistent with the enol form for phloroglucinol. However, the dianion exists in the oxo form, VIII; its n.m.r. spectrum in water shows olefinic (τ 4·97) and methylenic (τ 7·0) but no aromatic resonances, in contrast to the single aromatic resonance (τ 3·95–3·98) of phloroglucinol and its mono-anion. This change in structure also produces a large increase in the second acid dissociation constant of phloroglucinol compared with those of di- and other tri-hydroxybenzenes. Phloroglucinol, like resorcinol, undergoes many of the reactions of ketones, especially in basic solution. For example, it gives hexa-methylcyclohexane-1,3,5-trione and related compounds with methyl iodide (*A. R. Stein,* Canad. J. Chem., 1965, **43,** 1493 and references cited therein), mono-, di- and tri-adducts with bisulphite (*W. Fuchs,* Ber., 1921, **54,** 245), a cyanohydrin (*W. T. Gradwell* and *A. McGookin,* Chem. and Ind., 1956, 377) and a trioxime (*V. Farmer* and *R. H. Thomson, ibid.,* 1956, 86).

(3) *Substitution.* Nucleophilic substitution of the hydroxyl group by hydride ion occurs on treatment with sodium borohydride (*G. I. Fray,* Tetrahedron, 1958, **3,** 316). With ammonia, 5-aminoresorcinol and 3,5-diaminophenol are formed (*J. Pollak,* Monatsh., 1893, **14,** 419) and with secondary amines in an autoclave, 3,5-bis-dialkylaminophenol and 1,3,5-tris-dialkylaminobenzene are produced (*E. Effenberger* and *R. Ness,* Ber., 1968, **101,** 3787). Hydrolysis of the cyanohydrin gives 3,5-dihydroxybenzoic acid.

Phenolic reactions of phloroglucinol include carboxylation by aqueous potassium bicarbonate at 60^0 (*R. Mayer* and *A. Melhorn,* Z. chem., 1963, **3,** 390), diazo coupling to give mono-, bis- and tris-azo-compounds (*A. G. Perkin,* J. chem. Soc., 1897, **71,** 1154) and nitrosation with nitrous acid to give trinitrosophloroglucinol (*R. Benedikt,* Ber., 1878, **11,** 1375); see also p. 416.

(4) *Condensation reactions.* Phloroglucinols form coumarins on condensation with sodium acetate and acetic anhydride, and with ethyl acetoacetate in the presence of acid (*e.g., F. M. Dean, E. Evans* and *A. Robertson,* J. chem. Soc., 1954, 4565), phloroglucinolphthalein with phthalic anhydride, and benzoxazines with formaldehyde and secondary amines (*W. J. Burke* and *C. Weatherbee,* J. Amer. chem. Soc., 1950, **72,** 4691).

(*iii*) *Ethers and esters*

Phloroglucinol mono-, di- and **tri-methyl ethers,** m.p. 79^0, b.p. 189^0/14 mm; m.p. 37^0, b.p. 170^0/14 mm; m.p. 50^0, b.p. 130^0/10 mm, respectively (*H. Bredereck, I. Hennig* and *W. Rau,* Ber., 1953, **86,** 1085). **Phloroglucinol di-** and **tri-acetates,** m.p. 104^0 and 105^0, *mono-, di-* and *tri-benzoates,* m.p. 196^0, 126^0 and 174^0, respectively (*Bredereck* and *H. Heckh, ibid.,* 1958, **91,** 1314).

(*iv*) *Halogenation and halogenophloroglucinols*

Chlorination of phloroglucinol in chloroform produces hexachlorocyclohexane-

1,3,5-trione, which can be reduced to **trichlorophloroglucinol,** m.p. 134⁰. Ring-opened chlorinated products can also be formed, *e.g.*, dichloroacetic acid and tetrachloro-acetone from chlorination in aqueous solution. Direct bromination in non-aqueous solvents gives analogous products; in aqueous solution, a penta-bromocyclohexenoldione was isolated. *Tribromophloroglucinol,* m.p. 152–153⁰ (*T. Zincke* and *O. Kegel, ibid.*, 1889, **22,** 1467; *A. Sonn* and *K. Winzer, ibid.*, 1928, **61,** 2303; *A. W. Francis* and *A. J. Hill,* J. Amer. chem. Soc., 1924, **46,** 2503). Sulphuryl chloride is also used to prepare trichlorophloroglucinol.

Chloro-, dichloro- and *trichloro-*1,3,5-*trimethoxybenzenes,* m.p. 93⁰, 129⁰ and 130⁰, respectively (*G. Lloyd* and *W. B. Whalley,* J. chem. Soc., 1956, 3209; *P. Bartolotti,* Gazz., 1897, **27,** 289).

Bromo-, dibromo- and *tribromo-*1,3,5-*trimethoxybenzenes,* m.p. 97⁰, 132⁰ and 145⁰, respectively (*H. Leuchs,* Ann., 1928, **460,** 1; *W. Will,* Ber., 1888, **21,** 602).

(*v*) *Nitrophloroglucinols*

Direct nitration of phloroglucinol gives the **nitro-, dinitro-** and **trinitro-phloroglucinols,** m.p. 183–185⁰, 205⁰ and 165⁰, respectively (*S. V. Dubiel* and *S. Zuffanti,* J. org. Chem., 1954, **19,** 1359). Trinitrophloroglucinol is also prepared by hydrolysis of trifluoro-1,3,5-trinitrobenzene (*G. C. Shaw* and *D. L. Seaton, ibid.*, 1961, **26,** 5227) and pentanitroaniline (*B. Flürscheim* and *E. L. Holmes,* J. chem. Soc., 1928, 3041) or by oxidation of trinitrosophloroglucinol by nitric acid (*K. Freudenberg, H. Fikentscher* and *W. Wenner,* Ann., 1925, 442, 309).

The i.r. spectra of nitrophloroglucinols indicate extensive chelation between the hydroxyl and nitro groups.

2-*Nitrophloroglucinol* 3- and 5-*methyl ether,* m.p. 187⁰ and 153⁰, respectively; 1,3- and 1,5-*dimethyl ethers,* m.p. 165⁰ and 131⁰; *trimethyl ether,* m.p. 152⁰ (*E. M. Kampouris,* J. chem. Soc., C, 1967, 2568).

2,4-*Dinitro-* and 2,4,6-*trinitro-phloroglucinol trimethyl ethers,* m.p. 164⁰ and 76⁰, respectively (*Dubiel* and *Zuffanti, loc. cit.*).

(*vi*) *Aminophloroglucinols*

These are produced by reduction of the nitro-compounds, *e.g.* 2-nitrophloroglucinol 1,3-dimethyl ether is reduced by dithionite to **2-aminophloroglucinol 1,3-dimethyl ether,** m.p. 184⁰ (*Kampouris,* J. chem. Soc., C, 1968, 2125).

(*vii*) *Homologues of phloroglucinol*

Methyl homologues have been prepared by reduction followed by hydrolyses of the appropriate trinitro- and aminonitro-benzenes, decarboxylation of methyl-trihydroxybenzoic acids, direct synthesis from the requisite aliphatic compounds and reduction of phloroglucinaldehydes. The last-mentioned are conveniently prepared by formylation of phloroglucinols (*Robertson* and *W. B. Whalley,* J. chem. Soc., 1951, 3355 and references cited therein).

2-Methylphloroglucinol, m.p. 214⁰; 3-*mono-*, 3,5-*di-* and *tri-methyl ethers,* m.p. 117⁰, 66⁰ and 27⁰, respectively. 2,4-*Dimethyl-* and 2,4,6-*trimethyl-phloroglucinols,* m.p. 162⁰ and 187⁰, respectively.

n-Alkyl derivatives are best prepared by Clemmenson reduction of acyl-derivatives (*T. Kariyone* and *I. Inagaki*, J. pharm. Soc. Japan, 1949, **69**, 431). 2-*Ethyl*-, 2-n-*propyl*-, 2-n-*butyl*-, and 2-n-*pentyl-phloroglucinol*, m.p. 185°, 171°, 101°, and 97°, respectively.

(d) Tetra-, penta- and hexa-hydroxybenzenes

1,2,3,4-**Tetrahydroxybenzene, apionol,** m.p. 161°, *tetra-acetate*, m.p. 142°, is obtained by hydrolysis of 4-aminopyrogallol hydrochloride (*Einhorn, Cobliner* and *Pfeiffer, loc. cit.*) or of 2,3-dimethoxyhydroquinone (*H. A. Anderson* and *R. H. Thomson*, J. chem. Soc., C, 1967, 2152). It gives a blue colour with ferric chloride and its alkaline solutions do not absorb oxygen. It is soluble in water and ethanol but only sparingly so in benzene.

Acidifying an alkaline solution of apionol at low temperatures leads to the oxo form, IX, which reverts to apionol in concentrated acid or on heating. No equilibrium exists, however (*W. Mayer* and *R. Weiss*, Angew. Chem., 1956, **68**, 680). *E. M. Terry* and *N. A. Milas* (J. Amer. chem. Soc., 1926, **48**, 2647) claimed that an oxo form was produced by oxidising hydroquinone with sodium chlorate and osmium tetroxide, but the product is a tricyclic "dimer" of 2,3-dihydro-2,3-dihydroxy-*p*-benzoquinone (*Anderson* and *Thomson, loc. cit.*).

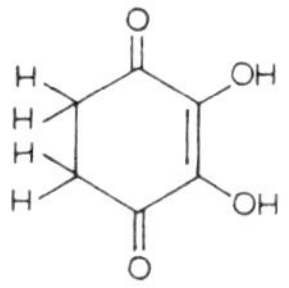

(IX)

Nitro and carboxylic acid derivatives are obtained on a preparative scale by γ-irradiation of an aqueous solution of the corresponding pyrogallols, especially if oxygen or hydrogen peroxide is present (*F. Merger* and *D. Graesslin*, G.P., 1,228,258/1966; Angew. Chem., 1964, **3**, 640). Aromatisation of inoses gives 1,2,3,4-*tetra-acetoxybenzene*, m.p. 106° (*T. Posternak* and *J. Deshusses*, Helv., 1961, 44, 2088).

1,3,4-**Trihydroxy-2-methoxybenzene**, m.p. 101° (*E. Spath et al.*, Ber., 1937, **70**, 1672); 2,3,4-*Trihydroxy*-1-*methoxybenzene*, m.p. 116° (*idem, ibid.*, 1938, **71**, 1831).

1,2-*Dihydroxy*-3,4-*dimethoxybenzene*, b.p. 160–170°/20 mm, *diacetate*, m.p. 85°, is prepared by the Dakin oxidation of galloacetophenone-3,4-dimethyl ether (*W. Baker, E. H. T. Dukes* and *C. A. Subrahmanyam*, J. chem. Soc., 1934, 1671). 1,4-*Dihydroxy*-2,3-*dimethoxybenzene*, 2,3-*dimethoxyhydroquinone*, m.p. 84°, *diacetate*, m.p. 54°, by oxidation of pyrogallol 1,2-dimethyl ether with persulphate or Fremy's salt and subsequent reduction and by decarboxylation of 2,5-dihydroxy-3,4-dimethoxybenzoic acid (*Baker* and *R. I. Savage, ibid.*, 1938, 1602; *F. Weygand, H. Weber* and *E. Maekawa*, Ber., 1957, **90**, 1879). 2,3-*Dihydroxy*-1,4-*dimethoxy*-

benzene, m.p. 105^0 (*G. Ciamician* and *P. Silber*, *ibid.*, 1889, **22**, 119, 2481). 1-*Hydroxy*-2,3,4-*trimethoxybenzene*, (*Spath et al.*, *ibid.*, 1940, **73**, 795). 1,2,3,4-*Tetramethoxybenzene*, m.p. 88·5^0 (*Einhorn, Cobliner* and *Pfeiffer*, *loc. cit.*).

1,2,3,5-**Tetrahydroxybenzene,** m.p. 168^0, is generally reported to be unstable and sensitive to light. However, *R. A. Baxter* and *J. P. Brown* (Chem. and Ind., 1967, 1171) who prepared it by the hydrogenation of 2,6-dibenzyloxy-*p*-benzoquinone, report that it does not darken on keeping in air for months. Other modes of preparation include the acid hydrolysis of 2,4,6-triaminophenol (*M. Nierenstein*, J. chem. Soc., 1917, **111**, 4) and, nitration of hydroquinone diacetate to 3,6-dinitrohydroquinone monoacetate followed by reduction and hydrolysis (*G. Zemplen* and *J. Schawartz*, C.A., 1955, **49**, 2350).

It is soluble in water and alcohol and insoluble in chloroform and benzene. In aqueous solution, it gives a red colour with ferric chloride.

1,2,3,5-**Tetrahydroxybenzene 2-methyl ether, iretol,** m.p. 186^0, is formed by decomposition of iregenin by aqueous alkali at 100^0 (*G. de Laire* and *F. Tiemann*, Ber., 1893, **26**, 2010) or by the reduction of 2,4,6-trinitroanisole and subsequent hydrolysis of the triamino compound (*E. Kohner*, Monatsh., 1899, **20**, 927). 2,5-*Dihydroxy*-1,3-*dimethoxybenzene*, m.p. 160^0, is obtained from 2,6-dimethoxy-1,4-benzoquinone. 1,5-*Dihydroxy*-2,3-*dimethoxybenzene*, m.p. 116^0, is produced by reducing 3,5-dinitroveratrole and subsequent hydrolysis (*E. Chapman, A. G. Perkin* and *R. Robinson*, J. chem. Soc., 1927, 3015). 1,3-*Dihydroxy*-2,5-*dimethoxybenzene*, anhydrous m.p. 86^0, $2H_2O$ m.p. 60^0 is synthesised from 2-nitrophloroglucinol 1,3-dibenzenesulphonate, and is used to prepare 2,3,5-*trimethoxyphenol*, m.p. 55^0 (*E. M. Kampouris*, J. chem. Soc., C, 1968, 2125). 3,4,5-*Trimethoxyphenol*, m.p. 146^0; 1,2,3,5-*tetramethoxybenzene*, m.p. 47^0 (*Chapman, Perkin* and *Robinson*, *loc. cit.*). 1,2,3,5-*Tetra-acetoxybenzene*, m.p. 106^0.

For amino and nitro derivatives, see *G. Zemplen, L. Mester* and *C. Szantay* (C.A., 1955, **49**, 12352). 2,3,4,6-Tetrahydroxybenzoic acid is formed by treatment of 1,2,3,5-tetrahydroxybenzene with potassium bicarbonate and carbon dioxide (*Nierenstein*, *loc. cit.*).

Tetrahydroxy-m-*xylene*, m.p. 189^0, is obtained from dihydroxy-*m*-xyloquinone (*H. Brunmayr*, Monatsh., 1900, **21**, 1), *tetra-acetate*, m.p. 161^0, from 2-hydroxy-3,5-dimethyl-*p*-benzoquinone and acetic anhydride (*Corbett*, J. chem. Soc., C, 1967, 2408).

1,2,4,5-**Tetrahydroxybenzene,** m.p. 232^0 with some decomposition at 200^0, is produced by reduction of 2,5-dihydroxy-*p*-benzoquinone by stannous chloride and hydrochloric acid or by catalytic hydrogenation. 2,5-Dihydroxy-*p*-benzoquinone is readily obtained by oxidation of hydroquinone by hydrogen peroxide in concentrated alkali (*R. Nietzki* and *F. Schmidt*, Ber., 1888, **21**, 2374; *W. K. Anslow* and *H. Raistrick*, J. chem. Soc., 1939, 1446). Poor yields have been obtained from the reaction of molten alkali-metals and carbon monoxide (*W. Buechner*, Ber., 1965, **98**, 3118; *Buechner* and *E. A. C. Lucken*, Helv., 1964, **47**, 2113); the semiquinone radical-anion was detected during this reaction.

It is soluble in water, ethanol, and ether. Its aqueous solution darkens on exposure to air. Oxidation by ferric chloride or autoxidation in alkaline solution produces 2,5-dihydroxy-*p*-benzoquinone. Halogeno and amino derivatives are prepared by reduction of the corresponding *p*-benzoquinones.

1,2,4-*Trihydroxy-5-methoxybenzene*, m.p. 133^0 (*Anslow* and *Raistrick, loc. cit.*) and 1,4-*dihydroxy-2,5-dimethoxybenzene*, m.p. 168^0 (*R. Scholl* and *P. Dahll*, Ber., 1924, **57**, 80) are obtained from the corresponding *p*-benzoquinones. 2,4,5-*Trimethoxyphenol*, m.p. 59^0, is obtained from 2,4,5-trimethoxybenzaldehyde (*Spath*, Ber., 1940, **73**, 795). 1,2,4,5-*Tetramethoxybenzene*, m.p. 101^0. 1,2,4,5-**Tetra-acetoxybenzene,** m.p. 228^0 (*A. H. Crosby* and *R. E. Lutz*, J. Amer. chem. Soc., 1956, **78**, 1233).

Pentahydroxybenzene is obtained by hydrolysing diaminopyrogallol in boiling water (*A. Einhorn, J. Cobliner* and *H. Pfeiffer*, Ber., 1904, **37**, 100). The *penta-acetate*, m.p. 165^0, is produced by the usual acetylation methods or in one step from *myo*-inositol (*A. J. Fatiadi*, J. chem. Eng. Data, 1969, **14**, 18).

Pentahydroxybenzene gives a red-brown colour with ferric chloride. Its ethers are prepared using the Dakin oxidation of the appropriate acetophenones, (1,2-*dihydroxy*-3,4,6- and -3,4,5-*trimethoxybenzenes*, m.p. 82^0 and 90^0, respectively) and by reduction of *p*-benzoquinones, (1,2,4-*trihydroxy*-3,6- and -5,6-*dimethoxybenzenes*, m.p. 144^0 and 157^0, respectively) (*Baker*, J. chem. Soc., 1941, 662; *Anslow* and *Raistrick, loc. cit.*). Further methylation produces *pentamethoxybenzene*, m.p. 58^0.

Hexahydroxybenzene is prepared by reduction of tetrahydroxy-*p*-benzoquinone, formed by aeration of a solution of glyoxal and bisulphite in aqueous sodium carbonate (*A. Fatiadi* and *W. F. Sager*, Org. Synth., 1962, **42**, 66, 90). Salts of hexahydroxybenzene are produced from carbon monoxide and alkali metals (*E. Nietzki* and *T. Benckiser*, Ber., 1885, **18**, 1833). Yields of 80% are obtained using temperatures of $\sim 300^0$ with either potassium (*W. Buechner* and *E. Weiss*, Helv., 1964, **47**, 1415) or sodium (*H. C. Miller*, U.S.P., 2,858,194/1958). Addition of acid gives the free phenol. Other preparations begin from hydroquinone, inositol and triquinol (*R. C. Anderson* and *E. S. Wallis*, J. Amer. chem. Soc., 1948, **70**, 2931).

Hexahydroxybenzene melts above 310^0 and is a white crystalline solid when freshly prepared. It is sparingly soluble in water, ethanol, and benzene. Its solutions rapidly become red-violet on exposure to air and turn violet with ferric chloride. In sodium carbonate solution, autoxidation gives tetrahydroxy-*p*-benzoquinone, while nitric acid oxidises it to triquinonyl. Potassium crotonate is obtained by evaporating an aqueous solution with potassium hydroxide or carbonate. Hydrogenation using a platinum or nickel catalyst produces a complex mixture of inositols and quercitols and other hydroxycyclohexanes (*S. J. Angyal* and *D. J. McHugh*, J. chem. Soc., 1957, 3682; *Anderson* and *Wallis, loc. cit.*). Phloroglucinol is obtained, when platinum is used as the catalyst at 55^0 (*R. Kuhn, G. Quadbeck* and *E. Rohm*, Ann., 1949, **565**, 1).

Hexa-esters, such as the *acetate*, m.p. 203^0, are prepared by standard means (*I. E. Neifert* and *E. Bartow*, J. Amer. chem. Soc., 1943, **65**, 1770); *hexamethoxybenzene*, m.p. 81^0 (*Anderson* and *Wallis, loc. cit.*).

Chapter 5

Mononuclear hydrocarbons carrying substituents attached through sulphur: Thiophenols, Sulphides, etc.

A. R. FORRESTER and J. L. WARDELL

1. Arenethiols or thiophenols

These are formally the sulphur analogues of the phenols, an SH group replacing the OH group of the phenols. Like the phenols they are acidic and form alkyl and aryl sulphides (thio-ethers), but are much more sensitive to oxidation. The SH group has less tendency than the OH group to undergo association (*M. J. Copley, C. S. Marvel* and *E. Ginsberg*, J. Amer. chem. Soc., 1939, **61**, 3161; *B. D. N. Rao et al.*, Canad. J. Chem., 1962, **40**, 963). Thiophenol* has a lower boiling point than phenol; the thiocresols have about the same boiling points as the cresols while the higher thiophenols boil at higher temperatures than the corresponding phenols.

A general review of thiophenols and aryl sulphides is given by *E. E. Reid*, in "Organic Chemistry of Bivalent Sulphur", Vols. I and II, Chemical Publishing Co., New York, 1958. Other relevant information is provided by *C. C. Price* and *S. Oae*, in "Sulphur Bonding", Ronald Press, New York, 1962, and *W. A. Pryor*, in "Mechanisms of Sulphur Reactions", McGraw-Hill, New York, 1962.

(a) Methods of preparation

(1) By vigorous reduction of arenesulphonyl chlorides (p. 449).

This is a useful method of preparation for thiophenols having no other

* Whilst still allowing the trivial name "thiophenol", the IUPAC Rules require that such names shall apply only to relatively simple compounds; the alternative terms "arenethiol" or "mercaptoarene" are preferred.

reducible group in the nucleus. Arenesulphinic acids are similarly reduced. Arenesulphonamides are reduced to thiophenols by heating with concentrated hydriodic acid and phosphonium iodide (*E. Fischer*, Ber., 1915, **48**, 93), zinc and hydrochloric acid, stannous chloride, or sodium and isoamyl alcohol (*D. Klamann* and *G. Hofbauer, ibid.*, 1953, **86**, 1246).

(2) From diaryl disulphides by reduction:

$$\text{ArS}\cdot\text{SAr} \xrightarrow{2\text{H}} 2\ \text{ArSH}$$

A wide range of reducing agents is available for this reaction. These include glucose and aqueous alkali (*M. Claasz, ibid.*, 1912, **45**, 2424); zinc and aqueous acids (*C. Vogt*, Ann., 1861, **119**, 142); electrolytic reduction in aqueous alcoholic alkali (*F. Taboury*, Ann. Chim., 1908, [viii], **15**, 49); triphenylphosphine in aqueous methanol (*R. E. Humphrey* and *J. M. Hawkins*, Anal. Chem., 1964, **36**, 1812); lithium aluminium hydride (*R. C. Arnold, A. P. Lien* and *R. M. Alm*, J. Amer. chem. Soc., 1950, **72**, 731); sodium in xylene (*H. Lecher*, Ber., 1915, **48**, 524) and catalytic hydrogenation (*Farbenfabriken Bayer A.-G.*, F. P. Appl., 2,008,330 and 2,008,331/1970).

(3) From phenols: The reagents which can bring about this conversion are: (a) thiocarbamoyl chlorides, $R_2N\overset{\overset{S}{\|}}{C}Cl$ (the best of these reagents) (*M. S. Newman* and *H. A. Karnes*, J. org. Chem., 1966, **31**, 3980):

$$\text{ArOH} \rightarrow \text{ArOCSNR}_2 \xrightarrow{\Delta} \text{ArSCONR}_2 \xrightarrow[\text{(b) H}_3\text{O}^\oplus]{\text{(a) OH}^\ominus} \text{ArSH}$$

(b) phosphorus pentasulphide (*E. O. Beckmann*, J. pr. Chem., 1878, **17**, 439); (c) hydrogen sulphide at 400–600° over a metal oxide catalyst (*S. A. Ballard* and *D. E. Winkler*, U.S.P., 2,438,838/1948).

(4) By hydrolysis of an *O*-alkyl *S*-aryl xanthate (*R. Leuchart*, J. pr. Chem., 1890, [ii], **41**, 187; *D. S. Tarbell* and *D. K. Fukushima*, Org. Synth., Coll. Vol. III, 1963, p. 809):

$$\text{ArS}\cdot\text{CS}\cdot\text{OAlk} \xrightarrow{\text{KOH}} \text{ArSK} \xrightarrow{\text{H}_3\text{O}^\oplus} \text{ArSH}$$

(The *O*-alkyl *S*-aryl xanthates are formed by the reaction of a diazo compound and potassium xanthate.)

(5) By reaction between sulphur and a Grignard reagent (*Taboury*, Compt. rend., 1904, **138**, 982; *M. S. Kharasch* and *D. Reinmuth*, "Grignard Reactions of Non-metallic Substances", Constable and Co., London, 1954, 1954, p. 1274):

$$\text{ArMgBr} + \text{S} \rightarrow \text{ArSMgBr} \xrightarrow{\text{H}_3\text{O}^\oplus} \text{ArSH}$$

or an aryl-lithium (*H. Gilman* and *L. Fullhart*, J. Amer. chem. Soc., 1949, 74, 1478):

$$ArLi \xrightarrow{S} ArSLi \xrightarrow{H_3O^{\oplus}} ArSH$$

Aryl-lithiums also react with epithioalkanes to give on hydrolysis thiophenols (*F. G. Bordwell, H. M. Andersen* and *B. M. Pitt, ibid.*, 1954, 76, 1082):

$$ArLi + \underset{\diagdown S \diagup}{CH_2{-}CH_2} \rightarrow ArSLi + CH_2{=}CH_2$$

(6) By hydrolysis of an aryl 2,4-dinitrophenyl sulphide.

The aryl 2,4-dinitrophenyl sulphide can be prepared from 2,4-dinitrobenzenesulphenyl chloride and an arene in the presence of aluminium chloride at low temperature:

$$2,4\text{-}(NO_2)_2C_6H_3SCl + C_6H_5CH_3 \xrightarrow{AlCl_3} 2,4\text{-}(NO_2)_2C_6H_3{-}S{-}C_6H_4CH_3$$

Hydrolysis is by methanolic potassium hydroxide (*N. Kharasch* and *R. Swidler*, J. org. Chem., 1954, 19, 1704):

$$2,4\text{-}(NO_2)_2C_6H_3{-}S{-}C_6H_4Me \xrightarrow[MeOH]{KOH} 2,4\text{-}(NO_2)_2C_6H_3OMe + MeC_6H_4S^-K^+$$

(7) Reactive aryl halides, such as 2,4-dinitrochlorobenzene give the corresponding thiols by the methods used for the alkyl halides (Vol. IB, p. 74).

(b) Properties and reactions

Thiophenols are generally very unpleasant-smelling compounds. *Thiophenol* itself is a mobile liquid, b.p. 169°, $d_4^{20°}$ 1·078, $n_D^{20°}$ 1·5888 (*I. N. Tits-Skvortsova et al.*, J. gen. Chem. USSR, 1952, 22, 135). Other thiophenols are listed in Table 1.

Methods for the determination of thiophenols have been reviewed (*F. Pellerin*, Bull. Soc. chim. Fr., 1962, 2319).

(i) Spectra

The u.v. absorption spectra of thiophenols suggest a first order conjugation between the unshared electron pairs of the sulphur atom and the aromatic ring.

In iso-octane solution, thiophenol has a well-defined maximum at 235 nm and a region of relatively weak absorption from 265 to 295 nm showing fine structure (*R. C. Passerini*, in "Organic Sulphur Compounds", Vol. I, Ed. *N. Kharasch*, Pergamon, Oxford, 1961).

The SH stretching absorption appears as a sharp, easily recognised band in the 2600–2550 cm^{-1} region. The intensity is between a tenth and a twentieth of that of the OH stretch in phenols. In solvents, such as pyridine, which can strongly hydrogen bond with the thiophenol, shifts of about 80 cm^{-1} in the SH stretching frequency occur (*A. Wagner, H. J. Becher* and *K. G. Kottenhahn*, Ber., 1956, **89**, 1708).

TABLE 1

THIOPHENOLS

Compound	M.p.°	B.p.°	Compound	B.p. °/mm
o-thiocresol[1]	15	188	2,3-thioxylenol[6]	132·2/50
m-thiocresol[2]		200	2,4-thioxylenol[6]	127·0/50
p-thiocresol[3]	43	195	2,5-thioxylenol[6]	126·3/50
o-ethylthiophenol[4]		210	2,6-thioxylenol[6]	122·0/50
p-ethylthiophenol[5]		91/12 mm	3,4-thioxylenol[6]	132·5/50
p-(*tert*-butyl)thiophenol[7]		120/20 mm	3,5-thioxylenol[6]	127·5/50

References

1 *R. Leuckart*, J. pr. Chem., 1890, [ii], **41**, 179.
2 *D. S. Tarbell* and *D. K. Fukushima*, Org. Synth., Coll. Vol. III, 1963, p. 809.
3 *L. Field* and *F. A. Grunwald*, J. org. Chem., 1951, **16**, 946.
4 *R. Fricke* and *G. Spilker*, Ber., 1925, **58**, 24.
5 *J. Pollak, J. von Fiedler* and *H. Pott*, Monatsh., 1918, **39**, 179.
6 *E. A. Bartkus, E. B. Hotelling* and *M. B. Neuworth*, J. org. Chem., 1957, **22**, 1185.
7 *Idem, ibid.*, 1960, **25**, 232.

Other spectral studies of thiophenols have been made; u.v. and i.r. spectra (*S. I. Miller* and *G. S. Krishnamurthy*, J. org. Chem., 1962, **27**, 645; *J. G. David* and *H. E. Hallum*, Trans. Faraday Soc., 1964, **60**, 2013); mass spectra (*S-O. Lawesson, J. O. Madsen* and *G. Schroll*, Acta Chem. Scand., 1966, **20**, 2325); proton magnetic resonance spectra (*S. H. Marcus* and *Miller*, J. phys. Chem., 1964, **68**, 331).

(*ii*) *Acidity*

Thiophenols are more strongly acidic than the corresponding phenols. They can be titrated in alcohol with alkali (*P. Klason* and *T. Carlson*, Ber., 1906, **39**, 741) and they readily form salts, including heavy metal covalent arenethiolates*. These heavy metal "salts" are their most distinctive derivatives and

* The terms "*mercaptan*" and "*mercaptide*", for the thiophenol and its metal derivatives, respectively, are no longer acceptable in IUPAC nomenclature.

are formed rapidly on mixing solutions of the thiophenols and heavy metal ions, especially in the presence of a base. They are generally insoluble in aqueous media. The most extensively studied are the lead, mercury, zinc and copper compounds. The mercuric salt, $(PhS)_2Hg$, obtained by shaking mercuric oxide with thiophenol in pyridine (*Lecher, ibid.*, 1915, **48**, 1425; 1920, **53**, 575), decomposes on heating into mercury and diphenyl disulphide, whereas the lead salt $(PhS)_2Pb$, decomposes on heating to give lead sulphide and diphenyl sulphide (*R. Otto, ibid.*, 1880, **13**, 1289). Aryl chloromercury sulphides, *e.g.* PhSHgCl, and aryl alkyl-(aryl)mercury sulphides, *e.g.* PhSHgR, are also known (see *E. E. Reid*, "Organic Chemistry of Bivalent Sulphur", Vol. I, Chemical Publishing Co., New York, N.Y., 1958, p. 141).

TABLE 2

ACIDITIES OF THIOPHENOLS

p*K*a in 48% methanol at 25°		p*K*a in water at 25°	
p-$NO_2C_6H_4SH$[1]	5·11	*p*-$NO_2C_6H_4OH$[2]	7·16
C_6H_5SH[1]	7·76	C_6H_5OH[2]	9·98
p-$CH_3 \cdot C_6H_4SH$[1]	8·03	*p*-$CH_3 \cdot C_6H_4OH$[2]	10·25

References

1 *F. G. Bordwell* and *H. M. Andersen*, J. Amer. chem. Soc., 1953, **75**, 6019.
2 *Bordwell* and *G. D. Cooper, ibid.*, 1952, **74**, 1058.

(*iii*) *Oxidation of thiophenols*

Thiophenols are very readily oxidised by a large variety of reagents. The products depend on the conditions used and include disulphides, sulphonic acids and thianthrenes.

Thiophenols are oxidised to *diaryl disulphides*, ArS·SAr, by air and oxygen especially in the presence of aqueous ammonia (*Leuckart*, J. pr. Chem., 1890, [ii], **41**, 179; see also *J. D. Hopton, C. J. Swan* and *D. L. Trimm*, "Oxidation of Organic Compounds", Vol. I, American Chemical Society, 1968, p. 216). Other reagents (and methods) available, include dimethyl sulphoxide (*T. J. Wallace*, J. Amer. chem. Soc., 1964, **86**, 2018), ferric salts (*T. Zincke* and *W. Frohneburg*, Ber., 1910, **43**, 837; *Wallace*, J. org. Chem., 1966, **31**, 3071), transition metal oxides (*Wallace*, J. org. Chem., 1966, **31**, 1217), X-ray radiation (*M. S. Simonidze* and *E. M. Nanobashvili*, C.A., 1965, **63**, 4198), lead tetra-acetate (*L. Field* and *J. E. Lawson*, J. Amer. chem. Soc., 1958, **80**, 838), thiol oxidase from *Piricularia oxyzae* (*S. M. Bocks*, Biochem. J.,

1966, 98, 9C) and iodine in alkaline solution (*A. I. Vogel*, "A Textbook of Practical Organic Chemistry", Longmans, London, 1959).

The products of oxidation by chlorine and bromine depend on the solvent used; in dry carbon tetrachloride the disulphide (*Zincke* and *Frohneburg*, *loc. cit.*), or the *sulphenyl halide*, ArSX (*Lecher et al.*, Ber., 1925, **58**, 409; *F. Kumar* and *J. R. Powell*, Org. Synth., Coll. Vol. IV, 1963, p. 934) is formed; in dry acetic acid, the *sulphonyl halide*, $ArSO_2X$ (*Zincke* and *Frohneburg*, *loc. cit.*), or, in moist acetic acid and in aqueous solution, the *sulphonic acid* results. Ring halogenation may also occur (*E. Gebauer-Fülnegg*, J. Amer. chem. Soc., 1927, **49**, 2270).

Other oxidants which give products beyond the disulphide stage include nitric acid, permanganate and hydrogen peroxide (*Reid*, *loc. cit.*, p. 118). High yields of sulphonic acids are also produced from thiophenols with oxygen in potassium hydroxide, hexamethylphosphoramide or dimethylformamide solutions (*Wallace* and *A. Schriesheim*, Tetrahedron, 1965, **21**, 2271).

Oxidation of a thiophenol to a *thianthrene* (I) can be effected in concentrated sulphuric acid solution (*K. Fries* and *W. Volk*, Ber., 1909, **42**, 1170) or in the vapour phase by passage over a silica-alumina catalyst at 300°C (*Tits-Skvortsova et al.*, Zhur. obshchei Khim., 1953, **23**, 303).

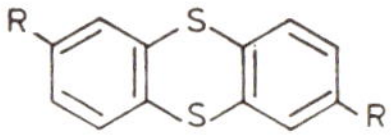

R S S R

(I)

(*iv*) *Desulphurisation*

Thiophenols like many other sulphur-containing compounds can be desulphurised by heating with Raney nickel (*G. R. Pettit* and *E. E. van Tamelen*, Org. Reactions, 1952, **12**, 356; *H. Hauptmann* and *W. G. Walter*, Chem. Reviews, 1962, **62**, 347):

$$ArSH \rightarrow ArH$$

(*v*) *Substitution*

As a rule the sulphur atom of thiophenols must be protected prior to carrying out electrophilic substitutions in the benzene ring. A good protecting group for this purpose is the carboxymethyl group which can be conveniently removed after the substitution has taken place by hydrogen peroxide in boiling mineral acid. Bromination, chlorination and acylation have been achieved in the following manner (*D. Walker* and *J. Leib*, J. org. Chem., 1962, **27**, 4455; 1963, **28**, 3077):

$$PhSH + ClCH_2{\cdot}CO_2H \xrightarrow{\text{base}} PhSCH_2{\cdot}CO_2H$$

$$PhSCH_2{\cdot}CO_2H + X^{\oplus} \xrightarrow{-H^{\oplus}} XC_6H_4SCH_2{\cdot}CO_2H$$

$$XC_6H_4SCH_2{\cdot}CO_2H \xrightarrow{H_2O_2/H^{\oplus}} XC_6H_4SH$$

C-Alkylation with olefins in the presence of boron trifluoride or aluminium trichloride has been reported (*E. A. Bartkus, E. B. Hotelling* and *M. B. Neuworth, ibid.*, 1960, **25**, 232; *K. L. Kreuz*, U.S.P., 2,753,378/1956):

$$(CH_3)_2C{=}CH_2 + PhSH \xrightarrow{BF_3/\Delta} p\text{-}(t\text{-}C_4H_9)C_6H_4SH$$

Alkylation can also be brought about by the disproportionation of an alkyl aryl sulphide with the parent thiophenol in the presence of boron trifluoride (*Bartkus, Hotelling* and *Neuworth, loc. cit.*).

(*vi*) *Condensations*

Thiophenols condense with aldehydes and ketones to give mono- and di-thioacetals. The products are more stable and are formed more quickly than the corresponding hemi-acetals and acetals (*E. Campaigne*, in "Organic Sulphur Compounds", Ed. *N. Kharasch*, Pergamon Press, Oxford, Vol. I, 1961). Thiophenols combine additively with hydrogen cyanide in the presence of hydrogen chloride to give *S*-arylthiolformimidates, ArSCH=NH (*W. Authenrieth* and *A. Brüning*, Ber., 1903, **36**, 3464); nitriles behave similarly. Thiophenols and isocyanides form thioformimidates, ArSCH:NR (*T. Saegusa, S. Kobayashi* and *Y. Ito*, J. org. Chem., 1970, **35**, 2119).

(*vii*) *Thioester formation*

With acyl chlorides and acid anhydrides (with a catalytic amount of triethylamine), *S*-aryl esters of carbothioic acids, ArSCOR, are formed (*W. Michler*, Ann., 1875, **176**, 177; *R. Schiller* and *Otto*, Ber., 1876, **9**, 1635; *A. A. Schleppink*, U.S.P., 3,402,194/1968). *S*-Aryl carbothioic esters are also formed by reaction with carbon monoxide in the presence of cobalt carbonyl (*H. E. Holmqvist* and *J. E. Carnahan*, J. org. Chem., 1960, **25**, 2240):

$$2\,ArSH + CO \xrightarrow[CO_2(CO)_8]{1000\text{ atm., }250^0} ArCOSAr$$

Further reaction of carbothioic esters with thiophenols in the presence of boron trifluoride gives trithio-orthocarboxylates, *e.g.*, (*Tarbell* and *A. H. Herz*, J. Amer. chem. Soc., 1953, **75**, 1668):

$$MeCOSPh + 2\ HSPh \rightarrow MeC(SPh)_3$$

(viii) Other reactions

With Grignard reagents, arylthiomagnesium halides are formed (*Gilman* and *W. B. King*, J. Amer. chem. Soc., 1925, 47, 1136).

Many reactive halogeno compounds react with thiophenols. Thus phosgene and thiophosgene give *S,S'*-diaryl dithiocarbonates, $(ArS)_2CO$, and diaryl trithiocarbonates, $(ArS)_2CS$ (*W. R. Waldron* and *Reid*, *ibid.*, 1923, 45, 2399; *F. Runge, Z. El-Hewehi* and *E. Taeger*, J. pr. Chem., 1959 [iv], 7, 279); or acid chlorides of the types ArSCOCl and ArSCSCl; sulphenyl halides give trisulphides, ArS·S·SAr (*K. Tsutsui*, Japan. P., 10,321/1961); thionyl chloride gives ArS·SO·SAr (*L. Field* and *W. B. Lacefield*, J. org. Chem., 1966, 31, 3555); oxalyl chloride gives ArSCO·COCl, which can cyclise to thianaphthenequinones (*D. Paps, E. Schwenk* and *H. F. Ginsberg*, *ibid.*, 1949, 14, 723; *O. Bothner Machinenfabrik.*, G.P., 291,793/1914).

2. Sulphides: thioethers

(a) Methods of preparation

*(i) Alkyl aryl sulphides or thioethers**

(1) Thiophenols form sulphides by reaction of their alkali metal salts with alkyl halides or dialkyl sulphates (*R. Leuckart*, J. pr. Chem., 1890, [ii], 41, 179):

$$ArSNa + RI \rightarrow ArSR + NaI$$

The alkyl esters of arenesulphonic acids can be similarly used (*H. Gilman* and *N. J. Beaber*, J. Amer. chem. Soc., 1925, 47, 1449; *D. A. Shirley* and *W. H. Reedy*, *ibid.*, 1951, 73, 4885).

(2) Alkyl aryl sulphides are very readily obtained by addition of thiophenols to olefins (*E. N. Prilezhaeva* and *M. F. Shostakovskii*, Russ. Chem. Reviews, 1963, 32, 399; *K. Griesbaum*, Angew. Chem., intern. Edn., 1970, 9, 273). The addition proceeds readily under homolytic conditions, with air, light or peroxide as catalyst, to give the anti-Markownikov adduct (*M. S. Kharasch, A. T. Read* and *F. R. Mayo*, Chem. and Ind., 1938, 752):

$$RCH{=}CH_2 + ArSH \rightarrow RCH_2{\cdot}CH_2SAr$$

or *via* a heterolytic route with acid catalysts (boron trifluoride, sulphuric

* The term "thioether" is no longer recommended in IUPAC rules.

acid, etc.) to produce the Markownikov product, RCH(SAr)Me (*V. N. Ipatieff, H. Pines* and *B. S. Friedman*, J. Amer. chem. Soc., 1938, **60**, 2731). α,β-Unsaturated acids, aldehydes, ketones, nitriles and nitro compounds react with thiophenols particularly readily under basic conditions (*B. H. Nicolet, ibid.*, 1931, **53**, 3066; 1935, **57**, 1098; *R. N. Ross, ibid.*, 1949, **71**, 3458):

$$ArSH + RCH{=}CHX \rightarrow RCH(SAr){\cdot}CH_2X$$

Thiophenols also add to acetylenes and dienes (*Shostakovskii, A. V. Bogdanova* and *G. I. Plotnikova*, Russ. Chem. Reviews, 1964, **33**, 66; *A. A. Oswald* and *K. Griesbaum*, in "Organosulphur Compounds", Ed. *N. Kharasch*, Vol. II, Pergamon Press, Oxford, 1966):

$$HC{\equiv}COEt + ArSH \rightarrow cis\text{-}ArSCH{=}CHOEt$$

$$CH_2{=}CH{-}CH{=}CH_2 + ArSH \rightarrow ArSCH_2{-}CH{=}CH{-}CH_3$$

Co-oxidation of a thiophenol with an acetylene in the presence of air occurs (*K. Griesbaum, A. A. Oswald* and *B. E. Hudson*, J. Amer. chem. Soc., 1963, **85**, 1969):

$$PhSH + PhC{\equiv}CH \xrightarrow[-75^{\circ}]{O_2} PhS{-}CH{=}C(Ph)OOH \xrightarrow{-10^{\circ}} PhSCH(OH){\cdot}COPh$$

(3) Treatment of an *O*-alkyl *S*-aryl xanthate with triethylamine at room temperature gives a sulphide (*H. Yoshida, S. Inokawa* and *I. Ogata*, Bull. chem. Soc. Japan, 1966, **39**, 411):

$$ArS{\cdot}CS{\cdot}OR \xrightarrow{Et_3N} ArSR + COS$$

(4) Benzyne and dialkyl sulphides yield alkyl phenyl sulphides (*H. Hellmann* and *D. Eberle*, Ann., 1963, **662**, 188). At low temperatures the intermediate ylide II can be trapped:

(ii) Other substituted alkyl aryl sulphides

(1) *β-Hydroxyalkyl sulphides* are formed from epoxyalkanes and thiophenols (*R. D. Schuetz*, J. Amer. chem. Soc., 1951, **73**, 1881; *J. P. Danehy* and *C. J. Noel, ibid.*, 1960, **82**, 2511):

$$ArSH + CH_2\text{—}CH_2 \text{ (with O bridge)} \rightarrow ArSCH_2{\cdot}CH_2OH$$

(2) *Aryl β-halogenoalkyl sulphides* are obtained from sulphenyl halides and olefins (see p. 481).

(3) *Aryl thioethers of α- and β-mercaptocarboxylic acids* may be prepared from thiophenols and the appropriate halogeno acids (*F. Arndt*, Ber., 1923, **56**, 1269; *O. Behaghel*, J. pr. Chem., 1926, [ii], **114**, 287; *A. Bistrzycki* and *J. Risi*, Helv., 1925, **8**, 582):

$$ArSH + ClCH_2{\cdot}CO_2H \xrightarrow{\text{Base}} ArSCH_2{\cdot}CO_2H$$

(iii) Diaryl sulphides

(1) Since aromatic halides, except in special cases such as the halogenonitrobenzenes, are unreactive toward nucleophiles, diaryl sulphides are best prepared in ways other than the simple reaction of a thiophenol with an aryl halide. However, the copper-catalysed reaction between a thiophenol and an aryl iodide is useful (*M. T. Bogert* and *M. R. Mandelbaum*, J. Amer. chem. Soc., 1923, **45**, 3045) especially using solvents such as dimethylformamide and dimethylacetamide (*J. R. Campbell*, J. org. Chem., 1964, **29**, 1830; *R. G. R. Bacon* and *H. A. O. Hill.*, J. chem. Soc., 1964, 1108).

(2) Another method is the copper catalysed decomposition of an aryldiazothiophenol (*J. H. Ziegler*, Ber., 1890, **23**, 2469; *G. E. Hilbert* and *T. B. Johnson*, J. Amer. chem. Soc., 1929, **51**, 1526):

$$ArN_2SAr' \xrightarrow{\text{Cu}} ArSAr'$$

(3) The Lewis acid catalysed reaction between sulphenyl halides and aromatic hydrocarbons is also used (*N. Kharasch* and *R. Swidler*, J. org. Chem., 1954, **19**, 1704; *C. M. Buess* and *Kharasch*, J. Amer. chem. Soc., 1950, **72**, 3529):

$$ArSX + Ar'H \xrightarrow{AlCl_3} ArSAr'$$

(b) Properties and reactions

(i) Oxidation

Alkyl aryl and diaryl sulphides, are oxidised successively to sulphoxides, ArSOR, and sulphones, $ArSO_2R$. At moderate temperatures, hydrogen peroxide and hydroperoxides normally yield only sulphoxides whereas peracids readily produce sulphones (*D. Barnard*, *L. Bateman* and *J. I. Cunneen*, in "Organic Sulphur Compounds", Ed. *N. Kharasch*, Vol. I, Pergamon Press, Oxford, 1961).

(ii) Reactions with electrophiles

(1) *Halogens.* Either oxidation or electrophilic substitution can occur. Bromine and chlorine add initially to sulphides to form dibromides and dichlorides, *e.g.*, $Ph_2S:Cl_2$, which are hydrolysed by water to give sulphoxides (*K. Fries* and *W. Vogt*, Ann., 1911, **381**, 337; *H. Szmant*, in "Organic Sulphur Compounds", Vol. 1, p. 154). The crystal structure of bis-(*p*-chlorophenyl)-sulphur di-chloride, $(p\text{-ClC}_6H_4)_2S:Cl_2$, has been determined (*N. C. Baeniger et al.*, J. Amer. chem. Soc., 1969, **91**, 5749). For the mechanism of the oxidation, of alkyl aryl sulphides by bromine in aqueous methanol, see *U. Miotti, G. Modena* and *L. Sedea*, J. chem. Soc., B, 1970, 802.

The halogen can also migrate into the ring, *e.g.*:

$$PhMeS:Br_2 \rightarrow p\text{-BrC}_6H_4SMe + HBr$$

(*E. Bourgeois* and *A. Abraham*, Rev. Trav. chim., 1911, **30**, 414). Whereas bromination of alkyl aryl sulphides in acetic acid (*S. Clemanti* and *P. Linda*, Tetrahedron, 1970, **26**, 2869) and in trifluoroacetic acid (*H. M. Gilow, R. B. Camp* and *E. C. Clifton*, J. org. Chem., 1968, **33**, 230) gives ring-substituted products, chlorination in acetic acid gives oxidised products, including sulphonyl chlorides, and chloroalkyl products (*Gilow, Camp* and *Clifton, loc. cit.*; *S. W. Lee* and *G. Dougherty*, *ibid.*, 1940, **5**, 81).

(2) *Other electrophiles.* Acetylation of aryl sulphides proceeds readily to give predominately *para*-substituted products (*Clemanti* and *Linda*, *loc. cit.*). Mercuration is inhibited by the complexation of the sulphur atom with the mercurating agent. Nitric acid gives oxidised products (*Gilow* and *G. L. Walker*, Tetrahedron Letters, 1965, 4295).

(iii) Other reactions

Alkyl aryl sulphides, such as methyl phenyl sulphide, unlike anisole, are metalated in the alkyl group, by butyl-lithium (*Gilman et al.*, J. Amer. chem. Soc., 1940, **62**, 987; 1953, **75**, 3760):

$$PhSMe + BuLi \rightarrow PhSCH_2Li \xrightarrow[\text{(b) } H_2O]{\text{(a) } CO_2} PhSCH_2.CO_2H$$

$$(PhOMe + BuLi \rightarrow o\text{-MeOC}_6H_4Li \xrightarrow[\text{(b) } H_2O]{\text{(a) } CO_2} o\text{-MeOC}_6H_4\cdot CO_2H)$$

A further difference is shown by the cleavage of anisole to methyl halide and phenol by hydrobromic or hydriodic acids under which conditions (130° for 2 hours) thioanisole is stable (*C. K. Hughes* and *E. O. P. Thompson*, J. proc. Roy. Soc., N.S. Wales, 1959, **83**, 269). Thioethers are cleaved by alkali metals (*R. Gerdil* and *E. A. C. Lucken*, J. chem. Soc., 1963, 2857):

$$PhSMe + 2\ K \rightarrow MeK + PhSK$$

A sulphonium salt is formed by the addition of an alkyl halide to a sulphide (see, *E. E. Reid*, "Organic Chemistry of Bivalent Sulphur", Vol. **II**, Chemical Publishing Co., N.Y., 1960).

Alkyl aryl sulphides form coordination compounds with covalent metal halides, *e.g.*, palladium and aluminium chlorides. The dihalide adducts mentioned above are related.

The Thio-Claisen rearrangement of allyl phenyl sulphide in quinoline gives 2-methyl-1-thiacoumaran (III) and 1-thiachroman (IV). The intermediacy of *o*-allylthiophenol has been confirmed (*H. Kwart et al.*, Chem. Comm., 1969, 44; J. org. Chem., 1967, **32**, 3135; 1966, **31**, 413; *C. Y. Meyers, C. Rinaldi* and *L. Bonoli*, *ibid.*, 1963, **28**, 2440):

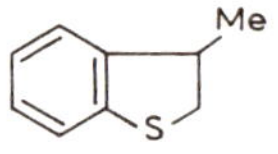

Me
S
(III) 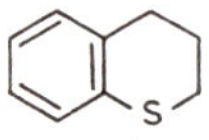
S
(IV)

(c) *Individual sulphides, sulphoxides and sulphones*

(*i*) *Alkyl aryl sulphides, sulphoxides and sulphones*

The following **alkyl aryl sulphides** are formed by standard means.

	B.p.°	n_D	*d*
MeSPh[1]	187–190	1·5869/20°	1·058/20°
EtSPh[2]	204	1·5662/22·5°	1·024/15°
n-PrSPh[3]	219	1·5571/20°	0·9995/20°
n-BuSPh[3]	94/4 mm	1·5463/20°	0·9852/20°
p-$MeSC_6H_4Me$[1]	209	1·573/20°	1·026/20°

References

1 *K. Brand* and *K. W. Kranz*, J. pr. Chem., 1927, [ii], **115**, 143.
2 *F. Taboury*, Ann. chim. phys., 1908, [8], **15**, 5.
3 *V. N. Ipatieff, H. Pines* and *B. S. Friedman*, J. Amer. chem. Soc., 1938, **60**, 2731.

See, *Reid, loc. cit.*, p. 113, for other sulphides.

Methyl phenyl sulphoxide, PhSOMe, m.p. 33°, 264°: 104°/7 mm, is obtained by the controlled oxidation of methyl phenyl sulphide by periodate (*C. R. Johnson* and *J. E. Kaiser*, Org. Synth., 1966, **46**, 78), ozone (*L. Horner, H. S. Schaefer* and *W. Ludwig*, Ber., 1958, **91**, 75) and hydrogen peroxide in glacial acetic acid (*C. C. Price* and *J. J. Hydock*, J. Amer. chem. Soc., 1952, **74**, 1943).

Like all sulphoxides, it can be further oxidised to the sulphone or reduced to the sulphide (*H. H. Szmant*, in "Organic Sulphur Compounds", Ed. *N. Kharasch*, Vol. I, Pergamon Press, Oxford, 1961).

Optically active sulphoxides are formed from the reaction of Grignard reagents with optically active menthyl arene- or alkane-sulphinates (*K. K. Andersen*, J. org. Chem., 1964, **29**, 1953):

O, S, Me, O-menthyl —(a) PhMgBr, (b) H_2O→ O, S, Ph, Me

Rates of racemisation and oxygen-exchange of optically active sulphoxides have been reported (*S. Oae* and *M. Kise*, Bull. Soc. chim. Japan, 1970, **43**, 1416).

Methyl phenyl sulphoxide reacts with active methylene compounds, such as acetylacetone in acetic anhydride at 100°, to give methylphenylsulphonium diacetylmethylide:

$$\text{MeSOPh} + \text{H}_2\text{C(COMe)}_2 \xrightarrow{(\text{MeCO})_2\text{O}} \text{MePhS}^{\oplus}\text{—O}^{\ominus}\text{(COMe)}_2$$

Methyl sulphoxides react with acetic anhydride to give acetoxymethyl sulphides, which on hydrolysis form thiolphenols (Pummerer reaction) (*Kise* and *Oae*, *ibid.*, p. 1426 and refs. cited therein):

$$\text{PhSOMe} + (\text{CH}_3\cdot\text{CO})_2\text{O} \rightleftharpoons \text{PhSCH}_2\text{OCOMe} + \text{CH}_3\cdot\text{CO}_2\text{H}$$

$$\text{PhSCH}_2\text{OCOMe} \rightarrow \text{PhSH} + \text{CH}_2\text{O} + \text{CH}_3\cdot\text{CO}_2\text{H}$$

Ethyl phenyl sulphoxide, PhSOEt, b.p. 68°/0·05 mm, is prepared by oxidation of ethyl phenyl sulphide at —78° with *tert*-butyl hypochlorite (*L. Skattebol*, *B. Boulette* and *S. Soloman*, *ibid.*, 1967, **32**, 3111; see also *Johnson* and *Kaiser*, *loc. cit.*).

Methyl phenyl sulphone, $PhSO_2Me$, m.p. 88°, is formed by reaction of methanesulphonyl chloride with benzene and aluminium chloride (*W. E. Truce* and *C. W. Vriesen*, J. Amer. chem. Soc., 1953, **75**, 5032; *F. R. Jensen* and *G. Goldman*, in "Friedel-Crafts and Related Reactions", Ed. *G. A. Olah*, Vol. III, pt. II, John Wiley and Sons, N.Y., 1964), or by oxidation of the sulphide, *e.g.* by ozone (*H. Bohme* and *H. Fischer*, Ber., 1942, **75**, 1310).

The methylene protons in methyl phenyl sulphone are activated (*L. Field*, J. Amer. chem. Soc., 1952, **74**, 3919):

$$\text{EtMgBr} + \text{PhSO}_2\text{Me} \rightarrow \text{PhSO}_2\text{CH}_2\text{MgBr} + \text{C}_2\text{H}_6$$

$$\text{PhSO}_2\text{CH}_2\text{MgBr} \xrightarrow{\text{PhCHO}} \text{PhSO}_2\text{CH}_2\cdot\text{CH(OH)Ph}$$

Sulphones are resistant to reduction.

(*ii*) *Diaryl sulphides, sulphoxides and sulphones*

Diphenyl sulphide, b.p. 296^0:162^0/18 mm, is prepared by reaction of benzene with sulphur monochloride, in the presence of aluminium chloride (*W. W. Hartman, L. A. Smith* and *J. B. Dickey,* Org. Synth., Coll. Vol. II, 1966, p. 242). It is also formed (1) by the copper-catalysed decomposition of benzene diazothiophenol (*Ziegler, loc. cit.*; *Hilbert* and *Johnson, loc. cit.*) and (2) by heating sodium benzenethiolate with iodobenzene at 240^0 (*F. Mauthner,* Ber., 1906, **39,** 3593).

Diphenyl sulphoxide, Ph_2SO, m.p. 70^0, is formed by gently boiling diphenyl sulphide with dilute nitric acid (*d* 1·1) (*F. Krafft* and *R. E. Lyons, ibid.*, 1896, **29,** 435). Other oxidising agents which have been used are hydrogen peroxide in acetic acid, perbenzoic acid, chromic oxide in acetic acid and permanganate (*O. Hinsberg, ibid.*, 1890, **43,** 289; *J. Böeseken* and *H. I. Waterman,* Rec. Trav. chim., 1909, **29,** 321). It is also formed by the action of thionyl chloride on benzene in the presence of aluminium chloride (*R. L. Shriner, H. C. Struck* and *W. J. Jorison,* J. Amer. chem. Soc., 1930, **52,** 2060).

Diphenyl sulphone, Ph_2SO_2, m.p. 128^0, is produced by the reaction of benzene with a preformed benzenesulphonyl chloride-aluminium trichloride complex at room temperature (*G. Holt* and *B. Pagdin,* J. chem. Soc., 1960, 2508):

$$AlCl_3 + PhSO_2Cl \rightarrow PhSO_2Cl{:}AlCl_3$$

$$PhH + PhSO_2Cl{:}AlCl_3 \rightarrow Ph_2SO_2{:}AlCl_3 + HCl$$

Polyphosphoric acid has been used to catalyse the reaction between benzenesulphonic acid and benzene (*B. M. Graybill,* J. org. Chem., 1967, **32,** 2931) (see also, *Jensen* and *Goldman, loc. cit.*). Oxidation of diphenyl sulphide or diphenyl sulphoxide with nitric acid, perbenzoic acid or permanganate is also used (*Barnard, Bateman* and *Cunneen, loc. cit.*).

Nitration gives the 3,3′-dinitrophenyl sulphone (*J. Martinet* and *A. Haedl,* Compt. rend., 1921, **173,** 775) and chlorosulphonic acid gives the 3-mono- and 3,3′-di-sulphonic acids (*R. Otto,* Ber., 1886, **19,** 2417).

Diphenyl disulphide, Ph_2S_2, m.p. 62^0, is formed by mild oxidation of thiophenol (p. 425). It is cleaved fairly easily by reduction or other means into thiophenol or its salts (p. 422). Chlorine brings about fission into two molecules of benzenesulphenyl chloride, PhSCl, and other products (*T. Zincke,* Ann., 1914, **406,** 103; *H. Lecher,* Ber., 1925, **58,** 409). Diphenyl disulphide is very readily cleaved by nucleophiles, *e.g.*, (*R. Schiller* and *Otto,* Ber., 1876, **9,** 1637).

$$2\,Ph_2S_2 + 4\,OH^{\ominus} \rightarrow 3\,PhS^{\ominus} + PhSO_2^{\ominus} + 2\,H_2O$$

Diphenyl disulphide can be readily oxidised, *e.g.*, hot nitric acid gives benzenesulphonic acid (*C. Vogt,* Ann., 1861, **119,** 142; for other oxidants, see *Reid,*

"Organic Chemistry of Bivalent Sulphur, Vol. III, Chemical Publishing Co., 1960, p. 374). Oxidation need not break the S–S bond. The following types of compounds are known: thiolsulphinate esters, ArSO·SAr (*H. J. Backer* and *H. Kloosterziji*, Rec. Trav. chim., 1954, **73**, 129), thiolsulphonate esters, $ArSO_2$·SAr (*J. Cymerman* and *J. B. Willis*, J. chem. Soc., 1951, 1332), sulphinylsulphonate esters, $ArSO_2$·SOAr (*H. Bredereck et al.*, Ber., 1960, **93**, 2736), disulphones, $ArSO_2$·SO_2Ar (*T. P. Hilditch*, J. chem. Soc., 1908, **93**, 1524).

(*iii*) *Substituted thiophenols and their derivatives*

(1) *Halogenothiophenols.* Hydrolysis of xanthates, formed from diazotised aromatic amines, has been widely used to produce halogenoarenethiols. o-**Chlorothiophenol,** ClC_6H_4SH, b.p. 205–206°, m-*chlorothiophenol*, b.p. 205–207°, and p-*chlorothiophenol*, m.p. 54° (*G. Daccomo, ibid.*, 1892, **62**, 307), o-*bromothiophenol*, b.p. 117°/18 mm, and m-*bromothiophenol*, b.p. 119°/20 mm (*H. F. Wilson* and *D. S. Tarbell*, J. Amer. chem. Soc., 1950, **72**, 5200) have all been prepared by this method. Reduction of the appropriate arenesulphonyl chloride has been used for p-*bromothiophenol*, m.p. 74° (*Wilson* and *Tarbell, loc. cit.*). p-*Fluorothiophenol*, b.p. 162°: 60°/15 mm, is obtained from *p*-fluorophenylmagnesium bromide and sulphur (*M. Seyhan*, Ber., 1939, **72**, 594) and also by reduction of the sulphonyl chloride; similarly for p-*iodothiophenol*, m.p. 85° (*M. Rajsner, V. Seidlova* and *M. Protiva*, C.A., 1963, **59**, 2773).

To illustrate the methods of preparation of halogen-substituted sulphides, sulphoxides and disulphides, some derivatives of *p*-chlorothiophenol are chosen as examples.

p-**Chlorophenyl methyl sulphide,** ClC_6H_4SMe, b.p. 170°: 108°/12 mm, is obtained from *p*-chlorothiophenol and dimethyl sulphate in alkaline solution (*G. Kresze, E. Ropte* and *B. Schrader*, Spectrochim. Acta, 1965, **21**, 1633).

p-**Chlorophenyl methyl sulphoxide,** ClC_6H_4SOMe, m.p. 46–48°, is formed by oxidation of *p*-chlorophenyl methyl sulphide either by *tert*-butyl hypochlorite at —70° (*K. K. Andersen et al.*, J. org. Chem., 1966, 31, 2859) or by periodate (*Kresze, Ropte* and *Schrader, loc. cit.*).

p-**Chlorophenyl methyl sulphone,** $ClC_6H_4SO_2Me$, m.p. 97°, is prepared by oxidation of the corresponding sulphide in acetic acid (*idem*) and by methylation of sodium *p*-chlorobenzenesulphinate with dimethyl sulphate (*C. J. Miller* and *S. Smiles*, J. chem. Soc., 1925, **127**, 224).

Di-(p-chlorophenyl) sulphide, $ClC_6H_4SC_6H_4Cl$, m.p. 93°, is formed by the chlorination of diphenyl sulphide with chlorine (*Krafft*, Ber., 1874, **7**, 1164) or by reduction of di-(*p*-chlorophenyl) sulphoxide with zinc and acetic acid (*C. W. N. Cumper, J. F. Read* and *A. I. Vogel*, J. chem. Soc., 1965, 5860).

Di-(p-chlorophenyl) sulphoxide, $ClC_6H_4SOC_6H_4Cl$, m.p. 143°, is prepared from chlorobenzene and thionyl chloride in the presence of aluminium chloride at 0°C (*G. C. Hampson, H. Farmer* and *L. E. Sutton*, Proc. roy. Soc., 1933, A, **143**, 146; *Cumper, Read* and *Vogel, loc. cit.*).

Di-(p-chlorophenyl) sulphone, $ClC_6H_4SO_2C_6H_4Cl$, m.p. 148°, is obtained by oxidation of the corresponding sulphide with permanganate (*Cumper, Read* and *Vogel, loc. cit.*). Chlorobenzene and concentrated sulphuric acid (*H. Heymann* and *L. F. Fieser,* J. Amer. chem. Soc., 1945, **67,** 1979) or an excess of diethyl sulphate and sulphur trioxide (*M. J. Keogh* and *A. K. Ingbermann,* U.S.P., 3,415,887/1968) also produce this compound.

Di-(p-chlorophenyl) disulphide, $ClC_6H_4S \cdot SC_6H_4Cl$, m.p. 73°, is produced by oxidation of *p*-chlorothiophenol by dimethyl sulphoxide (*T. J. Wallace,* J. Amer. chem. Soc., 1964, **86,** 2018) or by iodine in alkaline solution (*Vogel,* "A Textbook of Practical Organic Chemistry", Longmans, London, 1959).

(2) *Nitrothiophenols.* o-**Nitrothiophenol,** $NO_2C_6H_4SH$, m.p. 61°, is formed by the action of sodium sulphide on *o*-chloronitrobenzene, followed by addition of acid (*H. H. Hodgson* and *J. H. Wilson,* J. chem. Soc., 1925, **127,** 440). p-*Nitrothiophenol,* m.p. 75–76°, is produced similarly or by alkaline hydrolysis of the disulphide, formed from *p*-chloronitrobenzene and sodium disulphide (*C. C. Price* and *G. W. Stacy,* J. Amer. chem. Soc., 1946, **68,** 498). m-*Nitrothiophenol* is obtained by the xanthate route from *m*-nitroaniline (*F. G. Bordwell* and *H. M. Anderson, ibid.,* 1953, **75,** 6019). All three isomers are prepared from the corresponding phenols by the thiocarbamyl chloride reaction (p. 422).

2,4-**Dinitrothiophenol,** $(NO_2)_2C_6H_3SH$, m.p. 131°, is obtained by heating 2,4-dinitrochlorobenzene with thiourea or by decomposition of *S*-(2,4-dinitrophenyl)-isothiouronium chloride with alkali (*J. Taylor* and *A. E. Dixon,* J. chem. Soc., 1924, **125,** 243; *M. Guia* and *A. Ruggeri,* Gazz., 1923, **53,** 341). The corresponding disulphides are formed by reaction of the appropriate chloronitrobenzene with sodium disulphide (*M. I. Bogert* and *A. Stull,* Org. Synth., Coll. Vol. I, 1964, p. 220; *Price* and *Stacy, loc. cit.*) or by reduction of the appropriate nitrobenzenesulphonyl chloride with hydriodic acid (*W. A. Sheppard,* Org. Synth., 1960, **40,** 80).

p-**Nitrophenyl methyl sulphide,** m.p. 71°; p-**nitrophenyl methyl sulphoxide,** m.p. 149°; p-**nitrophenyl methyl sulphone,** m.p. 141° (*Kresze, Ropte* and *Schrader, loc. cit.*).

Di-(p-nitrophenyl) sulphide, m.p. 160°, is obtained from *p*-chloronitrobenzene and potassium ethyl xanthate (*Price* and *Stacy,* Org. Synth., Coll. Vol. III, 1963, p. 667).

(3) *Aminothiophenols.* o-**Aminothiophenol,** $NH_2C_6H_4SH$, m.p. 28°, b.p. 234°, is prepared by the reduction of di-(*o*-nitrophenyl) disulphide with zinc and acetic acid (*M. I. Bogert* and *F. D. Snell,* J. Amer. chem. Soc., 1924, **46,** 1308) or by reduction of *o*-nitrobenzenesulphonyl chloride with tin and hydrochloric acid. It resembles the *o*-aminophenols and *o*-phenylenediamines in readily forming cyclic condensation products, *e.g.,* 2-phenylbenzothiazole from benzoyl chloride (*Bogert* and *Snell, loc. cit.*).:

$$PhCOCl + C_6H_4(NH_2)(SH) \rightarrow C_6H_4{<}^{N}_{S}{>}C\text{—}Ph$$

2-Phenylbenzothiazole

o-Aminothiophenols, in general, are prepared by hydrolysing the condensation products of aromatic amines and sulphur monochloride, (Herz reaction) (*R. Herz,* G.P., 360,690/1922; *W. K. Warburton,* Chem. Reviews, 1957, **57,** 1011):

$$p\text{-}ClC_6H_4NH_2 + S_2Cl_2 \longrightarrow \text{(Cl, N, S, C–Cl)} \xrightarrow{NaOH} \text{(Cl, } NH_2\text{, } S^-Na^+\text{)}$$

m-*Aminothiophenol,* b.p. 180°/16 mm (*T. Zincke* and *J. Müller,* Ber., 1913, **46,** 775). It is prepared by reduction of *m*-nitrobenzenesulphonyl chloride by zinc and hydrochloric acid (*A. M. Kuliev* and *A. N. Agaev,* Zhur. org. Khim, 1970, **6,** 809). p-*Aminothiophenol,* m.p. 46°, b.p. 143°/17 mm, is obtained by refluxing *p*-nitrochlorobenzene and an excess of sodium sulphide in water (*Gilman* and *G. C. Grainer,* J. Amer. chem. Soc., 1949, **71,** 1747) and also by the reduction of *p*-acetamidobenzenesulphonyl chloride with zinc and hydrochloric acid (*Zincke* and *P. Jörg,* Ber., 1909, **42,** 3362).

p-**Aminophenyl methyl sulphide,** $NH_2C_6H_4SMe$, b.p. 140°, (*K. Brand* and *K. W. Kranz,* J. pr. Chem., 1927, [ii], **115,** 143).

Di-(p-**aminophenyl**) **sulphide,** $NH_2C_6H_4SC_6H_4NH_2$, m.p. 108°, is produced by reduction of di-(*p*-nitrophenyl) sulphide (*R. Nietzki* and *H. Bothof,* Ber., 1894, **27,** 3261).

Di-(p-**aminophenyl**) **sulphoxide,** $NH_2C_6H_4SOC_6H_4NH_2$, m.p. 175°, is formed by oxidation of the sulphide with hydrogen peroxide (*M. Gazdar* and *Smiles,* J. chem. Soc., 1908, **93,** 1833).

Di-(p-**aminophenyl**) **sulphone,** $NH_2C_6H_4SO_2C_6H_4NH_2$, m.p. 176°, is obtained by reducing di-(*p*-nitrophenyl) sulphone with stannous chloride and hydrochloric acid (*H. G. Biswas,* Sci. Cult. Calcutta, 1962, **28,** 586; C.A., 1963, **59,** 3806) and by reduction of 4′-acetamido-4-nitrodiphenyl sulphone [prepared from *p*-acetamidobenzenesulphinic acid and *p*-nitrochlorobenzene (*C. W. Ferry, J. S. Buck* and *R. Baltzly,* Org. Synth., Coll. Vol. III, 1963, p. 239)]. This compound has bactericidal activity comparable with that of sulphanilamide. It is used in the treatment of leprosy and other bacterial infections.

Di-(p-**aminophenyl**) **disulphide,** $NH_2C_6H_4S\cdot SC_6H_4NH_2$, m.p. 75–76°, is prepared by oxidation of the thiol in alkaline solution by air (*Hinsberg,* Ber., 1905, **38,** 1130). A preparation starting from *p*-nitrochlorobenzene is given in Organic Syntheses (*Price* and *Stacy,* Coll. Vol. III, 1963, p. 86).

3. Thio analogues of polyhydric phenols

(a) Thio analogues of catechol

Monothiocatechol, o-*mercaptophenol,* *o*-HOC_6H_4SH, b.p. 217°, is produced from *o*-aminophenol by the diazo reaction (*D. Greenwood* and *H. A. Stevenson,* J. chem. Soc., 1953, 1514). The use of lithium aluminium hydride is recommended to decompose the intermediate xanthate (*C. Djerassi et al.,* J. Amer. chem. Soc.,

1955, **77**, 568). A poorer yield of the thiol is obtained by reducing 2,2′-dihydroxydiphenyl disulphide, a product of the reaction between sodium phenoxide and sulphur at 200° (*L. Haitinger*, Monatsh., 1883, **4**, 166; *C. Le Fevre* and *C. Degrez*, Compt. rend., 1934, **198**, 1432). It forms a series of cyclic compounds on reaction with ketones, aldehydes, phosgene and ethylene dibromide (*Greenwood* and *Stevenson, loc. cit.*).

Alkyl *o*-hydroxyphenyl sulphides, RSC_6H_4OH, and *o*-alkoxythiophenols, ROC_6H_4SH, are prepared from the diazonium salts of the corresponding amines (*H. S. Holt* and *E. E. Reid*, J. Amer. chem. Soc., 1924, **46**, 2329) by reduction of alkoxybenzenesulphonyl chlorides (*L. Gatterman*, Ber., 1897, **32**, 1136) and also by alkylation of *o*-mercaptophenol by alkyl halides in methanolic sodium hydroxide (*A. O. Padersen et al.*, Tetrahedron, 1970, **24**, 449).

o-*Hydroxyphenyl methyl sulphide*, b.p. 105°/22 mm; o-*methoxythiophenol*, b.p. 218–219°; o-*hydroxyphenyl methyl sulphoxide*, HOC_6H_4SOMe, m.p. 127°; o-*hydroxyphenyl methyl sulphone*, $HOC_6H_4SO_2Me$, m.p. 80–83° (for these and other alkyl *o*-hydroxyphenyl sulphides, sulphoxides and sulphones, see *Pedersen et al., loc. cit.*).

Monothiocatechol dimethyl ether, o-*methoxyphenyl methyl sulphide*, $MeOC_6H_4SMe$, b.p. 238° (*E. L. Holmes, C. K. Ingold* and *E. W. Ingold, ibid.*, 1926, 1684); o-*methoxyphenyl methyl sulphoxide*, $MeOC_6H_4SOMe$, m.p. 44°; o-*methoxyphenyl methyl sulphone*, $MeOC_6H_4SO_2Me$, m.p. 91° (*A. Pollard* and *R. Robinson, ibid.*, 1926, 3090).

Dithiocatechol. *Benzene*-1,2-*dithiol*, *o*-$(HS)_2C_6H_4$, m.p. 27–28°, b.p. 238°, is synthesised by dealkylation of *o*-bisalkylthiobenzenes with sodium in liquid ammonia (*R. Adams* and *A. Ferretti*, J. Amer. chem. Soc., 1959, **81**, 4927, 4939; *Ferretti*, Org. Synth., 1962, **42**, 54). The thiolethers are prepared from *o*-dibromobenzene and alkyl cuprous sulphides. This method of preparing benzenedithiol is superior to the reduction of benzenedisulphonyl chloride (*W. R. H. Hurtley* and *Smiles*, J. chem. Soc., 1926, 1821, 2263; 1927, 534). Alternatively, diazotisation of *o*-aminophenol in glacial acetic acid to give 1,2,3-benzothiadiazole, which is then decomposed by carbon disulphide, can be used (*S. Hunig* and *E. Fleckenstein*, Ann., 1970, **738**, 192). The dithiol also forms cyclic compounds by condensation with aldehydes and ketones.

(b) Thio analogues of resorcinol

Monothioresorcinol, m-*mercaptophenol*, m.p. 16–17°, b.p. 168°/35 mm, is obtained by reduction of *m*-phenolsulphonyl chloride (*L. von Szathmary*, Ber., 1910, **43**, 2485) or by the diazo reaction on *m*-aminophenol (*E. R. Watson* and *S. Dutt*, J. chem. Soc., 1922, **121**, 2414).

m-*Methoxythiophenol*, b.p. 224°, is obtained either from *m*-anisidine (*F. Mauthner*, Ber., 1906, **39**, 3593) or the corresponding sulphonyl chloride (*K. Fries* and *E. Engelbertz*, Ann., 1915, **407**, 194). m-*Hydroxyphenyl methyl sulphide*, m.p. 15°, b.p. 225°, *sulphone*, m.p. 84° (*T. Zincke* and *C. Ebel*, Ber., 1914, **47**, 923).

Dithioresorcinol, *benzene*-1,3-*dithiol*, m.p. 27°, b.p. 243°, is obtained from benzene-*m*-disulphonyl chloride (*T. Zincke* and *O. Kruger*, *ibid.*, 1912, **45**, 3468).

1,3-*Bismethylthiobenzene, dithioresorcinol dimethyl ether*, b.p. 149°/17 mm, is oxidised by hydrogen peroxide to two *bis-sulphoxides*, m.p. 102° and 147°, and a *bis-sulphone*, m.p. 196° (*E. V. Bell* and *G. M. Bennett*, J. chem. Soc., 1928, 3189; *Zincke* and *Kruger*, *loc. cit.*). Dithioresorcinol and bis-acyl chlorides of the type $(CH_2)_n(COCl)_2$ condense to give polymers (*G. B. Gechele et al.*, European Polymer J., 1966, **2**, 1).

(c) Thio analogues of hydroquinone

Monothiohydroquinone, p-*mercaptophenol*, m.p. 29°, b.p. 167°/45 mm, is prepared from *p*-aminophenol by the diazo reaction or by reduction of (i) phenol-*p*-sulphonyl chloride (*Zincke* and *Ebel*, 1914, **47**, 1100), (ii), *p*-thiocyanophenol by sodium in liquid ammonia (*R. J. Laufer*, U.S.P., 3,129,262/1964) and (iii) the products of the reaction of phenol and sulphur chloride (*E. B. Hotelling et al.*, J. org. Chem., 1959, **24**, 1598) or by reaction of *p*-chlorophenol and sodium hydrogen sulphide (*K. W. Palmer*, B.P., 381,237/1932).

p-*Methoxybenzenethiol*, b.p. 228° (*C. M. Suter* and *H. L. Hansen*, J. Amer. chem. Soc., 1934, **54**, 4100). p-*Hydroxyphenyl methyl sulphide*, m.p. 84°, b.p. 113°/6 mm (*Zincke* and *Ebel*, *loc. cit.*). p-*Anisyl methyl sulphide*, m.p. 25°, b.p. 99°/4 mm (*Suter* and *Hansen*, *loc. cit.*). p-*Hydroxylphenyl methyl sulphoxide* and *sulphone*, m.p. 90° and 94°, respectively (*Zincke* and *Ebel*, *loc. cit.*).

Benzene-1,4-dithiol is obtained by reducing *p*-benzenedisulphonyl chloride (*Gechele et al.*, *loc. cit.*), by the diazo reaction (*R. Leuckart*, J. pr. Chem., 1890, [ii], **41**, 205) or by alkaline hydrolysis of *p*-dithiocyanobenzene (*F. Challenger* and *A. T. Peters*, J. chem. Soc., 1928, 1364); *monomethyl thioether*, m.p. 29°, b.p. 116°/3 mm (*A. Buraway*, *J. P. Critchley* and *A. R. Thompson*, Tetrahedron, 1958, **4**, 403); *monosulphone*, $HSC_6H_4SO_2Me$, m.p. 66° (*F. G. Bordwell* and *H. M. Anderson*, J. Amer. chem. Soc., 1953, **75**, 6019); *dimethyl thioether*, m.p. 85°, on oxidation gives two *bis-sulphoxides*, m.p. 136° and 183°, and a *bis-sulphone*, m.p. 260° (*Zincke* and *W. Frohneberg*, Ber., 1909, **42**, 2721; *Bell* and *Bennett*, *loc. cit.*).

(d) Thio analogues of trihydric phenols

2-Mercaptohydroquinone, m.p. 118°, results from the reaction of zinc and hydrochloric acid on 2,5-dihydroxybenzenethiosulphonic acid, $C_6H_3(OH)_2S \cdot SO_2H$; it is oxidised by ferric chloride to dibenzoquinone-2,2′-*disulphide*, m.p. 178° (*W. Alcalay*, Helv., 1947, **30**, 578):

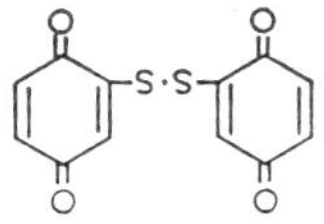

4-**Mercaptocatechol,** m.p. 57°, is prepared by condensation of *o*-benzoquinone and thiourea, followed by hydrolysis (*J. Daneke et al.*, Tetrahedron Letters, 1970, 271).

Monothiophloroglucinol, 5-*mercaptoresorcinol,* m.p. 88°, is produced by reduction of resorcinol-5-sulphonyl chloride, obtained from benzene-1,3,5-trisulphonic acid (*Suter* and *G. A. Harrington*, J. Amer. chem. Soc., 1937, **59**, 2575). 3,5-*Dihydroxyphenyl methyl sulphide,* m.p. 78°.

Trithiophloroglucinol, *benzene*-1,3,5-*trithiol,* m.p. 57–58°, is formed by reduction of benzene-1,3,5-trisulphonyl chloride (*J. Pollack* and *J. Carniol*, Ber., 1909, **42**, 3253); 1,3,5-*tris*(*methylthio*)*benzene,* m.p. 67°.

Chapter 6

Mononuclear hydrocarbons carrying nuclear substituents containing sulphur: Sulphonic Acids, Sulphinic Acids and Sulphenyl Compounds of the Benzene Series

D. R. HOGG

1. Nuclear sulphonic acids of benzene and its homologues; arenesulphonic acids

By contrast with the aliphatic hydrocarbons those of the aromatic series can be readily converted into sulphonic acids by the action of concentrated sulphuric acid and other sulphonating agents, nuclear hydrogen being replaced by the group SO_3H (cf. Vol. I B, p. 90). Sulphonic acids have played an important part in the development of organic chemistry, more especially in the naphthalene series. The sulphonic acids of aromatic amines and phenols provide essential intermediates for synthetic dyes, the alkali metal salts of long chain (10–15 carbon atoms) alkylbenzenesulphonic acids are the most widely used synthetic detergents and derivatives of sulphonic acids are used as drugs, sweetening agents, and pesticides. Aeruginosin B, 2-amino 6 carboxy-10-methyl-8-sulphophenazinium betaine, a red pigment

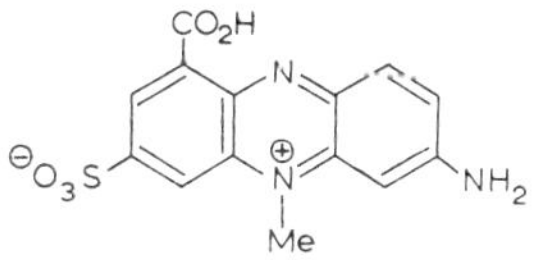

Aeruginosin

produced by *Pseudomonas aeruginosa*, is the only known example of an aromatic sulphonic acid arising from natural sources (*R. B. Herbert* and *F. G. Holliman*, Proc. chem. Soc., 1964, 19). The chemistry of aromatic sulphonic acids and derivatives up to 1954 has been reviewed (*F. Muth*, "Methoden der Organischen Chemie [Houben-Weyl]", Vol. 9, *E. Müller*, Ed., G. Thieme Verlag, Stuttgart, 1955, p. 435). The general methods of formation and general properties given below apply to most types of sulphonic acids, not only to those of the hydrocarbons.

(a) General methods of formation

The preparation of sulphonic acids has been reviewed (*E. E. Gilbert*, "Sulphonation and Related Reactions", Interscience, New York, 1965; Synthesis, 1969, 1, 3).

(i) Direct sulphonation

Arenesulphonic acids are prepared by treating the substance to be sulphonated with (1) concentrated sulphuric acid; (2) chlorosulphonic acid; (3) sulphur trioxide dissolved in sulphuric acid, dioxan, or, with more reactive substrates, in dimethylformamide or pyridine; or (4) with sulphur trioxide in the vapour phase.

The direct sulphonation of aromatic compounds is regarded as being an S_E2 reaction with monomeric sulphur trioxide, $HSO_3^{\oplus}$ or one of their solvated forms as the effective electrophilic species (*K. L. Nelson*, "Friedel-Crafts and related Reactions", Vol. III, Part 2, *G. A. Olah*, Ed., Interscience, New York, 1963, p. 1355; *H. Cerfontain*, "Mechanistic Aspects of Aromatic Sulfonation and Desulfonation", Interscience, New York, 1968; *Cerfontain* and *C. W. F. Kort*, Mechanism of reactions of sulphur compounds, 1968, 3, 23). In aqueous sulphuric acid solution the mechanism involving $H_3SO_4^{\oplus}$ is considered to predominate at low acid concentrations, reactions [1], [3] and [4], and the mechanism involving $H_2S_2O_7$ at higher acid concentrations, or with less reactive substrates, reactions [2], [3] and [4]:

$$ArH + H_3SO_4^{\oplus} \rightleftharpoons Ar^{\oplus}\begin{matrix}\diagup SO_3H\\ \diagdown H\end{matrix} + H_2O \qquad [1]$$

$$ArH + H_2S_2O_7 \rightleftharpoons Ar^{\oplus}\begin{matrix}\diagup SO_3H\\ \diagdown H\end{matrix} + HSO_4^{\ominus} \qquad [2]$$

$$Ar^{\oplus}\begin{matrix}\diagup SO_3H\\ \diagdown H\end{matrix} + HSO_4^{\ominus} \rightleftharpoons Ar^{\oplus}\begin{matrix}\diagup SO_3^{\ominus}\\ \diagdown H\end{matrix} + H_2SO_4 \qquad [3]$$

$$Ar^{\oplus}\begin{matrix}\diagup SO_3^{\ominus}\\ \diagdown H\end{matrix} + HSO_4^{\ominus} \rightleftharpoons ArSO_3^{\ominus} + H_2SO_4 \qquad [4]$$

In mildly fuming acid (100–103% H_2SO_4) the predominant electrophile is suggested to be $H_3S_2O_7^{\oplus}$ and the base to be H_2SO_4, while in more strongly fuming acid the species are $H_2S_4O_{13}$ and $HS_2O_7^{\ominus}$, respectively. At acid concentrations above 95% H_2SO_4 removal of the proton bound to the carbon atom becomes partially rate-limiting and reaction [5] is considered to be increasingly important:

$$Ar^{\oplus}\begin{matrix}\diagup SO_3H \\ \diagdown H\end{matrix} + \text{base} \rightleftharpoons ArSO_3H + H(\text{base})^{\oplus} \qquad [5]$$

(*A. J. Prinsen* and *Cerfontain,* Rec. Trav. chim., 1969, **88**, 833; and the preceding papers in this series). The overall reaction is reversible and with concentrated sulphuric acid the rate decreases as the concentration of water increases:

$$RH + H_2SO_4 \rightleftharpoons RSO_3H + H_2O$$

With benzene, under the usual conditions, equilibrium is reached at about 78% sulphuric acid. Diaryl sulphones are also formed particularly with chlorosulphonic acid (*J. I. Carr,* U.S.P. 2,000,061/1935, C.A., 1935, **29**, 4027) and vapour phase sulphonation (*P. M. Heertjes, H. C. A. van Beek* and *G. I. Grimmon,* Rec. Trav. chim., 1961, **80**, 82). Acetic acid may be used as a sulphone inhibitor (*W. H. C. Rueggeberg, T. W. Sauls* and *S. L. Norwood,* J. org. Chem., 1955, **20**, 455). Direct sulphonation of aromatic hydrocarbons and their halogen derivatives has been reviewed (*C. M. Suter* and *A. W. Weston,* Org. Reactions, 1946, **3**, 141; 406 references).

(*ii*) *Indirect sulphonation*

(1) Sulphonic acids are obtained by oxidation of thiophenols, disulphides or sulphinic acids with nitric acid, or with chlorine in aqueous acetic acid and subsequent hydrolysis of the resulting sulphonyl chloride (*E. Wertheim,* Org. Synth., Coll. Vol. II, 1943, p. 471). Sulphinic acids may also be oxidised with hydrogen peroxide or with permanganate ion (*W. E. Truce* and *J. F. Lyons,* J. Amer. chem. Soc. 1951, **73**, 126). Sulphoxides dissolved in hexamethylphosphoramide containing potassium *tert*-butoxide are oxidised to sulphonic acids with oxygen (*T. J. Wallace,* J. org. Chem., 1965, **30**, 4017):

$$PhSOCH_3 \xrightarrow{O_2} PhSO_3H$$

(2) Substitution of an activated halogen or nitro group by a sulphonic group may be effected with sodium sulphite in aqueous ethanol, *e.g.* *o*-chloro-

benzaldehyde gives benzaldehyde-*o*-sulphonic acid, 1-chloro-3,4-dinitrobenzene gives 5-chloro-2-nitrobenzene-sulphonic acid (*A. Laubenheimer*, Ber., 1882, **15**, 597) and 1-chloro-2,4-dinitrobenzene gives 2,4-dinitrobenzenesulphonic acid. Indirect sulphonation is of particular value for the preparation of compounds in which the orientation of the sulphonate group is not attainable by direct methods.

(b) General properties and reactions

The sulphonic acids are generally readily soluble in water from which they may crystallise with water of crystallisation. They are associated in relatively non-polar solvents *e.g.* benzene (*H. V. Kehiaian* and *C. D. Nenitzescu*, Ber., 1957, **90**, 685). The arenesulphonic acids are strong acids comparable with sulphuric acid (*L. P. Hammett* and *A. J. Deyrup*, J. Amer. chem. Soc., 1932, **54**, 4239; *R. J. Gillespie*, J. chem. Soc., 1950, 2542; *Cerfontain*, *loc. cit.* p. 186), and *p*-toluenesulphonic acid (tosic acid) is frequently used as an acid catalyst in reactions such as esterification, enol etherification, acetylation, enamine formation and dehydration. It is generally preferred to sulphuric acid, which has a similar reactivity, because it is less damaging to reactants and because it is a solid. The hydrocarbon sulphonic acids decompose on heating, but can be distilled unchanged under high vacuum (*F. Krafft* and *W. Wilke*, Ber., 1900, **33**, 3207). The sulphonic group is stable to reducing agents and to alkali except under drastic conditions. Sulphonic acids react with aromatic compounds on heating in polyphosphoric acid to give diaryl sulphones (*B. M. Graybill*, J. org. Chem., 1967, **32**, 2931). The sulphonic anhydride may be an intermediate:

$$RSO_3H + C_6H_6 \xrightarrow[\text{acid}]{\text{polyphosphoric}} RSO_2C_6H_5$$

Aromatic compounds containing electron-withdrawing substituents are very resistant to sulphone formation.

Sulphonic acids may be characterised by conversion to sulphonyl halides (p. 447) and thence to sulphonamides, by conversion to the *p*-bromophenacyl ester with *p*-bromo-α-diazoacetophenone (*T. Sumner*, *L. E. Ball* and *J. Platner*, *ibid.*, 1959, **24**, 2017), as the *S*-benzylisothiuronium salt (*E. Chambers* and *G. W. Watt*, *ibid.*, 1941, **6**, 376), or as the benzylamine salt (*G. W. H. Cheeseman* and *R. C. Poller*, Analyst, 1961, **86**, 256). The infrared spectra of anhydrous and hydrated sulphonic acids show the the following main bands with small deviations, ν_{OH} 2900, 2350, $\nu_{\text{a(SO}_2)}$ 1350, $\nu_{\text{s(SO}_2)}$ 1160, $\nu_{\text{(SO)}}$ 900 cm^{-1} (*S. Detoni* and *D. Hadži*, Spectrochim. Acta, 1957, Suppl. 601).

(i) Replacement of the sulphonyl group

(1) *By hydrogen.* Aromatic sulphonic acids are hydrolysed by heating in an aqueous medium in the presence of a mineral acid catalyst, generally sulphuric or phosphoric acid, the sulphonic groups being replaced by hydrogen. The reaction is the reverse of sulphonation:

$$RSO_3H + H_2O \rightleftharpoons RH + H_2SO_4$$

A frequently convenient procedure is to heat the sulphonic acid with diluted sulphuric acid in a current of steam (*R. C. Huston* and *M. M. Ballard*, Org. Synth., Coll. Vol. II, 1943, p. 97).

The ease of hydrolysis for the toluenesulphonic acids decreases in the order *ortho-para-meta* the respective velocity constants in 60 % sulphuric acid at 152° being 27, 13 and $0{\cdot}37 \times 10^{-5}$ sec^{-1}, respectively (*L. Vollbracht, Cerfontain* and *F. L. J. Sixma*, Rec. Trav. chim., 1961, **80**, 11). Since similar results have been reported for the amino-, hydroxy- and chloro-benzenesulphonic acids, whereas for carboxylic acid and sulphonic acid substituents the corresponding order is *ortho-meta-para*, it has been suggested that hydrolysis is facilitated by electron-donating substituents and by *ortho* substituents regardless of their electronic properties (*Gilbert, loc. cit.*, ch. 8, and references therein). The rates of hydrolysis of the isomeric sulphonic acids are of importance in determining the isomer distribution in direct sulphonation when the system begins to approach equilibrium.

The relatively easy removal of the sulphonic group by hydrolysis makes it a useful blocking group in aromatic synthesis (see *inter alia*, *H. P. Schultz*, Org. Synth., Coll. Vol. IV, 1963, p. 364). Sulphonation followed by hydrolysis has been used on a commercial scale for the separation of *m*-xylene (*H. P. Hetzner* and *R. J. Miller*, U.S.P., 2,511,711/1950; C.A. 1950, **44**, 8368d).

(2) *By hydroxyl.* On fusing an arenesulphonic acid salt with sodium or potassium hydroxide a phenol is produced. This reaction provides a convenient method for the introduction of the hydroxyl group into the aromatic nucleus:

$$PhSO_3Na + NaOH \rightarrow PhOH + Na_2SO_3$$

The monosulphonic acids give phenols (*S. W. Englund, R. S. Aries* and *D. F. Othmer*, Ind. Eng. Chem., 1953, **45**, 189). *m*-Benzenedisulphonic acid gives resorcinol on fusion with potassium hydroxide, but the *ortho*- and *para*-isomers give the appropriate phenolsulphonic acid together with a little phenol; no rearrangement is observed (*L. R. Buzbee*, J. org. Chem., 1966, **31**, 3289).

m-Phenolsulphonic acid is obtained by heating the disulphonic acid with 10 % aqueous sodium hydroxide at 250° (*F. Wilson* and *K. H. Meyer*, Ber., 1914, **47**, 3160). On fusion with potassium hydrogen sulphide thiophenols are produced (*O. Stadler*, *ibid.*, 1884, **17**, 2075) but the yields appear to be unsatisfactory.

(3) *By nitrile.* Heating sulphonic acid salts with potassium cyanide or ferrocyanide gives nitriles (*V. Merz* and *H. Muhlhauser*, *ibid.*, 1870, **3**, 709):

$$PhSO_3K + KCN \rightarrow PhCN$$

(4) *By carboxyl and by amino groups.* Heating the potassium salts of sulphonic acids with sodium formate gives arylcarboxylic acids in poor yield (*V. Meyer*, *ibid.*, 1870, **3**, 112; Ann., 1870, **156**, 265; *S. Coffey*, Rec. Trav. chim., 1923, **42**, 1029); similarly, by heating with sodamide an amine is obtained. In the latter case the addition of naphthalene improves the yield (*F. Sachs*, *ibid.*, 1906, **39**, 3006). For details of side-reactions and of the limitations of these reactions see *Suter*, "The Organic Chemistry of Sulphur", Wiley, London, 1944, p. 436.

Activated sulphonic groups may be replaced by amino groups by heating with ammonia in a sealed tube, *e.g.*, 2,4-dinitrobenzenesulphonic acid gives 2,4-dinitroaniline (*C. Willgerodt* and *P. Mohr*, J. pr. Chem., 1886, [2], **34**, 113).

(5) *By halogen.* Chlorination or bromination of sulphonic acids in aqueous solution frequently leads to the replacement of the sulphonic group by halogen.

p-Bromobenzenesulphonic acid gives *p*-bromochlorobenzene with chlorine in warm water (*R. L. Datta* and *J. C. Bhoumik*, J. Amer. chem. Soc., 1921, **43**, 303). The replacement is facilitated by electron-releasing groups situated *ortho* or *para* to the sulphonic group. The reaction of bromine with an aqueous solution of sodium 3,5-dibromo-4-hydroxybenzenesulphonate to give 2,4,6-tribromophenol is postulated to proceed through the cyclohexadienone intermediate shown below (*L. G. Cannell*, *ibid.*, 1957, **79**, 2927, 2932):

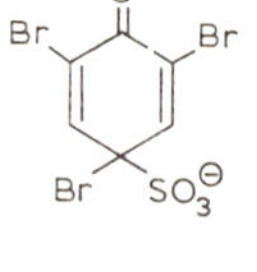

The sulphonic group may similarly be replaced by a nitro group during nitration (see *Suter*, *loc. cit.*, p. 406 *et seq.*).

(c) *Sulphonyl halides, $ArSO_2X$ ($X = F, Cl, Br$ or I)*

(*i*) *Preparation*

(1) Sulphonyl fluorides and chlorides can be prepared by reaction of the aromatic compound with an excess of fluoro- or chloro-sulphonic acid. The reaction involves two steps:

$$RH + ClSO_3H \rightarrow RSO_3H + HCl$$

$$RSO_3H + ClSO_3H \rightarrow RSO_2Cl + H_2SO_4$$

At higher temperatures more than one group may be introduced. Ester, amide, nitro, hydroxy and olefinic substituents are unaffected (*A. A. Spryskov* and *N. V. Apar'eva*, Zhur. obshchei Khim., 1950, **20**, 1818; C.A., 1951, **45**, 2434a).

(2) Sulphonyl fluorides are obtained by heating the corresponding chlorides under reflux with a saturated aqueous solution of potassium fluoride (*W. Davies* and *J. H. Dick*, J. chem. Soc., 1931, 2104) or by treating the sulphonic acid anhydrides with anhydrous hydrogen fluoride (*Olah* and *S. J. Kuhn*, J. org. Chem., 1962, **27**, 2667):

$$(RSO_2)_2O + HF \rightarrow RSO_2F + RSO_2OH$$

(3) Sulphonyl chlorides and bromides can be prepared by the oxidation of thiophenols, disulphides or sulphinic acids with chlorine or bromine in aqueous acetic acid solution.

(4) Sulphonyl chlorides are obtained by treatment of sulphonic acids or their sodium salts with phosphorus pentachloride or oxychloride (see *e.g.*, *R. Adams* and *C. S. Marvel*, Org. Synth., Coll. Vol. I, 1932, p. 84); or with thionyl chloride in dimethylformamide (*H. H. Bosshard et al.*, Helv., 1959, **42**, 1653). Phosphorus pentabromide similarly, but less satisfactorily, gives the sulphonyl bromides.

(5) Sulphonyl chlorides may be obtained by treating arylmagnesium chlorides in tetrahydrofuran with sulphuryl chloride (*S. N. Bhattacharaya*, *C. Eaborn* and *D. R. M. Walton*, J. chem. Soc. [C], 1968, 1265):

$$RMgCl + SO_2Cl_2 \rightarrow RSO_2Cl + MgCl_2$$

The use of arylmagnesium bromides gives a mixture of sulphonyl chlorides and bromides which is difficult to separate.

(6) Sulphonyl chlorides and bromides may be prepared from the correspond-

ing amine by diazotisation in concentrated acid and treatment of the appropriate diazonium salt with sulphur dioxide in the presence of a cupric halide catalyst (*A. J. Prinsen* and *H. Cerfontain*, Rec. Trav. chim., 1965, 84, 24).

(7) Sulphonyl bromides and iodides are obtained by treating the sulphonyl hydrazide with bromine or iodine (*L. M. Litvinenko et al.*, J. gen. Chem. U.S.S.R., 1964, **34**, 3730):

$$RSO_2Cl + N_2H_4 \rightarrow RSO_2NHNH_2 + HCl$$

$$RSO_2NHNH_2 + 2Br_2 \rightarrow RSO_2Br + 3HBr + N_2$$

(8) Sulphonyl iodides are obtained by the reaction of iodine with the sodium salts of sulphinic acids (*R. Otto* and *J. Tröger*, Ber., 1891, **24**, 478).

(ii) Properties and reactions

Arenesulphonyl fluorides are, in general, less reactive than the corresponding sulphonyl chlorides and bromides; the sulphonyl iodides have not been so extensively studied. They are, however, comparatively unstable and decompose after standing for a few days, or on exposure to light. *p*-Toluenesulphonyl iodide is reported to give *p*-toluenesulphonyl radicals on heating in the presence of copper powder (*C. M. M. da Silva Corrêa* and *W. A. Waters*, J. chem. Soc. [C], 1968, 1880). The main bands in the infra-red spectra of sulphonyl chlorides have been assigned as follows, $\nu_{s\,(SO_2)}$ 1170–1190, $\nu_{as\,(SO_2)}$ 1360–1390, $\delta_{(SO_2)}$ 570–590, $\nu_{S\text{-}Cl}$ 373 cm^{-1}; the corresponding assignments for sulphonyl fluorides are 1200–1215, 1395–1420, 585–600, 755–790 cm^{-1}, respectively (*N. S. Ham, A. N. Hambly* and *R. H. Laby*, Austral. J. chem., 1960, **13**, 443; *E. A. Robinson*, Canad. J. chem., 1961, **39**, 247).

(1) *Hydrolysis.* Sulphonyl chlorides and bromides are slowly hydrolysed by water, but sulphonyl fluorides are stable to hot water. Hydrolysis occurs more readily in aqueous alkali, and sulphonyl chlorides are smoothly hydrolysed in acidic media. Benzenesulphonyl chloride is considered to hydrolyse in aqueous media by an "S_N2 type" mechanism and details of the solvolysis of arenesulphonyl chlorides in various solvent mixtures have been reported (*H. K. Hall Jr.*, J. Amer. chem. Soc., 1956, **78**, 1450; *F. E. Jenkins* and *Hambly*, Austral. J. chem., 1961, **14**, 190, 205; *R. V. Vizgert* and *E. V. Savchuk*, Zhur. obshchei Khim., 1956, **26**, 2261; C.A., 1957, **51**, 4984d).

Mechanisms for the hydrolysis of sulphonyl chlorides have been reviewed (*Vizgert*, Russ. chem. Reviews, 1963, **32**, 1). Sulphonyl chlorides and bromides readily react with alcohols or phenols in the presence of pyridine or alkali to give sulphonic esters (see p. 454).

(2) *Ammonolysis.* Sulphonyl halides react with ammonia and with primary

or secondary amines to give sulphonamides; basic catalysts are frequently used. Arylamines do not appear to react with sulphonyl fluorides. The reaction is used in the Hinsberg method for the separation of primary, secondary and tertiary amines, and, with *p*-toluenesulphonyl chloride, for the protection of the amine group in peptide synthesis:

$$RSO_2Cl + RNH_2 \rightarrow RSO_2NHR + HCl$$

(3) *Reduction.* Reduction of the sulphonyl fluoride group is effected only with difficulty but the sulphonyl chloride group is readily reduced. Reduction with lithium aluminium hydride gives either sulphinic acid or thiol depending on the conditions:

$$RSO_2H \leftarrow RSO_2Cl \rightarrow RSH$$

inverse addition of hydride at low temperatures gives sulphinic acid, whereas the use of excess hydride under reflux in ether gives the thiol (*L. Field* and *F. A. Grunwald,* J. org. Chem., 1951, **16**, 946). Sodium borohydride (3 moles) and aluminium chloride in "diglyme" at 25° also gives the thiol (*H. C. Brown* and *B. C. Subba Rao,* J. Amer. chem. Soc., 1956, **78**, 2582). In general, reduction with metals in an alkaline or neutral medium yields sulphinates, but in a strongly acidic medium reduction to the thiol occurs. Thus zinc dust and water, or sodium (*K. Otto* and *O. von Gruber,* Ann., 1867, **142**, 92), react with an ethereal solution of a sulphonyl chloride to give the sulphinate. Zinc and aqueous mineral acids give the thiol (*Adams* and *Marvel, loc. cit.*, p. 504). Sodium sulphite in alkaline solution gives the sulphinate (see *e.g.*, *W. L. F. Armarego* and *E. E. Turner,* J. chem. Soc., 1965, 1665), and in acid solution gives the corresponding disulphide (*F. Challenger, S. A. Miller* and *G. M. Gibson, ibid.*, 1948, 769). The disulphide is also obtained by reduction with hydriodic acid under reflux (*A. Ekbom,* Ber., 1891, **24**, 335; see also *W. A. Sheppard,* Org. Synth., 1960, **40**, 80). The reduction of sulphonyl chlorides to sulphinic acids has been reviewed (*W. E. Truce* and *A. M. Murphy,* Chem. Reviews, 1951, **48**, 69).

(4) *Friedel-Crafts reaction.* In the presence of a suitable catalyst, sulphonyl chlorides react with aromatic compounds which do not contain strongly electron-withdrawing substituents to give sulphones:

$$RSO_2Cl + C_6H_6 \xrightarrow{AlCl_3} RSO_2C_6H_5$$

Sulphonyl chlorides substituted with electron-withdrawing substituents simultaneously undergo reduction to the sulphinic acid and under some conditions this is the only observable reaction (*E. C. Dart, G. Holt* and *K. D. Jeffreys*, J. chem. Soc., 1964, 5663; *ibid.*, [C], 1966, 1284). Sulphonyl fluorides do not appear to react. This reaction has been reviewed (*F. R. Jensen* and *G. Goldman*, "Friedel-Crafts and Related Reactions", Vol. III, Pt. 2, *Olah*, Ed., Interscience, New York, 1964, p. 1319).

(5) *Reaction with organometallic compounds. p*-Toluenesulphonyl fluoride reacts with Grignard reagents or with alkyl-lithiums under reflux in ether to give *β*-disulphones (*H. Fukuda, F. J. Frank* and *Truce*, J. org. Chem., 1963, **28**, 1420):

$$ArSO_2F + {R \atop R}{>}CHLi \longrightarrow {R \atop R}{>}C{<}{SO_2Ar \atop SO_2Ar}$$

Inverse addition of diarylcadmiums to sulphonyl chlorides gives diaryl sulphones (*H. R. Henze* and *N. E. Artman*, *ibid.*, 1957, **22**, 1410):

$$ArSO_2Cl + Ar'_2Cd \rightarrow ArSO_2Ar'$$

The reaction of sulphonyl chlorides with Grignard reagents is generally complex.

(6) *Reaction with enamines.* Sulphonyl chlorides react with enamines in basic media to give, on hydrolysis, the appropriate arenesulphonyl ketone (*M. E. Kuehne*, J. org. Chem., 1963, **28**, 2124):

$$+ \; ArSO_2Cl \xrightarrow{Et_3N} \quad SO_2Ar \quad \xrightarrow{H^{\oplus}/H_2O} \quad SO_2Ar$$

(7) *Phenylation.* Benzenesulphonyl chloride phenylates aromatic compounds such as naphthalene or biphenyl on heating the reactants above 200°. Sulphur dioxide and hydrogen chloride are evolved and the reaction is considered to follow a free radical mechanism. *p*-Nitrobenzenesulphonyl chloride and *p*-toluenesulphonyl chloride react similarly (*P. J. Bain et al.*, Proc. chem. Soc., 1962, 186):

e.g., $PhSO_2Cl$ + → (64%) + (16%) + SO_2 + HCl

3-Carboxybenzenesulphonyl chloride also gives ketonic sulphonic acids presumably formed by electrophilic substitution (*E. K. Fields*, J. chem. Soc., 1965, 5766).

(8) *Replacement by halogen.* The sulphonyl chloride group is replaced by chlorine on heating with phosphorus pentachloride at 200–210° (*G. A. Barbaglia* and *A. Kekulé*, Ber., 1872, **5**, 876) or better with thionyl chloride at 160–200° (*H. Meyer*, Monatsh., 1915, **36**, 721).

(9) *Addition of acetylenes and allenes.* These react with arenesulphonyl iodides to give $\alpha\beta$-unsaturated β-iodoalkenyl arylsulphones in high yield (*Truce* and *G. C. Wolf*, Chem. Comm., 1969, 150):

$$CH_3\cdot C_6H_4\cdot SO_2I + RC{\equiv}CR \xrightarrow[h\nu]{\text{ether}} CH_3\cdot C_6H_4\cdot SO_2CR{=}CRI$$

(d) Sulphonic anhydrides, $RSO_2\cdot O\cdot SO_2R$

(i) Preparation

Arenesulphonic anhydrides are obtained: (1) in high yield by reacting sulphonic acids with carbodiimides in benzene at room temperature (*H. G. Khorana*, Canad. J. chem., 1953, **31**, 585):

$$RSO_2OH + R'N{=}C{=}NR' \rightarrow (RSO_2)_2O + (R'NH)_2CO$$
$$R' = p\text{–}CH_3\cdot C_6H_4\text{–}$$

(2) by heating sulphonic acids with phosphorus pentoxide and an inert support of asbestos and kieselguhr (*L. Field*, J. Amer. chem. Soc., 1952, **74**, 394), or with phosphorus oxychloride and pentachloride (*V. O. Lukashevich*, Doklady Akad. Nauk S.S.S.R., 1957, **114**, 1025; C.A., 1958, **52**, 3717e); (3) by treating the aromatic compound with sulphur trioxide in nitromethane at 0° (*N. H. Christensen*, Acta chem. Scand., 1961, **15**, 1507).

(ii) Properties and reactions

The arenesulphonic anhydrides are little affected by cold water but hydrolyse on heating. They resemble the arenesulphonyl chlorides in many of their reactions and in some instances are superior as sulphonylating agents, *e.g.*, *p*-toluenesulphonic anhydride reacts with toluene-*p*-thiol in the presence of pyridine to give the thiolsulphonate in high yield:

$$(p\text{-}CH_3C_6H_4SO_2)_2O + p\text{-}CH_3\cdot C_6H_4SH \xrightarrow{\text{pyridine}} p\text{-}CH_3\cdot C_6H_4SO_2\cdot SC_6H_4\cdot CH_3\text{-}p$$

and benzenesulphonic anhydride gives a quantitative yield of diphenyl sulphone on treatment with aluminium chloride and benzene (*Field, loc. cit.*):

$$(C_6H_5SO_2)_2O + C_6H_6 \xrightarrow{AlCl_3} (C_6H_5)_2SO_2$$

Arenesulphonic anhydrides show strong infra-red absorption at around 1375 and 1175 cm^{-1} (*Detoni* and *Hadži, loc. cit.*).

(e) *Sulphonamides, $ArSO_2NR_2$*

(i) *Preparation*

(1) Sulphonamides are obtained by reacting sulphonyl chlorides with ammonia or primary or secondary amines (see *e.g., H. J. Scheifele Jr.,* and *D. F. DeTar,* Org. Synth., Coll. Vol. IV, 1963, p. 34; *Th. J. de Boer* and *H. J. Backer, ibid.,* p. 943).

(2) *N*-Alkylsulphonamides may be obtained by reacting a sulphonamide having a replaceable hydrogen with an alkyl halide, or a similar alkylating agent, in the presence of alkali (*R. Lukeŝ* and *J. Prêucil,* Coll. Czech. chem. Comm., 1938, **10**, 384):

$$Me\cdot C_6H_4SO_2NHMe + BuBr \xrightarrow[\text{NaOH}]{\text{reflux}} Me\cdot C_6H_4SO_2N(Bu)Me$$

(ii) *Properties and reactions*

Sulphonamides have been extensively studied, especially as regards their chemotherapeutic activities which, have been reviewed by *E. H. Northey* (Chem. Reviews, 1940, **27**, 85).

The amides of aromatic sulphonic acids are, in general, colourless crystalline solids which are sparingly soluble in cold water. With the exception of some higher *N*-alkyl-, *N*-cycloalkyl- and *N*-benzyl-sulphonamides (*P. E. Fanta* and *C. S. Wang,* J. chem. Educ., 1964, **41**, 280), sulphonamides with a hydrogen attached to the nitrogen of the amide group are soluble in alkali. This property has been used to separate these sulphonamides from those formed from secondary amines. The infra-red spectra of benzene- and *p*-toluene-sulphonamide show strong absorption at around 1358 and 1167 cm^{-1} (*Robinson, loc. cit.*).

(1) *Hydrolysis.* The sulphonamide group is resistant to hydrolysis in alkaline solution, but may be hydrolysed by heating with 25 % hydrochloric

acid for 10–36 h, or at more moderate temperatures with sulphuric acid in acetic acid. Methods for the cleavage of sulphonamides under milder conditions include:

(a) The use of 48 % hydrobromic acid together with phenol as a bromine-acceptor (*H. R. Snyder* and *R. E. Heckert*, J. Amer. chem. Soc., 1952, 74, 2006):

$$C_6H_5SO_2NHC_6H_5 \xrightarrow[\text{20 min.}]{\text{HBr, } C_6H_5OH} C_6H_5NH_2$$

(b) Treatment with sodium in liquid ammonia (*J. Kovacs* and *U. R. Ghatak*, J. org. Chem., 1966, 31, 119; *J. M. Swan* and *V. du Vigneaud*, J. Amer. chem. Soc., 1954, 76, 3110):

$$p\text{-Me}\cdot C_6H_4SO_2NHC_6H_5 \xrightarrow{\text{Na/liq.}NH_3} C_6H_5NH_2 + p\text{-Me}\cdot C_6H_4SH + C_6H_5\cdot \text{Me} + SO_2$$

p-*N*,*N*-Dialkyltoluenesulphonamides give the substituted benzyl carbanion with sodamide in liquid ammonia. This anion undergoes substitution reactions with alkyl halides and condensation reactions with aromatic aldehydes and ketones (*F. H. Rash* and *C. R. Hauser*, J. org. Chem., 1967, 32, 3379).

(c) Treatment with excess naphthalenesodium in 1,2-dimethoxyethane at room temperature under nitrogen (*W. D. Clossen et al.*, J. Amer. chem. Soc., 1967, 89, 5311):

$$p\text{-Me}\cdot C_6H_4SO_2NHC_8H_{17} \rightarrow C_8H_{17}NH_2 + C_6H_5\cdot \text{Me} + p\text{-Me}\cdot C_6H_4SO_2^{\ominus}$$

(d) Treatment with nitrosonium tetrafluoroborate in acetonitrile at -15^0 gives the sulphonic acid (*G. A. Olah* and *J. A. Olah*, J. org. Chem., 1965, 30, 2386). The cleavage of sulphonamides has been reviewed (*S. Searles* and *S. Nukina*, Chem. Reviews, 1959, 59, 1077).

(2) *Halogenation*. Sulphonamides react readily with halogen in alkaline solution to give *N*-halogeno- and *N*,*N*-dihalogeno-sulphonamides (*F. D. Chattaway*, J. chem. Soc., 1905, 87, 145):

$$RSO_2NH_2 \xrightarrow{Cl_2/\text{NaOH}} RSO_2NCl_2 + RSO_2NHCl$$

N,*N*-Dichlorosulphonamides react with olefines by addition giving *N*-chloro--*N*-(β-chloroalkyl)sulphonamides which have predominantly the anti-Markownikoff orientation. Addition to 1,3-dienes gives 1,4-adducts (*F. A. Daniher* and *P. E. Butler*, J. org. Chem., 1968, 33, 4336).

(3) *Hydroxymethylation.* Sulphonamides and *N*-alkylsulphonamides react with aldehydes in the presence of acid or base to give a variety of products including plasticizers and resins. The initial product appears to be the substituted hydroxymethylamide:

$$e.g.,\ p\text{-Me}\cdot C_6H_4SO_2NH_2 + HCHO \rightarrow p\text{-Me}\cdot C_6H_4SO_2NHCH_2OH$$

which may then dimerise or trimerise (*L. McMaster*, J. Amer. chem. Soc., 1934, **56**, 204).

(4) *Metalation.* *N*-Methyl- and *N*-phenyl-benzene- and -*p*-toluene-sulphonamides undergo *ortho* metalation as well as *N*-metalation with excess *n*-butyl-lithium. The resulting dilithiosulphonamides condense with carbonyl compounds to give *tert*-hydroxyalkylsulphonamides and hence sultams (*H. Watanabe, R. L. Gay* and *Hauser*, J. org. Chem., 1968, **33**, 900):

SO_2NHR^1 2 LiBu / THF → Li, SO_2NR^1, Li → (1) R^2R^3CO / (2) $H_2O/H^{\oplus}$ → SO_2NHR^1, OH, C, R^2, R^3 → heat → SO_2, $N\cdot R^1$, R^2, R^3

Similar condensations with nitriles give iminosulphonamides which are in equilibrium with the cyclic amine (*Hauser et al., ibid.*, 1969, **34**, 919).

(f) *Esters of arenesulphonic acids, $ArSO_2OR$ (R = alkyl, aryl)*

(i) *Preparation*

(1) Primary alkyl sulphonates are obtained by reacting the sulphonyl chloride with the appropriate sodium alkoxide (*D. J. Chadbourne* and *A. J. Nunn*, J. chem. Soc., 1965, 4458) or with the alcohol in the presence of pyridine at room temperature (*V. C. Sekera* and *C. S. Marvel*, J. Amer. chem. Soc., 1933, **55**, 345; Org. Synth., Coll. Vol. III, 1955, p. 366). Secondary alcohols react less readily and few *tert*-alkyl esters can be prepared using these conditions. Reaction with the alcohol alone is slow and frequently gives alkyl halides, ethers, alkenes and the sulphonic acid.

(2) Sulphonate esters may be prepared by treating the silver salt of the sulphonic acid with an alkyl halide in acetonitrile solution (*W. D. Emmons* and *A. F. Ferris*, J. Amer. chem. Soc., 1953, **75**, 2257). *tert*-Alkyl sulphonates have been obtained by careful reaction at low temperatures (*H. M. R. Hoff-*

mann, J. chem. Soc., 1965, 6748, 6753), but at normal temperatures elimination occurs with secondary and tertiary alkyl halides:

$$RSO_2OAg + R'X \rightarrow RSO_2OR' + AgX$$

(3) Unstable *sec*-alkyl sulphonates have been prepared by the reaction of diazoalkanes with the sulphonic acid (*A. Ledwith* and *D. G. Morris*, *ibid.*, 1964, 508):

$$R_2CN_2 + ArSO_2OH \rightarrow R_2CHOSO_2Ar + N_2$$

(4) Aryl esters of sulphonic acids are obtained by reaction of a phenol with a sulphonyl chloride in the presence of sodium hydroxide or carbonate (see, *e.g.*, *G. Leandri*, *G. Monaco* and *D. Spinelli*, Ann. chim. [Italy], 1959, 49, 407), or by reacting the thallium(I) salt of the phenol with the sulphonyl chloride in an inert medium (*E. C. Taylor*, *G. W. McLay* and *A. McKillop*, J. Amer. chem. Soc., 1968, 90, 2422).

(5) *p*-Toluenesulphonates are obtained from *p*-toluenesulphinates by oxidation with *m*-chloroperbenzoic acid in methylene chloride at 0^0. The method is suitable for reactive esters and for esters of hindered alcohols (*R. M. Coates* and *J. P. Chen*, Tetrahedron Letters, 1969, 2705).

(*ii*) *Properties and reactions*

The sulphonate esters of benzyl, alkyl, some secondary alcohols, and tertiary alcohols decompose on standing to give the sulphonic acid and when possible the olefin. The stability is increased by the absence of traces of water or acid (*W. J. Hickinbottom* and *N. W. Rodgers*, J. chem. Soc., 1957, 4131). *tert*-Butyl *p*-toluenesulphonate has a half-life of 63 min at 0^0 in acetonitrile (*Hoffmann*, *loc. cit.*). Arenesulphonate esters, especially *p*-toluenesulphonate esters, have been used extensively as protecting groups in steroid and in carbohydrate chemistry, and as leaving groups in solvolysis reactions. Alkyl arenesulphonates absorb in the infra-red at 1365, 1190, 565 and near 800 cm^{-1}; certain aryl esters have a band at 1095–1100 cm^{-1} (*D. E. Freeman* and *A. N. Hambly*, Austral. J. Chem., 1957, **10**, 239).

(1) *Hydrolysis*. Esters of sulphonic acids are hydrolysed by water and more readily by dilute alkali. In general alkyl esters hydrolyse more rapidly than aryl esters:

$$RSO_2OR' + H_2O \rightarrow RSO_2OH + R'OH$$

The hydrolysis of alkyl sulphonates involves carbon-oxygen bond fission, the nucleophile reacting at the alkyl carbon atom. In contrast the hydrolysis of simple aryl esters involves a bimolecular reaction at the sulphur atom (*C. A. Bunton* and *V. A. Welch*, J. chem. Soc., 1956, 3240). ^{18}O studies have confirmed the postulated sulphur-oxygen bond fission (*R. V. Vizgert, E. V. Savchuk* and *M. P. Ponomarchuk*, Doklady Akad. Nauk S.S.S.R., 1959, **125**, 1257) and shown that the mechanism is a one step process unlike the hydrolysis of carboxylic esters (*D. R. Christman* and *S. Oae*, Chem. and Ind., 1959, 1251). The rates of alkaline hydrolysis in aqueous dioxan increase as the electron-withdrawing power of the substituent in the arenesulphonyl moiety increases (*Vizgert* and *Savchuk*, Zhur. obshchei Khim., 1956, **26**, 2261, 2268; C.A., 1957, **51**, 4984[d,i]). The mechanisms for the hydrolysis of sulphonate esters have been reviewed (*Vizgert*, Russ. chem. Reviews, 1963, **32**, 1).

(2) *Substitution reactions*. Alkyl esters of sulphonic acids undergo substitution reactions in a similar manner to alkyl halides *e.g.*, the *p*-toluenesulphonates of primary alcohols give the alkyl iodide on treatment with sodium iodide in acetone (*R. S. Tipson, M. A. Clapp* and *L. H. Cretcher*, J. org. Chem., 1947, **12**, 133) and the fluoride on heating with potassium fluoride in diethylene glycol (*E. D. Bergmann* and *I. Shahak*, Chem. and Ind., 1958, 157):

$$ArSO_2OR + X^{\ominus} \rightarrow ArSO_2O^{\ominus} + RX$$

In basic media primary alkyl *p*-toluenesulphonates give a higher substitution/elimination ratio than the corresponding alkyl halides. Thus, with the exception of the 2-phenylethyl ester, primary alkyl *p*-toluenesulphonates react with potassium *tert*-butoxide in *tert*-butanol solution to give almost entirely the alkyl *tert*-butyl ethers (*P. Veeravagu, R. T. Arnold* and *E. W. Eigenmann*, J. Amer. chem. Soc., 1964, **86**, 3072) and in dimethyl sulphoxide solution to give 60–70 % of alkyl *tert*-butyl ethers together with 20–25 % of alkenes (*C. H. Snyder* and *A. R. Soto*, J. org. Chem., 1964, **29**, 742). The sulphonate esters of secondary and cyclic alcohols give ~80 % of alkenes under the latter conditions.

(3) *Formation of aldehydes*. Benzyl and *n*-alkyl *p*-toluenesulphonates give aldehydes on heating with dimethyl sulphoxide containing sodium bicarbonate (*N. Kornblum, W. J. Jones* and *G. J. Anderson*, J. Amer. chem. Soc., 1959, **81**, 4113):

$$Me_2SO + H_2C(C_7H_{15})\text{—}OTs \longrightarrow Me_2S^{\oplus}\text{—}O\text{—}CH_2C_7H_{15} + OTs^{\ominus} \longrightarrow Me_2S + O{=}CH\text{—}C_7H_{15} + H^{\oplus}$$

Under similar conditions *sec*-alkyl sulphonates give mainly alkenes (see *H. R. Nace, ibid.*, 1959, **81**, 5428).

(4) *As alkylating agents.* Alkyl esters of *p*-toluenesulphonic acid on heating with aromatic hydrocarbons, anisole or phenol, alkylate the aromatic compound (*Hickinbottom* and *Rogers*, J. chem. Soc., 1957, 4124; *C. Nenitzescu, V. Ioan* and *L. Teodorescu*, Ber., 1957, **90**, 585). *n*-Alkyl esters do not give rearranged products.

CH_3, CH_3 + p-$CH_3 \cdot C_6H_4SO_2OCH(CH_3) \cdot CH_2CH_3$ ⟶ CH_3, CH_3, $CH_3 \cdot CH \cdot CH_2 \cdot CH_3$ + $CH_3 \cdot CH{=}CH \cdot CH_3$ + p-$CH_3C_6H_4SO_3H$

(5) *Cleavage.* Alkyl *p*-toluenesulphonates may be cleaved to give the alcohol with naphthalenesodium in tetrahydrofuran (*W. D. Closson, P. Wriede* and *S. Bank*, J. Amer. chem. Soc., 1966, **88**, 1581), by alkali-metal ammonia reduction (*D. B. Denny* and *B. Goldstein*, J. org. Chem., 1956, **21**, 479), and by treatment with hydrogen and Raney nickel (*G. W. Kenner* and *M. A. Murray*, J. chem. Soc., 1949, S 178). Aryl esters on reduction with Raney nickel and hydrogen give the aromatic hydrocarbons.

(6) *Fries rearrangement.* Aryl esters of sulphonic acids undergo the Fries rearrangement on heating in the presence of Lewis acids to give phenolic sulphones (see *Jensen* and *Goldman, loc. cit.*, p. 1349):

OSO_2Ar, CH_3 $\xrightarrow[130°]{AlCl_3}$ OH, SO_2Ar, CH_3

(g) *Individual arenesulphonic acids*

(i) *Monosulphonic acids*

(1) **Benzenesulphonic acid,** $PhSO_2OH$, m.p. 52° (chloroform); + 2 H_2O, m.p. 43°, is a colourless, crystalline, deliquescent solid which is exceedingly soluble in water giving a highly acidic solution ($K_a = 2 \times 10^{-1}$ at 25°). It is sparingly soluble in benzene and virtually insoluble in ether and carbon disulphide. It is prepared on a large scale by mixing benzene with $2^1/_4$ times its weight of 100 %

sulphuric acid and gradually raising the temperature to 60–70°; alternatively a larger proportion of 96 % sulphuric acid is used under the same conditions. A variant of industrial value is to pass a continuous stream of benzene vapour through sulphuric acid at 120–185° (*G. F. Lisk*, Ind. Eng. Chem., 1948, **40**, 1671). After dilution the acid is neutralised with milk of lime, and the filtrate treated with sodium carbonate to give sodium benzenesulphonate.

An aqueous solution of benzenesulphonic acid is obtained by addition of the necessary amount of sulphuric acid to the barium salt, or by hydrolysis of benzenesulphonyl chloride. Concentration and drying in a desiccator gives the crystalline acid.

The salts of benzenesulphonic acid are generally readily soluble in water; the *sodium salt* crystallises with $1^1/_2H_2O$, the *potassium salt* is anhydrous. The solubility of the *calcium, strontium* and *barium salts* enables them to be separated from the sulphates. For quantitative information on the solubility and water of crystallisation of some of these salts see *F. Ephraim* and *A. Pfister*, Helv., 1925, **8**, 229. S-*benzylisothiuronium salt*, m.p. 148°; p-*bromophenacyl ester*, m.p. 110–111°.

Benzenesulphonyl fluoride, $PhSO_2F$, b.p. 207°, n_D^{20} 1·4922, is volatile in steam; it is prepared by heating benzenesulphonyl chloride under reflux with excess aqueous potassium fluoride solution (*W. Davis* and *J. H. Dick*, J. chem. Soc., 1931, 2104). **Benzenesulphonyl chloride,** $PhSO_2Cl$, m.p. 15°, b.p. 119°/15 mm, is prepared from benzene and chlorosulphonic acid (*H. T. Clarke, G. S. Babcock* and *T. F. Murray*, Org. Synth., Coll. Vol. I, 1941, p. 85), from phosphorus pentachloride or oxychloride and sodium benzenesulphonate (*R. Adams* and *C. S. Marvel*, *ibid.*, p. 84) or from phenylmagnesium chloride and sulphuryl chloride (*S. N. Battacharaya, C. Eaborn* and *D. R. M. Walton*, J. chem. Soc. [C], 1968, 1265). **Benzenesulphonyl bromide,** $PhSO_2Br$, m.p. 140–141°, is formed as an oil in high yield by the reaction of benzenesulphonylhydrazide with bromine in aqueous acidic solution (*A. C. Poshkus, J. E. Herweh* and *F. A. Magnotta*, J. org. Chem., 1963, **28**, 2766). **Benzenesulphonyl iodide,** $PhSO_2I$, m.p. 45°, is obtained by the action of iodine on benzenesulphinic acid (*R. Otto* and *J. Troeger*, Ber., 1891, **24**, 478) or on benzenesulphonhydrazide (*L. M. Litvinenko et al.*, J. gen. Chem. U.S.S.R., 1964, **34**, 3730. **Benzenesulphonic anhydride,** $(PhSO_2)_2O$, m.p. 90°, prepared by reacting benzenesulphonic acid with di-*p*-tolylcarbodiimide in benzene solution (*H. G. Khorana*, Canad. J. Chem., 1953, **31**, 585) reacts with ammonia to give a mixture of the amide and the ammonium salt. **Dibenzenesulphonyl sulphide,** $(PhSO_2)_2S$, m.p. 134°, is prepared by the reaction of iodine with an alkali salt of benzenethiolsulphonic acid, or of sulphur dichloride with an alkali salt of benzenesulphinic acid. *Dibenzenesulphonyl disulphide*, $(PhSO_2)_2S_2$, has m.p. 76°, the *trisulphide*, m.p. 103° and the *tetrasulphide*, m.p. 84°. **Benzoic benzenesulphonic anhydride,** PhCO·O·SO_2Ph, m.p. 75–78°, is formed by reacting benzoyl chloride with silver benzenesulphonate in acetonitrile (*C. G. Overberger* and *E. Sarlo*, J. Amer. chem. Soc., 1963, **85**, 2446); it disproportionates on standing to give a mixture of benzoic and benzenesulphonic anhydrides; *propionic benzenesulphonic anhydride*, m.p. 44–46°. These mixed anhydrides are acylating agents and readily cleave ethers (*M. H. Karger* and *Y. Mazur, ibid.*, 1968, **90**, 3878).

The lower alkyl esters of benzenesulphonic acid are liquids, the aryl esters are solids. **Methyl benzenesulphonate,** b.p. 154°/20 mm; *ethyl ester,* b.p. 160°/18 mm; *allyl ester,* b.p. 120°/1 mm; *phenyl ester,* m.p. 36°; *α-naphthyl ester,* m.p. 117°. Certain aryl esters show fungicidal and insecticidal activity. *tert*-**Butyl benzenepersulphonate,** $PhSO_2 \cdot O \cdot O \cdot CMe_3$, is obtained by the reaction of benzenesulphonyl chloride with *tert*-butyl hydroperoxide in the presence of pyridine at —25°. It decomposes exothermically and violently to acetone and benzenesulphonic acid (*P. D. Bartlett* and *B. T. Storey, ibid.*, 1958, **80**, 4954).

Benzenesulphonyl peroxide, $(PhSO_2)_2O_2$, m.p. 53–54° (decomp.), prepared by the action of sodium peroxide on benzenesulphonyl chloride (*R. F. Weinland* and *H. Lewkowitz,* Ber., 1903, **36**, 2702; *L. W. Crovatt* and *R. L. McKee,* J. org. Chem., 1959, **24**, 2031), decomposes in water forming phenol, sulphuric acid and presumably benzenesulphonic acid. It decomposes in benzene at room temperature yielding phenyl benzenesulphonate and benzenesulphonic acid.

Benzenesulphonamide, $PhSO_2NH_2$, m.p. 150°, formed by reacting benzenesulphonyl chloride with ammonia; N-*methylbenzenesulphonamide,* m.p. 31°; N,N-*dimethylbenzenesulphonamide,* m.p. 47°; *benzenesulphonanilide,* $PhSO_2NHPh$, m.p. 110°. **Benzenesulphonyldichloroamide,** $PhSO_2NCl_2$, m.p. 76°, is prepared by the action of aqueous hypochlorite on benzenesulphonamide (*F. D. Chattaway,* J. chem. Soc., 1905, **87**, 145). It liberates iodine from acidified potassium iodide solution, and is thereby reduced to the amide. It dissolves in aqueous alkali to give a salt of the monochloro-compound, $PhSO_2N(Na)Cl$. *Benzenesulphonyldibromoamide,* $PhSO_2NBr_2$, m.p. 116°, behaves similarly. N-**Nitrobenzenesulphonamide,** $PhSO_2NHNO_2$, m.p. 100° (decomp.), is a strong acid. It is prepared by the action of a mixture of nitric and sulphuric acids on benzenesulphonamide (*O. Hinsberg,* Ber., 1892, **25**, 1902; *B. R. Mathews,* J. phys. Chem., 1920, **24**, 108). N-**Hydroxybenzenesulphonamide** (*benzenesulphohydroxamic acid*), $PhSO_2NHOH$, m.p. ~126°, is formed by the reaction of benzenesulphonyl chloride with hydroxylamine (*O. Piloty,* Ber., 1896, **29**, 1559). It reacts with aldehydes in alkaline solution forming benzenesulphinic acid and the hydroxamic acid corresponding to the aldehyde (*A. Angeli,* Atti Accad. Lincei, 1911, **20**, [ii], 447; 1912, **21**, [i], 622; 1913, **22**, [i], 851), and is oxidised by ferric chloride or by hypochlorites to give nitrous acid and *N,N*-dibenzenesulphonylhydroxylamine, $(PhSO_2)_2NOH$. The *O*-methyl ester, $PhSO_2NHOCH_3$, is formed in low yield, together with the methyl ester of phenylsulphamic acid by u.v. irradiation of a solution of benzenesulphonyl azide in methanol (*W. Lwowski* and *E. Scheiffle,* J. Amer. chem. Soc., 1965, **87**, 4359).

Benzenesulphonohydrazide, $PhSO_2NHNH_2$, m.p. 104°, is obtained by the reaction of benzenesulphonyl chloride with hydrazine hydrate (*T. Curtius* and *F. Lorenzen,* J. pr. Chem., 1898, [ii], **58**, 160). It acts as a source of di-imide for reductions in alkaline solution (*S. Hünig, H. R. Müller* and *W. Thier,* Tetrahedron Letters, 1961, 353). The mechanism of this reaction has been studied (*J. F. Bunnett* and *H. Takayama,* J. Amer. chem. Soc., 1968, **90**, 5173). Benzenesulphonohydrazide reacts with aromatic esters to give the sulphonyl derivatives of the acid hydrazides, which are decomposed on heating with sodium ethoxide to give aldehydes (*J. S. McFadyen* and *T. S. Stevens*, J. chem. Soc., 1936, 584; *M. S. Newman* and *E. G. Caflisch Jr.*, J. Amer. chem. Soc., 1958, **80**, 862):

$$ArCO_2R + PhSO_2NHNH_2 \rightarrow ArCONHNHSO_2Ph \xrightarrow{OEt^{\ominus}} ArCHO + N_2 + PhSO_2^{\ominus}$$

Similarly 1,1-dibenzyl-2-benzenesulphonohydrazide and its derivatives are decomposed by alkali to give bibenzyl and substituted bibenzyls (*L. A. Carpino*, *ibid.*, 1957, **79**, 4427; *R. L. Hinman* and *K. L. Hamm*, *ibid.*, 1959, **81**, 3294):

$$ArCH_2(Ar'CH_2)N{\cdot}NHSO_2Ph \xrightarrow{OH^{\ominus}} ArCH_2{\cdot}CH_2Ar' + N_2 + PhSO_2^{\ominus}$$

Benzenesulphonohydrazide reacts with aldehydes and ketones to form benzenesulphonohydrazones, which decompose on treatment with base.

Benzenesulphonyl azide, $PhSO_2N_3$, m.p. 13–14°, is obtained by reacting benzenesulphonyl chloride with sodium azide in aqueous acetone at 0° (*J. E. Leffler* and *Y. Tsuno*, J. org. Chem., 1963, **28**, 902), or by the action of nitrous acid on benzenesulphonohydrazide. With strained olefins it undergoes cycloaddition reactions which are postulated to proceed through a reactive triazoline intermediate which can decompose to give either an aziridine or a sulphonimide derivative (*L. H. Zalkow et al.*, Chem. and Ind., 1964, 1556; *J. E. Franz, C. Osuch* and *M. W. Dietrich*, J. org. Chem., 1964, **29**, 2922; *A. C. Oehlschlager* and *Zalkow*, *ibid.*, 1965, **30**, 4205):

+ $PhSO_2N_3$ $\xrightarrow{-N_2}$ NSO$_2$Ph + =NSO$_2$Ph

Similar reactions occur with enamines and vinyl ethers (*Franz et al.*, J. org. Chem., 1966, **31**, 2847; *D. L. Rector* and *R. E. Harmon*, *ibid.*, 1966, **31**, 2837). Benzenesulphonyl azide decomposes on heating to 105–120° in aromatic solvents giving nitrogen and the substituted benzenesulphonanilide (*J. F. Heacock* and *M. T. Edmison*, J. Amer. chem. Soc., 1960, **82**, 3460):

$$PhSO_2N_3 + PhOMe \rightarrow PhSO_2NHC_6H_4OMe + N_2$$
$$(o, 71\%;\ m, 2\%;\ p, 27\%)$$

Benzenesulphonyl isocyanate, $PhSO_2NCO$, b.p. 69–72°/0·15 mm, is formed by the reaction of benzenesulphonamide with phosgene (*H. Krizkalla*, G.P., 817, 602/1951; C.A., 1953, **47**, 2206e), or, together with benzenesulphonic anhydride, by the reaction of benzenesulphonyl chloride with silver cyanate (*O. C. Billeter*, Ber., 1904, **37**, 690). It reacts readily with hindered alcohols and phenols on heating, and with tertiary alcohols at room temperature to give urethanes, which are used for characterisation. Triarylmethanols react abnormally to give *N*-triarylmethylbenzenesulphonamides, $PhSO_2NHCAr_3$, and carbon dioxide, while *tert*-butanol yields benzenesulphonamide (*J. W. McFarland* and *J. B. Howard*, J. org. Chem., 1965, **30**, 957; *McFarland, D. E. Lenz* and *D. J. Grosse ibid.*, 1966, **31**, 3798; *ibid.*, 1968, **33**, 3514; *McFarland* and *R. W. Hauser, ibid.*, 1968, **33**, 340). Reaction with *N,N*-dialkylamides gives *N,N*-dialkyl-*N'*-arylsulphonylamidines, $ArSO_2N{=}C(R')NR_2$; amines and thiols react to give substituted ureas and thiourethanes, respectively (*Billeter, loc. cit.*; *C. King*, J. org. Chem., 1960, **25**, 352). Benzenesulphonyl isocyanate is readily decomposed by water to give benzenesulphonamide. It undergoes cycloaddition reactions with double bonded substrates yielding acyclic 1:1 adducts, which then undergo ring closure to form four-membered ring cycloadducts, or react further with dipolarophiles to give cycloadducts with a six-membered ring (*H. Ulrich et al., ibid.*, 1969, **34**, 2250).

(2) **Toluenesulphonic acids.** Toluene, like all alkylbenzenes, reacts more readily than benzene with sulphuric acid. At 25° with 82·3% acid the relative rate of sulphonation, $k_{toluene}/k_{benzene}$, is 66·1, the partial rate factors being *para*: 258, *meta*: 5·7, *ortho*: 63·4 (*L. Vollbracht, H. Cerfontain* and *F. L. J. Sixma*, Rec. Trav. chim., 1961, **80**, 11). The relative rate of sulphonation and the isomer distribution varies with the acid strength and the reaction temperature (for details see *Cerfontain loc. cit.*, Chap. 3; *Cerfontain, Sixma* and *Vollbracht*, Rec. Trav. chim., 1963, **82**, 659). The toluenesulphonic acids may be determined by U.V. spectrophotometry (*Cerfontain, H. G. J. Duin* and *Vollbracht*, Analyt. Chem., 1963, **35**, 1005). *o*-**Toluenesulphonic acid,** $Me{\cdot}C_6H_4SO_3H,2H_2O$, m.p. 67·5°; *fluoride*, b.p. 100°/13 mm, n_D^{20} 1·5007; *chloride*, m.p. 67·5°, used in the manufacture of saccharin; *amide*, m.p. 156°. *m*-**Toluenesulphonic acid,** m.p. 23–24°, b.p. 160–170°/0·1 mm, best prepared by sulphonation of *o*- or *p*-toluidine and deamination (*A. A. Spryskov* and *T. I. Yakovleva*, Zhur. obshchei Khim., 1957, **27**, 239; C.A., 1957, **51**, 12847e); *chloride*, m.p. 12°; *amide*, m.p. 108°. *p*-**Toluenesulphonic acid,** $Me{\cdot}C_6H_4SO_3H,H_2O$, m.p. 105°, loses its water of crystallisation *in vacuo*; m.p. 35° anhydr.; manufactured by sulphonation of toluene by sulphur trioxide using sulphur dioxide as solvent; *anhydride*, m.p. 128°, prepared by dehydration of the acid with phosphorus pentoxide (*L. Field* and *J. W. McFarland*, Org. Synth., Coll. Vol. IV, 1963, p. 940); *fluoride*, m.p. 42°; *chloride* (*tosyl chloride*), m.p. 69°, manufactured by the action of chlorosulphonic acid on toluene, used as a reagent for Beckmann rearrangements and with pyridine as a catalyst for dehydration and esterification reactions, especially of *N*-acylated amino acids (*G. Blotny, J. R. Biernat* and *E. Taschner*, Ann., 1963, **663**, 194), and tertiary acetylenic alcohols (*G. F. Hennion* and *S. O. Barrett*, J. Amer. chem. Soc., 1957, **79**, 2146); *bromide*, m.p. 95–96°; *iodide*, m.p. 85° (decomp.) on heating with anthracene in the presence of copper powder gives 9,10-dihydro-9,10-di-*p*-toluenesulphonylanthracene (*C. M. M. da Silva Corrêa* and *W. A. Waters*, J. chem. Soc. [C], 1968, 1880); *amide*, m.p. 137°; *anilide*, m.p.

103°; *methyl ester*, m.p. 28°; n-*butyl ester*, b.p. 170–171°/10 mm (*A. T. Roos, H. Gilman* and *N. J. Beaber*, Org. Synth., Coll. Vol. I, 1932, p. 145); p-*bromophenacyl ester*, m.p. 128–129°; *S-benzylisothiuronium salt*, m.p. 181–182°. The N,N-*dichloroamide*, $Me{\cdot}C_6H_4SO_2NCl_2$, m.p. 83°, dissolves in aqueous alkali to give alkali salts of the *monochloroamide*. The sodium salt, $Me{\cdot}C_6H_4SO_2N(Na)Cl$, is used as an antiseptic under the name "Chloramine T", which chlorinates phenols in neutral aqueous solution. The chlorination step is considered to involve dichloroamine-T, produced by disproportionation, rather than hypochlorous acid (*T. Higuchi et al.*, J. chem. Soc. [B], 1967, 546, 549). Chloramine-T reacts with dialkyl and alkyl aryl sulphides to give *S,S*-dialkyl (or alkyl aryl)-*N*-*p*-toluenesulphonylsulphimides (sulphilimines) $Me{\cdot}C_6H_4SO_2N{:}SR_2$, and with thiolsulphonates to give sulphonic acids and *S*-alkyl-*S*-arenesulphonamidosulphinides, $Me{\cdot}C_6H_4SO_2N^{\ominus}RS^{\oplus}{=}NH{-}SO_2C_6H_4{\cdot}Me$ (*G. Leandri* and *D. Spinelli*, Ann. chim. [Italy], 1959, **49**, 964; 1960, **50**, 1616). *p-Toluenesulphonyl azide*, $Me{\cdot}C_6H_4SO_2N_3$, m.p. 22°, is obtained by the reaction of *p*-toluenesulphonyl chloride with sodium azide (*Rector* and *Harmon, loc. cit.*). It reacts with cyclopentadienyllithium to give diazocyclopentadiene and the *p*-toluenesulphonamide anion (*W. E. Doering* and *C. H. DePuy*, J. Amer. chem. Soc., 1953, **75**, 5955):

$$C_5H_5^{\ominus}Li^{\oplus} + Me{\cdot}C_6H_4SO_2N_3 \xrightarrow{\text{ether}} C_5H_4N_2 + Me{\cdot}C_6H_4SO_2NH^{\ominus}Li^{\oplus}$$

Diazo-compounds are also obtained by reaction with *β*-diketones in the presence of a base (*M. Regitz*, Ann., 1964, **676**, 101; *J. B. Hendrickson* and *W. Wolf*, J. org. Chem., 1968, **33**, 3610), and with the sodium salts of reactive phenols (*J. M. Tedder* and *B. Webster*, J. chem. Soc., 1960, 4417). In the latter case coupling occurs with the unreacted phenol. On reaction with the halogenomagnesium salt of aniline, phenyl azide and *p*-toluenesulphonamide are obtained (*W. Fischer* and *J. P. Anselme*, J. Amer. chem. Soc., 1967, **89**, 5284). With arylmagnesium halides in tetrahydrofuran the sulphonyltriazine salt is formed, which may be reduced *in situ* with Raney nickel alloy and aqueous sodium hydroxide to give the arylamine. Alternatively the salt may be converted to the aryl azide with aqueous tetrasodium pyrophosphate and the amine then obtained by reduction (*P. A. S. Smith et al.*, J. org. Chem., 1969, **34**, 3430):

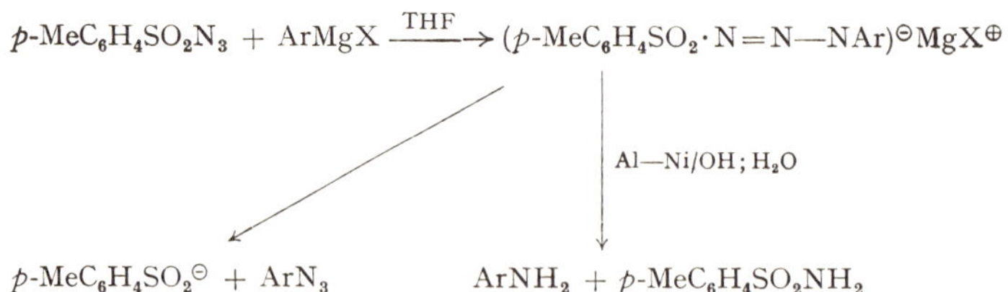

p-Toluenesulphonyl azide undergoes cycloaddition reactions with enamines, vinyl ethers, and strained olefins (see *R. Huisgen, R. Grashley* and *J. Sauer*, "The Chemistry of Alkenes", *S. Patai*, Ed., Wiley, New York, 1964, p. 835).

Thermal decomposition gives nitrogen and the sulphonylnitrene intermediate (*D. S. Breslow et al.*, J. Amer. chem. Soc., 1969, **91**, 2273).

p-*Toluenesulphonyl isocyanate*, $Me{\cdot}C_6H_4SO_2NCO$, b.p. 91–92°/0·5 mm, is prepared by reacting *p*-toluenesulphonamide with phosgene at 130° in the presence of butyl isocyanate as catalyst (*H. Ulrich, B. Tucker* and *A. A. R. Sayigh*, J. org. Chem., 1966, **31**, 2658). Reaction with salts of carboxylic acids gives *N*-acyl-*p*-toluenesulphonamides, *p*-$Me{\cdot}C_6H_4SO_2NHCOR$, and carbon dioxide (*R. C. Kerber, ibid.*, 1968, **33**, 4442).

p-*Toluenesulphonohydrazide*, $Me{\cdot}C_6H_4SO_2NHNH_2$, m.p. 104–107°, is prepared by reacting *p*-toluenesulphonyl chloride with hydrazine in tetrahydrofuran (*L. Friedman et al.*, Org. Synth., 1960, **40**, 93). On heating in "diglyme" under nitrogen it acts as a source of di-imide and reduces olefinic double bonds (*R. S. Dewey* and *E. E. van Tamelen*, J. Amer. chem. Soc., 1961, **83**, 3729). It reacts with aldehydes and ketones to give *p*-toluenesulphonohydrazones which decompose on treatment with a strong base, or on pyrolysing the dry salt, to give diazo-compounds (*M. P. Cava, R. L. Litle* and *D. R. Napier, ibid.*, 1958, **80**, 2257; *W. Ried* and *R. Dietrich*, Ber., 1961, **94**, 387):

$$e.g.,\ Me_2CH{\cdot}CH{=}N{\cdot}NHSO_2C_6H_4Me \xrightarrow{BuLi} Me_2CH{\cdot}CH{=}N{\cdot}\overset{\ominus}{N}SO_2C_6H_4{\cdot}Me,\ Li^{\oplus}$$

$$\xrightarrow[\text{at 0·2–3 mm}]{\text{pyrolysis}} Me_2CH{\cdot}CH{=}N_2 + Me{\cdot}C_6H_4SO_2Li$$

or products derived from the decomposition of the diazo-compound (*G. M. Kaufman et al.*, J. Amer. chem. Soc., 1965, **87**, 935). This decomposition appears to give a cationic intermediate in protic solvents, but a carbene in aprotic solvents or on pyrolysis of the dry salts (*J. A. Landgrebe* and *A. G. Kirk*, J. org. Chem., 1967, **32**, 3499 and references therein; *H. Babad, W. Flemon* and *J. B. Wood, ibid.*, 1967, **32**, 2871):

$$e.g.,\ Me{\cdot}C_6H_4SO_2NHN{=}C\ Me{\cdot}CMe_3 \xrightarrow[\text{diethylene glycol}]{Na} Me{\cdot}C_6H_4SO_2^{\ominus} + N_2 + Me_2C{=}CMe_2$$

(*W. R. Bamford* and *T. S. Stevens*, J. chem. Soc., 1952, 4735):

$$PhC(Me)_2{\cdot}C(Ph){=}N{\cdot}\overset{\ominus}{N}SO_2C_6H_4{\cdot}Me \xrightarrow{160\text{–}195^{\circ}}$$

$$Ph_2C{=}CMe_2 + Ph(Me)C{=}C(Ph)Me + Ph(Me)C\overset{CH_2}{\diagup\diagdown}CHPh + Me{\cdot}C_6H_4SO_2^{\ominus} + N_2$$

(*Landgrebe* and *Kirk, loc. cit.*).

p-Toluenesulphonohydrazones are reduced by excess of sodium borohydride in dioxan to give the hydrocarbon derived from the original aldehyde or ketone:

$$RCHO + Me{\cdot}C_6H_4SO_2NHNH_2 \rightarrow RCH{=}NNHSO_2C_6H_4{\cdot}Me \xrightarrow{NaBH_4} R{\cdot}CH_3$$

Lithium aluminium hydride reduction is preferred for aromatic compounds (*L. Cagliotti* and *P. Grasselli*, Chem. and Ind., 1964, 153).

Sulphonohydrazides are used as pneumatogens for producing expanded foams and sponges from rubber and plastics.

p-*Toluenesulphonyl cyanide*, Me·$C_6H_4SO_2CN$, m.p. 46–48°, is prepared by interaction of nitrosyl chloride and triphenyl (*p*-toluenesulphonylmethylene)-phosphorane in the presence of pyridine at —40°. It reacts with phenyl-magnesium bromide to give benzonitrile (*A. M. van Leusen, A. J. W. Iedema* and *J. Strating*, Chem. Comm., 1968, 440).

(3) **Xylenesulphonic acids.** The rates of sulphonation, relative to benzene, in 78·1 % acid at 25·0° are *p*-xylene, 265; *o*-xylene, 387; *m*-xylene, 1660. The isomer ratio, 3-sulphonic acid/4-sulphonic acid, obtained in the monosulphonation of *o*-xylene increases with increasing acid concentration. At 25° the percentage of the 3-sulphonic acid increases from 8 % to 41 % as the acid concentration is increased from 70·9 % to 98·9 %. For a given acid concentration the ratio decreases with increasing temperature. In the monosulphonation of *m*-xylene the isomer distribution is also markedly dependent on the acid concentration. At 25·0° with 78·6 % acid the isomer distribution is 2-sulphonic acid, 0·6 %; 4-sulphonic acid, 98·7 %; 5-sulphonic acid, 0·6 % (*A. J. Prinsen* and *Cerfontain*, Rec. Trav. chim., 1969, **88,** 833).

o-**Xylene**-4-**sulphonic acid**; *fluoride*, b.p. 150°/14 mm; *chloride*, m.p. 51–52°; *amide*, m.p. 144°. o-**Xylene**-3-**sulphonic acid**; *chloride*, m.p. 47°; *amide*, m.p. 167°. m-**Xylene**-4-**sulphonic acid** separates from dilute sulphuric acid with $2H_2O$. It may be prepared by the sulphonation of *m*-xylene with the addition complex of dioxan with two moles of sulphur trioxide (*C. M. Suter, P. B. Evans* and *J. M. Kiefer*, J. Amer. chem. Soc., 1938, **60,** 538); *fluoride*, b.p. 246°; *chloride*, m.p. 34°; *amide*, m.p. 137°. m-**Xylene**-2-**sulphonic acid** rearranges on heating in acid solution to the 4-sulphonic acid. It has been prepared by indirect methods; *chloride*, m.p. 39°; *amide*, m.p. 113°. p-**Xylene**-2-**sulphonic acid**, m.p. 86° ($2H_2O$) (*J. E. Weber*, J. chem. Educ., 1950, **27,** 384); *fluoride*, m.p. 24°; *chloride*, m.p. 26°; *amide*, m.p. 147°; S-*benzylisothiuronium salt*, m.p. 184°; p-*bromophenacyl ester*, m.p. 114°.

(4) **Trimethylbenzenesulphonic acids.** ψ-**Cumenesulphonic acid,** 1,2,4-*trimethylbenzene*-5-*sulphonic acid*, m.p. 111°; *fluoride*, b.p. 139°/20 mm; *chloride*, m.p. 62°; *amide*, m.p. 181°.

Mesitylenesulphonic acid, 1,3,5-*trimethylbenzene*-2-*sulphonic acid, dihydrate*, m.p. 77°; the sulphonic acid group is displaced by the action of an excess of aqueous chlorine or bromine (*R. L. Dhatta* and *J. C. Bhoumik*, J. Amer. chem. Soc., 1921, **43,** 303) and replaced to a very small extent during nitration (*C. S. Gibson*, J. chem. Soc., 1920, **117,** 948); *anhydride*, m.p. 225°; *fluoride*, m.p. 74°; *chloride*, m.p. 57°; *amide*, m.p. 141°.

(5) **Higher alkylbenzenesulphonic acids.** Sodium *n*-alkylbenzenesulphonates, which find use as detergents, are manufactured by the alkylation of benzene with higher olefins or propylene tetramers (averaging 10–15 carbon atoms), and

sulphonation with oleum. Neutralisation with sodium hydroxide gives the sodium salts (*N. V. Bataafsche Pet. Maats.*, B.P., 840,406/1960). Similar model compounds have been synthesised by *W. E. Truce* and *J. F. Lyons* (J. Amer. chem., Soc., 1951, **73**, 126), and by *F. W. Gray et al.* (J. org. Chem., 1955, **20**, 511; 1961, **26**, 209).

(ii) Halogen-substituted benzenesulphonic acids

The monohalogenobenzenesulphonic acids are formed by sulphonation of the halogenobenzenes, the *para*-compound being the major product. In 83–95 % sulphuric acid at 25° the percentage of the *para*-isomer formed is: fluorobenzene, 99 %; chlorobenzene, 98·8 %; bromobenzene, 98·7 %; iodobenzene, 96–98 %. The partial rate factors calculated for the *para*-positions are 3·3, 0·48, 0·29, and 0·57, respectively (*C. W. F. Kort* and *H. Cerfontain*, Rec. Trav. chim., 1969, **88**, 860; *ibid.*, 1967, **86**, 865). Sulphur trioxide in nitromethane at 0° reacts with aryl halides to give the sulphonic anhydrides, which may be hydrolysed to the acids (*N. H. Christensen*, Acta chem. Scand., 1961, **15**, 1507). Mono- or di-halogenobenzenes may be sulphonated by direct addition of liquid sulphur trioxide to the undiluted substrate at 25–120° (*E. E. Gilbert et al.*, Ind. Eng. Chem., 1953, **45**, 2065; *B. Veldhuis*, Analyt. Chem., 1960, **32**, 1681). When bromine or iodine is present in the substrate, side reactions may ensue, *e.g.*, diiodobenzene and benzenesulphonic acid are formed in the sulphonation of iodobenzene with 85 % sulphuric acid (*Kort* and *Cerfontain*, 1969, *loc. cit.*); *p*-diiodobenzene gives poly-iodo compounds on heating with concentrated sulphuric acid (*M. Boyle*, J. chem. Soc., 1909, **95**, 1683; *E. H. Huntress* and *F. H. Carten*, J. Amer. chem. Soc., 1940, **62**, 511; *J. Herzig*, Monatsh., 1881, **2**, 192). The mono- and di-halogenobenzenesulphonic acids are also readily obtained from the diazonium salts obtained from aminobenzenesulphonic or aminohalogenobenzenesulphonic acids. The *p*-halogenobenzenesulphonic acids as well as benzene- and *p*-toluene-sulphonic acids, form highly crystalline hydrated salts with divalent metals, *e.g.*, Mg, Zn, Fe, Co and Ni, of the general formula $(C_6H_4X\ SO_3)_2M,6H_2O$ (*H. E. Armstrong* and *E. H. Rodd*, Proc. roy. Soc., 1914, **A90**, 463).

p-**Fluorobenzenesulphonic acid,** m.p. 87°. p-**Chlorobenzenesulphonic acid,** m.p. 97–98°; *anhydride*, m.p. 150°; S-*benzylisothiuronium salt*, m.p. 175°. p-**Bromobenzenesulphonic acid,** m.p. 88–90°; *anhydride*, m.p. 170–178°; p-**Iodobenzenesulphonic acid,** m.p. 70°; *anhydride*, m.p. 220–221°, "*pipsan*", the anhydride labelled with ^{131}I or ^{35}S, is used in the determination of steroid hormones in sub-γ quantities (*Christensen*, Acta chem. Scand., 1961, **15**, 219); the *chloride*, "*pipsyl chloride*", is similarly used in the determination of amino acids in protein hydrolysates (*A. S. Keston, S. Udenfriend* and *R. K. Cannan*, J. Amer. chem.

TABLE 1

DERIVATIVES OF MONOHALOGENOBENZENESULPHONIC ACIDS

Benzene-sulphonic acid	*ortho*		*meta*		*para*	
	Chloride m.p.°	Amide m.p.°	Chloride m.p.°	Amide m.p.°	Chloride m.p.°	Amide m.p.°
Chloro-	28	188	[76–77/0·8]*	148	54	143
Bromo-	51	186	[95–95·5/0·8]	154	75	166
Iodo-	51	170	23	152	87	183
Fluoro-	36	123	[109–110/13]	—	35–36	126

* In brackets b.p.°/mm Hg.

TABLE 2

2,5-DIHALOGENOBENZENESULPHONIC ACIDS

Benzenesulphonic acid.	Chloride m.p.°	Bromide m.p.°	Anilide m.p.°
2,5-Dichloro-	38	74	182 (amide)
2-Bromo-5-chloro-	46	83	164
5-Bromo-2-chloro-	66	110	138
2,5-Dibromo-	71	114	143
5-Chloro-2-iodo-	69	102	168
5-Bromo-2-iodo-	97	—	150
2-Chloro-5-iodo-	88	143	—
2-Bromo-5-iodo-	91	—	—

Soc., 1949, **71**, 249). Table 1 gives the m.p.'s of chlorides and amides of the monohalogeno acids. 2,5-Dichloro- and -dibromo-benzenesulphonic acids form well crystallising salts with metals of the rare earths *e.g.* $(C_6H_3X_2 \cdot SO_3)_3Nd,9H_2O$ and $18H_2O$; their crystallographic constants have been recorded (*Armstrong* and *Rodd*, Proc. roy. Soc., 1912, **A87**, 204; *Rodd, ibid.*, 1913, **A89**, 292). The m.p.'s of halides and anilides of 2,5-dihalogenobenzenesulphonic acids are given in Table 2 (*R. T. Colgate* and *Rodd*, J. chem. Soc., 1910, **97**, 1585).

(*iii*) *Nitro-substituted benzenesulphonic acids*

The sulphonation of nitrobenzene gives almost entirely the *meta*-isomer, the relative rates of sulphonation, k_{PhNO_2}/k_{PhH}, being 0·016 (*F. J. Stubbs, C. D. Williams* and *C. N. Hinshelwood, ibid.*, 1948, 1065). *o*- and *p*-Nitrobenzenesulphonic acids are best prepared by oxidation of the corresponding thiols or disulphides (see *e.g. E. Wertheim*, Org. Synth., Coll. Vol. II, 1943, p. 471).

o-**Nitrobenzenesulphonic acid,** *chloride,* m.p. 69°; *amide,* m.p. 191°. m-**Nitrobenzenesulphonic acid,** *anhydride,* m.p. 160°; *chloride,* m.p. 64°; *amide,* m.p. 166°; p-*bromophenacyl ester,* m.p. 125–126°; S-*benzylisothiuronium salt,* m.p. 146°. p-**Nitrobenzenesulphonic acid,** *chloride,* m.p. 80°; *amide,* m.p. 178°; p-*bromophenacyl ester,* m.p. 173–174°.

Sulphonation of *o*-nitrotoluene gives the 4-sulphonic acid whilst *p*-nitrotoluene gives 4-**nitrotoluene-2-sulphonic acid** crystallising with H_2O (m.p. 130° anhydr.), which is manufactured as an intermediate in the production of stilbene dyes. The sulphonation of 4-chloronitrobenzene and 6-chloro-2-nitrotoluene with oleum can lead to violent explosions (*Gilbert* and *E. P. Jones,* Ind. Eng. Chem., 1951, **43,** 2034).

Nitrosulphonic acids can also be obtained by nitration of sulphonic acids or their functional derivatives. Thus nitration of 4-chloro-3-methylbenzenesulphonyl chloride gives 4-*chloro-3-methyl-5-nitrobenzenesulphonyl chloride,* m.p. 49–50°; *amide,* m.p. 201–202°; the *potassium salt* of the acid is very sparingly soluble in water (0·235 % at 0°) (*W. Davies,* J. chem. Soc., 1922, **121,** 785; 1923, **123,** 2976).

o-**Nitrobenzenesulphonyl peroxide,** $(o\text{-}NO_2C_6H_4SO_2)_2O_2$, m.p. 97° (decomp.); m-*nitrobenzenesulphonyl peroxide,* m.p. 112° (decomp.); p-*nitrobenzenesulphonyl peroxide,* m.p. 128° (decomp.), are prepared by reacting the sulphonyl chloride in chloroform solution with alkaline hydrogen peroxide in aqueous ethanol at −20° (*R. L. Dannley* and *G. E. Corbett,* J. org. Chem., 1966, **31,** 153). These compounds are stable at room temperature, but decompose thermally in aromatic solvents (hydrocarbons, ArH) to give aryl nitroarenesulphonates and the nitroarenesulphonic acids:

$$(NO_2C_6H_4SO_2)_2O_2 + ArH \rightarrow NO_2C_6H_4SO_3H + NO_2C_6H_4SO_2OAr$$

(*iv*) *Arene-di- and -tri-sulphonic acids*

Sulphonation of benzene with concentrated sulphuric acid (6 moles) gives largely disulphonation at around 115° and trisulphonation at around 250°. The *meta*-isomer is the principal disubstitution product. On increasing the temperature the percentage of *para*-isomer formed increases from 3·5 to 11 % but decreases when the sulphuric acid is replaced by oleum. 3,3′-Disulphodiphenyl sulphone is formed in increasing amounts as the sulphur trioxide concentration in the oleum is increased. Equilibrium is attained at 235° with 87 % sulphuric acid, the mixture comprising 66·3 % *meta*- and 33·7 % *para*-isomer (*A. A. Spryskov* and *S. P. Starkov,* Zhur. obshchei Khim., 1957, **27,** 2780; C.A., 1958, **52,** 8072e). Benzene-*p*-disulphonic acid is the isomer most resistant to hydrolysis and the *ortho*-isomer the least resistant (*idem, ibid.*, 1956, **26,** 2862; C.A., 1957, **51,** 8038h; 1957, **27,** 3067; C.A., 1958, **52,** 8072a). Sulphonation of *m*-toluenesulphonic acid with 100 % sulphuric acid at room temperature gives 38 % of toluene-2,5-disulphonic acid and 62 % of the more stable toluene-3,5-disulphonic acid. Chlorosulphonic acid gives a similar isomer ratio (*Spryskov* and *T. I. Yakovleva, ibid.*, 1957, **27,** 239; C.A., 1957, **51,** 128 47e).

o-**Benzenedisulphonic acid,** 1,2-$C_6H_4(SO_3H)_2$, is obtained from *o*-aminobenzenesulphonic acid by diazotisation, treatment with sodium disulphide, and oxidation of the resultant disulphide with potassium permanganate; *chloride,* m.p. 142–143° (*B. I. Karavaev* and *Starkov, ibid.*, 1957, **27,** 788; C.A., 1957, **51,** 16336g). The *imide,* $C_6H_4(SO_2)NH$, m.p. 195–196°, is prepared from the chloride and ammonia; it is a strong acid completely ionised in water; *ammonium salt,* m.p. 254°; N-*chloroimide,* m.p. 150–151°, is a powerful chlorinating agent; N-*hydroxyimide,* m.p. 128–130°, has pK_a ~1 (*J. B. Hendrickson et al.*, J. org. Chem., 1969, **34,** 3434). m-**Benzenedisulphonic acid,** can be prepared by deamination of aniline-2,4-disulphonic acid or by sulphonation of benzenesulphonic acid with oleum at 85° for 8 h. The sodium salt is used for the manufacture of resorcinol and also finds use in electroplating; *bromide,* m.p. 53–54°; S-*benzylisothiuronium salt,* m.p. 214°; *bis*-(p-*bromophenacyl ester*), m.p. 68–69°.

The melting points of the chlorides and amides of the benzenedisulphonic acids are as follows:

	ortho	*meta*	*para*
$C_6H_4(SO_2Cl)_2$	144°	63°	141°
$C_6H_4(SO_2NH_2)_2$	252°	229°	288°

Toluene-3,5-**disulphonic acid,** is obtained by deamination of 2-aminotoluene-3,5-disulphonic acid obtained by sulphonation of *o*-toluidine (*Spryskov* and *Yakovleva, loc. cit.*); *chloride,* m.p. 93°.

1,3,5-**Benzenetrisulphonic acid** is rapidly obtained by heating the *m*-disulphonic acid with 60 % oleum at 230°; the *ortho*- and *para*-isomers show little change under these conditions (*Spryskov* and *Starkov,* 1956, *loc. cit.*). On alkaline fusion, phloroglucinol is obtained, and on fusion with potassium cyanide 1,3,5-tricyanobenzene; *chloride,* m.p. 184°; *amide,* m.p. 310–315°; *anilide,* m.p. 234–236°. 1,2,4-**Benzenetrisulphonic acid,** has been prepared by reacting the sodium salt of bromobenzene-2,4-disulphonic acid with aqueous sodium sulphite in the presence of copper sulphate; *chloride,* m.p. 114–116°; S-*benzylisothiuronium salt,* m.p. 240–241° (decomp.) (*R. Lundquist,* Acta chem. Scand., 1957, **11,** 1421).

(h) Acids with the sulphonic group in the side chain

Phenylmethanesulphonic acid and its homologues may be prepared by many of the methods for indirect sulphonation previously outlined (see p. 443). The acids may be obtained directly by oxidation of thiols, disulphides or thioacetates with peracids or hydrogen peroxide (*C. J. Cavallitto* and *D. McK. Fruehauf*, J. Amer. chem. Soc., 1949, **71**, 2248). Salts of the acids are obtained by heating the appropriate benzyl halide under reflux with sodium sulphite in aqueous alcohol (*E. Fromm* and *J. de Seixas Palma*, Ber., 1906, **39**, 3308; *C. A. Bunton* and *E. A. Halevi*, J. chem. Soc., 1952, 4541):

$$Ph\cdot CH_2Cl + Na_2SO_3 \rightarrow Ph\cdot CH_2SO_3Na + NaCl$$

The corresponding sulphonyl chlorides are obtained by oxidation of the appropriate thiols, disulphides, thiocyanates and similar compounds with chlorine in acetic acid (*E. E. Gilbert* "Sulfonation and Related Reactions", Interscience, New York, 1965, p. 201).

When phenylmethanesulphonyl chloride is treated with triethylamine in an inert solvent, hydrogen chloride is eliminated and stilbene, oxythiobenzoyl chloride and sulphur dioxide are produced (*E. Wedekind* and *D. Schenk*, Ber., 1911, **44**, 198; *J. F. King* and *T. Durst*, Tetrahedron Letters, 1963, 585):

$$Ph\cdot CH_2SO_2Cl + Et_3N \rightarrow Ph\cdot CH{=}CH\cdot Ph + Ph\cdot C(Cl){=}SO + Et_3NH^{\oplus}Cl^{\ominus} + SO_2$$

This reaction was postulated to proceed through a sulphene intermediate, $Ph\cdot CH{=}SO_2$. Evidence for sulphene intermediates was provided by the rapid formation of monodeuterated sulphonates on treatment of the sulphonyl chloride with triethylamine and deuterated alcohols (*King* and *Durst*, J. Amer. chem. Soc., 1964, **86**, 287; 1965, **87**, 5684).

$$\textit{e.g.},\ R\cdot CH_2\cdot SO\cdot Cl + Et_3N + Me_2\cdot CHOD \rightarrow R\cdot CHDSO_2OCHMe_2 + Et_3NH^{\oplus}Cl^{\ominus}$$

In the absence of base, esterification proceeds more slowly and without exchange. Sulphenes undergo $(2+2)\pi$ cycloaddition to enamines and ketene acetals to give, in general, thietane 1,1-dioxides. In the absence of added base, but with excess ketene acetal, additional products are formed:

e.g. $Ph\cdot CH_2SO_2Cl + CH_2{=}C(OEt)_2 \xrightarrow{Et_3N}$ CH_2—$C(OEt)_2$ / SO_2—$CH\cdot Ph$ (thietane ring)

(without base, excess ketene acetal) → CH_2—$C(OEt)_2$ / SO_2—$CH\cdot Ph$ (thietane ring) + O=⟨ring, OEt, Ph, SO_2⟩ + EtO–⟨ring, OEt, Ph, SO_2⟩

(*W. E. Truce*, *D. J. Abraham* and *P. Son*, J. org. Chem., 1967, **32**, 990).

See also reviews by *T. J. Wallace* (Quart. Reviews, 1966, **20**, 67) and *G. Opitz* (Angew. Chem., internat. Edn., 1967, **6**, 107). Addition to ynamines similarly gives thieten dioxides, and reaction with the ylid (ethoxycarbonylmethylene)triphenylphosphorane, $Ph_3P{:}CH{\cdot}CO_2Et$, gave the stable ylid, $Ph_3P{:}C(SO_2CH_2Ph)CO_2Et$ (*A. M. Hamid* and *S. Trippett*, J. chem. Soc. [C], 1968, 1612).

Phenylmethanesulphonic acid, (benzylsulphonic acid), $Ph{\cdot}CH_2SO_2OH$, a hygroscopic solid, is prepared by oxidising toluene-ω-thiol with excess perbenzoic acid in acetonitrile (*Cavallito* and *Fruehauf, loc. cit.*). *Phenylmethanesulphonyl chloride*, m.p. 92–93°, decomposes on heating alone, or with Lewis acids, to give benzyl chloride and sulphur dioxide (*A. Reiche* and *E. Naumann*, J. pr. Chem., 1959, **281**, 108). It is reduced to the sulphinic acid by zinc dust and alcohol, and to the disulphide with red phosphorus and iodine. On treatment with sodium fluoride in tetramethylene sulphone it gives the *fluoride*, m.p. 91–92° (*D. E. Fahrney* and *A. M. Gold*, J. Amer. chem. Soc., 1963, **85**, 997); *amide*, m.p. 105°; *methylamide*, m.p. 110°; *dimethylamide*, m.p. 102°; *anilide*, m.p. 109°; *hydrazide*, m.p. 131°. The alkyl and aryl esters are rapidly reduced by hydrogen and Raney nickel to give the alcohol and phenol, respectively, the anilide similarly gives aniline (*G. W. Kenner and M. A. Murray*, J. chem. Soc., 1949, S178).

The sodium salt is chlorinated in aqueous solution to give a mixture of 2- and 4-chlorophenylmethanesulphonic acids (*Farbwerke vorm. Meister, Lucius & Brüning*, G.P., 146,946/1902). Nitration gives principally 4-nitrophenylmethanesulphonic acid together with the 2- and 3-isomers (*C. K. Ingold, E. H. Ingold* and *F. R. Shaw*, J. chem. Soc., 1927, 828; *A. C. Bottomley* and *R. Robinson*, *ibid.*, 1927, 2785). 2-*Chlorophenylmethanesulphonic acid*, $ClC_6H_4{\cdot}CH_2SO_2OH$, hygroscopic crystals; 4-*chlorophenylmethanesulphonic acid*, m.p. 108°; *chloride*, m.p. 95°. 2-*Nitrophenylmethanesulphonic acid*, hygroscopic needles, prepared by the reaction of 2-nitrobenzyl chloride with sodium sulphite (*W. Marckwald* and *H. H. Frahne*, Ber., 1898, **31**, 1855); *chloride*, m.p. 63–64°; *amide*, m.p. 137°. 3-*Nitrophenylmethanesulphonic acid monohydrate*, m.p. 137°; *methyl ester*, m.p. 100°; *amide*, m.p. 161°. 4-*Nitrophenylmethanesulphonic acid*, m.p. 71°; *chloride*, m.p. 92°; *methyl ester*, m.p. 113°; *amide*, m.p. 204° (*Ingold et al., loc. cit.*).

β-**Phenylethanesulphonic acid,** $Ph{\cdot}CH_2{\cdot}CH_2SO_2{\cdot}OH,5H_2O$, hygroscopic leaflets, m.p. 91°, prepared from β-phenylethyl chloride or bromide and sodium sulphite (*E. B. Evans, E. E. Mabbott* and *E. E. Turner*, J. chem. Soc., 1927, 1159); *chloride*, m.p. 34°; *fluoride*, m.p. 27°; *amide*, m.p. 122°.

α-*Phenylethanesulphonic acid*, prepared similarly from α-phenylethyl bromide, has been resolved by crystallisation of its strychnine salt; *chloride*, m.p. 79°. α-*Phenylpropane-* and α-*phenylbutane-sulphonic acids* have been prepared similarly (*Evans, Mabbott* and *Turner, loc. cit.*).

2-**Phenylethenesulphonic acid,** $Ph{\cdot}CH{=}CHSO_2{\cdot}OH$, is obtained by sulphonating styrene with sulphur trioxide in dioxan; *chloride*, m.p. 89°; S-*benzylisothiuronium salt*, m.p. 166–167° (*F. G. Bordwell et al.*, J. Amer. chem. Soc., 1946, **68**, 139; 1948, **70**, 2429).

γ-Phenylpropanesulphonyl chloride, b.p. 121–123°/1·5 mm, n_D^{20} 1·5401, prepared by oxidising the disulphide with chlorine in aqueous acetic acid, cyclises on treatment with aluminium chloride; *amide,* m.p. 61° (*Truce* and *J. P. Milionis, ibid.*, 1952, **74**, 974). Derivatives of α- and β-phenylpropanesulphonic acids have been isolated from spent sulphite liquor from the pulping of wattle wood (*J. R. Parrish,* J. chem. Soc. [C], 1967, 1145).

2. Arenesulphinic acids

Arenesulphinic acids react as tautomeric substances. Their formation by reduction of sulphonyl chlorides, and the conversion of their salts into sulphones by reaction with alkyl halides suggests the structure I.

$$\begin{matrix} R \\ H \end{matrix} > S \begin{matrix} \nearrow O \\ \searrow O \end{matrix} \qquad\qquad R\diagdown S \nearrow O \text{ (S–OH)}$$

(I) (II)

On the other hand, sulphinic acids give chlorides of the general formula RSO·Cl, and the esters, RSO·OR′ are dissymmetric and potentially resolvable, which suggests that the acids are best represented by structure II.

Dipole moment measurements support structure I (*E. N. Gur'yanova* and *Ya. K. Syrkin,* Zhur. fiz. Khim., 1949, **23**, 105; C.A., 1949, **43**, 5245f). The infra-red spectra of the acids show an intense band near 1090 cm^{-1} assigned to the S=O group as it is also present in sulphinate esters and in the acid chlorides, and is near to the band shown by sulphoxides (1050 cm^{-1}). Absorptions characteristic of the sulphonyl group are absent. On deuteration the band at 2900 cm^{-1} undergoes a large shift and is therefore assigned to an O–H stretching vibration (*S. Detoni* and *D. Hadži,* J. chem. Soc., 1955, 3163). The main features of the infra-red spectra therefore support structure II, although an alternative interpretation is that the spectra indicate an equilibrium mixture of structures I and II (*A. Hidalgo,* Anales real Soc. españ. Fís. Quím., 1958, **54**, 369). The ultra violet spectra support structure II (*H. Bredereck, G. Brod* and *G. Höschele,* Ber., 1955, **88**, 438).

The chemistry of sulphinic acids has been reviewed by *F. Muth* ("Methoden der Organischen Chemie [Houben-Weyl]" Vol. 9, *E. Müller,* Ed., G. Thieme Verlag, Stuttgart, 1955, p. 299) and the preparation of sulphinic acids by *W. E. Truce* and *A. M. Murphy* (Chem. Reviews, 1951, **48**, 69).

(a) Preparation

(1) By reduction of sulphonyl chlorides (see p. 449).

(2) By passing sulphur dioxide into a solution of a diazonium sulphate in the presence of clean copper powder or a cuprous salt which effects reduction (*L. Gattermann*, Ber., 1899, **32**, 1136).

(3) By the action of sulphur dioxide on a cooled ethereal solution of an aryllithium salt (*Truce* and *J. F. Lyons*, J. Amer. chem. Soc., 1951, **73**, 126) or an arylmagnesium halide (*A. Rosenheim* and *L. Singer*, Ber., 1904, **37**, 2152):

$$RMgBr + SO_2 \rightarrow RSO_2MgBr \xrightarrow[H_2O]{H^{\oplus}} RSO_2H$$

(4) Disulphides or thiophenols, dissolved in a mixture of chloroform and an alcohol, are oxidised by lead tetra-acetate to give the appropriate alkyl sulphinate. The method is not applicable to substrates containing electron-withdrawing substituents (*L. Field et al.*, J. Amer. chem. Soc., 1961, **83**, 1256; 1962, **84**, 847):

$$PhS\cdot SPh + 3\ Pb(OAc)_4 + 4\ MeOH \xrightarrow{\text{reflux}} 2\ PhSO\cdot OMe + 3\ Pb(OAc)_2 + 4\ AcOH + 2\ MeOAc$$

(5) By the action of sulphur dioxide and dry hydrogen chloride on an aromatic hydrocarbon in the presence of aluminium chloride (*E. Knoevenagel* and *J. Kenner*, Ber., 1908, **41**, 3315; *C. Friedel* and *J. M. Crafts*, Compt. rend., 1878, **86**, 1368; *S. Smiles* and *R. Le Rossignol*, J. chem. Soc., 1908, **93**, 745):

$$ArH + SO_2 \xrightarrow[(2)\ H_2O]{(1)\ AlCl_3,\ HCl} ArSO_2H$$

Smiles and *Le Rossignol* found that in the presence of *o,p*-directing groups (*e.g.* OEt) sulphination may proceed beyond the sulphinic acid stage to form the diaryl sulphone or a triarylsulphonium compound depending on the directive effect of the substituent groups and the accumulative steric effect of *ortho*-substituents.

(6) Sulphinic acids are obtained by the cleavage of sulphones with sodium (*F. Krafft* and *W. Vorster*, Ber., 1893, **26**, 2813), or with strong bases (*R. Otto*, J. pr. Chem., 1884, **30**, [2], 171, 321). Aryl 2,4-dinitrophenyl sulphones, prepared by oxidation of the sulphides obtained from reacting 2,4-dinitrobenzenesulphenyl chloride with aromatic compounds in the presence of aluminium chloride, are readily cleaved with sodium methoxide (*N. Kharasch* and *R. Swidler*, J. org. Chem., 1954, **19**, 1704):

$$2,4\text{-}(NO_2)_2C_6H_3SCl + ArH \xrightarrow{AlCl_3} 2,4\text{-}(NO_2)_2C_6H_3SAr \xrightarrow[HOAc]{H_2O_2}$$

$$2,4\text{-}(NO_2)_2C_6H_3SO_2Ar \xrightarrow{OMe^{\ominus}} ArSO_2^{\ominus} + 2,4\text{-}(NO_2)_2C_6H_3OMe$$

Another variant of this reaction is the cleavage of β-cyanoethyl sulphones, obtained by oxidation of the sulphides produced by the addition of thiophenols to acrylonitrile, with the sodium salt of a thiol. Overall yields of 70–75 % are recorded (*Truce* and *F. E. Roberts Jr., ibid.*, 1936, **28**, 593):

$$RSH + CH_2{=}CH{\cdot}CN \xrightarrow{base} RSCH_2{\cdot}CH_2{\cdot}CN \xrightarrow[HOAc]{H_2O_2}$$

$$RSO_2CH_2{\cdot}CH_2{\cdot}CN \xrightarrow{R'S^{\ominus}} RSO_2^{\ominus} + R'SCH_2{\cdot}CH_2{\cdot}CN$$

(7) Other methods include the reaction of thiolsulphonates with thiols (*C. J. Miller* and *Smiles*, J. chem. Soc., 1925, **127**, 224), sodium arsenite, or potassium cyanide (*A. Gutman*, Ber., 1908, **41**, 3351), and the decomposition of arylsulphonylhydroxylamines (*O. Piloty, ibid.*, 1896, **29**, 1559; *E. Divers, ibid.*, p. 2324) and arylsulphonohydrazides (see p. 449).

(b) Properties and reactions

The aromatic sulphinic acids are moderately strong acids with pK_a's between 2·0 and 3·0 (*R. K. Burkhard et al.*, J. org. Chem., 1959, **24**, 767; *D. de Fillipo* and *F. Momicchioli*, Tetrahedron, 1969, **25**, 5733). They are associated in dioxan (*Gur'yanova* and *Syrkin, loc. cit.*) and in nitrobenzene solution forming trimers and hexamers (*W. G. Wright*, J. chem. Soc., 1949, 683).

They may be characterised as the *S*-benzylisothiouronium salts (*F. Kurzer* and *J. R. Powell, ibid.*, 1952, 3728) and determined by titration with sodium nitrite (see, *e.g.*, *J. L. Kice* and *K. W. Bowers*, J. Amer. chem. Soc., 1962, **84**, 605).

(i) Disproportionation

Although the salts are comparatively stable, the free acids disproportionate to give a 1:1 mixture of sulphonic acid and thiolsulphonate (*L. Horner* and *O. H. Basedow*, Ann., 1958, **612**, 108; *Kice* and *Bowers, loc. cit.*):

$$3\ PhSO_2H \rightarrow PhSO_3H + PhSO_2SPh + H_2O$$

the rate of disproportionation increasing as the medium becomes more acidic (*H. Brederick et al.*, Angew. chem., 1958, **70**, 268). Heating sulphinic acids with dilute mineral acids provides a convenient method for the preparation of symmetrical thiolsulphonates (*Smiles* and *D. T. Gibson*, J. chem. Soc., 1924, **125**, 176; *T. P. Hilditch, ibid.*, 1910, **97**, 1091). Unsymmetrical thiolsulphonates may be obtained by reacting sulphinic acids with disulphides in acid solution (*Kice* and *E. H. Morkved*, J. Amer. chem. Soc., 1964, **86**, 2270):

$$RSO_2H + R'S{\cdot}SR' \xrightarrow[H_2SO_4]{HOAc} RSO_2SR'$$

(ii) Oxidation and reduction

Sulphinic acids are oxidised to sulphonic acids with air, peroxides or peracids. Sulphur and sodium polysulphides react similarly to give salts of thiolsulphonic acids (*Kurzer* and *Powell, loc. cit.*). Oxidation with potassium permanganate gives disulphones, $RSO_2{\cdot}SO_2R$, as by-products (*Hilditch*, J. chem. Soc., 1908, **93**, 1524). Reduction of sulphinic acids with zinc and sulphuric acid gives thiophenols (*Gatterman, loc. cit.*; *W. P. Winter*, Amer. chem. J., 1904, **31**, 572), and with lithium aluminium hydride gives disulphides (*J. Strating* and *H. J. Backer*, Rec. Trav. chim., 1950, **69**, 638). The sulphinyl group may be removed from aromatic sulphinic acids by treatment with aqueous mercuric chloride to give the arylmercuric chloride, which is subsequently hydrolysed with hydrochloric acid (*T. Okamoto* and *J. F. Bunnett*, J. Amer. chem. Soc., 1956, **78**, 5357):

$$ArSO_2H \xrightarrow{HgCl_2} ArHgCl \xrightarrow{HCl} ArH$$

(iii) Alkylation, etc.

Although the sulphinate anion is an ambident nucleophile, in the vast majority of cases studied sulphur is the predominant nucleophilic atom. Thus salts of sulphinic acids react with primary and secondary alkyl halides and reactive aryl halides to give sulphones as the major product:

$$RSO_2Na + MeI \rightarrow RSO_2Me + NaI$$

Similarly sulphinate anions react as S-nucleophiles with sulphenyl halides, sulphinyl halides and sulphonyl halides to give thiolsulphonates, sulphinyl sulphones and disulphones, respectively:

$$RSO_2^{\ominus} + RSCl \rightarrow RSO_2{\cdot}SR + Cl^{\ominus}$$

(*Stirling*, J. chem. Soc., 1957, 3597);

$$RSO_2^{\ominus} + RSOCl \rightarrow RSO_2\cdot SOR + Cl^{\ominus}$$

(*Bredereck et al.*, Ber., 1960, **93**, 2736);

$$RSO_2^{\ominus} + RSO_2Cl \rightarrow RSO_2\cdot SO_2\cdot R + Cl^{\ominus}$$

(*E. P. Kohler* and *M. B. MacDonald*, Amer. chem. J., 1899, **22**, 219). Alkylation with "hard" reagents, *e.g.* dimethyl sulphate, diazomethane, methyl tosylate, alkoxyphosphonium salts, gives predominantly the sulphinate ester (*J. S. Meek* and *J. S. Fowler*, J. org. Chem., 1968, **33**, 3421):

$$RSO_2Na + (MeO)_2SO_2 \xrightarrow[25^0]{DMF} \underset{(88\%)}{RSO\cdot OMe} + \underset{(12\%)}{RSO_2Me}$$

(iv) Reactions of halogens; addition compounds

Reaction of sodium sulphinates with halogens gives sulphonyl halides. With iodine an equilibrium occurs and sulphonyl iodide formation is favoured by electron-releasing substituents (*O. Foss*, Kgl. Norske Videnskab. Selskabs. Forh., 1946, **19**, No. 19, 68; C.A., 1948, **42**, 19f).

Sulphinic acids and their salts react additively with a large variety of compounds. They form sulphones by addition to $\alpha\beta$-unsaturated acids, ketones, esters, amides and nitriles (see *e.g. H. Gilman* and *L. F. Cason*, J. Amer. chem. Soc., 1950, **72**, 3469):

$$RCH{=}CH\cdot COR^1 + R^2SO_2H \rightarrow RCH(SO_2R^2)\cdot CH_2\cdot COR^1$$

aldehydes give addition compounds of the general type $RCH(OH)\cdot SO_2R$ (*Field* and *P. H. Settlage*, *ibid.*, 1951, **73**, 5870), and *p*-quinones give 2,5-dihydroxyaryl sulphones. Addition to *N,N*-dialkylquinone diimines gives a variety of products depending on the pH and the temperature (*K. T. Finley et al.*, J. org. Chem., 1969, **34**, 2083). Sulphinic acids also attack the nucleus of mono- and di-hydric phenols and arylamines with the formation of sulphones or sulphoxides, and sulphides (*O. Hinsberg*, Ber., 1903, **36**, 107).

(v) Sulphinyl chlorides

Thionyl chloride reacts with sulphinic acids or their sodium salts to give sulphinyl chlorides (*L. C. Raiford* and *S. E. Hazlet*, J. Amer. chem. Soc., 1935, **57**, 2172; *Kurzer*, Org. Synth., Coll. Vol. IV, 1963, p. 937). Sulphinyl chlorides react with alcohols, amines and thiols, preferably in the presence of a base, to give sulphinate esters, sulphinamides, RSONHR, and thiolsulphinates, RSO·SR, respectively. They react with zinc in ethereal solution to give thiolsulphonates (*D. Barnard*, J. chem. Soc., 1957, 4673), and are reduced by lithium aluminium hydride to give disulphides (*Strating* and *Backer*, *loc. cit.*).

(*vi*) *Sulphinyl sulphones*

Treatment of sulphinic acids with a mixture of acetic anhydride and sulphuric acid gives compounds originally formulated as sulphinic acid anhydrides, RSO·O·SOR (*E. Knovenagel* and *L. Polack*, Ber., 1908, **41**, 3323). These compounds have proved to be sulphinyl sulphones, RSO·SO_2R (*Bredereck et al.*, *loc. cit.*).

(*vii*) *Sulphinyl esters*

Sulphinate esters are also formed by reaction of the acid with alcohols containing dry hydrogen chloride or by the action of chloroformic esters on the salts of sulphinic acids (*Otto* and *A. Rossing*, Ber., 1885, **18**, 2493). The hydrolysis of sulphinate esters is subject to acid or base catalysis. Methyl *p*-toluenesulphinate hydrolyses by a bimolecular mechanism involving sulphur-oxygen bond fission (*C. A. Bunton* and *B. N. Hendy*, J. chem. Soc., 1962, 2562), but acetolysis of the (−)-1-phenylethyl ester gives a racemic acetate and presumably involves fission of the carbon–oxygen bond (*M. P. Balfe*, *J. Kenyon* and *A. N. Tárnoky*, *ibid.*, 1943, 446). Esters of the latter type, that is those capable of forming stable carbonium ions, rearrange on heating, particularly in acidic solvents, to form sulphones, with loss of optical activity (*C. L. Arcus*, *Balfe* and *Kenyon*, *ibid.*, 1938, 485; *A. H. Wragg*, *J. S. McFadyen* and *T. S. Stevens*, *ibid.*, 1958, 3603). Alkyl *p*-toluenesulphinates give unrearranged alcohols on heating in aprotic solvents like *N*-methyl-2-pyrrolidone (*J. W. Wilt*, *R. G. Stein* and *W. J. Wagner*, J. org. Chem., 1967, **32**, 2097). Thionyl chloride reacts with sulphinate esters to give sulphinyl chlorides and alkyl chlorosulphites (*H. F. Herbrandson*, *R. T. Dickerson Jr.* and *J. Winstein*, J. Amer. chem. Soc., 1956, **78**, 2576):

$$\text{RSO·OR}' \xrightarrow{\text{SOCl}_2} \text{RSO·Cl} + \text{R}'\text{O·SO·Cl}$$

Sulphinate esters are dissymmetric and configurationally stable under ordinary conditions. The asymmetric synthesis of menthyl sulphinates from (—)-menthol has been developed as a general method for use in the assignment of the absolute configuration of sulphinate esters and the optical rotatory dispersion of these compounds investigated (*K. Mislow et al.*, *ibid.*, 1965, **87**, 1958).

(viii) *Arenesulphinamides*

Arenesulphinamides are also obtained by the reaction of thionylarylamines with Grignard reagents or alkyl-lithiums (*D. Klamann, C. Sass* and *M. Zelenka*, Ber., 1959, **92**, 1910):

$$ArN{=}S{=}O + Ar'MgBr \rightarrow ArNHSOAr'$$

(ix) *Sulphines*

Hydrogen chloride can be eliminated from aralkylsulphinyl chlorides, *e.g.*, phenylmethanesulphinyl chloride, by treatment with triethylamine in ether to give a solution containing a sulphine (*Hamid* and *Trippett, loc. cit.*):

$$e.g.\quad PhCH_2SOCl + Et_3N \rightarrow PhCH{:}S{:}O + Et_3NH^{\oplus}Cl^{\ominus}$$

(x) *Individual arenesulphinic acids*

Benzenesulphinic acid, $PhSO_2H$, prisms, m.p. 84°, is prepared by reduction of benzenesulphonyl chloride with alkaline sodium sulphite (*Barnard, loc. cit.*). The free acid is obtained by acidification of a filtered solution of the sodium salt with 6*M*-hydrochloric acid and the precipitate dried under vacuum. The sodium, potassium, magnesium and barium salts are soluble in water, the ferric and zinc salts almost insoluble. The acid is easily soluble in hot water. *Benzenesulphinyl chloride*, m.p. 38°, b.p. 65°/0·012 mm, prepared by reacting the acid with redistilled thionyl chloride in ether at 0° (*Barnard, loc. cit.*), gives the *methyl ester*, b.p. 67–68°/0·04 mm, n_D^{25}, 1·5434, with methanol and pyridine; and the *sulphinamide*, m.p. 122°, with ammonia; it is rapidly hydrolysed by water. Methyl benzenesulphinate is best prepared by the oxidation of diphenyl disulphide with lead tetra-acetate in methanol (*L. Field* and *J. M. Locke*, Org. Synth., 1966, **46**, 62). *Phenyl ester*, m.p. 51–52°; *anilide*, m.p. 115°; *S-benzylisothiuronium salt*, m.p. 154–155°. *Benzenesulphinyl azide* is prepared by the reaction of benzenesulphinyl chloride with sodium azide in acetonitrile at —35°. It decomposes slowly at —20° to give benzenesulphonyl azide, diphenyl disulphide and a compound postulated to contain a trithiatriazine ring. At room temperatures it decomposes explosively (*T. J. Maricich*, J. Amer. chem. Soc., 1968, **90**, 7179).

o-**Toluenesulphinic acid,** m.p. 80°, is easily soluble in many organic solvents; the *methyl ester*, b.p. 66·5°/0·15 mm is obtained by the oxidation of *o*-tolyl disulphide with lead tetra-acetate in methanolic chloroform (*Field et al., loc. cit.*); S-*benzylisothiuronium salt*, m.p. 178–179°.

p-**Toluenesulphinic acid,** m.p. 85°, is prepared by the reduction of the sulphonyl chloride with zinc dust and water (*F. C. Whitmore* and *F. H. Hamilton*, Org. Synth., Coll. Vol. I, 1941, p. 492), or by sodium sulphite and bicarbonate (*Field* and *R. D. Clark*, Org. Synth., Coll. Vol. IV, 1963, p. 674). p-*Toluenesulphinyl chloride*, needles, m.p. 54–58° (*Hilditch* and *Smiles*, Ber., 1908, **41**, 4113), or a lemon yellow oil, b.p. 79°/0·012 mm, n_D^{20} 1·6007, is prepared by reacting thionyl chloride with sodium *p*-toluenesulphinate (*Kurzer*, Org. Synth., Coll. Vol. IV, 1963, p. 937; *amide*, m.p. 120°; *methyl ester*, b.p. 60–65°/0·1 mm; the (±)-*ethyl ester*, b.p. 99–104°/0·1 mm, d_4^{25} 1·114, n_D^{25} 1·5309 has been resolved, the (—)-ester has $[\alpha]_{5461}^{25}$ — 3·25; (—)-p-*toluenesulphinylanilide*, m.p. 134°, $[\alpha]_{5461}^{17}$ — 1·3 ($CHCl_3$) (*H. Phillips*, J. chem. Soc., 1925, **127**, 2552), the stereochemical course of the preparation from (—)-(*S*)-menthyl *p*-toluenesulphinate and lithium anilide in ether has been studied (*A. Nudelman* and *D. J. Cram*, J. Amer. chem. Soc., 1968, **90**, 3869); n-*butyl ester*, b.p. 90–95°/0·1 mm, n_D^{25} 1·5195, d_4^{25} 1·066; *phenyl ester*, m.p. 51–52°; S-*benzylisothiuronium salt*, m.p. 167–168°.

o-**Nitrobenzenesulphinic acid,** m.p. 143°, has been prepared by the reduction of *o*-nitrobenzenesulphonyl chloride with the calculated amount of stannous chloride and hydrochloric acid (*M. Claasz*, Ann., 1911, **380**, 303) and from *o*-nitroaniline by diazotisation and treatment of the diazonium salt with sulphur dioxide, ferrous sulphate and copper (*G. Wittig* and *R. W. Hoffmann*, Org. Synth., 1967, **47**, 4). Catalytic reduction and diazotisation gives 1,2,3-benzothiadiazole 1,1-dioxide, which decomposes smoothly at 10° to give dehydrobenzene ("benzyne"), nitrogen and sulphur dioxide. m-*Nitrobenzenesulphinic acid*, m.p. 97°, o-*ethoxybenzenesulphinic acid*, m.p. 92°, p-*chlorobenzenesulphinic acid*, m.p. 94°, p-*bromobenzenesulphinic acid*, m.p. 114°, have been obtained by reduction of the appropriate sulphonyl chloride with sodium sulphite (*Gur'yanov* and *Syrkin, loc. cit.*).

o-**Benzenedisulphinic acid,** m.p. 113–115°, is prepared by the reduction of the corresponding disulphonyl dichloride; it is stable at low temperatures (*J. B. Hendrickson et al.*, J. org. Chem., 1969, **34**, 3434).

Phenylmethanesulphinic acid, $PhCH_2 \cdot SOH_2$, from the reduction of the sulphonyl chloride, is resistant to disproportionation; *S-benzylisothiuronium salt*, m.p. 152°. Treatment of phenylmethanesulphinyl chloride with triethylamine in ether at room temperature gives a solution which is postulated to contain phenylsulphine, $C_6H_5CH{:}S{:}O$. Phenylsulphine, generated as above, does not react with aldehydes, phenyl isocyanate, benzoyl chloride or 1,3-dipolar systems. The solution reacts with 2,4-dinitrophenylhydrazine to give benzaldehyde 2,4-dinitrophenylhydrazone and benzaldehyde is also produced by irradiation of the solution with u.v. light. Enamines give a dipolar intermediate from which a benzylsulphinyl enamine is obtained:

e.g. [morpholino-cyclohexene] + $C_6H_5 \cdot CH:S:O$ ⟶ [morpholinium-ylidene cyclohexyl $S(O) \cdot \overset{\ominus}{C}H \cdot C_6H_5$] ⟶ [morpholino-cyclohexenyl $S(O) \cdot CH_2 \cdot C_6H_5$]

Phenylsulphine reacts with the ylid, $Ph_3P{:}CH \cdot CO_2Et$, to give the stable ylid, $Ph_3P{:}C(SO \cdot CH_2Ph)CO_2Et$. The solution of phenylsulphine slowly decomposes at room temperature to give a mixture of *cis-* and *trans*-stilbenes, *trans*-4,5-diphenyl-1,2,3-trithiolane 1,1-dioxide (III) and the 1,1- or 2,2-dioxide of the corresponding 5,6-diphenyltetrathiane (*Hamid* and *Trippett, loc. cit.*):

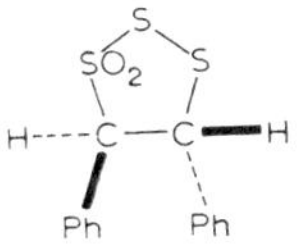

(III)

3. Derivatives of sulphenic acids

There is no authentic record of the isolation of sulphenic acids of the type RS·OH, in the benzene series, although compounds are known which can be regarded as being derivatives of such acids *e.g.*, sulphenyl halides RSBr and RSCl, sulphenamides, $RSNH_2$, alkyl and aryl sulphenates, RS·OAlk and RS·OAr, mixed anhydrides, *e.g.* acetic sulphenic anhydrides, $RS{\cdot}OCOR^1$, etc. The sulphenate anion, $RSO^{\ominus}$, like the sulphinate anion, is in principle an ambident nucleophile, but appears to react largely through the sulphur atom. Thus alkaline hydrolysis of azobenzene-2-sulphenyl bromide in the presence of dimethyl sulphate gives the aryl methyl sulphoxide in high yield (*A. Burawoy* and *A. Chaudhuri*, J. chem. Soc., 1956, 653).

(a) Preparation of sulphenic acid derivatives

Sulphenyl halides are prepared by reacting thiophenols or disulphides with dry bromine or chlorine in anhydrous solvents, frequently in the presence of a catalyst such as iodine or an aluminium halide (*T. Zincke et al.*, Ber., 1911, **44**, 769; Ann., 1912, **391**, 55; *H. Lecher* and *F. Holschneider*, Ber., 1924, **57**, 755):

$$RSH + Cl_2 \rightarrow RSCl + HCl$$

$$RS{\cdot}SR + Cl_2 \rightarrow 2\ RSCl$$

Sulphuryl chloride and pyridine may be used as an alternative to chlorine (*N. Kharasch*, U.S.P., 2,929,820/1960; C.A., 1960, **54**, 15318d; *D. D. Lawson* and *Kharasch*, J. org. Chem., 1959, **24**, 857). With bromine, side reactions frequently occur (see *e.g.*, *D. Spinelli*, Boll. Sci. Fac. Chim. Ind. Bologna, 1961, **19**, No. 1, 26; C.A., 1962, **56**, 2368c) and sulphenyl bromides are more conveniently obtained by shaking a benzene solution of the sulphenyl chloride with potassium bromide (*Kharasch, C. M. Buess* and *S. I. Strashun*, J. Amer. chem. Soc., 1952, **74**, 3422).

The other derivatives of sulphenic acids are best prepared from the sulphenyl halides. Reaction of sulphenyl halides with ammonia and with primary and secondary amines gives sulphenamides (*J. H. Billman et al.*, *ibid.*, 1941, **63**, 1920). Sulphenate esters are obtained by treating the sulphenyl halides with alcohols in the presence of pyridine (*Kharasch, D. P. McQuarrie* and *Buess, ibid.*, 1953, **75**, 2658) or with the sodium alkoxide or phenoxide (*Zincke* and *F. Farr*, Ann., 1912, **391**, 57). The lithium salts of allyl, crotyl and α-methallyl alcohols react with *p*-toluenesulphenyl chloride to give the rearranged sulphoxide. 2,4-Dinitrobenzenesulphenyl chloride gives the unrearranged sulphenate ester (*K. Mislow et al.*, J. Amer. chem. Soc., 1968, **90**, 4869). Sulphenyl acetates are formed by shaking a solution of the sulphenyl chloride in benzene with freshly fused sodium acetate (*A. J. Havlik* and *Kharasch, ibid.*, 1956, **78**, 1207).

(b) Reactions of sulphenyl compounds

In general the reactivity of the derivatives of sulphenic acids decreases in the order, sulphenyl halides > sulphenyl acetates > alkyl and aryl sulphenates > sulphenamides.

(1) Sulphenyl halides and acetates are hydrolysed readily by water, but with the sulphenate esters the uncatalysed hydrolysis is very slow. The base-catalysed hydrolysis of sulphenate esters is considered to involve an "S_N2-type" displacement at the sulphur atom (*C. Brown* and *D. R. Hogg*, Chem. Comm., 1967, 38). The free sulphenic acid has not been isolated from the hydrolysis. In neutral or weakly alkaline media a product is obtained which was formerly regarded as the sulphenic anhydride, RS·O·SR, but has now been shown to be the thiolsulphinate, RSO·SR, or in some cases a mixture of the disulphide, the thiolsulphonate, RSO_2·SR, and the thiolsulphinate (*E. Vinkler* and *F. Klivényi*, Acta chim. Acad. Sci. Hung., 1957, **11**, 15; C.A., 1958, **52**, 6242c; *S. Oae* and *S. Kawamura*, Ann. Rept. Radiation Centre Osaka Prefect., 1961, **2**, 120; C.A., 1964, **58**, 5556g):

$$2\,RSCl + H_2O \rightarrow RSO\cdot SR + 2\,HCl$$

$$2\,RSO\cdot SR \rightarrow RS\cdot SR + RSO_2\cdot SR$$

On hydrolysis with excess alkali the disulphide and sulphinic acid are produced under conditions in which the disulphide precipitates.

$$3\,RSCl + 4\,OH^{\ominus} \rightarrow RSO_2^{\ominus} + RS\cdot SR + 2\,H_2O + 3\,Cl^{\ominus}$$

When the reaction mixture remains homogeneous the disulphide is hydrolysed further to give the thiol (*D. R. Hogg* and *P. W. Vipond*, J. chem. Soc. [B], 1970, 1242):

$$2\ ArSCl + 4\ OH^{\ominus} \rightarrow ArSO_2 + ArS^{\ominus} + 2\ H_2O + 2\ Cl^{\ominus}$$

(2) The halogen in sulphenyl halides is reactive and is readily replaced by nucleophiles, thus reaction with potassium cyanide gives thiocyanates (*Zincke et al.*, Ann., 1913, **400**, 1), and potassium thiocyanate gives sulphenyl thiocyanates (*Kharasch, H. L. Wehrmeister* and *H. Tigerman*, J. Amer. chem. Soc., 1947, **69**, 1612):

$$RSCl + CNS^{\ominus} \rightarrow RS{\cdot}SCN + Cl^{\ominus}$$

Reaction with thiols gives disulphides (*I. Danielsson, J. E. Christian* and *G. L. Jenkins*, J. Amer. pharm. Assoc., 1947, **36**, 261), and with sodium sulphite gives Bunte salts, $RS{\cdot}SO_3^{\ominus}Na^{\oplus}$; the corresponding ammonium salts are obtained by treating sulphenamides with aqueous sulphurous acid (*H. Z. Lecher* and *E. M. Hardy*, J. org. Chem., 1955, **20**, 475):

$$RSCl + Na_2SO_3 \rightarrow RSSO_3Na + NaCl$$

Reactions with the silver salts of sulphinic acids give thiolsulphonates (*Zincke* and *Farr, loc. cit.*):

$$RSCl + R'SO_2Ag \rightarrow RS{\cdot}SO_2R' + AgCl$$

(3) Sulphenyl halides react with monosulphides to give disulphides together with the corresponding alkyl or aralkyl halides (*C. G. Moore* and *M. Porter*, Tetrahedron, 1960, **9**, 58):

$$R'SCl + RSR \xrightarrow[(20^{0})]{HOAc} R'S{\cdot}SR + RCl$$

and with disulphides exchange reactions occur also (*idem*, J. chem. Soc., 1958, 2890):

$$R'SCl + RS{\cdot}SR \xrightleftharpoons{HOAc} R'S{\cdot}SR + RSCl$$

(4) Sulphenyl halides readily add to olefins to give, in general, aryl β-chloroalkyl sulphides:

$$CH_2{=}CH_2 + ArSCl \rightarrow ClCH_2{\cdot}CH_2SAr$$

With aryl substituted olefins and with unsymmetrically substituted dialkylethenes the corresponding vinyl sulphides may also be produced, sometimes as the only product:

$$e.g., \ Ph_2C{=}CH_2 + 2{,}4\text{-}(NO_2)_2C_6H_3SCl \rightarrow Ph_2C{=}CHS{\cdot}C_6H_3(NO_2)_2 + HCl$$

The addition is *trans*-stereospecific (*D. J. Cram*, J. Amer. chem. Soc., 1949, **71**, 3883; *Kharasch* and *Havlik*, *ibid.*, 1953, **75**, 3734) and is postulated to proceed through an intermediate episulphonium cation, $\left[>C\overset{\oplus}{\underset{S-Ar}{\cdots}}C<\right] Cl^{\ominus}$ (*Kharasch* and *Buess*, J. Amer. chem. Soc., 1949, **71**, 2724). Under thermodynamically controlled conditions the addition follows Markownikoff's rule, but under kinetic control the orientation of the product changes to predominantly anti-Markownikoff as the substituents in the sulphenyl halide become increasingly electron-releasing (*G. M. Beverly* and *Hogg*, Chem. Comm., 1966, 138; *Beverly*, *Hogg* and *J. H. Smith*, Chem. and Ind., 1968, 1403; *Hogg* and *Smith* Mech. of Reactions of Sulphur Comps., 1968, **3**, 63).

This reaction has been used for the characterisation of olefins using 2,4-dinitrobenzenesulphenyl chloride; the m.p.'s of all derivatives prepared up to 1956 are given by *R. B. Langford* and *D. D. Lawson* (J. chem. Educ., 1957, **34**, 510). Sulphenyl acetates and sulphenate esters react similarly with olefins but require a much longer reaction time (*Havlik* and *Kharasch* 1956, *loc. cit.*; *Hogg*, *P. W. Vipond* and *Smith*, J. chem. Soc. [C], 1968, 2713).

(5) Sulphenyl halides add to acetylenes to give aryl β-chlorovinyl sulphides (*Kharasch* and *S. J. Assony*, J. Amer. chem. Soc., 1953, **75**, 1081):

$$RC{\equiv}CR + ArSCl \rightarrow \underset{\displaystyle Cl}{RC} = \overset{\displaystyle SAr}{CR}$$

The orientation of addition is dependent on the solvent and the substituents in the arenesulphenyl halide. The percentage of Markownikoff addition increases with an increase in the polarity of the solvent (*G. Modena et al.*, Tetrahedron Letters, 1965, 4399, 4405; Mech. of Reactions of Sulphur Comps., 1968, **3**, 115).

(6) In the presence of aluminium chloride, sulphenyl halides react with aromatic compounds which do not have strongly electron-withdrawing substituents, to give diaryl sulphides (*Kharasch* and *R. Swidler*, J. org. Chem., 1954, **19**, 1704; *Buess* and *Kharasch*, J. Amer. chem. Soc., 1950, **72**, 3529):

$$ArH + Ar'SCl \xrightarrow{AlCl_3} ArSAr'$$

With reactive aromatic compounds, *e.g.* amines and phenols, a catalyst is not necessary (*Zincke* and *Farr, loc. cit.*; *Buess* and *Kharasch*, 1950, *loc. cit.*).

(7) Sulphenyl chlorides react readily with ketones having an enolisable hydrogen to give aryl β-ketoalkyl sulphides (*T. Zincke* and *J. Baeumer*, Ann., 1918, **416**, 86; *J. A. Barltrop* and *K. J. Morgan*, J. chem. Soc., 1960, 4486):

$$RCH_2{\cdot}COR + ArSCl \rightarrow RCH(SAr){\cdot}COR + HCl$$

and with the sodium salts of nitroalkanes to give α-nitro sulphides (*Kharasch* and *J. L. Cameron*, J. Amer. chem. Soc., 1951, **73**, 3864):

$$R_2C^{\ominus}NO_2\ Na^{\oplus} + ArSCl \rightarrow R_2C(SAr)NO_2 + NaCl$$

(8) The photolysis of 2,4-dinitrobenzenesulphenyl derivatives of carboxylic acids, $2,4\text{-}(NO_2)_2C_6H_3S{\cdot}O{\cdot}CO{\cdot}R$, in benzene solution gives the carboxylic acid and 2,4-dinitrophenyl sulphide in high yield.

The reaction was shown to follow a polar mechanism; in contrast, photolysis of the corresponding *p*-toluenesulphenate gives products derived from the *p*-methylphenoxy radical together with the disulphide. In all these reactions minor products are obtained in which oxygen transfer from the *o*-nitro group to the sulphur occurs (*D. H. R. Barton et al.*, J. chem. Soc., 1965, 3571; [C], 1968, 322).

(9) Aryl esters of sulphenic acids and sulphenanilides rearrange on heating to form the corresponding hydroxy- or amino-diaryl sulphides (*M. L. Moore* and *T. B. Johnson*, J. Amer. chem. Soc., 1935, **57**, 1517; *Hogg, Vipond* and *Smith, loc. cit.*). Benzyl *p*-toluenesulphenate undergoes a thermal rearrangement by what appears to be an intramolecular concerted mechanism to give benzyl *p*-tolyl sulphoxide (*Mislow et al.*, J. Amer. chem. Soc., 1968, **90**, 4861). (*S*)-α-Methylallyl *p*-toluenesulphenate is postulated to rearrange at room temperature to give (*S*)-*trans*-crotyl *p*-tolyl sulphoxide by a cyclic concerted mechanism (*idem, ibid.*, p. 4869).

(10) With dry hydrogen chloride sulphenamides give the amine and the sulphenyl chloride, and consequently sulphenyl groups have been used as protecting groups in the syntheses of peptides (*L. Zervas, D. Borovas* and *E. Gazis, ibid.*, 1963, **85**, 3660; *P. H. Bentley et al.*, J. chem. Soc., 1964, 6130).

Sulphenate esters decompose similarly with dry hydrogen chloride (*Hogg, Vipond* and *Smith, loc. cit.*)

(11) *N*-Alkyl- or *N*-aralkyl-benzenesulphenamides on treatment with copper (II) carboxylates and triphenylphosphine in methylene chloride at room temperatures give carboxamides (*I. Mukaiyama*, J. Amer. chem. Soc., 1968, **90**, 4490):

$$(R'CO_2)_2Cu + 2\ RSNR''R''' + 2\ Ph_3P \rightarrow 2\ R'CO{\cdot}NR''R''' + 2\ Ph_3PO + (RS)_2Cu$$

(12) Treatment of an acetic acid solution of a sulphenyl halide with sodium iodide quantitatively gives the disulphide and iodine.

$$2\ ArSCl + 2\ I^{\ominus} \rightarrow ArS{\cdot}SAr + I_2 + 2\ Cl^{\ominus}$$

This reaction has been used for the determination of sulphenyl halides (*Kharasch* and *M. Wald*, Analyt. Chem., 1955, **27**, 996).

The chemistry of sulphenic acid derivatives has been reviewed by *Kharasch et al.* (Chem. Reviews, 1946, **39**, 269; J. chem. Educ., 1956, **33**, 585; Quart. Reports on Sulfur Chem., 1966, **1**, 93), by *E. E. Reid* ("Organic Chemistry of Bivalent Sulfur", Vol. I, Chem. Pub. Co. Inc., New York, 1958, p. 262) and by *A. Schöberl* and *A. Wagner*, "Methoden der Organischen Chemie [Houben-Weyl]", Vol. 9, *E. Müller*, Ed., G. Thieme Verlag, Stuttgart, 1955, p. 263).

(c) Individual sulphenyl compounds

Benzenesulphenyl chloride (*phenyl sulphur chloride*), PhSCl, is a deep red oil, b.p. 58°/3 mm, and has been reported to explode spontaneously after standing undisturbed for several months. With mercuric carboxylates at room temperature diphenyl disulphide, the acid anhydride and mercuric chloride are formed (*Mukaiyama et al.*, J. org. Chem., 1968, **33**, 2242). *Methyl benzenesulphenate*, PhSOMe, b.p. 88–89°/4 mm; *Benzenesulphen-anilide*, PhSNHPh, colourless crystals, m.p. 53–55°; *-dimethylamide*, $PhSNMe_2$, colourless oil, b.p. 65°/3 mm; *-diethylamide*, b.p. 90°/4 mm.

p-**Toluenesulphenyl chloride,** prepared by the action of chlorine on *p*-toluenethiol (*p*-thiocresol) in carbon tetrachloride solution in the presence of iodine (*Kurzer* and *J. R. Powell*, Org. Synth., Coll. Vol. IV, 1963, p. 934), red oil, b.p. 66–68°/0·8 mm, reacts with ammonia to form *di-p-toluenesulphenimide*, $(p\text{-}CH_3\text{-}C_6H_4S)_2NH$, m.p. 109–111° (*Kurzer*, J. chem. Soc., 1953, 3360); *anilide*, m.p. 81°; *benzyl ester*, b.p. 105/0·07 mm.

o-**Nitrobenzenesulphenyl chloride,** yellow needles, m.p. 75°, is prepared by the chlorinolysis of 2,2′-dinitrodiphenyl disulphide in dry carbon tetrachloride solution in the presence of iodine (*M. H. Hubacher,* Org. Synth., Coll. Vol. II., 1943, p. 455). Hydrolysis in aqueous methanol under reflux gives a mixture of the corresponding disulphide and sulphinic acid (*Zincke* and *Farr, loc. cit.*). Hydrolysis with cold water gives a product which is suggested to be a mixture of the corresponding thiolsulphinate, thiolsulphonate and disulphide (*Oae* and *Kawamura, loc. cit.*). Oxidation with concentrated nitric acid in acetic acid gives a mixture of 2-nitrobenzenesulphonic acid and the sulphonyl chloride. Reaction with anhydrous hydrogen fluoride gives bis(2,2′-fluorosulphonyl)-azobenzene (*D. L. Chamberlain, D. Peters* and *Kharasch,* J. org. Chem., 1958, **23,** 381). With cyclohexene in acetic acid it forms 2-*chlorocyclohexyl* 2-*nitrophenyl sulphide,* m.p. 103°, and with acetophenone forms ω-(o-*nitrophenylthio*)-*acetophenone,* m.p. 145–146°. 2-*Nitrobenzenesulphenyl bromide,* orange needles, m.p. 85°, obtained by shaking a benzene solution of the sulphenyl chloride with sodium bromide for several days, reacts with cyclohexene to give 2-*bromocyclohexyl* 2-*nitrophenyl sulphide,* m.p. 99°, but with acetophenone gives phenacyl bromide and 2,2′-dinitrodiphenyl disulphide. The *anilide,* m.p. 95°, rearranges on heating to give 2-nitro-4′-aminodiphenyl sulphide (*Moore* and *Johnson, loc. cit.*), and on hydrolysis with methanolic sodium hydroxide solution gives sodium azobenzene-2-sulphinate (*M. P. Cava* and *C. E. Blake,* J. Amer. chem. Soc., 1956, **78,** 5444; *C. Brown, ibid.,* 1969, **91,** 5832). *Amide,* m.p. 124°.

N-o-*Nitrobenzenesulphenylurea,* m.p. 235–238° (decomp.), and N,N′-*di*-o-*nitrobenzenesulphenylurea,* m.p. 299–300° (decomp. with explosive violence) are obtained from the sulphenyl chloride and urea in pyridine. Arylsulphenyl derivatives of arylureas can be obtained from arylsulphenamides and aryl isocyanates:

$$RSNH_2 + R'NCO \rightarrow RSNHCONHR'$$

N-o-*Nitrobenzenesulphenyl*-N′-*phenylurea,* pale yellow needles, m.p. 232–234° (decomp.) (*Kurzer, loc. cit.*).

The structure of the *methyl ester,* m.p. 54°, has been determined by X-ray diffraction (*W. C. Hamilton* and *S. J. La Placa,* J. Amer. chem. Soc., 1964, **86,** 2289); with the exception of the methyl group the molecule is planar, and the oxygen of the nitro-group exerts a strong non-bonding interaction on the sulphur atom. On heating under reflux for 7 days with a methanolic solution of cyclohexene the methyl ester gives 2-*methoxycyclohexyl* 2-*nitrophenyl sulphide,* m.p. 57–58°. The *phenyl ester,* m.p. 72°, rearranges on heating under reflux in carbon tetrachloride solution to give a mixture of 2-hydroxy- and 4-hydroxy- 2′-nitrodiphenyl sulphide (*Hogg, Vipond* and *Smith, loc. cit.*).

p-**Nitrobenzenesulphenyl chloride,** yellow scales, m.p. 52°; *amide,* deep yellow needles, m.p. 103°; *anilide,* m.p. 75°; *methyl ester,* m.p. 49°.

2,4-**Dinitrobenzenesulphenyl chloride,** m.p. 95–96°, obtained by treatment of benzyl 2,4-dinitrophenyl sulphide with sulphuryl chloride and pyridine (*Langford* and *Kharasch,* Org. Synth., 1964, **44,** 47), can be used to form derivatives with amines, epoxides, alcohols, thiols, aromatic compounds, olefins, acetylenes, ketones, and nitroalkanes; details of procedures and m.p.'s are given by *Langford* and *Lawson* (*loc. cit.*). On hydrolysis in acid solution it gives 2-amino-4-nitrobenzenesulphonic acid (*Kharasch, W. King* and *T. C. Bruice*, J. Amer. chem. Soc., 1955, **77**, 931). *Bromide*, m.p. 95–98^0; *amide* m.p. 120^0; *anilide*, m.p. 143^0; *methyl ester*, m.p. 125^0.

Pentachlorobenzenesulphenyl chloride, Cl_5C_6SCl, orange needles m.p. 99–101^0, prepared by chlorination of the thiol in carbon tetrachloride solution in the presence of iodine (*R. E. Putnam* and *W. H. Sharkey*, *ibid.*, 1957, **79**, 6526), reacts with excess of cyclohexane in diffused sunlight to give hydrogen chloride, cyclohexyl chloride, cyclohexyl pentachlorophenyl sulphide, and the disulphide. Similar light induced reactions are given with toluene, *p*-benzoquinone and diphenylacetylene (*Kharasch* and *Z. S. Ariyan*, Chem. and Ind., 1964, 929).

4. Arenethiolsulphonic acids and arenethiolsulphinic acids

(a) *Arenethiolsulphonic acids*

The aromatic thiolsulphonic acids, RSO_2SH, are unstable, but the salts and esters are known.

(*i*) *Salts*

The alkali salts are formed by heating the sodium salts of sulphinic acids with a saturated solution of sulphur in acetone (*F. Kurzer* and *J. R. Powell*, J. chem. Soc., 1952, 3728), or by the reaction of arylsulphonyl chlorides with alkali sulphides (*R. Otto et al.*, Ber., 1882, **15**, 121; 1892, **25**, 1477; *Ya. A. Stoyanovskaya* and *B. G. Boldyrev*, Zhur. org. Khim., 1969, **5**, 62). Acidification of the alkali salt with mineral acids gives a diarylsulphonyl trisulphide, an aryl arylthiolsulphonate and sulphur in approximately equal amounts:

$$4\ RSO_2SH \rightarrow (RSO_2)_2S_3 + RSO_2SR + S + 2\ H_2O$$

It is proposed that the reaction initially forms the sulphinic acid and sulphur which then react further yielding the above products (*Kurzer* and *Powell*, *loc. cit.*). Diarylsulphonyl di- and tri-sulphides are formed by the action of chlorine or iodine on the alkali salts in aqueous solution. The action of sulphur monochloride on potassium benzenethiolsulphonate yields dibenzenesulphonyl tetrasulphide (*J. Troeger* and *V. Hornung*, J. pr. Chem., 1899, [ii], 60, 113).

The preparation of salts of 4-chloro-, 4-bromo- and 4-iodo-benzenethiolsulphonic acids is described by *Troeger et al.* (*ibid.*, 1902, [ii], **65**, 82; 1904, [ii], **70**, 375).

S-**Benzylisothiuronium salts**: of *benzenethiolsulphonic acid*, m.p. 91–93°; o-*toluenethiolsulphonic acid*, m.p. 137–138°; p-*toluenethiolsulphonic acid*, m.p. 120–121°; p-*bromobenzenethiolsulphonic acid*, m.p. 146–147° (*Kurzer* and *Powell*, *loc. cit.*).

(ii) Esters of arenethiolsulphonic acids

(1) *Preparation.* The esters of arenethiolsulphonic acids may be prepared by (1) the oxidation of disulphides with hydrogen peroxide or peracids (*G. Leandri* and *A. Tundo*, Ann. Chim. [Italy], 1954, **44**, 74); (2) reacting sulphonic anhydrides with thiols (*L. Field*, J. Amer. chem. Soc., 1952, **74**, 394); (3) reacting arenesulphinic acids with thiols and ethyl nitrite (*G. Kresze* and *W. Wort*, Ber., 1961, **94**, 2624); (4) treating aryl disulphides with sulphinic acids in acid solution (*J. L. Kice* and *K. W. Bowers* J. Amer. chem. Soc., 1962, **84**, 2384); (5) reacting arenesulphonyl iodides with the silver salts of thiols (*D. T. Gibson, C. J. Miller* and *S. Smiles*, J. chem. Soc., 1925, **127**, 1821); (6) treating thiophenols or aryl disulphides with sulphuryl chloride in acetic acid and subsequent treatment with water (*Field et al.*, J. org. Chem., 1967, **32**, 1626); (7) the interaction of a sulphenyl halide with the silver salt of a sulphinic acid (*Leandri* and *Tundo*, Ann. Chim. [Italy], 1958, **47**, 575); (8) heating the sulphonyl chloride with an equimolar mixture of iron pentacarbonyl and boron trifluoride etherate in *N,N*-dimethylacetamide (*H. Alper*, Tetrahedron Letters, 1969, 1239). (9) Alkyl esters of benzenethiolsulphonic acid are prepared by reaction of alkyl halides with salts of the thiolsulphonic acid.

(2) *Reactions.* Thiolsulphonate esters react with alkali to give the disulphide and a salt of a sulphinic acid (*H. T. Hookway*, J. chem. Soc., 1950, 1932; *R. Otto* and *A. Rössing*, Ber., 1886, **19**, 1235):

$$3\,RS\cdot SO_2R + 4\,OH^{\ominus} \rightarrow RS\cdot SR + 4\,RSO_2^{\ominus} + 2\,H_2O$$

This reaction probably involves nucleophilic attack by hydroxide ion at the bivalent sulphur atom to give initially a sulphinate anion and a sulphenic acid, which yields the disulphide by further reaction. Thiolsulphonate esters react similarly with primary and secondary amines and with the sodium salts of thiols to give a salt of a sulphinic acid together with a sulphenamide and a disulphide respectively:

$$RSO_2\cdot SR + R'NH_2 \rightarrow RSO_2^{\ominus}\,R'NH_3^{\oplus} + RSNHR'$$

(*J. E. Dunbar* and *J. H. Rogers*, J. org. Chem., 1966, **31**, 2842);

$$RSO_2 \cdot SR + R'S^{\ominus} \rightarrow RSO_2^{\ominus} + RS \cdot SR'$$

(*Smiles* and *Gibson*, J. chem. Soc., 1924, **125**, 176; *Field et al.*, J. org. Chem., 1965, **30**, 1923).

Phenyl benzenethiolsulphonate (*"benzene disulphoxide"*), $PhS \cdot SO_2Ph$, m.p. 45°, is a product of the decomposition of benzenesulphinic acid. It is also formed by the action of benzenesulphenyl chloride on silver benzenesulphinate in dry ether (*H. Lecher et al.*, Ber., 1925, **58**, 409), and by the reaction of benzenesulphinyl chloride with zinc in dry ether (*D. Barnard*, J. chem. Soc., 1957, 4673). It reacts with bromine to give benzenesulphonyl bromide and 4,4′-dibromodiphenyl disulphide (*Leandri* and *Tundo*, Ann. Chim. [Italy], 1954, **44**, 255), and decomposes on heating to give diphenyl sulphide and sulphur dioxide (*F. Krafft* and *O. Steiner*, Ber., 1901, **34**, 560).

4-**Chlorophenyl** 4-**chlorobenzenethiolsulphonate,** m.p. 136°, is obtained from silver 4-chlorobenzenesulphinate and 4-chlorobenzenesulphenyl chloride (*Leandri* and *Tundo*, Ann. Chim. [Italy], 1958, **47**, 575). The following are similarly prepared: 4-**bromophenyl** 4-**bromobenzenethiolsulphonate,** m.p. 159–160° (*Alper, loc. cit.*); 2,5-**dichlorophenyl** 2-**nitrobenzenethiolsulphonate,** m.p. 129°; 2-**nitrophenyl** 4-**chlorobenzenethiolsulphonate,** m.p. 123° (*Miller* and *Smiles*, J. chem. Soc., 1925, **127**, 224); 2,5-**dichlorophenyl** 2,5-**dichlorobenzenethiolsulphonate,** m.p. 128° (*Smiles* and *Gibson, loc. cit.*); and 2-**aminophenyl benzenethiolsulphonate,** m.p. 83–85°, which on acetylation and treatment with nitrosyl chloride gives 1,2,3-benzothiadiazole and benzenesulphonic acid (*C. G. Overberger, M. P. Mazzeo* and *J. J. Godfrey*, J. org. Chem., 1959, **24**, 1407); 4-**methoxyphenyl** 4-**methoxybenzenethiolsulphonate**, m.p. 89°; 3-**nitrophenyl**-3-**nitrobenzenethiolsulphonate,** m.p. 124° (*Alper, loc. cit.*). For further details of unsymmetrical thiolsulphonates see *Kresze* and *Wort* (*loc. cit.*).

p-**Tolyl** p-**toluenethiosulphonate,** m.p. 78°, reacts with hydroxylamine hydrochloride and sodium acetate in ethanol to give di-*p*-tolyl disulphide and *p*-toluenesulphonamide, and with phenylhydrazine in aqueous alcoholic acetic acid to give the disulphide, *p*-toluenesulphonohydrazide and the phenylhydrazine salt of *p*-toluenesulphinic acid (*Hookway, loc. cit.*).

(b) Esters of arenethiolsulphinic acids

The thiol-esters of arenesulphinic acids, $RSO \cdot SR$, may be prepared by the oxidation of disulphides with peracids or by the reaction between thiols and arenesulphinyl chlorides (*H. J. Backer* and *H. Kloosterziel*, Rec. Trav. chim., 1954, **73**, 129). Arenethiolsulphinates have been prepared in an optically active form by oxidation of the appropriate disulphide with percamphoric acid at 0–10° in the dark. The optical stability depends upon the solvent (*W. E. Savige* and *A. Fava*, Chem. Comm., 1965, 417) and the mechanism of the racemisation has been studied (*P. Koch* and *Fava*, J.

Amer. chem. Soc., 1968, **90**, 3867; *J. L. Kice* and *G. B. Large*, *ibid.*, p. 4069). Thiolsulphinates are also obtained from the hydrolysis products of certain sulphenyl halides (cf. p. 480).

Thiolsulphinates are thermally unstable and disproportionate to give disulphides and thiolsulphonates (*Backer* and *Kloosterziel*, *loc. cit.*; *Barnard*, *loc. cit.*):

$$2\ RSO \cdot SR \rightarrow RSO_2 \cdot SR + RS \cdot SR$$

Oxidation with peracids gives thiolsulphonates, but treatment with ozone yields mixtures of thiolsulphonates and sulphonic anhydrides (*Barnard*, J. chem. Soc., 1957, 4547). Reaction with sulphinic acid in acid solution yields principally thiolsulphinates; this reaction is catalysed by sulphides (*J. L. Kice*, *C. G. Venier* and *L. Heasley*, J. Amer. chem. Soc., 1967, **89**, 3557):

$$ArSO \cdot SAr + Ar'SO_2H \xrightarrow{H^{\oplus}} 2\ Ar'SO_2SAr + H_2O$$

Phenyl benzenethiolsulphinate, m.p. 69·5–71°, is formed in the reduction of benzenesulphonyl chloride with sodium iodide in acetone. p-**Tolyl** p-**toluenethiolsulphinate,** m.p. 69·5–71°, is obtained by the hydrolysis of *p*-toluenesulphenyl chloride with sodium carbonate solution (*E. Vinkler* and *F. Klivényi*, Acta Chim. Acad. Sci. Hung., 1957, **11**, 15; C.A., 1958, **52**, 6242c); it reacts with benzylmagnesium chloride to give benzyl *p*-tolyl sulphoxide and with phenylmagnesium bromide to give phenyl *p*-tolyl sulphide (*Vinkler*, *Klivényi* and *E. Klivényi*, *ibid.*, 1958, **16**, 247; C.A., 1959, **53**, 8041b).

Chapter 7

Mononuclear hydrocarbons carrying nuclear substituents containing selenium or tellurium

A. G. DAVIES

1. Selenium compounds

The aromatic compounds of selenium have been reviewed (*T. W. Campbell, H. G. Walker* and *G. M. Coppinger*, Chem. Reviews, 1952, 50, 279; *H. Rheinboldt*, in Houben-Weyl's "Methoden der Organischen Chemie", Vol. IX, Thieme, Stuttgart, 1955, p. 916; *K. W. Bagnall*, "The Chemistry of

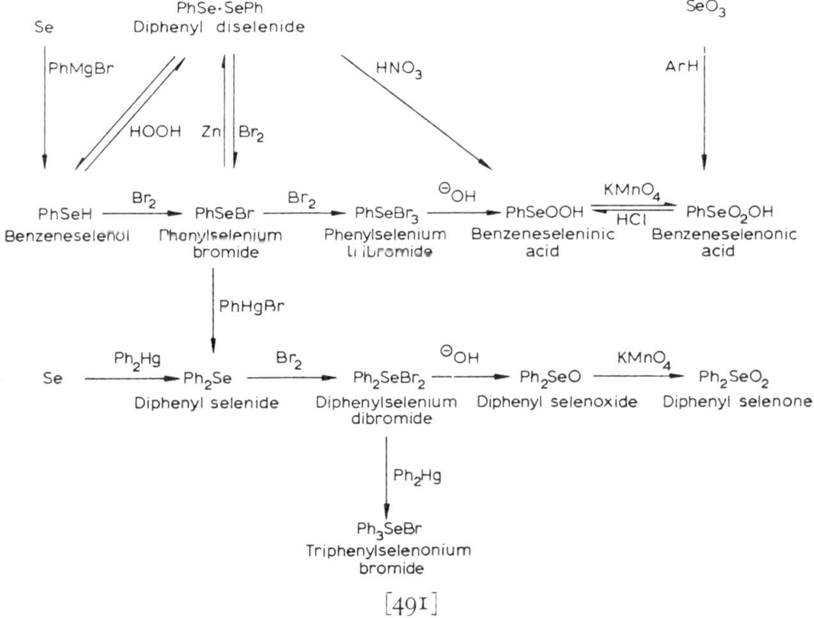

Selenium, Tellurium and Polonium", Elsevier, Amsterdam, 1966, Chap. 7). The nomenclature of the various types of compound, and their interrelation, are shown in the chart on p. 491. Many of these compounds are toxic and have an offensive odour, and may be affected by sunlight; they should be handled with due caution.

(a) Areneselenols, ArSeH

The aromatic selenols, or selenophenols, can be prepared in yields of 40–70 % by treating the appropriate Grignard reagent in ether with elementary selenium at room temperature (cf. the reaction of oxygen, sulphur, and tellurium) (*D. G. Foster* and *S. F. Brown*, J. Amer. chem. Soc., 1928, **50**, 1182; *Foster*, Org. Synth., 1944, **24**, 89):

$$\text{ArMgX} + \text{Se} \rightarrow \text{ArSeMgX} \xrightarrow{\text{HCl}} \text{ArSeH}$$

Alternatively, the aryl selenocyanate (from the arenediazonium chloride and potassium selenocyanate) or the diaryl diselenide may be reduced with zinc and acetic acid (*O. Behaghel* and *K. Hofmann*, Ber., 1939, **72**, 582):

$$\text{ArN}_2^{\oplus}\text{X}^{\ominus} + \text{KSeCN} \xrightarrow{-\text{N}_2} \text{ArSeCN} \xrightarrow{\text{Zn/H}^{\oplus}} \text{ArSeH}$$

The selenols have a repulsive odour, and are usually colourless liquids or solids soluble in organic solvents, but insoluble in water. In the air they are readily oxidised to the corresponding diselenides. They undergo the Zerewitinoff reaction with methylmagnesium iodide, and react with sodium or potassium in benzene to give the corresponding selenolates. **Selenophenol**, b.p. 57–59°/8 mm; p-*methylselenophenol*, m.p. 46–47°; p-*chloroselenophenol*, m.p. 57°; 1-*selenonaphthol*, b.p. 173–174°.

(b) Diaryl diselenides, ArSe·SeAr

The diaryl diselenides are obtained as yellow solids when the corresponding selenols are oxidized by air or dilute hydrogen peroxide, or when the selenocyanates are hydrolysed under alkaline conditions (*e.g. Behaghel* and *H. Seibert*, Ber., 1932, **65**, 812; *Rheinboldt*, *loc. cit.* pp. 1094–1098) or are treated with bromine (*Behaghel* and *Seibert*, Ber., 1933, **66**, 708). They can also be obtained by reducing areneseleninic acids with sodium hydrogen sulphite, or with iodide; this last reaction can be used for quantitative analysis (*J. D. McCullough*, *Campbell* and *N. J. Krilanovitch*, Ind. Eng. Chem. Anal. Edn., 1946, **18**, 638). **Diphenyl diselenide**, m.p. 63°; *di*-p-*tolyl* *diselenide*, m.p. 45–46°; 4,4′-*dinitrodiphenyl diselenide*, m.p. 179·5–180·5°; *di-2-naphthyl diselenide*, m.p. 138·5–139·5°.

Bis(pentafluorophenyl) diselenide, $C_6F_5Se \cdot SeC_6F_5$, m.p. 171–173°, has been prepared from pentafluorophenylmagnesium bromide and elementary selenium. It reacts with mercury to give $(C_6F_5Se)_2Hg$, and disproportionates with bis(pentafluorophenyl) disulphide giving bis(pentafluorophenyl) thioselenide $C_6F_5S \cdot SeC_6F_5$ (*E. Kostiner et al.*, J. organometallic Chem., 1968, **15**, 383).

(c) Arylselenium halides, ArSeX

The arylselenium chlorides or bromides (or areneselenenic halides) are formed when the corresponding selenols, diselenides, or selenocyanates are oxidised with the appropriate amount of halogen (*Behaghel* and *Seibert*, Ber., 1933, **66**, 714; *Rheinboldt* and *E. Giesbrecht*, Ann., 1950, **568**, 198; 1951, **574**, 227; *Rheinboldt* and *M. Perrier*, Bull. Soc. chim. Fr., 1953, 484):

$$ArSeH + X_2 \rightarrow ArSeX + HX$$

$$ArSe \cdot SeAr + X_2 \rightarrow 2\ ArSeX$$

$$ArSeCN + X_2 \rightarrow ArSeX + XCN$$

The last reaction must be carried out by adding the selenocyanate to the halogen, or the diselenide is formed, presumably by the reaction:

$$ArSeCN + ArSeX \rightarrow ArSe \cdot SeAr + XCN$$

Alternatively, the selenocyanate may be treated with sulphuryl chloride.

The chlorides are white to yellow solids, and the bromides orange to red, *e.g. phenylselenium chloride*, m.p. 59–60°; o-*nitrophenylselenium chloride*, m.p. 64°; *phenylselenium bromide*, m.p. 61–62°; p-*biphenylylselenium bromide*, m.p. 165–166°; o-*nitrophenylselenium bromide*, m.p. 64–65°; 2,4-*dinitrophenylselenium bromide*, m.p. 118°.

They smell of the free halogen. In water they are hydrolysed initially to the corresponding selenenic acid, ArSeOH, which disproportionates to the seleninic acid and selenol, and the selenol is oxidised in turn by the selenenic acid giving the diselenide (*Behaghel* and *W. Müller*, Ber., 1935, **68**, 1540):

$$2\ PhSeBr \xrightarrow{H_2O} 2\ PhSeOH \rightarrow PhSeH + PhSeOOH$$

$$PhSeOH + PhSeH \rightarrow PhSe \cdot SePh + H_2O$$

Zinc reduces the arylselenium halides back to the diselenides.

If the arylselenium chloride in benzene is shaken with finely powdered potassium thiocyanate or lead thiocyanate, the arylselenium thiocyanate is formed in quantitative yield (*Rheinboldt* and *Perrier*, Bull. Soc. chim. Fr., 1950, 245), *e.g.* o-*nitrophenylselenium thiocyanate*, *o*-$NO_2C_6H_4Se{\cdot}SCN$, m.p. 114°.

The corresponding arylselenium acetates, cyanides, selenolsulphonates and areneselenenic amides and anilides are prepared by treating arylselenium halides with the appropriate nucleophiles (*O. Foss*, J. Amer. chem. Soc., 1947, **69**, 2236; *W. S. Cook* and *R. A. Donia*, *ibid.*, 1951, **73**, 2275). Arylselenium halides react with ketones and add to carbon-carbon multiple bonds in a similar manner to the sulphenyl halides.

(d) Areneselenenic acids, ArSeOH

Most arylselenium halides undergo hydrolysis to give diselenides, but if the aryl group carries *ortho* or *para* electronegative substituents, the selenenic acids can be obtained. Thus *o*-nitrophenylselenium bromide gives *o*-nitrobenzeneselenenic acid, but *m*-nitrophenylselenium bromide gives the diselenide (*Behaghel* and *Müller*, *loc. cit.*). More generally, the selenenic acids can be reduced by hydrazine, thiols, selenols etc. (*Rheinboldt* and *Giesbrecht*, Ber., 1955, **88**, 666, 1037, 1974):

$$2\ ArSeOOH + N_2H_4 \rightarrow 2\ ArSeOH + N_2 + H_2O$$

Selenenic esters are obtained by treating the arylselenium bromides with the appropriate alcohol in the presence of silver acetate (*Cook* and *Donia*, *loc. cit.*).

The areneselenenic acids are usually dark orange or red crystalline solids, which give deep red to violet solutions in alcoholic alkali.

o-**Nitrobenzeneselenenic acid,** decomp. 165–170° (*thiocyanate*, m.p. 107°); 2,4-*dinitrobenzeneselenenic acid*, golden needles, m.p. 260° (*acetate*, m.p. 128–129°; *methyl ester*, m.p. 131–133°; *anilide*, m.p. 144–146°); 4-*chloro*-2-*nitrobenzeneselenenic acid*, m.p. 185°; 4-*methyl*-2-*nitrobenzeneselenenic acid*, m.p. 178–179° decomp.; 4-*methoxy*-2-*nitrobenzeneselenenic acid*, m.p. 169°; o-*benzoylbenzeneselenenic acid*, decomp. 140°.

(e) Diaryl selenides, Ar_2Se

The diaryl selenides are the most common arylselenium compounds. They can be prepared by treating an arenediazonium salt with an alkali selenide (*H. M. Leicester* and *F. W. Bergstrom*, J. Amer. chem. Soc., 1929, **51**, 3587; 1931, **53**, 4428; *Leicester*, Org. Synth., 1938, **18**, 27), or by heating a diarylmercury with selenium or selenium tetrabromide (*F. Krafft* and *R. E. Lyons*, Ber., 1894, **27**, 1768; *Leicester*, J. Amer. chem. Soc., 1938, **60**, 619), or a diaryl sulphoxide with selenium (*Krafft* and *W. Vorster*, Ber., 1893, **26**, 2813):

$$2\ ArN_2X + M_2Se \rightarrow Ar_2Se + 2\ N_2 + 2\ MX$$

$$Ar_2Hg + 2\ Se \rightarrow Ar_2Se + HgSe$$

$$3\ Ar_2Hg + 2\ SeBr_4 \rightarrow 2\ Ar_2Se + 2\ ArBr + 3\ HgBr_2$$

$$Ar_2SO_2 + Se \rightarrow Ar_2Se + SO_2$$

Bis(pentafluorophenyl) selenide is formed, together with bis(pentafluorophenyl) diselenide, when pentafluoroiodobenzene is heated with elementary selenium (*S. C. Cohen et al.*, J. organometallic Chem., 1968, **14**, 241).

Unsymmetrical diaryl selenides, ArSeAr′, are formed from an arenediazonium salt and the sodium salt of an areneselenol (*R. Lesser* and *R. Weiss*, Ber., 1914, **47**, 2522), from a diaryl diselenide and an arylmagnesium halide or from an arylselenium bromide and a diarylmercury (*Campbell* and *McCullough*, J. Amer. Chem. Soc., 1945, **67**, 1965; *McCullough* and *M. K. Barsh*, *ibid.*, 1949, **71**, 3029).

Mixed alkyl aryl selenides are usually prepared by alkylating sodium areneselenolates with alkyl halides or sulphates (*e.g. Foster*, Rec. Trav. chim., 1934, **53**, 409; 1935, **54**, 452; J. Amer. chem. Soc., 1941, **63**, 1361), or by treating arylselenium halides with aliphatic Grignard reagents (*Behaghel* and *Hofmann*, Ber., 1939, **72**, 710).

The diaryl selenides are usually colourless low-melting solids, without an unpleasant smell. Typical examples are as follows: **diphenyl selenide,** m.p. 2·5°, b.p. 301°; *di-1-naphthyl selenide*, m.p. 93–100°; *di-2-naphthyl selenide*, m.p. 110–114°; o-*bromodiphenyl selenide*, b.p. 156–158°/1·5 mm; p-*nitrodiphenyl selenide*, m.p. 57·2°.

X-ray analysis of di-*p*-tolyl selenide indicates a C-Se-C angle of 106° (*W. R. Blackmore* and *S. C. Abrahams*, Acta Cryst., 1955, **8**, 323); the dipole moment of di-*p*-anisyl selenide has been interpreted to imply that there is no significant conjugative interaction between the methoxy group and the selenium (*S. S. Krishnamurthy* and *S. Soundaravajan*, J. organometallic Chem., 1968, **15**, 367).

(f) *Arylselenium trihalides*, $ArSeX_3$

No fluorides or iodides appear to have been prepared. The trichlorides and tribromides are usually prepared by treating the diaryl diselenides, arylselenium halides, or arylselenocyanates with the appropriate halogen (*Behaghel* and *Seibert*, Ber., 1933, 66, 714):

$$\begin{array}{l} \text{ArSe·SeAr} \xrightarrow{X_2} \\ \qquad\qquad\qquad \text{ArSeX} \xrightarrow{X_2} \text{ArSeX}_3 \\ \text{ArSeCN} \xrightarrow{X_2} \end{array}$$

Like the corresponding monohalides, the trichlorides are colourless or pale yellow, and the tribromides are orange-yellow to deep red.

Phenylselenium trichloride, decomp. 133–134°; p-*nitrophenylselenium trichloride,* decomp. 154°; *phenylselenium tribromide,* decomp. 105°; o-*chlorophenylselenium tribromide,* decomp. 107°.

(g) *Areneseleninic acids, $ArSeO_2H$*

The arylselenium trihalides are hydrolysed by dilute aqueous sodium carbonate to the areneseleninic acids (*Behaghel* and *Hoffman*, Ber., 1939, 72, 582), which can also be prepared by oxidation of diaryl diselenides or aryl selenocyanates (*M. Stoecker* and *Krafft*, *ibid.*, 1906, 39, 2197; *McCullough* and *E. S. Gould*, J. Amer. chem. Soc., 1949, 71, 674) with nitric acid or hydrogen peroxide.

The products are colourless, odourless solids, and are amphoteric, being weaker acids than the corresponding carboxylic acids (*McCullough* and *Gould*, *loc. cit.*), and reacting with hydrogen halides to regenerate the arylselenium trihalides.

Benzeneseleninic acid, m.p. 120°, pK 4·79; *toluene*-p-*seleninic acid,* m.p. 170°, pK 4·88; p-*methoxybenzeneseleninic acid,* m.p. 101°, pK 5·05; p-*fluorobenzeneseleninic acid,* m.p. 132–140° (decomp.); p-*chlorobenzeneseleninic acid,* m.p. 170–185°; p-*bromobenzeneseleninic acid,* m.p. 187°, m-*nitrobenzeneseleninic acid,* m.p. 152°. The crystal structure of benzeneseleninic acid has been determined (*J. H. Bryden* and *McCullough,* Acta Cryst., 1954, 7, 833).

(h) *Areneselenonic acids, $ArSeO_3H$*

The selenonic acids are analogous to the sulphonic acids, and can be prepared by selenonation of the aromatic ring. Thus, under reflux benzene reacts with selenic acid in 100 h to give benzeneselenonic acid (*H. W. Doughty*, Amer. chem. J., 1909, 41, 326; *R. Anshütz* and *F. Teutenberg*, Ber., 1924, 57, 1019).

Alternatively the diaryl diselenides can be oxidised with chlorine (*Stoecker* and *Krafft*, Ber., 1906, **39**, 2187) or the seleninic acid with alkaline permanganate (*F. L. Pyman*, J. chem. Soc., 1919, **115**, 166).

The selenonic acids are colourless, odourless, crystalline solids. They are strong acids and also strong oxidising agents; for example they react with cold concentrated hydrochloric acid to give chlorine and a quantitative yield of the corresponding areneseleninic acid.

Benzeneselenonic acid, m.p. 142°; m-*nitrobenzeneselenonic acid,* m.p. 146°; 2,4-*dimethylbenzeneselenonic acid,* m.p. 130°; 2,5-*dimethylbenzeneselenonic acid,* m.p. 95–96°; 3,4-*dimethylbenzeneselenonic acid,* m.p. 108°.

(i) *Diarylselenium dihalides,* Ar_2SeX_2

If diaryl selenides are treated with an excess of halogen (or sulphuryl chloride), or diaryl selenoxides with hydrogen halides, the corresponding diarylselenium dihalides are formed (*Leicester*, Org. Synth., 1938, **18**, 30; *W. E. Bradt* and *J. F. Green*, J. org. Chem., 1937, **1**, 541; *Foster* and *Brown*, J. Amer. chem. Soc., 1928, **50**, 1185):

$$Ar_2Se \xrightarrow{X_2} Ar_2SeX_2 \xrightarrow{H_2O} Ar_2SeO$$

Alternatively, if the aromatic ring carries alkoxy groups, the $SeCl_2$ group can be introduced directly with selenium oxychloride (*F. Kunckell*, Ber., 1895, **28**, 609):

$$2\ ROC_6H_5 + OSeCl_2 \rightarrow (ROC_6H_4)_2SeCl_2 + H_2O$$

The chlorides are colourless or lemon yellow solids, the bromides orange, or red and the iodides, which are rather unstable, are violet-black. In solution, the dihalides are in equilibrium with the corresponding selenides and free halogen, the dissociation constants obeying the Hammett σ-ρ relationship (*McCullough* and *Barsh, loc. cit.*):

$$Ar_2SeX_2 \rightleftharpoons Ar_2Se + X_2$$

Diphenyl- and di-p-tolyl-selenium dichloride and dibromide show a slightly distorted pyramidal structure, with axial halides and equatorial aryl groups and unshared pair (*McCullough* and *G. Hamberger*, J. Amer. chem. Soc., 1941, **63**, 803; 1942, **64**, 508; *McCullough* and *R. E. Marsh*, Acta Cryst., 1950, **3**, 41).

Diphenylselenium dichloride, m.p. 187–188°; *di-(4-methoxyphenyl)selenium dichloride,* m.p. 163°.

(j) *Diaryl selenoxides, Ar_2SeO*

The diaryl selenoxides are formed when the corresponding dihalides are hydrolysed under basic conditions (*Krafft* and *Vorster, loc. cit.*; *W. R. Gaythwaite, J. Kenyon* and *H. Phillips*, J. chem. Soc., 1928, 2280; *C. K. Banks* and *C. S. Hamilton*, J. Amer. chem. Soc., 1939, **61**, 2306), or when the diaryl selenides are oxidised with hydrogen peroxide (*Rheinboldt* and *E. Geisbrecht, ibid.*, 1947, **69**, 644), peroxyacetic acid (*McCullough, Campbell* and *Gould, ibid.*, 1950, **72**, 5753), dichromate (*Krafft* and *Vorster, loc. cit.*), or permanganate (*Gaythwaite, Kenyon* and *Phillips, loc. cit.*).

The diaryl selenoxides are colourless, odourless, crystalline solids, which react with hydrogen halides to regenerate the dihalides, and add water to form dihydroxides. Several complexes of diphenyl selenoxides and metals have been prepared (*R. Paetzold* and *P. Vordank*, Z. Anorg. Chem., 1966, **347**, 294.). Presumably the selenoxides, which form mixed crystals with the sulphoxides, have pyramidal structures, but none has yet been optically resolved.

Diphenyl selenoxide, m.p. 113–114°; *methyl* p-*carboxyphenyl selenoxide,* m.p. 183–184°; *di-1-naphthyl selenoxide,* m.p. 105°; *bis*(3,4-*dimethoxyphenyl*)-*selenium dihydroxide,* m.p. 170–172°.

(k) *Diaryl selenones, Ar_2SeO_2*

With more powerful agents, such as peroxyacids or hot permanganate, the selenoxides can be oxidised further to the selenones. These are white, odourless, crystalline solids, with powerful oxidising properties. Thus concentrated hydrochloric acid is oxidised in the cold, to give chlorine and the diarylselenium dichloride, and at 190° sulphur reacts to form the diaryl selenide and sulphur dioxide.

Diphenyl selenone, m.p. 154–155°; *di*-p-*methoxyphenyl selenone,* m.p. 148·5–149·5°; *di-2-naphthyl selenone,* m.p. 160–161°; *ethyl 2-naphthyl selenone,* m.p. 40–41°.

(l) *Arylselenonium salts, $R_2ArSe^{\oplus}X^{\ominus}$ and $Ar_3Se^{\oplus}X^{\ominus}$*

The selenonium salts are ionic compounds like the sulphonium salts. The mixed dialkylaryl compounds can be prepared by alkylating the appropriate alkyl aryl selenide with an alkyl halide or dialkyl sulphate, (*e.g.*, *W. J. Pope* and *A. Neville*, Proc. chem. Soc., 1902, **81**, 1552; *J. W. Baker* and *W. G. Moffit*, J. chem. Soc., 1930, 1722):

$$\text{PhSeMe} + \text{BrCH}_2\cdot\text{CO}_2\text{H} \rightarrow \text{PhMeSe}^{\oplus}\text{CH}_2\cdot\text{CO}_2\text{H Br}^{\ominus}$$

$$p\text{-}NO_2\cdot C_6H_4\cdot SeMe + Me_2SO_4 \rightarrow p\text{-}NO_2\cdot C_6H_4Se^{\oplus}Me_2\ MeSO_4^{\ominus}$$

The triarylselenonium salts are obtained from a Friedel-Crafts reaction (*Leicester* and *Bergstrom*, J. Amer. chem. Soc., 1929, **51**, 3587; *Leicester*, Org. Synth., 1938, **18**, 31; *G. T. Morgan* and *F. H. Burstall*, J. chem. Soc., 1928, 3260):

$$Ph_2SeCl + C_6H_6 \xrightarrow{AlCl_3} Ph_3Se^{\oplus}Cl^{\ominus}$$

$$C_6H_5OH + SeOCl_2 \rightarrow (p\text{-}HOC_6H_4)_3Se^{\oplus}Cl^{\ominus} + HCl + H_2O$$

or from the arylation of a diarylselenium dichloride with an arylmercury compound (*Leicester* and *Bergstrom*, *loc. cit.*).

The selenonium salts are colourless, odourless, crystalline solids, soluble in water; they react with moist silver oxide to form the corresponding hydroxides. Some mixed alkylarylsulphonium compounds (*e.g.* $PhMeSe^{\oplus}CO_2H\ Br^{\ominus}$) have been resolved through their camphor sulphonates (*Pope* and *Neville*, Proc. chem. Soc., *loc. cit.*); no complete determination of the structure of triphenylselenonium chloride has been carried out, but the minimum Se-Cl separation of 3·60Å confirms the ionic nature (*McCullough* and *R. E. March*, J. Amer. chem. Soc., 1950, **72**, 4556).

Triphenylselenonium chloride, decomp. 230°; *tris-(4-hydroxyphenyl)-selenonium chloride,* decomp. 125°.

2. Tellurium compounds

The chemistry of organotellurium compounds is broadly parallel to that of organoselenium compounds, but has been studied much less thoroughly. Recent reviews are by *K. W. Bagnall* ("The Chemistry of Selenium, Tellurium, and Polonium", Elsevier, Amsterdam, 1966, Chap. 7) and by *H. Rheinboldt* (Houben-Weyl, "Methoden der Organischen Chemie", Vol. IX, Thieme, Stuttgart, 1955, p. 988).

(a) Diaryl ditellurides, ArTe·TeAr

No simple arenetellurols, ArTeH, appear to have been reported. Diphenyl ditelluride, PhTe·TePh is obtained together with diphenyl telluride, Ph_2Te, when tellurium(II) chloride, bromide, or iodide, is treated with phenylmagnesium bromide (*K. Lederer*, Ber., 1915, **48**, 1345), and a number of other diaryl ditellurides have been prepared by reducing aryltellurium trihalides with potassium pyrosulphite or sodium sulphide (*L. Reichel* and *E. Kirschbaum*, *ibid.*, 1943, **76**, 1105: Ann., 1936, **523**, 211):

$$ArTeCl_3 \rightarrow ArTeTeAr$$

Diphenyl ditelluride decomp. 66–67°; *di-*p*-anisyl ditelluride,* decomp. 60°; *di-2-naphthyl ditelluride,* decomp. 123°.

(b) Aryltellurium halides, ArTeX

Again, no simple aryltellurium halides appear to have been characterised, but phenyltellurium trichloride is reduced by thiourea to give derivatives of divalent tellurium (*O. Foss* and *S. Hauge,* Acta Chem. Scand., 1959, **13,** 2155):

$$PhTeCl_3 \xrightarrow{SC(NH_2)_2} PhTe(thiourea)Cl \rightarrow PhTe^{\oplus}(thiourea)_2\,Cl^{\ominus}$$

The three ligands in $PhTe^{\oplus}(thiourea)_2Cl^{\ominus}$ have been shown to occupy a square planar disposition about the metal, with a vacant site *trans* to the phenyl group (*Foss* and *K. Maroy, ibid.,* 1966, **20,** 123).

(c) Diaryl tellurides, Ar_2Te

Elementary tellurium reacts with a diarylmercury (*F. Krafft* and *R. E. Lyons,* Ber., 1894, **27,** 1768), with an aryl Grignard reagent or with phenyllithium (*N. Petragnani* and *M. de M. Campos, ibid.,* 1963, **96,** 249). Alternatively, tellurium dibromide, tetrabromide or tetrachloride may be treated with a arylmagnesium bromide in ether; the diaryl telluride is purified through the diaryl tellurium dibromide, (*Lederer, ibid.,* 1911, **44,** 2287; 1915, **48,** 1345; Compt. rend., 1910, **151,** 611):

$$Ar_2Te \xrightarrow{Br_2} Ar_2TeBr_2 \xrightarrow{NaHSO_3} Ar_2Te$$

The diaryl tellurides are colourless or lightly coloured, low melting solids, without an unpleasant odour.

Diphenyl telluride, m.p. 4°, b.p. 182°/14 mm; *phenyl* p*-tolyl telluride,* m.p. 63–64°.

(d) Aryltellurium trichlorides, $ArTeCl_3$

A few aryltellurium trichlorides have been prepared by treating aromatic ethers or phenols or arylmercuric chlorides with tellurium (*G. T. Morgan* and *R. E. Kellett,* J. chem. Soc., 1926, 1080), or tellurium tetrachloride (*I. G. M. Campbell* and *E. E. Turner, ibid.,* 1938, 37):

$$ROC_6H_5 + TeCl_4 \xrightarrow[CCl_4]{CHCl_3 \text{ or}} ROC_6H_4TeCl_3 + HCl$$

$$ArHgCl + TeCl_4 \rightarrow ArTeCl_3 + HgCl_2$$

Phenyltellurium trichloride, decomp. 215–218°; p-*methoxyphenyltellurium trichloride,* m.p. 194–195°.

A few of the trichlorides have been hydrolysed to the corresponding **arenetellurinic acids,** $ArTeO_2H$ (*L. Reichell* and *E. Kirschbaum,* Ann., 1936, **523,** 211) and *benzenetellurinic acid,* $C_6H_5TeO_2H$, m.p. 210–211°, has been prepared by the oxidation of diphenyl ditelluride with 65 % nitric acid (*Lederer,* Ber., 1915, **48,** 1345). No telluronic acids, $ArTeO_3H$, appear to have been reported.

(e) Diaryltellurium dihalides, Ar_2TeX_2

Above about 150°, aromatic ethers or phenols react with tellurium tetrachloride to give the corresponding diaryltellurium dichlorides, which can also be prepared by treating diaryltellurides with the halogen (*Morgan* and *Kellett, loc. cit.*; *Morgan* and *F. H. Burstall,* J. chem. Soc., 1930, 2599).

The difluorides and dichlorides are colourless, the dibromides yellow, and the diiodides orange to deep red. Hydrolysis gives sequentially, the hydroxyhalides, $Ar_2Te(X)OH$, or their anhydrides, $Ar_2Te(X)OTeAr_2X$, and then the dihydroxides $Ar_2Te(OH)_2$, or tellurones Ar_2TeO.

Diphenyltellurium dichloride, m.p. 160°. Reduction of the dihalides with sodium sulphite regenerates the diaryl tellurides.

(f) Aryltelluronium salts, $R_2ArTe^{\oplus}X^{\ominus}$, $RAr_2Te^{\oplus}X^{\ominus}$, and $Ar_3Te^{\oplus}X^{\ominus}$

Alkylaryltelluronium salts are usually prepared by alkylation of the appropriate telluride (*e.g. Lederer,* Ann., 1913, **399,** 260):

$$Ph_2Te + MeI \rightarrow MePh_2Te^{\oplus}I^{\ominus}$$

The triaryltelluronium salts are formed when tellurium tetrachloride or diaryltellurium dihalides are caused to react with arylmagnesium halides (*idem,* Compt. rend., 1910, **151,** 611; Ber., 1911, **44,** 2287):

$$TeX_4 \xrightarrow{ArMgX} Ar_2TeX_2 \xrightarrow{ArMgX} Ar_3Te^{\oplus}X^{\ominus}$$

Methyldiphenyltelluronium iodide, m.p. 123–124° (decomp.); *triphenyltelluronium iodide,* m.p. 247–249°.

GUIDE TO THE INDEX

This index is constructed in a similar manner to the volume indexes of the first edition of the Chemistry of Carbon Compounds. However, to make the index easier to use, more descriptive entries have been made for the commonly occurring individual, and groups of chemicals.

The indexes cover primarily the chemical compounds mentioned in the text, and also include reactions and techniques, where named, and some sources of chemical compounds such as plant and animal species, oils, etc.

Chemical compounds have been indexed alphabetically under the names used by authors, editing being restricted to ensuring uniformity of entries under the same heading. In view of the alternative nomenclature that can often be used, a limited amount of cross-referencing has been done where it is considered to be helpful, but attention is particularly drawn to Convention 2 below.

For this and the succeeding volumes, the indexing conventions listed below have been adopted.

1. *Alphabetisation*

(a) The following prefixes have not been counted for alphabetising:

n-	*o-*	*as-*	*meso-*	D	*C-*
sec-	*m-*	*sym-*	*cis-*	DL	*O-*
tert-	*p-*	*gem-*	*trans-*	L	*N-*
	vic-				*S-*
		lin-			*Bz-*
					Py-

Some prefixes and numbering have been omitted in the index, where they do not usefully contribute to the reference.

(b) The following prefixes have been alphabetised:

Allo	Epi	Neo
Anti	Hetero	Nor
Cyclo	Homo	Pseudo
	Iso	

(c) A letter by letter alphabetical sequence is followed for entries, firstly for the main entry, followed by the descriptive entry. The only exception to this sequence is the placing of plural entries in front of the corresponding individual entries to prevent these being overlooked by a strict alphabetical sequence which could lead to a considerable separation of plural from individual entries. Thus "butanes" will come before *n*-butane, "butenes" before 1-butene, and 2-butene, etc.

2. *Cross references*

In view of the many alternative trivial and systematic names for chemical compounds, the indexes should be searched under any alternative names which may be indicated in the main body of the text. Only a limited amount of cross-referencing has been carried out, where it is considered that it would be helpful to the user.

3. *Esters*

In the case of lower alcohols esters are indexed only under the acid, *e.g.* propionic methyl ester, not methyl propionate. Ethyl is normally omitted *e.g.* acetic ester.

4. *Derivatives*

Simple derivatives are not normally indexed if they follow in the same short section of the text.

5. *Collective and plural entries*

In place of "— derivatives" or "— compounds" the plural entry has normally been used. Plural entries have occasionally been used where compounds of the same name but differing numbering appear in the same section of the text.

6. *Main entries*

The main entry of the more common individual compounds is indicated by heavy type. Where entries relate to sections of three pages or more, the page number is followed by "ff".

INDEX